DESIGN AND ANALYSIS OF EXPERIMENTS

Fourth Edition

Douglas C. Montgomery

ARIZONA STATE UNIVERSITY

JOHN WILEY & SONS
New York   Chichester   Brisbane   Toronto   Singapore   Weinheim

COVER PHOTO Doreen L. Wynja

ACQUISITIONS EDITOR Charity Robey
MARKETING MANAGER Harper Mooy
PRODUCTION COORDINATION Elm Street Publishing Services, Inc.
MANUFACTURING MANAGER Mark Cirillo
ILLUSTRATION COORDINATOR Eugene Aiello

This book was set in 10/12 Times Ten by Bi-Comp, Inc., and printed and bound
by Courier/Westford. The cover was printed by Phoenix Color.

Recognizing the importance of preserving what has been written, it is a
policy of John Wiley & Sons, Inc. to have books of enduring value published
in the United States printed on acid-free paper, and we exert our best
efforts to that end.

The paper in this book was manufactured by a mill whose forest management programs include
sustained yield harvesting of its timberlands. Sustained yield harvesting principles ensure that
the number of trees cut each year does not exceed the amount of new growth.

Library of Congress Cataloging-in-Publication Data
Montgomery, Douglas C.
    Design and analysis of experiments/Douglas C. Montgomery.—4th
  ed.
       p. cm.
    Includes index.
    ISBN 0-471-15746-5 (cloth : alk. paper)
    1. Experimental design. I. Title.
  QA279.M66 1996
  001.4'34—dc20   96-18037

Printed in the United States of America

10 9 8 7 6 5 4 3 2 1

Preface

This is an introductory textbook dealing with the design and analysis of experiments. It is based on college-level courses in design of experiments that I have taught for the last 25 years at Arizona State University, the University of Washington, and the Georgia Institute of Technology. It also reflects the methods that I have found useful in my own professional practice as an engineering and statistical consultant in the general areas of product and process design, process improvement, and quality engineering.

The book is intended for students who have completed a first course in statistical methods. This background course should include at least some techniques of descriptive statistics, the normal distribution, and an introduction to basic concepts of confidence intervals and hypothesis testing for means and variances. Chapters 13 and 14 require some familiarity with matrix algebra.

Because the prerequisites are relatively modest, this book can be used in a second course on statistics focusing on statistical design of experiments for undergraduate students in engineering, the physical and chemical sciences, mathematics, and other fields of science. For many years I have taught a course from the book at the first-year graduate level in engineering. Students in this course come from all the traditional fields of engineering, physics, chemistry, mathematics, operations research, and statistics. I have also used this book as the basis of an industrial short course on design of experiments for practicing technical professionals with a wide variety of backgrounds. There are numerous examples illustrating all of the design and analysis techniques. These examples are based on real-world applications of experimental design and are drawn from many different fields of engineering and the sciences. This adds a strong applications flavor to an academic course for engineers and scientists and makes the book useful as a reference tool for experimenters in a variety of disciplines.

ABOUT THE BOOK

The fourth edition is a major revision of the book. I have tried to maintain the balance between design and analysis topics of previous editions; however, there are many new topics and examples, and I have reorganized much of the material. There is much more emphasis on the computer in this edition. During the last few years a number of excellent software products to assist experimenters in both the design and analysis phases of this subject have appeared. I have included output from several of these products at many points in the text. I urge all instructors who use this book to incorporate computer software into your course. (In my course, I bring a laptop computer and an overhead display panel to every lecture, and every design or analysis topic discussed in class is illustrated with the computer.)

Perhaps the most important change in the book is the increased emphasis on the connection between the experiment and the model that the experimenter can develop from the results of the experiment. Engineers (and physical and chemical scientists to a large extent) learn about physical mechanisms and their underlying mechanistic models early in their academic training, and throughout much of their professional careers they are involved with manipulation of these models. Statistically designed experiments offer the engineer a valid basis for developing an *empirical* model of the system being investigated. This empirical model can then be manipulated (perhaps through a response surface or contour plot, or perhaps mathematically) just as any other engineering model. I have discovered through many years of teaching that this viewpoint is very effective in creating enthusiasm in the engineering community for statistically designed experiments. Therefore, the notion of an underlying empirical model for the experiment and response surfaces appears much earlier in the book than in earlier editions and receives much more emphasis.

I have also made an effort to get the reader to factorial designs much faster. The material on blocking has been condensed into a single chapter (Chapter 5) and much of the material on incomplete block designs has been deleted. Chapter 4, which contains more topics concerning the analysis of variance, could be skipped on first reading, if desired. I have expanded and rearranged the material on factorial and fractional factorial designs in an effort to make the material flow more effectively from both the reader's and the instructor's viewpoint and to place more emphasis on the empirical model. There is a new chapter on experiments with random factors (Chapter 11). In previous editions, this material was scattered throughout the text; in this edition it is consolidated. Chapter 11 also has some new material on variance component estimation. Since nested and split-plot designs often involve random effects, and since they are of increasing importance in some industrial settings, they have received more emphasis in this edition of the book (see Chapter 12). The chapter on regression models (Chapter 13) focuses more sharply on the uses of regression analysis in experimental design problems. Chapter 14, on response surface methodology and other techniques for optimizing processes, contains some new response surface topics, evolutionary operation, as well as the discussion and critique of the Taguchi approach to parameter design. Since the response surface framework is the best way to implement Taguchi's philosophy, this chapter is the ideal place to present this material.

Throughout the book I have stressed the importance of experimental design as a tool for practicing engineers to use for product design and development as well as process development and improvement. The use of experimental design in developing products that are robust to environmental factors and other sources of variability is illustrated. I believe that the use of experimental design early in the product cycle can substantially reduce development lead time and cost, leading to processes and products that perform better in the field and have higher reliability than those developed using other approaches.

The book contains more material than can be covered comfortably in one course, and I hope that instructors will be able to either vary the content of each course offering or discuss some topics in greater depth, depending on class interest. There are problem sets at the end of each chapter (except Chapter 1). These problems vary in scope from computational exercises, designed to reinforce the fundamentals, to extensions or elaboration of basic principles.

My own course focuses extensively on factorial and fractional factorial designs, so I usually cover Chapter 1, Chapter 2 (very quickly), most of Chapter 3, Chapter 5 (excluding the material on incomplete blocks and only mentioning Latin squares briefly), and I discuss Chapters 6–9 on factorials and two-level factorial and fractional factorial designs in detail. To conclude the course, I introduce the concept of random effects models in Chapter 11, discuss nested and split-plot designs, and overview response surface methods, usually with an example. I always require the students to complete a term project that involves designing, conducting, and presenting the results of a statistically designed experiment. I require them to do this in teams, as this is the way that much industrial experimentation is conducted. They must present the results of this project, both orally and in written form.

ACKNOWLEDGMENTS

I express my appreciation to the many students, instructors, and colleagues who have used the three earlier editions of this book and who have made helpful suggestions for its revision. The contributions of Dr. Raymond H. Myers, Dr. G. Geoffrey Vining, Dr. Dennis Lin, Dr. John Ramberg, Dr. Joseph Pignatiello, Dr. Lloyd S. Nelson, Dr. Andre Khuri, Dr. Peter Nelson, Dr. John A. Cornell, Dr. Pat Spagon, Dr. William DuMouche, Dr. Bert Keats, Dr. Dwayne Rollier, Dr. Norma Hubele, Dr. Cynthia Lowry, Dr. Stephen R. Schmidt, Dr. Russell G. Heikes, Dr. Harrison M. Wadsworth, Dr. William W. Hines, Dr. Arvind Shah, Dr. Jane Ammons, Dr. Diane Schaub, and Mr. Pat Whitcomb were particularly valuable.

The contributions of the professional practitioners with whom I have worked have been invaluable. It is impossible to mention everyone, but some of the major contributors include Mr. Dan McCarville and Ms. Lisa Custer of Motorola; Mr. Tom Bingham, Mr. Dick Vaughn, Dr. Julian Anderson, Mr. Richard Alkire, and Mr. Chase Neilson of the Boeing Company; Mr. Mike Goza, Mr. Don Walton, Ms. Karen Madison, Mr. Jeff Stevens, and Mr. Bob Kohm of Alcoa; Dr. Jay Gardiner, Mr. John Butora, Mr. Dana Lesher, Mr. Lolly Marwah, Dr. Paul Tobias, and Mr. Leon Mason of IBM; Ms. Elizabeth A. Peck of the Coca-Cola Company; Dr. Sadri Khalessi and Mr. Franz Wagner of Signetics; Mr. Robert V. Baxley of Monsanto Chemicals; Mr. Harry Peterson-Nedry and Dr. Russell Boyles of Precision Castparts Corporation; Mr. Bill New and Mr. Randy Schmid of Allied-Signal Aerospace; Mr. John M. Fluke, Jr. of the John Fluke Manufacturing Company; Mr. Larry Newton and Mr. Kip Howlett of Georgia-Pacific; and Dr. Ernesto Ramos of BBN Software Products Corporation.

I am also indebted to Professor E. S. Pearson and the *Biometrika* Trustees, John Wiley & Sons, Prentice-Hall, The American Statistical Association, The Institute of Mathematical Statistics, and the editors of *Biometrics* for permission to use copyrighted material. I am grateful to the Office of Naval Research, the National Science Foundation, and the IBM Corporation for supporting much of my research in engineering statistics and experimental design.

Douglas C. Montgomery
Tempe, Arizona

Contents

Chapter 1
Introduction

1-1 STRATEGY OF EXPERIMENTATION

Experiments are performed by investigators in virtually all fields of inquiry, usually to discover something about a particular process or system. Literally, an experiment is a **test.** More formally, we can define an **experiment** as a test or series of tests in which purposeful changes are made to the input variables of a process or system so that we may observe and identify the reasons for changes in the output response.

This book is about planning and conducting experiments and about analyzing the resulting data so that valid and objective conclusions are obtained. Our focus is on experiments in the engineering, physical, and chemical sciences. In engineering, experimentation plays an important role in new product design, manufacturing process development, and process improvement. The objective in many cases may be to develop a **robust** process, that is, a process affected minimally by external sources of variability.

As an example of an experiment, suppose that a metallurgical engineer is interested in studying the effect of two different hardening processes, oil quenching and saltwater quenching, on an aluminum alloy. Here the objective of the experimenter is to determine which quenching solution produces the maximum hardness for this particular alloy. The engineer decides to subject a number of alloy specimens to each quenching medium and measure the hardness of the specimens after quenching. The average hardness of the specimens treated in each quenching solution will be used to determine which solution is best.

As we consider this simple experiment, a number of important questions come to mind:

1. Are these two solutions the only quenching media of potential interest?
2. Are there any other factors that might affect hardness that should be investigated or controlled in this experiment?
3. How many specimens of alloy should be tested in each quenching solution?

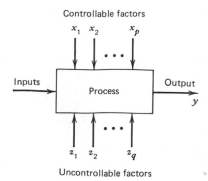

Controllable factors

Figure 1–1.   General model of a process or system.

4. How should the specimens be assigned to the quenching solutions, and in what order should the data be collected?

5. What method of data analysis should be used?

6. What difference in average observed hardness between the two quenching media will be considered important?

All of these questions, and perhaps many others, will have to be answered satisfactorily before the experiment is performed.

In any experiment, the results and conclusions that can be drawn depend to a large extent on the manner in which the data were collected. To illustrate this point, suppose that the metallurgical engineer in the above experiment used specimens from one heat in the oil quench and specimens from a second heat in the saltwater quench. Now, when the mean hardness is compared, the engineer is unable to say how much of the observed difference is the result of the quenching media and how much is the result of inherent differences between the heats.[1] Thus, the method of data collection has adversely affected the conclusions that can be drawn from the experiment.

In general, experiments are used to study the performance of processes and systems. The process or system can be represented by the model shown in Figure 1–1. We can usually visualize the process as a combination of machines, methods, people, and other resources that transforms some input (often a material) into an output that has one or more observable responses. Some of the process variables $x_1, x_2, \ldots, x_p$ are controllable, whereas other variables $z_1, z_2, \ldots, z_q$ are uncontrollable (although they may be controllable for purposes of a test). The objectives of the experiment may include the following:

1. Determining which variables are most influential on the response y.

2. Determining where to set the influential x's so that y is almost always near the desired nominal value.

---

[1] A specialist in experimental design would say that the effects of quenching media and heat were *confounded;* that is, the effects of these two factors cannot be separated.

3. Determining where to set the influential x's so that variability in y is small.
4. Determining where to set the influential x's so that the effects of the uncontrollable variables $z_1, z_2, \ldots, z_q$ are minimized.

As you can see from the foregoing discussion, experiments often involve several factors. Usually, an objective of the person conducting the experiment, called the **experimenter,** is to determine the influence that these factors have on the output response of the system. The general approach to planning and conducting the experiment is called the **strategy of experimentation.** There are several strategies that an experimenter could use. We will illustrate some of these with a very simple example.

The author really likes to play golf. Unfortunately, he does not enjoy practicing, so he is always looking for a simpler solution to lowering his score. Some of the factors that he thinks may be important, or that may influence his golf score, are as follows:

1. The type of driver used (oversized or regular-sized).
2. The type of ball used (balata or three-piece).
3. Walking and carrying the golf clubs or riding in a golf cart.
4. Drinking water or drinking beer while playing.
5. Playing in the morning or playing in the afternoon.

There are obviously many other factors that could be considered, but let's assume that these are the ones of primary interest. Furthermore, based on long experience with the game, he decides that factor 5 can be ignored; that is, this factor is not important because its effect is so small that it has no practical value. Engineers and scientists often must make these types of decisions about some of the factors they are considering in real experiments.

Now, let's consider how factors 1 through 4 could be experimentally tested to determine their effect on the author's golf score. Suppose that a maximum of 8 rounds of golf can be played over the course of the experiment. One approach would be to select an arbitrary combination of these factors, test them, and see what happens. For example, suppose the oversized driver, balata ball, golf cart, and water combination is selected, and the resulting score is 87. During the round, however, the author noticed several wayward shots with the big driver (long is not always good in golf), and, as a result, he decides to play another round with the regular-sized driver, holding the other factors at the same levels used previously. This approach could be continued almost indefinitely, switching the levels of one (or perhaps two) factors for the next test, based on the outcome of the current test. This strategy of experimentation, which we call the **best-guess approach,** is used frequently in practice by engineers and scientists. It often works reasonably well, too, because the experimenters often have a great deal of technical or theoretical knowledge of the system they are studying, as well as

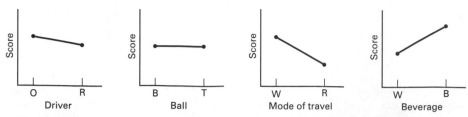

Figure 1–2.   Results of the one-factor-at-a-time strategy for the golf experiment.

considerable practical experience. There are, however, at least two disadvantages of the best-guess approach. First, suppose the initial "best-guess" does not produce the desired results. Now the experimenter has to take another "guess" at the correct combination of factor levels. This could continue for a long time, without any guarantee of success. Second, suppose the initial "best-guess" produces an acceptable result. Now the experimenter is tempted to stop testing, although there is no guarantee that the *best* solution has been found.

Another strategy of experimentation that is used extensively in practice is the **one-factor-at-a-time** approach. This method consists of selecting a starting point or **baseline** set of levels for each factor, then successively varying each factor over its range with the other factors held constant at the baseline level. After all tests are performed, a series of graphs are usually constructed showing how the response variable is affected by varying each factor with all other factors held constant. Figure 1–2 shows a set of these graphs for the golf experiment, using the oversized driver, balata ball, walking, and drinking water levels of the four factors as the baseline. The interpretation of this graph is straightforward; for example, since the slope of the mode of travel curve is negative, this would conclude that riding improves the score. Using these one-factor-at-a-time graphs, we would select the optimal combination to be the regular-sized driver, riding, and drinking water. The type of golf ball seems unimportant.

The major disadvantage of the one-factor-at-a-time strategy is that it fails to consider any possible **interaction** between the factors. An interaction is the failure of one factor to produce the same effect on the response at different levels of another factor. Figure 1–3 shows an interaction between the type of

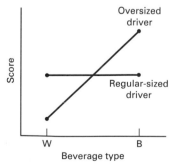

Figure 1–3.   Interaction between type of driver and type of beverage for the golf experiment.

driver and the beverage factors for the golf experiment. Notice that if the author uses the regular-sized driver, the type of beverage consumed has virtually no effect on his score, but if he uses the oversized driver, much better results are obtained by drinking water instead of beer. Interactions between factors are very common, and if they occur, the one-factor-at-a-time strategy will usually produce poor results. Many people do not recognize this, and, consequently, one-factor-at-a-time experiments are run frequently in practice. (Some individuals actually think that this strategy is related to the scientific method or that it is a "sound" engineering principle.) One-factor-at-a-time experiments are always less efficient than other methods based on a statistical approach to design. We will discuss this in more detail in Chapter 6.

The correct approach to dealing with several factors is to conduct a **factorial** experiment. This is an experimental strategy in which factors are varied *together,* instead of one at a time. The factorial experimental design concept is extremely important, and several chapters in this book are devoted to presenting basic factorial experiments and a number of useful variations and special cases.

To illustrate how a factorial experiment is conducted, consider the golf experiment and suppose that only two factors, type of driver and type of ball, are of interest. Figure 1–4 shows a two-factor factorial experiment for studying the joint effects of these two factors on the author's golf score. Notice that this factorial experiment has both factors at two levels and that all possible combinations of the two factors across their levels are used in the design. Geometrically, the four runs form the corners of a square. This particular type of factorial experiment is called a 2^2 **factorial design** (two factors, each at two levels). Since the author can reasonably expect to play eight rounds of golf to investigate these factors, a reasonable plan would be to play two rounds of golf at each combination of factor levels shown in Figure 1–4. An experimental designer would say that we have **replicated** the design twice. This experimental design would enable the experimenter to investigate the individual effects of each factor (or the **main** effects) and to determine whether the factors interact.

We can extend this concept to three factors. Suppose that the author wishes to study the effects of type of driver, type of ball, and the type of beverage consumed on his golf score. Assuming that all three factors have two levels, a

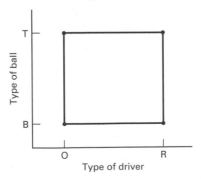

Figure 1–4. A two-factor factorial experiment involving type of driver and type of ball.

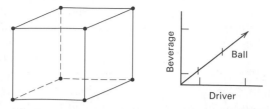

Figure 1–5.   A three-factor factorial experiment involving type of driver, type of ball, and type of beverage.

factorial design can be set up as shown in Figure 1–5. Notice that there are eight combinations of these three factors across the two levels of each and that these eight trials can be represented geometrically as the corners of a cube. This is an example of a **2^3 factorial design.** Since the author only wants to play eight rounds of golf, this experiment would require that one round be played at each combination of factors represented by the eight corners of the cube in Figure 1–5. However, if we compare this to the two-factor factorial in Figure 1–4, the 2^3 factorial design would provide the same information about the factor effects. For example, in both designs there are four tests that provide information about the regular-sized driver and four tests that provide information about the over-sized driver, assuming that each run in the two-factor design in Figure 1–4 is replicated twice.

Figure 1–6 illustrates how all four factors—driver, ball, beverage, and mode of travel (walking or riding)—could be investigated in a **2^4 factorial design.** As in any factorial design, all possible combinations of the levels of the factors are used. Since all four factors are at two levels, this experimental design can still be represented geometrically as a cube (actually a hypercube).

Generally, if there are k factors, each at two levels, the factorial design would require 2^k runs. For example, the experiment in Figure 1–6 requires 16 runs. Clearly, as the number of factors of interest increases, the number of runs required increases rapidly; for instance, a ten-factor experiment with all factors at two levels would require 1024 runs. This quickly becomes infeasible from a time and resource viewpoint. In the golf experiment, the author can only play eight rounds of golf, so even the experiment in Figure 1–6 is too large.

Fortunately, if there are four to five or more factors, it is usually unnecessary to run all possible combinations of factor levels. A **fractional factorial experiment**

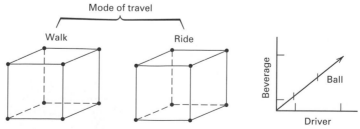

Figure 1–6.   A four-factor factorial experiment involving type of driver, type of ball, type of beverage, and mode of travel.

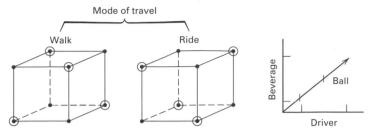

Figure 1–7. A four-factor fractional factorial experiment involving type of driver, type of ball, type of beverage, and mode of travel.

is a variation of the basic factorial design in which only a subset of the runs are made. Figure 1–7 shows a fractional factorial design for the four-factor version of the golf experiment. This design requires only 8 runs instead of the original 16 and would be called a **one-half fraction.** If the author can only play eight rounds of golf, this is an excellent design in which to study all four factors. It will provide good information about the main effects of the four factors as well as some information about how these factors interact.

Fractional factorial designs are used extensively in industrial research and development, and for process improvement. These designs will be discussed in Chapter 9.

1–2 SOME TYPICAL APPLICATIONS OF EXPERIMENTAL DESIGN

Experimental design methods have found broad application in many disciplines. In fact, we may view experimentation as part of the scientific process and as one of the ways we learn about how systems or processes work. Generally, we learn through a series of activities in which we make conjectures about a process, perform experiments to generate data from the process, and then use the information from the experiment to establish new conjectures, which lead to new experiments, and so on.

Experimental design is a critically important tool in the engineering world for improving the performance of a manufacturing process. It also has extensive application in the development of new processes. The application of experimental design techniques early in process development can result in

1. Improved process yields.
2. Reduced variability and closer conformance to nominal or target requirements.
3. Reduced development time.
4. Reduced overall costs.

Experimental design methods also play a major role in **engineering design** activities, where new products are developed and existing ones improved. Some applications of experimental design in engineering design include

1. Evaluation and comparison of basic design configurations.
2. Evaluation of material alternatives.
3. Selection of design parameters so that the product will work well under a wide variety of field conditions, that is, so that the product is **robust.**
4. Determination of key product design parameters that impact product performance.

The use of experimental design in these areas can result in products that are easier to manufacture, products that have enhanced field performance and reliability, lower product cost, and shorter product design and development time. We now present several examples that illustrate some of these ideas.

Example 1–1

Characterizing a Process

A flow solder machine is used in the manufacturing process for printed circuit boards. The machine cleans the boards in a flux, preheats the boards, and then moves them along a conveyor through a wave of molten solder. This solder process makes the electrical and mechanical connections for the leaded components on the board.

The process currently operates around the one percent defective level. That is, about one percent of the solder joints on a board are defective and require manual retouching. However, since the average printed circuit board contains over 2000 solder joints, even a one percent defective level results in far too many solder joints requiring rework. The process engineer responsible for this area would like to use a designed experiment to determine which machine parameters are influential in the occurrence of solder defects and which adjustments should be made to those variables to reduce solder defects.

The flow solder machine has several variables that can be controlled. They include

1. Solder temperature.
2. Preheat temperature.
3. Conveyor speed.
4. Flux type.
5. Flux specific gravity.
6. Solder wave depth.
7. Conveyor angle.

In addition to these controllable factors, there are several other factors that cannot be easily controlled during routine manufacturing, although they could be controlled for the purposes of a test. They are

1. Thickness of the printed circuit board.
2. Types of components used on the board.
3. Layout of the components on the board.
4. Operator.
5. Production rate.

In this situation, the engineer is interested in **characterizing** the flow solder machine; that is, he wants to determine which factors (both controllable and uncontrollable) affect the occurrence of defects on the printed circuit boards. To accomplish this, he can design an experiment that will enable him to estimate the magnitude and direction of the factor effects; that is, how much does the response variable (defects per unit) change when each factor is changed, and does changing the factors *together* produce different results than are obtained from individual factor adjustments, that is, do the factors interact. Sometimes we call an experiment such as this a **screening experiment.** Typically, screening or characterization experiments involve using fractional factorial designs, such as in the golf example in Figure 1–7.

The information from this screening or characterization experiment will be used to identify the critical process factors and to determine the direction of adjustment for these factors to reduce further the number of defects per unit. The experiment may also provide information about which factors should be more carefully controlled during routine manufacturing to prevent high defect levels and erratic process performance. Thus, one result of the experiment could be the application of techniques such as control charts to one or more **process variables** (such as solder temperature), in addition to control charts on process output. Over time, if the process is improved enough, it may be possible to base most of the process control plan on controlling process input variables instead of control charting the output.

∎

Example 1–2

Optimizing a Process

In a characterization experiment, we are usually interested in determining which process variables affect the response. A logical next step is to optimize, that is, to determine the region in the important factors that leads to the best possible response. For example, if the response is yield, we would look for a region of maximum yield, whereas if the response is variability in a critical product dimension, we would seek a region of minimum variability.

Suppose that we are interested in improving the yield of a chemical process. We know from the results of a characterization experiment that the two most important process variables that influence the yield are operating temperature and reaction time. The process currently runs at 155°F and 1.7 hours of reaction time, producing yields of around 75 percent. Figure 1–8 on page 10 shows a view of the time–temperature region from above. In this graph, the lines of constant yield are connected to form response **contours,** and we have shown the contour lines for yields

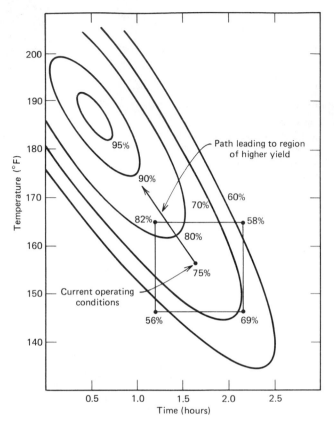

Figure 1–8. Contour plot of yield as a function of reaction time and reaction temperature, illustrating an optimization experiment.

of 60 percent, 70 percent, 80 percent, 90 percent, and 95 percent. These contours are projections on the time–temperature region of cross sections of the yield surface corresponding to the aforementioned percent yields. This surface is sometimes called a **response surface.** The true response surface in Figure 1–8 is unknown to the process personnel, so experimental methods will be required to optimize the yield with respect to time and temperature.

To locate the optimum, it is necessary to perform an experiment that varies time and temperature together, that is, a factorial experiment. The results of a factorial experiment with both time and temperature run at two levels is shown in Figure 1–8. The responses observed at the four corners of the square indicate that we should move in the general direction of increased temperature and decreased reaction time to increase yield. A few additional runs could be performed in this direction, and this additional experimentation would be sufficient to locate the region of maximum yield.

■

Example 1–3

A Product Design Example

Experimental design methods can often be applied in the product design process. To illustrate, suppose that a group of engineers are designing a door hinge for an automobile. The quality characteristic of interest is the check effort, or the holding ability of the door latch that prevents the door from swinging closed when the vehicle is parked on a hill. The check mechanism consists of a leaf spring and a roller. When the door is opened, the roller travels through an arc causing the leaf spring to be compressed. To close the door, the spring must be forced aside, and this creates the check effort. The engineering team thinks that check effort is a function of the following factors:

1. Roller travel distance.
2. Spring height from pivot to base.
3. Horizontal distance from pivot to spring.
4. Free height of the reinforcement spring.
5. Free height of the main spring.

The engineers can build a prototype hinge mechanism in which all of these factors can be varied over certain ranges. Once appropriate levels for these five factors have been identified, an experiment can be designed consisting of various combinations of the factor levels, and the prototype hinge can be tested at these combinations. This will produce information concerning which factors are most influential on the latch check effort, and through analysis of this information, the latch design can be improved.

∎

Example 1–4

Determining System and Component Tolerances

The Wheatstone bridge shown in Figure 1–9 is a device used for measuring an unknown resistance, Y. The adjustable resistor B is manipulated until a particular

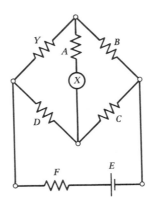

Figure 1–9.   A Wheatstone bridge.

current flow is obtained through the ammeter (usually $X = 0$), and then the unknown resistance is calculated as

$$Y = \frac{BD}{C} - \frac{X^2}{C^2E}[A(D + C) + D(B + C)][B(C + D) + F(B + C)] \qquad (1\text{–}1)$$

The engineer wants to design the circuit so that overall gage capability is good; that is, he would like for the standard deviation of the measurement error to be small. He has decided that $A = 20\ \Omega$, $C = 2\ \Omega$, $D = 50\ \Omega$, $E = 1.5$ V, and $F = 2\ \Omega$ are the best choices for the design parameters so far as gage capability is concerned, but the overall measurement error is still too high. This is likely due to the tolerances that have been specified on the circuit components. These tolerances are ±1 percent for each resistor (A, B, C, D, and F) and ±5 percent for the power supply E. These tolerance bands can be used to define high and low factor levels, and an experiment can be performed to determine which circuit components have the most critical tolerances and by how much these critical tolerances must be tightened to produce adequate gage capability. This will result in a design specification that tightens only the most critical tolerances the minimum amount possible consistent with desired measurement capability; consequently, a lower-cost design that is easier to manufacture will result.

Notice that in this experiment it is unnecessary to actually build the hardware since the response from the circuit can be calculated via Equation 1–1. The actual response variable for the experiment should be the standard deviation of Y. However, an equation for the standard deviation of Y for the circuit can be found by using a Taylor series expansion of Equation 1–1, as explained in many engineering statistics textbooks [for example, see Hines and Montgomery (1990), Chapter 5]. Therefore, the entire experiment can be performed using a computer model of the Wheatstone bridge.

■

1–3   BASIC PRINCIPLES

If an experiment such as the ones described in Examples 1–1 through 1–4 is to be performed most efficiently, then a scientific approach to planning the experiment must be employed. By the **statistical design of experiments,** we refer to the process of planning the experiment so that appropriate data that can be analyzed by statistical methods will be collected, resulting in valid and objective conclusions. The statistical approach to experimental design is necessary if we wish to draw meaningful conclusions from the data. When the problem involves data that are subject to experimental errors, statistical methodology is the only objective approach to analysis. Thus, there are two aspects to any experimental problem: the design of the experiment and the statistical analysis of the data. These two subjects are closely related since the method of analysis depends directly on the design employed. Both topics will be addressed in this book.

The three basic principles of experimental design are **replication, randomization,** and **blocking.** By replication we mean a repetition of the basic experiment.

In the metallurgical experiment discussed in Section 1–1, replication would consist of treating a specimen by oil quenching and treating a specimen by saltwater quenching. Thus, if five specimens are treated in each quenching medium, we say that five **replicates** have been obtained. Replication has two important properties. First, it allows the experimenter to obtain an estimate of the experimental error. This estimate of error becomes a basic unit of measurement for determining whether observed differences in the data are really *statistically* different. Second, if the sample mean (e.g., $\bar{y}$) is used to estimate the effect of a factor in the experiment, then replication permits the experimenter to obtain a more precise estimate of this effect. For example; if σ^2 is the variance of an individual observation and there are n replicates, then the variance of the sample mean is

$$\sigma_{\bar{y}}^2 = \frac{\sigma^2}{n}$$

The practical implication of this is that if we had $n = 1$ replicates and observed $y_1 = 145$ (oil quench) and $y_2 = 147$ (saltwater quench), we would probably be unable to make satisfactory inferences about the effect of the quenching medium—that is, the observed difference could be the result of experimental error. On the other hand, if n was reasonably large, and the experimental error was sufficiently small, then if we observed $\bar{y}_1 < \bar{y}_2$, we would be reasonably safe in concluding that saltwater quenching produces a higher hardness in this particular aluminum alloy than does oil quenching.

Randomization is the cornerstone underlying the use of statistical methods in experimental design. By randomization we mean that both the allocation of the experimental material and the order in which the individual runs or trials of the experiment are to be performed are randomly determined. Statistical methods require that the observations (or errors) be independently distributed random variables. Randomization usually makes this assumption valid. By properly randomizing the experiment, we also assist in "averaging out" the effects of extraneous factors that may be present. For example, suppose that the specimens in the above experiment are of slightly different thicknesses and that the effectiveness of the quenching medium may be affected by specimen thickness. If all the specimens subjected to the oil quench are thicker than those subjected to the saltwater quench, then we may be continually handicapping one quenching medium over the other. Randomly assigning the specimens to the quenching media alleviates this problem.

Blocking is a technique used to increase the precision of an experiment. A block is a portion of the experimental material that should be more homogeneous than the entire set of material. Blocking involves making comparisons among the conditions of interest in the experiment within each block. A simple example of the blocking principle is given in Section 2–5.1 of Chapter 2.

These basic principles of experimental design are an important part of every experiment. We will illustrate and emphasize them repeatedly throughout this book.

1-4   GUIDELINES FOR DESIGNING EXPERIMENTS

To use the statistical approach in designing and analyzing an experiment, it is necessary that everyone involved in the experiment have a clear idea in advance of exactly what is to be studied, how the data are to be collected, and at least a qualitative understanding of how these data are to be analyzed. An outline of the recommended procedure is shown in Table 1–1. We now give a brief discussion of this outline and elaborate on some of the key points. For more details, see Coleman and Montgomery (1993), and the references therein.

1. Recognition of and statement of the problem. This may seem to be a rather obvious point, but in practice it is often not simple to realize that a problem requiring experimentation exists, nor is it simple to develop a clear and generally accepted statement of this problem. It is necessary to develop all ideas about the objectives of the experiment. Usually, it is important to solicit input from all concerned parties: engineering, quality assurance, manufacturing, marketing, management, the customer, and operating personnel (who usually have much insight and who are too often ignored). A clear statement of the problem often contributes substantially to a better understanding of the phenomena and the final solution of the problem. For this reason, a **team approach** to designing experiments is recommended.

2. Choice of factors, levels, and ranges. (As noted in Table 1–1, steps 2 and 3 are often done simultaneously, or in the reverse order.) The experimenter must choose the factors to be varied in the experiment, the ranges over which these factors will be varied, and the specific levels at which runs will be made. Thought must also be given to how these factors are to be controlled at the desired values and how they are to be measured. For instance, in the flow solder experiment, the engineer has defined 12 variables that may affect the occurrence of solder defects. The engineer will also have to decide on a region of interest for each variable (that is, the range over which each factor will be varied) and on how many levels of each variable to use. Process knowledge is required to do this.

Table 1–1   Guidelines for Designing an Experiment

1. Recognition of and statement of the problem
2. Choice of factors, levels, and ranges[a]
3. Selection of the response variable[a]
4. Choice of experimental design
5. Performing the experiment
6. Statistical analysis of the data
7. Conclusions and recommendations

[a] In practice, steps 2 and 3 are often done simultaneously, or in reverse order.

This process knowledge is usually a combination of practical experience and theoretical understanding. It is important to investigate all factors that may be of importance and to not be overly influenced by past experience, particularly when we are in the early stages of experimentation or when the process is not very mature.

When the objective of the experiment is factor screening or process characterization, it is usually best to keep the number of factor levels low. Generally, two levels work very well in factor screening studies. Choosing the region of interest is also important. In factor screening, the region of interest should be relatively large—that is, the range over which the factors are varied should be broad. As we learn more about which variables are important and which levels produce the best results, the region of interest will usually become more narrow.

3. Selection of the response variable. In selecting the response variable, the experimenter should be certain that this variable really provides useful information about the process under study. Most often, the average or standard deviation (or both) of the measured characteristic will be the response variable. Multiple responses are not unusual. Gauge capability (or measurement error) is also an important factor. If gauge capability is inadequate, then only relatively large factor effects will be detected by the experiment or perhaps additional replication will be required. In some situations where gauge capability is poor, the experimenter may decide to measure each experimental unit several times and use the average of the repeated measurements as the observed response. It is usually critically important to identify issues related to defining the responses of interest and how they are to be measured *before* conducting the experiment.

We reiterate how crucial it is to bring out all points of view and process information in steps 1 through 3 above. We refer to this as **pre-experimental planning.** It is unlikely that one person has all the knowledge required to do this adequately in many situations. Therefore, we argue strongly for a team effort in planning the experiment. Most of your success will hinge on how well the pre-experimental planning is done.

4. Choice of experimental design. If the pre-experimental planning activities above are done correctly, this step is relatively easy. Choice of design involves the consideration of sample size (number of replicates), the selection of a suitable run order for the experimental trials, and the determination of whether or not blocking or other randomization restrictions are involved. This book discusses some of the more important types of experimental designs, and it can ultimately be used as a catalog for selecting an appropriate experimental design for a wide variety of problems.

There are also several interactive statistical software packages that support this phase of experimental design. The experimenter can enter information about the number of factors, levels, and ranges, and these programs will either present a selection of designs for consideration or recommend a particular design. (We prefer to see several alternatives instead of relying on a computer recommendation in most cases.) These programs will usually also provide a worksheet (with the order of the runs randomized) for use in conducting the experiment.

In selecting the design, it is important to keep the experimental objectives in mind. In many engineering experiments, we already know at the outset that some of the factor levels will result in different values for the response. Consequently, we are interested in identifying *which* factors cause this difference and in estimating the *magnitude* of the response change. In other situations, we may be more interested in verifying uniformity. For example, two production conditions *A* and *B* may be compared, *A* being the standard and *B* being a more cost-effective alternative. The experimenter will then be interested in demonstrating that, say, there is no difference in yield between the two conditions.

5. *Performing the experiment.* When running the experiment, it is vital to monitor the process carefully to ensure that everything is being done according to plan. Errors in experimental procedure at this stage will usually destroy experimental validity. Up-front planning is crucial to success. It is easy to underestimate the logistical and planning aspects of running a designed experiment in a complex manufacturing or research and development environment.

6. *Statistical analysis of the data.* Statistical methods should be used to analyze the data so that results and conclusions are *objective* rather than judgmental in nature. If the experiment has been designed correctly and if it has been performed according to the design, then the statistical methods required are not elaborate. There are many excellent software packages designed to assist in data analysis, and many of the programs used in step 4 to select the design provide a seamless, direct interface to the statistical analysis. Often we find that simple graphical methods play an important role in data analysis and interpretation. Residual analysis and model adequacy checking are also important analysis techniques. We will discuss these issues in detail later.

Remember that statistical methods cannot prove that a factor (or factors) has a particular effect. They only provide guidelines as to the reliability and validity of results. Properly applied, statistical methods do not allow anything to be proved, experimentally, but they do allow us to measure the likely error in a conclusion or to attach a level of confidence to a statement. The primary advantage of statistical methods is that they add objectivity to the decision-making process. Statistical techniques coupled with good engineering or process knowledge and common sense will usually lead to sound conclusions.

7. *Conclusions and recommendations.* Once the data have been analyzed, the experimenter must draw *practical* conclusions about the results and recommend a course of action. Graphical methods are often useful in this stage, particularly in presenting the results to others. **Follow-up runs** and **confirmation testing** should also be performed to validate the conclusions from the experiment.

Throughout this entire process, it is important to keep in mind that experimentation is an important part of the learning process, where we tentatively formulate hypotheses about a system, perform experiments to investigate these hypotheses, and on the basis of the results formulate new hypotheses, and so on. This suggests that experimentation is **iterative.** It is usually a major mistake to design a single, large, comprehensive experiment at the start of a study. A successful experiment requires knowledge of the important factors, the ranges over which these factors should be varied, the appropriate number of levels to use, and the proper units of measurement for these variables. Generally, we do not perfectly know the answers to these questions, but we learn about them as we go along. As an experimental program progresses, we often drop some input variables, add others, change the region of exploration for some factors, or add new response variables. Consequently, we usually experiment *sequentially,* and as a general rule, no more than about 25 percent of the available resources should be invested in the first experiment. This will ensure that sufficient resources are available to perform confirmation runs and ultimately accomplish the final objective of the experiment.

1–5 HISTORICAL PERSPECTIVE

The late Sir Ronald A. Fisher was the innovator in the use of statistical methods in experimental design. For several years he was responsible for statistics and data analysis at the Rothamsted Agricultural Experiment Station in London, England. Fisher developed and first used the analysis of variance as the primary method of statistical analysis in experimental design. In 1933, Fisher took a professorship at the University of London. He later was on the faculty of Cambridge University and held visiting professorships at several universities throughout the world. For an excellent biography of Fisher, see J. F. Box (1978). While Fisher was clearly the pioneer, there have been many other significant contributors to the literature of experimental design, including F. Yates, R. C. Bose, O. Kempthorne, W. G. Cochran, R. H. Myers, J. S. Hunter, W. G. Hunter, and G. E. P. Box. The bibliography at the end of the book contains several works by these authors.

Many of the early applications of experimental design methods were in the agricultural and biological sciences, and as a result, much of the terminology of the field is derived from this heritage. However, the first industrial applications of experimental design began to appear in the 1930s, initially in the British textile and woolen industry. After World War II, experimental design methods were introduced to the chemical and process industries in the United States and Western Europe. These industry groups are still very fertile areas for using experimental design in product and process development. The semiconductor and electronics industry has also used experimental design methods for many years with considerable success.

In recent years, there has been a revival of interest in experimental design in the United States because many industries discovered that their off-shore competitors have been using designed experiments for many years and that this has been an important factor in their competitive success. The day is approaching (hopefully rapidly) when all engineers will receive formal training in experimental design as part of their undergraduate education. The successful integration of experimental design into the engineering profession is a key factor in the future competitiveness of the industrial base of the United States.

1–6 SUMMARY: USING STATISTICAL TECHNIQUES IN EXPERIMENTATION

Much of the research in engineering, science, and industry is empirical and makes extensive use of experimentation. Statistical methods can greatly increase the efficiency of these experiments and often strengthens the conclusions so obtained. The proper use of statistical techniques in experimentation requires that the experimenter keep the following points in mind:

1. Use your nonstatistical knowledge of the problem. Experimenters are usually highly knowledgeable in their fields. For example, a civil engineer working on a problem in hydrology typically has considerable practical experience and formal academic training in this area. In some fields there is a large body of physical theory on which to draw in explaining relationships between factors and responses. This type of nonstatistical knowledge is invaluable in choosing factors, determining factor levels, deciding how many replicates to run, interpreting the results of the analysis, and so forth. Using statistics is no substitute for thinking about the problem.

2. Keep the design and analysis as simple as possible. Don't be overzealous in the use of complex, sophisticated statistical techniques. Relatively simple design and analysis methods are almost always best. This is a good place to reemphasize step 4 of the procedure recommended in Section 1–4. If you do the design carefully and correctly, the analysis will almost always be relatively straightforward. However, if you botch the design badly, it is unlikely that even the most complex and elegant statistics can save the situation.

3. Recognize the difference between practical and statistical significance. Just because two experimental conditions produce mean responses that are statistically different, there is no assurance that this difference is large enough to have any practical value. For example, an engineer may determine that a modification to an automobile fuel injection system may produce a true mean improvement in gasoline mileage of 0.1 mi/gal. This is a statistically significant result. However, if the cost of the modification is $1000, then the 0.1 mi/gal difference is probably too small to be of any practical value.

4. Experiments are usually iterative. Remember that, in most situations, it is unwise to design too comprehensive an experiment at the start of a study. Successful design requires knowledge of the important factors, the ranges over which these factors are varied, the appropriate number of levels for each factor, and the proper units of measurement for each factor and response. Generally, we are not well-equipped to answer these questions at the beginning of the experiment, but we learn the answers as we go along. This argues in favor of the *iterative* or *sequential* approach discussed previously. Of course, there are situations where comprehensive experiments are entirely appropriate, but as a general rule, most experiments should be iterative. Consequently, we usually should not invest more than about 25 percent of the resources of experimentation (runs, budget, time, etc.) in the initial design. Often these first efforts are just learning experiences, and some resources must be available to accomplish the final objectives of the experiment.

Chapter 2
Simple Comparative Experiments

In this chapter, we consider experiments to compare two **conditions** (sometimes called **treatments**). These are often called **simple comparative experiments.** We begin with an example of an experiment performed to determine whether two different formulations of a product give equivalent results. The discussion leads to a review of several basic statistical concepts, such as random variables, probability distributions, random samples, sampling distributions, and tests of hypotheses.

2–1 INTRODUCTION

The tension bond strength of portland cement mortar is an important characteristic of the product. An engineer is interested in comparing the strength of a modified formulation in which polymer latex emulsions have been added during mixing to the strength of the unmodified mortar. The experimenter has collected 10 observations on strength for the modified formulation and another 10 observations for the unmodified formulation. The data are shown in Table 2–1. We could refer to the two different formulations as two **treatments** or as two **levels** of the **factor** formulations.

The data from this experiment are plotted in Figure 2–1. This display is called a **dot diagram.** Visual examination of these data give the immediate impression that the strength of the unmodified mortar is greater than the strength of the modified mortar. This impression is supported by comparing the *average* tension bond strengths, $\bar{y}_1 = 16.76$ kgf/cm^2 for the modified mortar and $\bar{y}_2 = 17.92$ kgf/cm^2 for the unmodified mortar. The average tension bond strengths in these two samples differ by what seems to be a nontrivial amount. However, it is not obvious that this difference is large enough to imply that the two formulations really *are* different. Perhaps this observed difference in average strengths is the result of sampling fluctuation and the two formulations are really identical.

20

Table 2–1 Tension Bond Strength
Data for the Portland Cement
Formulation Experiment

j	Modified Mortar y_{1j}	Unmodified Mortar y_{2j}
1	16.85	17.50
2	16.40	17.63
3	17.21	18.25
4	16.35	18.00
5	16.52	17.86
6	17.04	17.75
7	16.96	18.22
8	17.15	17.90
9	16.59	17.96
10	16.57	18.15

Possibly another two samples would give opposite results, with the strength of the modified mortar exceeding that of the unmodified formulation.

A technique of statistical inference called **hypothesis testing** (some prefer **significance testing**) can be used to assist the experimenter in comparing these two formulations. Hypothesis testing allows the comparison of the two formulations to be made on *objective* terms, with a knowledge of the risks associated with reaching the wrong conclusion. Before presenting procedures for hypothesis testing in simple comparative experiments, we will briefly review some elementary statistical concepts.

2–2 BASIC STATISTICAL CONCEPTS

Each of the observations in the portland cement experiment described above would be called a **run.** Notice that the individual runs differ, so there is fluctuation

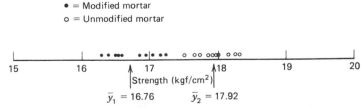

Figure 2–1. Dot diagram for the tension bond strength data in Table 2–1.

or **noise** in the results. This noise is usually called **experimental error** or simply **error.** It is a **statistical error,** meaning that it arises from variation that is uncontrolled and generally unavoidable. The presence of error or noise implies that the response variable, tension bond strength, is a **random variable.** A random variable may be either **discrete** or **continuous.** If the set of all possible values of the random variable is either finite or countably infinite, then the random variable is discrete, whereas if the set of all possible values of the random variable is an interval, then the random variable is continuous.

Graphical Description of Variability ▪ We often use simple graphical methods to assist in analyzing the data from an experiment. The **dot diagram,** illustrated in Figure 2–1, is a very useful device for displaying a small body of data (say up to about 20 observations). The dot diagram enables the experimenter to see quickly the general **location** or **central tendency** of the observations and their **spread.** For example, in the portland cement tension bond experiment, the dot diagram reveals that the two formulations probably differ in mean strength but that both formulations produce about the same variation in strength.

If the data are fairly numerous, then the dots in a dot diagram become difficult to distinguish, and a histogram may be preferable. Figure 2–2 presents a histogram for 200 observations on the metal recovery or yield from a smelting process. The histogram shows the central tendency, spread, and general shape of the distribution of the data. Recall that a histogram is constructed by dividing the horizontal axis into intervals (usually of equal length) and drawing a rectangle over the jth interval with the area of the rectangle proportional to n_j, the number of observations that fall in that interval.

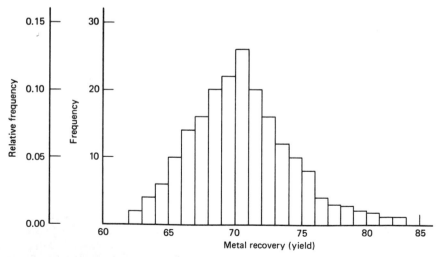

Figure 2–2. Histogram for 200 observations on metal recovery (yield) from a smelting process.

The **box plot** (or **box and whisker plot**) is a very useful way to display data. A box plot displays the minimum, the maximum, the lower and upper quartiles (the 25th percentile and the 75th percentiles, respectively), and the median (the 50th percentile) on a rectangular box aligned either horizontally or vertically. The box extends from the lower quartile to the upper quartile, and a line is drawn through the box at the median. Lines (or whiskers) extend from the ends of the box to (typically) the minimum and maximum values.

Figure 2–3 presents the box plots for the two samples of tension bond strength in the portland cement mortar experiment. This display clearly reveals the difference in mean strength between the two formulations. It also indicates that both formulations produce reasonably symmetric distributions of strength with similar variability or spread.

Dot diagrams, histograms, and box plots are useful for summarizing the information in a **sample** of data. To describe the observations that might occur in a sample more completely, we use the concept of the probability distribution.

Probability Distributions ▪ The probability structure of a random variable, say y, is described by its **probability distribution.** If y is discrete, we often call the probability distribution of y, say $p(y)$, the probability function of y. If y is continuous, the probability distribution of y, say $f(y)$, is often called the probability density function for y.

Figure 2–4 (see page 24) illustrates hypothetical discrete and continuous probability distributions. Notice that in the discrete probability distribution, it is the height of the function $p(y_j)$ that represents probability, whereas in the continuous case, it is the area under the curve $f(y)$ associated with a given interval that represents probability. The properties of probability distributions

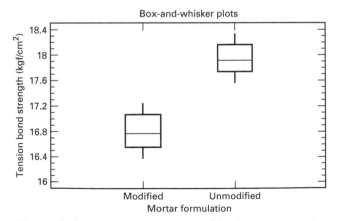

Figure 2–3.   Box plots for the portland cement tension bond strength experiment.

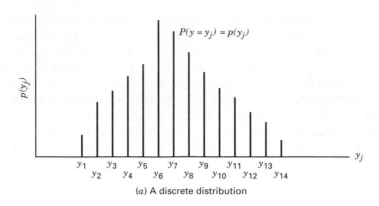

(a) A discrete distribution

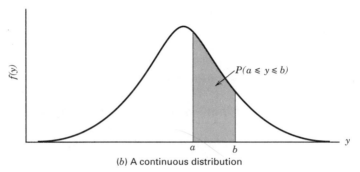

(b) A continuous distribution

Figure 2–4.   Discrete and continuous probability distributions.

may be summarized quantitatively as follows:

y discrete:	$0 \leqslant p(y_j) \leqslant 1$	all values of y_j

$$P(y = y_j) = p(y_j) \qquad \text{all values of } y_j$$

$$\sum_{\substack{\text{all values} \\ \text{of } y_j}} p(y_j) = 1$$

y continuous: $\qquad 0 \leqslant f(y)$

$$P(a \leqslant y \leqslant b) = \int_a^b f(y)\, dy$$

$$\int_{-\infty}^{\infty} f(y)\, dy = 1$$

Mean, Variance, and Expected Values ▪ The **mean** of a probability distribution is a measure of its central tendency or location. Mathematically, we define the mean (e.g., μ) as

$$\mu = \begin{cases} \displaystyle\int_{-\infty}^{\infty} y f(y)\, dy & y \text{ continuous} \\[2ex] \displaystyle\sum_{\text{all } y} y p(y) & y \text{ discrete} \end{cases} \qquad (2\text{–}1)$$

We may also express the mean in terms of the **expected value** or the long-run average value of the random variable y as

$$\mu = E(y) = \begin{cases} \int_{-\infty}^{\infty} y f(y) \, dy & y \text{ continuous} \\ \sum_{\text{all } y} y p(y) & y \text{ discrete} \end{cases} \tag{2-2}$$

where E denotes the **expected value operator.**

The spread or dispersion of a probability distribution can be measured by the **variance,** defined as

$$\sigma^2 = \begin{cases} \int_{-\infty}^{\infty} (y - \mu)^2 f(y) \, dy & y \text{ continuous} \\ \sum_{\text{all } y} (y - \mu)^2 p(y) & y \text{ discrete} \end{cases} \tag{2-3}$$

Note that the variance can be expressed entirely in terms of expectation since

$$\sigma^2 = E[(y - \mu)^2] \tag{2-4}$$

Finally, the variance is used so extensively that it is convenient to define a **variance operator** V such that

$$V(y) \equiv E[(y - \mu)^2] = \sigma^2 \tag{2-5}$$

The concepts of expected value and variance are used extensively throughout this book, and it may be helpful to review several elementary results concerning these operators. If y is a random variable with mean μ and variance σ^2 and c is a constant, then

1. $E(c) = c$
2. $E(y) = \mu$
3. $E(cy) = cE(y) = c\mu$
4. $V(c) = 0$
5. $V(y) = \sigma^2$
6. $V(cy) = c^2 V(y) = c^2 \sigma^2$

If there are two random variables, for example, y_1 with $E(y_1) = \mu_1$ and $V(y_1) = \sigma_1^2$ and y_2 with $E(y_2) = \mu_2$ and $V(y_2) = \sigma_2^2$, then we have

7. $E(y_1 + y_2) = E(y_1) + E(y_2) = \mu_1 + \mu_2$

It is possible to show that

8. $V(y_1 + y_2) = V(y_1) + V(y_2) + 2 \text{Cov}(y_1, y_2)$

where

$$\text{Cov}(y_1, y_2) = E[(y_1 - \mu_1)(y_2 - \mu_2)] \tag{2-6}$$

is the **covariance** of the random variables y_1 and y_2. The covariance is a measure of the linear association between y_1 and y_2. More specifically, we may show that if y_1 and y_2 are independent,[1] then $\text{Cov}(y_1, y_2) = 0$. We may also show that

9. $V(y_1 - y_2) = V(y_1) + V(y_2) - 2\,\text{Cov}(y_1, y_2)$

If y_1 and y_2 are **independent,** then we have

10. $V(y_1 \pm y_2) = V(y_1) + V(y_2) = \sigma_1^2 + \sigma_2^2$

and

11. $E(y_1 \cdot y_2) = E(y_1) \cdot E(y_2) = \mu_1 \cdot \mu_2$

However, note that, in general,

12. $E\left(\dfrac{y_1}{y_2}\right) \neq \dfrac{E(y_1)}{E(y_2)}$

regardless of whether or not y_1 and y_2 are independent.

2-3   SAMPLING AND SAMPLING DISTRIBUTIONS

Random Samples, Sample Mean, and Sample Variance ▪ The objective of statistical inference is to draw conclusions about a population using a sample from that population. Most of the methods that we will study assume that random samples are used. That is, if the population contains N elements and a sample of n of them is to be selected, then if each of the $N!/(N - n)!n!$ possible samples has an equal probability of being chosen, the procedure employed is called **random sampling.** In practice, it is sometimes difficult to obtain random samples, and tables of random numbers, such as Table XI in the Appendix, may be helpful.

Statistical inference makes considerable use of quantities computed from

---

[1] Note that the converse of this is not necessarily so; that is, we may have $\text{Cov}(y_1, y_2) = 0$ and yet this does not imply independence. For an example, see Hines and Montgomery (1990, pp. 128–129).

the observations in the sample. We define a **statistic** as any function of the observations in a sample that does not contain unknown parameters. For example, suppose that $y_1, y_2, \ldots, y_n$ represents a sample. Then the **sample mean**

$$\bar{y} = \frac{\sum\limits_{i=1}^{n} y_i}{n} \tag{2-7}$$

and the **sample variance**

$$S^2 = \frac{\sum\limits_{i=1}^{n} (y_i - \bar{y})^2}{n - 1} \tag{2-8}$$

are both statistics. These quantities are measures of the central tendency and dispersion of the sample, respectively. Sometimes $S = \sqrt{S^2}$, called the **sample standard deviation,** is used as a measure of dispersion. Engineers often prefer to use the standard deviation to measure dispersion because its units are the same as those for the variable of interest y.

Properties of the Sample Mean and Variance ▪ The sample mean $\bar{y}$ is a point estimator of the population mean μ, and the sample variance S^2 is a point estimator of the population variance σ^2. In general, an **estimator** of an unknown parameter is a statistic that corresponds to that parameter. Note that a point estimator is a random variable. A particular numerical value of an estimator, computed from sample data, is called an **estimate.** For example, suppose we wish to estimate the mean and variance of the breaking strength of a particular type of textile fiber. A random sample of $n = 25$ fiber specimens is tested and the breaking strength is recorded for each. The sample mean and variance are computed according to Equations 2-7 and 2-8, respectively, and are $\bar{y} = 18.6$ and $S^2 = 1.20$. Therefore, the estimate of μ is $\bar{y} = 18.6$ and the estimate of σ^2 is $S^2 = 1.20$.

There are several properties required of good point estimators. Two of the most important are the following:

1. The point estimator should be **unbiased.** That is, the long-run average or expected value of the point estimator should be the parameter that is being estimated. Although unbiasedness is desirable, this property alone does not always make an estimator a good one.

2. An unbiased estimator should have **minimum variance.** This property states that the minimum variance point estimator has a variance that is smaller than the variance of any other estimator of that parameter.

We may easily show that $\bar{y}$ and S^2 are unbiased estimators of μ and σ^2, respectively. First consider $\bar{y}$. Using the properties of expectation, we have

$$E(\bar{y}) = E\left(\frac{\sum_{i=1}^{n} y_i}{n}\right)$$

$$= \frac{1}{n} E\left(\sum_{i=1}^{n} y_i\right)$$

$$= \frac{1}{n} \sum_{i=1}^{n} E(y_i)$$

$$= \frac{1}{n} \sum_{i=1}^{n} \mu$$

$$= \mu$$

since the expected value of each observation y_i is μ. Thus, $\bar{y}$ is an unbiased estimator of μ.

Now consider the sample variance S^2. We have

$$E(S^2) = E\left[\frac{\sum_{i=1}^{n}(y_i - \bar{y})^2}{n-1}\right]$$

$$= \frac{1}{n-1} E\left[\sum_{i=1}^{n}(y_i - \bar{y})^2\right]$$

$$= \frac{1}{n-1} E(SS)$$

where $SS = \sum_{i=1}^{n}(y_i - \bar{y})^2$ is the **corrected sum of squares** of the observations y_i. Now

$$E(SS) = E\left[\sum_{i=1}^{n}(y_i - \bar{y})^2\right] \qquad (2\text{-}9)$$

$$= E\left[\sum_{i=1}^{n} y_i^2 - n\bar{y}^2\right]$$

$$= \sum_{i=1}^{n}(\mu^2 + \sigma^2) - n(\mu^2 + \sigma^2/n)$$

$$= (n-1)\sigma^2 \qquad (2\text{-}10)$$

Therefore,

$$E(S^2) = \frac{1}{n-1} E(SS)$$

$$= \sigma^2$$

and we see that S^2 is an unbiased estimator of σ^2.

Degrees of Freedom ▪ The quantity $n - 1$ in Equation 2–10 is called the **number of degrees of freedom** of the sum of squares SS. This is a very general result; that is, if y is a random variable with variance σ^2 and $SS = \Sigma(y_i - \bar{y})^2$ has ν degrees of freedom, then

$$E\left(\frac{SS}{\nu}\right) = \sigma^2 \tag{2–11}$$

The number of degrees of freedom of a sum of squares is equal to the number of independent elements in that sum of squares. For example, $SS = \Sigma_{i=1}^{n}(y_i - \bar{y})^2$ in Equation 2–9 consists of the sum of squares of the n elements $y_1 - \bar{y}$, $y_2 - \bar{y}, \ldots, y_n - \bar{y}$. These elements are not all independent since $\Sigma_{i=1}^{n}(y_i - \bar{y}) = 0$; in fact, only $n - 1$ of them are independent, implying that SS has $n - 1$ degrees of freedom.

The Normal and Other Sampling Distributions ▪ Often we are able to determine the probability distribution of a particular statistic if we know the probability distribution of the population from which the sample was drawn. The probability distribution of a statistic is called a **sampling distribution.** We will now briefly discuss several useful sampling distributions.

One of the most important sampling distributions is the **normal distribution.** If y is a normal random variable, then the probability distribution of y is

$$f(y) = \frac{1}{\sigma\sqrt{2\pi}}e^{-(1/2)[(y-\mu)/\sigma]^2} \qquad -\infty < y < \infty \tag{2–12}$$

where $-\infty < \mu < \infty$ is the mean of the distribution and $\sigma^2 > 0$ is the variance. The normal distribution is shown in Figure 2–5.

Because sample runs that differ as a result of experimental error often are well described by the normal distribution, the normal plays a central role in the analysis of data from designed experiments. Many important sampling distributions may also be defined in terms of normal random variables. We often use the notation $y \sim N(\mu, \sigma^2)$ to denote that y is distributed normally with mean μ and variance σ^2.

An important special case of the normal distribution is the **standard normal**

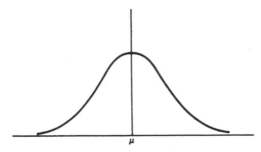

Figure 2–5. The normal distribution.

distribution; that is, $\mu = 0$ and $\sigma^2 = 1$. We see that if $y \sim N(\mu, \sigma^2)$, then the random variable

$$z = \frac{y - \mu}{\sigma} \tag{2-13}$$

follows the standard normal distribution, denoted $z \sim N(0, 1)$. The operation demonstrated in Equation 2–13 is often called **standardizing** the normal random variable y. A table of the cumulative standard normal distribution is given in Appendix Table I.

Many statistical techniques assume that the random variable is normally distributed. The central limit theorem is often a justification of approximate normality.

THEOREM 2–1   THE CENTRAL LIMIT THEOREM   If $y_1, y_2, \ldots, y_n$ is a sequence of n independent and identically distributed random variables with $E(y_i) = \mu$ and $V(y_i) = \sigma^2$ (both finite) and $x = y_1 + y_2 + \cdots + y_n$, then

$$z_n = \frac{x - n\mu}{\sqrt{n\sigma^2}}$$

has an approximate $N(0, 1)$ distribution in the sense that, if $F_n(z)$ is the distribution function of z_n and $\Phi(z)$ is the distribution function of the $N(0, 1)$ random variable, then $\lim_{n \to \infty} [F_n(z)/\Phi(z)] = 1$.

■ ■

This result states essentially that the sum of n independent and identically distributed random variables is approximately normally distributed. In many cases this approximation is good for very small n, say $n < 10$, whereas in other cases large n is required, say $n > 100$. Frequently, we think of the error in an experiment as arising in an additive manner from several independent sources; consequently, the normal distribution becomes a plausible model for the combined experimental error.

An important sampling distribution that can be defined in terms of normal random variables is the **chi-square** or χ^2 **distribution.** If $z_1, z_2, \ldots, z_k$ are normally and independently distributed random variables with mean 0 and variance 1, abbreviated NID(0, 1), then the random variable

$$x = z_1^2 + z_2^2 + \cdots + z_k^2$$

follows the chi-square distribution with k degrees of freedom. The density function of chi-square is

$$f(x) = \frac{1}{2^{k/2}\Gamma\left(\dfrac{k}{2}\right)} x^{(k/2)-1}e^{-x/2} \qquad x > 0 \tag{2-14}$$

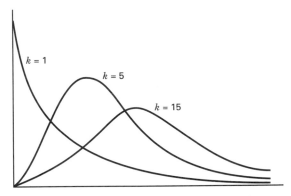

Figure 2–6.   Several chi-square distributions.

Several chi-square distributions are shown in Figure 2–6. The distribution is asymmetric or **skewed**, with mean and variance

$$\mu = k$$
$$\sigma^2 = 2k$$

respectively. A table of percentage points of the chi-square distribution is given in Table III of the Appendix.

As an example of a random variable that follows the chi-square distribution, suppose that $y_1, y_2, \ldots, y_n$ is a random sample from an $N(\mu, \sigma^2)$ distribution. Then

$$\frac{SS}{\sigma^2} = \frac{\sum_{i=1}^{n} (y_i - \bar{y})^2}{\sigma^2} \sim \chi^2_{n-1} \qquad (2\text{–}15)$$

That is, SS/σ^2 is distributed as chi-square with $n - 1$ degrees of freedom.

Many of the techniques used in this book involve the computation and manipulation of sums of squares. The result given in Equation 2–15 is extremely important and occurs repeatedly; a sum of squares in normal random variables when divided by σ^2 follows the chi-square distribution.

Examining Equation 2–8, we see that the sample variance can be written as

$$S^2 = \frac{SS}{n - 1} \qquad (2\text{–}16)$$

If the observations in the sample are $NID(\mu, \sigma^2)$, then the distribution of S^2 is $[\sigma^2/(n - 1)]\chi^2_{n-1}$. Thus, the sampling distribution of the sample variance is a constant times the chi-square distribution if the population is normally distributed.

If z and χ^2_k are independent standard normal and chi-square random variables, respectively, then the random variable

$$t_k = \frac{z}{\sqrt{\chi^2_k/k}} \qquad (2\text{–}17)$$

follows the *t* **distribution with** *k* **degrees of freedom,** denoted t_k. The density function of *t* is

$$f(t) = \frac{\Gamma[(k+1)/2]}{\sqrt{k\pi}\,\Gamma(k/2)} \frac{1}{[(t^2/k) + 1]^{(k+1)/2}} \qquad -\infty < t < \infty \qquad (2\text{–}18)$$

and the mean and variance of *t* are $\mu = 0$ and $\sigma^2 = k/(k-2)$ for $k > 2$, respectively. Several *t* distributions are shown in Figure 2–7. Note that if $k = \infty$, the *t* distribution becomes the standard normal distribution. A table of percentage points of the *t* distribution is given in Table II of the Appendix. If $y_1, y_2, \ldots, y_n$ is a random sample from the $N(\mu, \sigma^2)$ distribution, then the quantity

$$t = \frac{\bar{y} - \mu}{S/\sqrt{n}} \qquad (2\text{–}19)$$

is distributed as *t* with $n - 1$ degrees of freedom.

The final sampling distribution that we will consider is the *F* **distribution.** If χ_u^2 and χ_v^2 are two independent chi-square random variables with *u* and *v* degrees of freedom, respectively, then the ratio

$$F_{u,v} = \frac{\chi_u^2/u}{\chi_v^2/v} \qquad (2\text{–}20)$$

follows the *F* distribution with *u numerator* degrees of freedom and *v denominator* degrees of freedom. If *x* is an *F* random variable with $\Gamma(u/2)$ numerator and *v* denominator degrees of freedom, then the probability distribution of *x* is

$$h(x) = \frac{\Gamma\left(\dfrac{u+v}{2}\right)\left(\dfrac{u}{v}\right)^{u/2} x^{(u/2)-1}}{\Gamma\left(\dfrac{u}{2}\right)\Gamma\left(\dfrac{v}{2}\right)\left[\left(\dfrac{u}{v}\right)x + 1\right]^{(u+v)/2}} \qquad 0 < x < \infty \qquad (2\text{–}21)$$

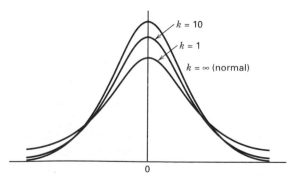

Figure 2–7. Several *t* distributions.

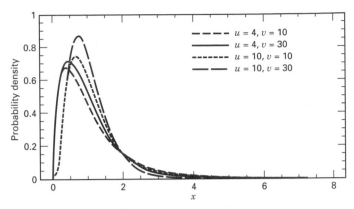

Figure 2–8. Several F distributions.

Several F distributions are shown in Figure 2–8. This distribution is very important in the statistical analysis of designed experiments. A table of percentage points of the F distribution is given in Table IV of the Appendix.

As an example of a statistic that is distributed as F, suppose we have two independent normal populations with common variance σ^2. If y_{11}, y_{12}, . . . , y_{1n_1} is a random sample of n_1 observations from the first population, and if y_{21}, y_{22}, . . . , y_{2n_2} is a random sample of n_2 observations from the second, then

$$\frac{S_1^2}{S_2^2} \sim F_{n_1-1,\,n_2-1} \tag{2-22}$$

where S_1^2 and S_2^2 are the two sample variances. This result follows directly from Equations 2–15 and 2–20.

2–4 INFERENCES ABOUT THE DIFFERENCES IN MEANS, RANDOMIZED DESIGNS

We are now ready to return to the portland cement mortar problem posed in Section 2–1. Recall that two different formulations of mortar were being investigated to determine if they differ in tension bond strength. In this section we discuss how the data from this simple comparative experiment can be analyzed using **hypothesis testing** and **confidence interval** procedures for comparing two treatment means.

Throughout this section we assume that a **completely randomized experimental design** is used. In such a design, the data are viewed as if they were a random sample from a normal distribution.

2–4.1 Hypothesis Testing

A **statistical hypothesis** is a statement about the parameters of a probability distribution. For example, in the portland cement experiment, we may think that the mean tension bond strengths of the two mortar formulations are equal. This may be stated formally as

$$H_0: \mu_1 = \mu_2$$
$$H_1: \mu_1 \neq \mu_2$$

where μ_1 is the mean tension bond strength of the modified mortar and μ_2 is the mean tension bond strength of the unmodified mortar. The statement $H_0: \mu_1 = \mu_2$ is called the **null hypothesis** and $H_1: \mu_1 \neq \mu_2$ is called the **alternative hypothesis.** The alternative hypothesis specified here is called a **two-sided alternative hypothesis** since it would be true either if $\mu_1 < \mu_2$ or if $\mu_1 > \mu_2$.

To test a hypothesis we devise a procedure for taking a random sample, computing an appropriate **test statistic,** and then rejecting or failing to reject the null hypothesis H_0. Part of this procedure is specifying the set of values for the test statistic that leads to rejection of H_0. This set of values is called the **critical region** or **rejection region** for the test.

Two kinds of errors may be committed when testing hypotheses. If the null hypothesis is rejected when it is true, then a type I error has occurred. If the null hypothesis is *not* rejected when it is false, then a type II error has been made. The probabilities of these two errors are given special symbols:

$$\alpha = P(\text{type I error}) = P(\text{reject } H_0 | H_0 \text{ is true})$$
$$\beta = P(\text{type II error}) = P(\text{fail to reject } H_0 | H_0 \text{ is false})$$

Sometimes it is more convenient to work with the **power** of the test, where

$$\text{Power} = 1 - \beta = P(\text{reject } H_0 | H_0 \text{ is false})$$

The general procedure in hypothesis testing is to specify a value of the probability of type I error α, often called the **significance level** of the test, and then design the test procedure so that the probability of type II error β has a suitably small value.

Suppose that we could assume that the variances of tension bond strengths were identical for both mortar formulations. Then the appropriate test statistic to use for comparing two treatment means in the completely randomized design is

$$t_0 = \frac{\bar{y}_1 - \bar{y}_2}{S_p \sqrt{\dfrac{1}{n_1} + \dfrac{1}{n_2}}} \tag{2–23}$$

where $\bar{y}_1$ and $\bar{y}_2$ are the sample means, n_1 and n_2 are the sample sizes, S_p^2 is an estimate of the common variance $\sigma_1^2 = \sigma_2^2 = \sigma^2$ computed from

$$S_p^2 = \frac{(n_1 - 1)S_1^2 + (n_2 - 1)S_2^2}{n_1 + n_2 - 2} \tag{2-24}$$

and S_1^2 and S_2^2 are the two individual sample variances. To determine whether to reject $H_0 : \mu_1 = \mu_2$, we would compare t_0 to the t distribution with $n_1 + n_2 - 2$ degrees of freedom. If $|t_0| > t_{\alpha/2, n_1+n_2-2}$, where $t_{\alpha/2, n_1+n_2-2}$ is the upper $\alpha/2$ percentage point of the t distribution with $n_1 + n_2 - 2$ degrees of freedom, we would *reject H_0* and conclude that the mean strengths of the two formulations of portland cement mortar differ.

This procedure may be justified as follows. If we are sampling from independent normal distributions, then the distribution of $\bar{y}_1 - \bar{y}_2$ is $N[\mu_1 - \mu_2, \sigma^2(1/n_1 + 1/n_2)]$. Thus, if σ^2 were known, and if $H_0 : \mu_1 = \mu_2$ were true, the distribution of

$$Z_0 = \frac{\bar{y}_1 - \bar{y}_2}{\sigma \sqrt{\dfrac{1}{n_1} + \dfrac{1}{n_2}}} \tag{2-25}$$

would be $N(0, 1)$. However, in replacing σ in Equation 2–25 by S_p, the distribution of Z_0 changes from standard normal to t with $n_1 + n_2 - 2$ degrees of freedom. Now if H_0 is true, t_0 in Equation 2–23 is distributed as $t_{n_1+n_2-2}$ and, consequently, we would expect $100(1 - \alpha)$ percent of the values of t_0 to fall between $-t_{\alpha/2, n_1+n_2-2}$ and $t_{\alpha/2, n_1+n_2-2}$. A sample producing a value of t_0 outside these limits would be unusual if the null hypothesis were true and is evidence that H_0 should be rejected. Thus the t distribution with $n_1 + n_2 - 2$ degrees of freedom is the appropriate **reference distribution** for the test statistic t_0. That is, it describes the behavior of t_0 when the null hypothesis is true. Note that α is the probability of type I error for the test.

In some problems, one may wish to reject H_0 only if one mean is larger than the other. Thus, one would specify a **one-sided alternative hypothesis** $H_1 : \mu_1 > \mu_2$ and would reject H_0 only if $t_0 > t_{\alpha, n_1+n_2-2}$. If one wants to reject H_0 only if μ_1 is less than μ_2, then the alternative hypothesis is $H_1 : \mu_1 < \mu_2$, and one would reject H_0 if $t_0 < -t_{\alpha, n_1+n_2-2}$.

To illustrate the procedure, consider the portland cement data in Table 2–1. For these data, we find that

Modified Mortar	Unmodified Mortar
$\bar{y}_1 = 16.76 \text{ kgf/cm}^2$	$\bar{y}_2 = 17.92 \text{ kgf/cm}^2$
$S_1^2 = 0.100$	$S_2^2 = 0.061$
$S_1 = 0.316$	$S_2 = 0.247$
$n_1 = 10$	$n_2 = 10$

Since the sample standard deviations are reasonably similar, it is not unreasonable to conclude that the population standard deviations (or variances) are equal. Therefore, we can use Equation 2–23 to test the hypotheses

$$H_0 : \mu_1 = \mu_2$$
$$H_1 : \mu_1 \neq \mu_2$$

Furthermore, $n_1 + n_2 - 2 = 10 + 10 - 2 = 18$, and if we choose $\alpha = 0.05$, then we would reject $H_0 : \mu_1 = \mu_2$ if the numerical value of the test statistic $t_0 > t_{0.025,18} = 2.101$, or if $t_0 < -t_{0.025,18} = -2.101$. These boundaries of the critical region are shown on the reference distribution (t with 18 degrees of freedom) in Figure 2–9.

Using Equation 2–24 we find that

$$S_p^2 = \frac{(n_1 - 1)S_1^2 + (n_2 - 1)S_2^2}{n_1 + n_2 - 2}$$
$$= \frac{9(0.100) + 9(0.061)}{10 + 10 - 2}$$
$$= 0.081$$
$$S_p = 0.284$$

and the test statistic is

$$t_0 = \frac{\bar{y}_1 - \bar{y}_2}{S_p \sqrt{\dfrac{1}{n_1} + \dfrac{1}{n_2}}}$$
$$= \frac{16.76 - 17.92}{0.284 \sqrt{\frac{1}{10} + \frac{1}{10}}}$$
$$= -9.13$$

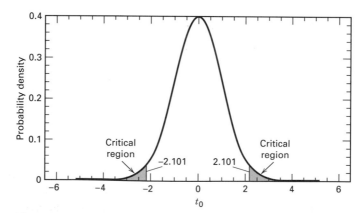

Figure 2–9.   The t distribution with 18 degrees of freedom with the critical region $\pm t_{0.025,18} = \pm 2.101$.

Since $t_0 = -9.13 < -t_{0.025,18} = -2.101$, we would reject H_0 and conclude that the mean tension bond strengths of the two formulations of portland cement mortar are different.

The Use of *P*-Values in Hypothesis Testing ▪ One way to report the results of a hypothesis test is to state that the null hypothesis was or was not rejected at a specified α-value or **level of significance.** For example, in the portland cement mortar formulation above, we can say that $H_0 : \mu_1 = \mu_2$ was rejected at the 0.05 level of significance. This statement of conclusions is often inadequate, because it gives the decision maker no idea about whether the computed value of the test statistic was just barely in the rejection region or whether it was very far into this region. Furthermore, stating the results this way imposes the predefined level of significance on other users of the information. This approach may be unsatisfactory, as some decision makers might be uncomfortable with the risks implied by $\alpha = 0.05$.

To avoid these difficulties, the ***P*-value approach** has been adopted widely in practice. The *P*-value is the probability that the test statistic will take on a value that is at least as extreme as the observed value of the statistic when the null hypothesis H_0 is true. Thus, a *P*-value conveys much information about the weight of evidence against H_0, and so a decision maker can draw a conclusion at *any* specified level of significance. More formally, we define the ***P*-value** as the smallest level of significance that would lead to rejection of the null hypothesis H_0.

It is customary to call the test statistic (and the data) significant when the null hypothesis H_0 is rejected; therefore, we may think of the *P*-value as the smallest level α at which the data are significant. Once the *P*-value is known, the decision maker can determine for himself or herself how significant the data are without the data analyst formally imposing a preselected level of significance.

It is not always easy to compute the exact *P*-value for a test. However, most modern computer programs for statistical analysis report *P*-values, and they can be obtained on some handheld calculators. We will show how to approximate the *P*-value for the portland cement mortar experiment. From Appendix Table II, for a t distribution with 18 degrees of freedom, the smallest tail area probability is 0.0005, for which $t_{0.0005,18} = 3.922$. Now $|t_0| = 9.13 > 3.922$, so since the alternative hypothesis is two-sided, we know that the *P*-value must be less than $2(0.0005) = 0.001$. Some handheld calculators have the capability to calculate *P*-values. One such calculator is the HP-48. From this calculator, we obtain the *P*-value for the value $t_0 = -9.13$ in the portland cement mortar formulation experiment as $P = 3.68 \times 10^{-8}$. Thus the null hypothesis $H_0 : \mu_1 = \mu_2$ would be rejected at any level of significance $\alpha \geq 3.68 \times 10^{-8}$.

Computer Solution ▪ There are many statistical software packages that have capability for statistical hypothesis testing. The output from Statgraphics Two-Sample Analysis applied to the portland cement mortar formulation experiment is shown in Table 2–2. Notice that the output includes some summary statistics about the two samples as well as some information about confidence intervals

Table 2–2   Statgraphics Two-Sample Analysis for the Portland Cement Mortar Experiment

		Sample 1	Sample 2	Pooled
Sample Statistics:	Number of Obs.	10	10	20
	Average	16.764	17.922	17.343
	Variance	0.100138	0.0614622	0.0808
	Std. Deviation	0.316446	0.247916	0.284253
	Median	16.72	17.93	17.355

Difference between Means = −1.158
Conf. Interval For Diff. in Means: 95   Percent
  (Equal Vars.)   Sample 1 − Sample 2   −1.42514 −0.890861   18 D.F.
  (Unequal Vars.)   Sample 1 − Sample 2   −1.42624 −0.88976   17.0 D.F.

Ratio of Variances = 1.62926
Conf. Interval for Ratio of Variances: 95   Percent
                   Sample 1 − Sample 2   0.404684 6.55938   9   9 D.F.

Hypothesis Test for H0: Diff = 0   Computed t statistic = −9.10936
                     vs Alt: NE   Sig. Level = 3.67808E−8
                 at Alpha = 0.05   so reject H0.

(which we will discuss in Sections 2–4.3 and 2–6). The program also tests the hypothesis of interest, allowing the analyst to specify the difference in the two means of interest in H_0 (in this case 0), the nature of the alternative hypothesis ("NE" implies $H_1: \mu_1 \neq \mu_2$), and the choice of α (here $\alpha = 0.05$).

The output includes the computed value of t_0, the P-value (called the significance level), and the decision that should be made given the specified value of α. Notice that $H_0: \mu_1 = \mu_2$ should be rejected in favor of $H_1: \mu_1 \neq \mu_2$, which is the identical conclusion we reached previously.

Checking Assumptions in the t Test ▪ In using the t test procedure we make the assumptions that both samples are drawn from independent populations that can be described by a normal distribution, that the standard deviation or variances of both populations are equal, and that the observations are independent random variables. The assumption of independence is critical, and if the run order is randomized (and, if appropriate, other experimental units and materials are selected at random), this assumption will be satisfied. The equal-variance and normality assumptions are easy to check using a **normal probability plot.**

Generally, probability plotting is a graphical technique for determining whether sample data conform to a hypothesized distribution based on a subjective visual examination of the data. The general procedure is very simple and can be performed quickly. Probability plotting typically uses special graph paper, known as **probability paper,** that has been designed for the hypothesized distribution. Probability paper is widely available for the normal distribution.

To construct a probability plot, the observations in the sample are first ranked from smallest to largest. That is, the sample $y_1, y_2, \ldots, y_n$ is arranged as $y_{(1)}$, $y_{(2)}$, . . . , $y_{(n)}$ where $y_{(1)}$ is the smallest observation, $y_{(2)}$ is the second smallest observation, and so forth, with $y_{(n)}$ the largest. The ordered observations $y_{(j)}$ are then plotted against their observed cumulative frequency $(j - 0.5)/n$ on the appropriate probability paper. If the hypothesized distribution adequately describes the data, the plotted points will fall approximately along a straight line; if the plotted points deviate significantly from a straight line, then the hypothesized model is not appropriate. Usually, the determination of whether or not the data plot as a straight line is subjective.

To illustrate the procedure, suppose that we wish to check the assumption that tension bond strength in the portland cement mortar formulation experiment is normally distributed. We will consider only the observations from the unmodified mortar formulation. We arrange the observations in ascending order and calculate their cumulative frequencies $(j - 0.5)/10$ as follows.

j	$y_{(j)}$	$(j - 0.5)/10$
1	17.50	0.05
2	17.63	0.15
3	17.75	0.25
4	17.86	0.35
5	17.90	0.45
6	17.96	0.55
7	18.00	0.65
8	18.15	0.75
9	18.22	0.85
10	18.25	0.95

The pairs of values $y_{(j)}$ and $(j - 0.5)/10$ are now plotted on normal probability paper. This plot is shown in Figure 2–10a. Most normal probability paper plots $100(j - 0.5)/n$ on the left vertical scale (and occasionally $100[1 - (j - 0.5)/n]$ is plotted on the right vertical scale), with the variable value plotted on the horizontal scale. A straight line, chosen subjectively, has been drawn through the plotted points. In drawing the straight line, you should be influenced more by the points near the middle of the plot than by the extreme points. A good rule of thumb is to draw the line approximately between the 25th and 75th percentile points. This is how the line in Figure 2–10a was determined. In assessing the "closeness" of the points to the straight line, imagine a "fat pencil" lying along the line. If all the points are covered by this imaginary pencil, then a normal distribution adequately describes the data. Since the points in Figure 2–10a would pass the "fat pencil" test, we conclude that the normal distribution is an appropriate model for tension bond strength for the unmodified mortar. Figure 2–10b presents the normal probability plot for the 10 observations on tension bond strength for the modified mortar. Once again, we would conclude that the assumption of a normal distribution is reasonable.

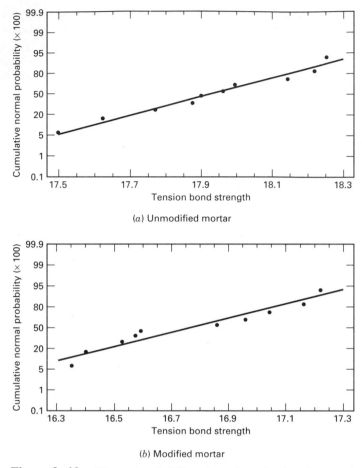

Figure 2–10.   Normal probability plots of tension bond strength in the portland cement experiment.

We can obtain an estimate of the mean and standard deviation directly from the normal probability plot. The mean is estimated as the 50th percentile on the probability plot, and the standard deviation is estimated as the difference between the 84th and 50th percentiles. This means that we can verify the assumption of equal population variances in the portland cement experiment by simply comparing the slopes of the two straight lines in Figure 2–10a and 2–10b. Both lines have very similar slopes, and so the assumption of equal variances is a reasonable one. If this assumption is violated, then you should use the version of the t test described in Section 2–4.4.

When assumptions are badly violated, the performance of the t test will be affected. Generally, small to moderate violations of assumptions are not a major concern, but *any* failure of the independence assumption and strong indications of nonnormality should not be ignored. Both the significance level of the test and the ability to detect differences between the means will be adversely affected by departures from assumptions. **Transformations** are one approach to dealing with this problem. We will discuss this in more detail in Chapter 3. Nonparametric hypothesis testing procedures can also be used if the observations come from nonnormal populations. Refer to Montgomery and Runger (1994) for more details.

An Alternate Justification to the t Test ▪ The two-sample t test we have just presented depends in theory on the underlying assumption that the two populations from which the samples were randomly selected are normal. Although the normality assumption is required to develop the test procedure formally, as we discussed above, moderate departures from normality will not seriously affect the results. It can be argued [e.g., see Box, Hunter, and Hunter (1978)] that the use of a randomized design enables one to test hypotheses without *any* assumptions regarding the form of the distribution. Briefly, the reasoning is as follows. If the treatments have no effect, then all $[20!/(10!10!)] = 184,756$ possible ways that the 20 observations could occur are equally likely. Corresponding to each of these 184,756 possible arrangements is a value of t_0. If the value of t_0 actually obtained from the data is unusually large or unusually small with reference to the set of 184,756 possible values, then it is an indication that $\mu_1 \neq \mu_2$.

This type of procedure is called a **randomization test.** It can be shown that the t test is a good approximation of the randomization test. Thus, we will use t tests (and other procedures that can be regarded as approximations of randomization tests) without extensive concern about the assumption of normality. This is one reason a simple procedure such as normal probability plotting is adequate to check the assumption of normality.

2-4.2 Choice of Sample Size

Selection of an appropriate sample size is one of the most important aspects of any experimental design problem. The choice of sample size and the probability of type II error β are closely connected. Suppose that we are testing the hypotheses

$$H_0 : \mu_1 = \mu_2$$
$$H_1 : \mu_1 \neq \mu_2$$

and that the means are *not* equal so that $\delta = \mu_1 - \mu_2$. Since $H_0 : \mu_1 = \mu_2$ is not true, we are concerned about wrongly failing to reject H_0. The probability of type II error depends on the true difference in means δ. A graph of β versus δ for a particular sample size is called the **operating characteristic curve** or **O.C. curve** for the test. The β error is also a function of sample size. Generally, for a given value of δ, the β error decreases as the sample size increases. That is, a specified difference in means is easier to detect for larger sample sizes than for smaller ones.

A set of operating characteristic curves for the hypotheses

$$H_0: \mu_1 = \mu_2$$
$$H_1: \mu_1 \neq \mu_2$$

for the case where the two population variances σ_1^2 and σ_2^2 are unknown but equal ($\sigma_1^2 = \sigma_2^2 = \sigma^2$) and for a level of significance of $\alpha = 0.05$ is shown in Figure 2–11. The curves also assume that the sample sizes from the two populations are equal; that is, $n_1 = n_2 = n$. The parameter on the horizontal axis in Figure 2–11 is

$$d = \frac{|\mu_1 - \mu_2|}{2\sigma} = \frac{|\delta|}{2\sigma}$$

Dividing $|\delta|$ by 2σ allows the experimenter to use the same set of curves, regardless of the value of the variance (the difference in means is expressed in standard deviation units). Furthermore, the sample size used to construct the curves is actually $n^* = 2n - 1$.

From examining these curves, we note the following:

1. The greater the difference in means, $\mu_1 - \mu_2$, the smaller the probability of type II error for a given sample size and α. That is, for a specified sample size and α, the test will detect large differences more easily than small ones.

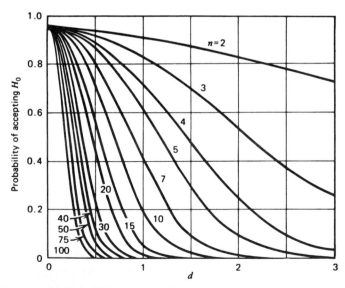

Figure 2–11. Operating characteristic curves for the two-sided t test with $\alpha = 0.05$. (Reproduced with permission from "Operating Characteristics for the Common Statistical Tests of Significance," C. L. Ferris, F. E. Grubbs, and C. L. Weaver, *Annals of Mathematical Statistics*, June 1946.)

2. As the sample size gets larger, the probability of type II error gets smaller for a given difference in means and α. That is, to detect a specified difference δ, we may make the test more powerful by increasing the sample size.

Operating characteristic curves are often helpful in selecting a sample size to use in an experiment. For example, consider the portland cement mortar problem discussed previously. Suppose that if the two formulations differ in mean strength by as much as 0.5 kgf/cm^2 we would like to detect it with a high probability. Thus, since $\mu_1 - \mu_2 = 0.5$ kgf/cm^2 is the "critical" difference in means we wish to detect, we find that d, the parameter on the horizontal axis of the operating characteristic curve in Figure 2–11, is

$$d = \frac{|\mu_1 - \mu_2|}{2\sigma} = \frac{0.5}{2\sigma} = \frac{0.25}{\sigma}$$

Unfortunately, d involves the unknown parameter σ. However, suppose that we think on the basis of prior experience that it is very unlikely that the standard deviation of any observation on strength would exceed 0.25 kgf/cm^2. Then using $\sigma = 0.25$ in the above expression for d yields $d = 1$. If we wish to reject the null hypothesis 95 percent of the time when $\mu_1 - \mu_2 = 0.5$, then $\beta = 0.05$, and Figure 2–11 with $\beta = 0.05$ and $d = 1$ yields $n^* = 16$, approximately. Therefore, since $n^* = 2n - 1$, the required sample size is

$$n = \frac{n^* + 1}{2} = \frac{16 + 1}{2} = 8.5 \approx 9$$

and we would use sample sizes of $n_1 = n_2 = n = 9$.

In our example, the experimenter actually used a sample size of 10. Perhaps the experimenter elected to increase the sample size slightly to guard against the possibility that the prior estimate of the common standard deviation σ was too conservative and was likely to be somewhat larger than 0.25.

Operating characteristic curves often play an important role in the choice of sample size in experimental design problems. Their use in this respect is discussed in subsequent chapters. For a discussion of the uses of operating characteristic curves for other simple comparative experiments similar to the two-sample t test see Montgomery and Runger (1994).

2–4.3 Confidence Intervals

Although hypothesis testing is a useful procedure, it sometimes does not tell the entire story. It is often preferable to provide an interval within which the value of the parameter or parameters in question would be expected to lie. These interval statements are called **confidence intervals.** In many engineering and industrial experiments, the experimenter already knows that the means μ_1 and μ_2 differ; consequently, hypothesis testing on $\mu_1 = \mu_2$ is of little interest. The experimenter would usually be more interested in a confidence interval on the difference in means $\mu_1 - \mu_2$.

To define a confidence interval, suppose that θ is an unknown parameter. To obtain an interval estimate of θ, we need to find two statistics L and U such that the probability statement

$$P(L \leqslant \theta \leqslant U) = 1 - \alpha \qquad (2\text{-}26)$$

is true. The interval

$$L \leqslant \theta \leqslant U \qquad (2\text{-}27)$$

is called a **100(1 − α) percent confidence interval** for the parameter θ. The interpretation of this interval is that if, in repeated random samplings, a large number of such intervals are constructed, $100(1 - \alpha)$ percent of them will contain the true value of θ. The statistics L and U are called the **lower** and **upper confidence limits,** respectively, and $1 - \alpha$ is called the **confidence coefficient.** If $\alpha = 0.05$, then Equation 2–27 is called a 95 percent confidence interval for θ. Note that confidence intervals have a frequency interpretation; that is, we do not know if the statement is true for this specific sample, but we do know that the *method* used to produce the confidence interval yields correct statements $100(1 - \alpha)$ percent of the time.

Suppose that we wish to find a $100(1 - \alpha)$ percent confidence interval on the true difference in means $\mu_1 - \mu_2$ for the portland cement problem. The interval can be derived in the following way. The statistic

$$\frac{\bar{y}_1 - \bar{y}_2 - (\mu_1 - \mu_2)}{S_p \sqrt{\dfrac{1}{n_1} + \dfrac{1}{n_2}}}$$

is distributed as $t_{n_1 + n_2 - 2}$. Thus,

$$P\left(-t_{\alpha/2, n_1 + n_2 - 2} \leqslant \frac{\bar{y}_1 - \bar{y}_2 - (\mu_1 - \mu_2)}{S_p \sqrt{\dfrac{1}{n_1} + \dfrac{1}{n_2}}} \leqslant t_{\alpha/2, n_1 + n_2 - 2}\right) = 1 - \alpha$$

or

$$P\left(\bar{y}_1 - \bar{y}_2 - t_{\alpha/2, n_1 + n_2 - 2}\, S_p \sqrt{\dfrac{1}{n_1} + \dfrac{1}{n_2}} \leqslant \mu_1 - \mu_2\right.$$

$$\left. \leqslant \bar{y}_1 - \bar{y}_2 + t_{\alpha/2, n_1 + n_2 - 2}\, S_p \sqrt{\dfrac{1}{n_1} + \dfrac{1}{n_2}}\right) = 1 - \alpha \qquad (2\text{-}28)$$

Comparing Equations 2–28 and 2–26, we see that

$$\bar{y}_1 - \bar{y}_2 - t_{\alpha/2, n_1 + n_2 - 2}\, S_p \sqrt{\dfrac{1}{n_1} + \dfrac{1}{n_2}} \leqslant \mu_1 - \mu_2$$

$$\leqslant \bar{y}_1 - \bar{y}_2 + t_{\alpha/2, n_1 + n_2 - 2}\, S_p \sqrt{\dfrac{1}{n_1} + \dfrac{1}{n_2}} \qquad (2\text{-}29)$$

is a $100(1 - \alpha)$ percent confidence interval for $\mu_1 - \mu_2$.

The actual 95 percent confidence interval estimate for the difference in mean tension bond strength for the formulations of portland cement mortar is found by substituting in Equation 2-29 as follows:

$$16.76 - 17.92 - (2.101)0.284 \sqrt{\tfrac{1}{10} + \tfrac{1}{10}} \leq \mu_1 - \mu_2$$
$$\leq 16.76 - 17.92 + (2.101)0.284 \sqrt{\tfrac{1}{10} + \tfrac{1}{10}}$$
$$-1.16 - 0.27 \leq \mu_1 - \mu_2 \leq -1.16 + 0.27$$
$$-1.43 \leq \mu_1 - \mu_2 \leq -0.89$$

Thus, the 95 percent confidence interval estimate on the difference in means extends from -1.43 kgf/cm^2 to -0.89 kgf/cm^2. Put another way, the confidence interval is $\mu_1 - \mu_2 = -1.16$ kgf/cm^2 $\pm$ 0.27 kgf/cm^2, or the difference in mean strengths is -1.16 kgf/cm^2, and the accuracy of this estimate is ± 0.27 kgf/cm^2. Note that since $\mu_1 - \mu_2 = 0$ is *not* included in this interval, the data do not support the hypothesis that $\mu_1 = \mu_2$ at the 5 percent level of significance. It is likely that the mean strength of the unmodified formulation exceeds the mean strength of the modified formulation. Notice from Table 2-2 that Statgraphics also reported this confidence interval when the hypothesis testing procedure was conducted.

2-4.4   The Case Where $\sigma_1^2 \neq \sigma_2^2$

If we are testing

$$H_0 : \mu_1 = \mu_2$$
$$H_1 : \mu_1 \neq \mu_2$$

and cannot reasonably assume that the variances σ_1^2 and σ_2^2 are equal, then the two-sample t test must be modified slightly. The test statistic becomes

$$t_0 = \frac{\bar{y}_1 - \bar{y}_2}{\sqrt{\dfrac{S_1^2}{n_1} + \dfrac{S_2^2}{n_2}}} \qquad (2\text{-}30)$$

This statistic is not distributed exactly as t. However, the distribution of t_0 is well-approximated by t if we use

$$\nu = \frac{\left(\dfrac{S_1^2}{n_1} + \dfrac{S_2^2}{n_2}\right)^2}{\dfrac{(S_1^2/n_1)^2}{n_1 - 1} + \dfrac{(S_2^2/n_2)^2}{n_2 - 1}} \qquad (2\text{-}31)$$

as the degrees of freedom. A strong indication of unequal variances on a normal probability plot would be a situation calling for this version of the t test. Note that in Table 2-2, Statgraphics reported a confidence interval on $\mu_1 - \mu_2$ assuming unequal variances. You should be able to develop the equation for finding that confidence interval easily.

2–4.5 The Case Where σ_1^2 and σ_2^2 Are Known

If the variances of both populations are *known,* then the hypotheses

$$H_0: \mu_1 = \mu_2$$
$$H_1: \mu_1 \neq \mu_2$$

may be tested using the statistic

$$Z_0 = \frac{\bar{y}_1 - \bar{y}_2}{\sqrt{\dfrac{\sigma_1^2}{n_1} + \dfrac{\sigma_2^2}{n_2}}} \tag{2–32}$$

If both populations are normal, or if the sample sizes are large enough so that the central limit theorem applies, the distribution of Z_0 is $N(0, 1)$ if the null hypothesis is true. Thus, the critical region would be found using the normal distribution rather than the t. Specifically, we would reject H_0 if $|Z_0| > Z_{\alpha/2}$, where $Z_{\alpha/2}$ is the upper $\alpha/2$ percentage point of the standard normal distribution.

   Unlike the t test of the previous sections, the test on means with known variances does not require the assumption of sampling from normal populations. One can use the central limit theorem to justify an approximate normal distribution for the difference in sample means $\bar{y}_1 - \bar{y}_2$.

   The $100(1 - \alpha)$ percent confidence interval on $\mu_1 - \mu_2$ where the variances are known is

$$\bar{y}_1 - \bar{y}_2 - Z_{\alpha/2}\sqrt{\frac{\sigma_1^2}{n_1} + \frac{\sigma_2^2}{n_2}} \leq \mu_1 - \mu_2 \leq \bar{y}_1 - \bar{y}_2 + Z_{\alpha/2}\sqrt{\frac{\sigma_1^2}{n_1} + \frac{\sigma_2^2}{n_2}} \tag{2–33}$$

As noted previously, the confidence interval is often a useful supplement to the hypothesis testing procedure.

2–4.6 Comparing a Single Mean to a Specified Value

Some experiments involve comparing only one population mean μ to a specified value, say, μ_0. The hypotheses are

$$H_0: \mu = \mu_0$$
$$H_1: \mu \neq \mu_0$$

If the population is normal with known variance, or if the population is nonnormal but the sample size is large enough so that the central limit theorem applies, then the hypothesis may be tested using a direct application of the normal distribution. The test statistic is

$$Z_0 = \frac{\bar{y} - \mu_0}{\sigma/\sqrt{n}} \tag{2–34}$$

If $H_0 : \mu = \mu_0$ is true, then the distribution of Z_0 is $N(0, 1)$. Therefore, the decision rule for $H_0 : \mu = \mu_0$ is to reject the null hypothesis if $|Z_0| > Z_{\alpha/2}$. The value of the mean μ_0 specified in the null hypothesis is usually determined in one of three ways. It may result from past evidence, knowledge, or experimentation. It may be the result of some theory or model describing the situation under study. Finally, it may be the result of contractual specifications.

The $100(1 - \alpha)$ percent confidence interval on the true population mean is

$$\bar{y} - Z_{\alpha/2}\sigma/\sqrt{n} \leqslant \mu \leqslant \bar{y} + Z_{\alpha/2}\sigma/\sqrt{n} \qquad (2\text{--}35)$$

Example 2–1

A vendor submits lots of fabric to a textile manufacturer. The manufacturer wants to know if the lot average breaking strength exceeds 200 psi. If so, she wants to accept the lot. Past experience indicates that a reasonable value for the variance of breaking strength is $100(psi)^2$. The hypotheses to be tested are

$$H_0 : \mu = 200$$
$$H_1 : \mu > 200$$

Note that this is a one-sided alternative hypothesis. Thus, we would accept the lot only if the null hypothesis $H_0 : \mu = 200$ could be rejected (i.e., if $Z_0 > Z_\alpha$).

Four specimens are randomly selected, and the average breaking strength observed is $\bar{y} = 214$ psi. The value of the test statistic is

$$Z_0 = \frac{\bar{y} - \mu_0}{\sigma/\sqrt{n}} = \frac{214 - 200}{10/\sqrt{4}} = 2.80$$

If a type I error of $\alpha = 0.05$ is specified, we find $Z_\alpha = Z_{0.05} = 1.645$ from Appendix Table I. Thus H_0 is rejected, and we conclude that the lot average breaking strength exceeds 200 psi.

■

If the variance of the population is unknown, we must make the additional assumption that the population is normally distributed, although moderate departures from normality will not seriously affect the results.

To test $H_0 : \mu = \mu_0$ in the variance unknown case, the sample variance S^2 is used to estimate σ^2. Replacing σ with S in Equation 2–34, we have the test statistic

$$t_0 = \frac{\bar{y} - \mu_0}{S/\sqrt{n}} \qquad (2\text{--}36)$$

The null hypothesis $H_0 : \mu = \mu_0$ would be rejected if $|t_0| > t_{\alpha/2,n-1}$, where $t_{\alpha/2,n-1}$ denotes the upper $\alpha/2$ percentage point of the t distribution with $n - 1$ degrees of freedom. The $100(1 - \alpha)$ percent confidence interval in this case is

$$\bar{y} - t_{\alpha/2,n-1}S/\sqrt{n} \leqslant \mu \leqslant \bar{y} + t_{\alpha/2,n-1}S/\sqrt{n} \qquad (2\text{--}37)$$

Table 2–3   Tests on Means with Variance Known

Hypothesis	Test Statistic	Criteria for Rejection
$H_0: \mu = \mu_0$ $H_1: \mu \neq \mu_0$		$\lvert Z_0 \rvert > Z_{\alpha/2}$
$H_0: \mu = \mu_0$ $H_1: \mu < \mu_0$	$Z_0 = \dfrac{\bar{y} - \mu_0}{\sigma/\sqrt{n}}$	$Z_0 < -Z_\alpha$
$H_0: \mu = \mu_0$ $H_1: \mu > \mu_0$		$Z_0 > Z_\alpha$
$H_0: \mu_1 = \mu_2$ $H_1: \mu_1 \neq \mu_2$		$\lvert Z_0 \rvert > Z_{\alpha/2}$
$H_0: \mu_1 = \mu_2$ $H_1: \mu_1 < \mu_2$	$Z_0 = \dfrac{\bar{y}_1 - \bar{y}_2}{\sqrt{\dfrac{\sigma_1^2}{n_1} + \dfrac{\sigma_2^2}{n_2}}}$	$Z_0 < -Z_\alpha$
$H_0: \mu_1 = \mu_2$ $H_1: \mu_1 > \mu_2$		$Z_0 > Z_\alpha$

2–4.7   Summary

Tables 2–3 and 2–4 summarize the test procedures discussed above for sample means. Critical regions are shown for both two-sided and one-sided alternative hypotheses.

2–5   INFERENCES ABOUT THE DIFFERENCES IN MEANS, PAIRED COMPARISON DESIGNS

2–5.1   The Paired Comparison Problem

In some simple comparative experiments we can greatly improve the precision by making comparisons within matched pairs of experimental material. For example, consider a hardness testing machine that presses a rod with a pointed tip into a metal specimen with a known force. By measuring the depth of the depression caused by the tip, the hardness of the specimen is determined. Two different tips are available for this machine, and although the precision (variability) of the measurements made by the two tips seems to be the same, it is suspected that one tip produces different hardness readings than the other.

An experiment could be performed as follows. A number of metal specimens (e.g., 20) could be randomly selected. Half of these specimens could be tested by tip 1 and the other half by tip 2. The exact assignment of specimens to tips

Table 2–4 Tests on Means of Normal Distributions, Variance Unknown

Hypothesis	Test Statistic	Criteria for Rejection
$H_0: \mu = \mu_0$ $H_1: \mu \neq \mu_0$		$\lvert t_0 \rvert > t_{\alpha/2, n-1}$
$H_0: \mu = \mu_0$ $H_1: \mu < \mu_0$	$t_0 = \dfrac{\bar{y} - \mu_0}{S/\sqrt{n}}$	$t_0 < -t_{\alpha, n-1}$
$H_0: \mu = \mu_0$ $H_1: \mu > \mu_0$		$t_0 > t_{\alpha, n-1}$
	if $\sigma_1^2 = \sigma_2^2$	
$H_0: \mu_1 = \mu_2$ $H_1: \mu_1 \neq \mu_2$	$t_0 = \dfrac{\bar{y}_1 - \bar{y}_2}{S_p \sqrt{\dfrac{1}{n_1} + \dfrac{1}{n_2}}}$ $\nu = n_1 + n_2 - 2$	$\lvert t_0 \rvert > t_{\alpha/2, \nu}$
	if $\sigma_1^2 \neq \sigma_2^2$	
$H_0: \mu_1 = \mu_2$ $H_1: \mu_1 < \mu_2$	$t_0 = \dfrac{\bar{y}_1 - \bar{y}_2}{\sqrt{\dfrac{S_1^2}{n_1} + \dfrac{S_2^2}{n_2}}}$	$t_0 < -t_{\alpha, \nu}$
$H_0: \mu_1 = \mu_2$ $H_1: \mu_1 > \mu_2$	$\nu = \dfrac{\left(\dfrac{S_1^2}{n_1} + \dfrac{S_2^2}{n_2}\right)^2}{\dfrac{(S_1^2/n_1)^2}{n_1 - 1} + \dfrac{(S_2^2/n_2)^2}{n_2 - 1}}$	$t_0 > t_{\alpha, \nu}$

would be randomly determined. Since this is a completely randomized design, the average hardness of the two samples could be compared using the t test described in Section 2–4.

A little reflection will reveal a serious disadvantage in the completely randomized design for this problem. Suppose the metal specimens were cut from different bar stock that were produced in different heats or that were not exactly homogeneous in some other way that might affect the hardness. This lack of homogeneity between specimens will contribute to the variability of the hardness measurements and will tend to inflate the experimental error, thus making a true difference between tips harder to detect.

To protect against this possibility, consider an alternate experimental design. Assume that each specimen is large enough so that *two* hardness determinations may be made on it. This alternative design would consist of dividing each specimen into two parts, then randomly assigning one tip to one-half of each specimen and the other tip to the remaining half. The order in which the tips are tested

Table 2–5   Data for the Hardness
Testing Experiment

Specimen	Tip 1	Tip 2
1	7	6
2	3	3
3	3	5
4	4	3
5	8	8
6	3	2
7	2	4
8	9	9
9	5	4
10	4	5

for a particular specimen would also be randomly selected. The experiment, when performed according to this design with 10 specimens, produced the (coded) data shown in Table 2–5.

We may write a statistical model that describes the data from this experiment as

$$y_{ij} = \mu_i + \beta_j + \epsilon_{ij} \begin{cases} i = 1, 2 \\ j = 1, 2, \ldots, 10 \end{cases} \tag{2–38}$$

where y_{ij} is the observation on hardness for tip i on specimen j, μ_i is the true mean hardness of the ith tip, β_j is an effect on hardness due to the jth specimen, and ϵ_{ij} is a random experimental error with mean zero and variance σ_i^2. That is, σ_1^2 is the variance of the hardness measurements from tip 1 and σ_2^2 is the variance of the hardness measurements from tip 2. We will frequently use models such as this one to describe the results of a designed experiment.

Note that if we compute the jth paired difference

$$d_j = y_{1j} - y_{2j} \qquad j = 1, 2, \ldots, 10 \tag{2–39}$$

the expected value of this difference is

$$\begin{aligned} \mu_d &= E(d_j) \\ &= E(y_{1j} - y_{2j}) \\ &= E(y_{1j}) - E(y_{2j}) \\ &= \mu_1 + \beta_j - (\mu_2 + \beta_j) \\ &= \mu_1 - \mu_2 \end{aligned}$$

That is, we may make inferences about the difference in the mean hardness readings of the two tips $\mu_1 - \mu_2$ by making inferences about the mean of the differences μ_d. Notice that the additive effect of the specimens β_j cancels out when the observations are paired in this manner.

Testing $H_0: \mu_1 = \mu_2$ is equivalent to testing

$$H_0: \mu_d = 0$$
$$H_1: \mu_d \neq 0$$

The test statistic for this hypothesis is

$$t_0 = \frac{\bar{d}}{S_d/\sqrt{n}} \tag{2–40}$$

where

$$\bar{d} = \frac{1}{n}\sum_{j=1}^{n} d_j \tag{2–41}$$

is the sample mean of the differences and

$$S_d = \left[\frac{\sum_{j=1}^{n}(d_j - \bar{d})^2}{n-1}\right]^{1/2} = \left[\frac{\sum_{j=1}^{n} d_j^2 - \frac{1}{n}\left(\sum_{j=1}^{n} d_j\right)^2}{n-1}\right]^{1/2} \tag{2–42}$$

is the sample standard deviation of the differences. $H_0: \mu_d = 0$ would be rejected if $|t_0| > t_{\alpha/2,n-1}$. For the data in Table 2–5, we find

$$
\begin{array}{ll}
d_1 = 7 - 6 = 1 & d_6 = 3 - 2 = 1 \\
d_2 = 3 - 3 = 0 & d_7 = 2 - 4 = -2 \\
d_3 = 3 - 5 = -2 & d_8 = 9 - 9 = 0 \\
d_4 = 4 - 3 = 1 & d_9 = 5 - 4 = 1 \\
d_5 = 8 - 8 = 0 & d_{10} = 4 - 5 = -1
\end{array}
$$

Thus,

$$\bar{d} = \frac{1}{n}\sum_{j=1}^{n} d_j = \frac{1}{10}(-1) = -0.10$$

$$S_d = \left[\frac{\sum_{j=1}^{n} d_j^2 - \frac{1}{n}\left(\sum_{j=1}^{n} d_j\right)^2}{n-1}\right]^{1/2} = \left[\frac{13 - \frac{1}{10}(-1)^2}{10-1}\right]^{1/2} = 1.20$$

Suppose we choose $\alpha = 0.05$. Now to make a decision, we would compute t_0 and reject H_0 if $|t_0| > t_{0.025,9} = 2.262$.

The computed value of the test statistic is

$$t_0 = \frac{\bar{d}}{S_d/\sqrt{n}}$$

$$= \frac{-0.10}{1.20/\sqrt{10}}$$

$$= -0.26$$

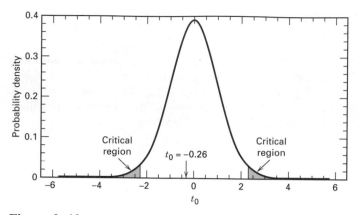

Figure 2–12. The reference distribution (t with 9 degrees of freedom) for the hardness testing problem.

and since $|t_0| = 0.26 > t_{0.025,9} = 2.262$, we cannot reject the hypothesis $H_0 : \mu_d = 0$. That is, there is no evidence to indicate that the two tips produce different hardness readings. Figure 2–12 shows the t_0 distribution with 9 degrees of freedom, the reference distribution for this test, with the value of t_0 shown relative to the critical region.

Table 2–6 shows the computer output from the One-Sample Analysis Program in Statgraphics for this problem. Notice that the P-value for this test is $P \approx 0.80$, implying that we cannot reject the null hypothesis at *any* reasonable level of significance.

Table 2–6 Statgraphics One-Sample Analysis Results for the Hardness Testing Example

		Tip1-Tip2	
Sample Statistics:	Number of Obs.	10	
	Average	−0.1	
	Variance	1.43333	
	Std. Deviation	1.19722	
	Median	0	
Confidence Interval for Mean:		95 Percent	
	Sample 1	−0.956672 0.756672	9 D.F.
Confidence Interval for Variance:		95 Percent	
	Sample 1	0.678135 4.77734	9 D.F.
Hypothesis Test for H0: Mean = 0		Computed t statistic = −0.264135	
	vs Alt: NE	Sig. Level = 0.797625	
	at Alpha = 0.05	so do not reject H0.	

2–5.2 Advantages of the Paired Comparison Design

The design actually used for this experiment is called the **paired comparison design** and it illustrates the blocking principle discussed in Section 1–3. Actually, it is a special case of a more general type of design called the **randomized block design.** The term *block* refers to a relatively homogeneous experimental unit (in our case, the metal specimens are the blocks), and the block represents a restriction on complete randomization because the treatment combinations are only randomized within the block. We look at designs of this type in Chapter 5. In that chapter the mathematical model for the design, Equation 2–38, is written in a slightly different form.

Before leaving this experiment, several points should be made. Note that, although $2n = 2(10) = 20$ observations have been taken, only $n - 1 = 9$ degrees of freedom are available for the t statistic. (We know that as the degrees of freedom for t increase the test becomes more sensitive.) By blocking or pairing, we have effectively "lost" $n - 1$ degrees of freedom, but we hope we have gained a better knowledge of the situation by eliminating an additional source of variability (the difference between specimens). We may obtain an indication of the quality of information produced from the paired design by comparing the standard deviation of the differences S_d with the pooled standard deviation S_p that would have resulted had the experiment been conducted in a completely randomized manner and the data of Table 2–5 been obtained. Using the data in Table 2–5 as two independent samples, we compute the pooled standard deviation from Equation 2–24 to be $S_p = 2.32$. Comparing this value to $S_d = 1.20$, we see that blocking or pairing has reduced the estimate of variability by nearly 50 percent. We may also express this information in terms of a confidence interval on $\mu_1 - \mu_2$. Using the paired data, a 95 percent confidence interval on $\mu_1 - \mu_2$ is

$$\bar{d} \pm t_{0.025,9}S_d/\sqrt{n}$$
$$-0.10 \pm (2.262)(1.20)/\sqrt{10}$$
$$-0.10 \pm 0.86$$

Conversely, using the pooled or independent analysis, a 95 percent confidence interval on $\mu_1 - \mu_2$ is

$$\bar{y}_1 - \bar{y}_2 \pm t_{0.025,18}S_p \sqrt{\frac{1}{n_1} + \frac{1}{n_2}}$$
$$4.80 - 4.90 \pm (2.101)(2.32)\sqrt{\tfrac{1}{10} + \tfrac{1}{10}}$$
$$-0.10 \pm 2.18$$

The confidence interval based on the paired analysis is much narrower than the confidence interval from the independent analysis. This illustrates the **noise reduction** property of blocking.

Blocking is not always the best design strategy. If the within-block variability is the same as the between-block variability, then the variance of $\bar{y}_1 - \bar{y}_2$ will be the same regardless of which design is used. Actually, blocking in this situation would be a poor choice of design because blocking results in the loss of $n - 1$ degrees of freedom and will actually lead to a wider confidence interval on $\mu_1 - \mu_2$. A further discussion of blocking is given in Chapter 5.

2–6 INFERENCES ABOUT THE VARIANCES OF NORMAL DISTRIBUTIONS

In many experiments, we are interested in possible differences in the mean response for two treatments. However, in some experiments it is the comparison of variability in the data that is important. In the food and beverage industry, for example, it is important that the variability of filling equipment be small so that all packages have close to the nominal net weight or volume of content. In chemical laboratories, we may wish to compare the variability of two analytical methods. We now briefly examine tests of hypotheses and confidence intervals for variances of normal distributions. Unlike the tests on means, the procedures for tests on variances are rather sensitive to the normality assumption. A good discussion of the normality assumption is in Appendix 2A of Davies (1956).

Suppose we wish to test the hypothesis that the variance of a normal population equals a constant, for example, σ_0^2. Stated formally, we wish to test

$$H_0: \sigma^2 = \sigma_0^2$$
$$H_1: \sigma^2 \neq \sigma_0^2$$

(2–43)

The test statistic for Equation 2–43 is

$$\chi_0^2 = \frac{SS}{\sigma_0^2} = \frac{(n-1)S^2}{\sigma_0^2}$$

(2–44)

where $SS = \sum_{i=1}^{n}(y_i - y)^2$ is the corrected sum of squares of the sample observations. The appropriate reference distribution for χ_0^2 is the chi-square distribution with $n - 1$ degrees of freedom. The null hypothesis is rejected if $\chi_0^2 > \chi_{\alpha/2,n-1}^2$ or if $\chi_0^2 < \chi_{1-(\alpha/2),n-1}^2$, where $\chi_{\alpha/2,n-1}^2$ and $\chi_{1-(\alpha/2),n-1}^2$ are the upper $\alpha/2$ and lower $1 - (\alpha/2)$ percentage points of the chi-square distribution with $n - 1$ degrees of freedom, respectively. Table 2–7 gives the critical regions for the one-sided alternative hypotheses. The $100(1 - \alpha)$ percent confidence interval on σ^2 is

$$\frac{(n-1)S^2}{\chi_{\alpha/2,n-1}^2} \leq \sigma^2 \leq \frac{(n-1)S^2}{\chi_{1-(\alpha/2),n-1}^2}$$

(2–45)

This equation was used in computing the confidence interval on the variance in the Statgraphics output in Table 2–6.

Table 2–7   Tests on Variances of Normal Distributions

Hypothesis	Test Statistic	Criteria for Rejection
$H_0: \sigma^2 = \sigma_0^2$ $H_1: \sigma^2 \neq \sigma_0^2$		$\chi_0^2 > \chi_{\alpha/2, n-1}^2$ or $\chi_0^2 < \chi_{1-\alpha/2, n-1}^2$
$H_0: \sigma^2 = \sigma_0^2$ $H_1: \sigma^2 < \sigma_0^2$	$\chi_0^2 = \dfrac{(n-1)S^2}{\sigma_0^2}$	$\chi_0^2 < \chi_{1-\alpha, n-1}^2$
$H_0: \sigma^2 = \sigma_0^2$ $H_1: \sigma^2 > \sigma_0^2$		$\chi_0^2 > \chi_{\alpha, n-1}^2$
$H_0: \sigma_1^2 = \sigma_2^2$ $H_1: \sigma_1^2 \neq \sigma_2^2$	$F_0 = \dfrac{S_1^2}{S_2^2}$	$F_0 > F_{\alpha/2, n_1-1, n_2-1}$ or $F_0 < F_{1-\alpha/2, n_1-1, n_2-1}$
$H_0: \sigma_1^2 = \sigma_2^2$ $H_1: \sigma_1^2 < \sigma_2^2$	$F_0 = \dfrac{S_2^2}{S_1^2}$	$F_0 > F_{\alpha, n_2-1, n_1-1}$
$H_0: \sigma_1^2 = \sigma_2^2$ $H_1: \sigma_1^2 > \sigma_2^2$	$F_0 = \dfrac{S_1^2}{S_2^2}$	$F_0 > F_{\alpha, n_1-1, n_2-1}$

Now consider testing the equality of the variances of two normal populations. If independent random samples of size n_1 and n_2 are taken from populations 1 and 2, respectively, then the test statistic for

$$H_0: \sigma_1^2 = \sigma_2^2$$
$$H_1: \sigma_1^2 \neq \sigma_2^2$$
$$(2\text{--}46)$$

is the ratio of the sample variances

$$F_0 = \frac{S_1^2}{S_2^2} \qquad (2\text{--}47)$$

The appropriate reference distribution for F_0 is the F distribution with $n_1 - 1$ numerator degrees of freedom and $n_2 - 1$ denominator degrees of freedom. The null hypothesis would be rejected if $F_0 > F_{\alpha/2, n_1-1, n_2-1}$ or if $F_0 < F_{1-(\alpha/2), n_1-1, n_2-1}$, where $F_{\alpha/2, n_1-1, n_2-1}$ and $F_{1-(\alpha/2), n_1-1, n_2-1}$ denote the upper $\alpha/2$ and lower $1 - (\alpha/2)$ percentage points of the F distribution with $n_1 - 1$ and $n_2 - 1$ degrees of freedom. Table IV of the Appendix gives only upper-tail percentage points of F; however, the upper- and lower-tail points are related by

$$F_{1-\alpha, \nu_1, \nu_2} = \frac{1}{F_{\alpha, \nu_2, \nu_1}} \qquad (2\text{--}48)$$

Test procedures for more than two variances are discussed in Chapter 3, Section 3–4.3. We will also discuss the use of the variance or standard deviation as a response variable in more general experimental settings.

Example 2-2

A chemical engineer is investigating the inherent variability of two types of test equipment that can be used to monitor the output of a production process. He suspects that the old equipment, type 1, has a larger variance than the new one. Thus, he wishes to test the hypothesis

$$H_0: \sigma_1^2 = \sigma_2^2$$
$$H_1: \sigma_1^2 > \sigma_2^2$$

Two random samples of $n_1 = 12$ and $n_2 = 10$ observations are taken, and the sample variances are $S_1^2 = 14.5$ and $S_2^2 = 10.8$. The test statistic is

$$F_0 = \frac{S_1^2}{S_2^2} = \frac{14.5}{10.8} = 1.34$$

From Appendix Table IV we find that $F_{0.05,11,9} = 3.10$, so the null hypothesis cannot be rejected. That is, we have found insufficient statistical evidence to conclude that the variance of the old equipment is greater than the variance of the new equipment.

∎

The $100(1 - \alpha)$ confidence interval for the ratio of the population variances σ_1^2/σ_2^2 is

$$\frac{S_1^2}{S_2^2} F_{1-\alpha/2, n_2-1, n_1-1} \leq \frac{\sigma_1^2}{\sigma_2^2} \leq \frac{S_1^2}{S_2^2} F_{\alpha/2, n_2-1, n_1-1} \qquad (2-49)$$

This equation was used in finding the confidence interval on the ratio of the two variances in the Statgraphics output for the portland cement experiment in Table 2–2.

To illustrate the use of Equation 2–49, the 95 percent confidence interval for the ratio of variances σ_1^2/σ_2^2 in Example 2–2 is, using $F_{0.025,9,11} = 3.59$ and $F_{0.975,9,11} = 1/F_{0.025,11,9} = 1/3.92 = 0.255$,

$$\frac{14.5}{10.8}(0.255) \leq \frac{\sigma_1^2}{\sigma_2^2} \leq \frac{14.5}{10.8}(3.59)$$

$$0.34 \leq \frac{\sigma_1^2}{\sigma_2^2} \leq 4.81$$

2-7 PROBLEMS

2-1 The breaking strength of a fiber is required to be at least 150 psi. Past experience has indicated that the standard deviation of breaking strength is $\sigma = 3$ psi. A random sample of four specimens is tested, and the results are $y_1 = 145$, $y_2 = 153$, $y_3 = 150$, and $y_4 = 147$.

(a) State the hypotheses that you think should be tested in this experiment.

(b) Test these hypotheses using $\alpha = 0.05$. What are your conclusions?

(c) Find the P-value for the test in part (b).

(d) Construct a 95 percent confidence interval on the mean breaking strength.

2-2 The viscosity of a liquid detergent is supposed to average 800 centistokes at 25°C. A random sample of 16 batches of detergent is collected, and the average viscosity is 812. Suppose we know that the standard deviation of viscosity is $\sigma = 25$ centistokes.

(a) State the hypotheses that should be tested.

(b) Test these hypotheses using $\alpha = 0.05$. What are your conclusions?

(c) What is the P-value for the test?

(d) Find a 95 percent confidence interval on the mean.

2-3 The diameters of steel shafts produced by a certain manufacturing process should have a mean diameter of 0.255 inches. The diameter is known to have a standard deviation of $\sigma = 0.0001$ inch. A random sample of 10 shafts has an average diameter of 0.2545 inch.

(a) Set up appropriate hypotheses on the mean μ.

(b) Test these hypotheses using $\alpha = 0.05$. What are your conclusions?

(c) Find the P-value for this test.

(d) Construct a 95 percent confidence interval on the mean shaft diameter.

2-4 A normally distributed random variable has an unknown mean μ and a known variance $\sigma^2 = 9$. Find the sample size required to construct a 95 percent confidence interval on the mean that has total width of 1.0.

2-5 The shelf life of a carbonated beverage is of interest. Ten bottles are randomly selected and tested, and the following results are obtained:

Days	
108	138
124	163
124	159
106	134
115	139

(a) We would like to demonstrate that the mean shelf life exceeds 120 days. Set up appropriate hypotheses for investigating this claim.

(b) Test these hypotheses using $\alpha = 0.01$. What are your conclusions?

(c) Find the P-value for the test in part (b).

(d) Construct a 99 percent confidence interval on the mean shelf life.

2-6 Consider the shelf life data in Problem 2–5. Can shelf life be described or modeled adequately by a normal distribution? What effect would violation of this assumption have on the test procedure you used in solving Problem 2–5?

2-7 The time to repair an electronic instrument is a normally distributed random variable measured in hours. The repair times for 16 such instruments chosen at random are as follows:

	Hours		
159	280	101	212
224	379	179	264
222	362	168	250
149	260	485	170

(a) You wish to know if the mean repair time exceeds 225 hours. Set up appropriate hypotheses for investigating this issue.

(b) Test the hypotheses you formulated on part (a). What are your conclusions? Use $\alpha = 0.05$.

(c) Find the P-value for the test.

(d) Construct a 95 percent confidence interval on mean repair time.

2-8 Reconsider the repair time data in Problem 2–7. Can repair time, in your opinion, be adequately modeled by a normal distribution?

2-9 Two machines are used for filling plastic bottles with a net volume of 16.0 ounces. The filling processes can be assumed to be normal, with standard deviations of $\sigma_1 = 0.015$ and $\sigma_2 = 0.018$. The quality engineering department suspects that both machines fill to the same net volume, whether or not this volume is 16.0 ounces. An experiment is performed by taking a random sample from the output of each machine.

Machine 1		Machine 2	
16.03	16.01	16.02	16.03
16.04	15.96	15.97	16.04
16.05	15.98	15.96	16.02
16.05	16.02	16.01	16.01
16.02	15.99	15.99	16.00

(a) State the hypotheses that should be tested in this experiment.

(b) Test these hypotheses using $\alpha = 0.05$. What are your conclusions?

(c) Find the P-value for this test.

(d) Find a 95 percent confidence interval on the difference in mean fill volume for the two machines.

2-10 Two types of plastic are suitable for use by an electronic calculator manufacturer. The breaking strength of this plastic is important. It is known that $\sigma_1 = \sigma_2 = 1.0$ psi. From random samples of $n_1 = 10$ and $n_2 = 12$ we obtain $\bar{y}_1 = 162.5$ and $\bar{y}_2 = 155.0$. The company will not adopt plastic 1 unless its breaking strength exceeds that of plastic 2 by at least 10 psi. Based on the sample information, should they use plastic 1? In answering this question, set up and test appropriate hypotheses using $\alpha = 0.01$. Construct a 99 percent confidence interval on the true mean difference in breaking strength.

2-11 The following are the burning times of chemical flares of two different formulations. The design engineers are interested in both the mean and variance of the burning times.

Type 1		Type 2	
65	82	64	56
81	67	71	69
57	59	83	74
66	75	59	82
82	70	65	79

(a) Test the hypothesis that the two variances are equal. Use $\alpha = 0.05$.

(b) Using the results of (a), test the hypothesis that the mean burning times are equal. Use $\alpha = 0.05$. What is the P-value for this test?

(c) Discuss the role of the normality assumption in this problem. Check the assumption of normality for both types of flares.

2-12 An article in Solid State Technology, "Orthogonal Design for Process Optimization and Its Application to Plasma Etching" by G. Z. Yin and D. W. Jillie (May, 1987) describes an experiment to determine the effect of the C_2F_6 flow rate on the uniformity of the etch on a silicon wafer used in integrated circuit manufacturing. Data for two flow rates are as follows:

C_2F_6 Flow (SCCM)	Uniformity Observation					
	1	2	3	4	5	6
125	2.7	4.6	2.6	3.0	3.2	3.8
200	4.6	3.4	2.9	3.5	4.1	5.1

(a) Does the C_2F_6 flow rate affect average etch uniformity? Use $\alpha = 0.05$.

(b) What is the P-value for the test in part (a)?

(c) Does the C_2F_6 flow rate affect the wafer-to-wafer variability in etch uniformity? Use $\alpha = 0.05$.

(d) Draw box plots to assist in the interpretation of the data from this experiment.

2–13   A new filtering device is installed in a chemical unit. Before its installation, a random sample yielded the following information about the percentage of impurity: $\bar{y}_1 = 12.5$, $S_1^2 = 101.17$, and $n_1 = 8$. After installation, a random sample yielded $\bar{y}_2 = 10.2$, $S_2^2 = 94.73$, $n_2 = 9$.

(a) Can you conclude that the two variances are equal? Use $\alpha = 0.05$.

(b) Has the filtering device reduced the percentage of impurity significantly? Use $\alpha = 0.05$.

2–14   Twenty observations on etch uniformity on silicon wafers are taken during a qualification experiment for a plasma etcher. The data are as follows:

5.34	6.65	4.76	5.98	7.25
6.00	7.55	5.54	5.62	6.21
5.97	7.35	5.44	4.39	4.98
5.25	6.35	4.61	6.00	5.32

(a) Construct a 95 percent confidence interval estimate of σ^2.

(b) Test the hypothesis that $\sigma^2 = 1.0$. Use $\alpha = 0.05$. What are your conclusions?

(c) Discuss the normality assumption and its role in this problem.

(d) Check normality by constructing a normal probability plot. What are your conclusions?

2–15   The diameter of a ball bearing was measured by 12 inspectors, each using two different kinds of calipers. The results were

Inspector	Caliper 1	Caliper 2
1	0.265	0.264
2	0.265	0.265
3	0.266	0.264
4	0.267	0.266
5	0.267	0.267
6	0.265	0.268
7	0.267	0.264
8	0.267	0.265
9	0.265	0.265
10	0.268	0.267
11	0.268	0.268
12	0.265	0.269

(a) Is there a significant difference between the means of the population of measurements represented by the two samples? Use $\alpha = 0.05$.

(b) Find the P-value for the test in part (a).

(c) Construct a 95 percent confidence interval on the difference in mean diameter measurements for the two types of calipers.

2–16   An article in the *Journal of Strain Analysis* (vol. 18, no. 2, 1983) compares several procedures for predicting the shear strength for steel plate girders. Data for nine girders in the form of the ratio of predicted to observed load for two of these procedures, the Karlsruhe and Lehigh methods, are as follows:

Girder	Karlsruhe Method	Lehigh Method
S1/1	1.186	1.061
S2/1	1.151	0.992
S3/1	1.322	1.063
S4/1	1.339	1.062
S5/1	1.200	1.065
S2/1	1.402	1.178
S2/2	1.365	1.037
S2/3	1.537	1.086
S2/4	1.559	1.052

(a) Is there any evidence to support a claim that there is a difference in mean performance between the two methods? Use $\alpha = 0.05$.

(b) What is the P-value for the test in part (a)?

(c) Construct a 95 percent confidence interval for the difference in mean predicted to observed load.

2–17   Two popular pain medications are being compared on the basis of the speed of absorption by the body. Specifically, tablet 1 is claimed to be absorbed twice as fast as tablet 2. Assume that σ_1^2 and σ_2^2 are known. Develop a test statistic for

$$H_0 : 2\mu_1 = \mu_2$$
$$H_1 : 2\mu_1 \neq \mu_2$$

2–18   Suppose we are testing

$$H_0 : \mu_1 = \mu_2$$
$$H_1 : \mu_1 \neq \mu_2$$

where σ_1^2 and σ_2^2 are known. Our sampling resources are constrained such that $n_1 + n_2 = N$. How should we allocate the N observations between the two populations to obtain the most powerful test?

2–19   Develop Equation 2–45 for a $100(1 - \alpha)$ percent confidence interval for the variance of a normal distribution.

2–20   Develop Equation 2–49 for a $100(1 - \alpha)$ percent confidence interval for the ratio σ_1^2/σ_2^2, where σ_1^2 and σ_2^2 are the variances of two normal distributions.

2–21   Develop an equation for finding a $100(1 - \alpha)$ percent confidence interval on the difference in the means of two normal distributions where $\sigma_1^2 \neq \sigma_2^2$. Show that your equation produces the results in Table 2–2 for the portland cement experiment data.

Chapter 3
Experiments with a Single Factor: The Analysis of Variance

In Chapter 2 we discussed methods for comparing two conditions or treatments. For example, the portland cement tension bond experiment involved two different mortar formulations. Another way to describe this experiment is as a single-factor experiment with two levels of the factor, where the factor is mortar formulation and the two levels are the two different formulation methods. Many experiments of this type involve more than two levels of the factor. In this chapter and the next, we present methods for the design and analysis of single-factor experiments with *a* levels of the factor (or *a* treatments). We will assume that the experiment has been completely randomized.

3-1 AN EXAMPLE

A product development engineer is interested in investigating the tensile strength of a new synthetic fiber that will be used to make cloth for men's shirts. The engineer knows from previous experience that the strength is affected by the weight percent of cotton used in the blend of materials for the fiber. Furthermore, he suspects that increasing the cotton content will increase the strength, at least initially. He also knows that cotton content should range beween about 10 percent and 40 percent if the final cloth is to have other quality characteristics that are desired (such as the ability to take a permanent-press finishing treatment). The engineer decides to test specimens at five levels of cotton weight percent: 15 percent, 20 percent, 25 percent, 30 percent, and 35 percent. He also decides to test five specimens at each level of cotton content.

This is an example of a single-factor experiment with $a = 5$ **levels** of the factor and $n = 5$ **replicates.** The 25 runs should be made in random order. To illustrate how the run order may be randomized, suppose that we number the runs as follows:

Cotton Weight Percent	Experimental Run Number				
15	1	2	3	4	5
20	6	7	8	9	10
25	11	12	13	14	15
30	16	17	18	19	20
35	21	22	23	24	25

Now we select a random number between 1 and 25. Suppose this number is 8. Then the number 8 observation (20 percent cotton) is run first. This process would be repeated until all 25 observations have been assigned a position in the test sequence.[1]

Suppose that the test sequence obtained is

Test Sequence	Run Number	Cotton Weight Percent
1	8	20
2	18	30
3	10	20
4	23	35
5	17	30
6	5	15
7	14	25
8	6	20
9	15	25
10	20	30
11	9	20
12	4	15
13	12	25
14	7	20
15	1	15
16	24	35
17	21	35
18	11	25
19	2	15
20	13	25

---

[1] The only restriction on randomization here is that if the same number (e.g., 8) is drawn again it is discarded. This is a minor restriction and is ignored.

(*continued*)

Test Sequence	Run Number	Cotton Weight Percent
21	22	35
22	16	30
23	25	35
24	19	30
25	3	15

This randomized test sequence is necessary to prevent the effects of unknown nuisance variables, perhaps varying out of control during the experiment, from contaminating the results. To illustrate, suppose that we were to run the 25 test specimens in the original nonrandomized order (that is, all five 15 percent cotton specimens are tested first, all five 20 percent cotton specimens are tested next, and so on). If the tensile strength testing machine exhibits a warm-up effect such that the longer it is on, the lower the observed tensile strength readings will be, then the warm-up effect will potentially contaminate the tensile strength data and destroy the validity of the experiment.

Suppose that the engineer runs the test in the random order we have determined. The observations that he obtains on tensile strength are shown in Table 3–1.

It is always a good idea to examine experimental data graphically. Figure 3–1 presents box plots for tensile strength at each level of cotton weight percent, and Figure 3–2 is a scatter diagram of tensile strength versus cotton weight percent. (See page 66.) In Figure 3–2, the solid dots are the individual observations and the open circles are the average observed tensile strengths. Both graphs indicate that tensile strength increases as cotton content increases, up to about 30 percent cotton. Beyond 30 percent cotton, there is a marked decrease in tensile strength. There is no strong evidence to suggest that the variability in tensile strength around the average depends on the cotton weight percent. Based on this simple graphical analysis, we strongly suspect that (1) cotton content affects tensile strength and (2) around 30 percent cotton would result in maximum strength.

Table 3–1   Data (in lb/in^2) from the Tensile Strength Experiment

Cotton Weight Percent	Observations					Total	Average
	1	2	3	4	5		
15	7	7	15	11	9	49	9.8
20	12	17	12	18	18	77	15.4
25	14	18	18	19	19	88	17.6
30	19	25	22	19	23	108	21.6
35	7	10	11	15	11	54	10.8
						376	15.04

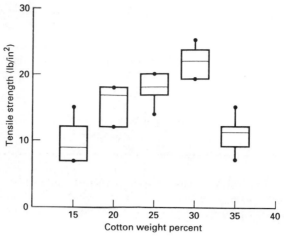

Figure 3–1. Box plots of tensile strength versus cotton weight percent.

Suppose that we wish to be more objective in our analysis of the data. Specifically, suppose that we wish to test for differences between the mean strengths at all $a = 5$ levels of cotton weight percent. Thus, we are interested in testing the equality of all five means. It might seem that this problem could be solved by performing a *t* test on all the possible pairs of means. However, this is not the best solution to this problem, since it would lead to considerable distortion in the type I error. For example, suppose we wish to test the equality of the five means using pairwise comparisons. There are 10 possible pairs, and if the probability of correctly accepting the null hypothesis for each individual test is $1 - \alpha = .95$, then the probability of correctly accepting the null hypothesis for all 10 tests is $(.95)^{10} = .60$ if the tests are independent. Thus, a substantial increase in the type I error has occurred.

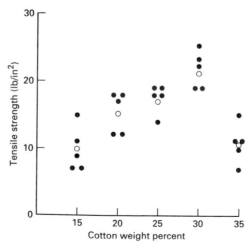

Figure 3–2. Scatter diagram of tensile strength versus cotton weight percent.

The appropriate procedure for testing the equality of several means is the **analysis of variance.** However, the analysis of variance has a much wider application than the problem above. It is probably the most useful technique in the field of statistical inference.

3-2 THE ANALYSIS OF VARIANCE

Suppose we have a treatments or different levels of a single factor that we wish to compare. The observed response from each of the a treatments is a random variable. The data would appear as in Table 3-2. An entry in Table 3-2 (e.g., y_{ij}) represents the jth observation taken under treatment i. There will be, in general, n observations under the ith treatment. Notice that Table 3-2 is the general case of the data from the tensile strength experiment in Table 3-1.

We will find it useful to describe the observations with a **linear statistical model:**

$$y_{ij} = \mu + \tau_i + \epsilon_{ij} \begin{cases} i = 1, 2, \ldots, a \\ j = 1, 2, \ldots, n \end{cases} \tag{3-1}$$

where y_{ij} is the (ij)th observation, μ is a parameter common to all treatments called the **overall mean,** τ_i is a parameter unique to the ith treatment called the **ith treatment effect,** and ϵ_{ij} is a **random error** component. Our objectives will be to test appropriate hypotheses about the treatment effects and to estimate them. For hypothesis testing, the model errors are assumed to be normally and independently distributed random variables with mean zero and variance σ^2. The variance σ^2 is assumed to be constant for all levels of the factor.

This model is called the **one-way** or **single-factor analysis of variance** because only one factor is investigated. Furthermore, we will require that the experiment be performed in random order so that the environment in which the treatments are used (often called the **experimental units**) is as uniform as possible. Thus, the experimental design is a **completely randomized design.**

The statistical model, Equation 3-1, describes two different situations with

Table 3-2 Typical Data for a Single-Factor Experiment

Treatment (level)	Observations				Totals	Averages
1	y_{11}	y_{12}	$\cdots$	y_{1n}	$y_{1.}$	$\bar{y}_{1.}$
2	y_{21}	y_{22}	$\cdots$	y_{2n}	$y_{2.}$	$\bar{y}_{2.}$
$\vdots$	$\vdots$	$\vdots$	$\cdots$	$\vdots$	$\vdots$	$\vdots$
a	y_{a1}	y_{a2}	$\cdots$	y_{an}	$y_{a.}$	$\bar{y}_{a.}$
					$y_{..}$	$\bar{y}_{..}$

respect to the treatment effects. First, the a treatments could have been specifically chosen by the experimenter. In this situation we wish to test hypotheses about the treatment means, and our conclusions will apply only to the factor levels considered in the analysis. The conclusions cannot be extended to similar treatments that were not explicitly considered. We may also wish to estimate the model parameters (μ, τ_i, σ^2). This is called the **fixed effects model.** Alternatively, the a treatments could be a **random sample** from a larger population of treatments. In this situation we should like to be able to extend the conclusions (which are based on the sample of treatments) to all treatments in the population, whether they were explicitly considered in the analysis or not. Here the τ_i are random variables, and knowledge about the particular ones investigated is relatively useless. Instead, we test hypotheses about the variability of the τ_i and try to estimate this variability. This is called the **random effects model** or **components of variance model.**

3–3   ANALYSIS OF THE FIXED EFFECTS MODEL

In this section, we develop the single-factor analysis of variance for the fixed effects model. In the fixed effects model, the treatment effects τ_i are usually defined as deviations from the overall mean, so

$$\sum_{i=1}^{a} \tau_i = 0 \tag{3–2}$$

Let $y_{i.}$ represent the total of the observations under the ith treatment and $\bar{y}_{i.}$ represent the average of the observations under the ith treatment. Similarly, let $y_{..}$ represent the grand total of all the observations and $\bar{y}_{..}$ represent the grand average of all the observations. Expressed symbolically,

$$y_{i.} = \sum_{j=1}^{n} y_{ij}, \qquad \bar{y}_{i.} = y_{i.}/n \qquad i = 1, 2, \ldots, a$$

$$y_{..} = \sum_{i=1}^{a} \sum_{j=1}^{n} y_{ij} \qquad \bar{y}_{..} = y_{..}/N \tag{3–3}$$

where $N = an$ is the total number of observations. We see that the "dot" subscript notation implies summation over the subscript that it replaces.

   The mean of the ith treatment is $E(y_{ij}) \equiv \mu_i = \mu + \tau_i$, $i = 1, 2, \ldots, a$. Thus, the mean of the ith treatment consists of the overall mean plus the ith treatment effect. We are interested in testing the equality of the a treatment means; that is,

$$H_0: \mu_1 = \mu_2 = \cdots = \mu_a$$
$$H_1: \mu_i \neq \mu_j \qquad \text{for at least one pair } (i, j)$$

Note that if H_0 is true, all treatments have a common mean μ. An equivalent way to write the above hypotheses is in terms of the treatment effects τ_i, say

$$H_0: \tau_1 = \tau_2 = \cdots \tau_a = 0$$
$$H_1: \tau_i \neq 0 \qquad \text{for at least one } i$$

Thus, we may speak of testing the equality of treatment means or testing that the treatment effects (the τ_i) are zero. The appropriate procedure for testing the equality of a treatment means is the analysis of variance.

3–3.1 Decomposition of the Total Sum of Squares

The name **analysis of variance** is derived from a partitioning of total variability into its component parts. The total corrected sum of squares

$$SS_T = \sum_{i=1}^{a} \sum_{j=1}^{n} (y_{ij} - \bar{y}_{..})^2$$

is used as a measure of overall variability in the data. Intuitively, this is reasonable since, if we were to divide SS_T by the appropriate number of degrees of freedom (in this case, $an - 1 = N - 1$), we would have the **sample variance** of the y's. The sample variance is, of course, a standard measure of variability.

Note that the total corrected sum of squares SS_T may be written as

$$\sum_{i=1}^{a} \sum_{j=1}^{n} (y_{ij} - \bar{y}_{..})^2 = \sum_{i=1}^{a} \sum_{j=1}^{n} [(\bar{y}_{i.} - \bar{y}_{..}) + (y_{ij} - \bar{y}_{i.})]^2 \tag{3-4}$$

or

$$\sum_{i=1}^{a} \sum_{j=1}^{n} (y_{ij} - \bar{y}_{..})^2 = n \sum_{i=1}^{a} (\bar{y}_{i.} - \bar{y}_{..})^2 + \sum_{i=1}^{a} \sum_{j=1}^{n} (y_{ij} - \bar{y}_{i.})^2$$
$$+ 2 \sum_{i=1}^{a} \sum_{j=1}^{n} (\bar{y}_{i.} - \bar{y}_{..})(y_{ij} - \bar{y}_{i.}) \tag{3-5}$$

However, the cross-product term in Equation 3–5 is zero, since

$$\sum_{j=1}^{n} (y_{ij} - \bar{y}_{i.}) = y_{i.} - n\bar{y}_{i.} = y_{i.} - n(y_{i.}/n) = 0$$

Therefore, we have

$$\sum_{i=1}^{a} \sum_{j=1}^{n} (y_{ij} - \bar{y}_{..})^2 = n \sum_{i=1}^{a} (\bar{y}_{i.} - \bar{y}_{..})^2 + \sum_{i=1}^{a} \sum_{j=1}^{n} (y_{ij} - \bar{y}_{i.})^2 \tag{3-6}$$

Equation 3–6 states that the total variability in the data, as measured by the total corrected sum of squares, can be partitioned into a sum of squares of the differences **between** the treatment averages and the grand average, plus a sum of squares of the differences of observations **within** treatments from the treatment average. Now, the difference between the observed treatment averages and the grand average is a measure of the differences between treatment means, whereas the differences of observations within a treatment from the treatment average can be due only to random error. Thus, we may write Equation 3–6 symbolically as

$$SS_T = SS_{\text{Treatments}} + SS_E$$

where $SS_{\text{Treatments}}$ is called the sum of squares due to treatments (i.e., between treatments), and SS_E is called the sum of squares due to error (i.e., within treatments). There are $an = N$ total observations; thus, SS_T has $N - 1$ degrees of freedom. There are a levels of the factor (and a treatment means), so $SS_{\text{Treatments}}$ has $a - 1$ degrees of freedom. Finally, within any treatment there are n replicates providing $n - 1$ degrees of freedom with which to estimate the experimental error. Since there are a treatments, we have $a(n - 1) = an - a = N - a$ degrees of freedom for error.

It is helpful to examine explicitly the two terms on the right-hand side of the fundamental analysis of variance identity (Equation 3–6). Consider the error sum of squares

$$SS_E = \sum_{i=1}^{a} \sum_{j=1}^{n} (y_{ij} - \bar{y}_{i.})^2 = \sum_{i=1}^{a} \left[\sum_{j=1}^{n} (y_{ij} - \bar{y}_{i.})^2 \right]$$

In this form it is easy to see that the term within square brackets, if divided by $n - 1$, is the sample variance in the ith treatment, or

$$S_i^2 = \frac{\sum_{j=1}^{n} (y_{ij} - \bar{y}_{i.})^2}{n - 1} \qquad i = 1, 2, \ldots, a$$

Now a sample variances may be combined to give a single estimate of the common population variance as follows:

$$\frac{(n - 1)S_1^2 + (n - 1)S_2^2 + \cdots + (n - 1)S_a^2}{(n - 1) + (n - 1) + \cdots + (n - 1)} = \frac{\sum_{i=1}^{a} \left[\sum_{j=1}^{n} (y_{ij} - \bar{y}_{i.})^2 \right]}{\sum_{i=1}^{a} (n - 1)}$$

$$= \frac{SS_E}{(N - a)}$$

Thus, $SS_E/(N - a)$ is an estimate of the common variance within each of the a treatments.

Similarly, if there were no differences between the a treatment means, we could use the variation of the treatment averages from the grand average to estimate σ^2. Specifically,

$$\frac{SS_{\text{Treatments}}}{a-1} = \frac{n\sum_{i=1}^{a}(\bar{y}_{i.} - \bar{y}_{..})^2}{a-1}$$

is an estimate of σ^2 if the treatment means are equal. The reason for this may be intuitively seen as follows: The quantity $\sum_{i=1}^{a}(\bar{y}_{i.} - \bar{y}_{..})^2/(a-1)$ estimates σ^2/n, the variance of the treatment averages, so $n\sum_{i=1}^{a}(\bar{y}_{i.} - \bar{y}_{..})^2/(a-1)$ must estimate σ^2 if there are no differences in treatment means.

We see that the analysis of variance identity (Equation 3–6) provides us with two estimates of σ^2—one based on the inherent variability within treatments and one based on the variability between treatments. If there are no differences in the treatment means, these two estimates should be very similar, and if they are not, we suspect that the observed difference must be caused by differences in the treatment means. Although we have used an intuitive argument to develop this result, a somewhat more formal approach can be taken.

The quantities

$$MS_{\text{Treatments}} = \frac{SS_{\text{Treatments}}}{a-1}$$

and

$$MS_E = \frac{SS_E}{N-a}$$

are called **mean squares.** We now examine the **expected values** of these mean squares. Consider

$$E(MS_E) = E\left(\frac{SS_E}{N-a}\right) = \frac{1}{N-a}E\left[\sum_{i=1}^{a}\sum_{j=1}^{n}(y_{ij} - \bar{y}_{i.})^2\right]$$

$$= \frac{1}{N-a}E\left[\sum_{i=1}^{a}\sum_{j=1}^{n}(y_{ij}^2 - 2y_{ij}\bar{y}_{i.} + \bar{y}_{i.}^2)\right]$$

$$= \frac{1}{N-a}E\left[\sum_{i=1}^{a}\sum_{j=1}^{n}y_{ij}^2 - 2n\sum_{i=1}^{a}\bar{y}_{i.}^2 + n\sum_{i=1}^{a}\bar{y}_{i.}^2\right]$$

$$= \frac{1}{N-a}E\left[\sum_{i=1}^{a}\sum_{j=1}^{n}y_{ij}^2 - \frac{1}{n}\sum_{i=1}^{a}y_{i.}^2\right]$$

Substituting the model (Equation 3–1) into this equation, we obtain

$$E(MS_E) = \frac{1}{N-a}E\left[\sum_{i=1}^{a}\sum_{j=1}^{n}(\mu + \tau_i + \epsilon_{ij})^2 - \frac{1}{n}\sum_{i=1}^{a}\left(\sum_{j=1}^{n}\mu + \tau_i + \epsilon_{ij}\right)^2\right]$$

Now when squaring and taking expectation of the quantity within the brackets, we see that terms involving ϵ_{ij}^2 and $\epsilon_{i.}^2$ are replaced by σ^2 and $n\sigma^2$, respectively, because $E(\epsilon_{ij}) = 0$. Furthermore, all cross-products involving ϵ_{ij} have zero expectation. Therefore, after squaring and taking expectation, the last equation becomes

$$E(MS_E) = \frac{1}{N-a}\left[N\mu^2 + n\sum_{i=1}^{a} \tau_i^2 + N\sigma^2 - N\mu^2 - n\sum_{i=1}^{a} \tau_i^2 - a\sigma^2 \right]$$

or

$$E(MS_E) = \sigma^2$$

By a similar approach, we may also show that

$$E(MS_{\text{Treatments}}) = \sigma^2 + \frac{n\sum_{i=1}^{a} \tau_i^2}{a-1}$$

Thus, as we argued heuristically, $MS_E = SS_E/(N-a)$ estimates σ^2, and, if there are no differences in treatment means (which implies that $\tau_i = 0$), $MS_{\text{Treatments}} = SS_{\text{Treatments}}/(a-1)$ also estimates σ^2. However, note that if treatment means do differ, the expected value of the treatment mean square is greater than σ^2.

It seems clear that a test of the hypothesis of no difference in treatment means can be performed by comparing $MS_{\text{Treatments}}$ and MS_E. We now consider how this comparison may be made.

3–3.2   Statistical Analysis

We now investigate how a formal test of the hypothesis of no differences in treatment means ($H_0: \mu_1 = \mu_2 = \cdots = \mu_a$, or equivalently, $H_0: \tau_1 = \tau_2 = \cdots \tau_a = 0$) can be performed. Since we have assumed that the errors ϵ_{ij} are normally and independently distributed with mean zero and variance σ^2, the observations y_{ij} are normally and independently distributed with mean $\mu + \tau_i$ and variance σ^2. Thus, SS_T is a sum of squares in normally distributed random variables; consequently, it can be shown that SS_T/σ^2 is distributed as chi-square with $N - 1$ degrees of freedom. Furthermore, we can show that SS_E/σ^2 is chi-square with $N - a$ degrees of freedom and that $SS_{\text{Treatments}}/\sigma^2$ is chi-square with $a - 1$ degrees of freedom if the null hypothesis $H_0: \tau_i = 0$ is true. However, all three sums of squares are not independent since $SS_{\text{Treatments}}$ and SS_E add to SS_T. The following theorem, which is a special form of one attributed to William Cochran, is useful in establishing the independence of SS_E and $SS_{\text{Treatments}}$.

THEOREM 3–1   COCHRAN'S THEOREM   Let Z_i be NID$(0, 1)$ for $i = 1$, $2, \ldots, \nu$ and

$$\sum_{i=1}^{\nu} Z_i^2 = Q_1 + Q_2 + \cdots + Q_s$$

where $s \leq \nu$, and Q_i has ν_i degrees of freedom ($i = 1, 2, \ldots, s$). Then $Q_1, Q_2, \ldots, Q_s$ are independent chi-square random variables with $\nu_1, \nu_2, \ldots, \nu_s$ degrees of freedom, respectively, if and only if

$$\nu = \nu_1 + \nu_2 + \cdots + \nu_s$$

■ ■

Since the degrees of freedom for $SS_{\text{Treatments}}$ and SS_E add to $N - 1$, the total number of degrees of freedom, Cochran's theorem implies that $SS_{\text{Treatments}}/\sigma^2$ and SS_E/σ^2 are independently distributed chi-square random variables. Therefore, if the null hypothesis of no difference in treatment means is true, the ratio

$$F_0 = \frac{SS_{\text{Treatments}}/(a - 1)}{SS_E/(N - a)} = \frac{MS_{\text{Treatments}}}{MS_E} \tag{3-7}$$

is distributed as F with $a - 1$ and $N - a$ degrees of freedom. Equation 3-7 is the **test statistic** for the hypothesis of no differences in treatment means.

From the expected mean squares we see that, in general, MS_E is an unbiased estimator of σ^2. Also, under the null hypothesis, $MS_{\text{Treatments}}$ is an unbiased estimator of σ^2. However, if the null hypothesis is false, then the expected value of $MS_{\text{Treatments}}$ is greater than σ^2. Therefore, under the alternative hypothesis, the expected value of the numerator of the test statistic (Equation 3-7) is greater than the expected value of the denominator, and we should reject H_0 on values of the test statistic that are too large. This implies an upper-tail, one-tail critical region. Therefore, we should reject H_0 and conclude that there are differences in the treatment means if

$$F_0 > F_{\alpha, a-1, N-a}$$

where F_0 is computed from Equation 3-7. Alternatively, we could use the P-value approach for decision making.

Computing formulas for the sums of squares may be obtained by rewriting and simplifying the definitions of $MS_{\text{Treatments}}$ and SS_T in Equation 3-6. This yields

$$SS_T = \sum_{i=1}^{a} \sum_{j=1}^{n} y_{ij}^2 - \frac{y_{..}^2}{N} \tag{3-8}$$

and

$$SS_{\text{Treatments}} = \frac{1}{n} \sum_{i=1}^{a} y_{i.}^2 - \frac{y_{..}^2}{N} \tag{3-9}$$

The error sum of squares is obtained by subtraction as

$$SS_E = SS_T - SS_{\text{Treatments}} \tag{3-10}$$

The test procedure is summarized in Table 3-3. This is called an **analysis of variance table.**

Table 3–3   The Analysis of Variance Table for the Single-Factor, Fixed Effects Model

Source of Variation	Sum of Squares	Degrees of Freedom	Mean Square	F_0
Between treatments	$SS_{\text{Treatments}}$	$a - 1$	$MS_{\text{Treatments}}$	$F_0 = \dfrac{MS_{\text{Treatments}}}{MS_E}$
Error (within treatments)	SS_E	$N - a$	MS_E	
Total	SS_1	$N - 1$		

Example 3–1

The Tensile Strength Experiment

To illustrate the analysis of variance, return to the example first discussed in Section 3–1. Recall that the development engineer is interested in determining if the cotton weight percent in a synthetic fiber affects the tensile strength, and he has run a completely randomized experiment with five levels of cotton weight percent and five replicates. For convenience, we repeat the data from Table 3–1 here:

Weight Percent of Cotton	Observed Tensile Strength (lb/in²)					Totals y_i	Averages $\bar{y}_{i.}$
	1	2	3	4	5		
15	7	7	15	11	9	49	9.8
20	12	17	12	18	18	77	15.4
25	14	18	18	19	19	88	17.6
30	19	25	22	19	23	108	21.6
35	7	10	11	15	11	54	10.8
						$y_{..} = 376$	$\bar{y}_{..} = 15.04$

We will use the analysis of variance to test $H_0 : \mu_1 = \mu_2 = \mu_3 = \mu_4 = \mu_5$ against the alternative $H_1 :$ some means are different. The sums of squares required are computed as follows:

$$SS_T = \sum_{i=1}^{5} \sum_{j=1}^{5} y_{ij}^2 - \frac{y_{..}^2}{N}$$

$$= (7)^2 + (7)^2 + (15)^2 + \cdots + (15)^2 + (11)^2 - \frac{(376)^2}{25} = 636.96$$

$$SS_{\text{Treatments}} = \frac{1}{n} \sum_{i=1}^{5} y_{i.}^2 - \frac{y_{..}^2}{N}$$

$$= \frac{1}{5} [(49)^2 + \cdots + (54)^2] - \frac{(376)^2}{25} = 475.76$$

$$SS_E = SS_T - SS_{\text{Treatments}}$$

$$= 636.96 - 475.76 = 161.20$$

Table 3–4   Analysis of Variance for the Tensile Strength Data

Source of Variation	Sum of Squares	Degrees of Freedom	Mean Square	F_0	P-Value
Cotton weight percent	475.76	4	118.94	$F_0 = 14.76$	<0.01
Error	161.20	20	8.06		
Total	636.96	24			

Usually, these calculations would be performed on a computer, using a software package with the capability to analyze data from designed experiments.

The analysis of variance is summarized in Table 3–4. Note that the between-treatment mean square (118.94) is many times larger than the within-treatment or error mean square (8.06). This indicates that it is unlikely that the treatment means are equal. More formally, we can compute the F ratio $F_0 = 118.94/8.06 = 14.76$ and compare this to an appropriate upper-tail percentage point of the $F_{4,20}$ distribution. Suppose that the experimenter has selected $\alpha = 0.05$. From Appendix Table IV we find that $F_{0.05,4,20} = 2.87$. Since $F_0 = 14.76 > 2.87$, we reject H_0 and conclude that the treatment means differ; that is, the cotton weight percent in the fiber significantly affects the mean tensile strength. We could also compute a P-value for this test statistic. Figure 3–3 shows the reference distribution ($F_{4,20}$) for the test statistic F_0. Clearly, the P-value is very small in this case. Since $F_{0.01,4,20} = 4.43$ and $F_0 > 4.43$, we can conclude that an upper bound for the P-value is 0.01; that is, $P < 0.01$ (the exact P-value is $P = 9.11 \times 10^{-6}$).

◼

Computations ▪ The reader has likely noted that we defined sum of squares in terms of averages; that is, from Equation 3–6,

$$SS_{\text{Treatments}} = n \sum_{i=1}^{a} (\bar{y}_{i.} - \bar{y}_{..})^2$$

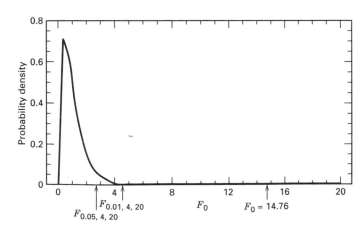

Figure 3–3.   The reference distribution ($F_{4,20}$) for the test statistic F_0 in Example 3–1.

but we developed the computing formulas using totals. For example, to compute $SS_{\text{Treatments}}$, we would use Equation 3–9:

$$SS_{\text{Treatments}} = \frac{1}{n} \sum_{i=1}^{a} y_{i.}^2 - \frac{y_{..}^2}{N}$$

The primary reason for this is convenience; furthermore, the totals $y_{i.}$ and $y_{..}$ are not as subject to rounding error as are the averages $\bar{y}_{i.}$ and $\bar{y}_{..}$.

Generally, we need not be too concerned with computing, as there are many widely available computer programs for performing the calculations. These computer programs are also helpful in performing many other analyses associated with experimental design (such as residual analysis and model adequacy checking). In many cases, these programs will also assist the experimenter in setting up the design.

When hand calculations are necessary, it is sometimes helpful to code the observations. This is illustrated in the next example.

Example 3–2

Coding the Observations

The calculations in the analysis of variance may often be made more accurate or simplified by coding the observations. For example, consider the tensile strength data in Example 3–1. Suppose we subtract 15 from each observation. The coded data are shown in Table 3–5. It is easy to verify that

$$SS_T = (-8)^2 + (-8)^2 + \cdots + (-4)^2 - \frac{(1)^2}{25} = 636.96$$

$$SS_{\text{Treatments}} = \frac{(-26)^2 + (2)^2 + \cdots + (-21)^2}{5} - \frac{(1)^2}{25} = 475.76$$

and

$$SS_E = 161.20$$

Comparing these sums of squares to those obtained in Example 3–1, we see that subtracting a constant from the original data does not change the sums of squares.

Table 3–5   Coded Tensile Strength Data for Example 3–2

Cotton Weight Percent	Observations					Totals y_i
	1	2	3	4	5	
15	−8	−8	0	−4	−6	−26
20	−3	2	−3	3	3	2
25	−1	3	3	4	4	13
30	4	10	7	4	8	33
35	−8	−5	−4	0	−4	−21
						$1 = y_{..}$

Now suppose that we multiply each observation in Example 3-1 by 2. It is easy to verify that the sums of squares for the transformed data are $SS_T = 2547.84$, $SS_{\text{Treatments}} = 1903.04$, and $SS_E = 644.80$. These sums of squares appear to differ considerably from those obtained in Example 3-1. However, if they are divided by 4 (i.e., 2^2), then the results are identical. For example, for the treatment sum of squares $1903.04/4 = 475.76$. Also, for the coded data, the F ratio is $F = (1903.04/4)/(644.80/20) = 14.76$, which is identical to the F ratio for the original data. Thus, the analyses of variance are equivalent.

∎

Randomization Tests and Analysis of Variance ▪ In our development of the F test analysis of variance, we have used the assumption that the random errors ϵ_{ij} are normally and independently distributed random variables. The F test can also be justified as an approximation to a **randomization test.** To illustrate, suppose that we have five observations on each of two treatments and that we wish to test the equality of treatment means. The data would look like this:

Treatment 1	Treatment 2
y_{11}	y_{21}
y_{12}	y_{22}
y_{13}	y_{23}
y_{14}	y_{24}
y_{15}	y_{25}

We could use the F test analysis of variance to test $H_0: \mu_1 = \mu_2$. Alternatively, we could use a somewhat different approach. Suppose we consider all the possible ways of allocating the 10 numbers in the sample above to the two treatments. There are $10!/5!5! = 252$ possible arrangements of the 10 observations. If there is no difference in treatment means, all 252 arrangements are equally likely. For each of the 252 arrangements, we calculate the value of the F statistic using Equation 3-7. The distribution of these F values is called a **randomization distribution,** and a large value of F indicates that the data are not consistent with the hypothesis $H_0: \mu_1 = \mu_2$. For example, if the value of F actually observed was exceeded by only 5 of the values of the randomization distribution, then this would correspond to rejection of $H_0: \mu_1 = \mu_2$ at a significance level of $\alpha = 5/252 = .0198$ (or 1.98 percent). Notice that no normality assumption is required in this approach.

The difficulty with this approach is that, even for relatively small problems, it is computationally prohibitive to enumerate the exact randomization distribution. However, numerous studies have shown that the exact randomization distribution is well-approximated by the usual normal-theory F distribution. Thus, even without the normality assumption, the F test can be viewed as an approximation to the randomization test. For further reading on randomization tests in the analysis of variance, see Box, Hunter, and Hunter (1978).

3–3.3 Estimation of the Model Parameters

We now present estimators for the parameters in the single-factor model

$$y_{ij} = \mu + \tau_i + \epsilon_{ij}$$

and confidence intervals on the treatment means. We will prove later that reasonable estimates of the overall mean and the treatment effects are given by

$$\hat{\mu} = \bar{y}_{..}$$
$$\hat{\tau}_i = \bar{y}_{i.} - \bar{y}_{..}, \qquad i = 1, 2, \ldots, a \tag{3–11}$$

These estimators have considerable intuitive appeal; note that the overall mean is estimated by the grand average of the observations and that any treatment effect is just the difference between the treatment average and the grand average.

A confidence interval estimate of the ith treatment mean may be easily determined. The mean of the ith treatment is

$$\mu_i = \mu + \tau_i$$

A point estimator of μ_i would be $\hat{\mu}_i = \hat{\mu} + \hat{\tau}_i = \bar{y}_{i.}$. Now, if we assume that the errors are normally distributed, each $\bar{y}_{i.}$ is NID$(\mu_i, \sigma^2/n)$. Thus, if σ^2 were known, we could use the normal distribution to define the confidence interval. Using the MS_E as an estimator of σ^2, we would base the confidence interval on the t distribution. Therefore, a $100(1 - \alpha)$ percent confidence interval on the ith treatment mean μ_i is

$$[\bar{y}_{i.} \pm t_{\alpha/2,N-a} \sqrt{MS_E/n}] \tag{3–12}$$

A $100(1 - \alpha)$ percent confidence interval on the difference in any two treatments means, say $\mu_i - \mu_j$, would be

$$[\bar{y}_{i.} - \bar{y}_{j.} \pm t_{\alpha/2,N-a} \sqrt{2MS_E/n}] \tag{3–13}$$

Example 3–3

Using the data in Example 3–1, we may find the estimates of the overall mean and the treatment effects as $\hat{\mu} = 376/25 = 15.04$ and

$$\hat{\tau}_1 = \bar{y}_{1.} - \bar{y}_{..} = 9.80 - 15.04 = -5.24$$
$$\hat{\tau}_2 = \bar{y}_{2.} - \bar{y}_{..} = 15.40 - 15.04 = +0.36$$
$$\hat{\tau}_3 = \bar{y}_{3.} - \bar{y}_{..} = 17.60 - 15.04 = -2.56$$
$$\hat{\tau}_4 = \bar{y}_{4.} - \bar{y}_{..} = 21.60 - 15.04 = +6.56$$
$$\hat{\tau}_5 = \bar{y}_{5.} - \bar{y}_{..} = 10.80 - 15.04 = -4.24$$

A 95 percent confidence interval on the mean of treatment 4 (30 percent cotton) is computed from Equation 3–12 as

$$[21.60 \pm (2.086)\sqrt{8.06/5}]$$

or

$$[21.60 \pm 2.65]$$

Thus, the desired confidence interval is $18.95 \leq \mu_4 \leq 25.25$.

■

3–3.4 Unbalanced Data

In some single-factor experiments the number of observations taken within each treatment may be different. We then say that the design is **unbalanced.** The analysis of variance described above may still be used, but slight modifications must be made in the sum of squares formulas. Let n_i observations be taken under treatment i ($i = 1, 2, \ldots, a$) and $N = \sum_{i=1}^{a} n_i$. The manual computational formulas for SS_T and $SS_{\text{Treatments}}$ become

$$SS_T = \sum_{i=1}^{a} \sum_{j=1}^{n_i} y_{ij}^2 - \frac{y_{..}^2}{N} \tag{3–14}$$

and

$$SS_{\text{Treatments}} = \sum_{i=1}^{a} \frac{y_{i.}^2}{n_i} - \frac{y_{..}^2}{N} \tag{3–15}$$

No other changes are required in the analysis of variance.

There are two advantages in choosing a balanced design. First, the test statistic is relatively insensitive to small departures from the assumption of equal variances for the a treatments if the sample sizes are equal. This is not the case for unequal sample sizes. Second, the power of the test is maximized if the samples are of equal size.

3–4 MODEL ADEQUACY CHECKING

The decomposition of the variability in observations through an analysis of variance identity (Equation 3–6) is a purely algebraic relationship. However, the use of the partitioning to test formally for no differences in treatment means requires that certain assumptions be satisfied. Specifically, these assumptions are that the observations are adequately described by the model

$$y_{ij} = \mu + \tau_i + \epsilon_{ij}$$

and that the errors are normally and independently distributed with mean zero and constant but unknown variance σ^2. If these assumptions are valid, then the analysis of variance procedure is an exact test of the hypothesis of no difference in treatment means.

In practice, however, these assumptions will usually not hold exactly. Consequently, it is usually unwise to rely on the analysis of variance until the validity of these assumptions has been checked. Violations of the basic assumptions and model adequacy can be easily investigated by the examination of **residuals.** We define the residual for observation j in treatment i as

$$e_{ij} = y_{ij} - \hat{y}_{ij} \tag{3-16}$$

where $\hat{y}_{ij}$ is an estimate of the corresponding observation y_{ij} obtained as follows:

$$\begin{aligned} \hat{y}_{ij} &= \hat{\mu} + \hat{\tau}_i \\ &= \bar{y}_{..} + (\bar{y}_{i.} - \bar{y}_{..}) \\ &= \bar{y}_{i.} \end{aligned} \tag{3-17}$$

Equation 3–17 gives the intuitively appealing result that the estimate of any observation in the ith treatment is just the corresponding treatment average.

   Examination of the residuals should be an automatic part of any analysis of variance. If the model is adequate, the residuals should be **structureless;** that is, they should contain no obvious patterns. Through a study of residuals, many types of model inadequacies and violations of the underlying assumptions can be discovered. In this section, we show how model diagnostic checking can be done easily by graphical analysis of residuals and how to deal with several commonly occurring abnormalities.

3–4.1   The Normality Assumption

A check of the normality assumption could be made by plotting a histogram of the residuals. If the NID($0, \sigma^2$) assumption on the errors is satisfied, then this plot should look like a sample from a normal distribution centered at zero. Unfortunately, with small samples, considerable fluctuation often occurs, so the appearance of a moderate departure from normality does not necessarily imply a serious violation of the assumptions. Gross deviations from normality are potentially serious and require further analysis.

   An extremely useful procedure is to construct a normal probability plot of the residuals. Recall that in Chapter 2 we used a normal probability plot of the raw data to check the assumption of normality when using the t test. In the analysis of variance, it is usually more effective (and straightforward) to do this with the residuals. If the underlying error distribution is normal, this plot will resemble a straight line. In visualizing the straight line, place more emphasis on the central values of the plot than on the extremes.

   Table 3–6 shows the original data and the residuals for the tensile strength data in Example 3–1. In Table 3–7, the residuals are ranked in ascending order and their cumulative probability points $P_k = (k - \frac{1}{2})/n$ are calculated. The normal probability plot is shown in Figure 3–4 with the residuals plotted versus $P_k \times$ 100 on the right vertical scale. (See page 82.) Note that the bottom of this figure also gives a dot diagram of the residuals. The general impression from examining

Table 3–6 Data and Residuals from Example 3–1[a]

Weight Percent of Cotton	Observations (j)										$\hat{y}_{ij} = \bar{y}_{i.}$
	1		2		3		4		5		
15	7	$\boxed{-2.8}$ (15)	7	$\boxed{-2.8}$ (19)	15	$\boxed{5.2}$ (25)	11	$\boxed{1.2}$ (12)	9	$\boxed{-0.8}$ (6)	9.8
20	12	$\boxed{-3.4}$ (8)	17	$\boxed{1.6}$ (14)	12	$\boxed{-3.4}$ (1)	18	$\boxed{2.6}$ (11)	19	$\boxed{2.6}$ (3)	15.4
25	14	$\boxed{-3.6}$ (18)	18	$\boxed{0.4}$ (13)	18	$\boxed{0.4}$ (20)	19	$\boxed{1.4}$ (7)	19	$\boxed{1.4}$ (9)	17.6
30	19	$\boxed{-2.6}$ (22)	25	$\boxed{3.4}$ (5)	22	$\boxed{0.4}$ (2)	19	$\boxed{-2.6}$ (24)	23	$\boxed{1.4}$ (10)	21.6
35	7	$\boxed{-3.8}$ (17)	10	$\boxed{-0.8}$ (21)	11	$\boxed{0.2}$ (4)	15	$\boxed{4.2}$ (16)	11	$\boxed{0.2}$ (23)	10.8

[a] The residuals are shown in the box in each cell. The numbers in parentheses indicate the order of data collection.

this display is that the error distribution may be slightly skewed, with the right tail being longer than the left. The tendency of the normal probability plot to bend down slightly on the left side implies that the left tail of the error distribution is somewhat *thinner* than would be anticipated in a normal distribution; that is, the negative residuals are not quite as large (in absolute value) as expected. This plot is not grossly nonnormal, however.

In general, moderate departures from normality are of little concern in the

Table 3–7 Ordered Residuals and Probability Points for the Tensile Strength Data

Order k	Residual e_{ij}	$P_k = (k - \frac{1}{2})/25$	Order k	Residual e_{ij}	$P_k = (k - \frac{1}{2})/25$
1	−3.8	.0200	14	0.4	.5400
2	−3.6	.0600	15	0.4	.5800
3	−3.4	.1000	16	1.2	.6200
4	−3.4	.1400	17	1.4	.6600
5	−2.8	.1800	18	1.4	.7000
6	−2.8	.2200	19	1.4	.7400
7	−2.8	.2600	20	1.6	.7800
8	−2.6	.3000	21	2.6	.8200
9	−0.8	.3400	22	2.6	.8600
10	−0.8	.3800	23	3.4	.9000
11	0.2	.4200	24	4.2	.9400
12	0.2	.4600	25	5.2	.9800
13	0.4	.5000			

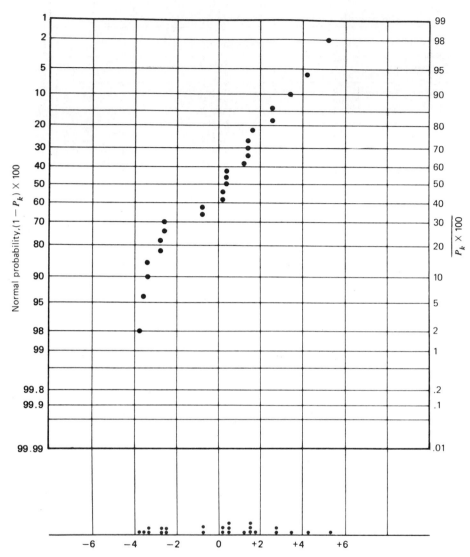

Figure 3–4.   Normal probability plot and dot diagram of residuals for Example 3–1.

fixed effects analysis of variance (recall our discussion of randomization tests in Section 3–3.2). An error distribution that has considerably thicker or thinner tails than the normal is of more concern than a skewed distribution. Since the F test is only slightly affected, we say that the analysis of variance (and related procedures such as multiple comparisons) is **robust** to the normality assumption. Departures from normality usually cause both the true significance level and the power to differ slightly from the advertised values, with the power generally being lower. The random effects model is more severely impacted by nonnormality. In particular, the true confidence levels on interval estimates of variance components may differ greatly from the advertised values.

A very common defect that often shows up on normal probability plots is one residual that is very much larger than any of the others. Such a residual is often called an **outlier.** The presence of one or more outliers can seriously distort the analysis of variance, so when a potential outlier is located, careful investigation is called for. Frequently, the cause of the outlier is a mistake in calculations or a data coding or copying error. If this is not the cause, then the experimental circumstances surrounding this run must be carefully studied. If the outlying response is a particularly desirable value (high strength, low cost, etc.), then the outlier may be more informative than the rest of the data. We should be careful not to reject or discard an outlying observation unless we have reasonable nonstatistical grounds for doing so. At worst, you may end up with two analyses; one with the outlier and one without.

There are several formal statistical procedures for detecting outliers [e.g., see Barnett and Lewis (1978), John and Prescott (1975), and Stefansky (1972)]. A rough check for outliers may be made by examining the standardized residuals

$$d_{ij} = \frac{e_{ij}}{\sqrt{MS_E}} \tag{3–18}$$

If the errors ϵ_{ij} are $N(0, \sigma^2)$, then the standardized residuals should be approximately normal with mean zero and unit variance. Thus, about 68 percent of the standardized residuals should fall within the limits ± 1, about 95 percent of them should fall within ± 2, and virtually all of them should fall within ± 3. A residual bigger than 3 or 4 standard deviations from zero is a potential outlier.

For the tensile strength data of Example 3–1, the normal probability plot and dot diagram give no indication of outliers. Furthermore, the largest standardized residual is

$$d_{13} = \frac{e_{13}}{\sqrt{MS_E}} = \frac{5.2}{\sqrt{8.06}} = \frac{5.2}{2.84} = 1.83$$

which should cause no concern.

3–4.2   Plot of Residuals in Time Sequence

Plotting the residuals in time order of data collection is helpful in detecting **correlation** between the residuals. A tendency to have runs of positive and negative residuals indicates positive correlation. This would imply that the **independence assumption** on the errors has been violated. This is a potentially serious problem and one that is difficult to correct, so it is important to prevent the problem if possible when the data are collected. Proper randomization of the experiment is an important step in obtaining independence.

Sometimes the skill of the experimenter (or the subjects) may change as the

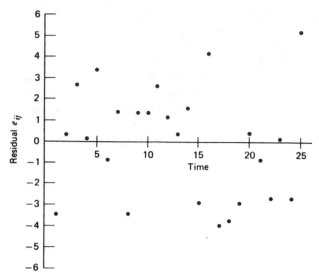

Figure 3-5.   Plot of residuals versus time.

experiment progresses, or the process being studied may "drift" or become more erratic. This will often result in a change in the error variance over time. This condition often leads to a plot of residuals versus time that exhibits more spread at one end than at the other. Nonconstant variance is a potentially serious problem. We will have more to say on the subject in Sections 3–4.3 and 3–4.4.

Table 3–6 displays the residuals and the time sequence of data collection for the tensile strength data. A plot of these residuals versus time is shown in Figure 3–5. There is no reason to suspect any violation of the independence or constant variance assumptions.

3–4.3   Plot of Residuals Versus Fitted Values

If the model is correct and if the assumptions are satisfied, the residuals should be structureless; in particular, they should be unrelated to any other variable including the predicted response. A simple check is to plot the residuals versus the fitted values $\hat{y}_{ij}$. (For the one-way model, remember that $\hat{y}_{ij} = \bar{y}_{i.}$, the ith treatment average.) This plot should not reveal any obvious pattern. Figure 3–6 plots the residuals versus the fitted values for the tensile strength data of Example 3–1. No unusual structure is apparent.

A defect that occasionally shows up on this plot is **nonconstant variance.** Sometimes the variance of the observations increases as the magnitude of the observation increases. This would be the case if the error or background noise in the experiment was a constant percentage of the size of the observation. (This

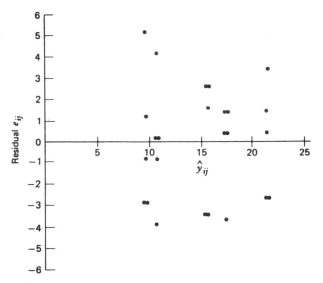

Figure 3–6. Plot of residuals versus fitted values.

commonly happens with many measuring instruments—error is a percent of the scale reading.) If this were the case, the residuals would get larger as y_{ij} gets larger, and the plot of residuals versus $\hat{y}_{ij}$ would look like an outward-opening funnel or megaphone. Nonconstant variance also arises in cases where the data follow a nonnormal, skewed distribution because in skewed distributions the variance tends to be a function of the mean.

If the assumption of homogeneity of variances is violated, the F test is only slightly affected in the balanced fixed effects model. However, in unbalanced designs or in cases where one variance is very much larger than the others, the problem is more serious. For the random effects model, unequal error variances can significantly disturb inferences on variance components even if balanced designs are used.

The usual approach to dealing with nonconstant variance when it occurs for the above reasons is to apply a **variance-stabilizing transformation** and then to run the analysis of variance on the transformed data. In this approach, one should note that the conclusions of the analysis of variance apply to the *transformed* populations.

Considerable research has been devoted to the selection of an appropriate transformation. If experimenters know the theoretical distribution of the observations, they may utilize this information in choosing a transformation. For example, if the observations follow the Poisson distribution, then the square root transformation $y_{ij}^* = \sqrt{y_{ij}}$ or $y_{ij}^* = \sqrt{1 + y_{ij}}$ would be used. If the data follow the log-normal distribution, then the logarithmic transformation $y_{ij}^* = \log y_{ij}$ is appropriate. For binomial data expressed as fractions, the arcsin transformation $y_{ij}^* = \arcsin\sqrt{y_{ij}}$ is useful. When there is no obvious transformation, the experimenter usually empirically seeks a transformation that equalizes the variance regardless of the value of the mean. In factorial experiments, which we discuss in Chapter 6, another approach is to select a transformation that minimizes the interaction mean square, resulting in an experiment that is easier to interpret. For more discussion of transformations, refer to Bartlett (1947), Box and Cox (1964), Dolby (1963), and Draper and Hunter (1969). In Section 3–4.4, we discuss in more detail methods for analytically selecting the form of the transformation. Transformations made for inequality of variance also affect the form of the error distribution. In most cases, the transformation brings the error distribution closer to normal.

Statistical Tests for Equality of Variance ▪ Although residual plots are frequently used to diagnose inequality of variance, several statistical tests have also been proposed. These tests may be viewed as formal tests of the hypotheses

$$H_0: \sigma_1^2 = \sigma_2^2 = \cdots = \sigma_a^2$$

$$H_1: \text{above not true for at least one } \sigma_i^2$$

A widely used procedure is **Bartlett's test.** The procedure involves computing a statistic whose sampling distribution is closely approximated by the chi-square distribution with $a - 1$ degrees of freedom when the a random samples are from independent normal populations. The test statistic is

$$\chi_0^2 = 2.3026 \frac{q}{c} \tag{3–19}$$

where

$$q = (N - a) \log_{10} S_p^2 - \sum_{i=1}^{a} (n_i - 1) \log_{10} S_i^2$$

$$c = 1 + \frac{1}{3(a - 1)} \left(\sum_{i=1}^{a} (n_i - 1)^{-1} - (N - a)^{-1} \right)$$

$$S_p^2 = \frac{\sum_{i=1}^{a} (n_i - 1) S_i^2}{N - a}$$

and S_i^2 is the sample variance of the ith population.

The quantity q is large when the sample variances S_i^2 differ greatly and is equal to zero when all S_i^2 are equal. Therefore, we should reject H_0 on values of χ_0^2 that are too large; that is, we reject H_0 only when

$$\chi_0^2 > \chi_{\alpha, a-1}^2$$

where $\chi_{\alpha, a-1}^2$ is the upper α percentage point of the chi-square distribution with $a - 1$ degrees of freedom. The P-value approach to decision making could also be used.

Various studies have indicated that Bartlett's test is very sensitive to the normality assumption and should not be applied when the normality assumption is doubtful. Other tests for equality of variance are reviewed by Anderson and McLean (1974).

Example 3-4

We can apply Bartlett's test to the tensile strength data from the cotton weight percent experiment in Example 3-1. We first compute the sample variances in each treatment and find that $S_1^2 = 11.2$, $S_2^2 = 9.8$, $S_3^2 = 4.3$, $S_4^2 = 6.8$, and $S_5^2 = 8.2$. Then

$$S_p^2 = \frac{4(11.2) + 4(9.8) + 4(4.3) + 4(6.8) + 4(8.2)}{20} = 8.06$$

$$q = 20 \log_{10}(8.06) - 4[\log_{10}11.2 + \log_{10}9.8 + \log_{10}4.3 + \log_{10}6.8 + \log_{10}8.2] = 0.45$$

$$c = 1 + \frac{1}{3(4)} \left(\frac{5}{4} - \frac{1}{20} \right) = 1.10$$

and the test statistic is

$$\chi_0^2 = 2.3026 \frac{(0.45)}{(1.10)} = 0.93$$

Since $\chi_{0.05,4}^2 = 9.49$, we cannot reject the null hypothesis and conclude that all five variances are the same. This is the same conclusion reached by analyzing the plot of residuals versus fitted values.

∎

3-4.4 Selecting a Variance-Stabilizing Transformation

In the previous section we noted that, if experimenters knew the relationship between the variance of the observations and the mean, they could use this information to guide them in selecting the form of the transformation. We now elaborate on this point and show how it is possible to estimate the form of the required transformation from the data.

Let $E(y) = \mu$ be the mean of y, and suppose that the standard deviation of y is proportional to a power of the mean of y such that

$$\sigma_y \propto \mu^\alpha$$

We want to find a transformation on y that yields a constant variance. Suppose that the transformation is a power of the original data, say

$$y^* = y^\lambda \tag{3-20}$$

Then it can be shown that

$$\sigma_{y^*} \propto \mu^{\lambda + \alpha - 1} \tag{3-21}$$

Table 3–8   Variance-Stabilizing Transformations

Relationship Between σ_y and μ	α	$\lambda = 1 - \alpha$	Transformation	Comment
$\sigma_y \propto$ constant	0	1	No transformation	
$\sigma_y \propto \mu^{1/2}$	1/2	1/2	Square root	Poisson (count) data
$\sigma_y \propto \mu$	1	0	Log	
$\sigma_y \propto \mu^{3/2}$	3/2	−1/2	Reciprocal square root	
$\sigma_y \propto \mu^2$	2	−1	Reciprocal	

Clearly, if we set $\lambda = 1 - \alpha$, then the variance of the transformed data $y*$ is constant.

Several of the common transformations discussed in Section 3–4.3 are summarized in Table 3–8. Note that $\lambda = 0$ implies the log transformation. These transformations are arranged in order of increasing **strength.** By the strength of a transformation we mean the amount of curvature it induces. A mild transformation applied to data spanning a narrow range has little effect on the analysis, whereas a strong transformation applied over a large range may have dramatic results. Transformations often have little effect unless the ratio y_{max}/y_{min} is larger than two or three.

Empirical Selection of α ▪ In many experimental design situations where there is replication, we can empirically estimate α from the data. Since in the ith treatment combination $\sigma_{y_i} \propto \mu_i^{\alpha} = \theta \mu_i^{\alpha}$, where θ is a constant of proportionality, we may take logs to obtain

$$\log \sigma_{y_i} = \log \theta + \alpha \log \mu_i \tag{3–22}$$

Therefore, a plot of $\log \sigma_{y_i}$ versus $\log \mu_i$ would be a straight line with slope α. Since we don't know σ_{y_i} and μ_i, we may substitute reasonable estimates of them in Equation 3–22 and use the slope of the resulting straight line fit as an estimate of α. Typically, we would use the standard deviation S_i and the average $\bar{y}_{i.}$ of the ith treatment (or, more generally, the ith treatment combination or set of experimental conditions) to estimate σ_{y_i} and μ_i.

Table 3–9   Peak Discharge Data

Estimation Method	Observations						$\bar{y}_{i.}$	S_i
1	0.34	0.12	1.23	0.70	1.75	0.12	0.71	0.66
2	0.91	2.94	2.14	2.36	2.86	4.55	2.63	1.09
3	6.31	8.37	9.75	6.09	9.82	7.24	7.93	1.66
4	17.15	11.82	10.95	17.20	14.35	16.82	14.72	2.77

Table 3–10 Analysis of Variance for Peak Discharge Data

Source of Variation	Sum of Squares	Degrees of Freedom	Mean Square	F_0	P-Value
Methods	708.3471	3	236.1157	76.07	<0.001
Error	62.0811	20	3.1041		
Total	770.4282	23			

Example 3–5

A civil engineer is interested in determining whether four different methods of estimating flood flow frequency produce equivalent estimates of peak discharge when applied to the same watershed. Each procedure is used six times on the watershed, and the resulting discharge data (in cubic feet per second) are shown in Table 3–9. The analysis of variance for the data, summarized in Table 3–10, implies that there is a difference in mean peak discharge estimates given by the four procedures. The plot of residuals versus fitted values, shown in Figure 3–7, is disturbing since the outward-opening funnel shape indicates that the constant variance assumption is not satisfied.

To investigate the possibility of using a variance-stabilizing transformation on the data, we plot log S_i versus log $\overline{y}_{i.}$ in Figure 3–8. The slope of a straight line passing through these four points is close to 1/2 and, from Table 3–8, this implies that the square root transformation may be appropriate. The analysis of variance for the transformed data $y^* = \sqrt{y}$ is presented in Table 3–11, and a plot of residuals is shown in Figure 3–9. This residual plot is much improved in comparison to Figure 3–7, so we conclude that the square root transformation has been helpful. Note that in Table 3–11 we have reduced the error degrees of freedom by one to account for the use of the data to estimate the transformation parameter α. ■

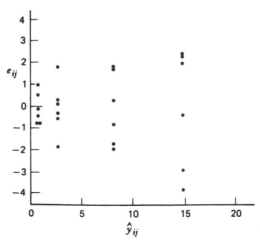

Figure 3–7. Plot of residuals versus $\hat{y}_{ij}$ for Example 3–5.

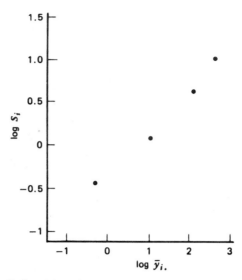

Figure 3–8.   Plot of log S_i versus log $\bar{y}_{i.}$ for Example 3–5.

Analytical Determination of the Transformation ▪ Box and Cox (1964) have shown how the transformation parameter λ in $y^* = y^\lambda$ may be estimated simultaneously with the other model parameters (overall mean and treatment effects) using the method of maximum likelihood. The procedure consists of performing, for various values of λ, a standard analysis of variance on

$$y^{(\lambda)} = \begin{cases} \dfrac{y^\lambda - 1}{\lambda \dot{y}^{\lambda-1}} & \lambda \neq 0 \\[2ex] \dot{y} \ln y & \lambda = 0 \end{cases} \tag{3–23}$$

where $\dot{y} = \ln^{-1}[(1/n) \sum \ln y]$ is the geometric mean of the observations. The maximum likelihood estimate of λ is the value for which the error sum of squares, say $SS_E(\lambda)$, is a minimum. This value of λ is usually found by plotting a graph of $SS_E(\lambda)$ versus λ and then reading the value of λ that minimizes $SS_E(\lambda)$ from the graph. Usually between 10 and 20 values of λ are sufficient for estimation of the optimum value. A second iteration using a finer mesh of values could be performed if a more accurate estimate of λ is necessary. Notice that we *cannot* select the value of λ by *directly* comparing the error sums of squares from analyses of variance on y^λ because for each value of λ the error sum of squares is measured

Table 3–11   Analysis of Variance for Transformed Peak Discharge Data, $y^* = \sqrt{y}$

Source of Variation	Sum of Squares	Degrees of Freedom	Mean Square	F_0	P-Value
Methods	32.6842	3	10.8947	76.99	<0.001
Error	2.6884	19	0.1415		
Total	35.3726	22			

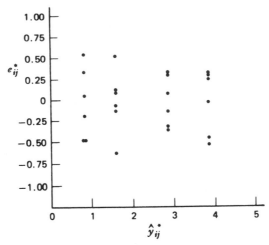

Figure 3-9.   Plot of residuals from transformed data versus $\hat{y}_{ij}^{*}$ for Example 3-5.

on a different scale. Equation 3-23 rescales the responses so that the error sums of squares are directly comparable. We recommend that the experimenter use simple choices for λ because the practical difference between $\lambda = 0.5$ and $\lambda = 0.55$ is likely to be small, but the former is much easier to interpret.

An approximate $100(1 - \alpha)$ percent confidence interval for λ can be found by computing

$$SS^{*} = SS_{E}(\lambda) \left(1 + \frac{t_{\alpha/2,\nu}^{2}}{\nu}\right) \tag{3-24}$$

where ν is the number of error degrees of freedom, and then by reading the corresponding confidence limits on λ directly from the graph. If this confidence interval includes the value $\lambda = 1$, this implies that the data do not support the need for transformation.

Example 3-6

Using the peak discharge data in Table 3-9, values of $SS_{E}(\lambda)$ for various values of λ are determined:

λ	$SS_{E}(\lambda)$
-1.00	7922.11
-0.50	687.10
-0.25	232.52
0.00	91.96
0.25	46.99
0.50	35.42
0.75	40.61
1.00	62.08
1.25	109.82
1.50	208.12

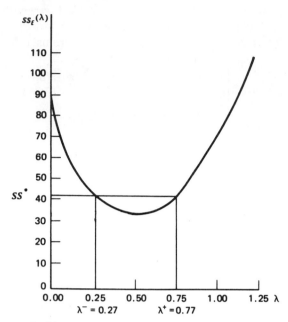

Figure 3–10. Plot of $SS_E(\lambda)$ versus λ for Example 3–6.

A graph of values close to the minimum is shown in Figure 3–10, from which it is seen that $\lambda = 0.52$ gives a minimum value of approximately $SS_E(\lambda) = 35.00$. An approximate 95 percent confidence interval on λ is found by calculating the quantity SS^* from Equation 3–24 as follows:

$$SS^* = SS_E(\lambda) \left(1 + \frac{t^2_{0.025,20}}{20}\right)$$

$$= 35.00 \left[1 + \frac{(2.086)^2}{20}\right]$$

$$= 42.61$$

By plotting SS^* on the graph in Figure 3–10 and by reading the points on the λ scale where this line intersects the curve, we obtain lower and upper confidence limits on λ of $\lambda^- = 0.27$ and $\lambda^+ = 0.77$. Since these confidence limits do not include the value 1, the use of a transformation is indicated, and the square root transformation ($\lambda = 0.50$) actually used is easily justified.

■

3–4.5 Plots of Residuals Versus Other Variables

If data have been collected on any other variables that might possibly affect the response, then the residuals should be plotted against these variables. For example, in the tensile strength experiment of Example 3–1, strength may be significantly affected by the thickness of the fiber, so the residuals should be plotted versus fiber thickness. If different testing machines were used to collect the data, then the residuals should be plotted against machines. Patterns in such residual plots imply that the variable affects the response. This suggests that the variable should be either controlled more carefully in future experiments or included in the analysis.

3–5 PRACTICAL INTERPRETATION OF RESULTS

After running the experiment, performing the statistical analysis, and investigating the underlying assumptions, the experimenter is ready to draw practical conclusions about the problem he or she is studying. Often this is realtively easy, and certainly in the simple experiments we have considered so far, this might be done somewhat informally, perhaps by inspection of graphical displays such as the box plots and scatter diagram in Figures 3–1 and 3–2. However, in some cases, more formal techniques need to be applied. We will present some of these techniques in this section.

3–5.1 A Regression Model

The factors involved in an experiment can be either **quantitative** or **qualitative.** A quantitative factor is one whose levels can be associated with points on a numerical scale, such as temperature, pressure, or time. Qualitative factors, on the other hand, are factors for which the levels cannot be arranged in order of magnitude. Operators, batches of raw material, and shifts are typical qualitative factors because there is no reason to rank them in any particular numerical order.

Insofar as the initial design and analysis of the experiment are concerned, both types of factors are treated identically. The experimenter is interested in determining the differences, if any, between the levels of the factors. If the factor is qualitative, such as operators, it is meaningless to consider the response for a subsequent run at an intermediate level of the factor. However, with a quantitative factor such as time, the experimenter is usually interested in the entire range of values used, particularly the response from a subsequent run at an intermediate factor level. That is, if the levels 1.0 hours, 2.0 hours, and 3.0 hours are used in the experiment, we may wish to predict the response at 2.5 hours. Thus, the experimenter is frequently interested in developing an interpolation equation for the response variable in the experiment. This equation is an **empirical model** of the process that has been studied.

The general approach to fitting empirical models is called **regression analysis,** which is discussed extensively in Chapter 13. This section briefly illustrates the technique using the tensile strength data of Example 3–1.

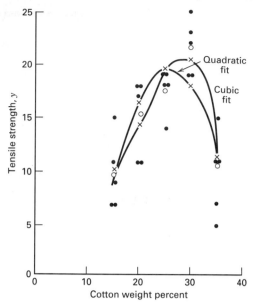

Figure 3–11. Scatter diagram for the tensile strength data of Example 3–1.

Figure 3–11 presents a scatter diagram of tensile strength y versus the weight percent of cotton in the fabric x for the experiment in Example 3–1. The open circles on the graph are the mean tensile strengths at each value of cotton weight percent x. From examining the scatter diagram, it is clear that the relationship between tensile strength and weight percent cotton is not linear. As a first approximation, we could try fitting a quadratic model to the data, say

$$y = \beta_0 + \beta_1 x + \beta_2 x^2 + \epsilon$$

where β_0, β_1, and β_2 are unknown parameters to be estimated and ϵ is a random error term. The method often used to estimate the parameters in a model such as this is the method of least squares. This consists of choosing estimates of the β's such that the sum of the squares of the errors (the ϵ's) are minimized. The least squares fit in our example is

$$\hat{y} = -39.9886 + 4.596x - 0.0886x^2$$

(If you are unfamiliar with regression methods, see Chapter 13.)

This quadratic model is shown in Figure 3–11. It does not appear to be very satisfactory because it drastically underestimates the responses at $x = 30$ percent cotton and overestimates the responses at $x = 25$ percent. Perhaps an improvement can be obtained by adding a cubic term in x. The resulting cubic fit is

$$\hat{y} = 62.6114 - 9.0114x + 0.4814x^2 - 0.0076x^3$$

This cubic fit is also shown in Figure 3–11. The cubic model appears to be superior to the quadratic because it provides a better fit at $x = 25$ and $x = 30$ percent cotton.

In general, we would like to fit the lowest-order polynomial that adequately describes the system or process. In this example, the cubic polynomial seems to fit better than the quadratic, so the extra complexity of the cubic model is justified. Selecting the order of the approximating polynomial is not always easy, however, and it is relatively easy to overfit, that is, to add high-order polynomial terms that do not really improve the fit but increase the complexity of the model and often damage its usefulness as a predictor or interpolation equation.

In this example, the empirical model could be used to predict mean tensile strength at values of cotton weight percent within the region of experimentation. In other cases, the empirical model could be used for **process optimization;** that is, finding the levels of the design variables that result in the best values of the response. We will discuss and illustrate these problems extensively later in the book.

3-5.2 Comparisons Among Treatment Means

Suppose that in conducting an analysis of variance for the fixed effects model the null hypothesis is rejected. Thus, there are differences between the treatment means, but exactly *which* means differ is not specified. Sometimes in this situation, further comparisons and analysis among **groups** of treatment means may be useful. The ith treatment mean is defined as $\mu_i = \mu + \tau_i$, and μ_i is estimated by $\bar{y}_{i.}$. Comparisons between treatment means are made in terms of either the treatment totals $\{y_{i.}\}$ or the treatment averages $\{\bar{y}_{i.}\}$. The procedures for making these comparisons are usually called **multiple comparison methods.** In the next several sections, we discuss methods for making comparisons among individual treatment means or groups of these means.

3-5.3 Graphical Comparisons of Means

It is very easy to develop a graphical procedure for the comparison of means following an analysis of variance. Suppose that the factor of interest has a levels and that $\bar{y}_{1.}, \bar{y}_{2.}, \ldots, \bar{y}_{a.}$ are the treatment averages. If we know σ, any treatment average would have a standard deviation $\sigma/\sqrt{n}$. Consequently, if all factor level means are identical, the observed sample means $\bar{y}_{i.}$ would behave as if they were a set of observations drawn at random from a normal distribution with mean $\bar{y}_{..}$ and standard deviation $\sigma/\sqrt{n}$. Visualize a normal distribution capable of being slid along an axis below which the $\bar{y}_{1.}, \bar{y}_{2.}, \ldots, \bar{y}_{a.}$ are plotted. If the treatment means are all equal, there should be some position for this distribution that makes it obvious that the $\bar{y}_{i.}$ values were drawn from the same distribution. If this is not the case, then the $\bar{y}_{i.}$ values that appear *not* to have been drawn from this distribution are associated with factor levels that produce different mean responses.

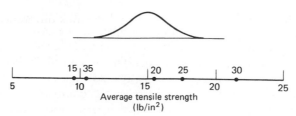

Figure 3–12. Tensile strength averages from the cotton weight percent experiment in relation to a t distribution with a scale factor $\sqrt{MS_E/n} = \sqrt{8.06/5} = 1.27$.

The only flaw in this logic is that σ is unknown. However, we can replace σ with $\sqrt{MS_E}$ from the analysis of variance and use a t distribution with a scale factor $\sqrt{MS_E/n}$ instead of the normal. Such an arrangement for the tensile strength data of Example 3–1 is shown in Figure 3–12.

To sketch the t distribution in Figure 3–12, simply multiply the abscissa t value by the scale factor

$$\sqrt{MS_E/n} = \sqrt{8.06/5} = 1.27$$

and plot this against the ordinate of t at that point. Since the t distribution looks much like the normal, except that it is a little flatter near the center and has longer tails, this sketch is usually easily constructed by eye. If you wish to be more precise, there is a table of abscissa t values and the corresponding ordinates in Box, Hunter, and Hunter (1978). The distribution can have an arbitrary origin, although it is usually best to choose one in the region of the $\bar{y}_{i\cdot}$ values to be compared. In Figure 3–12, the origin is 15 lb/in².

Now visualize sliding the t distribution in Figure 3–12 along the horizontal axis, and examine the five means plotted in the figure. Notice that there is no location for the distribution such that all five averages could be thought of as typical, randomly selected observations from the distribution. This implies that all five means are not equal; thus, the figure is a graphical display of the results of the analysis of variance. The figure does indicate that 30 percent cotton produces much higher tensile strengths than would 20 or 25 percent cotton (which are about the same) or that either 15 or 35 percent cotton (which are about the same) would result in even lower tensile strengths.

This simple procedure is a rough but effective technique for many multiple comparison problems. However, there are more formal methods. We now give a brief discussion of some of these procedures.

3–5.4 Contrasts

Many multiple comparison methods use the idea of a contrast. Consider the synthetic fiber testing problem of Example 3–1. Since the hypothesis $H_0: \tau_i = 0$ was rejected, we know that some cotton weight percents produce different tensile strengths than others, but which ones actually cause this difference? We might suspect at the outset of the experiment that levels 4 and 5 cotton weight percent (30 and 35 percent) produce the same tensile strength, implying that we would like to test the hypothesis

$$H_0: \mu_4 = \mu_5$$
$$H_1: \mu_4 \neq \mu_5$$

This hypothesis could be tested by investigating an appropriate linear combination of treatment totals, say

$$y_{4.} - y_{5.} = 0$$

If we had suspected that the *average* of levels 1 and 3 cotton weight percent did not differ from the *average* of levels 4 and 5 cotton weight percent, then the hypothesis would have been

$$H_0: \mu_1 + \mu_3 = \mu_4 + \mu_5$$
$$H_1: \mu_1 + \mu_3 \neq \mu_4 + \mu_5$$

which implies the linear combination

$$y_{1.} + y_{3.} - y_{4.} - y_{5.} = 0$$

In general, the comparison of treatment means of interest will imply a linear combination of treatment totals such as

$$C = \sum_{i=1}^{a} c_i y_{i.}$$

with the restriction that $\sum_{i=1}^{a} c_i = 0$. Such linear combinations are called **contrasts.** The sum of squares for any contrast is

$$SS_C = \frac{\left(\sum_{i=1}^{a} c_i y_{i.} \right)^2}{n \sum_{i=1}^{a} c_i^2} \tag{3-25}$$

and has a single degree of freedom. If the design is unbalanced, then the comparison of treatment means requires that $\sum_{i=1}^{a} n_i c_i = 0$, and Equation 3-25 becomes

$$SS_C = \frac{\left(\sum_{i=1}^{a} c_i y_{i.} \right)^2}{\sum_{i=1}^{a} n_i c_i^2} \tag{3-26}$$

A contrast is tested by comparing its sum of squares to the error mean square. The resulting statistic would be distributed as F with 1 and $N - a$ degrees of freedom. Many important comparisons between treatment means may be made using contrasts.

3–5.5 Orthogonal Contrasts

A very important special case of the above procedure is that of **orthogonal contrasts.** Two contrasts with coefficients $\{c_i\}$ and $\{d_i\}$ are orthogonal if

$$\sum_{i=1}^{a} c_i d_i = 0$$

or, for an unbalanced design, if

$$\sum_{i=1}^{a} n_i c_i d_i = 0$$

For a treatments the set of $a - 1$ orthogonal contrasts partition the sum of squares due to treatments into $a - 1$ independent single-degree-of-freedom components. Thus, tests performed on orthogonal contrasts are independent.

There are many ways to choose the orthogonal contrast coefficients for a set of treatments. Usually, something in the nature of the experiment should suggest which comparisons will be of interest. For example, if there are $a = 3$ treatments with treatment 1 a control and treatments 2 and 3 actual levels of the factor of interest to the experimenter, then appropriate orthogonal contrasts might be as follows:

Treatment	Coefficients for Orthogonal Contrasts	
1 (control)	−2	0
2 (level 1)	1	−1
3 (level 2)	1	1

Note that contrast 1 with $c_i = -2, 1, 1$ compares the average effect of the factor with the control, whereas contrast 2 with $d_i = 0, -1, 1$ compares the two levels of the factor of interest.

Contrast coefficients must be chosen prior to running the experiment and examining the data. The reason for this is that, if comparisons are selected after examining the data, most experimenters would construct tests that correspond to large observed differences in means. These large differences could be the result of the presence of real effects or they could be the result of random error. If experimenters consistently pick the largest differences to compare, they will inflate the type I error of the test since it is likely that, in an unusually high percentage of the comparisons selected, the observed differences will be the result of error.

Example 3–7

Consider the data in Example 3–1. There are five treatment means and four degrees of freedom between these treatments. The set of comparisons between these means and the associated orthogonal contrasts are

Hypothesis	**Contrast**
$H_0 : \mu_4 = \mu_5$	$C_1 = \qquad\qquad\qquad - y_{4.} + y_{5.}$
$H_0 : \mu_1 + \mu_3 = \mu_4 + \mu_5$	$C_2 = \quad y_{1.} \qquad + y_{3.} - y_{4.} - y_{5.}$
$H_0 : \mu_1 = \mu_3$	$C_3 = \quad y_{1.} \qquad\quad - y_{3.}$
$H_0 : 4\mu_2 = \mu_1 + \mu_3 + \mu_4 + \mu_5$	$C_4 = -y_{1.} + 4y_{2.} - y_{3.} - y_{4.} - y_{5.}$

Notice that the contrast coefficients are orthogonal. Using the data in Table 3–4, we find the numerical values of the contrasts and the sums of squares to be as follows.

$$C_1 = \qquad\qquad\qquad -1(108) + 1(54) = -54 \qquad SS_{C_1} = \frac{(-54)^2}{5(2)} = 291.60$$

$$C_2 = +1(49) \qquad +1(88) - 1(108) - 1(54) = -25 \qquad SS_{C_2} = \frac{(-25)^2}{5(4)} = 31.25$$

$$C_3 = +1(49) \qquad -1(88) \qquad\qquad = -39 \qquad SS_{C_3} = \frac{(-39)^2}{5(2)} = 152.10$$

$$C_4 = -1(49) + 4(77) - 1(88) - 1(108) - 1(54) = 9 \qquad SS_{C_4} = \frac{(9)^2}{5(20)} = 0.81$$

These contrast sums of squares completely partition the treatment sum of squares. The tests on the contrasts are usually incorporated in the analysis of variance, as shown in Table 3–12. We conclude from the P-values that there are significant differences between levels 4 and 5 and 1 and 3 of cotton weight percent but that the *average* of levels 1 and 3 does not differ from the average of levels 4 and 5 at the $\alpha = 0.05$ level, and that level 2 does not differ from the average of the other four levels. ∎

Table 3–12 Analysis of Variance for the Tensile Strength Data

Source of Variation	Sum of Squares	Degrees of Freedom	Mean Square	F_0	P-Value
Cotton Weight Percent	475.76	4	118.94	14.76	<0.001
orthogonal contrasts					
$C_1 : \mu_4 = \mu_5$	(291.60)	1	291.60	36.18	<0.001
$C_2 : \mu_1 + \mu_3 = \mu_4 + \mu_5$	(31.25)	1	31.25	3.88	0.06
$C_3 : \mu_1 = \mu_3$	(152.10)	1	152.10	18.87	<0.001
$C_4 : 4\mu_2 = \mu_1 + \mu_3 + \mu_4 + \mu_5$	(0.81)	1	0.81	0.10	0.76
Error	161.20	20	8.06		
Total	636.96	24			

3–5.6   Scheffé's Method for Comparing All Contrasts

In many situations, experimenters may not know in advance which contrasts they wish to compare, or they may be interested in more than $a - 1$ possible comparisons. In many exploratory experiments, the comparisons of interest are discovered only after preliminary examination of the data. Scheffé (1953) has proposed a method for comparing any and all possible contrasts between treatment means. In the Scheffé method, the type I error is at most α for any of the possible comparisons.

Suppose that a set of m contrasts in the treatment means

$$\Gamma_u = c_{1u}\mu_1 + c_{2u}\mu_2 + \cdots + c_{au}\mu_a \qquad u = 1, 2, \ldots, m \qquad (3\text{–}27)$$

of interest have been determined. The corresponding contrast in the treatment averages $\bar{y}_{i.}$ is

$$C_u = c_{1u}\bar{y}_{1.} + c_{2u}\bar{y}_{2.} + \cdots + c_{au}\bar{y}_{a.} \qquad u = 1, 2, \ldots, m \qquad (3\text{–}28)$$

and the standard error of this contrast is

$$S_{C_u} = \sqrt{MS_E \sum_{i=1}^{a} (c_{iu}^2/n_i)} \qquad (3\text{–}29)$$

where n_i is the number of observations in the ith treatment. It can be shown that the critical value against which C_u should be compared is

$$S_{\alpha,u} = S_{C_u} \sqrt{(a - 1)F_{\alpha,a-1,N-a}} \qquad (3\text{–}30)$$

To test the hypothesis that the contrast Γ_u differs significantly from zero, refer C_u to the critical value. If $|C_u| > S_{\alpha,u}$, then the hypothesis that the contrast Γ_u equals zero is rejected. The Scheffé procedure can also be used to form confidence intervals for all possible contrasts among treatment means. The resulting intervals, say $C_u - S_{\alpha,u} \leq \Gamma_u \leq C_u + S_{\alpha,u}$, are **simultaneous confidence intervals** in that the probability is at least $1 - \alpha$ that all of them are simultaneously true.

To illustrate the procedure, consider the data in Example 3–1 and suppose that the contrasts of interests are

$$\Gamma_1 = \mu_1 + \mu_3 - \mu_4 - \mu_5$$

and

$$\Gamma_2 = \mu_1 - \mu_4$$

The numerical values of these contrasts are

$$\begin{aligned}
C_1 &= \bar{y}_{1.} + \bar{y}_{3.} - \bar{y}_{4.} - \bar{y}_{5.} \\
&= 9.80 + 17.60 - 21.60 - 10.80 \\
&= 5.00
\end{aligned}$$

and

$$C_2 = \bar{y}_1. - \bar{y}_4.$$
$$= 9.80 - 21.60$$
$$= -11.80$$

and the standard errors are found from Equation 3-29 as

$$S_{C_1} = \sqrt{MS_E \sum_{i=1}^{5} (c_{i1}^2/n_i)} = \sqrt{8.06(1 + 1 + 1 + 1)/5} = 2.54$$

and

$$S_{C_2} = \sqrt{MS_E \sum_{i=1}^{5} (c_{i2}^2/n_i)} = \sqrt{8.06(1 + 1)/5} = 1.80$$

From Equation 3-30, the one percent critical values are

$$S_{0.01,1} = S_{C_1} \sqrt{(a - 1)F_{0.01,a-1,N-a}} = 2.54 \sqrt{4(4.43)} = 10.69$$

and

$$S_{0.01,2} = S_{C_2} \sqrt{(a - 1)F_{0.01,a-1,a(n-1)}} = 1.80 \sqrt{4(4.43)} = 7.58$$

Since $|C_1| < S_{0.01,1}$, we conclude that the contrast $\Gamma_1 = \mu_1 + \mu_3 - \mu_4 - \mu_5$ equals zero; that is, there is no strong evidence to conclude that the means of treatments 1 and 3 as a group differ from the means of treatments 4 and 5 as a group. However, since $|C_2| > S_{0.01,2}$, we conclude that the contrast $\Gamma_2 = \mu_1 - \mu_4$ does not equal zero; that is, the mean strengths of treatments 1 and 4 differ significantly.

In many practical situations, we will wish to compare only **pairs of means.** Frequently, we can determine which means differ by testing the differences between *all* pairs of treatment means. Thus, we are interested in contrasts of the form $\Gamma = \mu_i - \mu_j$ for all $i \neq j$. Although the Scheffé method could be easily applied to this problem, it is not the most sensitive procedure for such comparisons. We now turn to a consideration of methods specifically designed for pairwise comparisons between all a population means.

3-5.7 Comparing Pairs of Treatment Means

Suppose that we are interested in comparing all pairs of a treatment means and that the null hypotheses that we wish to test are $H_0: \mu_i = \mu_j$ for all $i \neq j$. We now present four methods for making such comparisons.

The Least Significant Difference (LSD) Method ▪ Suppose that, following an analysis of variance F test where the null hypothesis is rejected, we wish to test

$H_0: \mu_i = \mu_j$ for all $i \neq j$. This could be done by employing the t statistic

$$t_0 = \frac{\bar{y}_{i.} - \bar{y}_{j.}}{\sqrt{MS_E\left(\dfrac{1}{n_i} + \dfrac{1}{n_j}\right)}} \qquad (3\text{-}31)$$

Assuming a two-sided alternative, the pair of means μ_i and μ_j would be declared significantly different if $|\bar{y}_{i.} - \bar{y}_{j.}| > t_{\alpha/2,N-a} \sqrt{MS_E(1/n_i + 1/n_j)}$. The quantity

$$\text{LSD} = t_{\alpha/2,N-a} \sqrt{MS_E\left(\dfrac{1}{n_i} + \dfrac{1}{n_j}\right)} \qquad (3\text{-}32)$$

is called the **least significant difference.** If the design is balanced, then $n_1 = n_2 = \cdots = n_a = n$, and

$$\text{LSD} = t_{\alpha/2,N-a} \sqrt{\dfrac{2MS_E}{n}} \qquad (3\text{-}33)$$

To use the LSD procedure, we simply compare the observed difference between each pair of averages to the corresponding LSD. If $|\bar{y}_{i.} - \bar{y}_{j.}| > \text{LSD}$, we conclude that the population means μ_i and μ_j differ.

Example 3–8

To illustrate the procedure, if we use the data from the experiment in Example 3–1, the LSD at $\alpha = 0.05$ is

$$\text{LSD} = t_{.025,20} \sqrt{\dfrac{2MS_E}{n}} = 2.086 \sqrt{\dfrac{2(8.06)}{5}} = 3.75$$

Thus, any pair of treatment averages that differ in absolute value by more than 3.75 would imply that the corresponding pair of population means are significantly different. The five treatment averages are

$$\bar{y}_{1.} = 9.8 \qquad \bar{y}_{2.} = 15.4 \qquad \bar{y}_{3.} = 16.6 \qquad \bar{y}_{4.} = 21.6 \qquad \bar{y}_{5.} = 10.8$$

and the differences in averages are

$$
\begin{aligned}
\bar{y}_{1.} - \bar{y}_{2.} &= 9.8 - 15.4 = {-5.6}* \\
\bar{y}_{1.} - \bar{y}_{3.} &= 9.8 - 17.6 = {-7.8}* \\
\bar{y}_{1.} - \bar{y}_{4.} &= 9.8 - 21.6 = {-11.8}* \\
\bar{y}_{1.} - \bar{y}_{5.} &= 9.8 - 10.8 = {-1.0} \\
\bar{y}_{2.} - \bar{y}_{3.} &= 15.4 - 17.6 = {-2.2} \\
\bar{y}_{2.} - \bar{y}_{4.} &= 15.4 - 21.6 = {-6.2}* \\
\bar{y}_{2.} - \bar{y}_{5.} &= 15.4 - 10.8 = {4.6}* \\
\bar{y}_{3.} - \bar{y}_{4.} &= 17.6 - 21.6 = {-4.0}* \\
\bar{y}_{3.} - \bar{y}_{5.} &= 17.6 - 10.8 = {6.8}* \\
\bar{y}_{4.} - \bar{y}_{5.} &= 21.6 - 10.8 = {10.8}*
\end{aligned}
$$

$\bar{y}_{1.}$	$\bar{y}_{5.}$	$\bar{y}_{2.}$	$\bar{y}_{3.}$	$\bar{y}_{4.}$
9.8	10.8	15.4	17.6	21.6

Figure 3–13.   Results of the LSD procedure.

The starred values indicate pairs of means that are significantly different. It is helpful to draw a graph, such as Figure 3–13, underlining pairs of means that do not differ significantly. Clearly, the only pairs of means that do not differ significantly are 1 and 5 and 2 and 3, and treatment 4 produces a significantly greater tensile strength than the other treatments.

∎

Several comments should be made relative to this procedure. Note that the α risk may be considerably inflated using this method. Specifically, as a gets larger, the type I error of the experiment (the ratio of the number of experiments in which at least one type I error is made to the total number of experiments) becomes large. We also occasionally find that the overall F statistic in the analysis of variance is significant, but the LSD method fails to find any significant pairwise differences. This situation occurs because the F test is simultaneously considering all possible comparisons between the treatment means, not just pairwise comparisons. In the data at hand, the significant contrasts may not be of the form $\mu_i - \mu_j$.

Duncan's Multiple Range Test ▪ A widely used procedure for comparing all pairs of means is the **multiple range test** developed by Duncan (1955). To apply Duncan's multiple range test for equal sample sizes, the a treatment averages are arranged in ascending order, and the standard error of each average is determined as

$$S_{\bar{y}_{i.}} = \sqrt{\frac{MS_E}{n}} \qquad (3\text{--}34)$$

For unequal sample sizes, replace n in Equation 3–34 by the harmonic mean n_h of the $\{n_i\}$, where

$$n_h = \frac{a}{\sum_{i=1}^{a} (1/n_i)} \qquad (3\text{--}35)$$

Note that if $n_1 = n_2 = \cdots = n_a$, then $n_h = n$. From Duncan's table of significant ranges (Appendix Table VII), obtain the values $r_\alpha(p, f)$ for $p = 2, 3, \ldots, a$, where α is the significance level and f is the number of degrees of freedom for error. Convert these ranges into a set of $a - 1$ least significant ranges (e.g., R_p) for $p = 2, 3, \ldots, a$ by calculating

$$R_p = r_\alpha(p, f)S_{\bar{y}_{i.}} \qquad \text{for } p = 2, 3, \ldots, a \qquad (3\text{--}36)$$

Then, the observed differences between means are tested, beginning with largest versus smallest, which would be compared with the least significant range R_a. Next, the difference of the largest and the second-smallest is computed and compared with the least significant range R_{a-1}. These comparisons are continued until all means have been compared with the largest mean. Finally, the difference of the second-largest mean and the smallest is computed and compared against the least significant range R_{a-1}. This process is continued until the differences of all possible $a(a-1)/2$ pairs of means have been considered. If an observed difference is greater than the corresponding least significant range, then we conclude that the pair of means in question is significantly different. To prevent contradictions, no differences between a pair of means are considered significant if the two means involved fall between two other means that do not differ significantly.

Example 3–9

We can apply Duncan's multiple range test to the experiment of Example 3–1. Recall that $MS_E = 8.06$, $N = 25$, $n = 5$, and there are 20 error degrees of freedom. Ranking the treatment averages in ascending order, we have

$$\bar{y}_{1.} = 9.8$$
$$\bar{y}_{5.} = 10.8$$
$$\bar{y}_{2.} = 15.4$$
$$\bar{y}_{3.} = 17.6$$
$$\bar{y}_{4.} = 21.6$$

The standard error of each average is $S_{\bar{y}_{i.}} = \sqrt{8.06/5} = 1.27$. From the table of significant ranges in Appendix Table VII for 20 degrees of freedom and $\alpha = 0.05$, we obtain $r_{0.05}(2, 20) = 2.95$, $r_{0.05}(3, 20) = 3.10$, $r_{0.05}(4, 20) = 3.18$, and $r_{0.05}(5, 20) = 3.25$. Thus, the least significant ranges are

$$R_2 = r_{0.05}(2, 20) S_{\bar{y}_{i.}} = (2.95)(1.27) = 3.75$$
$$R_3 = r_{0.05}(3, 20) S_{\bar{y}_{i.}} = (3.10)(1.27) = 3.94$$
$$R_4 = r_{0.05}(4, 20) S_{\bar{y}_{i.}} = (3.18)(1.27) = 4.04$$
$$R_5 = r_{0.05}(5, 20) S_{\bar{y}_{i.}} = (3.25)(1.27) = 4.13$$

The comparisons would yield

$$\begin{aligned}
4 \text{ vs. } 1: \quad 21.6 - 9.8 &= 11.8 > 4.13(R_5) \\
4 \text{ vs. } 5: \quad 21.6 - 10.8 &= 10.8 > 4.04(R_4) \\
4 \text{ vs. } 2: \quad 21.6 - 15.4 &= 6.2 > 3.94(R_3) \\
4 \text{ vs. } 3: \quad 21.6 - 17.6 &= 4.0 > 3.75(R_2) \\
3 \text{ vs. } 1: \quad 17.6 - 9.8 &= 7.8 > 4.04(R_4) \\
3 \text{ vs. } 5: \quad 17.6 - 10.8 &= 6.8 > 3.95(R_3) \\
3 \text{ vs. } 2: \quad 17.6 - 15.4 &= 2.2 < 3.75(R_2)
\end{aligned}$$

$$2 \text{ vs. } 1: \quad 15.4 - 9.8 = 5.6 > 3.94(R_3)$$
$$2 \text{ vs. } 5: \quad 15.4 - 10.8 = 4.6 > 3.75(R_2)$$
$$5 \text{ vs. } 1: \quad 10.8 - 9.8 = 1.0 < 3.75(R_2)$$

From the analysis we see that there are significant differences between all pairs of means except 3 and 2 and 5 and 1. A graph underlining those means that are not significantly different is shown in Figure 3–14. Notice that, in this example, Duncan's multiple range test and the LSD method produce identical conclusions.

■

Duncan's test requires a greater observed difference to detect significantly different pairs of means as the number of means included in the group increases. For instance, in the above example $R_2 = 3.75$ (two means) whereas $R_3 = 3.94$ (three means). For two means, the critical value R_2 will always exactly equal the LSD value from the t test. The values $r_\alpha(p, f)$ in Appendix Table VII are chosen so that a specified **protection level** is obtained. That is, when two means that are p steps apart are compared, the protection level is $(1 - \alpha)^{p-1}$, where α is the level of significance specified for two *adjacent* means. Thus, the error rate is $1 - (1 - \alpha)^{p-1}$ of reporting at least one incorrect significant difference between two means when the group size is p. For example, if $\alpha = 0.05$, then $1 - (1 - 0.05)^1 = 0.05$ is the significance level for comparing the adjacent pair of means, $1 - (1 - 0.05)^2 = .10$ is the significance level for means that are one step apart, and so forth.

Generally, if the protection level is α, then tests on means have a significance level that is greater than or equal to α. Consequently, Duncan's procedure is quite powerful; that is, it is very effective at detecting differences between means when real differences exist. For this reason, Duncan's multiple range test is very popular.

The Newman–Keuls Test ▪ This test was devised by Newman (1939). Because new interest in Newman's test was generated by Keuls (1952), the procedure is usually called the Newman–Keuls test. Operationally, the procedure is similar to Duncan's multiple range test, except that the critical differences between means are calculated somewhat differently. Specifically, we compute a set of critical values

$$K_p = q_\alpha(p, f) S_{\bar{y}_{i.}} \qquad p = 2, 3, \dots, a \tag{3-37}$$

where $q_\alpha(p, f)$ is the upper α percentage point of the studentized range for groups of means of size p and f error degrees of freedom. The studentized range

$\bar{y}_{1.}$	$\bar{y}_{5.}$	$\bar{y}_{2.}$	$\bar{y}_{3.}$	$\bar{y}_{4.}$
9.8	10.8	15.4	17.6	21.6

Figure 3–14. Results of Duncan's multiple range test.

is defined as

$$q = \frac{\bar{y}_{max} - \bar{y}_{min}}{\sqrt{MS_E/n}}$$

where $\bar{y}_{max}$ and $\bar{y}_{min}$ are the largest and smallest sample means, respectively, out of a group of p sample means. Appendix Table VIII contains values of $q_\alpha(p, f)$, the upper α percentage point of q. Once the values K_p are computed from Equation 3–37, extreme pairs of means in groups of size p are compared with K_p exactly as in Duncan's multiple range test.

The Newman–Keuls test is more conservative than Duncan's test in that the type I error rate is smaller. Specifically, the type I error of the experiment is α for all tests involving the same number of means. Consequently, because α is generally lower, the power of the Newman–Keuls test is generally less than that of Duncan's test. To demonstrate that the Newman–Keuls procedure leads to a less powerful test than Duncan's multiple range test, we note from a comparison of Tables VII and VIII of the Appendix that for $p > 2$ we have $q_\alpha(p, f) > r_\alpha(p, f)$. That is, it is "harder" to declare a pair of means significantly different using the Newman–Keuls test than it is using Duncan's procedure. This is illustrated below for the case $\alpha = 0.01$, $a = 8$, and $f = 20$.

p	2	3	4	5	6	7	8
$r_{0.01}(p, 20)$	4.02	4.22	4.33	4.40	4.47	4.53	4.58
$q_{0.01}(p, 20)$	4.02	4.64	5.02	5.29	5.51	5.69	5.84

Tukey's Test ▪ Tukey (1953) proposed a multiple comparison procedure that is also based on the studentized range statistic. His procedure requires the use of $q_\alpha(a, f)$ to determine the critical value for *all* pairwise comparisons, regardless of how many means are in the group. Thus, Tukey's test declares two means significantly different if the absolute value of their sample differences exceeds

$$T_\alpha = q_\alpha(a, f) S_{\bar{y}_{i.}} \tag{3–38}$$

where $S_{\bar{y}_{i.}}$ is as defined in Equation 3–34. Note that a single critical value is used in all comparisons. Thus, in Example 3–9, we would use $q_{0.05}(5, 20) = 4.23$ as the basis of determining the critical value; consequently, no pair of means would be declared significantly different unless $|\bar{y}_{i.} - \bar{y}_{j.}|$ exceeded

$$T_{0.05} = q_{0.05}(5, 20) S_{\bar{y}_{i.}} = (4.23)(1.27) = 5.37$$

Using this critical value for the data in Example 3–9, we find now that $\bar{y}_{3.}$ and $\bar{y}_{4.}$ do not differ significantly; nor do $\bar{y}_{2.}$ and $\bar{y}_{5.}$. The graphical display of the results is

$\bar{y}_{1.}$	$\bar{y}_{5.}$	$\bar{y}_{2.}$	$\bar{y}_{3.}$	$\bar{y}_{4.}$
9.8	10.8	15.4	17.6	21.6

Note that these results are somewhat more difficult to interpret than those in either Figure 3–13 or Figure 3–14. Furthermore, Tukey's test has a type I error rate of α for all pairwise comparisons on an experimentwise basis. This is more conservative (it has a smaller type I error rate) than either the Newman–Keuls or Duncan tests. Consequently, Tukey's test has less power than either the Duncan or Newman–Keuls procedure.

Which Method Do I Use? ▪ Certainly a logical question at this point is, which one of these procedures should I use? Unfortunately, there is no clear-cut answer to this question, and professional statisticians often disagree over the utility of the various procedures. Carmer and Swanson (1973) have conducted Monte Carlo simulation studies of a number of multiple comparison procedures, including others not discussed here. They report that the least significant difference method is a very effective test for detecting true differences in means if it is applied *only after* the F test in the analysis of variance is significant at 5 percent. They also report good performance in detecting true differences with Duncan's multiple range test. This is not surprising, since these two methods are the most powerful of those we have discussed. Duncan's multiple range test is also available in many computer software packages for the analysis of variance. The Fisher LSD method is also very effective; however, we would want the analysis of variance F test to be significant with α no larger than 0.05 before using this test. These methods should be satisfactory for many general applications.

As indicated above, there are several other multiple comparison procedures. For articles describing these methods, see Miller (1977), O'Neill and Wetherill (1971), and Nelson (1989). The book by Miller (1981) is also recommended.

3–5.8 Comparing Treatment Means with a Control

In many experiments, one of the treatments is a **control,** and the analyst is interested in comparing each of the other $a - 1$ treatment means with the control. Thus, there are only $a - 1$ comparisons to be made. A procedure for making these comparisons has been developed by Dunnett (1964). Suppose that treatment a is the control. Then we wish to test the hypotheses

$$H_0 : \mu_i = \mu_a$$
$$H_1 : \mu_i \neq \mu_a$$

for $i = 1, 2, \ldots, a - 1$. Dunnett's procedure is a modification of the usual t test. For each hypothesis, we compute the observed differences in the sample means

$$|\bar{y}_{i.} - \bar{y}_{a.}| \qquad i = 1, 2, \ldots, a - 1$$

The null hypothesis $H_0: \mu_i = \mu_a$ is rejected using a type I error rate α if

$$|\bar{y}_{i.} - \bar{y}_{a.}| > d_\alpha(a - 1, f) \sqrt{MS_E \left(\frac{1}{n_i} + \frac{1}{n_a} \right)} \qquad (3\text{--}39)$$

where the constant $d_\alpha(a - 1, f)$ is given in Appendix Table IX. (Both two-sided and one-sided tests are possible.) Note that α is the **joint significance level** associated with all $a - 1$ tests.

Example 3–10

To illustrate Dunnett's test, consider the experiment from Example 3–1 with treatment 5 considered the control. In this example, $a = 5$, $a - 1 = 4$, $f = 20$, and $n_i = n = 5$. At the five percent level, we find from Appendix Table IX that $d_{0.05}(4, 20) = 2.65$. Thus, the critical difference becomes

$$d_{0.05}(4, 20) \sqrt{\frac{2MS_E}{n}} = 2.65 \sqrt{\frac{2(8.06)}{5}} = 4.76$$

(Note that this is a simplification of Equation 3–39 resulting from a balanced design.) Thus, any treatment mean that differs from the control by more than 4.76 would be declared significantly different. The observed differences are

$$
\begin{aligned}
1 \text{ vs. } 5: \quad & \bar{y}_{1.} - \bar{y}_{5.} = 9.8 - 10.8 = -1.0 \\
2 \text{ vs. } 5: \quad & \bar{y}_{2.} - \bar{y}_{5.} = 15.4 - 10.8 = 4.6 \\
3 \text{ vs. } 5: \quad & \bar{y}_{3.} - \bar{y}_{5.} = 17.6 - 10.8 = 6.8 \\
4 \text{ vs. } 5: \quad & \bar{y}_{4.} - \bar{y}_{5.} = 21.6 - 10.8 = 10.8
\end{aligned}
$$

Only the differences $\bar{y}_{3.} - \bar{y}_{5.}$ and $\bar{y}_{4.} - \bar{y}_{5.}$ indicate any significant difference when compared to the control; thus, we conclude that $\mu_3 \neq \mu_5$ and $\mu_4 \neq \mu_5$.

When comparing treatments with a control, it is a good idea to use more observations for the control treatment (say n_a) than for the other treatments (say n), assuming equal numbers of observations for the remaining $a - 1$ treatments. The ratio n_a/n should be chosen to be approximately equal to the square root of the total number of treatments. That is, choose $n_a/n = \sqrt{a}$.

■

3–6   SAMPLE COMPUTER OUTPUT

Computer programs for supporting experimental design and performing the analysis of variance are widely available. The output from one such program, the Statistical Analysis System (SAS) General Linear Models procedure, is shown in Figure 3–15, using the data in Example 3–1. The sum of squares corresponding

GENERAL LINEAR MODELS PROCEDURE

DEPENDENT VARIABLE Y

SOURCE	DF	SUM OF SQUARES	MEAN SQUARE	F VALUE	PR > F	R-SQUARE	C.V.
MODEL	4	475.76000000	118.94000000	14.76	0.0001	0.746923	18.8764
ERROR	20	161.20000000	8.06000000			STD DEV	Y MEAN
CORRECTED TOTAL	24	636.96000000				2.83901391	15.04000000

SOURCE	DF	TYPE I SS	F VALUE	PR > F	DF	TYPE III SS	F VALUE	PR > F
COTTON	4	475.76000000	14.76	0.0001	4	475.76000000	14.76	0.0001

OBSERVATION	OBSERVED VALUE	PREDICTED VALUE	RESIDUAL
1	7.00000000	9.80000000	-2.80000000
2	7.00000000	9.80000000	-2.80000000
3	15.00000000	9.80000000	5.20000000
4	11.00000000	9.80000000	1.20000000
5	9.00000000	9.80000000	-0.80000000
6	12.00000000	15.40000000	-3.40000000
7	17.00000000	15.40000000	1.60000000
8	12.00000000	15.40000000	-3.40000000
9	18.00000000	15.40000000	2.60000000
10	18.00000000	15.40000000	2.60000000
11	14.00000000	17.60000000	-3.60000000
12	18.00000000	17.60000000	0.40000000
13	18.00000000	17.60000000	0.40000000
14	19.00000000	17.60000000	1.40000000
15	19.00000000	17.60000000	1.40000000
16	19.00000000	21.60000000	-2.60000000
17	25.00000000	21.60000000	3.40000000
18	22.00000000	21.60000000	0.40000000
19	19.00000000	21.60000000	-2.60000000
20	23.00000000	21.60000000	1.40000000
21	7.00000000	10.80000000	-3.80000000
22	10.00000000	10.80000000	-0.80000000
23	11.00000000	10.80000000	0.20000000
24	15.00000000	10.80000000	4.20000000
25	11.00000000	10.80000000	0.20000000

SUM OF RESIDUALS   0.00000000
SUM OF SQUARED RESIDUALS   161.20000000
SUM OF SQUARED RESIDUALS - ERROR SS   -0.00000000
FIRST ORDER AUTOCORRELATION   -0.22555831
DURBIN-WATSON D   2.40223325

Figure 3–15. Sample computer output for the experiment in Example 3–1.

to the "model" is the usual $SS_{\text{Treatments}}$ for a single-factor design. That source is further identified as "cotton," and type I and type III sums of squares are displayed for that factor. These sums of squares are always identical for a balanced design, and in the case of a single-factor design, they are the same as the model sum of squares. Notice that the analysis of variance summary at the top of the computer output contains the usual sums of squares, degrees of freedom, mean squares, and test statistic F_0. The column "PR > F" is the P-value (actually, the upper bound on the P-value, because probabilities less than 0.0001 are defaulted to 0.0001.

In addition to the basic analysis of variance, the program displays some additional useful information. The quantity "R-SQUARE" is defined as

$$R^2 = \frac{SS_{\text{Model}}}{SS_{\text{Total}}} = \frac{475.76}{636.96} = 0.746923$$

and is loosely interpreted as the proportion of the variability in the data "explained" by the analysis of variance model. Thus, in the synthetic fiber strength testing data, the factor "cotton weight percent" explains about 74.69 percent of the variability in tensile strength. Clearly, we must have $0 \leq R^2 \leq 1$, with larger values being more desirable. "STD DEV" is the square root of the error mean square, $\sqrt{8.060} = 2.839$, and "C.V." is the coefficient of variation, defined as $(\sqrt{MS_E}/\bar{y})100$. The coefficient of variation measures the unexplained or residual variability in the data as a percentage of the mean of the response variable.

The computer program also calculates and displays the residuals, as defined in Equation 3–16. The program will also produce all of the residual plots that we discussed in Section 3–4. We will present several other examples of computer output from experimental design software packages throughout the book.

3-7 THE RANDOM EFFECTS MODEL

An experimenter is frequently interested in a factor that has a large number of possible levels. If the experimenter randomly selects a of these levels from the population of factor levels, then we say that the factor is **random.** Because the levels of the factor actually used in the experiment were chosen randomly, inferences are made about the entire population of factor levels. We assume that the population of factor levels is either of infinite size or is large enough to be considered infinite. Situations in which the population of factor levels is small enough to employ a finite population approach are not encountered frequently. Refer to Bennett and Franklin (1954) and Searle and Fawcett (1970) for a discussion of the finite population case.

The linear statistical model is

$$y_{ij} = \mu + \tau_i + \epsilon_{ij} \begin{cases} i = 1, 2, \ldots, a \\ j = 1, 2, \ldots, n \end{cases} \tag{3-40}$$

where both τ_i and ϵ_{ij} are random variables. If τ_i has variance σ_τ^2 and is independent of ϵ_{ij}, the variance of any observation is

$$V(y_{ij}) = \sigma_\tau^2 + \sigma^2$$

The variances σ_τ^2 and σ^2 are called **variance components,** and the model (Equation 3–40) is called the **components of variance** or **random effects model.** To test hypotheses in this model, we require that the $\{\epsilon_{ij}\}$ are NID(0, σ^2), that the $\{\tau_i\}$ are NID(0, σ_τ^2), and that τ_i and ϵ_{ij} are independent.[2]

The sum of squares identity

$$SS_T = SS_{\text{Treatments}} + SS_E \tag{3–41}$$

is still valid. That is, we partition the total variability in the observations into a component that measures the variation between treatments ($SS_{\text{Treatments}}$) and a component that measures the variation within treatments (SS_E). Testing hypotheses about individual treatment effects is meaningless, so instead we test the hypotheses

$$\begin{aligned} H_0&: \sigma_\tau^2 = 0 \\ H_1&: \sigma_\tau^2 > 0 \end{aligned} \tag{3–42}$$

If $\sigma_\tau^2 = 0$, all treatments are identical; but if $\sigma_\tau^2 > 0$, variability exists between treatments. As before, SS_E/σ^2 is distributed as chi-square with $N - a$ degrees of freedom, and under the null hypothesis, $SS_{\text{Treatments}}/\sigma^2$ is distributed as chi-square with $a - 1$ degrees of freedom. Both random variables are independent. Thus, under the null hypothesis $\sigma_\tau^2 = 0$, the ratio

$$F_0 = \dfrac{\dfrac{SS_{\text{Treatments}}}{a - 1}}{\dfrac{SS_E}{N - a}} = \dfrac{MS_{\text{Treatments}}}{MS_E} \tag{3–43}$$

is distributed as F with $a - 1$ and $N - a$ degrees of freedom. However, we need to examine the expected mean squares to fully describe the test procedure.

Consider

$$E(MS_{\text{Treatments}}) = \frac{1}{a - 1} E(SS_{\text{Treatments}}) = \frac{1}{a - 1} E\left[\sum_{i=1}^{a} \frac{y_{i.}^2}{n} - \frac{y_{..}^2}{N} \right]$$

$$= \frac{1}{a - 1} E\left[\frac{1}{n} \sum_{i=1}^{a} \left(\sum_{j=1}^{n} \mu + \tau_i + \epsilon_{ij} \right)^2 - \frac{1}{N} \left(\sum_{i=1}^{a} \sum_{j=1}^{n} \mu + \tau_i + \epsilon_{ij} \right)^2 \right]$$

When squaring and taking expectation of the quantities in brackets, we see that terms involving τ_i^2 are replaced by σ_τ^2 as $E(\tau_i) = 0$. Also, terms involving $\epsilon_{i.}^2$, $\epsilon_{..}^2$, and $\sum_{i=1}^{a} \sum_{j=1}^{n} \tau^2$ are replaced by $n\sigma^2$, $an\sigma^2$, and $an^2\sigma_\tau^2$, respectively. Further-

---

[2] The assumption that the $\{\tau_i\}$ are independent random variables implies that the usual assumption of $\sum_{i=1}^{a} \tau_i = 0$ from the fixed effects model does not apply to the random effects model.

more, all cross-product terms involving τ_i and ϵ_{ij} have zero expectation. This leads to

$$E(MS_{\text{Treatments}}) = \frac{1}{a-1}[N\mu^2 + N\sigma_\tau^2 + a\sigma^2 - N\mu^2 - n\sigma_\tau^2 - \sigma^2]$$

or

$$E(MS_{\text{Treatments}}) = \sigma^2 + n\sigma_\tau^2 \tag{3-44}$$

Similarly, we may show that

$$E(MS_E) = \sigma^2 \tag{3-45}$$

From the expected mean squares, we see that under H_0 both the numerator and denominator of the test statistic (Equation 3–43) are unbiased estimators of σ^2, whereas under H_1 the expected value of the numerator is greater than the expected value of the denominator. Therefore, we should reject H_0 for values of F_0 that are too large. This implies an upper-tail, one-tail critical region, so we reject H_0 if $F_0 > F_{\alpha,a-1,N-a}$.

The computational procedure and analysis of the variance table for the random effects model are identical to those for the fixed effects case. The conclusions, however, are quite different because they apply to the entire population of treatments.

We are usually interested in estimating the variance components (σ^2 and σ_τ^2) in the model. The procedure that we use to estimate σ^2 and σ_τ^2 is called the "analysis of variance method" because it makes use of the lines in the analysis of variance table. The procedure consists of equating the expected mean squares to their observed values in the analysis of variance table and solving for the variance components. In equating observed and expected mean squares in the single-factor random effects model, we obtain

$$MS_{\text{Treatments}} = \sigma^2 + n\sigma_\tau^2$$

and

$$MS_E = \sigma^2$$

Therefore, the estimators of the variance components are

$$\hat{\sigma}^2 = MS_E \tag{3-46}$$

and

$$\hat{\sigma}_\tau^2 = \frac{MS_{\text{Treatments}} - MS_E}{n} \tag{3-47}$$

For unequal sample sizes, replace n in Equation 3–47 by

$$n_0 = \frac{1}{a-1}\left[\sum_{i=1}^{a} n_i - \frac{\sum_{i=1}^{a} n_i^2}{\sum_{i=1}^{a} n_i}\right] \tag{3-48}$$

The analysis of variance method of variance component estimation does not require the normality assumption. It does yield estimators of σ^2 and σ_τ^2 that are best quadratic unbiased (i.e., of all unbiased quadratic functions of the observations, these estimators have minimum variance).

Occasionally, the analysis of variance method produces a negative estimate of a variance component. Clearly, variance components are by definition nonnegative, so a negative estimate of a variance component is viewed with some concern. One course of action is to accept the estimate and use it as evidence that the true value of the variance component is zero, assuming that sampling variation led to the negative estimate. This has intuitive appeal, but it suffers from some theoretical difficulties. For instance, using zero in place of the negative estimate can disturb the statistical properties of other estimates. Another alternative is to reestimate the negative variance component using a method that always yields nonnegative estimates. Still another alternative is to consider the negative estimate as evidence that the assumed linear model is incorrect and reexamine the problem. A good discussion of variance component estimation is given by Searle (1971a, 1971b).

Example 3–11

A textile company weaves a fabric on a large number of looms. They would like the looms to be homogeneous so that they obtain a fabric of uniform strength. The process engineer suspects that, in addition to the usual variation in strength within samples of fabric from the same loom, there may also be significant variations in strength between looms. To investigate this, she selects four looms at random and makes four strength determinations on the fabric manufactured on each loom. This experiment is run in random order, and the data obtained are shown in Table 3–13. The analysis of variance is conducted and is shown in Table 3–14. From the analysis of variance, we conclude that the looms in the plant differ significantly.

The variance components are estimated by $\hat{\sigma}^2 = 1.90$ and

$$\hat{\sigma}_\tau^2 = \frac{29.73 - 1.90}{4} = 6.96$$

Therefore, the variance of any observation on strength is estimated by $\hat{\sigma}^2 + \hat{\sigma}_\tau^2 = 1.90 + 6.96 = 8.86$. Most of this variability is attributable to differences *between* looms.

■

Table 3–13   Strength Data for Example 3–11

| Looms | Observations | | | | |
	1	2	3	4	$y_{i.}$
1	98	97	99	96	390
2	91	90	93	92	366
3	96	95	97	95	383
4	95	96	99	98	388

$1527 = y_{..}$

Table 3–14   Analysis of Variance for the Strength Data

Source of Variation	Sum of Squares	Degrees of Freedom	Mean Square	F_0	P-Value
Looms	89.19	3	29.73	15.68	<0.001
Error	22.75	12	1.90		
Total	111.94	15			

This example illustrates an important use of variance components—the isolating of different sources of variability that affect a product or system. The problem of product variability frequently arises in quality assurance, and it is often difficult to isolate the sources of variability. For example, this study may have been motivated by too much variability in the strength of the fabric, as illustrated in Figure 3–16a. This graph displays the process output (fiber strength) modeled as a normal distribution with variance $\hat{\sigma}_y^2 = 8.86$. (This is the estimate of the variance of any observation on strength from Example 3–11.) Upper and lower specifications on strength are also shown in Figure 3–16a, and it is relatively easy to see that a fairly large proportion of the process output is outside the specifications (the shaded tail areas in Figure 3–16a). The process engineer has asked why so much fabric is defective and must be scrapped, reworked, or downgraded to a lower quality product. The answer is that most of the product strength variability is the result of differences between looms. Different loom performance could be the result of faulty set-up, poor maintenance, ineffective supervision, poorly trained operators, defective input fiber, and so forth.

The process engineer must now try to isolate the specific causes of the difference in loom performance. If she could identify and eliminate these sources

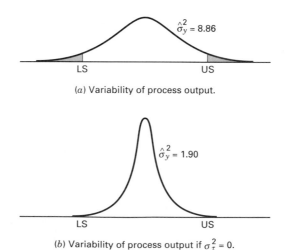

(a) Variability of process output.

(b) Variability of process output if $\sigma_\tau^2 = 0$.

Figure 3–16.   Process output in the fiber strength problem.

of between-loom variability, the variance of the process output could be reduced considerably, perhaps to as low as $\hat{\sigma}_y^2 = 1.90$, the estimate of the within-loom (error) variance component in Example 3–11. Figure 3–16b shows a normal distribution of fiber strength with $\hat{\sigma}_y^2 = 1.90$. Note that the proportion of defective product in the output has been dramatically reduced. Although it is unlikely that *all* of the between-loom variability can be eliminated, it is clear that a significant reduction in this variance component would greatly increase the quality of the fiber produced.

We may easily find a confidence interval for the variance component σ^2. If the observations are normally and independently distributed, then $(N - a)MS_E/\sigma^2$ is distributed as χ^2_{N-a}. Thus,

$$P\left(\chi^2_{1-(\alpha/2),N-a} \leq \frac{(N - a)MS_E}{\sigma^2} \leq \chi^2_{\alpha/2,N-a}\right) = 1 - \alpha$$

and a $100(1 - \alpha)$ percent confidence interval for σ^2 is

$$\frac{(N - a)MS_E}{\chi^2_{\alpha/2,N-a}} \leq \sigma^2 \leq \frac{(N - a)MS_E}{\chi^2_{1-(\alpha/2),N-a}} \tag{3-49}$$

Now consider the variance component σ_τ^2. The point estimator of σ_τ^2 is

$$\hat{\sigma}_\tau^2 = \frac{MS_{\text{Treatments}} - MS_E}{n}$$

The random variable $(a - 1)MS_{\text{Treatments}}/(\sigma^2 + n\sigma_\tau^2)$ is distributed as χ^2_{a-1}, and $(N - a)MS_E/\sigma^2$ is distributed as χ^2_{N-a}. Thus, the probability distribution of $\hat{\sigma}_\tau^2$ is a linear combination of two chi-square random variables, say

$$u_1\chi^2_{a-1} - u_2\chi^2_{N-a}$$

where

$$u_1 = \frac{\sigma^2 + n\sigma_\tau^2}{n(a - 1)} \quad \text{and} \quad u_2 = \frac{\sigma^2}{n(N - a)}$$

Unfortunately, a closed-form expression for the distribution of this linear combination of chi-square random variables cannot be obtained. Thus, an exact confidence interval for σ_τ^2 cannot be constructed. Approximate procedures are given in Graybill (1961) and Searle (1971a). Also see Chapter 11.

It is easy to find an exact expression for a confidence interval on the ratio $\sigma_\tau^2/(\sigma_\tau^2 + \sigma^2)$. This is a meaningful ratio since it reflects the *proportion* of the variance of an observation [recall that $V(y_{ij}) = \sigma_\tau^2 + \sigma^2$] that is the result of differences between treatments. To develop this confidence interval for the case of a balanced design, note that $MS_{\text{Treatments}}$ and MS_E are independent random variables and, furthermore, that it can be shown that

$$\frac{MS_{\text{Treatments}}/(n\sigma_\tau^2 + \sigma^2)}{MS_E/\sigma^2} \sim F_{a-1,N-a}$$

Thus,

$$\left(F_{1-\alpha/2,a-1,N-a} \leq \frac{MS_{\text{Treatments}}}{MS_E} \frac{\sigma^2}{n\sigma_\tau^2 + \sigma^2} \leq F_{\alpha/2,a-1,N-a} \right) = 1 - \alpha \qquad (3\text{-}50)$$

By rearranging Equation 3–48, we may obtain the following:

$$P\left(L \leq \frac{\sigma_\tau^2}{\sigma^2} \leq U \right) = 1 - \alpha \qquad (3\text{-}51)$$

where

$$L = \frac{1}{n}\left(\frac{MS_{\text{Treatments}}}{MS_E} \frac{1}{F_{\alpha/2,a-1,N-a}} - 1 \right) \qquad (3\text{-}52a)$$

and

$$U = \frac{1}{n}\left(\frac{MS_{\text{Treatments}}}{MS_E} \frac{1}{F_{1-\alpha/2,a-1,N-a}} - 1 \right) \qquad (3\text{-}52b)$$

Note that L and U are $100(1 - \alpha)$ percent lower and upper confidence limits, respectively, for the ratio σ_τ^2/σ^2. Therefore, a $100(1 - \alpha)$ percent confidence interval for $\sigma_\tau^2/(\sigma_\tau^2 + \sigma^2)$ is

$$\frac{L}{1 + L} \leq \frac{\sigma_\tau^2}{\sigma_\tau^2 + \sigma^2} \leq \frac{U}{1 + U} \qquad (3\text{-}53)$$

To illustrate this procedure, we find a 95 percent confidence interval on $\sigma_\tau^2/(\sigma_\tau^2 + \sigma^2)$ for the strength data in Example 3–11. Recall that $MS_{\text{Treatments}} = 29.73$, $MS_E = 1.90$, $a = 4$, $n = 4$, $F_{0.025,3,12} = 4.47$, and $F_{0.975,3,12} = 1/F_{0.025,12,3} = 1/14.34 = 0.070$. Therefore, from Equation 3–52,

$$L = \frac{1}{4}\left[\left(\frac{29.73}{1.90} \right)\left(\frac{1}{4.47} \right) - 1 \right] = 0.625$$

$$U = \frac{1}{4}\left[\left(\frac{29.73}{1.90} \right)\left(\frac{1}{0.070} \right) - 1 \right] = 54.883$$

and from Equation 3–51, the 95 percent confidence interval on $\sigma_\tau^2/(\sigma_\tau^2 + \sigma^2)$ is

$$\frac{0.625}{1.625} \leq \frac{\sigma_\tau^2}{\sigma_\tau^2 + \sigma^2} \leq \frac{54.883}{55.883}$$

or

$$0.39 \leq \frac{\sigma_\tau^2}{\sigma_\tau^2 + \sigma^2} \leq 0.98$$

We conclude that variability between looms accounts for between 39 and 98 percent of the variance in the observed strength of the fabric produced. This confidence interval is relatively wide because of the small sample size that was used in the experiment. Clearly, however, the variability between looms (σ_τ^2) is not negligible.

3–8 PROBLEMS

3–1 The tensile strength of portland cement is being studied. Four different mixing techniques can be used economically. The following data have been collected:

Mixing Technique	Tensile Strength (lb/in^2)			
1	3129	3000	2865	2890
2	3200	3300	2975	3150
3	2800	2900	2985	3050
4	2600	2700	2600	2765

(a) Test the hypothesis that mixing techniques affect the strength of the cement. Use $\alpha = 0.05$.

(b) Use Duncan's multiple range test with $\alpha = 0.05$ to make comparisons between pairs of means.

(c) Construct a graphical display as described in Section 3–5.3 to compare the mean tensile strengths for the four mixing techniques. What are your conclusions?

(d) Construct a normal probability plot of the residuals. What conclusion would you draw about the validity of the normality assumption?

(e) Plot the residuals versus the predicted tensile strength. Comment on the plot.

(f) Prepare box plots of the results to aid the interpretation of the results of this experiment.

3–2 Rework part (b) of Problem 3–1 using the Fisher LSD method. Does this make any difference in your conclusions?

3–3 Reconsider the experiment in Problem 3–1. Find a 95 percent confidence interval on the mean tensile strength of the portland cement produced by each of the four mixing techniques. Also find a 95 percent confidence interval on the difference in means for techniques 1 and 3. Does this aid you in interpreting the results of the experiment?

3–4 A textile mill has a large number of looms. Each loom is supposed to provide the same output of cloth per minute. To investigate this assumption, five looms are chosen at random and their output is noted at different times. The following data are obtained:

Loom	Output (lb/min)				
1	14.0	14.1	14.2	14.0	14.1
2	13.9	13.8	13.9	14.0	14.0
3	14.1	14.2	14.1	14.0	13.9
4	13.6	13.8	14.0	13.9	13.7
5	13.8	13.6	13.9	13.8	14.0

(a) Explain why this is a random effects experiment. Are the looms equal in output? Use $\alpha = 0.05$.

(b) Estimate the variability between looms.

(c) Estimate the experimental error variance.

(d) Find a 95 percent confidence interval for $\sigma_\tau^2/(\sigma_\tau^2 + \sigma^2)$.

(e) Analyze the residuals from this experiment. Do you think that the analysis of variance assumptions are satisfied?

3–5   An experiment was run to determine whether four specific firing temperatures affect the density of a certain type of brick. The experiment led to the following data:

Temperature	Density				
100	21.8	21.9	21.7	21.6	21.7
125	21.7	21.4	21.5	21.4	
150	21.9	21.8	21.8	21.6	21.5
175	21.9	21.7	21.8	21.4	

(a) Does the firing temperature affect the density of the bricks? Use $\alpha = 0.05$.

(b) Is it appropriate to compare the means using Duncan's multiple range test (for example) in this experiment?

(c) Analyze the residuals from this experiment. Are the analysis of variance assumptions satisfied?

(d) Construct a graphical display of the treatment as described in Section 3–5.3. Does this graph adequately summarize the results of the analysis of variance in part (a)?

3–6   A manufacturer of television sets is interested in the effect on tube conductivity of four different types of coating for color picture tubes. The following conductivity data are obtained:

Coating Type	Conductivity			
1	143	141	150	146
2	152	149	137	143
3	134	136	132	127
4	129	127	132	129

(a) Is there a difference in conductivity due to coating type? Use $\alpha = 0.05$.

(b) Estimate the overall mean and the treatment effects.

(c) Compute a 95 percent confidence interval estimate of the mean of coating type 4. Compute a 99 percent confidence interval estimate of the mean difference between coating types 1 and 4.

(d) Test all pairs of means using the Fisher LSD method with $\alpha = 0.05$.

(e) Use the graphical method discussed in Section 3–5.3 to compare the means. Which coating type produces the highest conductivity?

(f) Assuming that coating type 4 is currently in use, what are your recommendations to the manufacturer? We wish to minimize conductivity.

3-7 Reconsider the experiment from Problem 3–6. Analyze the residuals and draw conclusions about model adequacy.

3-8 An article in the *ACI Materials Journal* (Vol. 84, 1987, pp. 213–216) describes several experiments investigating the rodding of concrete to remove entrapped air. A 3″ × 6″ cylinder was used, and the number of times this rod was used is the design variable. The resulting compressive strength of the concrete specimen is the response. The data are shown in the following table.

Rodding Level	Compressive Strength		
10	1530	1530	1440
15	1610	1650	1500
20	1560	1730	1530
25	1500	1490	1510

(a) Is there any difference in compressive strength due to the rodding level? Use $\alpha = 0.05$.

(b) Find the *P*-value for the *F* statistic in part (a).

(c) Analyze the residuals from this experiment. What conclusions can you draw about the underlying model assumptions?

(d) Construct a graphical display to compare the treatment means as described in Section 3–5.3.

3–9 An article in *Environment International* (Vol. 18, No. 4, 1992) describes an experiment in which the amount of radon released in showers was investigated. Radon-enriched water was used in the experiment, and six different orifice diameters were tested in shower heads. The data from the experiment are shown in the following table.

Orifice Diameter	Radon Released (%)			
0.37	80	83	83	85
0.51	75	75	79	79
0.71	74	73	76	77
1.02	67	72	74	74
1.40	62	62	67	69
1.99	60	61	64	66

(a) Does the size of the orifice affect the mean percentage of radon released? Use $\alpha = 0.05$.

(b) Find the P-value for the F statistic in part (a).

(c) Analyze the residuals from this experiment.

(d) Find a 95 percent confidence interval on the mean percent of radon released when the orifice diameter is 1.40.

(e) Construct a graphical display to compare the treatment means as described in Section 3–5.3. What conclusions can you draw?

3–10 The response time in milliseconds was determined for three different types of circuits that could be used in an automatic valve shutoff mechanism. The results are shown in the following table.

Circuit Type	Response Time				
1	9	12	10	8	15
2	20	21	23	17	30
3	6	5	8	16	7

(a) Test the hypothesis that the three circuit types have the same response time. Use $\alpha = 0.01$.

(b) Use Tukey's test to compare pairs of treatment means. Use $\alpha = 0.01$.

(c) Use the graphical procedure in Section 3–5.3 to compare the treatment means. What conclusions can you draw? How do they compare with the conclusions from part (b)?

(d) Construct a set of orthogonal contrasts, assuming that at the outset of the experiment you suspected the response time of circuit type 2 to be different from the other two.

(e) If you were the design engineer and you wished to minimize the response time, which circuit type would you select?

(f) Analyze the residuals from this experiment. Are the basic analysis of variance assumptions satisfied?

3–11 The effective life of insulating fluids at an accelerated load of 35 kV is being studied. Test data have been obtained for four types of fluids. The results were as follows:

Fluid Type	Life (in h) at 35 kV Load					
1	17.6	18.9	16.3	17.4	20.1	21.6
2	16.9	15.3	18.6	17.1	19.5	20.3
3	21.4	23.6	19.4	18.5	20.5	22.3
4	19.3	21.1	16.9	17.5	18.3	19.8

(a) Is there any indication that the fluids differ? Use $\alpha = 0.05$.

(b) Which fluid would you select, given that the objective is long life?

(c) Analyze the residuals from this experiment. Are the basic analysis of variance assumptions satisfied?

3–12 Four different designs for a digital computer circuit are being studied in order to compare the amount of noise present. The following data have been obtained:

Circuit Design	Noise Observed				
1	19	20	19	30	8
2	80	61	73	56	80
3	47	26	25	35	50
4	95	46	83	78	97

(a) Is the amount of noise present the same for all four designs? Use $\alpha = 0.05$.

(b) Analyze the residuals from this experiment. Are the analysis of variance assumptions satisfied?

(c) Which circuit design would you select for use? Low noise is best.

3–13 A manufacturer suspects that the batches of raw material furnished by her supplier differ significantly in calcium content. There are a large number of batches currently in the warehouse. Five of these are randomly selected for study. A chemist makes five determinations on each batch and obtains the following data:

Batch 1	Batch 2	Batch 3	Batch 4	Batch 5
23.46	23.59	23.51	23.28	23.29
23.48	23.46	23.64	23.40	23.46
23.56	23.42	23.46	23.37	23.37
23.39	23.49	23.52	23.46	23.32
23.40	23.50	23.49	23.39	23.38

(a) Is there significant variation in calcium content from batch to batch? Use $\alpha = 0.05$.

(b) Estimate the components of variance.

(c) Find a 95 percent confidence interval for $\sigma_\tau^2/(\sigma_\tau^2 + \sigma^2)$.

(d) Analyze the residuals from this experiment. Are the analysis of variance assumptions satisfied?

3–14   Several ovens in a metal working shop are used to heat metal specimens. All the ovens are supposed to operate at the same temperature, although it is suspected that this may not be true. Three ovens are selected at random and their temperatures on successive heats are noted. The data collected are as follows:

Oven	Temperature				
1	491.50	498.30	498.10	493.50	493.60	
2	488.50	484.65	479.90	477.35		
3	490.10	484.80	488.25	473.00	471.85	478.65

(a) Is there significant variation in temperature between ovens? Use $\alpha = 0.05$.

(b) Estimate the components of variance for this model.

(c) Analyze the residuals from this experiment and draw conclusions about model adequacy.

3–15   Four chemists are asked to determine the percentage of methyl alcohol in a certain chemical compound. Each chemist makes three determinations, and the results are the following:

Chemist	Percentage of Methyl Alcohol		
1	84.99	84.04	84.38
2	85.15	85.13	84.88
3	84.72	84.48	85.16
4	84.20	84.10	84.55

(a) Do chemists differ significantly? Use $\alpha = 0.05$.

(b) Analyze the residuals from this experiment.

(c) If chemist 2 is a new employee, construct a meaningful set of orthogonal contrasts that might have been useful at the start of the experiment.

3–16 Three brands of batteries are under study. It is suspected that the lives (in weeks) of the three brands are different. Five batteries of each brand are tested with the following results:

Weeks of Life		
Brand 1	Brand 2	Brand 3
100	76	108
96	80	100
92	75	96
96	84	98
92	82	100

(a) Are the lives of these brands of batteries different?

(b) Analyze the residuals from this experiment.

(c) Construct a 95 percent confidence interval estimate on the mean life of battery brand 2. Construct a 99 percent confidence interval estimate on the mean difference between the lives of battery brands 2 and 3.

(d) Which brand would you select for use? If the manufacturer will replace without charge any battery that fails in less than 85 weeks, what percentage would the company expect to replace?

3–17 Four catalysts that may affect the concentration of one component in a three-component liquid mixture are being investigated. The following concentrations are obtained:

Catalyst			
1	2	3	4
58.2	56.3	50.1	52.9
57.2	54.5	54.2	49.9
58.4	57.0	55.4	50.0
55.8	55.3		51.7
54.9			

(a) Do the four catalysts have the same effect on the concentration?

(b) Analyze the residuals from this experiment.

(c) Construct a 99 percent confidence interval estimate of the mean response for catalyst 1.

3–18 An article in the *Journal of the Electrochemical Society* (Vol. 139, No. 2, 1992, pp. 524–532) describes an experiment to investigate the low-pressure vapor deposition of polysilicon. The experiment was carried out in a large-capacity reactor at Sematech in Austin, Texas. The reactor has several wafer positions, and four of these positions are selected at random. The response variable is film thickness uniformity. Three replicates of the experiment were run, and the data are as follows:

Wafer Position	Uniformity		
1	2.76	5.67	4.49
2	1.43	1.70	2.19
3	2.34	1.97	1.47
4	0.94	1.36	1.65

(a) Is there a difference in the wafer positions? Use $\alpha = 0.05$.

(b) Estimate the variability due to wafer positions.

(c) Estimate the random error component.

(d) Analyze the residuals from this experiment and comment on model adequacy.

3–19 Consider the vapor-deposition experiment described in Problem 3–18.

(a) Estimate the total variability in the uniformity response.

(b) How much of the total variability in the uniformity response is due to the difference between positions in the reactor?

(c) To what level could the variability in the uniformity response be reduced, if the position-to-position variability in the reactor could be eliminated? Do you believe this is a significant reduction?

3–20 An article in the *Journal of Quality Technology* (Vol. 13, No. 2, 1981, pp. 111–114) describes an experiment that investigates the effects of four bleaching chemicals on pulp brightness. These four chemicals were selected at random from a large population of potential bleaching agents. The data are as follows:

Chemical	Pulp Brightness				
1	77.199	74.466	92.746	76.208	82.876
2	80.522	79.306	81.914	80.346	73.385
3	79.417	78.017	91.596	80.802	80.626
4	78.001	78.358	77.544	77.364	77.386

(a) Is there a difference in the chemical types? Use $\alpha = 0.05$.

(b) Estimate the variability due to chemical types.

(c) Estimate the variability due to random error.

(d) Analyze the residuals from this experiment and comment on model adequacy.

3-21 Consider testing the equality of the means of two normal populations, where the variances are unknown but are assumed to be equal. The appropriate test procedure is the pooled t test. Show that the pooled t test is equivalent to the single-factor analysis of variance.

3-22 Show that the variance of the linear combination $\sum_{i=1}^{a} c_i y_{i.}$ is $\sigma^2 \sum_{i=1}^{a} n_i c_i^2$.

3-23 In a fixed effects experiment, suppose that there are n observations for each of four treatments. Let Q_1^2, Q_2^2, Q_3^2 be single-degree-of-freedom components for the orthogonal contrasts. Prove that $SS_{\text{Treatments}} = Q_1^2 + Q_2^2 + Q_3^2$.

3-24 Consider the one-way, balanced, random effects method. Develop a procedure for finding a $100(1 - \alpha)$ percent confidence interval for $\sigma^2/(\sigma_\tau^2 + \sigma^2)$.

3-25 Use Bartlett's test to determine if the assumption of equal variances is satisfied in Problem 3-16. Use $\alpha = 0.05$. Did you reach the same conclusion regarding equality of variance by examining residual plots?

Chapter 4
More About Single-Factor Experiments

This chapter discusses several additional topics concerning single-factor experiments and the analysis of variance. These include methods for choosing an appropriate sample size for comparing treatment means, using the analysis of variance to detect dispersion effects, more information about fitting a response curve to develop an interpolation equation for quantitative factors, a brief look at a more general method for developing the analysis of variance, a nonparametric alternative to the usual F test in the analysis of variance, an introduction to experiments with repeated measures, and an introduction to the analysis of covariance, a technique useful in dealing with certain types of nuisance variables in designed experiments.

4-1 CHOICE OF SAMPLE SIZE

In any experimental design problem, a critical decision is the choice of sample size—that is, determining the number of replicates to run. Generally, if the experimenter is interested in detecting small effects, more replicates are required than if the experimenter is interested in detecting large effects. In this section, we discuss several approaches to determining sample size. Although our discussion focuses on a single-factor design, most of the methods can be used in more complex experimental situations.

4-1.1 Operating Characteristic Curves

Recall that an **operating characteristic curve** is a plot of the type II error probability of a statistical test for a particular sample size versus a parameter that reflects the extent to which the null hypothesis is false. These curves can be used to guide the experimenter in selecting the number of replicates so that the design will be sensitive to important potential differences in the treatments.

We first consider the probability of type II error of the fixed effects model for the case of equal sample sizes per treatment, say

$$\beta = 1 - P\{\text{Reject } H_0 | H_0 \text{ is false}\}$$
$$= 1 - P\{F_0 > F_{\alpha, a-1, N-a} | H_0 \text{ is false}\} \tag{4-1}$$

To evaluate the probability statement in Equation 4-1, we need to know the distribution of the test statistic F_0 if the null hypothesis is false. It can be shown that, if H_0 is false, the statistic $F_0 = MS_{\text{Treatments}}/MS_E$ is distributed as a **noncentral** F random variable with $a - 1$ and $N - a$ degrees of freedom and the noncentrality parameter δ. If $\delta = 0$, then the noncentral F distribution becomes the usual (central) F distribution.

Operating characteristic curves given in Chart V of the Appendix are used to evaluate the probability statement in Equation 4-1. These curves plot the probability of type II error (β) against a parameter Φ, where

$$\Phi^2 = \frac{n \sum_{i=1}^{a} \tau_i^2}{a\sigma^2} \tag{4-2}$$

The quantity Φ^2 is related to the noncentrality parameter δ. Curves are available for $\alpha = 0.05$ and $\alpha = 0.01$ and a range of degrees of freedom for numerator and denominator.

In using the operating characteristic curves, the experimenter must specify the parameter Φ. This is often difficult to do in practice. One way to determine Φ is to choose the actual values of the treatment means for which we would like to reject the null hypothesis with high probability. Thus, if $\mu_1, \mu_2, \ldots, \mu_a$ are the specified treatment means, we find the τ_i in Equation 4-2 as $\tau_i = \mu_i - \overline{\mu}$, where $\overline{\mu} = (1/a) \sum_{i=1}^{a} \mu_i$ is the average of the individual treatment means. We also require an estimate of σ^2. Sometimes this is available from prior experience, a previous experiment, or a judgment estimate. When we are uncertain about the value of σ^2, sample sizes could be determined for a range of likely values of σ^2 to study the effect of this parameter on the required sample size before a final choice is made.

Example 4-1

Consider the tensile strength experiment described in Example 3-1. Suppose that the experimenter is interested in rejecting the null hypothesis with a probability of at least 0.90 if the five treatment means are

$$\mu_1 = 11 \qquad \mu_2 = 12 \qquad \mu_3 = 15 \qquad \mu_4 = 18 \qquad \text{and} \qquad \mu_5 = 19$$

She plans to use $\alpha = 0.01$. In this case, since $\sum_{i=1}^{5} \mu_i = 75$, we have $\overline{\mu} = (1/5)75 = 15$ and

$$\tau_1 = \mu_1 - \overline{\mu} = 11 - 15 = -4$$
$$\tau_2 = \mu_2 - \overline{\mu} = 12 - 15 = -3$$

$$\tau_3 = \mu_3 - \bar{\mu} = 15 - 15 = \quad 0$$
$$\tau_4 = \mu_4 - \bar{\mu} = 18 - 15 = \quad 3$$
$$\tau_5 = \mu_5 - \bar{\mu} = 19 - 15 = \quad 4$$

Thus, $\sum_{i=1}^{5} \tau_i^2 = 50$. Suppose the experimenter feels that the standard deviation of tensile strength at any particular level of cotton weight percent will be no larger than $\sigma = 3$ psi. Then, by using Equation 4–2, we have

$$\Phi^2 = \frac{n \sum_{i=1}^{5} \tau_i^2}{a\sigma^2} = \frac{n(50)}{5(3)^2} = 1.11n$$

We use the operating characteristic curve for $a - 1 = 5 - 1 = 4$ with $N - a = a(n - 1) = 5(n - 1)$ error degrees of freedom and $\alpha = 0.01$ (see Appendix Chart V). As a first guess at the required sample size, try $n = 4$ replicates. This yields $\Phi^2 = 1.11(4) = 4.44$, $\Phi = 2.11$, and $5(3) = 15$ error degrees of freedom. Consequently, from Chart V, we find that $\beta \approx 0.30$. Therefore, the power of the test is approximately $1 - \beta = 1 - 0.30 = 0.70$, which is less than the required 0.90, and so we conclude that $n = 4$ replicates are not sufficient. Proceeding in a similar manner, we can construct the following display:

n	Φ^2	Φ	$a(n - 1)$	β	Power $(1 - \beta)$
4	4.44	2.11	15	0.30	0.70
5	5.55	2.36	20	0.15	0.85
6	6.66	2.58	25	0.04	0.96

Thus, at least $n = 6$ replicates must be run in order to obtain a test with the required power.

∎

The only problem with this approach to using the operating characteristic curves is that it is usually difficult to select a set of treatment means on which the sample size decision should be based. An alternate approach is to select a sample size such that if the difference between any two treatment means exceeds a specified value the null hypothesis should be rejected. If the difference between any two treatment means is as large as D, then it can be shown that the minimum value of Φ^2 is

$$\Phi^2 = \frac{nD^2}{2a\sigma^2} \tag{4–3}$$

Since this is a minimum value of Φ^2, the corresponding sample size obtained from the operating characteristic curve is a conservative value; that is, it provides a power at least as great as that specified by the experimenter.

To illustrate this approach, suppose that in the tensile strength experiment from Example 3–1, the experimenter wished to reject the null hypothesis with probability at least 0.90 if any two treatment means differed by as much as 10 psi. Then, assuming that $\sigma = 3$ psi, we find the minimum value of Φ^2 to be

$$\Phi^2 = \frac{n(10)^2}{2(5)(3^2)} = 1.11n$$

and, from the analysis in Example 4–1, we conclude that $n = 6$ replicates are required to give the desired sensitivity when $\alpha = 0.01$.

Now consider the random effects model. The type II error probability for the random effects model is

$$\begin{aligned} \beta &= 1 - P\{\text{Reject } H_0 | H_0 \text{ is false}\} \\ &= 1 - P\{F_0 > F_{\alpha, a-1, N-a} | \sigma_\tau^2 > 0\} \end{aligned} \tag{4-4}$$

Once again, the distribution of the test statistic $F_0 = MS_{\text{Treatments}}/MS_E$ under the alternative hypothesis is needed. It can be shown that if H_1 is true ($\sigma_\tau^2 > 0$), the distribution of F_0 is central F with $a - 1$ and $N - a$ degrees of freedom.

Since the type II error probability of the random effects model is based on the usual central F distribution, we could use the tables of the F distribution in the Appendix to evaluate Equation 4–4. However, it is simpler to determine the sensitivity of the test through the use of operating characteristic curves. A set of these curves for various values of numerator degrees of freedom, denominator degrees of freedom, and α of 0.05 or 0.01 is provided in Chart VI of the Appendix. These curves plot the probability of type II error against the parameter λ, where

$$\lambda = \sqrt{1 + \frac{n\sigma_\tau^2}{\sigma^2}} \tag{4-5}$$

Note that λ involves two unknown parameters, σ^2 and σ_τ^2. We may be able to estimate σ_τ^2 if we have an idea about how much variability in the population of treatments it is important to detect. An estimate of σ^2 may be chosen using prior experience or judgment. Sometimes it is helpful to define the value of σ_τ^2 we are interested in detecting in terms of the ratio σ_τ^2/σ^2.

Example 4–2

Suppose we have five treatments selected at random with six observations per treatment and $\alpha = 0.05$, and we wish to determine the power of the test if σ_τ^2 is equal to σ^2. Since $a = 5$, $n = 6$, and $\sigma_\tau^2 = \sigma^2$, we may compute

$$\lambda = \sqrt{1 + 6(1)} = 2.646$$

From the operating characteristic curve with $a - 1 = 4$, $N - a = 25$ degrees of freedom, and $\alpha = 0.05$, we find that

$$\beta \simeq 0.20$$

and thus the power is approximately 0.80.

■

4–1.2   Specifying a Standard Deviation Increase

This approach is occasionally helpful in choosing the sample size. Consider first the fixed effects model. If the treatment means do not differ, then the standard deviation of an observation chosen at random is σ. If the treatment means are different, however, then the standard deviation of a randomly chosen observation is

$$\sqrt{\sigma^2 + \left(\sum_{i=1}^{a} \tau_i^2/a\right)}$$

If we choose a percentage P for the increase in the standard deviation of an observation beyond which we wish to reject the hypothesis that all treatment means are equal, then this is equivalent to choosing

$$\frac{\sqrt{\sigma^2 + \left(\sum_{i=1}^{a} \tau_i^2/a\right)}}{\sigma} = 1 + 0.01P \qquad (P = \text{percent})$$

or

$$\frac{\sqrt{\sum_{i=1}^{a} \tau_i^2/a}}{\sigma} = \sqrt{(1 + 0.01P)^2 - 1}$$

so that

$$\Phi = \frac{\sqrt{\sum_{i=1}^{a} \tau_i^2/a}}{\sigma/\sqrt{n}} = \sqrt{(1 + 0.01P)^2 - 1}(\sqrt{n}) \qquad (4\text{–}6)$$

Thus, for a specified value of P, we may compute Φ from Equation 4–6 and then use the operating characteristic curves in Appendix Chart V to determine the required sample size.

For example, in the tensile strength experiment from Example 3–1, suppose that we wish to detect a standard deviation increase of 20 percent with a probability of at least 0.90 and $\alpha = 0.05$. Then

$$\Phi = \sqrt{(1.2)^2 - 1}(\sqrt{n}) = 0.66\sqrt{n}$$

Reference to the operating characteristic curves shows that $n = 9$ is required to give the desired sensitivity.

A similar approach can be used for the random effects model. If the treatments are homogeneous, then the standard deviation of an observation selected at random is σ. However, if the treatments are different, then the standard deviation of a randomly chosen observation is

$$\sqrt{\sigma^2 + \sigma_\tau^2}$$

If P is the fixed percentage increase in the standard deviation of an observation beyond which rejection of the null hypothesis is desired, then

$$\frac{\sqrt{\sigma^2 + \sigma_\tau^2}}{\sigma} = 1 + 0.01P$$

or

$$\frac{\sigma_\tau^2}{\sigma^2} = (1 + 0.01P)^2 - 1$$

Therefore, using Equation 4–5, we find that

$$\lambda = \sqrt{1 + \frac{n\sigma_\tau^2}{\sigma^2}} = \sqrt{1 + n[(1 + 0.01P)^2 - 1]} \qquad (4\text{–}7)$$

For a given P, the operating characteristic curves in Appendix Chart VI can be used to find the desired sample size.

4–1.3 Confidence Interval Estimation Method

This approach assumes that the experimenter wishes to express the final results in terms of confidence intervals and is willing to specify in advance how wide he or she wants these confidence intervals to be. For example, suppose that in the tensile strength experiment from Example 3–1 we wanted a 95 percent confidence interval on the difference in mean tensile strength for any two cotton weight percents to be ± 5 psi and a prior estimate of σ is 3. Then, using Equation 3–13, we find that the accuracy of the confidence interval is

$$\pm t_{\alpha/2, N-a} \sqrt{\frac{2MS_E}{n}}$$

Suppose that we try $n = 5$ replicates. Then, using $\sigma^2 = 3^2 = 9$ as an estimate of MS_E, the accuracy of the confidence interval becomes

$$\pm 2.086 \sqrt{\frac{2(9)}{5}} = \pm 3.96$$

which is more accurate than the requirement. Trying $n = 4$ gives

$$\pm 2.132 \sqrt{\frac{2(9)}{4}} = \pm 4.52$$

Trying $n = 3$ gives

$$\pm 2.228 \sqrt{\frac{2(9)}{3}} = \pm 5.46$$

Clearly $n = 4$ is the smallest sample size that will lead to the desired accuracy.

The quoted level of significance in the above illustration applies only to one confidence interval. However, the same general approach can be used if the experimenter wishes to prespecify a *set* of confidence intervals about which a **joint** or **simultaneous confidence statement** is made. Furthermore, the confidence intervals could be constructed about more general contrasts in the treatment means than the pairwise comparison illustrated above. For further examples of this approach to determining sample sizes, see Neter, Wasserman, and Kutner (1990).

4–2   DISCOVERING DISPERSION EFFECTS

We have focused on using the analysis of variance and related methods to determine which factor levels result in differences among treatment or factor level means. It is customary to refer to these effects as **location effects.** If there was inequality of variance at the different factor levels, we used transformations to stabilize the variance to improve our inference on the location effects. In some problems, however, we are interested in discovering whether the different factor levels affect **variability;** that is, we are interested in discovering potential **dispersion effects.** This will occur whenever the standard deviation, variance, or some other measure of variability is used as a response variable.

To illustrate these ideas, consider the data in Table 4–1, which resulted from a designed experiment in an aluminum smelter. Aluminum is produced by combining alumina with other ingredients in a reaction cell and applying heat by passing electric current through the cell. Alumina is added continuously to the cell to maintain the proper ratio of alumina to other ingredients. Four different ratio control algorithms were investigated in this experiment. The response variables studied were related to cell voltage. Specifically, a sensor scans cell voltage several times each second, producing thousands of voltage measurements during each run of the experiment. The process engineers decided to use the average voltage and the standard deviation of cell voltage (shown in parentheses) over the run as the response variables. The average voltage is important because it impacts cell temperature, and the standard deviation of

Table 4–1   Data for the Smelting Experiment

Ratio Control Algorithm	Observations					
	1	2	3	4	5	6
1	4.93(0.05)	4.86(0.04)	4.75(0.05)	4.95(0.06)	4.79(0.03)	4.88(0.05)
2	4.85(0.04)	4.91(0.02)	4.79(0.03)	4.85(0.05)	4.75(0.03)	4.85(0.02)
3	4.83(0.09)	4.88(0.13)	4.90(0.11)	4.75(0.15)	4.82(0.08)	4.90(0.12)
4	4.89(0.03)	4.77(0.04)	4.94(0.05)	4.86(0.05)	4.79(0.03)	4.76(0.02)

Table 4-2 Analysis of Variance for the Natural Logarithm of Pot Noise

Source of Variation	Sum of Squares	Degrees of Freedom	Mean Square	F_0	P-Value
Ratio Control Algorithm	6.166	3	2.055	21.96	<0.001
Error	1.872	20	0.094		
Total	8.038	23			

voltage (called "pot noise" by the process engineers) is important because it impacts the overall cell efficiency.

An analysis of variance was performed to determine if the different ratio control algorithms affect average cell voltage. This revealed that the ratio control algorithm had no **location effect;** that is, changing the ratio control algorithms does not change the average cell voltage. (Refer to Problem 4–7.)

To investigate dispersion effects, it is usually best to use

$$\log (s)$$

as a response variable as the log transformation is effective in stabilizing variability in the distribution of the sample standard deviation. As all sample standard deviations of pot voltage are less than unity, we will use

$$y = -\ln (s)$$

as the response variable. Table 4–2 presents the analysis of variance for this response, the natural logarithm of "pot noise." Notice that the choice of a ratio control algorithm affects pot noise; that is, the ratio control algorithm has a **dispersion effect.** Standard tests of model adequacy, including normal probability plots of the residuals, indicate that there are no problems with experimental validity. (Refer to Problem 4–8.)

Figure 4–1 plots the average log pot noise for each ratio control algorithm and also presents a scaled t distribution for use as a **reference distribution** in discriminating between ratio control algorithms. This plot clearly reveals that ratio control algorithm 3 produces greater pot noise or greater cell voltage standard deviation than the other algorithms. There does not seem to be much difference between algorithms 1, 2, and 4.

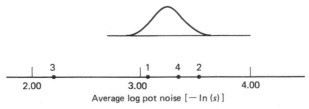

Figure 4–1. Average log pot noise $[-\ln (s)]$ for four ratio control algorithms relative to a scaled t distribution with scale factor $\sqrt{MS_E/n} = \sqrt{0.094/6} = 0.125$.

Table 4–3   Calculation of Polynomial Effects for Example 3–1

Weight Percent of Cotton	Treatment Totals $y_{i.}$	Orthogonal Contrast Coefficients (c_i)			
		Linear	Quadratic	Cubic	Quartic
15	49	−2	2	−1	1
20	77	−1	−1	2	−4
25	88	0	−2	0	6
30	108	1	−1	−2	−4
35	54	2	2	1	1
Effects: $\left(\sum_{i=1}^{a} c_i y_{i.}\right)$		41	−155	−57	−109
Sums of squares: $\left[\dfrac{\left(\sum_{i=1}^{a} c_i y_{i.}\right)^2}{n\sum_{i=1}^{a} c_i^2}\right]$		$\dfrac{(41)^2}{5(10)} =$ 33.62	$\dfrac{(-155)^2}{5(14)} =$ 343.21	$\dfrac{(-57)^2}{5(10)} =$ 64.98	$\dfrac{(-109)^2}{5(70)} =$ 33.95

4–3 FITTING RESPONSE CURVES IN THE SINGLE-FACTOR MODEL

In Section 3–5.1 we showed how an **empirical model** could be developed for the tensile strength experiment from Example 3–1. We used regression analysis as the technique to generate the fitted equation, which was the third-order polynomial

$$\hat{y} = 62.6114 - 9.0114x + 0.4814x^2 - 0.0076x^3$$

where $\hat{y}$ is the predicted tensile strength and x is the weight percent of cotton in the fiber. The technique used to estimate the parameters in this model is the **method of least squares.** Empirical models such as this are very useful in summarizing the results of many designed experiments involving quantitative factors.

In general, when the levels of a quantitative factor are equally spaced, there is a special technique for obtaining the least squares fit to the polynomial regression model. This technique is called the **method of orthogonal polynomials.**[1] In addition to producing the least squares fit to the model, this technique will also generate a linear, quadratic, cubic, and so forth, effect and the sum of squares for the factor. This allows the contribution of each term to the polynomial to be tested. It is possible to extract polynomial effects up through order $a - 1$ if there are a levels of the factor involved in the experiment.

The process is illustrated in Table 4–3 using the data of Example 3–1. In

---

[1] The method of orthogonal polynomials is not really necessary for model fitting. Modern computer software for regression analysis (see Chapter 13) will do this quite effectively. However, the method is useful in understanding some other aspects of designed experiments covered in Chapter 10.

Table 4-4   Analysis of Variance

Source of Variation	Sum of Squares	Degrees of Freedom	Mean Square	F_0	P-Value
Weight percent of cotton	475.76	4	118.94	14.76	<0.001
(Linear)	(33.62)	1	33.62	4.17	0.0546
(Quadratic)	(343.21)	1	343.21	42.58	<0.001
(Cubic)	(64.98)	1	64.98	8.06	0.0101
(Quartic)	(33.95)	1	33.95	4.21	0.0535
Error	161.20	20	8.06		
Total		24			

this problem the independent factor, weight percent of cotton, is equally spaced at five levels. The orthogonal contrast coefficients in Table 4–3 are found in Appendix Table X. The sums of squares for the linear, quadratic, cubic, and quartic effects of the factor form a partition of the treatment sum of squares and can be incorporated into the analysis of variance, as shown in Table 4–4. Each effect has one degree of freedom and can be tested by comparing its sum of squares to the mean square error.

From inspection of Table 4–4, we note that the quadratic and cubic effects of cotton weight percent are statistically significant at $\alpha = 0.05$. Therefore, we will fit a cubic polynomial to the data, say

$$y = \alpha_0 + \alpha_1 P_1(x) + \alpha_2 P_2(x) + \alpha_3 P_3(x) + \epsilon \qquad (4\text{--}8)$$

where $P_u(x)$ is the uth-order **orthogonal polynomial,** which implies that if there are a levels of x we have $\sum_{j=1}^{a} P_u(x_j)P_s(x_j) = 0$ for $u \neq s$. The first five orthogonal polynomials are

$$P_0(x) = 1$$
$$P_1(x) = \lambda_1 \left[\frac{(x - \bar{x})}{d} \right]$$
$$P_2(x) = \lambda_2 \left[\left(\frac{x - \bar{x}}{d} \right)^2 - \left(\frac{a^2 - 1}{12} \right) \right]$$
$$P_3(x) = \lambda_3 \left[\left(\frac{x - \bar{x}}{d} \right)^3 - \left(\frac{x - \bar{x}}{d} \right) \left(\frac{3a^2 - 7}{20} \right) \right]$$
$$P_4(x) = \lambda_4 \left[\left(\frac{x - \bar{x}}{d} \right)^4 - \left(\frac{x - \bar{x}}{d} \right)^2 \left(\frac{3a^2 - 13}{14} \right) + \frac{3(a^2 - 1)(a^2 - 9)}{560} \right]$$

where d is the distance between the levels of x, a is the total number of levels, and $\{\lambda_i\}$ are constants such that the polynomials have integer values. Appendix Table X lists coefficients of the orthogonal polynomials and values of the λ_i for $a \leq 10$. The least squares estimates of the parameters in the orthogonal polynomial model are

$$\hat{\alpha}_i = \frac{\sum yP_i(x)}{\sum [P_i(x)]^2} \qquad i = 0, 1, \ldots, a - 1$$

For the data in Example 3–1, we may estimate the model parameters as

$$\hat{\alpha}_0 = \frac{\sum yP_0(x)}{\sum [P_0(x)]^2} = \frac{\sum y}{25} = \frac{376}{5(5)} = 15.0400$$

$$\hat{\alpha}_1 = \frac{\sum yP_1(x)}{\sum [P_1(x)]^2} = \frac{41}{5(10)} = 0.8200$$

$$\hat{\alpha}_2 = \frac{\sum yP_2(x)}{\sum [P_2(x)]^2} = \frac{-155}{5(14)} = -2.2143$$

$$\hat{\alpha}_3 = \frac{\sum yP_3(x)}{\sum [P_3(x)]^2} = \frac{-57}{5(10)} = -1.1400$$

If we wish either to add or to delete terms from the model, it is not necessary to recompute the $\{\hat{\alpha}_i\}$ already in the model because of the orthogonality property of the polynomials $\{P_i(x)\}$.

Since we have $a = 5$ levels of x and the spacing between the levels is $d = 5$, the orthogonal polynomial model becomes

$$\hat{y} = 15.0400 + 0.8200(1)\left(\frac{x - 25}{5}\right) - 2.2143(1)\left[\left(\frac{x - 25}{5}\right)^2 - \left(\frac{5^2 - 1}{12}\right)\right]$$

$$- 1.1400(5/6)\left[\left(\frac{x - 25}{5}\right)^3 - \left(\frac{x - 25}{5}\right)\left(\frac{3(5)^2 - 7}{20}\right)\right]$$

where $\lambda_1 = \lambda_2 = 1$ and $\lambda_3 = 5/6$ have been obtained from Appendix Table X. This equation can be simplified to

$$\hat{y} = 62.6111 - 9.0100x + 0.4814x^2 - 0.00786x^3$$

which is essentially the same equation found earlier by more general regression methods. (The difference is due primarily to round-off errors in the regression computer program used earlier and the manual calculations above.)

Notice that we have fit a model that, according to the P-values in Table 4–4, contains statistically significant (at $\alpha = 0.05$) quadratic and cubic terms, and a nonsignificant (again at $\alpha = 0.05$) linear term. A logical question is, "why did we retain the nonsignificant linear term in Equation 4–8?" We left the linear term in Equation 4–8 to preserve **hierarchy** in the model. This is a model-building principle that suggests that when a particular polynomial term is included in a model, all lower-order polynomial terms should also be included. We feel that, in general, the hierarchy principle is a good one, as it promotes a certain kind of internal consistency in the model.

4-4 THE REGRESSION APPROACH TO THE ANALYSIS OF VARIANCE

We have given an intuitive or heuristic development of the analysis of variance. However, it is possible to give a more formal development. The method will be useful later in understanding the basis for the statistical analysis of more complex designs. Called the **general regression significance test,** the procedure essentially consists of finding the reduction in the total sum of squares for fitting the model with all parameters included and the reduction in sum of squares when the model is restricted to the null hypotheses. The difference between these two sums of squares is the treatment sum of squares with which a test of the null hypothesis can be conducted. The procedure requires the least squares estimators of the parameters in the analysis of variance model. We have given these parameter estimates previously (in Section 3–3.3); however, we now give a formal development.

4-4.1 Least Squares Estimation of the Model Parameters

We now develop estimators for the parameter in the single-factor model

$$y_{ij} = \mu + \tau_i + \epsilon_{ij}$$

using the method of least squares. To find the least squares estimators of μ and τ_i, we first form the sum of squares of the errors

$$L = \sum_{i=1}^{a} \sum_{j=1}^{n} \epsilon_{ij}^2 = \sum_{i=1}^{a} \sum_{j=1}^{n} (y_{ij} - \mu - \tau_i)^2 \tag{4-9}$$

and then choose values of μ and τ_i, say $\hat{\mu}$ and $\hat{\tau}_i$, that minimize L. The appropriate values would be the solutions to the $a + 1$ simultaneous equations

$$\left. \frac{\partial L}{\partial \mu} \right|_{\hat{\mu}, \hat{\tau}_i} = 0$$

$$\left. \frac{\partial L}{\partial \tau_i} \right|_{\hat{\mu}, \hat{\tau}_i} = 0 \qquad i = 1, 2, \ldots, a$$

Differentiating Equation 4–9 with respect to μ and τ_i and equating to zero, we obtain

$$-2 \sum_{i=1}^{a} \sum_{j=1}^{n} (y_{ij} - \hat{\mu} - \hat{\tau}_i) = 0$$

and

$$-2 \sum_{j=1}^{n} (y_{ij} - \hat{\mu} - \hat{\tau}_i) = 0 \qquad i = 1, 2, \ldots, a$$

which, after simplification, yield

$$
\begin{aligned}
N\hat{\mu} + n\hat{\tau}_1 + n\hat{\tau}_2 + \cdots + n\hat{\tau}_a &= y_{..} \\
n\hat{\mu} + n\hat{\tau}_1 \qquad\qquad\qquad &= y_{1.} \\
n\hat{\mu} \qquad + n\hat{\tau}_2 \qquad\qquad &= y_{2.} \\
&\vdots \\
n\hat{\mu} \qquad\qquad\qquad + n\hat{\tau}_a &= y_{a.}
\end{aligned}
\tag{4-10}
$$

The $a + 1$ equations (Equation 4–10) in $a + 1$ unknowns are called the **least squares normal equations.** Notice that if we add the last a normal equations, we obtain the first normal equation. Therefore, the normal equations are not linearly independent, and no unique solution for $\mu, \tau_1, \ldots, \tau_a$ exists. This difficulty can be overcome by several methods. Since we have defined the treatment effects as deviations from the overall mean, it seems reasonable to apply the **constraint**

$$
\sum_{i=1}^{a} \hat{\tau}_i = 0
\tag{4-11}
$$

Using this constraint, we obtain as the solution to the normal equations

$$
\begin{aligned}
\hat{\mu} &= \bar{y}_{..} \\
\hat{\tau}_i &= \bar{y}_{i.} - \bar{y}_{..} \qquad i = 1, 2, \ldots, a
\end{aligned}
\tag{4-12}
$$

This solution is obviously not unique and depends on the constraint (Equation 4–11) that we have chosen. At first this may seem unfortunate because two different experimenters could analyze the same data and obtain different results if they apply different constraints. However, certain **functions** of the model parameter *are* uniquely estimated, regardless of the constraint. Some examples are $\tau_i - \tau_j$, which would be estimated by $\hat{\tau}_i - \hat{\tau}_j = \bar{y}_{i.} - \bar{y}_{j.}$, and the ith treatment mean $\mu_i = \mu + \tau_i$, which would be estimated by $\hat{\mu}_i = \hat{\mu} + \hat{\tau}_i = \bar{y}_{i.}$.

Since we are usually interested in differences among the treatment effects rather than their actual values, it causes no concern that the τ_i cannot be uniquely estimated. In general, any function of the model parameters that is a linear combination of the left-hand side of the normal equations (Equations 4–10) can be uniquely estimated. Functions that are uniquely estimated regardless of which constraint is used are called **estimable functions.** We are now ready to use these parameter estimates in a general development of the analysis of variance.

4–4.2 The General Regression Significance Test

A fundamental part of this procedure is writing the normal equations for the model. These equations may always be obtained by forming the least squares function and differentiating it with respect to each unknown parameter, as we did in Section 4–4.1. However, an easier method is available. The following list of rules allow the normal equations for *any* experimental design model to be written directly:

RULE 1. There is one normal equation for each parameter in the model to be estimated.

RULE 2. The right-hand side of any normal equation is just the sum of all observations that contain the parameter associated with that particular normal equation.

To illustrate this rule, consider the single-factor model. The first normal equation is for the parameter μ; therefore, the right-hand side is $y_{..}$ because *all* observations contain μ.

RULE 3. The left-hand side of any normal equation is the sum of all model parameters, where each parameter is multiplied by the number of times it appears in the total on the right-hand side. The parameters are written with a circumflex ($\hat{}$) to indicate that they are *estimators* and not the true parameter values.

For example, consider the first normal equation in a single-factor experiment. According to the above rules, it would be

$$N\hat{\mu} + n\hat{\tau}_1 + n\hat{\tau}_2 + \cdots + n\hat{\tau}_a = y_{..}$$

since μ appears in all N observations, τ_1 appears only in the n observations taken under the first treatment, τ_2 appears only in the n observations taken under the second treatment, and so on. From Equation 4–10, we verify that the equation shown above is correct. The second normal equation would correspond to τ_1 and is

$$n\hat{\mu} + n\hat{\tau}_1 = y_{1.}$$

since only the observations in the first treatment contain τ_1 (this gives $y_{1.}$ as the right-hand side), μ and τ_1 appear exactly n times in $y_{1.}$, and all other τ_i appear zero times. In general, the left-hand side of any normal equation is the expected value of the right-hand side.

Now, consider finding the reduction in the sum of squares by fitting a particular model to the data. By fitting a model to the data, we "explain" some of the variability; that is, we reduce the unexplained variability by some amount. The reduction in the unexplained variability is always the sum of the parameter estimates, each multiplied by the right-hand side of the normal equation that corresponds to that parameter. For example, in a single-factor experiment, the reduction due to fitting the model $y_{ij} = \mu + \tau_i + \epsilon_{ij}$ is

$$R(\mu, \tau) = \hat{\mu}y_{..} + \hat{\tau}_1 y_{1.} + \hat{\tau}_2 y_{2.} + \cdots + \hat{\tau}_a y_{a.}$$
$$= \hat{\mu}y_{..} + \sum_{i=1}^{a} \hat{\tau}_i y_{i.} \tag{4–13}$$

The notation $R(\mu, \tau)$ means that reduction in the sum of squares from fitting the model containing μ and $\{\tau_i\}$. $R(\mu, \tau)$ is also sometimes called the "regression" sum of squares for the model $y_{ij} = \mu + \tau_i + \epsilon_{ij}$. The number of degrees of freedom associated with a reduction in the sum of squares, such as $R(\mu, \tau)$, is always equal to the number of linearly independent normal equations. The remaining variability unaccounted for by the model is found from

$$SS_E = \sum_{i=1}^{a} \sum_{j=1}^{n} y_{ij}^2 - R(\mu, \tau) \tag{4–14}$$

This quantity is used in the denominator of the test statistic for $H_0: \tau_1 = \tau_2 = \cdots = \tau_a = 0$.

We now illustrate the general regression significance test for a single-factor experiment and show that it yields the usual one-way analysis of variance. The model is $y_{ij} = \mu + \tau_i + \epsilon_{ij}$, and the normal equations are found from the above rules as

$$
\begin{aligned}
N\hat{\mu} + n\hat{\tau}_1 + n\hat{\tau}_2 + \cdots + n\hat{\tau}_a &= y_{..} \\
n\hat{\mu} + n\hat{\tau}_1 \qquad\qquad\qquad\quad &= y_{1.} \\
n\hat{\mu} \qquad + n\hat{\tau}_2 \qquad\qquad &= y_{2.} \\
&\;\;\vdots \\
n\hat{\mu} \qquad\qquad\qquad + n\hat{\tau}_a &= y_{a.}
\end{aligned}
$$

Compare these normal equations with those obtained in Equation 4–10.

Applying the constraint $\sum_{i=1}^{a} \hat{\tau}_i = 0$, the estimators for μ and τ_i are

$$\hat{\mu} = \bar{y}_{..} \qquad \hat{\tau}_i = \bar{y}_{i.} - \bar{y}_{..} \qquad i = 1, 2, \ldots, a$$

The reduction in the sum of squares due to fitting this model is found from Equation 4–13 as

$$
\begin{aligned}
R(\mu, \tau) &= \hat{\mu}y_{..} + \sum_{i=1}^{a} \hat{\tau}_i y_{i.} \\
&= (\bar{y}_{..})y_{..} + \sum_{i=1}^{a} (\bar{y}_{i.} - \bar{y}_{..})y_{i.} \\
&= \frac{y_{..}^2}{N} + \sum_{i=1}^{a} \bar{y}_{i.}y_{i.} - \bar{y}_{..}\sum_{i=1}^{a} y_{i.} \\
&= \sum_{i=1}^{a} \frac{y_{i.}^2}{n}
\end{aligned}
$$

which has a degrees of freedom because there are a linearly independent normal equations. The error sum of squares is, from Equation 4–14,

$$
\begin{aligned}
SS_E &= \sum_{i=1}^{a} \sum_{j=1}^{n} y_{ij}^2 - R(\mu, \tau) \\
&= \sum_{i=1}^{a} \sum_{j=1}^{n} y_{ij}^2 - \sum_{i=1}^{a} \frac{y_{i.}^2}{n}
\end{aligned}
$$

and has $N - a$ degrees of freedom.

To find the sum of squares resulting from the treatment effects (the $\{\tau_i\}$), we consider the model to be restricted to the null hypothesis; that is, $\tau_i = 0$ for all i. The reduced model is $y_{ij} = \mu + \epsilon_{ij}$. There is only one normal equation for this model:

$$N\hat{\mu} = y_{..}$$

and the estimator of μ is $\hat{\mu} = \bar{y}_{..}$. Thus, the reduction in the sum of squares that results from fitting only μ is

$$R(\mu) = (\bar{y}_{..})(y_{..}) = \frac{y_{..}^2}{N}$$

Since there is only one normal equation for this reduced model, $R(\mu)$ has one degree of freedom. The sum of squares due to the $\{\tau_i\}$, given that μ is already in the model, is the difference between $R(\mu, \tau)$ and $R(\mu)$, which is

$$R(\tau|\mu) = R(\mu, \tau) - R(\mu)$$
$$= \frac{1}{n}\sum_{i=1}^{a} y_{i.}^2 - \frac{y_{..}^2}{N}$$

with $a - 1$ degrees of freedom, which we recognize from Equation 3–9 as $SS_{\text{Treatments}}$. Making the usual normality assumption, the appropriate statistic for testing $H_0: \tau_1 = \tau_2 = \cdots = \tau_a = 0$ is

$$F_0 = \frac{R(\tau|\mu)/(a-1)}{\left[\sum_{i=1}^{a}\sum_{j=1}^{n} y_{ij}^2 - R(\mu, \tau)\right]/(N-a)}$$

which is distributed as $F_{a-1,N-a}$ under the null hypothesis. This is, of course, the test statistic for the single-factor analysis of variance.

By now you may suspect that there is a close connection between analysis of variance and regression. In fact, every analysis of variance model can be expressed in terms of a regression equation, and the general regression significance test described above can be used to develop a test for the hypothesis of interest. This will lead to the analysis of variance procedure we have used previously.

To illustrate this connection, suppose that we have a single-factor analysis of variance model with $a = 3$ treatments so that the model is

$$y_{ij} = \mu + \tau_i + \epsilon_{ij} \qquad \begin{cases} i = 1, 2, 3 \\ j = 1, 2, \ldots, n \end{cases}$$

The equivalent regression model is

$$y_{ij} = \beta_0 + \beta_1 x_{1j} + \beta_2 x_{2j} + \epsilon_{ij} \qquad \begin{cases} i = 1, 2, 3 \\ j = 1, 2, \ldots, n \end{cases} \qquad (4\text{--}15)$$

where x_{1j} and x_{2j} are defined as follows:

$$x_{1j} = \begin{cases} 1 & \text{if observation } j \text{ is from treatment 1} \\ 0 & \text{otherwise} \end{cases}$$

$$x_{2j} = \begin{cases} 1 & \text{if observation } j \text{ is from treatment 2} \\ 0 & \text{otherwise} \end{cases}$$

The relationships among the parameters β_0, β_1, and β_2 in the regression model and the parameters μ and τ_i ($i = 1, 2, 3$) in the analysis of variance model are easily determined. If the observations come from treatment 1, then

$$x_{1j} = 1 \quad \text{and} \quad x_{2j} = 0$$

and Equation 4–15 becomes

$$y_{1j} = \beta_0 + \beta_1(1) + \beta_2(0) + \epsilon_{1j}$$
$$= \beta_0 + \beta_1 + \epsilon_{1j}$$

Since in the analysis of variance model an observation from treatment 1 is represented by $y_{1j} = \mu + \tau_1 + \epsilon_{1j} = \mu_1 + \epsilon_{1j}$, this implies that

$$\beta_0 + \beta_1 = \mu_1 = \mu + \tau_1$$

Similarly, if the observations are from treatment 2, then $x_{1j} = 0$ and $x_{2j} = 1$ and

$$y_{2j} = \beta_0 + \beta_1(0) + \beta_2(1) + \epsilon_{2j}$$
$$= \beta_0 + \beta_2 + \epsilon_{2j}$$

Considering the corresponding analysis of variance model, $y_{2j} = \mu + \tau_2 + \epsilon_{2j} = \mu_2 + \epsilon_{2j}$, so

$$\beta_0 + \beta_2 = \mu_2 = \mu + \tau_2$$

Finally, consider observations from treatment 3, for which $x_{1j} = x_{2j} = 0$. The regression model becomes

$$y_{3j} = \beta_0 + \beta_1(0) + \beta_2(0) + \epsilon_{3j}$$
$$= \beta_0 + \epsilon_{3j}$$

The corresponding analysis of variance model is $y_{3j} = \mu + \tau_3 + \epsilon_{3j} = \mu_3 + \epsilon_{3j}$, so

$$\beta_0 = \mu_3 = \mu + \tau_3$$

Thus, in the regression model formulation of the one-way analysis of variance model, the regression coefficients describe comparisons of the first two treatment means with the third mean; that is,

$$\beta_0 = \mu_3$$
$$\beta_1 = \mu_1 - \mu_3$$
$$\beta_2 = \mu_2 - \mu_3$$

In general, if there are a treatments, the regression model will have $a - 1$ variables, say

$$y_{ij} = \beta_0 + \beta_1 x_{1j} + \beta_2 x_{2j} + \cdots + \beta_{a-1} x_{a-1,j} + \epsilon_{ij} \qquad \begin{cases} i = 1, 2, \ldots, a \\ j = 1, 2, \ldots, n \end{cases} \qquad (4\text{–}16)$$

where

$$x_{ij} = \begin{cases} 1 & \text{if observation } j \text{ is from treatment } i \\ 0 & \text{otherwise} \end{cases}$$

The relationship between the parameters in the regression and analysis of variance models is

$$\beta_0 = \mu_a$$
$$\beta_i = \mu_i - \mu_a \qquad i = 1, 2, \ldots, a - 1$$

Thus, β_0 always estimates the mean of the ath treatment and β_i estimates the difference between the means of treatment i and treatment a.

Now consider testing hypotheses. In the analysis of variance model, we want to test $H_0 : \mu_1 = \mu_2 = \mu_3$ (or equivalently, $H_0 : \tau_1 = \tau_2 = \tau_3 = 0$). If this null hypothesis is true, then the parameters in the regression model become

$$\beta_0 = \mu$$
$$\beta_1 = 0$$
$$\beta_2 = 0$$

Therefore, testing $H_0 : \beta_1 = \beta_2 = 0$ in the regression model provides a test of the equality of the three treatment means. The general regression significance test can be used to obtain the test procedure for $H_0 : \beta_1 = \beta_2 = 0$. The resulting test statistic is the F test in the single-factor analysis of variance.

4–5   NONPARAMETRIC METHODS IN THE ANALYSIS OF VARIANCE

4–5.1   The Kruskal–Wallis Test

In situations where the normality assumption is unjustified, the experimenter may wish to use an alternative procedure to the F test analysis of variance that does not depend on this assumption. Such a procedure has been developed by Kruskal and Wallis (1952). The Kruskal–Wallis test is used to test the null hypothesis that the a treatments are identical against the alternative hypothesis that some of the treatments generate observations that are larger than others.

Because the procedure is designed to be sensitive for testing differences in means, it is sometimes convenient to think of the Kruskal–Wallis test as a test for equality of treatment means. The Kruskal–Wallis test is a **nonparametric alternative** to the usual analysis of variance.

To perform a Kruskal–Wallis test, first rank the observations y_{ij} in ascending order and replace each observation by its rank, say R_{ij}, with the smallest observation having rank 1. In the case of ties (observations having the same value), assign the average rank to each of the tied observations. Let $R_{i.}$ be the sum of the ranks in the ith treatment. The test statistic is

$$H = \frac{1}{S^2}\left[\sum_{i=1}^{a}\frac{R_{i.}^2}{n_i} - \frac{N(N+1)^2}{4}\right]$$ (4–17)

where n_i is the number of observations in the ith treatment, N is the total number of observations, and

$$S^2 = \frac{1}{N-1}\left[\sum_{i=1}^{a}\sum_{j=1}^{n_i} R_{ij}^2 - \frac{N(N+1)^2}{4}\right]$$ (4–18)

Note that S^2 is just the variance of the ranks. If there are no ties, $S^2 = N(N+1)/12$, and the test statistic simplifies to

$$H = \frac{12}{N(N+1)}\sum_{i=1}^{a}\frac{R_{i.}^2}{n_i} - 3(N+1)$$ (4–19)

When the number of ties is moderate, there will be little difference between Equations 4–17 and 4–19, and the simpler form (Equation 4–19) may be used. If the n_i are reasonably large, say $n_i \geq 5$, then H is distributed approximately as χ_{a-1}^2 under the null hypothesis. Therefore, if

$$H > \chi_{\alpha,a-1}^2$$

the null hypothesis is rejected. The P-value approach could also be used.

Example 4–3

The data from Example 3–1 and their corresponding ranks are shown in Table 4–5. Since there is a fairly large number of ties, we use Equation 4–17 as the test statistic. From Equation 4–18 we find

$$S^2 = \frac{1}{N-1}\left[\sum_{i=1}^{a}\sum_{j=1}^{n_i} R_{ij}^2 - \frac{N(N+1)^2}{4}\right]$$

$$= \frac{1}{24}\left[5497.79 - \frac{25(26)^2}{4}\right]$$

$$= 53.03$$

Table 4–5   Data and Ranks for the Tensile Testing Experiment in Example 3–1

				Weight Percent of Cotton					
15		20		25		30		35	
y_{1j}	R_{1j}	y_{2j}	R_{2j}	y_{3j}	R_{3j}	y_{4j}	R_{4j}	y_{5j}	R_{5j}
7	2.0	12	9.5	14	11.0	19	20.5	7	2.0
7	2.0	17	14.0	18	16.5	25	25.0	10	5.0
15	12.5	12	9.5	18	16.5	22	23.0	11	7.0
11	7.0	18	16.5	19	20.5	19	20.5	15	12.5
9	4.0	18	16.5	19	20.5	23	24.0	11	7.0
$R_{i.}$	27.5		66.0		85.0		113.0		33.5

and the test statistic is

$$H = \frac{1}{S^2} \left[\sum_{i=1}^{a} \frac{R_{i.}^2}{n_i} - \frac{N(N+1)^2}{4} \right]$$

$$= \frac{1}{53.03} \left[5245.0 - \frac{25(26)^2}{4} \right]$$

$$= 19.25$$

Since $H > \chi_{0.01,4}^2 = 13.28$, we would reject the null hypothesis and conclude that the treatments differ. (The P-value for $H = 19.25$ is $P = 0.0002$.) This is the same conclusion given by the usual analysis of variance F test.

■

4–5.2   General Comments on the Rank Transformation

The procedure used in the previous section of replacing the observations by their ranks is called the **rank transformation.** It is a very powerful and widely useful technique. If we were to apply the ordinary F test to the ranks rather than to the original data, we would obtain

$$F_0 = \frac{H/(a-1)}{(N-1-H)/(N-a)} \tag{4–20}$$

as the test statistic [see Conover (1980), p. 337]. Note that, as the Kruskal–Wallis statistic H increases or decreases, F_0 also increases or decreases, so the Kruskal–Wallis test is equivalent to applying the usual analysis of variance to the ranks.

The rank transformation has wide applicability in experimental design problems for which no nonparametric alternative to the analysis of variance exists. This includes many of the designs in subsequent chapters of this book. If the data are ranked and the ordinary F test is applied, an approximate procedure that has good statistical properties results [see Conover and Iman (1976, 1981), and Conover (1980)]. When we are concerned about the normality assumption or the effect of outliers or "wild" values, we recommend that the usual analysis of variance be performed on both the original data and the ranks. When both procedures give similar results, the analysis of variance assumptions are probably satisfied reasonably well, and the standard analysis is satisfactory. When the two procedures differ, the rank transformation should be preferred since it is less likely to be distorted by nonnormality and unusual observations. In such cases, the experimenter may want to investigate the use of transformations for nonnormality and examine the data and the experimental procedure to determine if outliers are present and why they have occurred.

4–6　REPEATED MEASURES

In experimental work in the social and behavioral sciences and some aspects of engineering and the physical sciences, the experimental units are frequently people. Because of differences in experience, training, or background, the differences in the responses of different people to the same treatment may be very large in some experimental situations. Unless it is controlled, this variability between people would become part of the experimental error, and in some cases, it would significantly inflate the error mean square, making it more difficult to detect real differences between treatments.

It is possible to control this variability between people by using a design in which each of the a treatments is used on each person (or "subject"). Such a design is called a **repeated measures design.** In this section we give a brief introduction to repeated measures experiments with a single factor.

Suppose that an experiment involves a treatments and every treatment is to be used exactly once on each of n subjects. The data would appear as in Table 4–6. Note that the observation y_{ij} represents the response of subject j to treatment i and that only n subjects are used. The model that we use for this design is

$$y_{ij} = \mu + \tau_i + \beta_j + \epsilon_{ij} \tag{4–21}$$

where τ_i is the effect of the ith treatment and β_j is a parameter associated with the jth subject. We assume that treatments are fixed (so $\sum_{i=1}^{a} \tau_i = 0$) and that the subjects employed are a random sample of subjects from some larger population of potential subjects. Thus, the subjects collectively represent a random effect, so we assume that the mean of β_j is zero and that the variance of β_j is σ_β^2. Since the term β_j is common to all a measurements on the same subject, the covariance between y_{ij} and $y_{i'j}$ is not, in general, zero. It is customary to assume that the covariance between y_{ij} and $y_{i'j}$ is constant across all treatments and subjects.

Table 4-6 Data for a Single-Factor Repeated Measures Design

Treatment	Subject 1	2	$\cdots$	n	Treatment Totals
1	y_{11}	y_{12}	$\cdots$	y_{1n}	$y_{1.}$
2	y_{21}	y_{22}	$\cdots$	y_{2n}	$y_{2.}$
$\vdots$	$\vdots$	$\vdots$		$\vdots$	
a	y_{a1}	y_{a2}		y_{an}	$y_{a.}$
Subject Totals	$y_{.1}$	$y_{.2}$	$\cdots$	$y_{.n}$	$y_{..}$

Consider an analysis of variance partitioning of the total sum of squares, say

$$\sum_{i=1}^{a} \sum_{j=1}^{n} (y_{ij} - \bar{y}_{..})^2 = a \sum_{j=1}^{n} (\bar{y}_{.j} - \bar{y}_{..})^2 + \sum_{i=1}^{a} \sum_{j=1}^{n} (y_{ij} - \bar{y}_{.j})^2 \qquad (4\text{--}22)$$

We may view the first term on the right-hand side of Equation 4–22 as a sum of squares that results from differences *between subjects* and the second term as a sum of squares of differences *within subjects*. That is,

$$SS_T = SS_{\text{Between Subjects}} + SS_{\text{Within Subjects}}$$

The sums of squares $SS_{\text{Between Subjects}}$ and $SS_{\text{Within Subjects}}$ are statistically independent, with degrees of freedom

$$an - 1 = (n - 1) + n(a - 1)$$

The differences within subjects depend on both differences in treatment effects and uncontrolled variability (noise or error). Therefore, we may decompose the sum of squares resulting from differences within subjects as follows:

$$\sum_{i=1}^{a} \sum_{j=1}^{n} (y_{ij} - \bar{y}_{.j})^2 = n \sum_{i=1}^{a} (\bar{y}_{i.} - \bar{y}_{..})^2 + \sum_{i=1}^{a} \sum_{j=1}^{n} (y_{ij} - \bar{y}_{i.} - \bar{y}_{.j} + \bar{y}_{..})^2 \qquad (4\text{--}23)$$

The first term on the right-hand side of Equation 4–23 measures the contribution of the difference between treatment means to $SS_{\text{Within Subjects}}$, and the second term is the residual variation due to error. Both components of $SS_{\text{Within Subjects}}$ are independent. Thus,

$$SS_{\text{Within Subjects}} = SS_{\text{Treatments}} + SS_E$$

with degrees of freedom given by

$$n(a - 1) = (a - 1) + (a - 1)(n - 1)$$

respectively.

Table 4-7 Analysis of Variance for a Single-Factor Repeated Measures Design

Source of Variation	Sums of Squares	Degrees of Freedom	Mean Square	F_0
1. Between subjects	$\sum_{j=1}^{n} \dfrac{y_{.j}^2}{a} - \dfrac{y_{..}^2}{an}$	$n-1$		
2. Within subjects	$\sum_{i=1}^{a}\sum_{j=1}^{n} y_{ij}^2 - \sum_{j=1}^{n} \dfrac{y_{.j}^2}{a}$	$n(a-1)$		
3. (Treatments)	$\sum_{i=1}^{a} \dfrac{y_{i.}^2}{n} - \dfrac{y_{..}^2}{an}$	$a-1$	$MS_{\text{Treatment}} = \dfrac{SS_{\text{Treatment}}}{a-1}$	$\dfrac{MS_{\text{Treatment}}}{MS_E}$
4. (Error)	Subtraction: line (2) – line (3)	$(a-1)(n-1)$	$MS_E = \dfrac{SS_E}{(a-1)(n-1)}$	
5. Total	$\sum_{i=1}^{a}\sum_{j=1}^{n} y_{ij}^2 - \dfrac{y_{..}^2}{an}$	$an-1$		

To test the hypothesis of no treatment effect, that is,

$$H_0: \tau_1 = \tau_2 = \cdots = \tau_a = 0$$
$$H_1: \text{At least one } \tau_i \neq 0$$

we would use the ratio

$$F_0 = \frac{SS_{\text{Treatment}}/(a-1)}{SS_E/(a-1)(n-1)} = \frac{MS_{\text{Treatments}}}{MS_E} \tag{4-24}$$

If the model errors are normally distributed, then under the null hypothesis, $H_0: \tau_i = 0$, the statistic F_0 follows an $F_{a-1,(a-1)(n-1)}$ distribution. The null hypothesis would be rejected if $F_0 > F_{\alpha,a-1,(a-1)(n-1)}$.

The analysis of variance procedure is summarized in Table 4–7, which also gives convenient computing formulas for the sums of squares. Readers with prior exposure to experimental design will recognize the analysis of variance for a single-factor design with repeated measures as equivalent to the analysis for a randomized complete block design, with subjects considered to be the blocks. Randomized block designs are discussed in the next chapter.

4-7 THE ANALYSIS OF COVARIANCE

In Chapter Two we introduced the use of the blocking principle to improve the precision with which comparisons between treatments are made. The paired t test was the specific example illustrated. In general, the **blocking principle** can be used to eliminate the effect of controllable nuisance factors. We will discuss this in more detail in the next chapter. The **analysis of covariance** is another technique that is occasionally useful for improving the precision of an experiment. Suppose that in an experiment with a response variable y there is another variable, such as x, and that y is linearly related to x. Furthermore, suppose that x cannot be controlled by the experimenter but can be observed along with y. The variable x is called a **covariate** or **concomitant variable.** The analysis of covariance involves adjusting the observed response variable for the effect of the concomitant variable. If such an adjustment is not performed, the concomitant variable could inflate the error mean square and make true differences in the response due to treatments harder to detect. Thus, the analysis of covariance is a method of adjusting for the effects of an uncontrollable nuisance variable. As we will see, the procedure is a combination of analysis of variance and regression analysis.

As an example of an experiment in which the analysis of covariance may be employed, consider a study performed to determine if there is a difference in the strength of a monofilament fiber produced by three different machines. The data from this experiment are shown in Table 4–8. Figure 4–2 presents a scatter diagram of strength (y) versus the diameter (or thickness) of the sample. Clearly the strength of the fiber is also affected by its thickness; consequently,

Table 4–8   Breaking Strength Data (y = strength in pounds and x = diameter in 10^{-3} inches)

Machine 1		Machine 2		Machine 3	
y	x	y	x	y	x
36	20	40	22	35	21
41	25	48	28	37	23
39	24	39	22	42	26
42	25	45	30	34	21
49	32	44	28	32	15
207	126	216	130	180	106

a thicker fiber will generally be stronger than a thinner one. The analysis of covariance could be used to remove the effect of thickness (x) on strength (y) when testing for differences in strength between machines.

4–7.1   Description of the Procedure

The basic procedure for the analysis of covariance is now described and illustrated for a single-factor experiment with one covariate. Assuming that there is a linear relationship between the response and the covariate, an appropriate statistical model is

$$y_{ij} = \mu + \tau_i + \beta(x_{ij} - \bar{x}_{..}) + \epsilon_{ij} \begin{cases} i = 1, 2, \ldots, a \\ j = 1, 2, \ldots, n \end{cases} \tag{4–25}$$

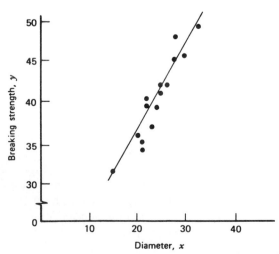

Figure 4–2.   Breaking strength (y) versus fiber diameter (x).

where y_{ij} is the jth observation on the response variable taken under the ith treatment or level of the single factor, x_{ij} is the measurement made on the covariate or concomitant variable corresponding to y_{ij} (i.e., the ijth run), $\bar{x}_{..}$ is the mean of the x_{ij} values, μ is an overall mean, τ_i is the effect of the ith treatment, β is a linear regression coefficient indicating the dependency of y_{ij} on x_{ij}, and ϵ_{ij} is a random error component. We assume that the errors ϵ_{ij} are NID$(0, \sigma^2)$, that the slope $\beta \neq 0$ and the true relationship between y_{ij} and x_{ij} is linear, that the regression coefficients for each treatment are identical, that the treatment effects sum to zero ($\sum_{i=1}^{a} \tau_i = 0$), and that the concomitant variable x_{ij} is not affected by the treatments.

Notice from Equation 4–25 that the analysis of covariance model is a combination of the linear models employed in analysis of variance and regression. That is, we have treatment effects $\{\tau_i\}$ as in a single-factor analysis of variance and a regression coefficient β as in a regression equation. The concomitant variable in Equation 4–25 is expressed as $(x_{ij} - \bar{x}_{..})$ instead of x_{ij} so that the parameter μ is preserved as the overall mean. The model could have been written as

$$y_{ij} = \mu' + \tau_i + \beta x_{ij} + \epsilon_{ij} \qquad \begin{cases} i = 1, 2, \ldots, a \\ j = 1, 2, \ldots, n \end{cases} \qquad (4\text{–}26)$$

where μ' is a constant not equal to the overall mean, which for this model is $\mu' + \beta \bar{x}_{..}$. Equation 4–25 is more widely found in the literature.

To describe the analysis, we introduce the following notation:

$$S_{yy} = \sum_{i=1}^{a} \sum_{j=1}^{n} (y_{ij} - \bar{y}_{..})^2 = \sum_{i=1}^{a} \sum_{j=1}^{n} y_{ij}^2 - \frac{y_{..}^2}{an} \qquad (4\text{–}27)$$

$$S_{xx} = \sum_{i=1}^{a} \sum_{j=1}^{n} (x_{ij} - \bar{x}_{..})^2 = \sum_{i=1}^{a} \sum_{j=1}^{n} x_{ij}^2 - \frac{x_{..}^2}{an} \qquad (4\text{–}28)$$

$$S_{xy} = \sum_{i=1}^{a} \sum_{j=1}^{n} (x_{ij} - \bar{x}_{..})(y_{ij} - \bar{y}_{..}) = \sum_{i=1}^{a} \sum_{j=1}^{n} x_{ij} y_{ij} - \frac{(x_{..})(y_{..})}{an} \qquad (4\text{–}29)$$

$$T_{yy} = \sum_{i=1}^{a} (\bar{y}_{i.} - \bar{y}_{..})^2 = \frac{1}{n} \sum_{i=1}^{a} y_{i.}^2 - \frac{y_{..}^2}{an} \qquad (4\text{–}30)$$

$$T_{xx} = \sum_{i=1}^{a} (\bar{x}_{i.} - \bar{x}_{..})^2 = \frac{1}{n} \sum_{i=1}^{a} x_{i.}^2 - \frac{x_{..}^2}{an} \qquad (4\text{–}31)$$

$$T_{xy} = \sum_{i=1}^{a} (\bar{x}_{i.} - \bar{x}_{..})(\bar{y}_{i.} - \bar{y}_{..}) = \frac{1}{n} \sum_{i=1}^{a} (x_{i.})(y_{i.}) - \frac{(x_{..})(y_{..})}{an} \qquad (4\text{–}32)$$

$$E_{yy} = \sum_{i=1}^{a} \sum_{j=1}^{n} (y_{ij} - \bar{y}_{i.})^2 = S_{yy} - T_{yy} \qquad (4\text{–}33)$$

$$E_{xx} = \sum_{i=1}^{a} \sum_{j=1}^{n} (x_{ij} - \bar{x}_{i.})^2 = S_{xx} - T_{xx} \qquad (4\text{–}34)$$

$$E_{xy} = \sum_{i=1}^{a} \sum_{j=1}^{n} (x_{ij} - \bar{x}_{i.})(y_{ij} - \bar{y}_{i.}) = S_{xy} - T_{xy} \qquad (4\text{–}35)$$

Note that, in general, $S = T + E$, where the symbols S, T, and E are used to denote sums of squares and cross-products for total, treatments, and error, respectively. The sums of squares for x and y must be nonnegative; however, the sums of cross-products (xy) may be negative.

We now show how the analysis of covariance adjusts the response variable for the effect of the covariate. Consider the full model (Equation 4–25). The least squares estimators of μ, τ_i, and β and $\hat{\mu} = \bar{y}_{..}$, $\hat{\tau}_i = \bar{y}_{i.} - \bar{y}_{..} - \hat{\beta}(\bar{x}_{i.} - \bar{x}_{..})$, and

$$\hat{\beta} = \frac{E_{xy}}{E_{xx}} \tag{4-36}$$

The error sum of squares in this model is

$$SS_E = E_{yy} - (E_{xy})^2 / E_{xx} \tag{4-37}$$

with $a(n - 1) - 1$ degrees of freedom. The experimental error variance is estimated by

$$MS_E = \frac{SS_E}{a(n - 1) - 1}$$

Now suppose that there is no treatment effect. The model (Equation 4–25) would then be

$$y_{ij} = \mu + \beta(x_{ij} - \bar{x}_{..}) + \epsilon_{ij} \tag{4-38}$$

and it can be shown that the least squares estimators of μ and β are $\hat{\mu} = \bar{y}_{..}$ and $\hat{\beta} = S_{xy}/S_{xx}$. The sum of squares for error in this reduced model is

$$SS'_E = S_{yy} - (S_{xy})^2 / S_{xx} \tag{4-39}$$

with $an - 2$ degrees of freedom. In Equation 4–39, the quantity $(S_{xy})^2/S_{xx}$ is the reduction in the sum of squares of y obtained through the linear regression of y on x. Furthermore, note that SS_E is smaller than SS'_E [because the model (Equation 4–25) contains additional parameters $\{\tau_i\}$] and that the quantity $SS'_E - SS_E$ is a reduction in sum of squares due to the $\{\tau_i\}$. Therefore, the difference between SS'_E and SS_E, that is, $SS'_E - SS_E$, provides a sum of squares with $a - 1$ degrees of freedom for testing the hypothesis of no treatment effects. Consequently, to test $H_0: \tau_i = 0$, compute

$$F_0 = \frac{(SS'_E - SS_E)/(a - 1)}{SS_E/[a(n - 1) - 1]} \tag{4-40}$$

which, if the null hypothesis is true, is distributed as $F_{a-1,a(n-1)-1}$. Thus, we reject $H_0: \tau_i = 0$ if $F_0 > F_{\alpha,a-1,a(n-1)-1}$. The P-value approach could also be used.

It is instructive to examine the display in Table 4–9. In this table we have presented the analysis of covariance as an "adjusted" analysis of variance. In the source of variation column, the total variability is measured by S_{yy} with $an - 1$ degrees of freedom. The source of variation "regression" has the sum

Table 4-9 Analysis of Covariance as an "Adjusted" Analysis of Variance

Source of Variation	Sum of Squares	Degrees of Freedom	Mean Square	F_0
Regression	$(S_{xy})^2/S_{xx}$	1		
Treatments	$SS'_E - SS_E =$ $S_{yy} - (S_{xy})^2/S_{xx} -$ $[E_{yy} - (E_{xy})^2/E_{xx}]$	$a - 1$	$\dfrac{SS'_E - SS_E}{a - 1}$	$\dfrac{(SS'_E - SS_E)/(a - 1)}{MS_E}$
Error	$SS_E = E_{yy} - (E_{xy})^2/E_{xx}$	$a(n - 1) - 1$	$MS_E = \dfrac{SS_E}{a(n - 1) - 1}$	
Total	S_{yy}	$an - 1$		

Table 4–10 Analysis of Covariance for a Single-Factor Experiment with One Covariate

Source of Variation	Degrees of Freedom	Sums of Squares and Products			Adjusted for Regression		
		x	xy	y	y	Degrees of Freedom	Mean Square
Treatments	$a - 1$	T_{xx}	T_{xy}	T_{yy}			
Error	$a(n-1)$	E_{xx}	E_{xy}	E_{yy}	$SS_E = E_{yy} - (E_{xy})^2/E_{xx}$	$a(n-1) - 1$	$MS_E = \dfrac{SS_E}{a(n-1)-1}$
Total	$an - 1$	S_{xx}	S_{xy}	S_{yy}	$SS'_E = S_{yy} - (S_{xy})^2/S_{xx}$	$an - 2$	
Adjusted treatments					$SS'_E - SS_E$	$a - 1$	$\dfrac{SS'_E - SS_E}{a-1}$

of squares $(S_{xy})^2/S_{xx}$ with one degree of freedom. If there were no concomitant variable, we would have $S_{xy} = S_{xx} = E_{xy} = E_{xx} = 0$. Then the sum of squares for error would be simply E_{yy} and the sum of squares for treatments would be $S_{yy} - E_{yy} = T_{yy}$. However, because of the presence of the concomitant variable, we must "adjust" S_{yy} and E_{yy} for the regression of y on x as shown in Table 4-9. The adjusted error sum of squares has $a(n - 1) - 1$ degrees of freedom instead of $a(n - 1)$ degrees of freedom because an additional parameter (the slope β) is fitted to the data.

The computations are usually displayed in an analysis of covariance table such as Table 4-10. This layout is employed because it conveniently summarizes all the required sums of squares and cross-products as well as the sums of squares for testing hypotheses about treatment effects. In addition to testing the hypothesis that there are no differences in the treatment effects, we frequently find it useful in interpreting the data to present the adjusted treatment means. These adjusted means are computed according to

$$\text{Adjusted } \bar{y}_{i.} = \bar{y}_{i.} - \hat{\beta}(\bar{x}_{i.} - \bar{x}_{..}) \qquad i = 1, 2, \ldots, a \tag{4-41}$$

where $\hat{\beta} = E_{xy}/E_{xx}$. This adjusted treatment mean is the least squares estimator of $\mu + \tau_i$, $i = 1, 2, \ldots, a$, in the model (Equation 4-25). The standard error of any adjusted treatment mean is

$$S_{\text{adj}\bar{y}_{i.}} = \left[MS_E \left(\frac{1}{n} + \frac{(\bar{x}_{i.} - \bar{x}_{..})^2}{E_{xx}} \right) \right]^{1/2} \tag{4-42}$$

Finally, we recall that the regression coefficient β in the model (Equation 4-25) has been assumed to be nonzero. We may test the hypothesis $H_0: \beta = 0$ by using the test statistic

$$F_0 = \frac{(E_{xy})^2/E_{xx}}{MS_E} \tag{4-43}$$

which under the null hypothesis is distributed as $F_{1,a(n-1)-1}$. Thus, we reject $H_0: \beta = 0$ if $F_0 > F_{\alpha,1,a(n-1)-1}$.

Example 4-4

Consider the experiment described at the beginning of Section 4-7. Three different machines produce a monofilament fiber for a textile company. The process engineer is interested in determining if there is a difference in the breaking strength of the fiber produced by the three machines. However, the strength of a fiber is related to its diameter, with thicker fibers being generally stronger than thinner ones. A random sample of five fiber specimens is selected from each machine. The fiber strength (y) and the corresponding diameter (x) for each specimen are shown in Table 4-8.

The scatter diagram of breaking strength versus the fiber diameter (Figure 4-2) shows a strong suggestion of a linear relationship between breaking strength and diameter, and it seems appropriate to remove the effect of diameter on strength by an analysis of covariance. Assuming that a linear relationship between breaking strength and diameter is appropriate, the model is

$$y_{ij} = \mu + \tau_i + \beta(x_{ij} - \bar{x}_{..}) + \epsilon_{ij} \qquad \begin{cases} i = 1, 2, 3 \\ j = 1, 2, \ldots, 5 \end{cases}$$

Using Equations 4–27 through 4–35, we may compute

$$S_{yy} = \sum_{i=1}^{3} \sum_{j=1}^{5} y_{ij}^2 - \frac{y_{..}^2}{an} = (36)^2 + (41)^2 + \cdots + (32)^2 - \frac{(603)^2}{(3)(5)} = 346.40$$

$$S_{xx} = \sum_{i=1}^{3} \sum_{j=1}^{5} x_{ij} - \frac{x_{..}^2}{an} = (20)^2 + (25)^2 + \cdots + (15)^2 - \frac{(362)^2}{(3)(5)} = 261.73$$

$$S_{xy} = \sum_{i=1}^{3} \sum_{j=1}^{5} x_{ij} y_{ij} - \frac{(x_{..})(y_{..})}{an} = (20)(36) + (25)(41) + \cdots + (15)(32)$$

$$-\frac{(362)(603)}{(3)(5)} = 282.60$$

$$T_{yy} = \frac{1}{n} \sum_{i=1}^{3} y_{i.}^2 - \frac{y_{..}^2}{an} = \frac{1}{5}[(207)^2 + (216)^2 + (180)^2] - \frac{(603)^2}{(3)(5)} = 140.40$$

$$T_{xx} = \frac{1}{n} \sum_{i=1}^{3} x_{i.}^2 - \frac{x_{..}^2}{an} = \frac{1}{5}[(126)^2 + (130)^2 + (106)^2] - \frac{(362)^2}{(3)(5)} = 66.13$$

$$T_{xy} = \frac{1}{n} \sum_{i=1}^{3} x_{i.} y_{i.} - \frac{(x_{..})(y_{..})}{an} = \frac{1}{5}[(126)(207) + (130)(216) + (106)(184)]$$

$$-\frac{(362)(603)}{(3)(5)} = 96.00$$

$$E_{yy} = S_{yy} - T_{yy} = 346.40 - 140.40 = 206.00$$
$$E_{xx} = S_{xx} - T_{xx} = 261.73 - 66.13 = 195.60$$
$$E_{xy} = S_{xy} - T_{xy} = 282.60 - 96.00 = 186.60$$

From Equation 4–39, we find

$$SS_E' = S_{yy} - (S_{xy})^2/S_{xx}$$
$$= 346.40 - (282.60)^2/261.73$$
$$= 41.27$$

with $an - 2 = (3)(5) - 2 = 13$ degrees of freedom; and from Equation 4–37, we find

$$SS_E = E_{yy} - (E_{xy})^2/E_{xx}$$
$$= 206.00 - (186.60)^2/195.60$$
$$= 27.99$$

with $a(n - 1) - 1 = 3(5 - 1) - 1 = 11$ degrees of freedom.
The sum of squares for testing $H_0: \tau_1 = \tau_2 = \tau_3 = 0$ is

$$SS_E' - SS_E = 41.27 - 27.99$$
$$= 13.28$$

with $a - 1 = 3 - 1 = 2$ degrees of freedom. These calculations are summarized in Table 4–11.

Table 4-11 Analysis of Covariance for the Breaking Strength Data

Source of Variation	Degrees of Freedom	Sums of Squares and Products			Adjusted for Regression				
		x	xy	y	y	Degrees of Freedom	Mean Square	F_0	P-Value
Machines	2	66.13	96.00	140.40					
Error	12	195.60	186.60	206.00	27.99	11	2.54		
Total	14	261.73	282.60	346.40	41.27	13			
Adjusted machines					13.28	2	6.64	2.61	0.1181

157

To test the hypothesis that machines differ in the breaking strength of fiber produced, that is, $H_0: \tau_i = 0$, we compute the test statistic from Equation 4–40 as

$$F_0 = \frac{(SS'_E - SS_E)/(a - 1)}{SS_E/[a(n - 1) - 1]}$$

$$= \frac{13.28/2}{27.99/11} = \frac{6.64}{2.54} = 2.61$$

Comparing this to $F_{0.10,2,11} = 2.86$, we find that the null hypothesis cannot be rejected. The P-value of this test statistic is $P = 0.1181$. Thus there is no strong evidence that the fibers produced by the three machines differ in breaking strength.

The estimate of the regression coefficient is computed from Equation 4–36 as

$$\hat{\beta} = \frac{E_{xy}}{E_{xx}} = \frac{186.60}{195.60} = 0.9540$$

We may test the hypothesis $H_0: \beta = 0$ by using Equation 4–43. The test statistic is

$$F_0 = \frac{(E_{xy})^2/E_{xx}}{MS_E} = \frac{(186.60)^2/195.60}{2.54} = 70.08$$

and since $F_{0.01,1,11} = 9.65$, we reject the hypothesis that $\beta = 0$. Therefore, there is a linear relationship between breaking strength and diameter, and the adjustment provided by the analysis of covariance was necessary.

The adjusted treatment means may be computed from Equation 4–41. These adjusted means are

$$\text{Adjusted } \bar{y}_{1.} = \bar{y}_{1.} - \hat{\beta}(\bar{x}_{1.} - \bar{x}_{..})$$
$$= 41.40 - (0.9540)(25.20 - 24.13) = 40.38$$
$$\text{Adjusted } \bar{y}_{2.} = \bar{y}_{2.} - \hat{\beta}(\bar{x}_{2.} - \bar{x}_{..})$$
$$= 43.20 - (0.9540)(26.00 - 24.13) = 41.42$$

and

$$\text{Adjusted } \bar{y}_{3.} = \bar{y}_{3.} - \hat{\beta}(\bar{x}_{3.} - \bar{x}_{..})$$
$$= 36.00 - (0.9540)(21.20 - 24.13) = 38.80$$

Comparing the adjusted treatment means with the unadjusted treatment means (the $\bar{y}_{i.}$), we note that the adjusted means are much closer together, another indication that the covariance analysis was necessary.

A basic assumption in the analysis of covariance is that the treatments do not influence the covariate x because the technique removes the effect of variations in the $\bar{x}_{i.}$. However, if the variability in the $\bar{x}_{i.}$ is due in part to the treatments, then analysis of covariance removes part of the treatment effect. Thus, we must be reasonably sure that the treatments do not affect the values x_{ij}. In some experiments this may be obvious from the nature of the covariate, whereas in others it may be more doubtful. In our example, there may be a difference in fiber diameter (x_{ij}) between the three machines. In such cases, Cochran and Cox (1957) suggest that an analysis of variance on the x_{ij} values may be helpful in determining the validity of this assumption. For our problem, this procedure yields

$$F_0 = \frac{66.13/2}{195.60/12} = \frac{33.07}{16.30} = 2.03$$

which is less than $F_{0.10,2,12} = 2.81$, so there is no reason to believe that machines produce fibers of different diameters.

■

Diagnostic checking of the covariance model is based on residual analysis. For the covariance model, the residuals are

$$e_{ij} = y_{ij} - \hat{y}_{ij}$$

where the fitted values are

$$\hat{y}_{ij} = \hat{\mu} + \hat{\tau}_i + \hat{\beta}(x_{ij} - \bar{x}_{..}) = \bar{y}_{..} + [\bar{y}_{i.} - \bar{y}_{..} - \hat{\beta}(\bar{x}_{i.} - \bar{x}_{..})]$$
$$+ \hat{\beta}(x_{ij} - \bar{x}_{..}) = \bar{y}_{i.} + \hat{\beta}(x_{ij} - \bar{x}_{i.})$$

Thus,

$$e_{ij} = y_{ij} - \bar{y}_{i.} - \hat{\beta}(x_{ij} - \bar{x}_{i.}) \tag{4-44}$$

To illustrate the use of Equation 4–44, the residual for the first observation from the first machine in Example 4–4 is

$$e_{11} = y_{11} - \bar{y}_{1.} - \hat{\beta}(x_{11} - \bar{x}_{1.}) = 36 - 41.4 - (0.9540)(20 - 25.2)$$
$$= 36 - 36.4392 = -0.4392$$

A complete listing of observations, fitted values, and residuals is given in the following table:

Observed Value y_{ij}	Fitted Value $\hat{y}_{ij}$	Residual $e_{ij} = y_{ij} - \hat{y}_{ij}$
36	36.4392	−0.4392
41	41.2092	−0.2092
39	40.2552	−1.2552
42	41.2092	0.7908
49	47.8871	1.1129
40	39.3840	0.6160
48	45.1079	2.8921
39	39.3840	−0.3840
45	47.0159	−2.0159
44	45.1079	−1.1079
35	35.8092	−0.8092
37	37.7171	−0.7171
42	40.5791	1.4209
34	35.8092	−1.8092
32	30.0852	1.9148

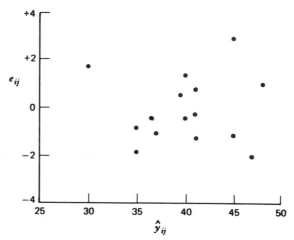

Figure 4–3.   Plot of residuals versus fitted values for
Example 4–4.

The residuals are plotted versus the fitted values $\hat{y}_{ij}$ in Figure 4–3, versus the
covariate x_{ij} in Figure 4–4, and versus the machines in Figure 4–5. A normal
probability plot of the residuals is shown in Figure 4–6 on page 162. These plots
do not reveal any major departures from the assumptions, so we conclude that the
covariance model (Equation 4–25) is appropriate for the breaking strength data.

It is interesting to note what would have happened in this experiment if an
analysis of covariance had not been performed, that is, if the breaking strength
data (y) had been analyzed as a single-factor experiment in which the covariate
x was ignored. The analysis of variance of the breaking strength data is shown

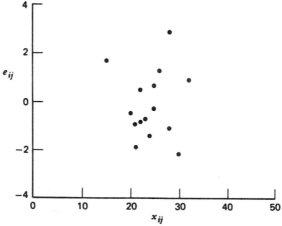

Figure 4–4.   Plot of residuals versus fiber diameter
x for Example 4–4.

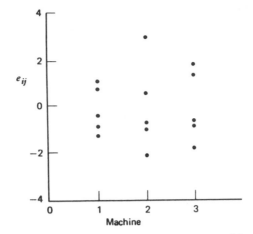

Figure 4–5. Plot of residuals versus machine.

in Table 4–12. We would conclude, based on this analysis, that machines differ significantly in the strength of fiber produced. This is exactly *opposite the conclusion* reached by the covariance analysis. If we suspected that the machines differed significantly in their effect on fiber strength, then we would try to equalize the strength output of the three machines. However, in this problem the machines do not differ in the strength of fiber produced after the linear effect of fiber diameter is removed. It would be helpful to reduce the within-machine fiber diameter variability since this would probably reduce the strength variability in the fiber.

4–7.2 Computer Solution

There are several computer software packages available that can perform the analysis of covariance. The output from the SAS General Linear Models procedure for the data in Example 4–4 is shown in Figure 4–7 (see page 163). This output is very similar to those presented previously. The model sum of squares contains both the sum of squares due to the machines and the sum of squares due to fiber diameter (x). The **type I sums of squares** correspond to a "sequential"

Table 4–12 Incorrect Analysis of the Breaking Strength Data as a
Single-Factor Experiment

Source of Variation	Sum of Squares	Degrees of Freedom	Mean Square	F_0	P-Value
Machines	140.40	2	70.20	4.09	0.0442
Error	206.00	12	17.17		
Total	346.40	14			

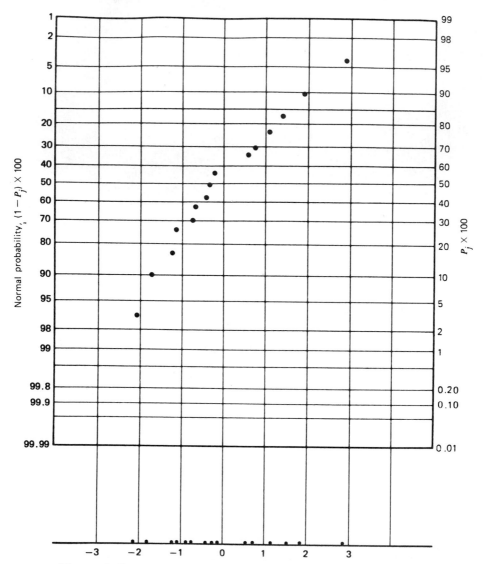

Figure 4–6. Normal probability plot of residuals for Example 4–4.

partitioning of the model sum of squares, say

$$SS \text{ (Model)} = SS \text{ (Machine)} + SS \text{ } (x|\text{Machine})$$
$$= 140.40 + 178.01$$
$$= 318.41$$

whereas the **type III sums of squares** correspond to the "extra" sum of squares for each factor, that is,

$$SS \text{ (Machine}|x) = 13.28$$

GENERAL LINEAR MODELS PROCEDURE

DEPENDENT VARIABLE: Y

SOURCE	DF	SUM OF SQUARES	MEAN SQUARE	F VALUE	PR > F	R-SQUARE	C.V.
MODEL	3	318.41411043	106.13803681	41.72	0.0001	0.919209	03.9678
ERROR	11	27.98588957	2.54417178			STD DEV	Y MEAN
CORRECTED TOTAL	14	346.40000000				1.59504601	40.20000000

SOURCE	DF	TYPE I SS	F VALUE	PR > F	DF	TYPE III SS	F VALUE	PR > F
MACHINE	2	140.40000000	27.59	0.0001	2	13.28385062	2.61	0.1181
X	1	178.01411043	69.97	0.0001	1	178.01411043	69.97	0.0001

OBSERVATION	OBSERVED VALUE	PREDICTED VALUE	RESIDUAL
1	36.00000000	36.43926380	-0.43926380
2	41.00000000	41.20920245	-0.20920245
3	39.00000000	40.25521472	-1.25521472
4	42.00000000	41.20920245	0.79079755
5	49.00000000	47.88711656	1.11288344
6	40.00000000	39.38404908	0.61595092
7	48.00000000	45.10797546	2.89202454
8	39.00000000	39.38404908	-0.38404908
9	45.00000000	47.01595092	-2.01595092
10	44.00000000	45.10797546	-1.10797546
11	35.00000000	35.80920245	-0.80920245
12	37.00000000	37.71717791	-0.71717791
13	42.00000000	40.57914110	1.42085890
14	34.00000000	35.80920245	-1.80920245
15	32.00000000	30.08527607	1.91472393

SUM OF RESIDUALS	0.00000000
SUM OF SQUARED RESIDUALS	27.98588957
SUM OF SQUARED RESIDUALS - ERROR SS	-0.00000000
FIRST ORDER AUTOCORRELATION	-0.03469267
DURBIN-WATSON D	1.93149012

LEAST SQUARES MEANS

MACHINE	Y LSMEAN	STD ERR LSMEAN	PROB > |T| H0:LSMEAN=0	HO: LSMEAN(I)-LSMEAN(J)		
				I/J	1	2	3
1	40.3824131	0.7236252	0.0001	1	.	0.3280	0.1803
2	41.4192229	0.7444169	0.0001	2	0.3280	.	0.0433
3	38.7983640	0.7878785	0.0001	3	0.1803	0.0433	.

NOTE: TO ENSURE OVERALL PROTECTION LEVEL, ONLY PROBABILITIES ASSOCIATED WITH PRE-PLANNED COMPARISONS SHOULD BE USED.

Figure 4-7 Sample computer output for Example 4-4.

and

$$SS\ (x|\text{Machine}) = 178.01$$

Note that SS (Machine$|x$) is the correct sum of squares to use for testing for no machine effect, and SS ($x|$Machine) is the correct sum of squares to use for testing the hypothesis that $\beta = 0$. The test statistics in Figure 4–7 differ slightly from those computed manually because of rounding.

The program also computes the adjusted treatment means from Equation 4–41 (SAS refers to those as least squares means or "LSMEAN" on the sample output) and the standard errors, and it constructs the test statistic for the hypothesis that each adjusted treatment mean equals zero. In our example, all these means differ from zero. The program also compares all pairs of treatment means and displays the significance level at which the pair of means would be different. We observe that means 2 and 3 differ at the 0.0433 level, whereas means 1 and 3 differ at the 0.1803 level, and means 1 and 2 differ at the 0.3280 level. It seems reasonable to conclude that only means 2 and 3 differ significantly.

4–7.3 Development by the General Regression Significance Test

It is possible to develop formally the procedure for testing $H_0: \tau_i = 0$ in the covariance model

$$y_{ij} = \mu + \tau_i + \beta(x_{ij} - \bar{x}_{..}) + \epsilon_{ij} \qquad \begin{cases} i = 1, 2, \ldots, a \\ j = 1, 2, \ldots, n \end{cases} \qquad (4\text{–}45)$$

using the general regression significance test. Consider estimating the parameters in the model (Equation 4–45) by least squares. The least squares function is

$$L = \sum_{i=1}^{a} \sum_{j=1}^{n} [y_{ij} - \mu - \tau_i - \beta(x_{ij} - \bar{x}_{..})]^2 \qquad (4\text{–}46)$$

and from $\partial L/\partial \mu = \partial L/\partial \tau_i = \partial L/\partial \beta = 0$, we obtain the normal equations

$$\mu: \quad an\hat{\mu} + n \sum_{i=1}^{a} \hat{\tau}_i = y_{..} \qquad (4\text{–}47a)$$

$$\tau_i: \quad n\hat{\mu} + n\hat{\tau}_i + \hat{\beta} \sum_{j=1}^{n} (x_{ij} - \bar{x}_{..}) = y_{i.} \qquad i = 1, 2, \ldots, a \qquad (4\text{–}47b)$$

$$\beta: \quad \sum_{i=1}^{a} \hat{\tau}_i \sum_{j=1}^{n} (x_{ij} - \bar{x}_{..}) + \hat{\beta} S_{xx} = S_{xy} \qquad (4\text{–}47c)$$

Adding the a equations in Equation 4–47b, we obtain Equation 4–47a since $\sum_{i=1}^{a} \sum_{j=1}^{n} (x_{ij} - \bar{x}_{..}) = 0$, so there is one linear dependency in the normal equations. Therefore, it is necessary to augment Equations 4–47 with a linearly independent equation in order to obtain a solution. A logical side condition is $\sum_{i=1}^{a} \hat{\tau}_i = 0$.

Using this condition, we obtain from Equation 4–47a

$$\hat{\mu} = \bar{y}_{..} \tag{4-48a}$$

and from Equation 4–47b

$$\hat{\tau}_i = \bar{y}_{i.} - \bar{y}_{..} - \hat{\beta}(\bar{x}_{i.} - \bar{x}_{..}) \tag{4-48b}$$

Equation 4–47c may be rewritten as

$$\sum_{i=1}^{a} (\bar{y}_{i.} - \bar{y}_{..}) \sum_{j=1}^{n} (x_{ij} - \bar{x}_{..}) - \hat{\beta} \sum_{i=1}^{a} (\bar{x}_{i.} - \bar{x}_{..}) \sum_{j=1}^{n} (x_{ij} - \bar{x}_{..}) + \hat{\beta} S_{xx} = S_{xy}$$

after substituting for $\hat{\tau}_i$. But we see that

$$\sum_{i=1}^{a} (\bar{y}_{i.} - \bar{y}_{..}) \sum_{j=1}^{n} (x_{ij} - \bar{x}_{..}) = T_{xy}$$

and

$$\sum_{i=1}^{a} (\bar{x}_{i.} - \bar{x}_{..}) \sum_{j=1}^{n} (x_{ij} - \bar{x}_{..}) = T_{xx}$$

Therefore, the solution to Equation 4–47c is

$$\hat{\beta} = \frac{S_{xy} - T_{xy}}{S_{xx} - T_{xx}} = \frac{E_{xy}}{E_{xx}}$$

which was the result given previously in Section 4–7.1, Equation 4–36.

We may express the reduction in the total sum of squares due to fitting the model (Equation 4–45) as

$$
\begin{aligned}
R(\mu, \tau, \beta) &= \hat{\mu}y_{..} + \sum_{i=1}^{a} \hat{\tau}_i y_{i.} + \hat{\beta}S_{xy} \\
&= (\bar{y}_{..})y_{..} + \sum_{i=1}^{a} [\bar{y}_{i.} - \bar{y}_{..} - (E_{xy}/E_{xx})(\bar{x}_{i.} - \bar{x}_{..})]y_{i.} + (E_{xy}/E_{xx})S_{xy} \\
&= y_{..}^2/an + \sum_{i=1}^{a} (\bar{y}_{i.} - \bar{y}_{..})y_{i.} - (E_{xy}/E_{xx}) \sum_{i=1}^{a} (\bar{x}_{i.} - \bar{x}_{..})y_{i.} + (E_{xy}/E_{xx})S_{xy} \\
&= y_{..}^2/an + T_{yy} - (E_{xy}/E_{xx})(T_{xy} - S_{xy}) \\
&= y_{..}^2/an + T_{yy} + (E_{xy})^2/E_{xx}
\end{aligned}
$$

This sum of squares has $a + 1$ degrees of freedom since the rank of the normal equations is $a + 1$. The error sum of squares for this model is

$$
\begin{aligned}
SS_E &= \sum_{i=1}^{a} \sum_{j=1}^{n} y_{ij}^2 - R(\mu, \tau, \beta) \\
&= \sum_{i=1}^{a} \sum_{j=1}^{n} y_{ij}^2 - y_{..}^2/an - T_{yy} - (E_{xy})^2/E_{xx} \tag{4-49} \\
&= S_{yy} - T_{yy} - (E_{xy})^2/E_{xx} \\
&= E_{yy} - (E_{xy})^2/E_{xx}
\end{aligned}
$$

with $an - (a + 1) = a(n - 1) - 1$ degrees of freedom. This quantity was obtained previously as Equation 4–37.

Now consider the model restricted to the null hypothesis, that is, to $H_0: \tau_1 = \tau_2 = \cdots = \tau_a = 0$. This reduced model is

$$y_{ij} = \mu + \beta(x_{ij} - \bar{x}_{..}) + \epsilon_{ij} \qquad \begin{cases} i = 1, 2, \ldots, a \\ j = 1, 2, \ldots, n \end{cases} \qquad (4\text{–}50)$$

This is a simple linear regression model, and the least squares normal equations for this model are

$$an\hat{\mu} = y_{..} \qquad (4\text{–}51\text{a})$$
$$\hat{\beta}S_{xx} = S_{xy} \qquad (4\text{–}51\text{b})$$

The solutions to these equations are $\hat{\mu} = \bar{y}_{..}$ and $\hat{\beta} = S_{xy}/S_{xx}$, and the reduction in the total sum of squares due to fitting the reduced model is

$$
\begin{aligned}
R(\mu, \beta) &= \hat{\mu}y_{..} + \hat{\beta}S_{xy} \\
&= (\bar{y}_{..})y_{..} + (S_{xy}/S_{xx})S_{xy} \qquad (4\text{–}52) \\
&= y_{..}^2/an + (S_{xy})^2/S_{xx}
\end{aligned}
$$

This sum of squares has 2 degrees of freedom.

We may find the appropriate sum of squares for testing $H_0: \tau_1 = \tau_2 = \cdots = \tau_a = 0$ as

$$
\begin{aligned}
R(\tau | \mu, \beta) &= R(\mu, \tau, \beta) - R(\mu, \beta) \\
&= y_{..}^2/an + T_{yy} + (E_{xy})^2/E_{xx} - y_{..}^2/an - (S_{xy})^2/S_{xx} \qquad (4\text{–}53) \\
&= S_{yy} - (S_{xy})^2/S_{xx} - [E_{yy} - (E_{xy})^2/E_{xx}]
\end{aligned}
$$

using $T_{yy} = S_{yy} - E_{yy}$. Note that $R(\tau | \mu, \beta)$ has $a + 1 - 2 = a - 1$ degrees of freedom and is identical to the sum of squares given by $SS'_E - SS_E$ in Section 4–7.1. Thus, the test statistic for $H_0: \tau_i = 0$ is

$$F_0 = \frac{R(\tau | \mu, \beta)/(a - 1)}{SS_E/[a(n - 1) - 1]} = \frac{(SS'_E - SS_E)/(a - 1)}{SS_E/[a(n - 1) - 1]} \qquad (4\text{–}54)$$

which we gave previously as Equation 4–40. Therefore, by using the general regression significance test, we have justified the heuristic development of the analysis of covariance in Section 4–7.1.

4–8   PROBLEMS

4–1   Refer to Problem 3–4.

(a) What is the probability of accepting H_0 if σ_τ^2 is four times the error variance σ^2?

(b) If the difference between looms is large enough to increase the standard deviation of an observation by 20 percent, we wish to detect this with a probability of at least 0.80. What sample size should be used?

4-2 Refer to Problem 3–10. If we wish to detect a maximum difference in mean response times of 10 milliseconds with a probability of at least 0.90, what sample size should be used? How would you obtain a preliminary estimate of σ^2?

4-3 Refer to Problem 3–13.

(a) If we wish to detect a maximum difference in mean calcium content of 0.5 percent with a probability of at least 0.90, what sample size should be used? Discuss how you would obtain a preliminary estimate of σ^2 for answering this question.

(b) If the difference between batches is great enough so that the standard deviation of an observation is increased by 25 percent, what sample size should be used if we wish to detect this with a probability of at least 0.90?

4-4 Consider the experiment in Problem 3–16. If we wish to construct a 95 percent confidence interval on the difference in two mean battery lives that has an accuracy of ± 2 weeks, how many batteries of each brand must be tested?

4-5 Suppose that four normal populations have means of $\mu_1 = 50$, $\mu_2 = 60$, $\mu_3 = 50$, and $\mu_4 = 60$. How many observations should be taken from each population so that the probability of rejecting the null hypothesis of equal population means is at least 0.90? Assume that $\alpha = 0.05$ and that a reasonable estimate of the error variance is $\sigma^2 = 25$.

4-6 Refer to Problem 4–5.

(a) How would your answer change if a reasonable estimate of the experimental error variance were $\sigma^2 = 36$?

(b) How would your answer change if a reasonable estimate of the experimental error variance were $\sigma^2 = 49$?

(c) Can you draw any conclusions about the sensitivity of your answer in this particular situation about how your estimate of σ affects the decision about sample size?

(d) Can you make any recommendations about how we should use this general approach to choosing n in practice?

4-7 Refer to the aluminum smelting experiment described in Section 4–2. Verify that ratio control methods do not affect average cell voltage. Construct a normal probability plot of the residuals. Plot the residuals versus the predicted values. Is there an indication that any underlying assumptions are violated?

4-8 Refer to the aluminum smelting experiment in Section 4–2. Verify the analysis of variance for pot noise summarized in Table 4–2. Examine the usual residual plots and comment on the experimental validity.

4-9 Four different feed rates were investigated in an experiment on a CNC machine producing a component part used in an aircraft auxiliary power unit. The manufacturing engineer in charge of the experiment knows that a critical part dimension of interest may be affected by the feed rate. However, prior experience has indicated that only dispersion effects are likely to be present. That is, changing the feed rate does not affect the *average* dimension, but it could affect dimensional variability. The engineer makes five production runs at each feed rate and obtains the standard deviation of the critical dimension (in 10^{-3} mm). The data are shown below. Assume that all runs were made in random order.

Feed Rate (in/min)	Production Run				
	1	2	3	4	5
10	0.09	0.10	0.13	0.08	0.07
12	0.06	0.09	0.12	0.07	0.12
14	0.11	0.08	0.08	0.05	0.06
16	0.19	0.13	0.15	0.20	0.11

(a) Does feed rate have any effect on the standard deviation of this critical dimension?

(b) Use the residuals from this experiment to investigate model adequacy. Are there any problems with experimental validity?

4–10 Consider the data shown in Problem 3–10.

(a) Write out the least squares normal equations for this problem, and solve them for $\hat{\mu}$ and $\hat{\tau}_i$, using the usual constraint ($\sum_{i=1}^{3} \hat{\tau}_i = 0$). Estimate $\tau_1 - \tau_2$.

(b) Solve the equations in (a) using the constraint $\hat{\tau}_3 = 0$. Are the estimators $\hat{\tau}_i$ and $\hat{\mu}$ the same as you found in (a)? Why? Now estimate $\tau_1 - \tau_2$ and compare your answer with that for (a). What statement can you make about estimating contrasts in the τ_i?

(c) Estimate $\mu + \tau_1$, $2\tau_1 - \tau_2 - \tau_3$, and $\mu + \tau_1 + \tau_2$ using the two solutions to the normal equations. Compare the results obtained in each case.

4–11 Apply the general regression significance test to the experiment in Example 3–1. Show that the procedure yields the same results as the usual analysis of variance.

4–12 Use the Kruskal–Wallis test for the experiment in Problem 3–12. Compare the conclusions obtained with those from the usual analysis of variance.

4–13 Use the Kruskal–Wallis test for the experiment in Problem 3–13. Are the results comparable to those found by the usual analysis of variance?

4–14 Consider the experiment in Example 3–1. Suppose that the largest observation on tensile strength is incorrectly recorded as 50. What effect does this have on the usual analysis of variance? What effect does it have on the Kruskal–Wallis test?

4–15 A soft drink distributor is studying the effectiveness of delivery methods. Three different types of hand trucks have been developed, and an experiment is performed in the company's methods engineering laboratory. The variable of interest is the delivery time in minutes (y); however, delivery time is also strongly related to the case volume delivered (x). Each hand truck is used four times and the data that follow are obtained. Analyze these data and draw appropriate conclusions. Use $\alpha = 0.05$.

	Hand Truck Type				
1		2		3	
y	x	y	x	y	x
27	24	25	26	40	38
44	40	35	32	22	26
33	35	46	42	53	50
41	40	26	25	18	20

4–16 Compute the adjusted treatment means and the standard errors of the adjusted treatment means for the data in Problem 4–15.

4–17 The sums of squares and products for a single-factor analysis of covariance follow. Complete the analysis and draw appropriate conclusions. Use $\alpha = 0.05$.

Source of Variation	Degrees of Freedom	Sums of Squares and Products		
		x	xy	y
Treatment	3	1500	1000	650
Error	12	6000	1200	550
Total	15	7500	2200	1200

4–18 Find the standard errors of the adjusted treatment means in Example 4–4.

4–19 Four different formulations of an industrial glue are being tested. The tensile strength of the glue when it is applied to join parts is also related to the application thickness. Five observations on strength (y) in pounds and thickness (x) in 0.01 inches are obtained for each formulation. The data are shown in the following table. Analyze these data and draw appropriate conclusions.

	Glue Formulation							
1		2		3		4	
y	x	y	x	y	x	y	x
46.5	13	48.7	12	46.3	15	44.7	16
45.9	14	49.0	10	47.1	14	43.0	15
49.8	12	50.1	11	48.9	11	51.0	10
46.1	12	48.5	12	48.2	11	48.1	12
44.3	14	45.2	14	50.3	10	48.6	11

4-20 Compute the adjusted treatment means and their standard errors using the data in Problem 4–19.

4-21 An engineer is studying the effect of cutting speed on the rate of metal removal in a machining operation. However, the rate of metal removal is also related to the hardness of the test specimen. Five observations are taken at each cutting speed. The amount of metal removed (y) and the hardness of the specimen (x) are shown in the following table. Analyze the data using an analysis of covariance. Use $\alpha = 0.05$.

Cutting Speed (rpm)					
1000		1200		1400	
y	x	y	x	y	x
68	120	112	165	118	175
90	140	94	140	82	132
98	150	65	120	73	124
77	125	74	125	92	141
88	136	85	133	80	130

4-22 Show that in a single-factor analysis of covariance with a single covariate a $100(1 - \alpha)$ percent confidence interval on the ith adjusted treatment mean is

$$\bar{y}_{i.} - \hat{\beta}(\bar{x}_{i.} - \bar{x}_{..}) \pm t_{\alpha/2, a(n-1)-1} \left[MS_E \left(\frac{1}{n} + \frac{(\bar{x}_{i.} - \bar{x}_{..})^2}{E_{xx}} \right) \right]^{1/2}$$

Using this formula, calculate a 95 percent confidence interval on the adjusted mean of machine 1 in Example 4–4.

4-23 Show that in a single-factor analysis of covariance with a single covariate, the standard error of the difference between any two adjusted treatment means is

$$S_{\text{Adj}\bar{y}_{i.} - \text{Adj}\bar{y}_{j.}} = \left[MS_E \left(\frac{2}{n} + \frac{(\bar{x}_{i.} - \bar{x}_{j.})^2}{E_{xx}} \right) \right]^{1/2}$$

4-24 Discuss how the operating characteristic curves for the analysis of variance can be used in the analysis of covariance.

Chapter 5
Randomized Blocks, Latin Squares, and Related Designs

5-1 THE RANDOMIZED COMPLETE BLOCK DESIGN

In any experiment, variability arising from a nuisance factor can impact the results. Generally, we define a **nuisance factor** as a design factor that probably has an effect on the response, but we are not interested in that effect. Sometimes a nuisance factor is **unknown and uncontrolled;** that is, we don't know that the factor exists and it may even be changing levels while we are conducting the experiment. **Randomization** is the design technique used to guard against such a "lurking" nuisance factor. In other cases, the nuisance factor is **known but uncontrollable.** If we can at least observe the value that the nuisance factor takes on at each run of the experiment, then we can compensate for it in the statistical analysis by using the **analysis of covariance,** as we discussed in Section 4–7. (In Example 4–4, the yarn thickness or the covariate was the nuisance factor.) When the nuisance source of variability is **known and controllable,** then a design technique called **blocking** can be used to systematically eliminate its effect on the statistical comparisons among treatments. Blocking is an extremely important design technique, used extensively in industrial experimentation, and is the subject of this chapter.

To illustrate the general idea, suppose we wish to determine whether or not four different tips produce different readings on a hardness testing machine. An experiment such as this might be part of a gage capability study. The machine operates by pressing the tip into a metal test coupon, and from the depth of the resulting depression, the hardness of the coupon can be determined. The experimenter has decided to obtain four observations for each tip. There is only one factor—tip type—and a completely randomized single-factor design would consist of randomly assigning each one of the $4 \times 4 = 16$ runs to an **experimental**

unit, that is, a metal coupon, and observing the hardness reading that results. Thus, 16 different metal test coupons would be required in this experiment, one for each run in the design.

There is a potentially serious problem with a completely randomized experiment in this design situation. If the metal coupons differ slightly in their hardness, as might happen if they are taken from ingots that are produced in different heats, then the experimental units (the coupons) will contribute to the variability observed in the hardness data. As a result, the experimental error will reflect *both* random error *and* variability between coupons.

We would like to make the experimental error as small as possible; that is, we would like to remove the variability between coupons from the experimental error. A design that would accomplish this requires the experimenter to test each tip once on each of four coupons. This design, shown in Table 5–1, is called a **randomized complete block design.** The observed response is the Rockwell C scale hardness minus 40. The word "complete" indicates that each block (coupon) contains all the treatments (tips). By using this design, the blocks or coupons form a more homogeneous experimental unit on which to compare the tips. Effectively, this design strategy improves the accuracy of the comparisons among tips by eliminating the variability among the coupons. Within a block, the order in which the four tips are tested is randomly determined. Notice the similarity of this design problem to the one of Section 2–5 in which the paired t test was discussed. The randomized complete block design is a generalization of that concept.

The randomized complete block design is one of the most widely used experimental designs. Situations for which the randomized complete block design is appropriate are numerous. Units of test equipment or machinery are often different in their operating characteristics and would be a typical blocking factor. Batches of raw material, people, and time are also common nuisance sources of variability in an experiment that can be systematically controlled through blocking.

Blocking may also be useful in situations that do not necessarily involve nuisance factors. For example, suppose that a chemical engineer is interested in the effect of catalyst feed rate on the viscosity of a polymer. He knows that there

Table 5–1   Randomized Complete Block Design for the Hardness Testing Experiment

	Test Coupon			
Type of Tip	1	2	3	4
1	9.3	9.4	9.6	10.0
2	9.4	9.3	9.8	9.9
3	9.2	9.4	9.5	9.7
4	9.7	9.6	10.0	10.2

are several factors, such as raw material source, temperature, operator, and raw material purity that are very difficult to control in the full-scale process. Therefore he decides to test the catalyst feed rate factor in blocks, where each block consists of some combination of these uncontrollable factors. In effect, he is using the blocks to test the **robustness** of his process variable (feed rate) to conditions he cannot easily control. For more discussion of this, see Coleman and Montgomery (1993).

5-1.1   Statistical Analysis

Suppose we have, in general, a treatments that are to be compared and b blocks. The randomized complete block design is shown in Figure 5-1. There is one observation per treatment in each block, and the order in which the treatments are run within each block is determined randomly. Because the only randomization of treatments is within the blocks, we often say that the blocks represent a **restriction on randomization.**

The **statistical model** for this design is

$$y_{ij} = \mu + \tau_i + \beta_j + \epsilon_{ij} \qquad \begin{cases} i = 1, 2, \ldots, a \\ j = 1, 2, \ldots, b \end{cases} \qquad (5\text{-}1)$$

where μ is an overall mean, τ_i is the effect of the ith treatment, β_j is the effect of the jth block, and ϵ_{ij} is the usual NID$(0, \sigma^2)$ random error term. Treatments and blocks are considered initially as fixed factors. Furthermore, the treatment and block effects are defined as deviations from the overall mean so that

$$\sum_{i=1}^{a} \tau_i = 0$$

and

$$\sum_{j=1}^{b} \beta_j = 0$$

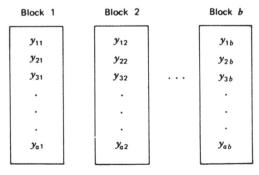

Figure 5-1.   The randomized complete block design.

We are interested in testing the equality of the treatment means. Thus, the hypotheses of interest are

$$H_0: \mu_1 = \mu_2 = \cdots = \mu_a$$
$$H_1: \text{at least one } \mu_i \neq \mu_j$$

Since the ith treatment mean $\mu_i = (1/b) \sum_{j=1}^{b}(\mu + \tau_i + \beta_j) = \mu + \tau_i$, an equivalent way to write the above hypotheses is in terms of the treatment effects, say

$$H_0: \tau_1 = \tau_2 = \cdots = \tau_a = 0$$
$$H_1: \tau_i \neq 0 \text{ at least one } i$$

Let $y_{i.}$ be the total of all observations taken under treatment i, $y_{.j}$ be the total of all observations in block j, $y_{..}$ be the grand total of all observations, and $N = ab$ be the total number of observations. Expressed mathematically,

$$y_{i.} = \sum_{j=1}^{b} y_{ij} \qquad i = 1, 2, \ldots, a \tag{5-2}$$

$$y_{.j} = \sum_{i=1}^{a} y_{ij} \qquad j = 1, 2, \ldots, b \tag{5-3}$$

and

$$y_{..} = \sum_{i=1}^{a} \sum_{j=1}^{b} y_{ij} = \sum_{i=1}^{a} y_{i.} = \sum_{j=1}^{b} y_{.j} \tag{5-4}$$

Similarly, $\bar{y}_{i.}$ is the average of the observations taken under treatment i, $\bar{y}_{.j}$ is the average of the observations in block j, and $\bar{y}_{..}$ is the grand average of all observations. That is,

$$\bar{y}_{i.} = y_{i.}/b \qquad \bar{y}_{.j} = y_{.j}/a \qquad \bar{y}_{..} = y_{..}/N \tag{5-5}$$

We may express the total corrected sum of squares as

$$\sum_{i=1}^{a} \sum_{j=1}^{b} (y_{ij} - \bar{y}_{..})^2 = \sum_{i=1}^{a} \sum_{j=1}^{b} [(\bar{y}_{i.} - \bar{y}_{..})$$
$$+ (\bar{y}_{.j} - \bar{y}_{..}) + (y_{ij} - \bar{y}_{i.} - \bar{y}_{.j} + \bar{y}_{..})]^2 \tag{5-6}$$

By expanding the right-hand side of Equation 5–6, we obtain

$$\sum_{i=1}^{a} \sum_{j=1}^{b} (y_{ij} - \bar{y}_{..})^2 = b \sum_{i=1}^{a} (\bar{y}_{i.} - \bar{y}_{..})^2 + a \sum_{j=1}^{b} (\bar{y}_{.j} - \bar{y}_{..})^2$$

$$+ \sum_{i=1}^{a} \sum_{j=1}^{b} (y_{ij} - \bar{y}_{i.} - \bar{y}_{.j} + \bar{y}_{..})^2 + 2 \sum_{i=1}^{a} \sum_{j=1}^{b} (\bar{y}_{i.} - \bar{y}_{..})(\bar{y}_{.j} - \bar{y}_{..})$$

$$+ 2 \sum_{i=1}^{a} \sum_{j=1}^{b} (\bar{y}_{.j} - \bar{y}_{..})(y_{ij} - \bar{y}_{i.} - \bar{y}_{.j} + \bar{y}_{..})$$

$$+ 2 \sum_{i=1}^{a} \sum_{j=1}^{b} (\bar{y}_{i.} - \bar{y}_{..})(y_{ij} - \bar{y}_{i.} - \bar{y}_{.j} + \bar{y}_{..})$$

Simple but tedious algebra proves that the three cross-products are zero. Therefore,

$$\sum_{i=1}^{a}\sum_{j=1}^{b}(y_{ij}-\bar{y}_{..})^2 = b\sum_{i=1}^{a}(\bar{y}_{i.}-\bar{y}_{..})^2 + a\sum_{j=1}^{b}(\bar{y}_{.j}-\bar{y}_{..})^2$$
$$+ \sum_{i=1}^{a}\sum_{j=1}^{b}(y_{ij}-\bar{y}_{.j}-\bar{y}_{i.}+\bar{y}_{..})^2 \tag{5-7}$$

represents a partition of the total sum of squares. Expressing the sums of squares in Equation 5-7 symbolically, we have

$$SS_T = SS_{\text{Treatments}} + SS_{\text{Blocks}} + SS_E \tag{5-8}$$

Since there are N observations, SS_T has $N-1$ degrees of freedom. There are a treatments and b blocks, so $SS_{\text{Treatments}}$ and SS_{Blocks} have $a-1$ and $b-1$ degrees of freedom, respectively. The error sum of squares is just a sum of squares between cells minus the sum of squares for treatments and blocks. There are ab cells with $ab-1$ degrees of freedom between them, so SS_E has $ab-1-(a-1)-(b-1) = (a-1)(b-1)$ degrees of freedom. Furthermore, the degrees of freedom on the right-hand side of Equation 5-8 add to the total on the left; therefore, making the usual normality assumptions on the errors, one may use Theorem 3-1 to show that $SS_{\text{Treatments}}/\sigma^2$, $SS_{\text{Blocks}}/\sigma^2$, and SS_E/σ^2 are independently distributed chi-square random variables. Each sum of squares divided by its degrees of freedom is a mean square. The expected value of the mean squares, if treatments and blocks are fixed, can be shown to be

$$E(MS_{\text{Treatments}}) = \sigma^2 + \frac{b\sum_{i=1}^{a}\tau_i^2}{a-1}$$

$$E(MS_{\text{Blocks}}) = \sigma^2 + \frac{a\sum_{j=1}^{b}\beta_j^2}{b-1}$$

$$E(MS_E) = \sigma^2$$

Therefore, to test the equality of treatment means, we would use the test statistic

$$F_0 = \frac{MS_{\text{Treatments}}}{MS_E}$$

which is distributed as $F_{a-1,(a-1)(b-1)}$ if the null hypothesis is true. The critical region is the upper tail of the F distribution, and we would reject H_0 if $F_0 > F_{\alpha,a-1,(a-1)(b-1)}$.

We may also be interested in comparing block means since, if these means do not differ greatly, blocking may not be necessary in future experiments. From the expected mean squares, it seems that the hypothesis $H_0: \beta_j = 0$ may be tested by comparing the statistic $F_0 = MS_{\text{Blocks}}/MS_E$ to $F_{\alpha,b-1,(a-1)(b-1)}$. However, recall that randomization has been applied only to treatments *within* blocks; that is, the blocks represent a **restriction on randomization.** What effect does this have

Table 5–2   Analysis of Variance for a Randomized Complete Block Design

Source of Variation	Sum of Squares	Degrees of Freedom	Mean Square	F_0
Treatments	$SS_{\text{Treatments}}$	$a - 1$	$\dfrac{SS_{\text{Treatments}}}{a - 1}$	$\dfrac{MS_{\text{Treatments}}}{MS_E}$
Blocks	SS_{Blocks}	$b - 1$	$\dfrac{SS_{\text{Blocks}}}{b - 1}$	
Error	SS_E	$(a - 1)(b - 1)$	$\dfrac{SS_E}{(a - 1)(b - 1)}$	
Total	SS_T	$N - 1$		

on the statistic $F_0 = MS_{\text{Blocks}}/MS_E$? Some difference in treatments of this question exists. For example, Box, Hunter, and Hunter (1978) point out that the usual analysis of variance F test can be justified on the basis of randomization only,[1] without direct use of the normality assumption. They further observe that the test to compare block means cannot appeal to such a justification because of the randomization restriction; but if the errors are NID(0, σ^2), then the statistic $F_0 = MS_{\text{Blocks}}/MS_E$ can be used to compare block means. On the other hand, Anderson and McLean (1974) argue that the randomization restriction prevents this statistic from being a meaningful test for comparing block means and that this F ratio really is a test for the equality of the block means plus the randomization restriction [which they call a restriction error; see Anderson and McLean (1974) for further details].

In practice, then, what do we do? Since the normality assumption is often questionable, to view $F_0 = MS_{\text{Blocks}}/MS_E$ as an exact F test on the equality of block means is not a good general practice. For that reason, we exclude this F test from the analysis of variance table. However, as an approximate procedure to investigate the effect of the blocking variable, examining the ratio of MS_{Blocks} to MS_E is certainly reasonable. If this ratio is large, it implies that the blocking factor has a large effect and that the noise reduction obtained by blocking was probably helpful in improving the precision of the comparison of treatment means.

The procedure is usually summarized in an analysis of variance table, such as the one shown in Table 5–2. The computing would usually be done with a statistical software package. However, manual computing formulas for the sums of squares may be obtained for the elements in Equation 5–7 by expressing them in terms of treatment and block totals. These computing formulas are

$$SS_T = \sum_{i=1}^{a} \sum_{j=1}^{b} y_{ij}^2 - \frac{y_{..}^2}{N} \tag{5–9}$$

---

[1] Actually, the normal-theory F distribution is an approximation to the randomization distribution generated by calculating F_0 from every possible assignment of the responses to the treatments.

Table 5-3   Randomized Complete Block Design for
the Hardness Testing Experiment

Type of Tip	Coupon (Block)			
	1	2	3	4
1	9.3	9.4	9.6	10.0
2	9.4	9.3	9.8	9.9
3	9.2	9.4	9.5	9.7
4	9.7	9.6	10.0	10.2

$$SS_{\text{Treatments}} = \frac{1}{b} \sum_{i=1}^{a} y_{i.}^2 - \frac{y_{..}^2}{N} \tag{5-10}$$

$$SS_{\text{Blocks}} = \frac{1}{a} \sum_{j=1}^{b} y_{.j}^2 - \frac{y_{..}^2}{N} \tag{5-11}$$

and the error sum of squares is obtained by subtraction as

$$SS_E = SS_T - SS_{\text{Treatments}} - SS_{\text{Blocks}} \tag{5-12}$$

Example 5-1

Consider the hardness testing experiment described in Section 5-1. There are four
tips and four available metal coupons. Each tip is tested once on each coupon,
resulting in a randomized complete block design. The data obtained are repeated
for convenience in Table 5-3. Remember that the *order* in which the tips were tested
on a particular coupon was determined randomly. To simplify the calculations, we
code the original data by subtracting 9.5 from each observation and multiplying the
result by 10. This yields the data in Table 5-4. The sums of squares are obtained
as follows:

$$SS_T = \sum_{i=1}^{4} \sum_{j=1}^{4} y_{ij}^2 - \frac{y_{..}^2}{N}$$

$$= 154.00 - \frac{(20)^2}{16} = 129.00$$

$$SS_{\text{Treatments}} = \frac{1}{b} \sum_{i=1}^{4} y_{i.}^2 - \frac{y_{..}^2}{N}$$

$$= \frac{1}{4} [(3)^2 + (4)^2 + (-2)^2 + (15)^2] - \frac{(20)^2}{16} = 38.50$$

$$SS_{\text{Blocks}} = \frac{1}{a} \sum_{j=1}^{4} y_{.j}^2 - \frac{y_{..}^2}{N}$$

$$= \frac{1}{4} [(-4)^2 + (-3)^2 + (9)^2 + (18)^2] - \frac{(20)^2}{16} = 82.50$$

$$SS_E = SS_T - SS_{\text{Treatments}} - SS_{\text{Blocks}}$$

$$= 129.00 - 38.50 - 82.50 = 8.00$$

Table 5-4   Coded Data for the Hardness Testing Experiment

Type of Tip	Coupon (Block)				$y_{i.}$
	1	2	3	4	
1	-2	-1	1	5	3
2	-1	-2	3	4	4
3	-3	-1	0	2	-2
4	2	1	5	7	15
$y_{.j}$	-4	-3	9	18	$20 = y_{..}$

The analysis of variance is shown in Table 5-5. Using $\alpha = 0.05$, the critical value of F is $F_{0.05,3,9} = 3.86$. Since $14.44 > 3.86$, we conclude that the type of tip affects the mean hardness reading. The P-value for the test is also quite small. Also, the coupons (blocks) seem to differ significantly, since the mean square for blocks is large relative to error.

It is interesting to observe the results we would have obtained had we not been aware of randomized block designs. Suppose we used four coupons, randomly assigned the tips to each, and (by chance) the same design resulted as in Table 5-3. The incorrect analysis of these data as a completely randomized single-factor design is shown in Table 5-6. (See page 180.) Since $F_{0.05,3,12} = 3.49$, the hypothesis of equal mean hardness measurements from the four tips cannot be rejected. Thus, the randomized block design reduces the amount of noise in the data sufficiently for differences among the four tips to be detected.

■

Sample Computer Output ▪ Computer output for the hardness testing data in Example 5-1, obtained from the GLM procedure in SAS, is shown in Figure 5-2. Recall that the original analysis in Table 5-5 used *coded* data. (The raw responses were coded by subtracting 9.5 and multiplying the result by 10.) The computer analysis used the raw responses. Consequently, the sums of squares in Figure 5-6 are equal to those in Table 5-5 divided by 100.

Table 5-5   Analysis of Variance for the Hardness Testing Experiment

Source of Variation	Sum of Squares	Degrees of Freedom	Mean Square	F_0	P-Value
Treatments (type of tip)	38.50	3	12.83	14.44	0.0009
Blocks (coupons)	82.50	3	27.50		
Error	8.00	9	0.89		
Total	129.00	15			

STATISTICAL ANALYSIS SYSTEM

13:16 WEDNESDAY, JANUARY 17, 1990

GENERAL LINEAR MODELS PROCEDURE

DEPENDENT VARIABLE: Y

SOURCE	DF	SUM OF SQUARES	MEAN SQUARE	F VALUE	PR > F	R-SQUARE	C.V.
MODEL	6	1.21000000	0.20166667	22.69	0.0001	0.937984	0.9795
ERROR	9	0.08000000	0.00888889		STD DEV		Y MEAN
CORRECTED TOTAL	15	1.29000000			0.09428090		9.62500000

SOURCE	DF	TYPE I SS	F VALUE	PR > F	DF	TYPE III SS	F VALUE	PR > F
TIPTYPE	3	0.38500000	14.44	0.0009	3	0.38500000	14.44	0.0009
BLOCK	3	0.82500000	30.94	0.0001	3	0.82500000	30.94	0.0001

OBSERVATION	OBSERVED VALUE	PREDICTED VALUE	RESIDUAL
1	9.30000000	9.35000000	-0.05000000
2	9.40000000	9.37500000	0.02500000
3	9.60000000	9.67500000	-0.07500000
4	10.00000000	9.90000000	0.10000000
5	9.40000000	9.37500000	0.02500000
6	9.30000000	9.40000000	-0.10000000
7	9.80000000	9.70000000	0.10000000
8	9.90000000	9.92500000	-0.02500000
9	9.20000000	9.22500000	-0.02500000
10	9.40000000	9.25000000	0.15000000
11	9.50000000	9.55000000	-0.50000000
12	9.70000000	9.77000000	-0.75000000
13	9.70000000	9.65000000	0.50000000
14	9.60000000	9.67500000	-0.07500000
15	10.00000000	9.97500000	0.25000000
16	10.20000000	10.20000000	0.00000000

SUM OF RESIDUALS   0.00000000
SUM OF SQUARED RESIDUALS   0.08000000
SUM OF SQUARED RESIDUALS - ERROR SS   -0.00000000
FIRST ORDER AUTOCORRELATION   -0.49218750
DURBIN-WATSON D   2.95312500

Figure 5-2. Sample computer output for Example 5-1.

179

Table 5–6 Incorrect Analysis of the Hardness Testing Experiment as a Completely Randomized Design

Source of Variation	Sum of Squares	Degrees of Freedom	Mean Square	F_0
Type of tip	38.50	3	12.83	1.70
Error	90.50	12	7.54	
Total	129.00	15		

The model sum of squares reported in the analysis of variance consists of the treatment sum of squares plus the block sum of squares; that is,

$$SS_{\text{Model}} = SS_{\text{Tip Type}} + SS_{\text{Block}}$$
$$= 0.3850 + 0.8250$$
$$= 1.21$$

The individual sources of variation resulting from types of tips and blocks are also identified in the computer output. Note that the type I and type III sums of squares for those two factors are identical, as will always be the case for balanced data.

The R^2 for this model, computed as

$$R^2 = \frac{SS_{\text{Model}}}{SS_T} = \frac{1.21}{1.29} = 0.937984$$

implies that about 94 percent of the variability in the data is explained by the model (i.e., the factors type of tip and blocks). This indicates that the model is a very good fit to the data.

The residuals are listed at the bottom of the computer output. They are calculated as

$$e_{ij} = y_{ij} - \hat{y}_{ij}$$

and, as we will later show, the fitted values are $\hat{y}_{ij} = \bar{y}_{i.} + \bar{y}_{.j} - \bar{y}_{..}$, so

$$e_{ij} = y_{ij} - \bar{y}_{i.} - \bar{y}_{.j} + \bar{y}_{..} \tag{5–13}$$

In the next section, we will show how the residuals are used in **model adequacy checking.**

Multiple Comparisons ▪ If the treatments in a randomized block design are fixed, and the analysis indicates a significant difference in treatment means, then the experimenter is usually interested in multiple comparisons to discover *which* treatment means differ. Any of the multiple comparison procedures discussed in Chapter 3 (Section 3–5) may be used for this purpose. In the formulas of Section 3–5, simply replace the number of replicates in the single-factor completely randomized design (n) by the number of blocks (b). Also, remember to use the number of error degrees of freedom for the randomized block [$(a - 1)$

$(b - 1)]$ instead of those for the completely randomized design $[a(n - 1)]$. For example, if we wish to use Duncan's multiple range test, the standard error of a treatment mean is

$$S_{\bar{y}_{i.}} = \sqrt{\frac{MS_E}{b}}$$

and Appendix Table VII would be entered with $(a - 1)(b - 1)$ degrees of freedom to find the significant ranges.

Example 5-2

Since we concluded in Example 5-1 that the mean hardness readings differ from tip to tip, it is of interest to discover the specific nature of the differences. The coded tip type or treatment means, in ascending order, are as follows:

$$\bar{y}_{3.} = -0.50 \qquad \bar{y}_{1.} = 0.75 \qquad \bar{y}_{2.} = 1.00 \qquad \bar{y}_{4.} = 3.75$$

and the standard error of a tip type or treatment mean is

$$S_{\bar{y}_{i.}} = \sqrt{\frac{MS_E}{b}} = \sqrt{\frac{0.89}{4}} = 0.47$$

Appendix Table VII provides the following significant ranges

$$r_{0.05}(2, 9) = 3.20 \qquad r_{0.05}(3, 9) = 3.34 \qquad \text{and} \qquad r_{0.05}(4, 9) = 3.41$$

Thus, the least significant ranges are

$$R_2 = r_{0.05}(2, 9)S_{\bar{y}_{i.}} = (3.20)(0.47) = 1.50$$
$$R_3 = r_{0.05}(3, 9)S_{\bar{y}_{i.}} = (3.34)(0.47) = 1.57$$
$$R_4 = r_{0.05}(4, 9)S_{\bar{y}_{i.}} = (3.41)(0.47) = 1.60$$

and the pairwise comparisons are

$$\begin{aligned}
4 \text{ vs. } 3: & \quad 3.75 - (-0.50) = 4.25 > 1.60(R_4) \\
4 \text{ vs. } 1: & \quad 3.75 - 0.75 = 3.00 > 1.57(R_3) \\
4 \text{ vs. } 2: & \quad 3.75 - 1.00 = 2.75 > 1.50(R_2) \\
2 \text{ vs. } 3: & \quad 1.00 - (-0.50) = 1.50 < 1.57(R_3)
\end{aligned}$$

Note that since the comparison of $\bar{y}_{2.}$ and $\bar{y}_{3.}$ spans the remaining means and this comparison indicates that $H_0: \mu_2 = \mu_3$ cannot be rejected, we can immediately conclude that all other pairs of means do not differ (i.e., $\mu_1 = \mu_2 = \mu_3$). Thus, tip type 4 produces a mean hardness reading that is significantly higher than the mean readings obtained from the other three types of tips.

∎

To illustrate further multiple comparisons in the randomized complete block design, Figure 5-3 plots the four tip type means from Example 5-1 relative to a scaled t distribution with a scale factor $\sqrt{MS_E/b} = \sqrt{0.89/4} = 0.47$. This plot indicates that tips 1, 2, and 3 probably produce identical average hardness mea-

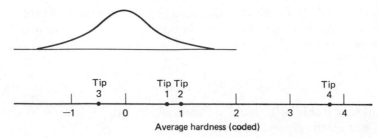

Figure 5–3. Tip type means relative to a scaled t distribution with a scale factor $\sqrt{MS_E/b} = \sqrt{0.89/4} = 0.47$.

surements but that tip 4 produces a much higher mean hardness. This figure confirms the results from the Duncan multiple range test in Example 5–2.

5–1.2   Model Adequacy Checking

We have previously discussed the importance of checking the adequacy of the assumed model. Generally, we should be alert for potential problems with the normality assumption, unequal error variance by treatment or block, and block–treatment interaction. As in the completely randomized design, residual analysis is the major tool used in this diagnostic checking. The residuals for the randomized block design in Example 5–1 are listed at the bottom of the SAS output in Figure 5–2. The coded residuals would be found by multiplying these residuals by 10. The observations, fitted values, and residuals for the coded hardness testing data in Example 5–1 are as follows:

y_{ij}	$\hat{y}_{ij}$	e_{ij}
−2.00	−1.50	−0.50
−1.00	−1.25	0.25
1.00	1.75	−0.75
5.00	4.00	1.00
−1.00	−1.25	0.25
−2.00	−1.00	−1.00
3.00	2.00	1.00
4.00	4.25	−0.25
−3.00	−2.75	−0.25
−1.00	−2.50	1.50
0.00	0.50	−0.50
2.00	2.75	−0.75
2.00	1.50	0.50
1.00	1.75	−0.75
5.00	4.75	0.25
7.00	7.00	0.00

A normal probability plot and a dot diagram of these residuals are shown in Figure 5–4. There is no severe indication of nonnormality, nor is there any evidence pointing to possible outliers. Figure 5–5 shows plots of the residuals by type of tip or treatment and by coupon or block. These plots could be, potentially, very informative. If there is more scatter in the residuals for a particular tip, that could indicate that this tip produces more erratic hardness readings than the others. More scatter in the residuals for a particular test coupon could indicate that the coupon is not of uniform hardness. However, in our example,

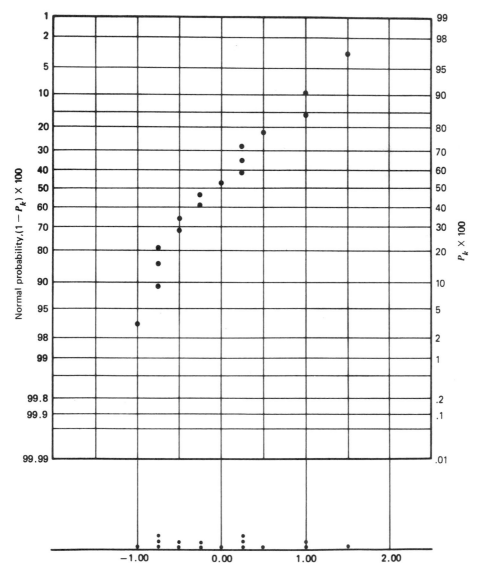

Figure 5–4. Normal probability plot and dot diagram of residuals for Example 5–1.

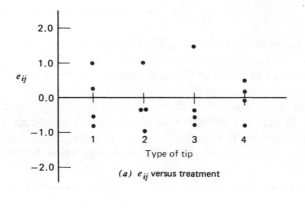

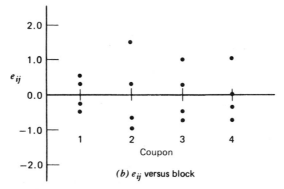

Figure 5–5. Plot of residuals by tip type (treatment) and by coupon (block) for Example 5–1.

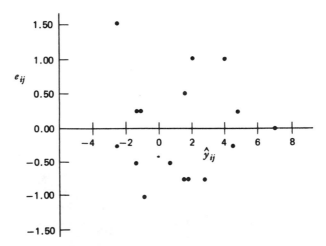

Figure 5–6. Plot of residuals versus $\hat{y}_{ij}$ for Example 5–1.

Figure 5–5 gives no indication of inequality of variance by treatment or by block. Figure 5–6 plots the residuals versus the fitted values $\hat{y}_{ij}$. There should be no relationship between the size of the residuals and the fitted values $\hat{y}_{ij}$. This plot reveals nothing of unusual interest.

Sometimes the plot of residuals versus $\hat{y}_{ij}$ has a curvilinear shape; for example, there may be a tendency for negative residuals to occur with low $\hat{y}_{ij}$ values, positive residuals with intermediate $\hat{y}_{ij}$ values, and negative residuals with high $\hat{y}_{ij}$ values. This type of pattern is suggestive of **interaction** between blocks and treatments. If this pattern occurs, a transformation should be used in an effort to eliminate or minimize the interaction. In Chapter 6 (Section 6–3.7), we describe a statistical test that can be used to detect the presence of interaction in a randomized block design.

5–1.3   Some Other Aspects of the Randomized Complete Block Design

Additivity of the Randomized Block Model ▪ The linear model that we have used for the randomized block design

$$y_{ij} = \mu + \tau_i + \beta_j + \epsilon_{ij}$$

is completely **additive.** This says that, for example, if the first treatment causes the expected response to increase by five units ($\tau_1 = 5$) and if the first block increases the expected response by 2 units ($\beta_1 = 2$), then the expected increase in response of *both* treatment 1 *and* block 1 together is $E(y_{11}) = \mu + \tau_1 + \beta_1 = \mu + 5 + 2 = \mu + 7$. In general, treatment 1 *always* increases the expected response by 5 units over the sum of the overall mean and the block effect.

Although this simple additive model is often useful, there are situations where it is inadequate. Suppose, for example, that we are comparing four formulations of a chemical product using six batches of raw material; the raw material batches are considered blocks. If an impurity in batch 2 affects formulation 2 adversely, resulting in an unusually low yield, but does not affect the other formulations, then an **interaction** between formulations (or treatments) and batches (or blocks) has occurred. Similarly, interactions between treatments and blocks can occur when the response is measured on the wrong scale. Thus, a relationship that is multiplicative in the original units, say

$$E(y_{ij}) = \mu \tau_i \beta_j$$

is linear or additive in a log scale since, for example,

$$\ln E(y_{ij}) = \ln \mu + \ln \tau_i + \ln \beta_j$$

or

$$E(y_{ij}^*) = \mu^* + \tau_i^* + \beta_j^*$$

Although this type of interaction can be eliminated by a transformation, not all interactions are so easily treated. For example, transformations do not eliminate the formulation–batch interaction discussed previously. Residual analysis and other diagnostic checking procedures can be helpful in detecting nonadditivity.

If interaction is present, it can seriously affect and possibly invalidate the analysis of variance. In general, the presence of interaction inflates the error mean square and may adversely affect the comparison of treatment means. In situations where both factors, as well as their possible interaction, are of interest, **factorial designs** must be used. These designs are discussed extensively in Chapters 6 through 10.

Random Treatments and Blocks ▪ Although we have described the test procedure considering treatments and blocks as fixed factors, the same analysis procedure is used if either the treatments or blocks (or both) are random. There are, of course, corresponding changes in the interpretation of the results. For example, if the blocks are random, then we expect the comparisons among the treatments to be the same throughout the **population of blocks** from which those used in the experiment were randomly selected. There are also corresponding changes in the expected mean squares. For example, if the blocks are independent random variables with common variance, then $E(MS_{\text{Blocks}}) = \sigma^2 + a\sigma_\beta^2$, where σ_β^2 is the variance component of the block effects. In any event, however, $E(MS_{\text{Treatments}})$ is always free of any block effect, and the test statistic for between-treatment variability is always $F_0 = MS_{\text{Treatments}}/MS_E$.

In situations where the blocks are random, if treatment–block interaction is present, then the test on treatment means is unaffected by the interaction. The reason for this is that the expected mean squares for treatments and error *both* contain the interaction effect; consequently, the test for differences in treatment means may be conducted as usual by comparing the treatment mean square to the error mean square. This procedure does not yield any information on the interaction.

Choice of Sample Size ▪ Choosing the **sample size,** or the **number of blocks** to run, is an important decision when using a randomized block design. Increasing the number of blocks increases the number of replicates and the number of error degrees of freedom, making the design more sensitive. Any of the techniques discussed in Chapter 4 (Section 4–1) for selecting the number of replicates to run in a completely randomized single-factor experiment may be applied directly to the randomized block design. For the case of a fixed factor, the operating characteristic curves in Appendix Chart V may be used with

$$\Phi^2 = \frac{b \sum_{i=1}^{a} \tau_i^2}{a\sigma^2}$$

where there are $a - 1$ numerator degrees of freedom and $(a - 1)(b - 1)$ denominator degrees of freedom. If the factor is random, then we use Appendix Chart VI with

$$\lambda = \sqrt{1 + \frac{b\sigma_\tau^2}{\sigma^2}}$$

where there are $a - 1$ numerator degrees of freedom and $(a - 1)(b - 1)$ denominator degrees of freedom.

Example 5-3

Consider the hardness testing problem described in Example 5-1. Suppose that we wish to determine the appropriate number of blocks to run if we are interested in detecting a true maximum difference in mean hardness readings of 0.4 with a high probability, and a reasonable estimate of the standard deviation of the errors is $\sigma = 0.1$. (These values are given in the *original* units; recall that the analysis of variance was performed using *coded* data.) From Equation 4-3, the minimum value of Φ^2 is (writing b, the number of blocks, for n)

$$\Phi^2 = \frac{bD^2}{2a\sigma^2}$$

where D is the maximum difference we wish to detect. Thus,

$$\Phi^2 = \frac{b(0.4)^2}{2(4)(0.1)^2} = 2.0b$$

If we use $b = 3$ blocks, then $\Phi = \sqrt{2.0b} = \sqrt{2.0(3)} = 2.45$, and there are $(a - 1)(b - 1) = 3(2) = 6$ error degrees of freedom. Appendix Chart V with $\nu_1 = a - 1 = 3$ and $\alpha = 0.05$ indicates that the β risk for this design is approximately 0.10 (power $= 1 - \beta = 0.90$). If we use $b = 4$ blocks, then $\Phi = \sqrt{2.0b} = \sqrt{2.0(4)} = 2.83$, with $(a - 1)(b - 1) = 3(3) = 9$ error degrees of freedom, and the corresponding β risk is approximately 0.03 (power $= 1 - \beta = 0.97$). Either three or four blocks will result in a design having a high probability of detecting the difference between the mean hardness readings considered important. Since coupons (blocks) are inexpensive and readily available and the cost of testing is low, the experimenter decides to use four blocks.

■

Relative Efficiency of the Randomized Block Design ▪ In Example 5-1 we illustrated the noise-reducing property of the randomized block design. If we look at the portion of the total sum of squares not accounted for by treatments (90.50; see Table 5-6), over 90 percent (82.50) is the result of differences between blocks. Thus, if we had run a completely randomized design, MS_E would have been much larger, and the resulting design would not have been as sensitive as the randomized block design.

It is often helpful to estimate the relative efficiency of the randomized block design compared to a completely randomized design. One way to define this relative efficiency is

$$R = \frac{(df_b + 1)(df_r + 3)}{(df_b + 3)(df_r + 1)} \cdot \frac{\sigma_r^2}{\sigma_b^2} \tag{5-14}$$

where σ_r^2 and σ_b^2 are the experimental error variances of the completely randomized and randomized block designs, respectively, and df_r and df_b are the corresponding error degrees of freedom. This statistic may be viewed as the increase in replications that is required if a completely randomized design is used as compared to a randomized block if the two designs are to have the same sensitivity. The ratio of degrees of freedom in R is an adjustment to reflect the different number of error degrees of freedom in the two designs.

To compute the relative efficiency, we must have estimates of σ_r^2 and σ_b^2. We can use MS_E from the randomized block design to estimate σ_b^2, and it may be shown [see Cochran and Cox (1957), pp. 112–114] that

$$\hat{\sigma}_r^2 = \frac{(b-1)MS_{\text{Blocks}} + b(a-1)MS_E}{ab - 1} \tag{5-15}$$

is an unbiased estimator of the error variance of a completely randomized design. To illustrate the procedure, consider the data in Example 5–1. Since $MS_E = 0.89$, we have

$$\hat{\sigma}_b^2 = 0.89$$

and, from Equation 5–15,

$$\begin{aligned}
\hat{\sigma}_r^2 &= \frac{(b-1)MS_{\text{Blocks}} + b(a-1)MS_E}{ab - 1} \\
&= \frac{(3)27.50 + 4(3)0.89}{(4)4 - 1} \\
&= 6.21
\end{aligned}$$

Therefore, using Equation 5–14, our estimate of the relative efficiency of the randomized block design in this example is

$$\begin{aligned}
R &= \frac{(df_b + 1)(df_r + 3)}{(df_b + 3)(df_r + 1)} \frac{\hat{\sigma}_r^2}{\hat{\sigma}_b^2} = \frac{(9+1)(12+3)}{(9+3)(12+1)} \frac{6.21}{0.89} \\
&= 6.71
\end{aligned}$$

This implies that we would have to use about seven times as many replicates with a completely randomized design to obtain the same sensitivity as is obtained by blocking on the metal coupons.

Clearly, blocking has paid off handsomely in this experiment. However, suppose that blocking was not really necessary. In such cases, if experimenters choose to block, what do they stand to lose? In general, the randomized block design has $(a-1)(b-1)$ error degrees of freedom. If blocking was unnecessary and the experiment was run as a completely randomized design with b replicates, we would have had $a(b - 1)$ degrees of freedom for error. Thus, incorrectly blocking has cost $a(b - 1) - (a - 1)(b - 1) = b - 1$ degrees of freedom for error, and the test on treatment means has been made less sensitive needlessly. However, if block effects really are large, then the experimental error may be so inflated that significant differences in treatment means would remain undetected. (Remember the incorrect analysis of Example 5–1.) As a general rule, when the importance of block effects is in doubt, the experimenter should block and gamble that the block means differ. If the experimenter is wrong, the slight loss in error degrees of freedom will have little effect on the outcome as long as a moderate number of degrees of freedom for error are available.

Estimating Missing Values ▪ When using the randomized complete block design, sometimes an observation in one of the blocks is missing. This may happen because of carelessness or error or for reasons beyond our control, such as unavoidable damage to an experimental unit. A missing observation introduces a new problem into the analysis since treatments are no longer **orthogonal to blocks;** that is, every treatment does not occur in every block. There are two general approaches to the missing value problem. The first is an **approximate analysis** in which the missing observation is estimated and the usual analysis of variance is performed just as if the estimated observation were real data, with the error degrees of freedom reduced by one. This approximate analysis is the subject of this section. The second is an **exact analysis,** which is discussed in Section 5–1.4.

Suppose the observation y_{ij} for treatment i in block j is missing. Denote the missing observation by x. As an illustration, suppose that in the hardness testing experiment of Example 5–1 specimen 3 was broken while tip 2 was being tested and that data point could not be obtained. The data might appear as in Table 5–7.

In general, we will let $y'_{..}$ represent the grand total with one missing observation, $y'_{i.}$ represent the total for the treatment with one missing observation, and $y'_{.j}$ be the total for the block with one missing observation. Suppose we wish to estimate the missing observation x so that x will have a minimum contribution to the error sum of squares. Since $SS_E = \sum_{i=1}^{a}\sum_{j=1}^{b}(y_{ij} - \bar{y}_{i.} - \bar{y}_{.j} + \bar{y}_{..})^2$, this is

Table 5–7 Randomized Complete Block Design for the Hardness Testing Experiment with One Missing Value

Type of Tip	Specimen (Block)			
	1	2	3	4
1	-2	-1	1	5
2	-1	-2	x	4
3	-3	-1	0	2
4	2	1	5	7

equivalent to choosing x in order to minimize

$$SS_E = \sum_{i=1}^{a} \sum_{j=1}^{b} y_{ij}^2 - \frac{1}{b} \sum_{i=1}^{a} \left(\sum_{j=1}^{b} y_{ij} \right)^2 - \frac{1}{a} \sum_{j=1}^{b} \left(\sum_{i=1}^{a} y_{ij} \right)^2 + \frac{1}{ab} \left(\sum_{i=1}^{a} \sum_{j=1}^{b} y_{ij} \right)^2$$

or

$$SS_E = x^2 - \frac{1}{b}(y_{i.}' + x)^2 - \frac{1}{a}(y_{.j}' + x)^2 + \frac{1}{ab}(y_{..}' + x)^2 + R \qquad (5\text{--}16)$$

where R includes all terms not involving x. From $dSS_E/dx = 0$, we obtain

$$x = \frac{ay_{i.}' + by_{.j}' - y_{..}'}{(a-1)(b-1)} \qquad (5\text{--}17)$$

as the estimate of the missing observation.

For the data in Table 5–7, we find that $y_{2.}' = 1$, $y_{.3}' = 6$, and $y_{..}' = 17$. Therefore, from Equation 5–17,

$$x \equiv y_{23} = \frac{4(1) + 4(6) - 17}{(3)(3)} = 1.22$$

The usual analysis of variance may now be performed using $y_{23} = 1.22$ and reducing the error degrees of freedom by one. The analysis of variance is shown in Table 5–8. Compare the results of this approximate analysis with the results obtained for the full data set (Table 5–5).

If several observations are missing, they may be estimated by writing the error sum of squares as a function of the missing values, differentiating with respect to each missing value, equating the results to zero, and solving the resulting equations. Alternatively, we may use Equation 5–17 iteratively to estimate the missing values. To illustrate the iterative approach, suppose that two values are missing. Arbitrarily estimate the first missing value, and then use this value along with the real data and Equation 5–17 to estimate the second. Now Equation 5–17 can be used to reestimate the first missing value, and following this, the second can be reestimated. This process is continued until convergence is obtained. In any missing value problem, the error degrees of freedom are reduced by one for each missing observation.

Table 5–8   Approximate Analysis of Variance for Example 5–1 with One Missing Value

Source of Variation	Sum of Squares	Degrees of Freedom	Mean Square	F_0	P-Value
Type of tip	39.98	3	13.33	17.12	0.0008
Specimens (blocks)	79.53	3	26.51		
Error	6.22	8	0.78		
Total	125.73	14			

5–1.4 Estimating Model Parameters and the General Regression Significance Test

If both treatments and blocks are fixed, we may estimate the parameters in the randomized complete block design model by least squares. Recall that the linear statistical model is

$$y_{ij} = \mu + \tau_i + \beta_j + \epsilon_{ij} \qquad \begin{cases} i = 1, 2, \ldots, a \\ j = 1, 2, \ldots, b \end{cases}$$

Applying the rules in Section 4–4.2 for finding the normal equations for an experimental design model, we obtain

$$
\begin{aligned}
\mu: \ & ab\hat{\mu} + b\hat{\tau}_1 + b\hat{\tau}_2 + \cdots + b\hat{\tau}_a + a\hat{\beta}_1 + a\hat{\beta}_2 + \cdots + a\hat{\beta}_b = y_{..} \\
\tau_1: \ & b\hat{\mu} + b\hat{\tau}_1 \qquad\qquad\qquad\quad\; + \hat{\beta}_1 + \hat{\beta}_2 + \cdots + \hat{\beta}_b = y_{1.} \\
\tau_2: \ & b\hat{\mu} \qquad + b\hat{\tau}_2 \qquad\qquad\quad\; + \hat{\beta}_1 + \hat{\beta}_2 + \cdots + \hat{\beta}_b = y_{2.} \\
& \;\vdots \qquad\qquad\qquad\qquad\qquad\quad\; \vdots \qquad\qquad\qquad\qquad \vdots \\
\tau_a: \ & b\hat{\mu} \qquad\qquad\qquad\quad + b\hat{\tau}_a + \hat{\beta}_1 + \hat{\beta}_2 + \cdots + \hat{\beta}_b = y_{a.} \\
\beta_1: \ & a\hat{\mu} + \hat{\tau}_1 + \hat{\tau}_2 + \cdots + \hat{\tau}_a + a\hat{\beta}_1 \qquad\qquad\qquad\quad = y_{.1} \\
\beta_2: \ & a\hat{\mu} + \hat{\tau}_1 + \hat{\tau}_2 + \cdots + \hat{\tau}_a \qquad + a\hat{\beta}_2 \qquad\qquad\quad = y_{.2} \\
& \;\vdots \qquad\qquad\qquad\qquad\quad\; \vdots \qquad\qquad\qquad\qquad \vdots \\
\beta_b: \ & a\hat{\mu} + \hat{\tau}_1 + \hat{\tau}_2 + \cdots + \hat{\tau}_a \qquad\qquad\qquad\qquad + a\hat{\beta}_b = y_{.b}
\end{aligned}
\tag{5–18}
$$

Notice that the second through the $(a + 1)$st equations in Equation 5–18 sum to the first normal equation, as do the last b equations. Thus, there are two linear dependencies in the normal equations, implying that two constraints must be imposed to solve Equation 5–18. The usual constraints are

$$\sum_{i=1}^{a} \hat{\tau}_i = 0 \qquad \sum_{j=1}^{b} \hat{\beta}_j = 0 \tag{5–19}$$

Using these constraints, the normal equations simplify considerably. In fact, they become

$$
\begin{aligned}
ab\hat{\mu} &= y_{..} \\
b\hat{\mu} + b\hat{\tau}_i &= y_{i.}, \qquad i = 1, 2, \ldots, a \\
a\hat{\mu} + a\hat{\beta}_j &= y_{.j} \qquad j = 1, 2, \ldots, b
\end{aligned}
\tag{5–20}
$$

whose solution is

$$
\begin{aligned}
\hat{\mu} &= \bar{y}_{..} \\
\hat{\tau}_i &= \bar{y}_{i.} - \bar{y}_{..} \qquad i = 1, 2, \ldots, a \\
\hat{\beta}_j &= \bar{y}_{.j} - \bar{y}_{..} \qquad j = 1, 2, \ldots, b
\end{aligned}
\tag{5–21}
$$

Using the solution to the normal equation in Equation 5–21, we may find the estimated or fitted values of y_{ij} as

$$\hat{y}_{ij} = \hat{\mu} + \hat{\tau}_i + \hat{\beta}_j$$
$$= \bar{y}_{..} + (\bar{y}_{i.} - \bar{y}_{..}) + (\bar{y}_{.j} - \bar{y}_{..})$$
$$= \bar{y}_{i.} + \bar{y}_{.j} - \bar{y}_{..}$$

This result was used previously in Equation 5–13 for computing the residuals from a randomized block design.

The general regression significance test can be used to develop the analysis of variance for the randomized complete block design. Using the solution to the normal equations given by Equation 5–21, the reduction in the sum of squares for fitting the complete model is

$$R(\mu, \tau, \beta) = \hat{\mu} y_{..} + \sum_{i=1}^{a} \hat{\tau}_i y_{i.} + \sum_{j=1}^{b} \hat{\beta}_j y_{.j}$$

$$= \bar{y}_{..} y_{..} + \sum_{i=1}^{a} (\bar{y}_{i.} - \bar{y}_{..}) y_{i.} + \sum_{j=1}^{b} (\bar{y}_{.j} - \bar{y}_{..}) y_{.j}$$

$$= \frac{y_{..}^2}{ab} + \sum_{i=1}^{a} \bar{y}_{i.} y_{i.} - \frac{y_{..}^2}{ab} + \sum_{j=1}^{b} \bar{y}_{.j} y_{.j} - \frac{y_{..}^2}{ab}$$

$$= \sum_{i=1}^{a} \frac{y_{i.}^2}{b} + \sum_{j=1}^{b} \frac{y_{.j}^2}{a} - \frac{y_{..}^2}{ab}$$

with $a + b - 1$ degrees of freedom, and the error sum of squares is

$$SS_E = \sum_{i=1}^{a} \sum_{j=1}^{b} y_{ij}^2 - R(\mu, \tau, \beta)$$

$$= \sum_{i=1}^{a} \sum_{j=1}^{b} y_{ij}^2 - \sum_{i=1}^{a} \frac{y_{i.}^2}{b} - \sum_{j=1}^{b} \frac{y_{.j}^2}{a} + \frac{y_{..}^2}{ab}$$

$$= \sum_{i=1}^{a} \sum_{j=1}^{b} (y_{ij} - \bar{y}_{i.} - \bar{y}_{.j} + \bar{y}_{..})^2$$

with $(a - 1)(b - 1)$ degrees of freedom. Compare this last equation with SS_E in Equation 5–7.

To test the hypothesis $H_0: \tau_i = 0$, the reduced model is

$$y_{ij} = \mu + \beta_j + \epsilon_{ij}$$

which is just a single-factor analysis of variance. By analogy with Equation 4–13, the reduction in the sum of squares for fitting this model is

$$R(\mu, \beta) = \sum_{j=1}^{b} \frac{y_{.j}^2}{a}$$

which has b degrees of freedom. Therefore, the sum of squares due to $\{\tau_i\}$ after fitting μ and $\{\beta_j\}$ is

$$R(\tau|\mu, \beta) = R(\mu, \tau, \beta) - R(\mu, \beta)$$

$$= \sum_{i=1}^{a} \frac{y_{i.}^2}{b} + \sum_{j=1}^{b} \frac{y_{.j}^2}{a} - \frac{y_{..}^2}{ab} - \sum_{j=1}^{b} \frac{y_{.j}^2}{a}$$

$$= \sum_{i=1}^{a} \frac{y_{i.}^2}{b} - \frac{y_{..}^2}{ab}$$

which we recognize as the treatment sum of squares with $a - 1$ degrees of freedom (Equation 5-10).

The block sum of squares is obtained by fitting the reduced model

$$y_{ij} = \mu + \tau_i + \epsilon_{ij}$$

which is also a single-factor analysis. Again, by analogy with Equation 4-13, the reduction in the sum of squares for fitting this model is

$$R(\mu, \tau) = \sum_{i=1}^{a} \frac{y_{i.}^2}{b}$$

with a degrees of freedom. The sum of squares for blocks $\{\beta_j\}$ after fitting μ and $\{\tau_i\}$ is

$$R(\beta|\mu, \tau) = R(\mu, \tau, \beta) - R(\mu, \tau)$$

$$= \sum_{i=1}^{a} \frac{y_{i.}^2}{b} + \sum_{j=1}^{b} \frac{y_{.j}^2}{a} - \frac{y_{..}^2}{ab} - \sum_{i=1}^{a} \frac{y_{i.}^2}{b}$$

$$= \sum_{j=1}^{b} \frac{y_{.j}^2}{a} - \frac{y_{..}^2}{ab}$$

with $b - 1$ degrees of freedom, which we have given previously as Equation 5-11.

We have developed the sums of squares for treatments, blocks, and error in the randomized complete block design using the general regression significance test. Although we would not ordinarily use the general regression significance test actually to analyze data in a randomized complete block, the procedure occasionally proves useful in more general randomized block designs, such as those discussed in Section 5-4.

Exact Analysis of the Missing Value Problem ▪ In Section 5-1.3 an approximate procedure for dealing with missing observations in a randomized block design was presented. This approximate analysis consists of estimating the missing value so that the error mean square is minimized. It can be shown that the approximate analysis produces a biased mean square for treatments in the sense that $E(MS_{\text{Treatments}})$ is larger than $E(MS_E)$ if the null hypothesis is true. Consequently, too many significant results are reported.

The missing value problem may be analyzed exactly by using the general regression significance test. The missing value causes the design to be **unbalanced**, and because all the treatments do not occur in all blocks, we say that the treatments and blocks are not **orthogonal**. This method of analysis is also used in more general types of randomized block designs; it is discussed further in Section 5–4. Problem 5–26 asks the reader to perform the exact analysis for a randomized complete block design with one missing value.

5–2   THE LATIN SQUARE DESIGN

In Section 5–1 we introduced the randomized complete block design as a design to reduce the residual error in an experiment by removing variability due to a known and controllable nuisance variable. There are several other types of designs that utilize the blocking principle. For example, suppose that an experimenter is studying the effects of five different formulations of a rocket propellant used in aircrew escape systems on the observed burning rate. Each formulation is mixed from a batch of raw material that is only large enough for five formulations to be tested. Furthermore, the formulations are prepared by several operators, and there may be substantial differences in the skills and experience of the operators. Thus, it would seem that there are two nuisance factors to be "averaged out" in the design: batches of raw material and operators. The appropriate design for this problem consists of testing each formulation exactly once in each batch of raw material and for each formulation to be prepared exactly once by each of five operators. The resulting design, shown in Table 5–9, is called a **Latin square design.** Notice that the design is a square arrangement, and that the five formulations (or treatments) are denoted by the Latin letters A, B, C, D, and E; hence the name Latin square. We see that both batches of raw material (rows) and operators (columns) are orthogonal to treatments.

The Latin square design is used to eliminate two nuisance sources of variability; that is, it systematically allows blocking in two directions. Thus, the rows and columns actually represent *two restrictions on randomization*. In general, a Latin square for p factors, or a p × p Latin square, is a square containing p rows

Table 5–9   Latin Square Design for the Rocket Propellant Problem

Batches of Raw Material	Operators				
	1	2	3	4	5
1	A = 24	B = 20	C = 19	D = 24	E = 24
2	B = 17	C = 24	D = 30	E = 27	A = 36
3	C = 18	D = 38	E = 26	A = 27	B = 21
4	D = 26	E = 31	A = 26	B = 23	C = 22
5	E = 22	A = 30	B = 20	C = 29	D = 31

and p columns. Each of the resulting p^2 cells contains one of the p letters that corresponds to the treatments, and each letter occurs once and only once in each row and column. Some examples of Latin squares are

4 × 4	5 × 5	6 × 6
A B D C	A D B E C	A D C E B F
B C A D	D A C B E	B A E C F D
C D B A	C B E D A	C E D F A B
D A C B	B E A C D	D C F B E A
	E C D A B	F B A D C E
		E F B A D C

The **statistical model** for a Latin square is

$$y_{ijk} = \mu + \alpha_i + \tau_j + \beta_k + \epsilon_{ijk} \qquad \begin{cases} i = 1, 2, \ldots, p \\ j = 1, 2, \ldots, p \\ k = 1, 2, \ldots, p \end{cases} \qquad (5\text{-}22)$$

where y_{ijk} is the observation in the ith row and kth column for the jth treatment, μ is the overall mean, α_i is the ith row effect, τ_j is the jth treatment effect, β_k is the kth column effect, and ϵ_{ijk} is the random error. The model is completely **additive;** that is, there is no interaction between rows, columns, and treatments. Since there is only one observation in each cell, only two of the three subscripts i, j, and k are needed to denote a particular observation. For example, referring to the rocket propellant problem in Table 5-9, if $i = 2$ and $k = 3$ we automatically find $j = 4$ (formulation D), and if $i = 1$ and $j = 3$ (formulation C) we find $k = 3$. This is a consequence of each treatment appearing exactly once in each row and column.

The analysis of variance consists of partitioning the total sum of squares of the $N = p^2$ observations into components for rows, columns, treatments, and error, for example,

$$SS_T = SS_{\text{Rows}} + SS_{\text{Columns}} + SS_{\text{Treatments}} + SS_E \qquad (5\text{-}23)$$

with respective degrees of freedom

$$p^2 - 1 = p - 1 + p - 1 + p - 1 + (p - 2)(p - 1)$$

Under the usual assumption that ϵ_{ijk} is NID $(0, \sigma^2)$, each sum of squares on the right-hand side of Equation 5-23 is, upon division by σ^2, an independently distributed chi-square random variable. The appropriate statistic for testing for no differences in treatment means is

$$F_0 = \frac{MS_{\text{Treatments}}}{MS_E}$$

Table 5–10   Analysis of Variance for the Latin Square Design

Source of Variation	Sum of Squares	Degrees of Freedom	Mean Square	F_0
Treatments	$SS_{\text{Treatments}}$ $= \dfrac{1}{p}\sum_{j=1}^{p} y_{.j.}^2 - \dfrac{y_{...}^2}{N}$	$p-1$	$\dfrac{SS_{\text{Treatments}}}{p-1}$	$F_0 = \dfrac{MS_{\text{Treatments}}}{MS_E}$
Rows	SS_{Rows} $= \dfrac{1}{p}\sum_{i=1}^{p} y_{i..}^2 - \dfrac{y_{...}^2}{N}$	$p-1$	$\dfrac{SS_{\text{Rows}}}{p-1}$	
Columns	SS_{Columns} $= \dfrac{1}{p}\sum_{k=1}^{p} y_{..k}^2 - \dfrac{y_{...}^2}{N}$	$p-1$	$\dfrac{SS_{\text{Columns}}}{p-1}$	
Error	SS_E (by subtraction)	$(p-2)(p-1)$	$\dfrac{SS_E}{(p-2)(p-1)}$	
Total	SS_T $= \sum_i \sum_j \sum_k y_{ijk}^2 - \dfrac{y_{...}^2}{N}$	p^2-1		

which is distributed as $F_{p-1,(p-2)(p-1)}$ under the null hypothesis. We may also test for no row effect and no column effect by forming the ratio of MS_{Rows} or MS_{Columns} to MS_E. However, since the rows and columns represent restrictions on randomization, these tests may not be appropriate.

The computational procedure for the analysis of variance is shown in Table 5–10. From the computational formulas for the sums of squares, we see that the analysis is a simple extension of the randomized block design, with the sum of squares resulting from rows obtained from the row totals.

Example 5–4

Consider the rocket propellant problem previously described, where both batches of raw material and operators represent randomization restrictions. The design for this experiment, shown in Table 5–9, is a 5×5 Latin square. After coding by subtracting 25 from each observation, we have the data in Table 5–11. The sums of squares for the total, batches (rows), and operators (columns) are computed as follows:

$$SS_T = \sum_i \sum_j \sum_k y_{ijk}^2 - \frac{y_{...}^2}{N}$$

$$= 680 - \frac{(10)^2}{25} = 676.00$$

$$SS_{Batches} = \frac{1}{p}\sum_{i=1}^{p} y_{i..}^2 - \frac{y_{...}^2}{N}$$

$$= \frac{1}{5}[(-14)^2 + 9^2 + 5^2 + 3^2 + 7^2] - \frac{(10)^2}{25} = 68.00$$

$$SS_{Operators} = \frac{1}{p}\sum_{k=1}^{p} y_{..k}^2 - \frac{y_{...}^2}{N}$$

$$= \frac{1}{5}[(-18)^2 + 18^2 + (-4)^2 + 5^2 + 9^2] - \frac{(10)^2}{25} = 150.00$$

The totals for the treatments (Latin letters) are

Latin Letter	Treatment Total	
A	$y_{.1.} =$	18
B	$y_{.2.} =$	-24
C	$y_{.3.} =$	-13
D	$y_{.4.} =$	24
E	$y_{.5.} =$	5

The sum of squares resulting from the formulations is computed from these totals as

$$SS_{Formulations} = \frac{1}{p}\sum_{j=1}^{p} y_{.j.}^2 - \frac{y_{...}^2}{N}$$

$$= \frac{18^2 + (-24)^2 + (-13)^2 + 24^2 + 5^2}{5} - \frac{(10)^2}{25} = 330.00$$

The error sum of squares is found by subtraction:

$$SS_E = SS_T - SS_{Batches} - SS_{Operators} - SS_{Formulations}$$
$$= 676.00 - 68.00 - 150.00 - 330.00 = 128.00$$

The analysis of variance is summarized in Table 5–12. We conclude that there is a significant difference in the mean burning rate generated by the different rocket propellant formulations. There is also an indication that there are differences between

Table 5–11 Coded Data for the Rocket Propellant Problem

| Batches of Raw Material | Operators | | | | | $y_{i..}$ |
	1	2	3	4	5	
1	$A = -1$	$B = -5$	$C = -6$	$D = -1$	$E = -1$	-14
2	$B = -8$	$C = -1$	$D = 5$	$E = 2$	$A = 11$	9
3	$C = -7$	$D = 13$	$E = 1$	$A = 2$	$B = -4$	5
4	$D = 1$	$E = 6$	$A = 1$	$B = -2$	$C = -3$	3
5	$E = -3$	$A = 5$	$B = -5$	$C = 4$	$D = 6$	7
$y_{..k}$	-18	18	-4	5	9	$10 = y_{...}$

Table 5–12   Analysis of Variance for the Rocket Propellant Experiment

Source of Variation	Sum of Squares	Degrees of Freedom	Mean Square	F_0	P-Value
Formulations	330.00	4	82.50	7.73	0.0025
Batches of raw material	68.00	4	17.00		
Operators	150.00	4	37.50		
Error	128.00	12	10.67		
Total	676.00	24			

operators, so blocking on this factor was a good precaution. There is no strong evidence of a difference between batches of raw material, so it seems that in this particular experiment we were unnecessarily concerned about this source of variability. However, blocking on batches of raw material is usually a good idea.

∎

As in any design problem, the experimenter should investigate the adequacy of the model by inspecting and plotting the residuals. For a Latin square, the residuals are given by

$$e_{ijk} = y_{ijk} - \hat{y}_{ijk}$$
$$= y_{ijk} - \bar{y}_{i..} - \bar{y}_{.j.} - \bar{y}_{..k} + 2\bar{y}_{...}$$

The reader should find the residuals for Example 5–4 and construct appropriate plots.

A Latin square in which the first row and column consists of the letters written in alphabetical order is called a **standard Latin square.** The design used in Example 5–4 is a standard Latin square. A standard Latin square can always be obtained by writing the first row in alphabetical order and then writing each successive row as the row of letters just above shifted one place to the left. Table 5–13 summarizes several important facts about Latin squares and standard Latin squares.

As with any experimental design, the observations in the Latin square should be taken in random order. The proper randomization procedure is to select the particular square employed at random. As we see in Table 5–13, there are a large number of Latin squares of a particular size, so it is impossible to enumerate all the squares and select one randomly. The usual procedure is to select a Latin square from a table of such designs, as in Fisher and Yates (1953), and then arrange the order of the rows, columns, and letters at random. This is discussed more completely in Fisher and Yates (1953).

Occasonally, one observation in a Latin square is missing. For a $p \times p$ Latin square, the missing value may be estimated by

$$y_{ijk} = \frac{p(y'_{i..} + y'_{.j.} + y'_{..k}) - 2y'_{...}}{(p-2)(p-1)} \tag{5-24}$$

Table 5-13 Standard Latin Squares and Number of Latin Squares of Various Sizes[a]

Size	3×3	4×4	5×5	6×6	7×7	$p \times p$
Examples of standard squares	A B C B C A C A B	A B C D B C D A C D A B D A B C	A B C D E B A E C D C D A E B D E B A C E C D B A	A B C D E F B C F A D E C F B E A D D E A B F C E A D F C B F D E C B A	A B C D E F G B C D E F G A C D E F G A B D E F G A B C E F G A B C D F G A B C D E G A B C D E F	A B C . . . P B C D . . . A C D E . . . B . . P A B . . . (P − 1)
Number of standard squares	1	4	56	9408	16,942,080	—
Total number of Latin squares	12	576	161,280	818,851,200	61,479,419,904,000	$p!(p - 1)! \times$ (number of standard squares)

[a] Some of the information in this table is found in *Statistical Tables for Biological, Agricultural and Medical Research*, 4th edition, by R. A. Fisher and F. Yates, Oliver and Boyd, Edinburgh, 1953. Little is known about the properties of Latin squares larger than 7×7.

where the primes indicate totals for the row, column, and treatment with the missing value, and $y'_{...}$ is the grand total with the missing value.

Latin squares can be useful in situations where the rows and columns represent factors the experimenter actually wishes to study and where there are no randomization restrictions. Thus, three factors (rows, columns, and letters), each at p levels, can be investigated in only p^2 runs. This design assumes that there is no interaction between the factors. More will be said later on the subject of interaction.

Replication of Latin Squares ▪ A disadvantage of small Latin squares is that they provide a relatively small number of error degrees of freedom. For example, a 3×3 Latin square has only 2 error degrees of freedom, a 4×4 Latin square has only 6 error degrees of freedom, and so forth. When small Latin squares are used, it is frequently desirable to replicate them to increase the error degrees of freedom.

There are several ways to replicate a Latin square. To illustrate, suppose that the 5×5 Latin square used in Example 5–4 is replicated n times. This could have been done as follows:

1.   Use the same batches and operators in each replicate.
2.   Use the same batches but different operators in each replicate (or, equivalently, use the same operators but different batches).
3.   Use different batches and different operators.

The analysis of variance depends on the method of replication.

Consider case 1, where the same levels of the row and column blocking factors are used in each replicate. Let y_{ijkl} be the observation in row i, treatment j, column k, and replicate l. There are $N = np^2$ total observations. The analysis of variance is summarized in Table 5–14.

Now consider case 2 and assume that new batches of raw material but the same operators are used in each replicate. Thus, there are now 5 new rows (in general, p new rows) within each replicate. The analysis of variance is summarized on page 202 in Table 5–15. Note that the source of variation for the rows really measures the variation between rows within the n replicates.

Finally, consider case 3, where new batches of raw material and new operators are used in each replicate. Now the variation that results from both the rows and columns measures the variation resulting from these factors within the replicates. The analysis of variance is summarized in Table 5–16. (See page 203.)

There are other approaches to analyzing replicated Latin squares that allow some interactions between treatments and squares. (Refer to Problem 5–19.)

Crossover Designs and Designs Balanced for Residual Effects ▪ Occasionally, one encounters a problem in which time periods are a factor in the experiment. In general, there are p treatments to be tested in p time periods using np

Table 5-14 Analysis of Variance for a Replicated Latin Square, Case 1

Source of Variation	Sum of Squares	Degrees of Freedom	Mean Square	F_0
Treatments	$\dfrac{1}{np}\sum_{j=1}^{p} y_{.j..}^2 - \dfrac{y_{....}^2}{N}$	$p - 1$	$\dfrac{SS_{\text{Treatments}}}{p-1}$	$\dfrac{MS_{\text{Treatments}}}{MS_E}$
Rows	$\dfrac{1}{np}\sum_{i=1}^{p} y_{i...}^2 - \dfrac{y_{....}^2}{N}$	$p - 1$	$\dfrac{SS_{\text{Rows}}}{p-1}$	
Columns	$\dfrac{1}{np}\sum_{k=1}^{p} y_{..k.}^2 - \dfrac{y_{....}^2}{N}$	$p - 1$	$\dfrac{SS_{\text{Columns}}}{p-1}$	
Replicates	$\dfrac{1}{p^2}\sum_{l=1}^{n} y_{...l}^2 - \dfrac{y_{....}^2}{N}$	$n - 1$	$\dfrac{SS_{\text{Replicates}}}{n-1}$	
Error	Subtraction	$(p-1)[n(p+1)-3]$	$\dfrac{SS_E}{(p-1)[n(p+1)-3]}$	
Total	$\sum\sum\sum\sum y_{ijkl}^2 - \dfrac{y_{....}^2}{N}$	$np^2 - 1$		

Table 5-15 Analysis of Variance for a Replicated Latin Square, Case 2

Source of Variation	Sum of Squares	Degrees of Freedom	Mean Square	F_0
Treatments	$\dfrac{1}{np}\sum\limits_{j=1}^{p} y_{.j..}^2 - \dfrac{y_{....}^2}{N}$	$p-1$	$\dfrac{SS_{\text{Treatments}}}{p-1}$	$\dfrac{MS_{\text{Treatments}}}{MS_E}$
Rows	$\dfrac{1}{p}\sum\limits_{l=1}^{n}\sum\limits_{i=1}^{p} y_{i.l}^2 - \sum\limits_{l=1}^{n}\dfrac{y_{...l}^2}{p^2}$	$n(p-1)$	$\dfrac{SS_{\text{Rows}}}{n(p-1)}$	
Columns	$\dfrac{1}{np}\sum\limits_{k=1}^{p} y_{..k.}^2 - \dfrac{y_{....}^2}{N}$	$p-1$	$\dfrac{SS_{\text{Columns}}}{p-1}$	
Replicates	$\dfrac{1}{p^2}\sum\limits_{l=1}^{n} y_{...l}^2 - \dfrac{y_{....}^2}{N}$	$n-1$	$\dfrac{SS_{\text{Replicates}}}{n-1}$	
Error	Subtraction	$(p-1)(np-1)$	$\dfrac{SS_E}{(p-1)(np-1)}$	
Total	$\sum\limits_{i}\sum\limits_{j}\sum\limits_{k}\sum\limits_{l} y_{ijkl}^2 - \dfrac{y_{....}^2}{N}$	np^2-1		

Table 5-16 Analysis of Variance for a Replicated Latin Square, Case 3

Source of Variation	Sum of Squares	Degrees of Freedom	Mean Square	F_0
Treatments	$\dfrac{1}{np}\sum_{j=1}^{p} y_{.j..}^2 - \dfrac{y_{....}^2}{N}$	$p-1$	$\dfrac{SS_{\text{Treatments}}}{p-1}$	$\dfrac{MS_{\text{Treatments}}}{MS_E}$
Rows	$\dfrac{1}{p}\sum_{l=1}^{n}\sum_{i=1}^{p} y_{i..l}^2 - \sum_{l=1}^{n}\dfrac{y_{...l}^2}{p^2}$	$n(p-1)$	$\dfrac{SS_{\text{Rows}}}{n(p-1)}$	
Columns	$\dfrac{1}{p}\sum_{l=1}^{n}\sum_{k=1}^{p} y_{..kl}^2 - \sum_{l=1}^{n}\dfrac{y_{...l}^2}{p^2}$	$n(p-1)$	$\dfrac{SS_{\text{Columns}}}{n(p-1)}$	
Replicates	$\dfrac{1}{p^2}\sum_{l=1}^{n} y_{...l}^2 - \dfrac{y_{....}^2}{N}$	$n-1$	$\dfrac{SS_{\text{Replicates}}}{n-1}$	
Error	Subtraction	$(p-1)[n(p-1)-1]$	$\dfrac{SS_E}{(p-1)[n(p-1)-1]}$	
Total	$\sum_i\sum_j\sum_k\sum_l y_{ijkl}^2 - \dfrac{y_{....}^2}{N}$	np^2-1		

203

		Latin Squares																		
	I		II		III		IV		V		VI		VII		VIII		IX		X	
Subject	1	2	3	4	5	6	7	8	9	10	11	12	13	14	15	16	17	18	19	20
Period 1	A	B	B	A	B	A	A	B	A	B	B	A	A	B	A	B	A	B	A	B
Period 2	B	A	A	B	A	B	B	A	B	A	A	B	B	A	B	A	B	A	B	A

Figure 5–7. A crossover design.

experimental units. For example, a human performance analyst is studying the effect of two replacement fluids on dehydration in 20 subjects. In the first period, half of the subjects (chosen at random) are given fluid A and the other half fluid B. At the end of the period, the response is measured, and a period of time is allowed to pass in which any physiological effect of the fluids is eliminated. Then the experimenter has the subjects who took fluid A take fluid B and those who took fluid B take fluid A. This design is called a **crossover design.** It is analyzed as a set of 10 Latin squares with two rows (time periods) and two treatments (fluid types). The two columns in each of the 10 squares correspond to subjects.

The layout of this design is shown in Figure 5–7. Notice that the rows in the Latin squares represent the time periods and the columns represent the subjects. The 10 subjects who received fluid A first (1, 4, 6, 7, 9, 12, 13, 15, 17, and 19) are randomly determined.

An abbreviated analysis of variance is summarized in Table 5–17. The subject sum of squares is computed as the corrected sum of squares among the 20 subject totals, the period sum of squares is the corrected sum of squares among the rows, and the fluid sum of squares is computed as the corrected sum of squares among the letter totals. For further details of the statistical analysis of these designs, see Cochran and Cox (1957), John (1971), and Anderson and McLean (1974).

It is also possible to employ Latin square type designs for experiments in which the treatments have a **residual effect**—that is, for example, if the data for fluid B in period 2 still reflected some effect of fluid A taken in period 1. Designs

Table 5–17 Analysis of Variance for the Crossover Design in Figure 5–7

Source of Validation	Degrees of Freedom
Subjects (columns)	19
Periods (rows)	1
Fluids (letters)	1
Error	18
Total	39

balanced for residual effects are discussed in detail by Cochran and Cox (1957) and John (1971).

5–3 THE GRAECO-LATIN SQUARE DESIGN

Consider a $p \times p$ Latin square, and superimpose on it a second $p \times p$ Latin square in which the treatments are denoted by Greek letters. If the two squares when superimposed have the property that each Greek letter appears once and only once with each Latin letter, the two Latin squares are said to be *orthogonal,* and the design obtained is called a **Graeco-Latin square.** An example of a 4 × 4 Graeco-Latin square is shown in Table 5–18.

The Graeco-Latin square design can be used to control systematically three sources of extraneous variability, that is, to block in *three* directions. The design allows investigation of four factors (rows, columns, Latin letters, and Greek letters), each at p levels in only p^2 runs. Graeco-Latin squares exist for all $p \geq 3$ except $p = 6$.

The statistical model for the Graeco-Latin square design is

$$y_{ijkl} = \mu + \theta_i + \tau_j + \omega_k + \Psi_l + \epsilon_{ijkl} \qquad \begin{cases} i = 1, 2, \ldots, p \\ j = 1, 2, \ldots, p \\ k = 1, 2, \ldots, p \\ l = 1, 2, \ldots, p \end{cases} \qquad (5\text{–}25)$$

where y_{ijkl} is the observation in row i and column l for Latin letter j and Greek letter k, θ_i is the effect of the ith row, τ_j is the effect of Latin letter treatment j, ω_k is the effect of Greek letter treatment k, Ψ_l is the effect of column l, and ϵ_{ijkl} is an NID $(0, \sigma^2)$ random error component. Only two of the four subscripts are necessary to completely identify an observation.

The analysis of variance is very similar to that of a Latin square. Since the Greek letters appear exactly once in each row and column and exactly once with each Latin letter, the factor represented by the Greek letters is orthogonal to

Table 5–18   4 × 4 Graeco-Latin Square Design

Row	Column			
	1	2	3	4
1	$A\alpha$	$B\beta$	$C\gamma$	$D\delta$
2	$B\delta$	$A\gamma$	$D\beta$	$C\alpha$
3	$C\beta$	$D\alpha$	$A\delta$	$B\gamma$
4	$D\gamma$	$C\delta$	$B\alpha$	$A\beta$

Table 5–19 Analysis of Variance for a Graeco-Latin Square Design

Source of Variation	Sum of Squares	Degrees of Freedom
Latin letter treatments	$SS_L = \dfrac{1}{p} \sum\limits_{j=1}^{p} y_{.j..}^2 - \dfrac{y_{....}^2}{N}$	$p - 1$
Greek letter treatments	$SS_G = \dfrac{1}{p} \sum\limits_{k=1}^{p} y_{..k.}^2 - \dfrac{y_{....}^2}{N}$	$p - 1$
Rows	$SS_{\text{Rows}} = \dfrac{1}{p} \sum\limits_{i=1}^{p} y_{i...}^2 - \dfrac{y_{....}^2}{N}$	$p - 1$
Columns	$SS_{\text{Columns}} = \dfrac{1}{p} \sum\limits_{l=1}^{p} y_{...l}^2 - \dfrac{y_{....}^2}{N}$	$p - 1$
Error	SS_E (by subtraction)	$(p - 3)(p - 1)$
Total	$SS_T = \sum\limits_{i}\sum\limits_{j}\sum\limits_{k}\sum\limits_{l} y_{ijkl}^2 - \dfrac{y_{....}^2}{N}$	$p^2 - 1$

rows, columns, and Latin letter treatments. Therefore, a sum of squares due to the Greek letter factor may be computed from the Greek letter totals and the experimental error is further reduced by this amount. The computational details are illustrated in Table 5–19. The null hypotheses of equal row, column, Latin letter, and Greek letter treatments would be tested by dividing the corresponding mean square by mean square error. The rejection region is the upper-tail point of the $F_{p-1,(p-3)(p-1)}$ distribution.

Example 5–5

Suppose that in the rocket propellant experiment of Example 5–4 an additional factor, test assemblies, could be of importance. Let there be five test assemblies denoted by the Greek letters α, β, γ, δ, and ϵ. The resulting 5×5 Graeco-Latin square design is shown in Table 5–20.

Table 5–20 Graeco-Latin Square Design for the Rocket Propellant Problem

Batches of Raw Material	Operators					$y_{i...}$
	1	2	3	4	5	
1	$A\alpha = -1$	$B\gamma = -5$	$C\epsilon = -6$	$D\beta = -1$	$E\delta = -1$	-14
2	$B\beta = -8$	$C\delta = -1$	$D\alpha = 5$	$E\gamma = 2$	$A\epsilon = 11$	9
3	$C\gamma = -7$	$D\epsilon = 13$	$E\beta = 1$	$A\delta = 2$	$B\alpha = -4$	5
4	$D\delta = 1$	$E\alpha = 6$	$A\gamma = 1$	$B\epsilon = -2$	$C\beta = -3$	3
5	$E\epsilon = -3$	$A\beta = 5$	$B\delta = -5$	$C\alpha = 4$	$D\gamma = 6$	7
$y_{...l}$	-18	18	-4	5	9	$10 = y_{....}$

Notice that, since the totals for batches of raw material (rows), operators (columns) and formulations (Latin letters) are identical to those in Example 5–4, we have

$$SS_{\text{Batches}} = 68.00 \qquad SS_{\text{Operators}} = 150.00 \qquad \text{and} \qquad SS_{\text{Formulations}} = 330.00$$

The totals for the test assemblies (Greek letters) are

Greek Letter	Test Assembly Total
α	$y_{..1.} = 10$
β	$y_{..2.} = -6$
γ	$y_{..3.} = -3$
δ	$y_{..4.} = -4$
ϵ	$y_{..5.} = 13$

Thus, the sum of squares due to the test assemblies is

$$SS_{\text{Assemblies}} = \frac{1}{p} \sum_{k=1}^{p} y_{..k.}^2 - \frac{y_{....}^2}{N}$$

$$= \frac{1}{5} [10^2 + (-6)^2 + (-3)^2 + (-4)^2 + 13^2] - \frac{(10)^2}{25} = 62.00$$

The complete analysis of variance is summarized in Table 5–21. Formulations are significantly different at 1 percent. In comparing Tables 5–21 and 5–12, we observe that removing the variability due to test assemblies has decreased the experimental error. However, in decreasing the experimental error, we have also reduced the error degrees of freedom from 12 (in the Latin square design of Example 5–4) to 8. Thus, our estimate of error has fewer degrees of freedom, and the test may be less sensitive.

∎

The concept of orthogonal pairs of Latin squares forming a Graeco-Latin square can be extended somewhat. A $p \times p$ **hypersquare** is a design in which three or more orthogonal $p \times p$ Latin squares are superimposed. In general, up to $p + 1$ factors could be studied if a complete set of $p - 1$ orthogonal Latin squares is available. Such a design would utilize all $(p + 1)(p - 1) = p^2 - 1$ degrees of freedom, so an independent estimate of the error variance is necessary.

Table 5–21   Analysis of Variance for the Rocket Propellant Problem

Source of Variation	Sum of Squares	Degrees of Freedom	Mean Square	F_0	P-Value
Formulations	330.00	4	82.50	10.00	0.0033
Batches of raw material	68.00	4	17.00		
Operators	150.00	4	37.50		
Test assemblies	62.00	4	15.50		
Error	66.00	8	8.25		
Total	676.00	24			

Of course, there must be no interactions between the factors when using hypersquares.

5–4 BALANCED INCOMPLETE BLOCK DESIGNS

In certain experiments using randomized block designs, we may not be able to run all the treatment combinations in each block. Situations like this usually occur because of shortages of experimental apparatus or facilities or the physical size of the block. For example, in the hardness testing experiment (Example 5–1), suppose that because of their size each coupon can be used only for testing three tips. Therefore, each tip cannot be tested on each coupon. For this type of problem it is possible to use randomized block designs in which every treatment is not present in every block. These designs are known as **randomized incomplete block designs.**

When all treatment comparisons are equally important, the treatment combinations used in each block should be selected in a balanced manner, that is, so that any pair of treatments occur together the same number of times as any other pair. Thus, a **balanced incomplete block design** is an incomplete block design in which any two treatments appear together an equal number of times. Suppose that there are a treatments and that each block can hold exactly k ($k < a$) treatments. A balanced incomplete block design may be constructed by taking $\binom{a}{k}$ blocks and assigning a different combination of treatments to each block. Frequently, however, balance can be obtained with fewer than $\binom{a}{k}$ blocks. Tables of balanced incomplete block designs are given in Fisher and Yates (1953), Davies (1956), and Cochran and Cox (1957).

As an example, suppose that a chemical engineer thinks that the time of reaction for a chemical process is a function of the type of catalyst employed. Four catalysts are currently being investigated. The experimental procedure consists of selecting a batch of raw material, loading the pilot plant, applying each catalyst in a separate run of the pilot plant, and observing the reaction time. Since variations in the batches of raw material may affect the performance of the catalysts, the engineer decides to use batches of raw material as blocks. However, each batch is only large enough to permit three catalysts to be run. Therefore, a randomized incomplete block design must be used. The balanced incomplete block design for this experiment, along with the observations recorded, is shown in Table 5–22. The order in which the catalysts are run in each block is randomized.

5–4.1 Statistical Analysis

As usual, we assume that there are a treatments and b blocks. In addition, we assume that each block contains k treatments, that each treatment occurs r times

Table 5–22   Balanced Incomplete Block Design for
Catalyst Experiment

Treatment (Catalyst)	Block (Batch of Raw Material)				$y_{i.}$
	1	2	3	4	
1	73	74	—	71	218
2	—	75	67	72	214
3	73	75	68	—	216
4	75	—	72	75	222
$y_{.j}$	221	224	207	218	$870 = y_{..}$

in the design (or is replicated r times), and that there are $N = ar = bk$ total observations. Furthermore, the number of times each pair of treatments appears in the same block is

$$\lambda = \frac{r(k-1)}{a-1}$$

If $a = b$, the design is said to be **symmetric.**

The parameter λ must be an integer. To derive the relationship for λ, consider any treatment, say treatment 1. Since treatment 1 appears in r blocks and there are $k - 1$ other treatments in each of those blocks, there are $r(k - 1)$ observations in a block containing treatment 1. These $r(k - 1)$ observations also have to represent the remaining $a - 1$ treatments λ times. Therefore, $\lambda(a - 1) = r(k - 1)$.

The **statistical model** is

$$y_{ij} = \mu + \tau_i + \beta_j + \epsilon_{ij} \tag{5–26}$$

where y_{ij} is the ith observation in the jth block, μ is the overall mean, τ_i is the effect of the ith treatment, β_j is the effect of the jth block, and ϵ_{ij} is the $NID(0, \sigma^2)$ random error component. The total variability in the data is expressed by the total corrected sum of squares:

$$SS_T = \sum_i \sum_j y_{ij}^2 - \frac{y_{..}^2}{N} \tag{5–27}$$

Total variability may be partitioned into

$$SS_T = SS_{\text{Treatments (adjusted)}} + SS_{\text{Blocks}} + SS_E$$

where the sum of squares for treatments is **adjusted** to separate the treatment and the block effects. This adjustment is necessary because each treatment is represented in a different set of r blocks. Thus, differences between unadjusted treatment totals $y_{1.}, y_{2.}, \ldots, y_{a.}$ are also affected by differences between blocks.

Table 5-23 Analysis of Variance for the Balanced Incomplete Block Design

Source of Variation	Sum of Squares	Degrees of Freedom	Mean Square	F_0
Treatments (adjusted)	$\dfrac{k \sum Q_i^2}{\lambda a}$	$a - 1$	$\dfrac{SS_{\text{Treatments (adjusted)}}}{a - 1}$	$F_0 = \dfrac{MS_{\text{Treatments (adjusted)}}}{MS_E}$
Blocks	$\dfrac{1}{k} \sum y_{\cdot j}^2 - \dfrac{y_{\cdot\cdot}^2}{N}$	$b - 1$	$\dfrac{SS_{\text{Blocks}}}{b - 1}$	
Error	SS_E (by subtraction)	$N - a - b + 1$	$\dfrac{SS_E}{N - a - b + 1}$	
Total	$\sum \sum y_{ij}^2 - \dfrac{y_{\cdot\cdot}^2}{N}$	$N - 1$		

The block sum of squares is

$$SS_{Blocks} = \frac{1}{k} \sum_{j=1}^{b} y_{.j}^2 - \frac{y_{..}^2}{N} \tag{5-28}$$

where $y_{.j}$ is the total in the jth block. SS_{Blocks} has $b - 1$ degrees of freedom. The adjusted treatment sum of squares is

$$SS_{Treatments\,(adjusted)} = \frac{k \sum_{i=1}^{a} Q_i^2}{\lambda a} \tag{5-29}$$

where Q_i is the adjusted total for the ith treatment, which is computed as

$$Q_i = y_{i.} - \frac{1}{k} \sum_{j=1}^{b} n_{ij} y_{.j} \qquad i = 1, 2, \ldots, a \tag{5-30}$$

with $n_{ij} = 1$ if treatment i appears in block j and $n_{ij} = 0$ otherwise. Thus, $(1/k) \cdot \sum_{j=1}^{b} n_{ij} y_{.j}$ is the average of the block totals containing treatment i. The adjusted treatment totals will always sum to zero. $SS_{Treatments\,(adjusted)}$ has $a - 1$ degrees of freedom. The error sum of squares is computed by subtraction as

$$SS_E = SS_T - SS_{Treatments\,(adjusted)} - SS_{Blocks} \tag{5-31}$$

and has $N - a - b + 1$ degrees of freedom.

The appropriate statistic for testing the equality of the treatment effects is

$$F_0 = \frac{MS_{Treatments\,(adjusted)}}{MS_E}$$

The analysis of variance is summarized in Table 5-23.

Example 5-6

Consider the data in Table 5-22 for the catalyst experiment. This is a balanced incomplete block design with $a = 4$, $b = 4$, $k = 3$, $r = 3$, $\lambda = 2$, and $N = 12$. The analysis of this data is as follows. The total sum of squares is

$$SS_T = \sum_i \sum_j y_{ij}^2 - \frac{y_{..}^2}{12}$$

$$= 63{,}156 - \frac{(870)^2}{12} = 81.00$$

The block sum of squares is found from Equation 5-28 as

$$SS_{Blocks} = \frac{1}{3} \sum_{j=1}^{4} y_{.j}^2 - \frac{y_{..}^2}{12}$$

$$= \frac{1}{3}[(221)^2 + (207)^2 + (224)^2 + (218)^2] - \frac{(870)^2}{12} = 55.00$$

To compute the treatment sum of squares adjusted for blocks, we first determine the adjusted treatment totals using Equation 5–30 as

$$Q_1 = (218) - \tfrac{1}{3}(221 + 224 + 218) = -9/3$$
$$Q_2 = (214) - \tfrac{1}{3}(207 + 224 + 218) = -7/3$$
$$Q_3 = (216) - \tfrac{1}{3}(221 + 207 + 224) = -4/3$$
$$Q_4 = (222) - \tfrac{1}{3}(221 + 207 + 218) = 20/3$$

The adjusted sum of squares for treatments is computed from Equation 5–29 as

$$
SS_{\text{Treatments (adjusted)}} = \frac{k \sum\limits_{i=1}^{4} Q_i^2}{\lambda a}
$$

$$
= \frac{3[(-9/3)^2 + (-7/3)^2 + (-4/3)^2 + (20/3)^2]}{(2)(4)} = 22.75
$$

The error sum of squares is obtained by subtraction as

$$SS_E = SS_T - SS_{\text{Treatments (adjusted)}} - SS_{\text{Blocks}}$$
$$= 81.00 - 22.75 - 55.00 = 3.25$$

The analysis of variance is shown in Table 5–24. Since the P-value is small, we conclude that the catalyst employed has a significant effect on the time of reaction.

∎

If the factor under study is fixed, then tests on individual treatment means may be of interest. If orthogonal contrasts are employed, the contrasts must be made on the **adjusted treatment totals,** the $\{Q_i\}$ rather than the $\{y_{i.}\}$. The contrast sum of squares is

$$
SS_C = \frac{k \left(\sum\limits_{i=1}^{a} c_i Q_i \right)^2}{\lambda a \sum\limits_{i=1}^{a} c_i^2}
$$

where $\{c_i\}$ are the contrast coefficients. Other multiple comparison methods may be used to compare all the pairs of adjusted treatment effects, which we will

Table 5–24   Analysis of Variance for Example 5–6

Source of Variation	Sum of Squares	Degrees of Freedom	Mean Square	F_0	P-Value
Treatments (adjusted for blocks)	22.75	3	7.58	11.66	0.0107
Blocks	55.00	3	—		
Error	3.25	5	0.65		
Total	81.00	11			

find in Section 5–4.2 are estimated by $\hat{\tau}_i = kQ_i/(\lambda a)$. The standard error of an adjusted treatment effect is

$$S = \sqrt{\frac{kMS_E}{\lambda a}} \tag{5-32}$$

As an example, suppose we apply Duncan's multiple range test to the data in Example 5–6. The adjusted treatment effects, in ascending order, are

$$\hat{\tau}_1 = kQ_1/(\lambda a) = (3)(-9/3)/(2)(4) = -9/8$$
$$\hat{\tau}_2 = kQ_2/(\lambda a) = (3)(-7/3)/(2)(4) = -7/8$$
$$\hat{\tau}_3 = kQ_3/(\lambda a) = (3)(-4/3)/(2)(4) = -4/8$$
$$\hat{\tau}_4 = kQ_4/(\lambda a) = (3)(20/3)/(2)(4)\ = 20/8$$

From Appendix Table VII with $\alpha = 0.05$ and five error degrees of freedom, we obtain $r_{0.05}(2, 5) = 3.64$, $r_{0.05}(3, 5) = 3.74$, and $r_{0.05}(4, 5) = 3.79$. Substituting $MS_E = 0.65$ from Table 5–24 into Equation 5–32 yields

$$S = \sqrt{\frac{kMS_E}{\lambda a}} = \sqrt{\frac{3(0.65)}{2(4)}} = 0.49$$

Thus,

$$R_2 = r_{0.05}(2, 5)S = (3.64)(0.49) = 1.80$$
$$R_3 = r_{0.05}(3, 5)S = (3.74)(0.49) = 1.83$$
$$R_4 = r_{0.05}(4, 5)S = (3.79)(0.49) = 1.86$$

and the comparisons yield

$$
\begin{array}{lll}
\text{4 vs. 1:} & 20/8 - (-9/8) = 29/8 > 1.86(R_4) \\
\text{4 vs. 2:} & 20/8 - (-7/8) = 27/8 > 1.83(R_3) \\
\text{4 vs. 3:} & 20/8 - (-4/8) = 24/8 > 1.80(R_2) \\
\text{3 vs. 1:} & -4/8 - (-9/8) = 5/8 < 1.83(R_3) \\
\text{3 vs. 2:} & -4/8 - (-7/8) = 3/8 < 1.80(R_2) \\
\text{2 vs. 1:} & -7/8 - (-9/8) = 2/8 < 1.80(R_2)
\end{array}
$$

The test results are summarized in Figure 5–8. Catalyst 4 differs from the other three, whereas all other pairs of means do not differ.

In the analysis we have elected to partition the total sum of squares into an adjusted sum of squares for treatments, an unadjusted sum of squares for blocks,

$\hat{\tau}_1$	$\hat{\tau}_2$	$\hat{\tau}_3$	$\hat{\tau}_4$
-9/8	-7/8	-4/8	20/8

Figure 5–8.   Results of Duncan's multiple range test.

and an error sum of squares. Sometimes we would like to assess the block effects. To do this, we require an alternate partitioning of SS_T, that is,

$$SS_T = SS_{\text{Treatments}} + SS_{\text{Blocks (adjusted)}} + SS_E$$

Here $SS_{\text{Treatments}}$ is unadjusted. If the design is symmetric, that is, if $a = b$, then a simple formula may be obtained for $SS_{\text{Blocks (adjusted)}}$. The adjusted block totals are

$$Q'_j = y_{.j} - \frac{1}{r} \sum_{i=1}^{a} n_{ij} y_{i.} \qquad j = 1, 2, \ldots, b \qquad (5\text{-}33)$$

and

$$SS_{\text{Blocks (adjusted)}} = \frac{r \sum_{j=1}^{b} (Q'_j)^2}{\lambda b} \qquad (5\text{-}34)$$

The balanced incomplete block design in Example 5–6 is symmetric since $a = b = 4$. Therefore,

$$Q'_1 = (221) - \tfrac{1}{3}(218 + 216 + 222) = 7/3$$
$$Q'_2 = (224) - \tfrac{1}{3}(218 + 214 + 216) = 24/3$$
$$Q'_3 = (207) - \tfrac{1}{3}(214 + 216 + 222) = -31/3$$
$$Q'_4 = (218) - \tfrac{1}{3}(218 + 214 + 222) = 0$$

and

$$SS_{\text{Blocks (adjusted)}} = \frac{3[(7/3)^2 + (24/3)^2 + (-31/3)^2 + (0)^2]}{(2)(4)} = 66.08$$

Also,

$$SS_{\text{Treatments}} = \frac{(218)^2 + (214)^2 + (216)^2 + (222)^2}{3} - \frac{(870)^2}{12} = 11.67$$

A summary of the analysis of variance for the symmetric balanced incomplete block design is given in Table 5–25. Notice that the sums of squares associated with the mean squares in Table 5–25 do not add to the total sum of squares, that is,

$$SS_T \neq SS_{\text{Treatments (adjusted)}} + SS_{\text{Blocks (adjusted)}} + SS_E$$

This is a consequence of the nonorthogonality of treatments and blocks.

Computer Output ▪ There are several computer packages that will perform the analysis for a balanced incomplete block design. The SAS General Linear Models procedure illustrated earlier is one of these, and Statgraphics, a widely used PC statistics package, is another. The upper portion of Table 5–26 is the Statgraphics analysis of variance output for Example 5–6. Comparing Tables 5–26 and 5–25, we see that Statgraphics has computed the adjusted treatment sum of squares

Table 5-25 Analysis of Variance for Example 5-6, Including Both Treatments and Blocks

Source of Variation	Sum of Squares	Degrees of Freedom	Mean Square	F_0	P-Value
Treatments (adjusted)	22.75	3	7.58	11.66	0.0107
Treatments (unadjusted)	11.67	3			
Blocks (unadjusted)	55.00	3			
Blocks (adjusted)	66.08	3	22.03	33.90	0.0010
Error	3.25	5	0.65		
Total	81.00	11			

and the adjusted block sum of squares (they are identified as Type III sums of squares, the same terminology used in SAS).

The lower portion of Table 5-26 is a multiple comparison analysis, using the Fisher LSD method. The least squares means are the overall mean ($\bar{y}_{..} = 72.5$) plus the adjusted treatment effects (the $\hat{\tau}_i$). Notice that the LSD method would

Table 5-26 Statgraphics Analysis for Example 5-6

Analysis of Variance for DXCOURSE.react_time - Type III Sums of Squares

Source of Variation	Sum of Squares	d.f.	Mean Square	F-Ratio	Sig. Level
MAIN EFFECTS					
A:DXCOURSE.catalyst	22.750000	3	7.583333	11.667	.0107
B:DXCOURSE.Block	66.083333	3	22.027778	33.889	.0010
RESIDUAL	3.2500000	5	.6500000		
TOTAL (CORRECTED)	81.00000	11			

4 missing values have been excluded.
All F-ratios are based on the residual mean square error.

Multiple Range Analysis for DXCOURSE.react_time by DXCOURSE.catalyst

Method: 95 Percent LSD

Level	Count	LS Mean	Homogeneous Groups
1	3	71.375000	X
2	3	71.625000	X
3	3	72.000000	X
4	3	75.000000	X

contrast	difference	+/-	limits	
1 - 2	-0.25000		1.79539	
1 - 3	-0.62500		1.79539	
1 - 4	-3.62500		1.79539	*
2 - 3	-0.37500		1.79539	
2 - 4	-3.37500		1.79539	*
3 - 4	-3.00000		1.79539	*

* Denotes a statistically significant difference.

lead us to conclude that catalyst 4 is different from the other three, the identical conclusion we reached earlier with Duncan's method.

5–4.2 Least Squares Estimation of the Parameters

Consider estimating the treatment effects for the balanced incomplete block model. The least squares normal equations are

$$\mu: N\hat{\mu} + r\sum_{i=1}^{a} \hat{\tau}_i + k\sum_{j=1}^{b} \hat{\beta}_j = y_{..}$$

$$\tau_i: r\hat{\mu} + r\hat{\tau}_i + \sum_{j=1}^{b} n_{ij}\hat{\beta}_j = y_{i.} \qquad i = 1, 2, \ldots, a \qquad (5\text{–}35)$$

$$\beta_j: k\hat{\mu} + \sum_{i=1}^{a} n_{ij}\hat{\tau}_i + k\hat{\beta}_j = y_{.j} \qquad j = 1, 2, \ldots, b$$

Imposing $\Sigma\hat{\tau}_i = \Sigma\hat{\beta}_j = 0$, we find that $\hat{\mu} = \bar{y}_{..}$. Furthermore, using the equations for $\{\beta_j\}$ to eliminate the block effects from the equations for $\{\tau_i\}$, we obtain

$$rk\hat{\tau}_i - r\hat{\tau}_i - \sum_{j=1}^{b}\sum_{\substack{p=1 \\ p\neq i}}^{a} n_{ij}n_{pj}\hat{\tau}_p = ky_{i.} - \sum_{j=1}^{b} n_{ij}y_{.j} \qquad (5\text{–}36)$$

Note that the right-hand side of Equation 5–36 is kQ_i, where Q_i is the ith adjusted treatment total (see Equation 5–29). Now, since $\Sigma_{j=1}^{b}n_{ij}n_{pj} = \lambda$ if $p \neq i$ and $n_{pj}^2 = n_{pj}$ (because $n_{pj} = 0$ or 1), we may rewrite Equation 5–36 as

$$r(k-1)\hat{\tau}_i - \lambda\sum_{\substack{p=1 \\ p\neq i}}^{a} \hat{\tau}_p = kQ_i \qquad i = 1, 2, \ldots, a \qquad (5\text{–}37)$$

Finally, note that the constraint $\Sigma_{i=1}^{a} \hat{\tau}_i = 0$ implies that $\sum_{\substack{p=1 \\ p\neq i}}^{a} \hat{\tau}_p = -\hat{\tau}_i$ and recall that $r(k-1) = \lambda(a-1)$ to obtain

$$\lambda a\hat{\tau}_i = kQ_i \qquad i = 1, 2, \ldots, a \qquad (5\text{–}38)$$

Therefore, the least squares estimators of the treatment effects in the balanced incomplete block model are

$$\hat{\tau}_i = \frac{kQ_i}{\lambda a} \qquad i = 1, 2, \ldots, a \qquad (5\text{–}39)$$

As an illustration, consider the balanced incomplete block design in Example 5–6. Since $Q_1 = -9/3$, $Q_2 = -7/3$, $Q_3 = -4/3$, and $Q_4 = 20/3$, we obtain

$$\hat{\tau}_1 = \frac{3(-9/3)}{(2)(4)} = -9/8 \qquad \hat{\tau}_2 = \frac{3(-7/3)}{(2)(4)} = -7/8$$

$$\hat{\tau}_3 = \frac{3(-4/3)}{(2)(4)} = -4/8 \qquad \hat{\tau}_4 = \frac{3(20/3)}{(2)(4)} = 20/8$$

as we found in Section 5–4.1.

5–4.3   Recovery of Interblock Information in the Balanced Incomplete Block Design

The analysis of the balanced incomplete block design given in Section 5–4.1 is usually called the **intrablock analysis** because block differences are eliminated and all contrasts in the treatment effects can be expressed as comparisons between observations in the same block. This analysis is appropriate regardless of whether the blocks are fixed or random. Yates (1940) noted that, if the block effects are uncorrelated random variables with zero means and variance σ_β^2, then one may obtain additional information about the treatment effects τ_i. Yates called the method of obtaining this additional information the **interblock analysis.**

Consider the block totals $y_{.j}$ as a collection of b observations. The model for these observations [following John (1971)] is

$$y_{.j} = k\mu + \sum_{i=1}^{a} n_{ij}\tau_i + \left(k\beta_j + \sum_{i=1}^{a} \epsilon_{ij} \right) \tag{5–40}$$

where the term in parentheses may be regarded as error. The interblock estimators of μ and τ_i are found by minimizing the least squares function

$$L = \sum_{j=1}^{b} \left(y_{.j} - k\mu - \sum_{i=1}^{a} n_{ij}\tau_i \right)^2$$

This yields the following least squares normal equations:

$$\mu: N\tilde{\mu} + r \sum_{i=1}^{a} \tilde{\tau}_i = y_{..}$$

$$\tau_i: kr\tilde{\mu} + r\tilde{\tau}_i + \lambda \sum_{\substack{p=1 \\ p \neq i}}^{a} \tilde{\tau}_p = \sum_{j=1}^{b} n_{ij}y_{.j} \qquad i = 1, 2, \ldots, a \tag{5–41}$$

where $\tilde{\mu}$ and $\tilde{\tau}_i$ denote the **interblock estimators.** Imposing the constraint $\sum_{i=1}^{a} \tilde{\tau}_i = 0$, we obtain the solutions to Equations 5–41 as

$$\tilde{\mu} = \bar{y}_{..} \tag{5–42}$$

$$\tilde{\tau}_i = \frac{\sum_{j=1}^{b} n_{ij}y_{.j} - kr\bar{y}_{..}}{r - \lambda} \qquad i = 1, 2, \ldots, a \tag{5–43}$$

It is possible to show that the interblock estimators $\{\tilde{\tau}_i\}$ and the intrablock estimators $\{\hat{\tau}_i\}$ are uncorrelated.

The interblock estimators $\{\tilde{\tau}_i\}$ can differ from the intrablock estimators $\{\hat{\tau}_i\}$. For example, the interblock estimators for the balanced incomplete block design in Example 5–6 are computed as follows:

$$\tilde{\tau}_1 = \frac{663 - (3)(3)(72.50)}{3 - 2} = 10.50$$

$$\tilde{\tau}_2 = \frac{649 - (3)(3)(72.50)}{3 - 2} = -3.50$$

$$\tilde{\tau}_3 = \frac{652 - (3)(3)(72.50)}{3 - 2} = -0.50$$

$$\tilde{\tau}_4 = \frac{646 - (3)(3)(72.50)}{3 - 2} = -6.50$$

Note that the values of $\sum_{j=1}^{b} n_{ij} y_{.j}$ were used previously on page 212 in computing the adjusted treatment totals in the intrablock analysis.

Now suppose we wish to combine the interblock and intrablock estimators to obtain a single, unbiased, minimum variance estimate of each τ_i. It is possible to show that both $\hat{\tau}_i$ and $\tilde{\tau}_i$ are unbiased and also that

$$V(\hat{\tau}_i) = \frac{k(a - 1)}{\lambda a^2} \sigma^2 \qquad \text{(intrablock)}$$

and

$$V(\tilde{\tau}_i) = \frac{k(a - 1)}{a(r - \lambda)} (\sigma^2 + k\sigma_\beta^2) \qquad \text{(interblock)}$$

We use a linear combination of the two estimators, say

$$\tau_i^* = \alpha_1 \hat{\tau}_i + \alpha_2 \tilde{\tau}_i \qquad (5\text{--}44)$$

to estimate τ_i. For this estimation method, the minimum variance unbiased combined estimator τ_i^* should have weights $\alpha_1 = u_1/(u_1 + u_2)$ and $\alpha_2 = u_2/(u_1 + u_2)$, where $u_1 = 1/V(\hat{\tau}_i)$ and $u_2 = 1/V(\tilde{\tau}_i)$. Thus, the optimal weights are inversely proportional to the variances of $\hat{\tau}_i$ and $\tilde{\tau}_i$. This implies that the best combined estimator is

$$\tau_i^* = \frac{\hat{\tau}_i \dfrac{k(a - 1)}{a(r - \lambda)} (\sigma^2 + k\sigma_\beta^2) + \tilde{\tau}_i \dfrac{k(a - 1)}{\lambda a^2} \sigma^2}{\dfrac{k(a - 1)}{\lambda a^2} \sigma^2 + \dfrac{k(a - 1)}{a(r - \lambda)} (\sigma^2 + k\sigma_\beta^2)} \qquad i = 1, 2, \ldots, a$$

which can be simplified to

$$\tau_i^* = \frac{kQ_i(\sigma^2 + k\sigma_\beta^2) + \left(\sum_{j=1}^{b} n_{ij} y_{.j} - kr\bar{y}_{..} \right) \sigma^2}{(r - \lambda)\sigma^2 + \lambda a(\sigma^2 + k\sigma_\beta^2)} \qquad i = 1, 2, \ldots, a \quad (5\text{--}45)$$

Unfortunately, Equation 5–45 cannot be used to estimate the τ_i because the variances σ^2 and σ_β^2 are unknown. The usual approach is to estimate σ^2 and σ_β^2 from the data and replace these parameters in Equation 5–45 by the estimates. The estimate usually taken for σ^2 is the error mean square from the intrablock analysis of variance, or the **intrablock error.** Thus,

$$\hat{\sigma}^2 = MS_E$$

The estimate of σ_β^2 is found from the mean square for blocks adjusted for treatments. In general, for a balanced incomplete block design, this mean square is

$$MS_{\text{Blocks (adjusted)}} = \left(\frac{k \sum_{i=1}^{a} Q_i^2}{\lambda a} + \sum_{j=1}^{b} \frac{y_{.j}^2}{k} - \sum_{i=1}^{a} \frac{y_{i.}^2}{r} \right) \bigg/ (b - 1) \qquad (5\text{-}46)$$

and its expected value [which is derived in Graybill (1961)] is

$$E[(MS_{\text{Blocks (adjusted)}}] = \sigma^2 + \frac{a(r - 1)}{b - 1} \sigma_\beta^2$$

Thus, if $MS_{\text{Blocks (adjusted)}} > MS_E$, the estimate of $\hat{\sigma}_\beta^2$ is

$$\hat{\sigma}_\beta^2 = \frac{[MS_{\text{Blocks (adjusted)}} - MS_E](b - 1)}{a(r - 1)} \qquad (5\text{-}47)$$

and if $MS_{\text{Blocks (adjusted)}} \leq MS_E$, we set $\hat{\sigma}_\beta^2 = 0$. This results in the combined estimator

$$\tau_i^* = \begin{cases} \dfrac{kQ_i(\hat{\sigma}^2 + k\hat{\sigma}_\beta^2) + \left(\sum_{j=1}^{b} n_{ij}y_{.j} - kr\bar{y}_{..} \right) \hat{\sigma}^2}{(r - \lambda)\hat{\sigma}^2 + \lambda a(\hat{\sigma}^2 + k\hat{\sigma}_\beta^2)}, & \hat{\sigma}_\beta^2 > 0 \quad (5\text{-}48a) \\[4mm] \dfrac{y_{i.} - (1/a)y_{..}}{r}, & \hat{\sigma}_\beta^2 = 0 \quad (5\text{-}48b) \end{cases}$$

We now compute the combined estimates for the data in Example 5-6. From Table 5-25 we obtain $\hat{\sigma}^2 = MS_E = 0.65$ and $MS_{\text{Blocks (adjusted)}} = 22.03$. (Note that in computing $MS_{\text{Blocks (adjusted)}}$ we make use of the fact that this is a symmetric design. In general, we must use Equation 5-46.) Since $MS_{\text{Blocks (adjusted)}} > MS_E$, we use Equation 5-47 to estimate σ_β^2 as

$$\hat{\sigma}_\beta^2 = \frac{(22.03 - 0.65)(3)}{4(3 - 1)} = 8.02$$

Therefore, we may substitute $\hat{\sigma}^2 = 0.65$ and $\hat{\sigma}_\beta^2 = 8.02$ into Equation 5-48a to obtain the combined estimates listed below. For convenience, the intrablock and interblock estimates are also given. In this example, the combined estimates are close to the intrablock estimates because the variance of the interblock estimates is relatively large.

Parameter	Intrablock Estimate	Interblock Estimate	Combined Estimate
τ_1	-1.12	10.50	-1.09
τ_2	-0.88	-3.50	-0.88
τ_3	-0.50	-0.50	-0.50
τ_4	2.50	-6.50	2.47

5-5   PROBLEMS

5-1   A chemist wishes to test the effect of four chemical agents on the strength of a particular type of cloth. Because there might be variability from one bolt to another, the chemist decides to use a randomized block design, with the bolts of cloth considered as blocks. She selects five bolts and applies all four chemicals in random order to each bolt. The resulting tensile strengths follow. Analyze the data from this experiment (use $\alpha = 0.05$) and draw appropriate conclusions.

	Bolt				
Chemical	1	2	3	4	5
1	73	68	74	71	67
2	73	67	75	72	70
3	75	68	78	73	68
4	73	71	75	75	69

5-2   Three different washing solutions are being compared to study their effectiveness in retarding bacteria growth in five-gallon milk containers. The analysis is done in a laboratory, and only three trials can be run on any day. Because days could represent a potential source of variability, the experimenter decides to use a randomized block design. Observations are taken for four days, and the data are shown here. Analyze the data from this experiment (use $\alpha = 0.05$) and draw conclusions.

	Days			
Solution	1	2	3	4
1	13	22	18	39
2	16	24	17	44
3	5	4	1	22

5-3   Plot the mean tensile strengths observed for each chemical type in Problem 5–1 and compare them to an appropriately scaled t distribution. What conclusions would you draw from this display?

5-4   Plot the average bacteria counts for each solution in Problem 5–2 and compare them to a scaled t distribution. What conclusions can you draw?

5-5   An article in the *Fire Safety Journal* ("The Effect of Nozzle Design on the Stability and Performance of Turbulent Water Jets," Vol. 4, August 1981) describes an experiment in which a shape factor was determined for several different nozzle designs at six levels of jet efflux velocity. Interest focused on potential differences between nozzle designs, with velocity considered as a nuisance variable. The data are shown below:

Nozzle Design	Jet Efflux Velocity (m/s)					
	11.73	14.37	16.59	20.43	23.46	28.74
1	0.78	0.80	0.81	0.75	0.77	0.78
2	0.85	0.85	0.92	0.86	0.81	0.83
3	0.93	0.92	0.95	0.89	0.89	0.83
4	1.14	0.97	0.98	0.88	0.86	0.83
5	0.97	0.86	0.78	0.76	0.76	0.75

(a) Does nozzle design affect the shape factor? Compare the nozzles with box plots and with an analysis of variance, using $\alpha = 0.05$.

(b) Analyze the residuals from this experiment.

(c) Which nozzle designs are different with respect to shape factor? Draw a graph of the average shape factor for each nozzle type and compare this to a scaled t distribution. Compare the conclusions that you draw from this plot to those from Duncan's multiple range test.

5-6 Consider the ratio control algorithm experiment described in Chapter 4 (p. 132). The experiment was actually conducted as a randomized block design, where six time periods were selected as the blocks, and all four ratio control algorithms were tested in each time period. The average cell voltage and the standard deviation of voltage (shown in parentheses) for each cell are as follows:

Ratio Control Algorithm	Time Period					
	1	2	3	4	5	6
1	4.93 (0.05)	4.86 (0.04)	4.75 (0.05)	4.95 (0.06)	4.79 (0.03)	4.88 (0.05)
2	4.85 (0.04)	4.91 (0.02)	4.79 (0.03)	4.85 (0.05)	4.75 (0.03)	4.85 (0.02)
3	4.83 (0.09)	4.88 (0.13)	4.90 (0.11)	4.75 (0.15)	4.82 (0.08)	4.90 (0.12)
4	4.89 (0.03)	4.77 (0.04)	4.94 (0.05)	4.86 (0.05)	4.79 (0.03)	4.76 (0.02)

(a) Analyze the average cell voltage data. (Use $\alpha = 0.05$.) Does the choice of ratio control algorithm affect the average cell voltage?

(b) Perform an appropriate analysis on the standard deviation of voltage. (Recall that this is called "pot noise.") Does the choice of ratio control algorithm affect the pot noise?

(c) Conduct any residual analyses that seem appropriate.

(d) Which ratio control algorithm would you select if your objective is to reduce both the average cell voltage *and* the pot noise?

5-7   An aluminum master alloy manufacturer produces grain refiners in ingot form. The company produces the product in four furnaces. Each furnace is known to have its own unique operating characteristics, so any experiment run in the foundry that involves more than one furnace will consider furnaces as a nuisance variable. The process engineers suspect that stirring rate impacts the grain size of the product. Each furnace can be run at four different stirring rates. A randomized block design is run for a particular refiner and the resulting grain size data is shown below.

	Furnace			
Stirring Rate (rpm)	1	2	3	4
5	8	4	5	6
10	14	5	6	9
15	14	6	9	2
20	17	9	3	6

(a) Is there any evidence that stirring rate impacts grain size?

(b) Graph the residuals from this experiment on a normal probability plot. Interpret this plot.

(c) Plot the residuals versus furnace and stirring rate. Does this plot convey any useful information?

(d) What should the process engineers recommend concerning the choice of stirring rate and furnace for this particular grain refiner if small grain size is desirable?

5-8   Analyze the data in Problem 5-2 using the general regression significance test.

5-9   Assuming that chemical types and bolts are fixed, estimate the model parameters τ_i and β_j in Problem 5-1.

5-10   Draw an operating characteristic curve for the design in Problem 5-2. Does the test seem to be sensitive to small differences in the treatment effects?

5-11   Suppose that the observation for chemical type 2 and bolt 3 is missing in Problem 5-1. Analyze the problem by estimating the missing value. Perform the exact analysis and compare the results.

5-12   *Two missing values in a randomized block.* Suppose that in Problem 5-1 the observations for chemical type 2 and bolt 3 and chemical type 4 and bolt 4 are missing.

(a) Analyze the design by iteratively estimating the missing values, as described in Section 5-1.3.

(b) Differentiate SS_E with respect to the two missing values, equate the results to zero, and solve for estimates of the missing values. Analyze the design using these two estimates of the missing values.

(c) Derive general formulas for estimating two missing values when the observations are in *different* blocks.

(d) Derive general formulas for estimating two missing values when the observations are in the *same* block.

5-13 An industrial engineer is conducting an experiment on eye focus time. He is interested in the effect of the distance of the object from the eye on the focus time. Four different distances are of interest. He has five subjects available for the experiment. Because there may be differences among individuals, he decides to conduct the experiment in a randomized block design. The data obtained follow. Analyze the data from this experiment (use $\alpha = 0.05$) and draw appropriate conclusions.

	Subject				
Distance (ft)	1	2	3	4	5
4	10	6	6	6	6
6	7	6	6	1	6
8	5	3	3	2	5
10	6	4	4	2	3

5-14 The effect of five different ingredients (A, B, C, D, E) on the reaction time of a chemical process is being studied. Each batch of new material is only large enough to permit five runs to be made. Furthermore, each run requires approximately $1\frac{1}{2}$ hours, so only five runs can be made in one day. The experimenter decides to run the experiment as a Latin square so that day and batch effects may be systematically controlled. She obtains the data that follow. Analyze the data from this experiment (use $\alpha = 0.05$) and draw conclusions.

	Day				
Batch	1	2	3	4	5
1	$A = 8$	$B = 7$	$D = 1$	$C = 7$	$E = 3$
2	$C = 11$	$E = 2$	$A = 7$	$D = 3$	$B = 8$
3	$B = 4$	$A = 9$	$C = 10$	$E = 1$	$D = 5$
4	$D = 6$	$C = 8$	$E = 6$	$B = 6$	$A = 10$
5	$E = 4$	$D = 2$	$B = 3$	$A = 8$	$C = 8$

5-15 An industrial engineer is investigating the effect of four assembly methods (A, B, C, D) on the assembly time for a color television component. Four operators are selected for the study. Furthermore, the engineer knows that each assembly method produces such fatigue that the time required for the last assembly may be greater than the time required for the first, regardless of the method. That is, a trend develops in the required assembly time. To account for this source of variability, the engineer uses the Latin square design shown below. Analyze the data from this experiment ($\alpha = 0.05$) and draw appropriate conclusions.

Order of	Operator			
Assembly	1	2	3	4
1	$C = 10$	$D = 14$	$A = 7$	$B = 8$
2	$B = 7$	$C = 18$	$D = 11$	$A = 8$
3	$A = 5$	$B = 10$	$C = 11$	$D = 9$
4	$D = 10$	$A = 10$	$B = 12$	$C = 14$

5–16   Suppose that in Problem 5–14 the observation from batch 3 on day 4 is missing. Estimate the missing value from Equation 5–24, and perform the analysis using the value.

5–17   Consider a $p \times p$ Latin square with rows (α_i), columns (β_k), and treatments (τ_j) fixed. Obtain least squares estimates of the model parameters α_i, β_k, and τ_j.

5–18   Derive the missing value formula (Equation 5–24) for the Latin square design.

5–19   *Designs involving several Latin squares.* [See Cochran and Cox (1957), John (1971).] The $p \times p$ Latin square contains only p observations for each treatment. To obtain more replications the experimenter may use several squares, say n. It is immaterial whether the squares used are the same or different. The appropriate model is

$$y_{ijkh} = \mu + \rho_h + \alpha_{i(h)} + \tau_j + \beta_{k(h)} + (\tau\rho)_{jh} + \epsilon_{ijkh} \qquad \begin{cases} i = 1, 2, \ldots, p \\ j = 1, 2, \ldots, p \\ k = 1, 2, \ldots, p \\ h = 1, 2, \ldots, n \end{cases}$$

where y_{ijkh} is the observation on treatment j in row i and column k of the hth square. Note that $\alpha_{i(h)}$ and $\beta_{k(h)}$ are the row and column effects in the hth square, ρ_h is the effect of the hth square, and $(\tau\rho)_{jh}$ is the interaction between treatments and squares.

(a) Set up the normal equations for this model, and solve for estimates of the model parameters. Assume that appropriate side conditions on the parameters are $\Sigma_h \hat{\rho}_h = 0$, $\Sigma_i \hat{\alpha}_{i(h)} = 0$, and $\Sigma_k \hat{\beta}_{k(h)} = 0$ for each h, $\Sigma_j \hat{\tau}_j = 0$, $\Sigma_j (\hat{\tau\rho})_{jh} = 0$ for each h, and $\Sigma_h (\hat{\tau\rho})_{jh} = 0$ for each j.

(b) Write down the analysis of variance table for this design.

5–20   Discuss how the operating characteristics curves in the Appendix may be used with the Latin square design.

5–21   Suppose that in Problem 5–14 the data taken on day 5 were incorrectly analyzed and had to be discarded. Develop an appropriate analysis for the remaining data.

5-22 The yield of a chemical process was measured using five batches of raw material, five acid concentrations, five standing times (A, B, C, D, E), and five catalyst concentrations (α, β, γ, δ, ϵ). The Graeco-Latin square that follows was used. Analyze the data from this experiment (use $\alpha = 0.05$) and draw conclusions.

Batch	\multicolumn{5}{c}{Acid Concentration}				

Batch	1	2	3	4	5
1	$A\alpha = 26$	$B\beta = 16$	$C\gamma = 19$	$D\delta = 16$	$E\epsilon = 13$
2	$B\gamma = 18$	$C\delta = 21$	$D\epsilon = 18$	$E\alpha = 11$	$A\beta = 21$
3	$C\epsilon = 20$	$D\alpha = 12$	$E\beta = 16$	$A\gamma = 25$	$B\delta = 13$
4	$D\beta = 15$	$E\gamma = 15$	$A\delta = 22$	$B\epsilon = 14$	$C\alpha = 17$
5	$E\delta = 10$	$A\epsilon = 24$	$B\alpha = 17$	$C\beta = 17$	$D\gamma = 14$

5-23 Suppose that in Problem 5–15 the engineer suspects that the workplaces used by the four operators may represent an additional source of variation. A fourth factor, workplace (α, β, γ, δ) may be introduced and another experiment conducted, yielding the Graeco-Latin square that follows. Analyze the data from this experiment (use $\alpha = 0.05$) and draw conclusions.

Order of Assembly	\multicolumn{4}{c}{Operator}			

Order of Assembly	1	2	3	4
1	$C\beta = 11$	$B\gamma = 10$	$D\delta = 14$	$A\alpha = 8$
2	$B\alpha = 8$	$C\delta = 12$	$A\gamma = 10$	$D\beta = 12$
3	$A\delta = 9$	$D\alpha = 11$	$B\beta = 7$	$C\gamma = 15$
4	$D\gamma = 9$	$A\beta = 8$	$C\alpha = 18$	$B\delta = 6$

5-24 Construct a 5×5 hypersquare for studying the effects of five factors. Exhibit the analysis of variance table for this design.

5-25 Consider the data in Problems 5–15 and 5–23. Suppressing the Greek letters in 5–23, analyze the data using the method developed in Problem 5–19.

5-26 Consider the randomized block design with one missing value in Table 5–7. Analyze this data by using the exact analysis of the missing value problem discussed in Section 5–1.4. Compare your results to the approximate analysis of these data given in Table 5–8.

5-27 An engineer is studying the mileage performance characteristics of five types of gasoline additives. In the road test he wishes to use cars as blocks; however, because of a time constraint, he must use an incomplete block design. He runs the balanced design with the five blocks that follow. Analyze the data from this experiment (use $\alpha = 0.05$) and draw conclusions.

			Car		
Additive	1	2	3	4	5
1		17	14	13	12
2	14	14		13	10
3	12		13	12	9
4	13	11	11	12	
5	11	12	10		8

5–28   Construct orthogonal contrasts for the data in Problem 5–27. Compute the sum of squares for each contrast.

5–29   Seven different hardwood concentrations are being studied to determine their effect on the strength of the paper produced. However the pilot plant can only produce three runs each day. As days may differ, the analyst uses the balanced incomplete block design that follows. Analyze the data from this experiment (use $\alpha = 0.05$) and draw conclusions.

Hardwood Concentration (%)			Days				
	1	2	3	4	5	6	7
2	114				120		117
4	126	120				119	
6		137	117				134
8	141		129	149			
10		145		150	143		
12			120		118	123	
14				136		130	127

5–30   Analyze the data in Example 5–6 using the general regression significance test.

5–31   Prove that $k\Sigma_{i=1}^{a}Q_i^2/(\lambda a)$ is the adjusted sum of squares for treatments in a balanced incomplete block design.

5–32   An experimenter wishes to compare four treatments in blocks of two runs. Find a balanced incomplete block design for this experiment with six blocks.

5–33   An experimenter wishes to compare eight treatments in blocks of four runs. Find a balanced incomplete block design with 14 blocks and $\lambda = 3$.

5–34   Perform the interblock analysis for the design in Problem 5–27.

5–35   Perform the interblock analysis for the design in Problem 5–29.

5–36   Verify that a balanced incomplete block design with the parameters $a = 8$, $r = 8$, $k = 4$, and $b = 16$ does not exist.

5-37 Show that the variance of the intrablock estimators $\{\hat{\tau}_i\}$ is $k(a-1)\sigma^2/(\lambda a^2)$.

5-38 *Extended incomplete block designs.* Occasionally the block size obeys the relationship $a < k < 2a$. An extended incomplete block design consists of a single replicate of each treatment in each block along with an incomplete block design with $k* = k - a$. In the balanced case, the incomplete block design will have parameters $k* = k - a$, $r* = r - b$, and $\lambda*$. Write out the statistical analysis. (*Hint:* In the extended incomplete block design, we have $\lambda = 2r - b + \lambda*$.)

Chapter 6
Introduction to Factorial Designs

6–1 BASIC DEFINITIONS AND PRINCIPLES

Many experiments involve the study of the effects of two or more factors. In general, **factorial designs** are most efficient for this type of experiment. By a factorial design, we mean that in each complete trial or replication of the experiment all possible combinations of the levels of the factors are investigated. For example, if there are a levels of factor A and b levels of factor B, then each replicate contains all ab treatment combinations. When factors are arranged in a factorial design, they are often said to be **crossed.**

The effect of a factor is defined to be the change in response produced by a change in the level of the factor. This is frequently called a **main effect** because it refers to the primary factors of interest in the experiment. For example, consider the simple experiment in Figure 6–1. This is a two-factor factorial experiment with both design factors at two levels. We have called these levels "low" and "high," and denoted them "$-$" and "$+$," respectively. The main effect of factor A in this two-level design can be thought of as the difference between the average response at the low level of A and the average response at the high level of A. Numerically, this is

$$A = \frac{40 + 52}{2} - \frac{20 + 30}{2} = 21$$

That is, increasing factor A from the low level to the high level causes an **average response increase** of 21 units. Similarly, the main effect of B is

$$B = \frac{30 + 52}{2} - \frac{20 + 40}{2} = 11$$

If the factors appear at more than two levels, the above procedure must be modified since there are other ways to define the effect of a factor. This point is discussed more completely later.

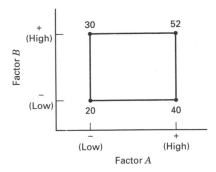

Figure 6–1.   A two-factor factorial experiment, with the response (y) shown at the corners.

In some experiments, we may find that the difference in response between the levels of one factor is not the same at all levels of the other factors. When this occurs, there is an **interaction** between the factors. For example, consider the two-factor factorial experiment shown in Figure 6–2. At the low level of factor B (or B^-), the A effect is

$$A = 50 - 20 = 30$$

and at the high level of factor B (or B^+), the A effect is

$$A = 12 - 40 = -28$$

Since the effect of A depends on the level chosen for factor B, we see that there is interaction between A and B. The magnitude of the interaction effect is the average *difference* in these two A effects, or $AB = (-28 - 30)/2 = -29$. Clearly, the interaction is large in this experiment.

These ideas may be illustrated graphically. Figure 6–3 plots the response data in Figure 6–1 against factor A for both levels of factor B. Note that the B^- and B^+ lines are approximately parallel, indicating a lack of interaction between factors A and B. Similarly, Figure 6–4 plots the response data in Figure 6–2. Here we see that the B^- and B^+ lines are not parallel. This indicates an interaction between factors A and B. Graphs such as these are frequently very useful in interpreting significant interactions and in reporting results to nonstatistically

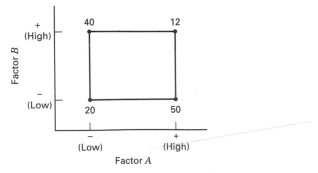

Figure 6–2.   A two-factor factorial experiment with interaction.

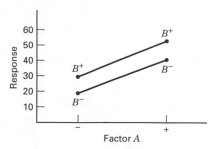

Figure 6–3.   A factorial experiment without interaction.

trained management. However, they should not be utilized as the sole technique of data analysis because their interpretation is subjective and their appearance is often misleading.

There is another way to illustrate the concept of interaction. Suppose that both of our design factors are **quantitative** (such as temperature, pressure, time, etc.). Then a **regression model representation** of the two-factor factorial experiment could be written as

$$y = \beta_0 + \beta_1 x_1 + \beta_2 x_2 + \beta_{12} x_1 x_2 + \epsilon$$

where y is the response, the β's are parameters whose values are to be determined, x_1 is a variable that represents factor A, x_2 is a variable that represents factor B, and ϵ is a random error term. The variables x_1 and x_2 are defined on a **coded scale** from -1 to $+1$ (the low and high levels of A and B), and $x_1 x_2$ represents the interaction between x_1 and x_2.

The parameter estimates in this regression model turn out to be related to the effect estimates. For the experiment shown in Figure 6–1 we found the main effects of A and B to be $A = 21$ and $B = 11$. The estimates of β_1 and β_2 are one-half the value of the corresponding main effect; therefore, $\hat{\beta}_1 = 21/2 = 10.5$ and $\hat{\beta}_2 = 11/2 = 5.5$. The interaction effect in Figure 6–1 is $AB = 1$, so the value of the interaction coefficient in the regression model is $\hat{\beta}_{12} = 1/2 = 0.5$. The parameter β_0 is estimated by the average of all four responses, or $\hat{\beta}_0 = (20 + 40 + 30 + 52)/4 = 35.5$. Therefore, the fitted regression model is

$$\hat{y} = 35.5 + 10.5 x_1 + 5.5 x_2 + 0.5 x_1 x_2$$

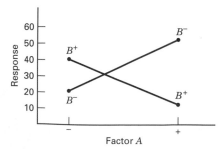

Figure 6–4.   A factorial experiment with interaction.

The parameter estimates obtained in the manner for the factorial design with all factors at two levels ($-$ and $+$) turn out to be *least squares* estimates (more on this later).

The interaction coefficient ($\hat{\beta}_{12} = 0.5$) is small relative to the main effect coefficients $\hat{\beta}_1$ and $\hat{\beta}_2$. We will take this to mean that interaction is small and can be ignored. Therefore, dropping the term $0.5x_1x_2$ gives us the model

$$\hat{y} = 35.5 + 10.5x_1 + 5.5x_2$$

Figure 6–5 presents graphical representations of this model. In Figure 6–5a we have a plot of the plane of y-values generated by the various combinations of x_1 and x_2. This three-dimensional graph is called a **response surface plot.** Figure 6–5b shows the contour lines of constant response y in the x_1, x_2 plane. Notice

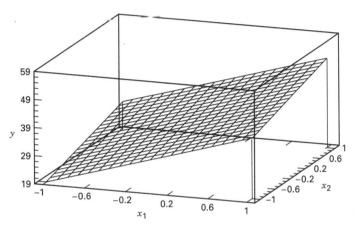

(a) The response surface

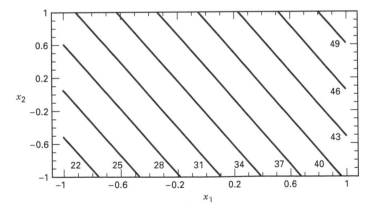

(b) The contour plot

Figure 6–5. Response surface and contour plot for the model $\hat{y} = 35.5 + 10.5x_1 + 5.5x_2$.

that since the response surface is a plane, the contour plot contains parallel straight lines.

Now suppose that the interaction contribution to this experiment was not negligible; that is, the coefficient β_{12} was not small. Figure 6–6 presents the response surface and contour plot for the model

$$\hat{y} = 35.5 + 10.5x_1 + 5.5x_2 + 8x_1x_2$$

(We have let the interaction effect be the average of the two main effects.) Notice that the significant interaction effect "twists" the plane in Figure 6–6a. This twisting of the response surface results in curved contour lines of constant

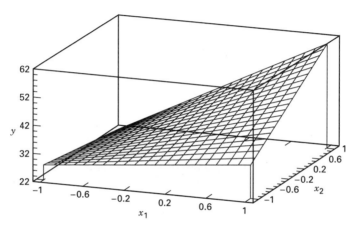

(a) The response surface

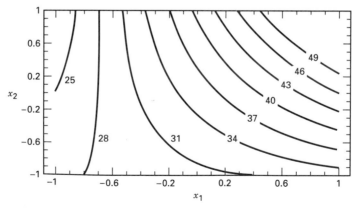

(b) The contour plot

Figure 6–6.   Response surface and contour plot for the model $\hat{y} = 35.5 + 10.5x_1 + 5.5x_2 + 8x_1x_2$.

response in the x_1, x_2 plane, as shown in Figure 6–6b. Thus, *interaction is a form of curvature* in the underlying response surface model for the experiment.

The response surface model for an experiment is extremely important and useful. We will say more about it in Section 6–5 and in subsequent chapters.

Generally, when an interaction is large, the corresponding main effects have little practical meaning. For the experiment in Figure 6–2, we would estimate the main effect of A to be

$$A = \frac{50 + 12}{2} - \frac{20 + 40}{2} = 1$$

which is very small, and we are tempted to conclude that there is no effect due to A. However, when we examine the effects of A at *different levels of factor B,* we see that this is not the case. Factor A has an effect, but it *depends on the level of factor B*. That is, knowledge of the AB interaction is more useful than knowledge of the main effect. A significant interaction will often *mask* the significance of main effects. These points are clearly indicated by the interaction plot in Figure 6–4. In the presence of significant interaction, the experimenter must usually examine the levels of one factor, say A, with levels of the other factors fixed to draw conclusions about the main effect of A.

6–2 THE ADVANTAGE OF FACTORIALS

The advantage of factorial designs can be easily illustrated. Suppose we have two factors A and B, each at two levels. We denote the levels of the factors by A^-, A^+, B^-, and B^+. Information on both factors could be obtained by varying the factors one at a time, as shown in Figure 6–7. The effect of changing factor A is given by $A^+B^- - A^-B^-$ and the effect of changing factor B is given by $A^-B^+ - A^-B^-$. Since experimental error is present, it is desirable to take two observations, say, at each treatment combination and estimate the effects of the factors using average responses. Thus, a total of six observations are required.

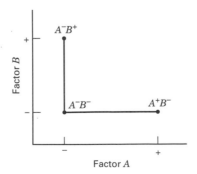

Figure 6–7. A one-factor-at-a-time experiment.

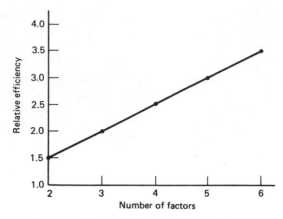

Figure 6–8.   Relative efficiency of a factorial design to a one-factor-at-a-time experiment (two factor levels).

If a factorial experiment had been performed, an additional treatment combination, A^+B^+, would have been taken. Now, using just *four* observations, two estimates of the A effect can be made: $A^+B^- - A^-B^-$ and $A^+B^+ - A^-B^+$. Similarly, two estimates of the B effect can be made. These two estimates of each main effect could be averaged to produce average main effects that are *just as precise* as those from the single-factor experiment, but only four total observations are required and we would say that the relative efficiency of the factorial design to the one-factor-at-a-time experiment is $(6/4) = 1.5$. Generally, this relative efficiency will increase as the number of factors increases, as shown in Figure 6–8.

Now suppose interaction is present. If the one-factor-at-a-time design indicated that A^-B^+ and A^+B^- gave better responses than A^-B^-, a logical conclusion would be that A^+B^+ would be even better. However, if interaction is present, this conclusion may be *seriously in error*. For an example, refer to the experiment in Figure 6–2.

In summary, note that factorial designs have several advantages. They are more efficient than one-factor-at-a-time experiments. Furthermore, a factorial design is necessary when interactions may be present to avoid misleading conclusions. Finally, factorial designs allow the effects of a factor to be estimated at several levels of the other factors, yielding conclusions that are valid over a range of experimental conditions.

6–3 THE TWO-FACTOR FACTORIAL DESIGN

6–3.1 An Example

The simplest types of factorial designs involve only two factors or sets of treatments. There are a levels of factor A and b levels of factor B, and these are arranged in a factorial design; that is, each replicate of the experiment contains all *ab* treatment combinations. In general, there are *n* replicates.

As an example of a factorial design involving two factors, an engineer is designing a battery for use in a device that will be subjected to some extreme variations in temperature. The only design parameter that he can select at this point is the plate material for the battery, and he has three possible choices. When the device is manufactured and is shipped to the field, the engineer has no control over the temperature extremes that the device will encounter, and he knows from experience that temperature will probably impact the effective battery life. However, temperature can be controlled in the product development laboratory for the purposes of a test.

The engineer decides to test all three plate materials at three temperature levels—15°F, 70°F, and 125°F—as these temperature levels are consistent with the product end-use environment. Four batteries are tested at each combination of plate material and temperature, and all 36 tests are run in random order. The experiment and the resulting observed battery life data are given in Table 6–1.

In this problem the engineer wants to answer the following questions:

1. What effects do material type and temperature have on the life of the battery?

2. Is there a choice of material that would give *uniformly long life regardless of temperature?*

This last question is particularly important. It may be possible to find a material alternative that is not greatly affected by temperature. If this is so, the engineer can make the battery **robust** to temperature variation in the field. This is an example of using statistical experimental design for **robust product design,** a very important engineering problem.

This design is a specific example of the general case of a two-factor factorial. To pass to the general case, let y_{ijk} be the observed response when factor A is at the ith level ($i = 1, 2, \ldots, a$) and factor B is at the jth level ($j = 1, 2, \ldots, b$) for the kth replicate ($k = 1, 2, \ldots, n$). In general, a two-factor factorial

Table 6–1   Life (in hours) Data for the Battery Design Example

Material Type	Temperature (°F)					
	15		70		125	
1	130	155	34	40	20	70
	74	180	80	75	82	58
2	150	188	136	122	25	70
	159	126	106	115	58	45
3	138	110	174	120	96	104
	168	160	150	139	82	60

Table 6-2   General Arrangement for a Two-Factor Factorial Design

		Factor B			
		1	2	$\cdots$	b
Factor A	1	$y_{111}, y_{112},$ $\ldots, y_{11n}$	$y_{121}, y_{122},$ $\ldots, y_{12n}$		$y_{1b1}, y_{1b2},$ $\ldots, y_{1bn}$
	2	$y_{211}, y_{212},$ $\ldots, y_{21n}$	$y_{221}, y_{222},$ $\ldots, y_{22n}$		$y_{2b1}, y_{2b2},$ $\ldots, y_{2bn}$
	$\vdots$				
	a	$y_{a11}, y_{a12},$ $\ldots, y_{a1n}$	$y_{a21}, y_{a22},$ $\ldots, y_{a2n}$		$y_{ab1}, y_{ab2},$ $\ldots, y_{abn}$

experiment will appear as in Table 6-2. The order in which the *abn* observations are taken is selected at random so that this design is a **completely randomized design.**

The observations may be described by the **linear statistical model**

$$y_{ijk} = \mu + \tau_i + \beta_j + (\tau\beta)_{ij} + \epsilon_{ijk} \qquad \begin{cases} i = 1, 2, \ldots, a \\ j = 1, 2, \ldots, b \\ k = 1, 2, \ldots, n \end{cases} \qquad (6\text{-}1)$$

where μ is the overall mean effect, τ_i is the effect of the *i*th level of the row factor A, β_j is the effect of the *j*th level of column factor B, $(\tau\beta)_{ij}$ is the effect of the interaction between τ_i and β_j, and ϵ_{ijk} is a random error component. Both factors are assumed to be **fixed,** and the treatment effects are defined as deviations from the overall mean, so $\sum_{i=1}^{a} \tau_i = 0$ and $\sum_{j=1}^{b} \beta_j = 0$. Similarly, the interaction effects are fixed and are defined such that $\sum_{i=1}^{a}(\tau\beta)_{ij} = \sum_{j=1}^{b}(\tau\beta)_{ij} = 0$. Since there are *n* replicates of the experiment, there are *abn* total observations.

In the two-factor factorial, both row and column factors (or treatments), A and B, are of equal interest. Specifically, we are interested in *testing hypotheses* about the equality of row treatment effects, say

$$H_0: \tau_1 = \tau_2 = \cdots = \tau_a = 0$$
$$H_1: \text{at least one } \tau_i \neq 0 \qquad (6\text{-}2a)$$

and the equality of column treatment effects, say

$$H_0: \beta_1 = \beta_2 = \cdots = \beta_b = 0$$
$$H_1: \text{at least one } \beta_j \neq 0 \qquad (6\text{-}2b)$$

We are also interested in determining whether row and column treatments *interact*. Thus, we also wish to test

$$H_0 : (\tau\beta)_{ij} = 0 \qquad \text{for all } i, j$$
$$H_1 : \text{at least one } (\tau\beta)_{ij} \neq 0 \qquad (6\text{–}2c)$$

We now discuss how these hypotheses are tested using a **two-factor analysis of variance.**

6–3.2 Statistical Analysis of the Fixed Effects Model

Let $y_{i..}$ denote the total of all observations under the ith level of factor A, $y_{.j.}$ denote the total of all observations under the jth level of factor B, $y_{ij.}$ denote the total of all observations in the ijth cell, and $y_{...}$ denote the grand total of all the observations. Define $\bar{y}_{i..}$, $\bar{y}_{.j.}$, $\bar{y}_{ij.}$, and $\bar{y}_{...}$ as the corresponding row, column, cell, and grand averages. Expressed mathematically,

$$
\begin{aligned}
y_{i..} = \sum_{j=1}^{b} \sum_{k=1}^{n} y_{ijk} \qquad & \bar{y}_{i..} = \frac{y_{i..}}{bn} \qquad i = 1, 2, \ldots, a \\[2mm]
y_{.j.} = \sum_{i=1}^{a} \sum_{k=1}^{n} y_{ijk} \qquad & \bar{y}_{.j.} = \frac{y_{.j.}}{an} \qquad j = 1, 2, \ldots, b \\[2mm]
y_{ij.} = \sum_{k=1}^{n} y_{ijk} \qquad & \bar{y}_{ij.} = \frac{y_{ij.}}{n} \qquad \begin{array}{l} i = 1, 2, \ldots, a \\ j = 1, 2, \ldots, b \end{array} \\[2mm]
y_{...} = \sum_{i=1}^{a} \sum_{j=1}^{b} \sum_{k=1}^{n} y_{ijk} \qquad & \bar{y}_{...} = \frac{y_{...}}{abn}
\end{aligned}
\qquad (6\text{–}3)
$$

The **total corrected sum of squares** may be written as

$$
\begin{aligned}
\sum_{i=1}^{a} \sum_{j=1}^{b} \sum_{k=1}^{n} (y_{ijk} - \bar{y}_{...})^2 &= \sum_{i=1}^{a} \sum_{j=1}^{b} \sum_{k=1}^{n} [(\bar{y}_{i..} - \bar{y}_{...}) + (\bar{y}_{.j.} - \bar{y}_{...}) \\
&\qquad + (\bar{y}_{ij.} - \bar{y}_{i..} - \bar{y}_{.j.} + \bar{y}_{...}) + (y_{ijk} - \bar{y}_{ij.})]^2 \\
&= bn \sum_{i=1}^{a} (\bar{y}_{i..} - \bar{y}_{...})^2 + an \sum_{j=1}^{b} (\bar{y}_{.j.} - \bar{y}_{...})^2 \qquad (6\text{–}4) \\
&\quad + n \sum_{i=1}^{a} \sum_{j=1}^{b} (\bar{y}_{ij.} - \bar{y}_{i..} - \bar{y}_{.j.} + \bar{y}_{...})^2 \\
&\quad + \sum_{i=1}^{a} \sum_{j-1}^{b} \sum_{k-1}^{n} (y_{ijk} - \bar{y}_{ij.})^2
\end{aligned}
$$

since the six cross-products on the right-hand side are zero. Notice that the total sum of squares has been partitioned into a sum of squares due to "rows" or factor A, (SS_A); a sum of squares due to "columns" or factor B, (SS_B); a sum of squares due to the interaction between A and B, (SS_{AB}); and a sum of squares due to error, (SS_E). From the last component on the right-hand side of Equation 6–4, we see that there must be at least two replicates $(n \geqslant 2)$ to obtain an error sum of squares.

We may write Equation 6–4 symbolically as

$$SS_T = SS_A + SS_B + SS_{AB} + SS_E \tag{6-5}$$

The number of degrees of freedom associated with each sum of squares is

Effect	Degrees of Freedom
A	$a - 1$
B	$b - 1$
AB interaction	$(a - 1)(b - 1)$
Error	$ab(n - 1)$
Total	$abn - 1$

We may justify this allocation of the $abn - 1$ total degrees of freedom to the sums of squares as follows: The main effects A and B have a and b levels, respectively; therefore they have $a - 1$ and $b - 1$ degrees of freedom as shown. The interaction degrees of freedom are simply the number of degrees of freedom for cells (which is $ab - 1$) minus the number of degrees of freedom for the two main effects A and B; that is, $ab - 1 - (a - 1) - (b - 1) = (a - 1)(b - 1)$. Within each of the ab cells, there are $n - 1$ degrees of freedom between the n replicates; thus there are $ab(n - 1)$ degrees of freedom for error. Note that the number of degrees of freedom on the right-hand side of Equation 6–5 adds to the total number of degrees of freedom.

Each sum of squares divided by its degrees of freedom is a **mean square.** The expected values of the mean squares are

$$E(MS_A) = E\left(\frac{SS_A}{a - 1}\right) = \sigma^2 + \frac{bn \sum_{i=1}^{a} \tau_i^2}{a - 1}$$

$$E(MS_B) = E\left(\frac{SS_B}{b - 1}\right) = \sigma^2 + \frac{an \sum_{j=1}^{b} \beta_j^2}{b - 1}$$

$$E(MS_{AB}) = E\left(\frac{SS_{AB}}{(a - 1)(b - 1)}\right) = \sigma^2 + \frac{n \sum_{i=1}^{a} \sum_{j=1}^{b} (\tau\beta)_{ij}^2}{(a - 1)(b - 1)}$$

and

$$E(MS_E) = E\left(\frac{SS_E}{ab(n - 1)}\right) = \sigma^2$$

Table 6-3 The Analysis of Variance Table for the Two-Factor Factorial, Fixed Effects Model

Source of Variation	Sum of Squares	Degrees of Freedom	Mean Square	F_0
A treatments	SS_A	$a - 1$	$MS_A = \dfrac{SS_A}{a - 1}$	$F_0 = \dfrac{MS_A}{MS_E}$
B treatments	SS_B	$b - 1$	$MS_B = \dfrac{SS_B}{b - 1}$	$F_0 = \dfrac{MS_B}{MS_E}$
Interaction	SS_{AB}	$(a - 1)(b - 1)$	$MS_{AB} = \dfrac{SS_{AB}}{(a - 1)(b - 1)}$	$F_0 = \dfrac{MS_{AB}}{MS_E}$
Error	SS_E	$ab(n - 1)$	$MS_E = \dfrac{SS_E}{ab(n - 1)}$	
Total	SS_T	$abn - 1$		

Notice that if the null hypotheses of no row treatment effects, no column treatment effects, and no interaction are true, then MS_A, MS_B, MS_{AB}, and MS_E all estimate σ^2. However, if there are differences between row treatment effects, say, then MS_A will be larger than MS_E. Similarly, if there are column treatment effects or interaction present, then the corresponding mean squares will be larger than MS_E. Therefore, to test the significance of both main effects and their interaction, simply divide the corresponding mean square by the error mean square. Large values of this ratio imply that the data do not support the null hypothesis.

If we assume that the model (Equation 6-1) is adequate and that the error terms ϵ_{ijk} are normally and independently distributed with constant variance σ^2, then each of the ratios of mean squares MS_A/MS_E, MS_B/MS_E, and MS_{AB}/MS_E are distributed as F with $a - 1$, $b - 1$, and $(a - 1)(b - 1)$ numerator degrees of freedom, respectively, and $ab(n - 1)$ denominator degrees of freedom,[1] and the critical region would be the upper tail of the F distribution. The test procedure is usually summarized in an **analysis of variance table,** as shown in Table 6-3.

Computationally, we usually employ a statistical software package to conduct an analysis of variance. However, manual computing formulas for the sums of squares in Equation 6-5 may be obtained easily. The total sum of squares is computed as usual by

$$SS_T = \sum_{i=1}^{a} \sum_{j=1}^{b} \sum_{k=1}^{n} y_{ijk}^2 - \frac{y_{...}^2}{abn} \tag{6-6}$$

The sums of squares for the main effects are

$$SS_A = \frac{1}{bn} \sum_{i=1}^{a} y_{i..}^2 - \frac{y_{...}^2}{abn} \tag{6-7}$$

---

[1] The F test may also be viewed as an approximation to a randomization test, as noted previously.

and

$$SS_B = \frac{1}{an} \sum_{j=1}^{b} y_{.j.}^2 - \frac{y_{...}^2}{abn} \tag{6-8}$$

It is convenient to obtain the SS_{AB} in two stages. First we compute the sum of squares between the ab cell totals, which is called the sum of squares due to "subtotals":

$$SS_{\text{Subtotals}} = \frac{1}{n} \sum_{i=1}^{a} \sum_{j=1}^{b} y_{ij.}^2 - \frac{y_{...}^2}{abn}$$

This sum of squares also contains SS_A and SS_B. Therefore, the second step is to compute SS_{AB} as

$$SS_{AB} = SS_{\text{Subtotals}} - SS_A - SS_B \tag{6-9}$$

We may compute SS_E by subtraction as

$$SS_E = SS_T - SS_{AB} - SS_A - SS_B \tag{6-10}$$

or

$$SS_E = SS_T - SS_{\text{Subtotals}}$$

Example 6-1

The Battery Design Experiment

Table 6–4 presents the effective life (in hours) observed in the battery design example described in Section 6–3.1. The row and column totals are shown in the margins of the table and the circled numbers are the cell totals.

The sums of squares are computed as follows:

$$SS_T = \sum_{i=1}^{a} \sum_{j=1}^{b} \sum_{k=1}^{n} y_{ijk}^2 - \frac{y_{...}^2}{abn}$$

$$= (130)^2 + (155)^2 + (74)^2 + \cdots + (60)^2 - \frac{(3799)^2}{36} = 77,646.97$$

Table 6-4   Life Data (in hours) for the Battery Design Experiment

Material Type	Temperature (°F)						$y_{i..}$
	15		70		125		
1	130 74	155 180 (539)	34 80	40 75 (229)	20 82	70 58 (230)	998
2	150 159	188 126 (623)	136 106	122 115 (479)	25 58	70 45 (198)	1300
3	138 168	110 160 (576)	174 150	120 139 (583)	96 82	104 60 (342)	1501
$y_{.j.}$	1738		1291		770		$3799 = y_{...}$

$$SS_{\text{Material}} = \frac{1}{bn} \sum_{i=1}^{a} y_{i..}^2 - \frac{y_{...}^2}{abn}$$

$$= \frac{1}{(3)(4)} [(998)^2 + (1300)^2 + (1501)^2] - \frac{(3799)^2}{36} = 10{,}683.72$$

$$SS_{\text{Temperature}} = \frac{1}{an} \sum_{j=1}^{b} y_{.j.}^2 - \frac{y_{...}^2}{abn}$$

$$= \frac{1}{(3)(4)} [(1738)^2 + (1291)^2 + (770)^2] - \frac{(3799)^2}{36} = 39{,}118.72$$

$$SS_{\text{Interaction}} = \frac{1}{n} \sum_{i=1}^{a} \sum_{j=1}^{b} y_{ij.}^2 - \frac{y_{...}^2}{abn} - SS_{\text{Material}} - SS_{\text{Temperature}}$$

$$= \frac{1}{4} [(539)^2 + (229)^2 + \cdots + (342)^2] - \frac{(3799)^2}{36} - 10{,}683.72$$

$$- 39{,}118.72 = 9613.78$$

and

$$SS_E = SS_T - SS_{\text{Material}} - SS_{\text{Temperature}} - SS_{\text{Interaction}}$$
$$= 77{,}646.97 - 10{,}683.72 - 39{,}118.72 - 9613.78 = 18{,}230.75$$

The analysis of variance is shown in Table 6–5. Since $F_{0.05,4,27} = 2.73$, we conclude that there is a significant interaction between material types and temperature. Furthermore, $F_{0.05,2,27} = 3.35$, so the main effects of material type and temperature are also significant. Table 6–5 also shows the P-values for the test statistics.

To assist in interpreting the results of this experiment, it is helpful to construct a graph of the average responses at each treatment combination. This graph is shown in Figure 6–9 on page 242. The significant interaction is indicated by the lack of parallelism of the lines. In general, longer life is attained at low temperature, regardless of material type. Changing from low to intermediate temperature, battery life with material type 3 actually *increases,* whereas it decreases for types 1 and 2. From intermediate to high temperature, battery life decreases for material types 2 and 3 and is essentially unchanged for type 1. Material type 3 seems to give the best results if we want less loss of effective life as the temperature changes.

■

Table 6–5   Analysis of Variance for Battery Life Data

Source of Variation	Sum of Squares	Degrees of Freedom	Mean Square	F_0	P-Value
Material types	10,683.72	2	5,341.86	7.91	0.0020
Temperature	39,118.72	2	19,558.36	28.97	<0.0001
Interaction	9,613.78	4	2,403.44	3.56	0.0186
Error	18,230.75	27	675.21		
Total	77,646.97	35			

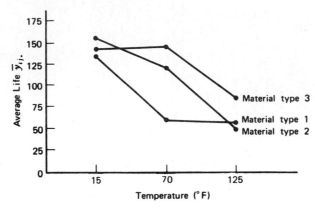

Figure 6–9. Material type–temperature plot for Example 6–1.

Multiple Comparisons ▪ When the analysis of variance indicates that row or column means differ, it is usually of interest to make comparisons between the individual row or column means to discover the specific differences. The multiple comparison methods discussed in Chapter 3 are useful in this regard.

We now illustrate the use of Duncan's multiple range test on the battery life data in Example 6–1. Note that in this experiment, interaction is significant. When interaction is significant, comparisons between the means of one factor (e.g., A) may be obscured by the AB interaction. One approach to this situation is to fix factor B at a specific level and apply Duncan's multiple range test to the means of factor A at that level. To illustrate, suppose that in Example 6–1 we are interested in detecting differences among the means of the three material types. Since interaction is significant, we make this comparison at just one level of temperature, say level 2 (70 degrees). We assume that the best estimate of the error variance is the MS_E from the analysis of variance table, utilizing the assumption that the experimental error variance is the same over all treatment combinations.

The three material type means arranged in ascending order are

$$\bar{y}_{12.} = 57.25 \quad \text{(material type 1)}$$
$$\bar{y}_{22.} = 119.75 \quad \text{(material type 2)}$$
$$\bar{y}_{32.} = 145.75 \quad \text{(material type 3)}$$

The standard error of these treatment means is

$$S_{\bar{y}_{i2.}} = \sqrt{\frac{MS_E}{n}} = \sqrt{\frac{675.21}{4}} = 12.99$$

since each mean contains $n = 4$ observations. From Appendix Table VII we obtain the values $r_{0.05}(2, 27) \simeq 2.91$ and $r_{0.05}(3, 27) \simeq 3.06$. The least significant ranges are

$$R_2 = r_{0.05}(2, 27)S_{\bar{y}_{i2.}} = (2.91)(12.99) = 37.80$$

$$R_3 = r_{0.05}(3, 27)S_{\bar{y}_{i2.}} = (3.06)(12.99) = 39.75$$

and the comparisons yield

$$3 \text{ vs. } 1: \quad 145.75 - 57.25 = 88.50 > 39.75(R_3)$$
$$3 \text{ vs. } 2: \quad 145.75 - 119.75 = 26.00 < 37.80(R_2)$$
$$2 \text{ vs. } 1: \quad 119.75 - 57.25 = 62.50 > 37.80(R_2)$$

This analysis indicates that at the temperature level 70 degrees, the mean battery life is the same for material types 2 and 3, and that the mean battery life for material type 1 is significantly lower in comparison to both types 2 and 3.

If interaction is significant, the experimenter could compare *all ab* cell means to determine which ones differ significantly. In this analysis, differences between cell means include interaction effects as well as both main effects. In Example 6–1, this would give 36 comparisons between all possible pairs of the nine cell means.

Computer Output ▪ Figure 6–10 (page 244) presents computer output from the SAS General Linear Models Procedure for the battery life data in Example 6–1. Note that

$$SS_{\text{Model}} = SS_{\text{Material}} + SS_{\text{Temperature}} + SS_{\text{Interaction}}$$
$$= 10{,}683.72 + 39{,}118{,}72 + 9613.78$$
$$= 59{,}416.22$$

and that

$$R^2 = \frac{SS_{\text{Model}}}{SS_{\text{Total}}} = \frac{59{,}416.22}{77{,}646.97} = 0.765210$$

That is, about 77 percent of the variability in the battery life is explained by the plate material in the battery, the temperature, and the material type—temperature interaction. The program also computes the material type, temperature, and material type—temperature interaction sums of squares. (Recall that both type I and type III sums of squares are always identical for balanced data.) The residuals from the fitted model are also displayed on the computer output. We now show how to use these residuals in model adequacy checking.

6-3.3 Model Adequacy Checking

Before the conclusions from the analysis of variance are adopted, the adequacy of the underlying model should be checked. As before, the primary diagnostic

GENERAL LINEAR MODELS PROCEDURE

DEPENDENT VARIABLE: Y

SOURCE	DF	SUM OF SQUARES	MEAN SQUARE	F VALUE	PR > F	R-SQUARE	C.V.
MODEL	8	59416.22222222	7427.02777778	11.00	0.0001	0.765210	24.6237
ERROR	27	18230.75000000	675.21296296			STD DEV	Y MEAN
CORRECTED TOTAL	35	77646.97222222				25.9486026	105.52777778

SOURCE	DF	TYPE I SS	F VALUE	PR > F		DF	TYPE III SS	F VALUE	PR > F
MATERIAL	2	10683.72222222	7.91	0.0020		2	10683.72222222	7.91	0.0020
TEMP	2	39118.72222222	28.97	0.0001		2	39118.72222222	28.97	0.0001
MATERIAL*TEMP	4	9613.77777778	3.56	0.0186		4	9613.77777778	3.56	0.0186

OBSERVATION	OBSERVED VALUE	PREDICTED VALUE	RESIDUAL
1	130.00000000	134.75000000	-4.75000000
2	155.00000000	134.75000000	20.25000000
3	74.00000000	134.75000000	-60.75000000
4	180.00000000	134.75000000	45.25000000
5	34.00000000	57.25000000	-23.25000000
6	40.00000000	57.25000000	-17.25000000
7	80.00000000	57.25000000	22.75000000
8	75.00000000	57.25000000	17.75000000
9	20.00000000	57.50000000	-37.50000000
10	70.00000000	57.50000000	12.50000000
11	82.00000000	57.50000000	24.50000000
12	58.00000000	57.50000000	0.50000000
13	150.00000000	155.75000000	-5.75000000
14	188.00000000	155.75000000	32.25000000
15	159.00000000	155.75000000	3.25000000
16	126.00000000	155.75000000	-29.75000000
17	136.00000000	119.75000000	16.25000000
18	122.00000000	119.75000000	2.25000000
19	106.00000000	119.75000000	-13.75000000
20	115.00000000	119.75000000	-4.75000000
21	25.00000000	49.50000000	-24.50000000
22	70.00000000	49.50000000	20.50000000
23	58.00000000	49.50000000	8.50000000
24	45.00000000	49.50000000	-4.50000000
25	138.00000000	144.00000000	-6.00000000
26	110.00000000	144.00000000	-34.00000000
27	168.00000000	144.00000000	24.00000000
28	160.00000000	144.00000000	16.00000000
29	174.00000000	145.75000000	28.25000000
30	120.00000000	145.75000000	-25.75000000
31	150.00000000	145.75000000	4.25000000
32	139.00000000	145.75000000	-6.75000000
33	96.00000000	85.50000000	10.50000000
34	104.00000000	85.50000000	18.50000000
35	82.00000000	85.50000000	-3.50000000
36	60.00000000	85.50000000	-25.50000000

SUM OF RESIDUALS	0.00000000
SUM OF SQUARED RESIDUALS	18230.75000000
SUM OF SQUARED RESIDUALS - ERROR SS	-0.00000000
FIRST ORDER AUTOCORRELATION	-0.37519370
DURBIN-WATSON D	2.71348203

Figure 6–10. Sample computer output for Example 6–1.

Table 6–6   Residuals for Example 6–1

Material Type	Temperature (°F)					
	15		70		125	
1	−4.75	20.25	−23.25	−17.25	−37.50	12.50
	−60.75	45.25	22.75	17.75	24.50	0.50
2	−5.75	32.25	16.25	2.25	−24.50	20.50
	3.25	−29.75	−13.75	−4.75	8.50	−4.50
3	−6.00	−34.00	28.25	−25.75	10.50	18.50
	24.00	16.00	4.25	−6.75	−3.50	−25.50

tool is residual analysis. The residuals for the two-factor factorial model are

$$e_{ijk} = y_{ijk} - \hat{y}_{ijk} \tag{6–11}$$

and since the fitted value $\hat{y}_{ijk} = \bar{y}_{ij.}$ (the average of the observations in the ijth cell), Equation 6–11 becomes

$$e_{ijk} = y_{ijk} - \bar{y}_{ij.} \tag{6–12}$$

The residuals from the battery life data in Example 6–1 are shown in Table 6–6. The normal probability plot and dot diagram of these residuals (Figure 6–11 page 246) do not reveal anything particularly troublesome, although the largest negative residual (−60.75 at 15°F for material type 1) does stand out somewhat from the others. The standardized value of this residual is $-60.75/\sqrt{675.21} = -2.34$, and this is the only residual whose absolute value is larger than 2.

Figure 6–12 (page 247) plots the residuals versus the fitted values $\hat{y}_{ijk}$. This plot indicates a mild tendency for the variance of the residuals to increase as the battery life increases. Figures 6–13 (page 247) and 6–14 (page 248) plot the residuals versus material types and temperature, respectively. Both plots indicate mild inequality of variance, with the treatment combination of 15°F and material type 1 possibly having larger variance than the others.

The 70°F–material type 1 cell contains both extreme residuals (−60.75 and 45.25). These two residuals are primarily responsible for the inequality of variance detected in Figures 6–12, 6–13, and 6–14. Reexamination of the data does not reveal any obvious problem, such as an error in recording, so we accept these responses as legitimate. It is possible that this particular treatment combination produces slightly more erratic battery life than the others. The problem, however, is not severe enough to have a dramatic impact on the analysis and conclusions.

6–3.4   Estimating the Model Parameters

The parameters in the two-factor analysis of variance model

$$y_{ijk} = \mu + \tau_i + \beta_j + (\tau\beta)_{ij} + \epsilon_{ijk} \tag{6–13}$$

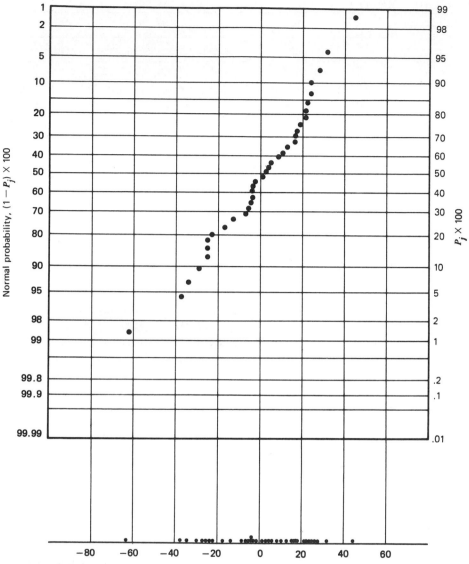

Figure 6–11. Normal probability plot and dot diagram of residuals for Example 6–1.

may be estimated by least squares. Since the model has $1 + a + b + ab$ parameters to be estimated, there are $1 + a + b + ab$ normal equations. Using the method of Section 4–4, it is not difficult to show that the normal equations are

$$\mu: \quad abn\hat{\mu} + bn\sum_{i=1}^{a}\hat{\tau}_i + an\sum_{j=1}^{b}\hat{\beta}_j + n\sum_{i=1}^{a}\sum_{j=1}^{b}\widehat{(\tau\beta)}_{ij} = y_{...} \qquad (6\text{–}14a)$$

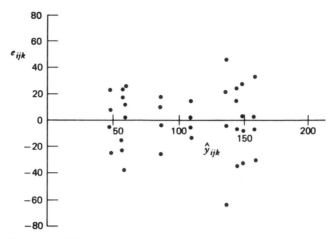

Figure 6–12.   Plot of residuals versus $\hat{y}_{ijk}$ for Example 6–1.

$$\tau_i: \quad bn\hat{\mu} + bn\hat{\tau}_i + n\sum_{j=1}^{b}\hat{\beta}_j + n\sum_{j=1}^{b}\widehat{(\tau\beta)}_{ij} = y_{i..} \qquad i = 1, 2, \ldots, a \quad \text{(6–14b)}$$

$$\beta_j: \quad an\hat{\mu} + n\sum_{i=1}^{a}\hat{\tau}_i + an\hat{\beta}_j + n\sum_{i=1}^{a}\widehat{(\tau\beta)}_{ij} = y_{.j.} \qquad j = 1, 2, \ldots, b \quad \text{(6–14c)}$$

$$(\tau\beta)_{ij}: \quad n\hat{\mu} + n\hat{\tau}_i + n\hat{\beta}_j + n\widehat{(\tau\beta)}_{ij} = y_{ij.} \qquad \begin{cases} i = 1, 2, \ldots, a \\ j = 1, 2, \ldots, b \end{cases} \quad \text{(6–14d)}$$

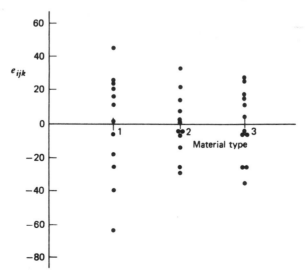

Figure 6–13.   Plot of residuals versus material type for Example 6–1.

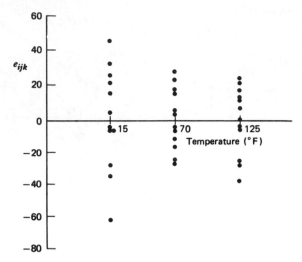

Figure 6–14. Plot of residuals versus temperature for Example 6–1.

For convenience, we have shown the parameter corresponding to each normal equation on the left in Equations 6–14.

As we attempt to solve the normal equations, we notice that the a equations in Equation 6–14b sum to Equation 6–14a and that the b equations of Equation 6–14c sum to Equation 6–14a. Also summing Equation 6–14d over j for a particular i will give Equation 6–14b, and summing Equation 6–14d over i for a particular j will give Equation 6–14c. Therefore, there are $a + b + 1$ **linear dependencies** in this system of equations and no unique solution will exist. In order to obtain a solution, we impose the constraints

$$\sum_{i=1}^{a} \hat{\tau}_i = 0 \tag{6–15a}$$

$$\sum_{j=1}^{b} \hat{\beta}_j = 0 \tag{6–15b}$$

$$\sum_{i=1}^{a} \widehat{(\tau\beta)}_{ij} = 0 \qquad j = 1, 2, \ldots, b \tag{6–15c}$$

and

$$\sum_{j=1}^{b} \widehat{(\tau\beta)}_{ij} = 0 \qquad i = 1, 2, \ldots, a \tag{6–15d}$$

Equations 6–15a and 6–15b constitute two constraints, whereas Equations 6–15c and 6–15d form $a + b - 1$ independent constraints. Therefore, we have $a + b + 1$ total constraints, the number needed.

Applying these constraints, the normal equations (Equations 6–14) simplify considerably, and we obtain the solution

$$\hat{\mu} = \bar{y}_{...}$$
$$\hat{\tau}_i = \bar{y}_{i..} - \bar{y}_{...} \qquad i = 1, 2, \ldots, a$$
$$\hat{\beta}_j = \bar{y}_{.j.} - \bar{y}_{...} \qquad j = 1, 2, \ldots, b \qquad\qquad (6\text{--}16)$$
$$\widehat{(\tau\beta)}_{ij} = \bar{y}_{ij.} - \bar{y}_{i..} - \bar{y}_{.j.} + \bar{y}_{...} \qquad \begin{cases} i = 1, 2, \ldots, a \\ j = 1, 2, \ldots, b \end{cases}$$

Notice the considerable intuitive appeal of this solution to the normal equations. Row treatment effects are estimated by the row average minus the grand average; column treatments are estimated by the column average minus the grand average; and the ijth interaction is estimated by the ijth cell average minus the grand average, the ith row effect, and the jth column effect.

Using Equation 6–16, we may find the **fitted value** y_{ijk} as

$$\hat{y}_{ijk} = \hat{\mu} + \hat{\tau}_i + \hat{\beta}_j + \widehat{(\tau\beta)}_{ij}$$
$$= \bar{y}_{...} + (\bar{y}_{i..} - \bar{y}_{...}) + (\bar{y}_{.j.} - \bar{y}_{...})$$
$$\qquad (\bar{y}_{ij.} - \bar{y}_{i..} - \bar{y}_{.j.} + \bar{y}_{...})$$
$$= \bar{y}_{ij.}$$

That is, the kth observation in the ijth cell is estimated by the average of the n observations in that cell. This result was used in Equation 6–12 to obtain the residuals for the two-factor factorial model.

Because constraints (Equations 6–15) have been used to solve the normal equations, the model parameters are not uniquely estimated. However, certain important **functions** of the model parameters *are* estimable, that is, uniquely estimated regardless of the constraint chosen. An example is $\tau_i - \tau_u + (\tau\beta)_{i.} - (\tau\beta)_{u.}$, which might be thought of as the "true" difference between the ith and uth levels of factor A. Notice that the true difference between the levels of any main effect includes an "average" interaction effect. It is this result that disturbs the tests on main effects in the presence of interaction, as noted earlier. In general, any function of the model parameters that is a linear combination of the left-hand side of the normal equations is **estimable.** This property was also noted in Chapter 4 when we were discussing the single-factor model.

6-3.5 Choice of Sample Size

The operating characteristic curves in Appendix Chart V can be used to assist the experimenter in determining an appropriate sample size (number of replicates, n) for a two-factor factorial design. The appropriate value of the parameter Φ^2 and the numerator and denominator degrees of freedom are shown in Table 6–7.

A very effective way to use these curves is to find the smallest value of Φ^2 corresponding to a specified difference between any two treatment means. For

Table 6–7   Operating Characteristic Curve Parameters for Chart V of the Appendix for the Two-Factor Factorial, Fixed Effects Model

Factor	Φ^2	Numerator Degrees of Freedom	Denominator Degrees of Freedom
A	$\dfrac{bn \sum_{i=1}^{a} \tau_i^2}{a\sigma^2}$	$a - 1$	$ab(n - 1)$
B	$\dfrac{an \sum_{j=1}^{b} \beta_j^2}{b\sigma^2}$	$b - 1$	$ab(n - 1)$
AB	$\dfrac{n \sum_{i=1}^{a} \sum_{j=1}^{b} (\tau\beta)_{ij}^2}{\sigma^2[(a - 1)(b - 1) + 1]}$	$(a - 1)(b - 1)$	$ab(n - 1)$

example, if the difference in any two row means is D, then the minimum value of Φ^2 is

$$\Phi^2 = \frac{nbD^2}{2a\sigma^2} \tag{6–17}$$

whereas if the difference in any two column means is D, then the minimum value of Φ^2 is

$$\Phi^2 = \frac{naD^2}{2b\sigma^2} \tag{6–18}$$

Finally, the minimum value of Φ^2 corresponding to a difference of D between any two interaction effects is

$$\Phi^2 = \frac{nD^2}{2\sigma^2[(a - 1)(b - 1) + 1]} \tag{6–19}$$

To illustrate the use of these equations, consider the battery life data in Example 6–1. Suppose that before running the experiment we decide that the null hypothesis should be rejected with a high probability if the difference in mean battery life between any two temperatures is as great as 40 hours. Thus $D = 40$, and if we assume that the standard deviation of battery life is approximately 25, then Equation 6–18 gives

$$\Phi^2 = \frac{naD^2}{2b\sigma^2}$$

$$= \frac{n(3)(40)^2}{2(3)(25)^2}$$

$$= 1.28n$$

as the minimum value of Φ^2. Assuming that $\alpha = 0.05$, we can now use Appendix Table V to construct the following display:

n	Φ^2	Φ	ν_1 = Numerator Degrees of Freedom	ν_2 = Error Degrees of Freedom	β
2	2.56	1.60	2	9	0.45
3	3.84	1.96	2	18	0.18
4	5.12	2.26	2	27	0.06

Note that $n = 4$ replicates give a β risk of about 0.06 or approximately a 94 percent chance of rejecting the null hypothesis if the difference in mean battery life at any two temperature levels is as large as 40 hours. Thus, we conclude that four replicates are enough to provide the desired sensitivity as long as our estimate of the standard deviation of battery life is not seriously in error. If in doubt, the experimenter could repeat the above procedure with other values of σ to determine the effect of misestimating this parameter on the sensitivity of the design.

6-3.6   The Assumption of No Interaction in a Two-Factor Model

Occasionally, an experimenter feels that a **two-factor model without interaction** is appropriate, say

$$y_{ijk} = \mu + \tau_i + \beta_j + \epsilon_{ijk} \qquad \begin{cases} i = 1, 2, \ldots, a \\ j = 1, 2, \ldots, b \\ k = 1, 2, \ldots, n \end{cases} \qquad (6\text{-}20)$$

We should be very careful in dispensing with the interaction terms, however, since the presence of significant interaction can have a dramatic impact on the interpretation of the data.

The statistical analysis of a two-factor factorial model without interaction is straightforward. Table 6-8 presents the analysis of the battery life data from Example 6-1, assuming that the no-interaction model (Equation 6-20) applies. As noted previously, both main effects are significant. However, as soon as a

Table 6-8   Analysis of Variance for Battery Life Data Assuming No Interaction

Source of Variation	Sum of Squares	Degrees of Freedom	Mean Square	F_0
Material types	10,683.72	2	5,341.86	5.95
Temperature	39,118.72	2	19,558.36	21.78
Error	27,844.52	31	898.21	
Total	77,646.96	35		

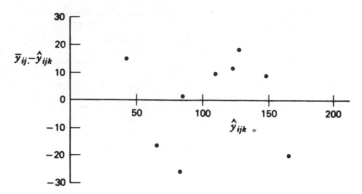

Figure 6–15. Plot of $\bar{y}_{ij.} - \hat{y}_{ijk}$ versus $\hat{y}_{ijk}$, battery life data.

residual analysis is performed for these data, it becomes clear that the no-interaction model is inadequate. For the two-factor model without interaction, the fitted values are $\hat{y}_{ijk} = \bar{y}_{i..} + \bar{y}_{.j.} - \bar{y}_{...}$. A plot of $\bar{y}_{ij.} - \hat{y}_{ijk}$ (the cell averages minus the fitted value for that cell) versus the fitted value $\hat{y}_{ijk}$ is shown in Figure 6–15. Now the quantities $\bar{y}_{ij.} - \hat{y}_{ijk}$ may be viewed as the differences between the observed cell means and the estimated cell means assuming no interaction. Any pattern in these quantities is suggestive of the presence of interaction. Figure 6–15 shows a distinct pattern as the quantities $\bar{y}_{ij.} - \hat{y}_{ijk}$ move from positive to negative to positive to negative again. This structure is the result of interaction between material types and temperature.

6–3.7   One Observation per Cell

Occasionally, one encounters a two-factor experiment with only a **single replicate,** that is, only one observation per cell. If there are two factors and only one observation per cell, the linear statistical model is

$$y_{ij} = \mu + \tau_i + \beta_j + (\tau\beta)_{ij} + \epsilon_{ij} \qquad \begin{cases} i = 1, 2, \ldots, a \\ j = 1, 2, \ldots, b \end{cases} \qquad (6\text{–}21)$$

The analysis of variance for this situation is shown in Table 6–9, assuming that both factors are fixed.

From examining the expected mean squares, we see that the error variance σ^2 is *not estimable;* that is, the two-factor interaction effect $(\tau\beta)_{ij}$ and the experimental error cannot be separated in any obvious manner. Consequently, there are no tests on main effects unless the interaction effect is zero. If there is no interaction present, then $(\tau\beta)_{ij} = 0$ for all i and j, and a plausible model is

$$y_{ij} = \mu + \tau_i + \beta_j + \epsilon_{ij} \qquad \begin{cases} i = 1, 2, \ldots, a \\ j = 1, 2, \ldots, b \end{cases} \qquad (6\text{–}22)$$

Table 6–9 Analysis of Variance for a Two-Factor Model, One Observation per Cell

Source of Variation	Sum of Squares	Degrees of Freedom	Mean Square	Expected Mean Square
Rows (A)	$\displaystyle\sum_{i=1}^{a} \frac{y_{i.}^2}{b} - \frac{y_{..}^2}{ab}$	$a - 1$	MS_A	$\sigma^2 + \dfrac{b \sum \tau_i^2}{a - 1}$
Columns (B)	$\displaystyle\sum_{j=1}^{b} \frac{y_{.j}^2}{a} - \frac{y_{..}^2}{ab}$	$b - 1$	MS_B	$\sigma^2 + \dfrac{a \sum \beta_j^2}{b - 1}$
Residual or AB	Subtraction	$(a - 1)(b - 1)$	MS_{Residual}	$\sigma^2 + \dfrac{\sum\sum (\tau\beta)_{ij}^2}{(a - 1)(b - 1)}$
Total	$\displaystyle\sum_{i=1}^{a}\sum_{j=1}^{b} y_{ij}^2 - \frac{y_{..}^2}{ab}$	$ab - 1$		

If the model (Equation 6–22) is appropriate, then the residual mean square in Table 6–9 is an unbiased estimator of σ^2, and the main effects may be tested by comparing MS_A and MS_B to MS_{Residual}.

A test developed by Tukey (1949a) is helpful in determining if interaction is present. The procedure assumes that the interaction term is of a particularly simple form; namely,

$$(\tau\beta)_{ij} = \gamma\tau_i\beta_j$$

where γ is an unknown constant. By defining the interaction term this way, we may use a regression approach to test the significance of the interaction term. The test partitions the residual sum of squares into a single-degree-of-freedom component due to nonadditivity (interaction) and a component for error with $(a - 1)(b - 1) - 1$ degrees of freedom. Computationally, we have

$$SS_N = \frac{\left[\displaystyle\sum_{i=1}^{a}\sum_{j=1}^{b} y_{ij}y_{i.}y_{.j} - y_{..}\left(SS_A + SS_B + \frac{y_{..}^2}{ab}\right)\right]^2}{ab\,SS_A\,SS_B} \tag{6-23}$$

with one degree of freedom, and

$$SS_{\text{Error}} = SS_{\text{Residual}} - SS_N \tag{6-24}$$

with $(a - 1)(b - 1) - 1$ degrees of freedom. To test for the presence of interaction, we compute

$$F_0 = \frac{SS_N}{SS_{\text{Error}}/[(a - 1)(b - 1) - 1]} \tag{6-25}$$

If $F_0 > F_{\alpha,1,(a-1)(b-1)-1}$, the hypothesis of no interaction must be rejected.

Example 6–2

The impurity present in a chemical product is affected by two factors—pressure and temperature. The data from a single replicate of a factorial experiment are shown in Table 6–10. The sums of squares are

$$SS_A = \frac{1}{b} \sum_{i=1}^{a} y_{i.}^2 - \frac{y_{..}^2}{ab}$$

$$= \frac{1}{5} [23^2 + 13^2 + 8^2] - \frac{44^2}{(3)(5)} = 23.33$$

$$SS_B = \frac{1}{a} \sum_{j=1}^{b} y_{.j}^2 - \frac{y_{..}^2}{ab}$$

$$= \frac{1}{3} [9^2 + 6^2 + 13^2 + 6^2 + 10^2] - \frac{44^2}{(3)(5)} = 11.60$$

$$SS_T = \sum_{i=1}^{a} \sum_{j=1}^{b} y_{ij}^2 - \frac{y_{..}^2}{ab}$$

$$= 166 - 129.07 = 36.93$$

and

$$SS_{\text{Residual}} = SS_T - SS_A - SS_B$$
$$= 36.93 - 23.33 - 11.60 = 2.00$$

The sum of squares for nonadditivity is computed from Equation 6–23 as follows:

$$\sum_{i=1}^{a} \sum_{j=1}^{b} y_{ij} y_{i.} y_{.j} = (5)(23)(9) + (4)(23)(6) + \cdots + (2)(8)(10) = 7236$$

$$SS_N = \frac{\left[\sum_{i=1}^{a} \sum_{j=1}^{b} y_{ij} y_{i.} y_{.j} - y_{..} \left(SS_A + SS_B + \frac{y_{..}^2}{ab} \right) \right]^2}{ab SS_A SS_B}$$

$$= \frac{[7236 - (44)(23.33 + 11.60 + 129.07)]^2}{(3)(5)(23.33)(11.60)}$$

$$= \frac{[20.00]^2}{4059.42} = 0.0985$$

Table 6–10 Impurity Data for Example 6–2

Temperature (°F)	Pressure					$y_{i.}$
	25	30	35	40	45	
100	5	4	6	3	5	23
125	3	1	4	2	3	13
150	1	1	3	1	2	8
$y_{.j}$	9	6	13	6	10	44 = $y_{..}$

Table 6–11   Analysis of Variance for Example 6–2

Source of Variation	Sum of Squares	Degrees of Freedom	Mean Square	F_0	P-Value
Temperature	23.33	2	11.67	42.97	0.0001
Pressure	11.60	4	2.90	10.68	0.0042
Nonadditivity	0.0985	1	0.0985	0.36	0.5674
Error	1.9015	7	0.2716		
Total	36.93	14			

and the error sum of squares is, from Equation 6–24,

$$SS_{\text{Error}} = SS_{\text{Residual}} - SS_N$$
$$= 2.00 - 0.0985 = 1.9015$$

The complete analysis of variance is summarized in Table 6–11. The test statistic for nonadditivity is $F_0 = 0.0985/0.2716 = 0.36$, so we conclude that there is no evidence of interaction in these data. The main effects of temperature and pressure are significant.

■

In concluding this section, we note that the two-factor factorial model with one observation per cell (Equation 6–22) looks exactly like the randomized complete block model (Equation 5–1). In fact, the Tukey single-degree-of-freedom test for nonadditivity can be directly applied to test for interaction in the randomized block model. However, remember that the **experimental situations** that lead to the randomized block and factorial models are very different. In the factorial model, *all ab* runs have been made in random order, whereas in the randomized block model, randomization occurs only *within the block*. The blocks are a randomization restriction. Hence, the manner in which the experiments are run and the interpretation of the two models are quite different.

6-4   THE GENERAL FACTORIAL DESIGN

The results for the two-factor factorial design may be extended to the general case where there are a levels of factor A, b levels of factor B, c levels of factor C, and so on, arranged in a factorial experiment. In general, there will be $abc \cdots n$ total observations if there are n replicates of the complete experiment. Once again, note that we must have at least two replicates ($n \geq 2$) in order to determine a sum of squares due to error if all possible interactions are included in the model.

If all factors in the experiment are fixed, we may easily formulate and test hypotheses about the main effects and interactions. For a fixed effects model, test statistics for each main effect and interaction may be constructed by dividing the corresponding mean square for the effect or interaction by the mean square error. All of these F tests will be upper-tail, one-tail tests. The number of degrees of freedom for any main effect is the number of levels of the factor minus one, and the number of degrees of freedom for an interaction is the product of the number of degrees of freedom associated with the individual components of the interaction.

For example, consider the **three-factor analysis of variance model:**

$$y_{ijkl} = \mu + \tau_i + \beta_j + \gamma_k + (\tau\beta)_{ij} + (\tau\gamma)_{ik} + (\beta\gamma)_{jk}$$

$$+ (\tau\beta\gamma)_{ijk} + \epsilon_{ijkl} \quad \begin{cases} i = 1, 2, \ldots, a \\ j = 1, 2, \ldots, b \\ k = 1, 2, \ldots, c \\ l = 1, 2, \ldots, n \end{cases} \quad (6\text{--}26)$$

Assuming that A, B, and C are fixed, the **analysis of variance table** is shown in Table 6–12. The F tests on main effects and interactions follow directly from the expected mean squares.

Usually, the analysis of variance computations would be done using a statistics software package. However, the manual computing formulas for the sums of squares in Table 6–12 are occasionally useful. The total sum of squares is found in the usual way as

$$SS_T = \sum_{i=1}^{a} \sum_{j=1}^{b} \sum_{k=1}^{c} \sum_{l=1}^{n} y_{ijkl}^2 - \frac{y_{....}^2}{abcn} \quad (6\text{--}27)$$

The sums of squares for the main effects are found from the totals for factors $A(y_{i...})$, $B(y_{.j..})$, and $C(y_{..k.})$ as follows:

$$SS_A = \frac{1}{bcn} \sum_{i=1}^{a} y_{i...}^2 - \frac{y_{....}^2}{abcn} \quad (6\text{--}28)$$

$$SS_B = \frac{1}{acn} \sum_{j=1}^{b} y_{.j..}^2 - \frac{y_{....}^2}{abcn} \quad (6\text{--}29)$$

$$SS_C = \frac{1}{abn} \sum_{k=1}^{c} y_{..k.}^2 - \frac{y_{....}^2}{abcn} \quad (6\text{--}30)$$

To compute the two-factor interaction sums of squares, the totals for the $A \times B$, $A \times C$, and $B \times C$ cells are needed. It is frequently helpful to collapse the original data table into three two-way tables in order to compute these quantities. The sums of squares are found from

$$SS_{AB} = \frac{1}{cn} \sum_{i=1}^{a} \sum_{j=1}^{b} y_{ij..}^2 - \frac{y_{....}^2}{abcn} - SS_A - SS_B$$

$$= SS_{\text{Subtotals}(AB)} - SS_A - SS_B \quad (6\text{--}31)$$

Table 6–12 The Analysis of Variance Table for the Three-Factor Fixed Effects Model

Source of Variation	Sum of Squares	Degrees of Freedom	Mean Square	Expected Mean Square	F_0
A	SS_A	$a - 1$	MS_A	$\sigma^2 + \dfrac{bcn \sum \tau_i^2}{a - 1}$	$F_0 = \dfrac{MS_A}{MS_E}$
B	SS_B	$b - 1$	MS_B	$\sigma^2 + \dfrac{acn \sum \beta_j^2}{b - 1}$	$F_0 = \dfrac{MS_B}{MS_E}$
C	SS_C	$c - 1$	MS_C	$\sigma^2 + \dfrac{abn \sum \gamma_k^2}{c - 1}$	$F_0 = \dfrac{MS_C}{MS_E}$
AB	SS_{AB}	$(a - 1)(b - 1)$	MS_{AB}	$\sigma^2 + \dfrac{cn \sum \sum (\tau\beta)_{ij}^2}{(a - 1)(b - 1)}$	$F_0 = \dfrac{MS_{AB}}{MS_E}$
AC	SS_{AC}	$(a - 1)(c - 1)$	MS_{AC}	$\sigma^2 + \dfrac{bn \sum \sum (\tau\gamma)_{ik}^2}{(a - 1)(c - 1)}$	$F_0 = \dfrac{MS_{AC}}{MS_E}$
BC	SS_{BC}	$(b - 1)(c - 1)$	MS_{BC}	$\sigma^2 + \dfrac{an \sum \sum (\beta\gamma)_{jk}^2}{(b - 1)(c - 1)}$	$F_0 = \dfrac{MS_{BC}}{MS_E}$
ABC	SS_{ABC}	$(a - 1)(b - 1)(c - 1)$	MS_{ABC}	$\sigma^2 + \dfrac{n \sum \sum \sum (\tau\beta\gamma)_{ijk}^2}{(a - 1)(b - 1)(c - 1)}$	$F_0 = \dfrac{MS_{ABC}}{MS_E}$
Error	SS_E	$abc(n - 1)$	MS_E	σ^2	
Total	SS_T	$abcn - 1$			

$$SS_{AC} = \frac{1}{bn}\sum_{i=1}^{a}\sum_{k=1}^{c} y_{i.k.}^2 - \frac{y_{....}^2}{abcn} - SS_A - SS_C$$

$$= SS_{\text{Subtotals}(AC)} - SS_A - SS_C \qquad (6\text{--}32)$$

and

$$SS_{BC} = \frac{1}{an}\sum_{j=1}^{b}\sum_{k=1}^{c} y_{.jk.}^2 - \frac{y_{....}^2}{abcn} - SS_B - SS_C$$

$$= SS_{\text{Subtotals}(BC)} - SS_B - SS_C \qquad (6\text{--}33)$$

Note that the sums of squares for the two-factor subtotals are found from the totals in each two-way table. The three-factor interaction sum of squares is computed from the three-way cell totals $\{y_{ijk.}\}$ as

$$SS_{ABC} = \frac{1}{n}\sum_{i=1}^{a}\sum_{j=1}^{b}\sum_{k=1}^{c} y_{ijk.}^2 - \frac{y_{....}^2}{abcn} - SS_A - SS_B - SS_C - SS_{AB} - SS_{AC} - SS_{BC} \qquad (6\text{--}34a)$$

$$= SS_{\text{Subtotals}(ABC)} - SS_A - SS_B - SS_C - SS_{AB} - SS_{AC} - SS_{BC} \qquad (6\text{--}34b)$$

The error sum of squares may be found by subtracting the sum of squares for each main effect and interaction from the total sum of squares or by

$$SS_E = SS_T - SS_{\text{Subtotals}(ABC)} \qquad (6\text{--}35)$$

Example 6–3

The Soft Drink Bottling Problem

A soft drink bottler is interested in obtaining more uniform fill heights in the bottles produced by his manufacturing process. The filling machine theoretically fills each bottle to the correct target height, but in practice, there is variation around this target, and the bottler would like to understand better the sources of this variability and eventually reduce it.

The process engineer can control three variables during the filling process: the percent carbonation (A), the operating pressure in the filler (B), and the bottles produced per minute or the line speed (C). The pressure and speed are easy to control, but the percent carbonation is more difficult to control during actual manufacturing because it varies with product temperature. However, for purposes of an experiment, the engineer can control carbonation at three levels: 10, 12, and 14 percent. He chooses two levels for pressure (25 and 30 psi) and two levels for line speed (200 and 250 bpm). He decides to run two replicates of a factorial design in these three factors, with all 24 runs taken in random order. The response variable observed is the average deviation from the target fill height observed in a production run of bottles at each set of conditions. The data that resulted from this experiment are shown in Table 6–13. Positive deviations are fill heights above the target, whereas negative deviations are fill heights below the target. The circled numbers in Table 6–13 are the three-way cell totals $y_{ijk.}$.

Table 6-13   Fill Height Deviation Data for Example 6-3

Percent Carbonation (A)	Operating Pressure (B)				$y_{i...}$
	25 psi		30 psi		
	Line Speed (C)		Line Speed (C)		
	200	250	200	250	
10	-3, -1 (-4)	-1, 0 (-1)	-1, 0 (-1)	1, 1 (2)	-4
12	0, 1 (1)	2, 1 (3)	2, 3 (5)	6, 5 (11)	20
14	5, 4 (9)	7, 6 (13)	7, 9 (16)	10, 11 (21)	59
B × C Totals $y_{.jk.}$	6	15	20	34	75 = $y_{....}$
$y_{.j..}$	21		54		

A × B Totals $y_{ij..}$				A × C Totals $y_{i.k.}$		
	B				C	
A	25	30		A	200	250
10	-5	1		10	-5	1
12	4	16		12	6	14
14	22	37		14	25	34

The total corrected sum of squares is found from Equation 6-27 as

$$SS_T = \sum_{i=1}^{a} \sum_{j=1}^{b} \sum_{k=1}^{c} \sum_{l=1}^{n} y_{ijkl}^2 - \frac{y_{....}^2}{abcn}$$

$$= 571 - \frac{(75)^2}{24} = 336.625$$

and the sums of squares for the main effects are calculated from Equations 6-28, 6-29, and 6-30 as

$$SS_{\text{Carbonation}} = \frac{1}{bcn} \sum_{i=1}^{a} y_{i...}^2 - \frac{y_{....}^2}{abcn}$$

$$= \frac{1}{8}[(-4)^2 + (20)^2 + (59)^2] - \frac{(75)^2}{24} = 252.750$$

$$SS_{\text{Pressure}} = \frac{1}{acn} \sum_{j=1}^{b} y_{.j..}^2 - \frac{y_{....}^2}{abcn}$$

$$= \frac{1}{12}[(21)^2 + (54)^2] - \frac{(75)^2}{24} = 45.375$$

and

$$SS_{\text{Speed}} = \frac{1}{abn} \sum_{k=1}^{c} y_{..k.}^2 - \frac{y_{....}^2}{abcn}$$

$$= \frac{1}{12} [(26)^2 + (49)^2] - \frac{(75)^2}{24} = 22.042$$

To calculate the sums of squares for the two-factor interactions, we must find the two-way cell totals. For example, to find the carbonation–pressure or AB interaction, we need the totals for the $A \times B$ cells $\{y_{ij..}\}$ shown in Table 6–13. Using Equation 6–31, we find the sums of squares as

$$SS_{AB} = \frac{1}{cn} \sum_{i=1}^{a} \sum_{j=1}^{b} y_{ij..}^2 - \frac{y_{....}^2}{abcn} - SS_A - SS_B$$

$$= \frac{1}{4} [(-5)^2 + (1)^2 + (4)^2 + (16)^2 + (22)^2 + (37)^2] - \frac{(75)^2}{24}$$

$$- 252.750 - 45.375$$

$$= 5.250$$

The carbonation–speed or AC interaction uses the $A \times C$ cell totals $\{y_{i.k.}\}$ shown in Table 6–13 and Equation 6–32:

$$SS_{AC} = \frac{1}{bn} \sum_{i=1}^{a} \sum_{k=1}^{c} y_{i.k.}^2 - \frac{y_{....}^2}{abcn} - SS_A - SS_C$$

$$= \frac{1}{4} [(-5)^2 + (1)^2 + (6)^2 + (14)^2 + (25)^2 + (34)^2] - \frac{(75)^2}{24}$$

$$- 252.750 - 22.042$$

$$= 0.583$$

The pressure–speed or BC interaction is found from the $B \times C$ cell totals $\{y_{.jk.}\}$ shown in Table 6–13 and Equation 6–33:

$$SS_{BC} = \frac{1}{an} \sum_{j=1}^{b} \sum_{k=1}^{c} y_{.jk.}^2 - \frac{y_{....}^2}{abcn} - SS_B - SS_C$$

$$= \frac{1}{6} [(6)^2 + (15)^2 + (20)^2 + (34)^2] - \frac{(75)^2}{24} - 45.375 - 22.042$$

$$= 1.042$$

The three-factor interaction sum of squares is found from the $A \times B \times C$ cell totals $\{y_{ijk.}\}$, which are circled in Table 6–13. From Equation 6–34a, we find

$$SS_{ABC} = \frac{1}{n} \sum_{i=1}^{a} \sum_{j=1}^{b} \sum_{k=1}^{c} y_{ijk.}^2 - \frac{y_{....}^2}{abcn} - SS_A - SS_B - SS_C - SS_{AB} - SS_{AC} - SS_{BC}$$

$$= \frac{1}{2} [(-4)^2 + (-1)^2 + (-1)^2 + \cdots + (16)^2 + (21)^2] - \frac{(75)^2}{24}$$

$$- 252.750 - 45.375 - 22.042 - 5.250 - 0.583 - 1.042$$

$$= 1.083$$

Finally, noting that

$$SS_{\text{Subtotals}(ABC)} = \frac{1}{n} \sum_{i=1}^{a} \sum_{j=1}^{b} \sum_{k=1}^{c} y_{ijk.}^2 - \frac{y_{....}^2}{abcn} = 328.125$$

we have

$$SS_E = SS_T - SS_{\text{Subtotals}(ABC)}$$
$$= 336.625 - 328.125$$
$$= 8.500$$

The analysis of variance is summarized in Table 6-14. We see that the percent carbonation, operating pressure, and line speed significantly affect the fill volume. The carbonation–pressure interaction F ratio has a P-value of 0.0558, indicating some interaction between these factors.

The next step should be an analysis of the residuals from this experiment. We leave this as an exercise for the reader but point out that a normal probability plot of the residuals and the other usual diagnostics do not indicate any major concerns.

To assist in the practical interpretation of this experiment, Figure 6-16 presents plots of the three main effects and the AB (carbonation–pressure) interaction. The main effect plots are just graphs of the marginal response averages at the levels of the three factors. Notice that all three variables have *positive* main effects; that is, increasing the variable moves the average deviation from the fill target upward. The interaction between carbonation and pressure is fairly small, as shown by the similar shape of the two curves in Figure 6-16d.

Since the company wants the average deviation from the fill target to be close to zero, the engineer decides to recommend the low level of operating pressure (25 psi) and the high level of line speed (250 bpm, which will maximize the production rate). Figure 6-17 (see page 263) plots the average observed deviation from the target fill height at the three different carbonation levels for this set of operating conditions. Now the carbonation level cannot presently be perfectly controlled in the manufacturing process, and the normal distribution shown with the solid curve in Figure 6-17 approximates the variability in the carbonation levels presently experienced. As the process is impacted by the values of the carbonation level drawn from

Table 6-14 Analysis of Variance for Example 6-3

Source of Variation	Sum of Squares	Degrees of Freedom	Mean Square	F_0	P-Value
Percent carbonation (A)	252.750	2	126.375	178.412	<0.0001
Operating pressure (B)	45.375	1	45.375	64.059	<0.0001
Line speed (C)	22.042	1	22.042	31.118	0.0001
AB	5.250	2	2.625	3.706	0.0558
AC	0.583	2	0.292	0.412	0.6713
BC	1.042	1	1.042	1.471	0.2485
ABC	1.083	2	0.542	0.765	0.4867
Error	8.500	12	0.708		
Total	336.625	23			

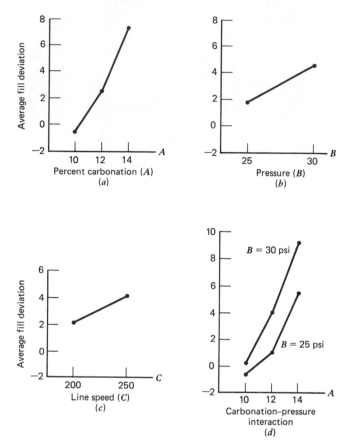

Figure 6–16. Main effects and interaction plots for Example 6–3. (*a*) Percent carbonation (*A*). (*b*) Pressure (*B*). (*c*) Line speed (*C*). (*d*) Carbonation-pressure interaction.

this distribution, the fill heights will fluctuate considerably. This variability in the fill heights could be reduced if the distribution of the carbonation level values followed the normal distribution shown with the dashed line in Figure 6–17. Reducing the standard deviation of the carbonation level distribution was ultimately achieved by improving temperature control during manufacturing.

■

We have indicated that if all the factors in a factorial experiment are fixed, then test statistic construction is straightforward. The statistic for testing any main effect or interaction is always formed by dividing the mean square for the main effect or interaction by the mean square error. However, if the factorial experiment involves one or more **random factors,** the test statistic construction is not always done this way. We must examine the expected mean squares to

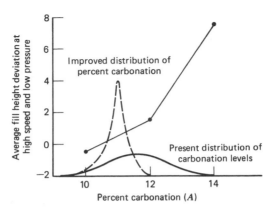

Figure 6–17. Average fill height deviation at high speed and low pressure for different carbonation levels.

determine the correct tests. We defer a complete discussion of experiments with random factors until Chapter 11.

6–5 FITTING RESPONSE CURVES AND SURFACES

We have observed that it is occasionally useful to fit a **response curve** to the levels of a quantitative factor so that the experimenter has an equation that relates the response to the factor. This equation might be used for interpolation, that is, for predicting the response at factor levels between those actually used in the experiment. When at least two factors are quantitative, we can fit a **response surface** for predicting y at various combinations of the design factors. The same principles discussed in Section 4–3 can be applied to factorial designs where at least one factor is quantitative. They are illustrated in the following two examples. In these examples, we use factor levels that are equally spaced and we use the method of **orthogonal polynomials.** General **regression methods** could also be used. In fact, if the factor levels are not equally spaced regression methods must be used. These procedures are illustrated in Chapter 14.

Example 6–4

Consider the experiment described in Example 6–1. The temperature is quantitative and material type is qualitative. Furthermore, the three levels of temperature are equally spaced. Consequently, we can compute a linear and a quadratic temperature effect to study how temperature affects the battery life. The calculation of the linear and quadratic effects of temperature, denoted T_L and T_Q, and their sums of squares are shown in Table 6–15.

The interaction sum of squares may be partitioned into an interaction component between the material type and the linear effect of temperature ($M \times T_L$) and an interaction component between the material type and the quadratic effect of tempera-

Table 6–15   Calculation of the Polynomial Effects of Temperature, Example 6–4

Temperature Levels (°F)	Treatment Totals $y_{.j.}$	Orthogonal Contrast Coefficients (c_j) Linear	Quadratic
15	1738	−1	+1
70	1291	0	−2
125	770	+1	+1
Effects: $\left(\sum\limits_{j=1}^{3} c_j y_{.j.} \right)$		$T_L = -968$	$T_Q = -74$
Sums of squares: $\left[\dfrac{\left(\sum\limits_{j=1}^{3} c_j y_{.j.} \right)^2}{an \sum\limits_{j=1}^{3} c_j^2} \right]$		$SS_{T_L} = \dfrac{(-968)^2}{(4)(3)(2)}$ $= 39{,}042.67$	$SS_{T_Q} = \dfrac{(-74)^2}{(4)(3)(6)}$ $= 76.05$

ture ($M \times T_Q$). To compute these interaction components, the linear and quadratic effects of temperature are determined for each material type, and then the sums of squares between these effects are calculated. The process is shown in Table 6–16.

Note that $SS_{MT} = SS_{M \times T_L} + SS_{M \times T_Q}$. The analysis of variance is displayed in Table 6–17. The linear temperature effect is evident, and there is no indication of significant quadratic response to temperature. Furthermore, the significant interaction is the result of the $M \times T_Q$ component.

Since the effect of temperature is significant, the experimenter could fit a **response curve** to the data so that the battery life may be predicted as a function of the temperature. However, the significant MT interaction implies that the battery life

Table 6–16   Polynomial Effects of the MT Interaction, Example 6–4

Temperature Levels (°F)	Totals for Material Types 1($y_{1j.}$)	2($y_{2j.}$)	3($y_{3j.}$)	Orthogonal Contrast Coefficients (c_j) Linear	Quadratic
15	539	623	576	−1	+1
70	229	479	583	0	−2
125	230	198	342	+1	+1

Effects: $M_1 \times T_L = -309,\ M_2 \times T_L = -425,\ M_3 \times T_L = -234$

$M_1 \times T_Q = 311,\ M_2 \times T_Q = -137,\ M_3 \times T_Q = -248$

Sums of squares:

$$SS_{M \times T_L} = \frac{(-309)^2 + (-425)^2 + (-234)^2}{4(2)} - \frac{(-968)^2}{12(2)} = 2315.08$$

$$SS_{M \times T_Q} = \frac{(311)^2 + (-137)^2 + (-248)^2}{4(6)} - \frac{(-74)^2}{12(6)} = 7298.70$$

Table 6-17 Analysis of Variance for Example 6-4

Source of Variation	Sum of Squares	Degrees of Freedom	Mean Square	F_0	P-Value
Material types (M)	10,683.72	2	5,341.86	7.91	0.0020
Temperature (T)	39,118.72	2	19,559.36	28.97	<0.0001
(Linear)	(39,042.67)	1	39,042.67	57.82	<0.0001
(Quadratic)	(76.05)	1	76.05	0.12	0.7317
MT interaction	9,613.78	4	2,403.44	3.56	0.0186
($M \times T_L$)	(2,315.08)	2	1157.54	1.71	0.1999
($M \times T_Q$)	(7,298.70)	2	3649.35	5.41	0.0106
Error	18,230.75	27	675.21		
Total	77,646.97	35			

response to temperature depends on which material type is used. Therefore, the experimenter would have to fit *three* response curves—one for each material type.

∎

If several factors in a factorial experiment are quantitative then a **response surface** may be used to model the relationship between y and the design factors. Furthermore, the factor effects may be decomposed into single-degree-of-freedom polynomial effects. Similarly, the interactions of quantitative factors can be partitioned into single-degree-of-freedom components of interaction. This is illustrated in the following example.

Example 6-5

The effective life of a cutting tool installed in a numerically controlled machine is thought to be affected by the cutting speed and the tool angle. Three speeds and three angles are selected, and a factorial experiment with two replicates is performed. The coded data are shown in Table 6-18. The circled numbers in the cells are the cell totals $\{y_{ij.}\}$.

Table 6-18 Data for Tool Life Experiment

Tool Angle (degrees)	Cutting Speed (in/min)			$y_{i..}$
	125	150	175	
15	-2 / -1 (-3)	-3 / 0 (-3)	2 / 3 (5)	-1
20	0 / 2 (2)	1 / 3 (4)	4 / 6 (10)	16
25	-1 / 0 (-1)	5 / 6 (11)	0 / -1 (-1)	9
$y_{.j.}$	-2	12	14	24 = $y_{....}$

Table 6–19   Analysis of Variance for Tool Life Data

Source of Variation	Sum of Squares	Degrees of Freedom	Mean Square	F_0	P-Value
Tool angle (T)	24.33	2	12.17	8.45	0.0086
Cutting speed (C)	25.33	2	12.67	8.80	0.0076
TC interaction	61.34	4	15.34	10.65	0.0018
Error	13.00	9	1.44		
Total	124.00	17			

Table 6–19 summarizes the analysis of variance. Both factors are treated as fixed effects. Cutting speed, tool angle, and their interaction are significant at the 5 percent level.

The linear and quadratic effects of both factors may be determined using the method presented in Example 6–4. The calculations are displayed in Table 6–20. Notice that the sum of squares for a factor is equal to the total sum of squares of the linear and quadratic components for that same factor.

Since both factors in this experiment are quantitative, it is possible to partition the TC interaction into four single-degree-of-freedom components, say

$$TC_{L \times L} \qquad TC_{L \times Q} \qquad TC_{Q \times L} \qquad \text{and} \qquad TC_{Q \times Q}$$

In general, any two-factor interaction between quantitative factors can be partitioned into $(a - 1)(b - 1)$ single-degree-of-freedom components. However, it may be difficult to interpret these interaction components. Often a particular component is significant because of atypical responses in one cell. Thus, it is relatively standard practice to compute only the lower-order components. In such a case, the remaining components are considered jointly as higher-order interaction effects. The statistical significance of these higher-order interaction components could indicate a very complex relationship between the factors.

Table 6–20   Polynomial Effects of Tool Angle (T) and Cutting Speed (C) for Example 6–5

Treatment Totals		Orthogonal Contrast Coefficients		
Tool Angle ($y_{i..}$)	Cutting speed ($y_{.j.}$)	Linear	Quadratic	
-1	-2	-1	$+1$	
16	12	0	-2	
9	14	$+1$	$+1$	
Effects:		$T_L = 10$	$T_Q = -24$	
		$C_L = 16$	$C_Q = -12$	
				Totals
Sums of squares:		$SS_{T_L} = 8.33$	$SS_{T_Q} = 16.00$	24.33
		$SS_{C_L} = 21.33$	$SS_{C_Q} = 4.00$	25.33

Table 6–21 $TC_{L \times L}$ Interaction
Coefficients

	C_L		
T_L	-1	0	$+1$
-1	$+1$	0	-1
0	0	0	0
$+1$	-1	0	$+1$

The technique for computing the single-degree-of-freedom interaction components requires the cell totals (recall that differences between cells define an interaction effect) and the orthogonal contrast coefficients. To illustrate the approach, consider the coefficients for the $TC_{L \times L}$ interaction component shown in Table 6–21. The entries in the body of the table (e.g., c_{ij}) are obtained by multiplying the coefficients of T_L by the coefficients of C_L.

To obtain the $TC_{L \times L}$ contrast, multiply the coefficient in the ijth cell in Table 6–21 by the ijth cell total (shown in circles in Table 6–18) and add, yielding

$$TC_{L \times L} = \sum_{i=1}^{3} \sum_{j=1}^{3} c_{ij} y_{ij.}$$

$$= +1(-3) + (0)(-3) + (-1)(5) + (0)(2) + (0)(4)$$
$$+ (0)(10) + (-1)(-1) + (0)(11) + 1(-1) = -8$$

The sum of squares for the $TC_{L \times L}$ contrast is

$$SS_{TC_{L \times L}} = \frac{(TC_{L \times L})^2}{n \sum \sum c_{ij}^2} = \frac{(-8)^2}{2[(1)^2 + (0)^2 + (-1)^2 + \cdots + (1)^2]} = \frac{(-8)^2}{2(4)} = 8.00$$

The coefficients for the $TC_{L \times Q}$, $TC_{Q \times L}$, and $TC_{Q \times Q}$ interaction components are derived similarly. They are shown in Table 6–22, along with the contrast and sum of squares for each interaction component.

■

It is easy to see that the sum of the four single-degree-of-freedom interaction components equals the interaction sum of squares; that is,

$$SS_{TC} = SS_{TC_{L \times L}} + SS_{TC_{L \times Q}} + SS_{TC_{Q \times L}} + SS_{TC_{Q \times Q}}$$
$$= 8.00 + 42.67 + 2.67 + 8.00 = 61.34$$

The expanded analysis of variance is summarized in Table 6–23. Both linear and quadratic components of tool angle are significant, whereas cutting speed has only a linear effect on tool life. The $TC_{L \times L}$, $TC_{L \times Q}$, and $TC_{Q \times Q}$ interactions are also significant, with the $TC_{L \times Q}$ component quite large relative to the others.

Figure 6–18 plots the cell averages against the tool angle for all three speeds and illustrates the quadratic effect of tool angle. Figure 6–19 plots cell averages versus cutting speed for the tool angles. See page 269.

Table 6–22 Computation of the $TC_{L \times Q}$, $TC_{Q \times L}$, and $TC_{Q \times Q}$ Interactions

	C_Q				C_L				C_Q		
T_L	+1	−2	+1	T_Q	−1	0	+1	T_Q	+1	−2	+1
−1	−1	+2	−1	+1	−1	0	+1	+1	+1	−2	+1
0	0	0	0	−2	+2	0	−2	−2	−2	+4	−2
+1	+1	−2	+1	+1	−1	0	+1	+1	+1	−2	+1

$$TC_{L \times Q} = \sum_{i=1}^{3} \sum_{j=1}^{3} c_{ij} y_{ij.}$$

$$= -32$$

$$SS_{TC_{L \times Q}} = \frac{(TC_{L \times Q})^2}{n \sum \sum c_{ij}^2}$$

$$= \frac{(-32)^2}{2(12)} = 42.67$$

$$TC_{Q \times L} = \sum_{i=1}^{3} \sum_{j=1}^{3} c_{ij} y_{ij.}$$

$$= -8$$

$$SS_{TC_{Q \times L}} = \frac{(TC_{Q \times L})^2}{n \sum \sum c_{ij}^2}$$

$$= \frac{(-8)^2}{2(12)} = 2.67$$

$$TC_{Q \times Q} = \sum_{i=1}^{3} \sum_{j=1}^{3} c_{ij} y_{ij.}$$

$$= -24$$

$$SS_{TC_{Q \times Q}} = \frac{(TC_{Q \times Q})^2}{n \sum \sum c_{ij}^2}$$

$$= \frac{(-24)^2}{2(36)} = 8.00$$

If we let x_1 represent tool angle and x_2 represent cutting speed, then Table 6–23 implies a polynomial approximation for tool life, for example,

$$y = \alpha_0 + \alpha_1 P_1(x_1) + \alpha_{11} P_2(x_1) + \alpha_2 P_1(x_2) + \alpha_{12} P_1(x_1)P_1(x_2)$$
$$+ \alpha_{122} P_1(x_1)P_2(x_2) + \alpha_{1122} P_2(x_1)P_2(x_2) + \epsilon \tag{6-36}$$

where α_1 denotes the linear coefficient of tool angle, α_{122} denotes the coefficient for linear tool angle and quadratic cutting speed, and so on, and $P_i(x_j)$ are the orthogonal polynomials. These coefficients and polynomials are evaluated exactly

Table 6–23 Analysis of Variance for Example 6–5

Source of Variance	Sum of Squares	Degrees of Freedom	Mean Square	F_0	P-Value
Tool angle, T	24.33	2	12.17	8.45	0.0086
(T_L)	(8.33)	1	(8.33)	5.78	0.0396
(T_Q)	(16.00)	1	(16.00)	11.11	0.0088
Cutting speed, C	25.33	2	12.67	8.80	0.0076
(C_L)	(21.33)	1	(21.33)	14.81	0.0039
(C_Q)	(4.00)	1	(4.00)	2.77	0.1304
TC interaction	61.34	4	15.34	10.65	0.0018
$(TC_{L \times L})$	(8.00)	1	(8.00)	5.56	0.0427
$(TC_{L \times Q})$	(42.67)	1	(42.67)	29.63	0.0004
$(TC_{Q \times L})$	(2.67)	1	(2.67)	1.85	0.2069
$(TC_{Q \times Q})$	(8.00)	1	(8.00)	5.56	0.0427
Error	13.00	9	1.44		
Total	124.00				

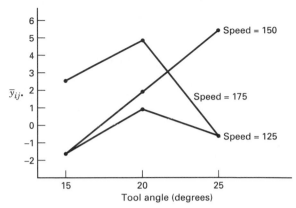

Figure 6–18. Cell averages versus tool angle for Example 6–5.

as in Section 4–3; for example,

$$\hat{\alpha}_{12} = \frac{\sum\limits_{i=1}^{3}\sum\limits_{j=1}^{3} y_{ij}P_1(x_{1i})P_1(x_{2j})}{\sum\limits_{i=1}^{3}\sum\limits_{j=1}^{3} [P_1(x_{1i})P_1(x_{2j})]^2} = \frac{-8}{8} = -1.00$$

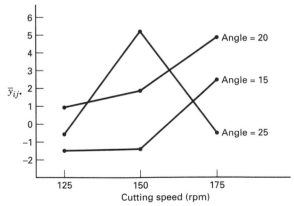

Figure 6–19. Cell averages versus cutting speed for Example 6–5.

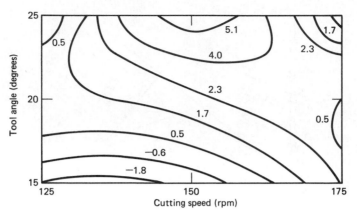

Figure 6–20. Two-dimensional tool life response surface for Example 6–5.

and

$$P_1(x_1)P_1(x_2) = \lambda_1\left(\frac{x_1 - \bar{x}_1}{d_1}\right)\lambda_1\left(\frac{x_2 - \bar{x}_2}{d_2}\right)$$

$$= (1)\left(\frac{x_1 - 20}{5}\right)(1)\left(\frac{x_2 - 150}{25}\right)$$

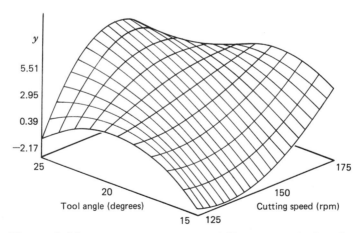

Figure 6–21. Three-dimensional tool life response surface for Example 6–5.

Evaluating each term in Equation 6–36 in a similar manner will yield the following **response surface model:**

$$\hat{y} = 2.667 + 0.700(x_1 - 20) - 0.0267(x_1 - 20)^2 + 0.0532(x_2 - 150)$$
$$- 0.008(x_1 - 20)(x_2 - 150) - 0.00128(x_1 - 20)(x_2 - 150)^2 \qquad (6\text{–}37)$$
$$- 0.000128(x_1 - 20)^2(x_2 - 150)^2$$

This response surface could be used to predict tool life at various values of tool angle and cutting speed or to assist in process optimization. Figure 6–20 presents a contour plot of the surface generated by Equation 6–37. Examination of the response surface reveals that maximum tool life is achieved at cutting speeds around 150 rpm and tool angles of 25 degrees. The three-dimensional response surface plot in Figure 6–21 provides essentially the same information, but it provides a different and sometimes more useful perspective of the tool life response surface. Exploration of response surfaces is a very important aspect of experimental design, which we will discuss in more detail in Chapter 14.

6–6   BLOCKING IN A FACTORIAL DESIGN

We have discussed factorial designs in the context of a **completely randomized** experiment. Sometimes, it is not feasible or practical to completely randomize all of the runs in a factorial. For example, the presence of a nuisance factor may require that the experiment be run in **blocks.** We discussed the basic concepts of blocking in the context of a single-factor experiment in Chapter 5. We now show how blocking can be incorporated in a factorial. Some other aspects of blocking in factorial designs are presented in Chapters 8, 9, and 12.

Consider a factorial experiment with two factors (A and B) and n replicates. The linear statistical model for this design is

$$y_{ijk} = \mu + \tau_i + \beta_j + (\tau\beta)_{ij} + \epsilon_{ijk} \qquad \begin{cases} i = 1, 2, \ldots, a \\ j = 1, 2, \ldots, b \\ k = 1, 2, \ldots, n \end{cases} \qquad (6\text{–}38)$$

where τ_i, β_j, and $(\tau\beta)_{ij}$ represent the effects of factors A, B, and the AB interaction, respectively. Now suppose that to run this experiment a particular raw material is required. This raw material is available in batches that are not large enough to allow *all abn* treatment combinations to be run from the *same* batch. However, if a batch contains enough material for *ab* observations, then an alternative design is to run each of the n replicates using a separate batch of raw material. Consequently, the batches of raw material represent a randomization restriction or a **block,** and a single replicate of a complete factorial experiment is run within each block. The model for this new design is

$$y_{ijk} = \mu + \tau_i + \beta_j + (\tau\beta)_{ij} + \delta_k + \epsilon_{ijk} \qquad \begin{cases} i = 1, 2, \ldots, a \\ j = 1, 2, \ldots, b \\ k = 1, 2, \ldots, n \end{cases} \qquad (6\text{-}39)$$

where δ_k is the effect of the kth block. Of course, within a block the order in which the treatment combinations are run is completely randomized.

The model (Equation 6–39) assumes that interaction between blocks and treatments is negligible. This was assumed previously in the analysis of randomized block designs. If these interactions do exist, they cannot be separated from the error component. In fact, the error term in this model really consists of the $(\tau\delta)_{ik}$, $(\beta\delta)_{jk}$, and $(\tau\beta\delta)_{ijk}$ interactions. The analysis of variance is outlined in Table 6–24. The layout closely resembles that of a factorial design, with the error sum of squares reduced by the sum of squares for blocks. Computationally, we find the sum of squares for blocks as the sum of squares between the n block totals $\{y_{..k}\}$.

In the previous example, the randomization was restricted to within a batch of raw material. In practice, a variety of phenomena may cause randomization restrictions, such as time, operators, and so on. For example, if we could not run the entire factorial experiment on one day, then the experimenter could run a complete replicate on day 1, a second replicate on day 2, and so on. Consequently, each day would be a block.

Table 6–24 Analysis of Variance for a Two-Factor Factorial in a Randomized Complete Block

Source of Variation	Sum of Squares	Degrees of Freedom	Expected Mean Square	F_0
Blocks	$\dfrac{1}{ab}\sum_k y_{..k}^2 - \dfrac{y_{...}^2}{abn}$	$n-1$	$\sigma^2 + ab\sigma_\delta^2$	
A	$\dfrac{1}{bn}\sum_i y_{i..}^2 - \dfrac{y_{...}^2}{abn}$	$a-1$	$\sigma^2 + \dfrac{bn\sum \tau_i^2}{a-1}$	$\dfrac{MS_A}{MS_E}$
B	$\dfrac{1}{an}\sum_j y_{.j.}^2 - \dfrac{y_{...}^2}{abn}$	$b-1$	$\sigma^2 + \dfrac{an\sum \beta_j^2}{b-1}$	$\dfrac{MS_B}{MS_E}$
AB	$\dfrac{1}{n}\sum_i \sum_j y_{ij.}^2 - \dfrac{y_{...}^2}{abn} - SS_A - SS_B$	$(a-1)(b-1)$	$\sigma^2 + \dfrac{n\sum\sum(\tau\beta)_{ij}^2}{(a-1)(b-1)}$	$\dfrac{MS_{AB}}{MS_E}$
Error	Subtraction	$(ab-1)(n-1)$	σ^2	
Total	$\sum_i \sum_j \sum_k y_{ijk}^2 - \dfrac{y_{...}^2}{abn}$	$abn-1$		

Example 6-6

An engineer is studying methods for improving the ability to detect targets on a radar scope. Two factors she considers to be important are the amount of background noise, or "ground clutter," on the scope and the type of filter placed over the screen. An experiment is designed using three levels of ground clutter and two filter types. We will consider these as fixed type factors. The experiment is performed by randomly selecting a treatment combination (ground clutter level and filter type) and then introducing a signal representing the target into the scope. The intensity of this target is increased until the operator observes it. The intensity level at detection is then measured as the response variable. Because of operator availability, it is convenient to select an operator and keep him or her at the scope until all the necessary runs have been made. Furthermore, operators differ in their skill and ability to use the scope. Consequently, it seems logical to use the operators as blocks. Four operators are randomly selected. Once an operator is chosen, the order in which the six treatment combinations are run is randomly determined. Thus, we have a 3 × 2 factorial experiment run in a randomized complete block. The data are shown in Table 6-25.

The linear model for this experiment is

$$y_{ijk} = \mu + \tau_i + \beta_j + (\tau\beta)_{ij} + \delta_k + \epsilon_{ijk} \qquad \begin{cases} i = 1, 2, 3 \\ j = 1, 2 \\ k = 1, 2, 3, 4 \end{cases}$$

where τ_i represents the ground clutter effect, β_j represents the filter type effect, $(\tau\beta)_{ij}$ is the interaction, δ_k is the block effect, and ϵ_{ijk} is the NID(0, σ^2) error component. The sums of squares for ground clutter, filter type, and their interaction are computed in the usual manner. The sum of squares due to blocks is found from the operator totals $\{y_{..k}\}$ as follows:

$$SS_{\text{Blocks}} = \frac{1}{ab} \sum_{k=1}^{n} y_{..k}^2 - \frac{y_{...}^2}{abn}$$

$$= \frac{1}{(3)(2)} [(572)^2 + (579)^2 + (597)^2 + (530)^2] - \frac{(2278)^2}{(3)(2)(4)}$$

$$= 402.17$$

The complete analysis of variance for this experiment is summarized in Table 6-26. The presentation in Table 6-26 indicates that all effects are tested by dividing their mean squares by the mean square error. Both ground clutter level and filter

Table 6-25 Intensity Level at Target Detection

Operators (blocks)	1		2		3		4	
Filter Type	1	2	1	2	1	2	1	2
Ground clutter								
Low	90	86	96	84	100	92	92	81
Medium	102	87	106	90	105	97	96	80
High	114	93	112	91	108	95	98	83

Table 6-26   Analysis of Variance for Example 6-6

Source of Variation	Sum of Squares	Degrees of Freedom	Mean Square	F_0	P-Value
Ground clutter (G)	335.58	2	167.79	15.13	0.0003
Filter type (F)	1066.67	1	1066.67	96.19	<0.0001
GF	77.08	2	38.54	3.48	0.0573
Blocks	402.17	3	134.06		
Error	166.33	15	11.09		
Total	2047.83	23			

type are significant at the 1 percent level, whereas their interaction is significant only at the 10 percent level. Thus, we conclude that both ground clutter level and the type of scope filter used affect the operator's ability to detect the target, and there is some evidence of mild interaction between these factors.

■

In the case of two randomization restrictions, each with p levels, if the number of treatment combinations in a k-factor factorial design exactly equals the number of restriction levels, that is, if $p = ab \ldots m$, then the factorial design may be run in a $p \times p$ Latin square. For example, consider a modification of the radar target detection experiment of Example 6-6. The factors in this experiment are filter type (two levels) and ground clutter (three levels), and operators are considered as blocks. Suppose now that because of the setup time required, only six runs can be made per day. Thus, days become a second randomization restriction, resulting in the 6×6 Latin square design, as shown in Table 6-27. In this table we have used the lowercase letters f_i and g_j to represent the ith and jth levels of filter type and ground clutter, respectively. That is, $f_1 g_2$ represents filter type 1 and medium ground clutter. Note that now six operators are required, rather than four as in the original experiment, so the number of treatment combinations in the 3×2 factorial design exactly equals the number of restriction levels. Furthermore, in this design, each operator would be used only once on each day. The Latin letters A, B, C, D, E, and F represent the $3 \times 2 = 6$ factorial treatment combinations as follows: $A = f_1 g_1$, $B = f_1 g_2$, $C = f_1 g_3$, $D = f_2 g_1$, $E = f_2 g_2$, and $F = f_2 g_3$.

The five degrees of freedom between the six Latin letters correspond to the main effects of filter type (one degree of freedom), ground clutter (two degrees of freedom), and their interaction (two degrees of freedom). The linear statistical model for this design is

$$y_{ijkl} = \mu + \alpha_i + \tau_j + \beta_k + (\tau\beta)_{jk} + \theta_l + \epsilon_{ijkl} \qquad \begin{cases} i = 1, 2, \ldots, 6 \\ j = 1, 2, 3 \\ k = 1, 2 \\ l = 1, 2, \ldots, 6 \end{cases} \qquad (6\text{-}40)$$

Table 6-27   Radar Detection Experiment Run in a 6 × 6 Latin Square

Day	Operator					
	1	2	3	4	5	6
1	$A(f_1g_1 = 90)$	$B(f_1g_2 = 106)$	$C(f_1g_3 = 108)$	$D(f_2g_1 = 81)$	$F(f_2g_3 = 90)$	$E(f_2g_2 = 88)$
2	$C(f_1g_3 = 114)$	$A(f_1g_1 = 96)$	$B(f_1g_2 = 105)$	$F(f_2g_3 = 83)$	$E(f_2g_2 = 86)$	$D(f_2g_1 = 84)$
3	$B(f_1g_2 = 102)$	$E(f_2g_2 = 90)$	$F(f_2g_3 = 95)$	$A(f_1g_1 = 92)$	$D(f_2g_1 = 85)$	$C(f_1g_3 = 104)$
4	$E(f_2g_2 = 87)$	$D(f_2g_1 = 84)$	$A(f_1g_1 = 100)$	$B(f_1g_2 = 96)$	$C(f_1g_3 = 110)$	$F(f_2g_3 = 91)$
5	$F(f_2g_3 = 93)$	$C(f_1g_3 = 112)$	$D(f_2g_1 = 92)$	$E(f_2g_2 = 80)$	$A(f_1g_1 = 90)$	$B(f_1g_2 = 98)$
6	$D(f_2g_1 = 86)$	$F(f_2g_3 = 91)$	$E(f_2g_2 = 97)$	$C(f_1g_3 = 98)$	$B(f_1g_2 = 100)$	$A(f_1g_1 = 92)$

Table 6–28   Analysis of Variance for the Radar Detection Experiment
Run as a 3 × 2 Factorial in a Latin Square

Source of Variation	Sum of Squares	Degrees of Freedom	General Formula for Degrees of Freedom	Mean Square	F_0	P-Value
Ground clutter, G	571.50	2	$a - 1$	285.75	28.86	<0.0001
Filter type, F	1469.44	1	$b - 1$	1469.44	148.43	<0.0001
GF	126.73	2	$(a - 1)(b - 1)$	63.37	6.40	0.0071
Days (rows)	4.33	5	$ab - 1$	0.87		
Operators (columns)	428.00	5	$ab - 1$	85.60		
Error	198.00	20	$(ab - 1)(ab - 2)$	9.90		
Total	2798.00	35	$(ab)^2 - 1$			

where τ_j and β_k are effects of ground clutter and filter type, respectively, and α_i and θ_l represent the randomization restrictions of days and operators, respectively. To compute the sums of squares, the following two-way table of treatment totals is helpful:

Ground Clutter	Filter Type 1	Filter Type 2	$y_{.j..}$
Low	560	512	1072
Medium	607	528	1135
High	646	543	1189
$y_{..k.}$	1813	1583	$3396 = y_{....}$

Furthermore, the row and column totals are

Rows ($y_{.jkl}$):   563   568   568   568   565   564

Columns ($y_{ijk.}$):   572   579   597   530   561   557

The analysis of variance is summarized in Table 6–28. We have added a column to this table indicating how the number of degrees of freedom for each sum of squares is determined.

6–7   UNBALANCED DATA IN A FACTORIAL DESIGN

The primary focus of this chapter has been the analysis of **balanced factorial designs,** that is, cases where there are an equal number of observations n in each cell. However, it is not unusual to encounter situations where the number of observations in the cells are unequal. These **unbalanced factorial designs** occur for various reasons. For example, the experimenter may have designed a balanced experiment initially, but because of unforeseen problems in running the experiment, resulting in the loss of some observations, she or he ends up with unbalanced data. On the other hand, some unbalanced experiments are deliberately designed that way. For instance, certain treatment combinations may be more expensive or more difficult to run than others, so fewer observations may be taken in those cells. Alternatively, some treatment combinations may be of greater interest to the experimenter because they represent new or unexplored conditions, so he or she may elect to obtain additional replication in those cells.

The orthogonality property of main effects and interactions present in balanced data do not carry over to the unbalanced case. This means that the usual analysis of variance techniques do not apply. Consequently, the analysis of unbalanced factorials is much more difficult than that for balanced designs.

In this section we give a brief overview of methods for dealing with unbalanced factorials, concentrating on the case of the two-factor fixed effects model. Suppose that the number of observations in the ijth cell is n_{ij}. Furthermore, let $n_{i.} = \sum_{j=1}^{b} n_{ij}$ be the number of observations in the ith row (the ith level of factor A), $n_{.j} = \sum_{i=1}^{a} n_{ij}$ be the number of observations in the jth column (the jth level of factor B), and $n_{..} = \sum_{i=1}^{a} \sum_{j=1}^{b} n_{ij}$ be the total number of observations.

6–7.1   Proportional Data: An Easy Case

One situation involving unbalanced data presents little difficulty in analysis; this is the case of **proportional data.** That is, the number of observations in the ijth cell is

$$n_{ij} = \frac{n_{i.} n_{.j}}{n_{..}} \tag{6–41}$$

This condition implies that the number of observations in any two rows or columns are proportional. When proportional data occur, the standard analysis of variance can be employed. Only minor modifications are required in the manual computing formulas for the sums of squares, which become

$$SS_T = \sum_{i=1}^{a} \sum_{j=1}^{b} \sum_{k=1}^{n_{ij}} y_{ijk}^2 - \frac{y_{...}^2}{n_{..}}$$

$$SS_A = \sum_{i=1}^{a} \frac{y_{i..}^2}{n_{i.}} - \frac{y_{...}^2}{n_{..}}$$

$$SS_B = \sum_{j=1}^{b} \frac{y_{.j.}^2}{n_{.j}} - \frac{y_{...}^2}{n_{..}}$$

Table 6-29   Battery Design Experiment with Proportional Data

Material Type	Temperature (°F) 15		Temperature (°F) 70		Temperature (°F) 125			
1	$n_{11} = 4$ 130 74	155 180	$n_{12} = 4$ 34 80	40 75	$n_{13} = 2$ 70	58	$n_{1.} = 10$ $y_{1..} = 896$	
2	$n_{21} = 2$ 159	126	$n_{22} = 2$ 136	115	$n_{23} = 1$ 45		$n_{2.} = 5$ $y_{2..} = 581$	
3	$n_{31} = 2$ 138	160	$n_{32} = 2$ 150	139	$n_{33} = 1$ 96		$n_{3.} = 5$ $y_{3..} = 683$	
	$n_{.1} = 8$ $y_{.1.} = 1122$		$n_{.2} = 8$ $y_{.2.} = 769$		$n_{.3} = 4$ $y_{.3.} = 269$		$n_{..} = 20$ $y_{...} = 2160$	

$$SS_{AB} = \sum_{i=1}^{a} \sum_{j=1}^{b} \frac{y_{ij.}^2}{n_{ij}} - \frac{y_{...}^2}{n_{..}} - SS_A - SS_B$$

$$SS_E = SS_T - SS_A - SS_B - SS_{AB}$$

$$= \sum_{i=1}^{a} \sum_{j=1}^{b} \sum_{k=1}^{n_{ij}} y_{ijk}^2 - \sum_{i=1}^{a} \sum_{j=1}^{b} \frac{y_{ij.}^2}{n_{ij}}$$

As an example of proportional data, consider the battery design experiment in Example 6-1. A modified version of the original data is shown in Table 6-29. Clearly, the data are proportional; for example, in cell 1,1 we have

$$n_{11} = \frac{n_{1.}n_{.1}}{n_{..}} = \frac{10(8)}{20} = 4$$

observations. The results of applying the usual analysis of variance to these data are shown in Table 6-30. Both material type and temperature are significant, which agrees with the analysis of the full data set in Example 6-1. However, the interaction noted in Example 6-1 is not present.

Table 6-30   Analysis of Variance for Battery Design Data in Table 6-29

Source of Variance	Sum of Squares	Degrees of Freedom	Mean Square	F_0
Material types	8,170.400	2	4,085.20	5.00
Temperature	16,090.875	2	8,045.44	9.85
Interaction	5,907.725	4	1,476.93	1.18
Error	8,981.000	11	816.45	
Total	39,150.000	19		

6–7.2 Approximate Methods

When unbalanced data are not too far from the balanced case, it is sometimes possible to use **approximate procedures** that convert the unbalanced problem into a balanced one. This of course makes the analysis only approximate, but the analysis of balanced data is so easy that we are frequently tempted to use this approach. In practice, we must decide when the data are not sufficiently different from the balanced case to make the degree of approximation introduced relatively unimportant. We now briefly describe some of these approximate methods. We assume that every cell has at least one observation (i.e., $n_{ij} \geq 1$).

Estimating Missing Observations ▪ If only a few n_{ij} are different, a reasonable procedure is to estimate the missing values. For example, consider the unbalanced design in Table 6–31. Clearly, estimating the single missing value in cell 2,2 is a reasonable approach. For a model with interaction, the estimate of the missing value in the ijth cell that minimizes the error sum of squares is $\bar{y}_{ij\cdot}$. That is, we estimate the missing value by taking the average of the observations that are available in that cell.

The estimated value is treated just like actual data. The only modification to the analysis of variance is to reduce the error degrees of freedom by the number of missing observations that have been estimated. For example, if we estimate the missing value in cell 2,2 in Table 6–31, we would use 26 error degrees of freedom instead of 27.

Setting Data Aside ▪ Consider the data in Table 6–32. Note that cell 2,2 has only one more observation than the others. Estimating missing values for the remaining 8 cells is probably not a good idea here because this would result in estimates constituting about 18 percent of the final data. An alternative is to set aside one of the observations in cell 2,2, giving a balanced design with $n = 4$ replicates.

The observation that is set aside should be chosen randomly. Furthermore, rather than completely discarding the observation, we could return it to the design, then randomly choose another observation to set aside and repeat the analysis. And, we hope, these two analyses will not lead to conflicting interpretations of the data. If they do, then we suspect that the observation that was set

Table 6–31 The n_{ij} Values
for an Unbalanced Design

Rows	Columns		
	1	2	3
1	4	4	4
2	4	3	4
3	4	4	4

Table 6–32   The n_{ij} Values
for an Unbalanced Design

	Columns		
Rows	1	2	3
1	4	4	4
2	4	5	4
3	4	4	4

aside is an outlier or a wild value and should be handled accordingly. In practice, this confusion is unlikely to occur when only small numbers of observations are set aside and the variability within the cells is small.

Method of Unweighted Means ▪ In this approach, introduced by Yates (1934), the cell averages are treated as data and are subjected to a standard balanced data analysis to obtain sums of squares for rows, columns, and interaction. The error mean square is found as

$$MS_E = \frac{\sum\limits_{i=1}^{a} \sum\limits_{j=1}^{b} \sum\limits_{k=1}^{n_{ij}} (y_{ijk} - \bar{y}_{ij.})^2}{n_{..} - ab} \tag{6–42}$$

Now MS_E estimates σ^2, the variance of y_{ijk}, an individual observation. However, we have done an analysis of variance on the cell *averages,* and since the variance of the average in the ijth cell is σ^2/n_{ij}, the error mean square actually used in the analysis of variance should be an estimate of the average variance of the $\bar{y}_{ij.}$, say

$$\bar{V}(\bar{y}_{ij.}) = \frac{\sum\limits_{i=1}^{a} \sum\limits_{j=1}^{b} \sigma^2/n_{ij}}{ab} = \frac{\sigma^2}{ab} \sum\limits_{i=1}^{a} \sum\limits_{j=1}^{b} \frac{1}{n_{ij}} \tag{6–43}$$

Using MS_E from Equation 6–42 to estimate σ^2 in Equation 6–43, we obtain

$$MS_E' = \frac{MS_E}{ab} \sum\limits_{i=1}^{a} \sum\limits_{j=1}^{b} \frac{1}{n_{ij}} \tag{6–44}$$

as the error mean square (with $n_{..} - ab$ degrees of freedom) to use in the analysis of variance.

The method of unweighted means is an approximate procedure because the sums of squares for rows, columns, and interaction are not distributed as chi-square random variables. The primary advantage of the method seems to be its computational simplicity. When the n_{ij} are not dramatically different, the method of unweighted means often works reasonably well.

A related technique is the **weighted squares of means method,** also proposed by Yates (1934). This technique is also based on the sums of squares of the cell means, but the terms in the sums of squares are weighted in inverse proportion to their variances. For further details of the procedure, see Searle (1971a) and Speed, Hocking, and Hackney (1978).

6–7.3 The Exact Method

In situations where approximate methods are inappropriate, such as when empty cells occur (some $n_{ij} = 0$) or when the n_{ij} are dramatically different, the experimenter must use an exact analysis. The approach used to develop sums of squares for testing main effects and interactions is to represent the analysis of variance model as a regression model, fit that model to the data, and use the general regression significance test approach. However, there are several ways that this may be done, and these methods may result in different values for the sums of squares. Furthermore, the hypotheses that are being tested are not always direct analogs of those for the balanced case, nor are they always easily interpretable. For further reading on the subject, see Searle (1971a); Speed and Hocking (1976); Hocking and Speed (1975); Hocking, Hackney, and Speed (1978); Speed, Hocking, and Hackney (1978); Searle, Speed, and Henderson (1981); Searle (1987); and Milliken and Johnson (1984). The SAS system of statistical software provides an excellent approach to the analysis of unbalanced data through PROC GLM.

6–8 PROBLEMS

6–1 The yield of a chemical process is being studied. The two most important variables are thought to be the pressure and the temperature. Three levels of each factor are selected, and a factorial experiment with two replicates is performed. The yield data follow:

Temperature	Pressure		
	200	215	230
Low	90.4	90.7	90.2
	90.2	90.6	90.4
Medium	90.1	90.5	89.9
	90.3	90.6	90.1
High	90.5	90.8	90.4
	90.7	90.9	90.1

(a) Analyze the data and draw conclusions. Use $\alpha = 0.05$.

(b) Prepare appropriate residual plots and comment on the model's adequacy.

(c) Under what conditions would you operate this process?

6–2   An engineer suspects that the surface finish of a metal part is influenced by the feed rate and the depth of cut. She selects three feed rates and four depths of cut. She then conducts a factorial experiment and obtains the following data:

	Depth of Cut (in)			
Feed Rate (in/min)	0.15	0.18	0.20	0.25
0.20	74	79	82	99
	64	68	88	104
	60	73	92	96
0.25	92	98	99	104
	86	104	108	110
	88	88	95	99
0.30	99	104	108	114
	98	99	110	111
	102	95	99	107

(a) Analyze the data and draw conclusions. Use $\alpha = 0.05$.

(b) Prepare appropriate residual plots and comment on the model's adequacy.

(c) Obtain point estimates of the mean surface finish at each feed rate.

(d) Find the P-values for the tests in part (a).

6–3   For the data in Problem 6–2, compute a 95 percent confidence interval estimate of the mean difference in response for feed rates of 0.20 and 0.25 in/min.

6–4   An article in *Industrial Quality Control* (1956, pp. 5–8) describes an experiment to investigate the effect of the type of glass and the type of phosphor on the brightness of a television tube. The response variable is the current necessary (in microamps) to obtain a specified brightness level. The data are as follows:

Glass Type	Phosphor Type		
	1	2	3
1	280	300	290
	290	310	285
	285	295	290
2	230	260	220
	235	240	225
	240	235	230

(a) Is there any indication that either factor influences brightness? Use $\alpha = 0.05$.

(b) Do the two factors interact? Use $\alpha = 0.05$.

(c) Analyze the residuals from this experiment.

6-5 Johnson and Leone (*Statistics and Experimental Design in Engineering and the Physical Sciences*, Wiley, 1977) describe an experiment to investigate warping of copper plates. The two factors studied were the temperature and the copper content of the plates. The response variable was a measure of the amount of warping. The data were as follows:

Temperature (°C)	Copper Content (%)			
	40	60	80	100
50	17, 20	16, 21	24, 22	28, 27
75	12, 9	18, 13	17, 12	27, 31
100	16, 12	18, 21	25, 23	30, 23
125	21, 17	23, 21	23, 22	29, 31

(a) Is there any indication that either factor affects the amount of warping? Is there any interaction between the factors? Use $\alpha = 0.05$.

(b) Analyze the residuals from this experiment.

(c) Plot the average warping at each level of copper content and compare them to an appropriately scaled t distribution. Describe the differences in the effects of the different levels of copper content on warping. If low warping is desirable, what level of copper content would you specify?

(d) Suppose that temperature cannot be easily controlled in the environment in which the copper plates are to be used. Does this change your answer for part (c)?

6-6 The factors that influence the breaking strength of a synthetic fiber are being studied. Four production machines and three operators are chosen and a factorial experiment is run using fiber from the same production batch. The results are as follows:

Operator	Machine			
	1	2	3	4
1	109	110	108	110
	110	115	109	108
2	110	110	111	114
	112	111	109	112
3	116	112	114	120
	114	115	119	117

(a) Analyze the data and draw conclusions. Use $\alpha = 0.05$.

(b) Prepare appropriate residual plots and comment on the model's adequacy.

6-7   A mechanical engineer is studying the thrust force developed by a drill press. He suspects that the drilling speed and the feed rate of the material are the most important factors. He selects four feed rates and uses a high and low drill speed chosen to represent the extreme operating conditions. He obtains the following results. Analyze the data and draw conclusions. Use $\alpha = 0.05$.

Drill Speed	Feed Rate			
	0.015	0.030	0.045	0.060
125	2.70	2.45	2.60	2.75
	2.78	2.49	2.72	2.86
200	2.83	2.85	2.86	2.94
	2.86	2.80	2.87	2.88

6-8   An experiment is conducted to study the influence of operating temperature and three types of face-plate glass in the light output of an oscilloscope tube. The following data are collected:

Glass Type	Temperature		
	100	125	150
1	580	1090	1392
	568	1087	1380
	570	1085	1386
2	550	1070	1328
	530	1035	1312
	579	1000	1299
3	546	1045	867
	575	1053	904
	599	1066	889

Use $\alpha = 0.05$ in the analysis. Is there a significant interaction effect? Does glass type or temperature affect the response? What conclusions can you draw? Use the method discussed in the text to partition the temperature effect into its linear and quadratic components. Break the interaction down into appropriate components.

6–9 Consider the data in Problem 6–1. Use the method described in the text to compute the linear and quadratic effects of pressure.

6–10 Use Duncan's multiple range test to determine which levels of the pressure factor are significantly different for the data in Problem 6–1.

6–11 An experiment was conducted to determine if either firing temperature or furnace position affects the baked density of a carbon anode. The data are shown below.

Position	Temperature (°C)		
	800	825	850
1	570	1063	565
	565	1080	510
	583	1043	590
2	528	988	526
	547	1026	538
	521	1004	532

Suppose we assume that no interaction exists. Write down the statistical model. Conduct the analysis of variance and test hypotheses on the main effects. What conclusions can be drawn? Comment on the model's adequacy.

6–12 Derive the expected mean squares for a two-factor analysis of variance with one observation per cell, assuming that both factors are fixed.

6–13 Consider the following data from a two-factor factorial experiment. Analyze the data and draw conclusions. Perform a test for nonadditivity. Use $\alpha = 0.05$.

Row Factor	Column Factor			
	1	2	3	4
1	36	39	36	32
2	18	20	22	20
3	30	37	33	34

6–14 The shear strength of an adhesive is thought to be affected by the application pressure and temperature. A factorial experiment is performed in which both factors are assumed to be fixed. Analyze the data and draw conclusions. Perform a test for nonadditivity.

	Temperature (°F)		
Pressure (lb/in²)	250	260	270
120	9.60	11.28	9.00
130	9.69	10.10	9.57
140	8.43	11.01	9.03
150	9.98	10.44	9.80

6–15   Consider the three-factor model

$$y_{ijk} = \mu + \tau_i + \beta_j + \gamma_k + (\tau\beta)_{ij} + (\beta\gamma)_{jk} + \epsilon_{ijk} \quad \begin{cases} i = 1, 2, \ldots, a \\ j = 1, 2, \ldots, b \\ k = 1, 2, \ldots, c \end{cases}$$

Notice that there is only one replicate. Assuming all the factors are fixed, write down the analysis of variance table, including the expected mean squares. What would you use as the "experimental error" in order to test hypotheses?

6–16   The percentage of hardwood concentration in raw pulp, the vat pressure, and the cooking time of the pulp are being investigated for their effects on the strength of paper. Three levels of hardwood concentration, three levels of pressure, and two cooking times are selected. A factorial experiment with two replicates is conducted, and the following data are obtained:

Percentage of Hardwood Concentration	Cooking Time 3.0 Hours			Cooking Time 4.0 Hours		
	Pressure			Pressure		
	400	500	650	400	500	650
2	196.6	197.7	199.8	198.4	199.6	200.6
	196.0	196.0	199.4	198.6	200.4	200.9
4	198.5	196.0	198.4	197.5	198.7	199.6
	197.2	196.9	197.6	198.1	198.0	199.0
8	197.5	195.6	197.4	197.6	197.0	198.5
	196.6	196.2	198.1	198.4	197.8	199.8

(a) Analyze the data and draw conclusions. Use $\alpha = 0.05$.

(b) Prepare appropriate residual plots and comment on the model's adequacy.

(c) Under what set of conditions would you operate this process? Why?

6–17   The quality control department of a fabric finishing plant is studying the effect of several factors on the dyeing of cotton–synthetic cloth used to manufacture men's shirts. Three operators, three cycle times, and two temperatures were selected, and three small specimens of cloth were dyed under each set of conditions. The finished cloth was compared to a standard, and a numerical score was assigned. The results follow. Analyze the data and draw conclusions. Comment on the model's adequacy.

	Temperature					
	300°			350°		
	Operator			Operator		
Cycle Time	1	2	3	1	2	3
40	23	27	31	24	38	34
	24	28	32	23	36	36
	25	26	29	28	35	39
50	36	34	33	37	34	34
	35	38	34	39	38	36
	36	39	35	35	36	31
60	28	35	26	26	36	28
	24	35	27	29	37	26
	27	34	25	25	34	24

6-18 In Problem 6-1, suppose that we wish to reject the null hypothesis with a high probability if the difference in the true mean yield at any two pressures is as great as 0.5. If a reasonable prior estimate of the standard deviation of yield is 0.1, how many replicates should be run?

6-19 Consider the data in Problem 6-2. Suppose that the first observations in cell 1,1 and cell 2,3 are missing. Analyze the data by estimating the missing values. Compare this analysis with that obtained in Problem 6-2. Comment on the adequacy of the approximate analysis in this case.

6-20 Repeat Problem 6-19 using the method of unweighted means.

6-21 Consider the data in Problem 6-2. Suppose that the first observation in every cell except cell 2,2 is missing. Analyze this data by setting aside one observation from cell 2,2. Compare the analysis with that obtained in Problem 6-2. Comment on the adequacy of the approximation in this case.

6-22 The yield of a chemical process is being studied. The two factors of interest are temperature and pressure. Three levels of each factor are selected; however, only 9 runs can be made in one day. The experimenter runs a complete replicate of the design on each day. The data are shown in the following table. Analyze the data, assuming that the days are blocks.

	Day 1 Pressure			Day 2 Pressure		
Temperature	250	260	270	250	260	270
Low	86.3	84.0	85.8	86.1	85.2	87.3
Medium	88.5	87.3	89.0	89.4	89.9	90.3
High	89.1	90.2	91.3	91.7	93.2	93.7

6–23 Consider the data in Problem 6–5. Analyze the data, assuming that replicates are blocks.

6–24 Consider the data in Problem 6–6. Analyze the data, assuming that replicates are blocks.

6–25 An article in the *Journal of Testing and Evaluation* (Vol. 16, no. 2, pp. 508–515) investigated the effects of cyclic loading and environmental conditions on fatigue crack growth at a constant 22 MPa stress for a particular material. The data from this experiment are shown below (the response is crack growth rate).

	Environment		
Frequency	Air	H_2O	Salt H_2O
10	2.29	2.06	1.90
	2.47	2.05	1.93
	2.48	2.23	1.75
	2.12	2.03	2.06
1	2.65	3.20	3.10
	2.68	3.18	3.24
	2.06	3.96	3.98
	2.38	3.64	3.24
0.1	2.24	11.00	9.96
	2.71	11.00	10.01
	2.81	9.06	9.36
	2.08	11.30	10.40

(a) Analyze the data from this experiment (use $\alpha = 0.05$).

(b) Analyze the residuals.

(c) Repeat the analyses from parts (a) and (b) using *ln* (y) as the response. Comment on the results.

6–26 An article in the *IEEE Transactions on Electron Devices* (Nov. 1986, pp. 1754) describes a study on polysilicon doping. The experiment shown below is a variation of their study. The response variable is base current.

Polysilicon	Anneal Temperature (°C)		
Doping (ions)	900	950	1000
1×10^{20}	4.60	10.15	11.01
	4.40	10.20	10.58
2×10^{20}	3.20	9.38	10.81
	3.50	10.02	10.60

(a) Is there evidence (with $\alpha = 0.05$) indicating that either polysilicon doping level or anneal temperature affect base current?

(b) Prepare graphical displays to assist in interpretation of this experiment.

(c) Analyze the residuals and comment on model adequacy.

(d) Is the model

$$y = \beta_0 + \beta_1 x_1 + \beta_2 x_2 + \beta_{22} x_2^2 + \beta_{12} x_1 x_2 + \epsilon$$

supported by this experiment ($x_1 =$ doping level, $x_2 =$ temperature)? Estimate the parameters in this model and plot the response surface.

Chapter 7
The 2^k Factorial Design

7–1 INTRODUCTION

Factorial designs are widely used in experiments involving several factors where it is necessary to study the joint effect of the factors on a response. Chapter 6 presented general methods for the analysis of factorial designs. However, there are several special cases of the general factorial design that are important because they are widely used in research work and also because they form the basis of other designs of considerable practical value.

The most important of these special cases is that of k factors, each at only two levels. These levels may be quantitative, such as two values of temperature, pressure, or time; or they may be qualitative, such as two machines, two operators, the "high" and "low" levels of a factor, or perhaps the presence and absence of a factor. A complete replicate of such a design requires $2 \times 2 \times \cdots \times 2 = 2^k$ observations and is called a 2^k **factorial design.**

This chapter focuses on this extremely important class of designs. Throughout this chapter we assume that (1) the factors are fixed, (2) the designs are completely randomized, and (3) the usual normality assumptions are satisfied.

The 2^k design is particularly useful in the early stages of experimental work, when there are likely to be many factors to be investigated. It provides the smallest number of runs with which k factors can be studied in a complete factorial design. Consequently, these designs are widely used in **factor screening experiments.**

Because there are only two levels for each factor, we assume that the response is approximately linear over the range of the factor levels chosen. In many factor screening experiments, when we are just starting to study the process or system, this is often a reasonable assumption. In Section 7–6, we will present a simple method for checking this assumption, and discuss what action to take if it is violated.

7-2 THE 2^2 DESIGN

The first design in the 2^k series is one with only two factors, say A and B, each run at two levels. This design is called a **2^2 factorial design.** The levels of the factors may be arbitrarily called "low" and "high." As an example, consider an investigation into the effect of the concentration of the reactant and the amount of the catalyst on the conversion (yield) in a chemical process. Let the reactant concentration be factor A, and let the two levels of interest be 15 percent and 25 percent. The catalyst is factor B, with the high level denoting the use of two pounds of the catalyst and the low level denoting the use of only one pound. The experiment is replicated three times, and the data are as follows:

Factor		Treatment	Replicate			
A	B	Combination	I	II	III	Total
−	−	A low, B low	28	25	27	80
+	−	A high, B low	36	32	32	100
−	+	A low, B high	18	19	23	60
+	+	A high, B high	31	30	29	90

The treatment combinations in this design are shown graphically in Figure 7-1. By convention, we denote the effect of a factor by a capital Latin letter. Thus "A" refers to the effect of factor A, "B" refers to the effect of factor B, and "AB" refers to the AB interaction. In the 2^2 design the low and high levels of A and B are denoted by "−" and "+," respectively, on the A and B axes. Thus, − on the A axis represents the low level of concentration (15%) whereas

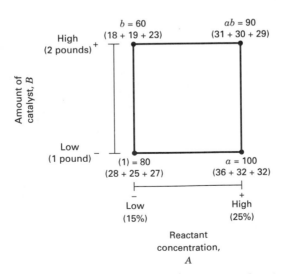

Figure 7-1. Treatment combinations in the 2^2 design.

+ represents the high level (25%), and − on the *B* axis represents the low level of catalyst whereas + denotes the high level.

The four treatment combinations in the design are usually represented by lowercase letters, as shown in Figure 7–1. We see from the figure that the high level of any factor in the treatment combination is denoted by the corresponding lowercase letter and that the low level of a factor in the treatment combination is denoted by the absence of the corresponding letter. Thus, *a* represents the treatment combination of *A* at the high level and *B* at the low level, *b* represents *A* at the low level and *B* at the high level, and *ab* represents both factors at the high level. By convention, (1) is used to denote both factors at the low level. This notation is used throughout the 2^k series.

In a two-level factorial design, we may define the average effect of a factor as the change in response produced by a change in the level of that factor averaged over the levels of the other factor. Also, the notations (1), *a*, *b*, and *ab* now represent the *total* of all *n* replicates taken at the treatment combination, as illustrated in Figure 7–1. Now the effect of *A* at the low level of *B* is $[a − (1)]/n$ and the effect of *A* at the high level of *B* is $[ab − b]/n$. Averaging these two quantities yields the **main effect** of *A*:

$$A = \frac{1}{2n}\{[ab − b] + [a − (1)]\}$$
$$= \frac{1}{2n}[ab + a − b − (1)] \tag{7–1}$$

The average main effect of *B* is found from the effect of *B* at the low level of *A* (i.e., $[b − (1)]/n$) and at the high level of *A* (i.e., $[ab − a]/n$) as

$$B = \frac{1}{2n}\{[ab − a] + [b − (1)]\}$$
$$= \frac{1}{2n}[ab + b − a − (1)] \tag{7–2}$$

We define the **interaction effect** *AB* as the average difference between the effect of *A* at the high level of *B* and the effect of *A* at the low level of *B*. Thus,

$$AB = \frac{1}{2n}\{[ab − b] − [a − (1)]\}$$
$$= \frac{1}{2n}[ab + (1) − a − b] \tag{7–3}$$

Alternatively, we may define *AB* as the average difference between the effect of *B* at the high level of *A* and the effect of *B* at the low level of *A*. This will also lead to Equation 7–3.

The formulas for the effects of *A*, *B*, and *AB* may be derived by another method. The effect of *A* can be found as the difference in the average response of the two treatment combinations on the right-hand side of the square in Figure

7-1 (call this average $\bar{y}_{A^+}$, since it is the average response at the treatment combinations where A is at the high level) and the two treatment combinations on the left-hand side (or $\bar{y}_{A^-}$). That is,

$$A = \bar{y}_{A^+} - \bar{y}_{A^-}$$

$$= \frac{ab + a}{2n} - \frac{b + (1)}{2n}$$

$$= \frac{1}{2n}[ab + a - b - (1)]$$

This is exactly the same result as in Equation 7-1. The effect of B, Equation 7-2, is found as the difference between the average of the two treatment combinations on the top of the square ($\bar{y}_{B^+}$) and the average of the two treatment combinations on the bottom ($\bar{y}_{B^-}$), or

$$B = \bar{y}_{B^+} - \bar{y}_{B^-}$$

$$= \frac{ab + b}{2n} - \frac{a + (1)}{2n}$$

$$= \frac{1}{2n}[ab + b - a - (1)]$$

Finally, the interaction effect AB is the average of the right-to-left diagonal treatment combinations in the square [ab and (1)] minus the average of the left-to-right diagonal treatment combinations (a and b), or

$$AB = \frac{ab + (1)}{2n} - \frac{a + b}{2n}$$

$$= \frac{1}{2n}[ab + (1) - a - b]$$

which is identical to Equation 7-3.

Using the experiment in Figure 7-1, we may estimate the average effects as

$$A = \frac{1}{2(3)}(90 + 100 - 60 - 80) = 8.33$$

$$B = \frac{1}{2(3)}(90 + 60 - 100 - 80) = -5.00$$

$$AB = \frac{1}{2(3)}(90 + 80 - 100 - 60) = 1.67$$

The effect of A (reactant concentration) is positive; this suggests that increasing A from the low level (15%) to the high level (25%) will increase the yield. The effect of B (catalyst) is negative; this suggests that increasing the amount of catalyst added to the process will decrease the yield. The interaction effect appears to be small relative to the two main effects.

In many experiments involving 2^k designs, we will examine the **magnitude** and **direction** of the factor effects to determine which variables are likely to be important. The analysis of variance can generally be used to confirm this interpretation. There are several excellent statistics software packages that are useful for setting up and analyzing 2^k designs. There are also special time-saving methods for performing the calculations manually.

Consider the sums of squares for A, B, and AB. Note from Equation 7–1 that a **contrast** is used in estimating A, namely

$$\text{Contrast}_A = ab + a - b - (1) \tag{7-4}$$

We usually call this contrast the **total effect** of A. From Equations 7–2 and 7–3, we see that contrasts are also used to estimate B and AB. Furthermore, these three contrasts are **orthogonal**. The sum of squares for any contrast can be computed from Equation 3–25, which states that the contrast sum of squares is equal to the contrast squared divided by the number of observations in each total in the contrast times the sum of the squares of the contrast coefficients. Consequently, we have

$$SS_A = \frac{[ab + a - b - (1)]^2}{4n} \tag{7-5}$$

$$SS_B = \frac{[ab + b - a - (1)]^2}{4n} \tag{7-6}$$

and

$$SS_{AB} = \frac{[ab + (1) - a - b]^2}{4n} \tag{7-7}$$

as the sums of squares for A, B, and AB.

Using the experiment in Figure 7–1, we may find the sums of squares from Equations 7–5, 7–6, and 7–7 as

$$SS_A = \frac{(50)^2}{4(3)} = 208.33$$

$$SS_B = \frac{(-30)^2}{4(3)} = 75.00 \tag{7-8}$$

and

$$SS_{AB} = \frac{(10)^2}{4(3)} = 8.33$$

The total sum of squares is found in the usual way, that is,

$$SS_T = \sum_{i=1}^{2}\sum_{j=1}^{2}\sum_{k=1}^{n} y_{ijk}^2 - \frac{y_{\cdots}^2}{4n} \tag{7-9}$$

In general, SS_T has $4n - 1$ degrees of freedom. The error sum of squares, with $4(n - 1)$ degrees of freedom, is usually computed by subtraction as

$$SS_E = SS_T - SS_A - SS_B - SS_{AB} \tag{7-10}$$

For the experiment in Figure 7–1, we obtain

$$SS_T = \sum_{i=1}^{2} \sum_{j=1}^{2} \sum_{k=1}^{3} y_{ijk}^2 - \frac{y_{...}^2}{4(3)}$$

$$= 9398.00 - 9075.00 = 323.00$$

and

$$SS_E = SS_T - SS_A - SS_B - SS_{AB}$$

$$= 323.00 - 208.33 - 75.00 - 8.33$$

$$= 31.34$$

using SS_A, SS_B, and SS_{AB} from Equations 7–8. The complete analysis of variance is summarized in Table 7–1. Based on the P-values, we conclude that the main effects are statistically significant and that there is no interaction between these factors. This confirms our initial interpretation of the data based on the magnitudes of the factor effects.

It is often convenient to write down the treatment combinations in the order (1), a, b, ab. This is referred to as **standard order.** Using this standard order, we see that the contrast coefficients used in estimating the effects are

Effects	(1)	a	b	ab
A:	-1	$+1$	-1	$+1$
B:	-1	-1	$+1$	$+1$
AB:	$+1$	-1	-1	$+1$

Note that the contrast coefficients for estimating the interaction effect are just the product of the corresponding coefficients for the two main effects. The

Table 7–1   Analysis of Variance for the Experiment in Figure 7–1

Source of Variation	Sum of Squares	Degrees of Freedom	Mean Square	F_0	P-Value
A	208.33	1	208.33	53.15	0.0001
B	75.00	1	75.00	19.13	0.0024
AB	8.33	1	8.33	2.13	0.1826
Error	31.34	8	3.92		
Total	323.00	11			

Table 7–2   Algebraic Signs for Calculating
Effects in the 2^2 Design

Treatment	Factorial Effect			
Combination	I	A	B	AB
(1)	+	−	−	+
a	+	+	−	−
b	+	−	+	−
ab	+	+	+	+

contrast coefficient is always either $+1$ or -1, and a **table of plus and minus signs** such as in Table 7–2 can be used to determine the proper sign for each treatment combination. The column headings in Table 7–2 are the main effects (A and B), the AB interaction, and I, which represents the total or average of the entire experiment. Notice that the column corresponding to I has only plus signs. The row designators are the treatment combinations. To find the contrast for estimating any effect, simply multiply the signs in the appropriate column of the table by the corresponding treatment combination and add. For example, to estimate A, the contrast is $-(1) + a - b + ab$, which agrees with Equation 7–1.

The Regression Model ▪ In a 2^k factorial design, it is easy to express the results of the experiment in terms of a **regression model.** For the chemical process experiment in Figure 7–1, the regression model is

$$y = \beta_0 + \beta_1 x_1 + \beta_2 x_2 + \epsilon$$

where x_1 is a **coded variable** that represents the reactant concentration and x_2 is a coded variable that represents the amount of catalyst and the β's are regression coefficients. The relationship between the **natural variables,** the reactant concentration and the amount of catalyst, and the coded variables is

$$x_1 = \frac{Conc - (Conc_{low} + Conc_{high})/2}{(Conc_{high} - Conc_{low})/2}$$

and

$$x_2 = \frac{Catalyst - (Catalyst_{low} + Catalyst_{high})/2}{(Catalyst_{high} - Catalyst_{low})/2}$$

When the natural variables have only two levels, this coding will produce the familiar ± 1 notation for the levels of the coded variables. To illustrate this for our example, note that

$$x_1 = \frac{Conc - (15 + 25)/2}{(25 - 15)/2}$$

$$= \frac{Conc - 20}{5}$$

Thus, if the concentration is at the high level (*Conc* = 25%), then x_1 = +1; if the concentration is at the low level (*Conc* = 15%), then x_1 = −1. Furthermore,

$$x_2 = \frac{Catalyst - (1+2)/2}{(2-1)/2}$$

$$= \frac{Catalyst - 1.5}{0.5}$$

Thus, if the catalyst is at the high level (*Catalyst* = 2 pounds), then x_2 = +1; if the catalyst is at the low level (*Catalyst* = 1 pound), then x_2 = −1.

The fitted regression model is

$$\hat{y} = 27.5 + \left(\frac{8.33}{2}\right)x_1 + \left(\frac{-5.00}{2}\right)x_2$$

where the intercept is the grand average of all 12 observations, and the regression coefficients $\hat{\beta}_1$ and $\hat{\beta}_2$ are one-half the corresponding factor effect estimates. The reason that the regression coefficient is one-half the effect estimate is that a regression coefficient measures the effect of a unit change in x on the mean of y, and the effect estimate is based on a two-unit change (from −1 to +1). We will show later that this simple method of estimating the regression coefficients is producing **least squares** parameter estimates.

Residuals and Model Adequacy ▪ The regression model can be used to obtain the predicted or fitted value of y at the four points in the design. The residuals are the differences between the observed and fitted values of y. For example, when the reactant concentration is at the low level (x_1 = −1) and the catalyst is at the low level (x_2 = −1), the predicted yield is

$$\hat{y} = 27.5 + \left(\frac{8.33}{2}\right)(-1) + \left(\frac{-5.00}{2}\right)(-1)$$

$$= 25.835$$

There are three observations at this treatment combination, and the residuals are

$$e_1 = 28 - 25.835 = 2.165$$
$$e_2 = 25 - 25.835 = -0.835$$
$$e_3 = 27 - 25.835 = 1.165$$

The remaining predicted values and residuals are calculated similarly. For the high level of the reactant concentration and the low level of the catalyst:

$$\hat{y} = 27.5 + \left(\frac{8.33}{2}\right)(+1) + \left(\frac{-5.00}{2}\right)(-1)$$

$$= 34.165$$

and

$$e_4 = 36 - 34.165 = 1.835$$
$$e_5 = 32 - 34.165 = -2.165$$
$$e_6 = 32 - 34.165 = -2.165$$

For the low level of the reactant concentration and the high level of the catalyst:

$$\hat{y} = 27.5 + \left(\frac{8.33}{2}\right)(-1) + \left(\frac{-5.00}{2}\right)(+1)$$
$$= 20.835$$

and

$$e_7 = 18 - 20.835 = -2.835$$
$$e_8 = 19 - 20.835 = -1.835$$
$$e_9 = 23 - 20.835 = 2.165$$

Finally, for the high level of both factors:

$$\hat{y} = 27.5 + \left(\frac{8.33}{2}\right)(+1) + \left(\frac{-5.00}{2}\right)(+1)$$
$$= 29.165$$

and

$$e_{10} = 31 - 29.165 = 1.835$$
$$e_{11} = 30 - 29.165 = 0.835$$
$$e_{12} = 29 - 29.165 = -0.165$$

Figure 7–2 presents a normal probability plot of these residuals and a plot of the residuals versus the predicted yield. These plots appear satisfactory, so we have no reason to suspect problems with the validity of our conclusions.

The Response Surface ▪ The regression model

$$\hat{y} = 27.5 + \left(\frac{8.33}{2}\right)x_1 + \left(\frac{-5.00}{2}\right)x_2$$

can be used to generate response surface plots. If it is desirable to construct these plots in terms of the **natural factor levels,** then we simply substitute the relationships between the natural and the coded variables that we showed earlier into the regression model, yielding

$$\hat{y} = 27.5 + \left(\frac{8.33}{2}\right)\left(\frac{Conc - 20}{5}\right) + \left(\frac{-5.00}{2}\right)\left(\frac{Catalyst - 1.5}{0.5}\right)$$
$$= 18.33 + 0.8333Conc - 5.00Catalyst$$

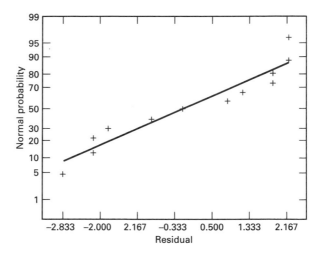

(a) Normal probability plot

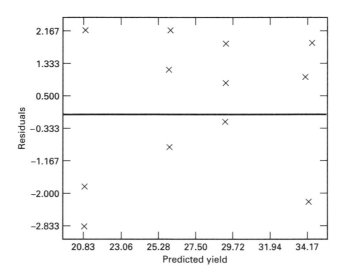

(b) Residuals versus predicted yield

Figure 7–2.   Residual plots for the chemical process experiment.

Figure 7–3*a* presents the three-dimensional response surface plot of yield from this model, and Figure 7–3*b* is the contour plot. Because the model is **first-order** (that is, it contains only the main effects), the fitted response surface is a plane. From examining the contour plot, we see that yield increases as reactant concentration increases and catalyst amount decreases. Often, we use a fitted surface such as this to find a **direction of potential improvement** for a process.

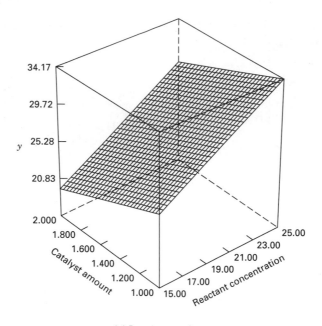

(a) Response surface

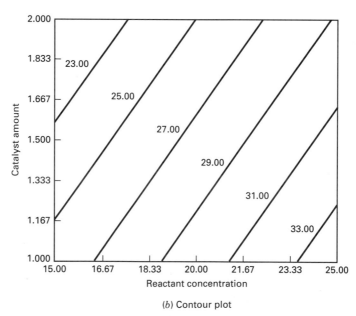

(b) Contour plot

Figure 7–3.   Response surface plot and contour plot of yield from the chemical process experiment.

A formal way to do this, called the **method of steepest ascent,** will be presented in Chapter 14 when we discuss methods for systematically exploring response surfaces.

7-3 THE 2³ DESIGN

Suppose that three factors, A, B, and C, each at two levels, are of interest. The design is called a **2³ factorial design** and the eight treatment combinations can now be displayed geometrically as a cube, as shown in Figure 7–4a on page 302. Using the "+ and –" notation to represent the low and high levels of the factors, we may list the eight runs in the 2³ design as in Figure 7–4b. This is sometimes called the **design matrix.** Extending the label notation discussed in Section 7–2, we write the treatment combinations in standard order as (1), a, b, ab, c, ac, bc, and abc. Remember that these symbols also represent the *total* of all n observations taken at that particular treatment combination.

There are actually three different notations that are widely used for the runs in the 2^k design. The first is the + and – notation, often called the **geometric notation.** The second is the use of lowercase letter labels to identify the treatment combinations. The final notation uses 1 and 0 to denote high and low factor levels, respectively, instead of + and –. These different notations are illustrated below for the 2³ design:

Run	A	B	C	Labels	A	B	C
1	–	–	–	(1)	0	0	0
2	+	–	–	a	1	0	0
3	–	+	–	b	0	1	0
4	+	+	–	ab	1	1	0
5	–	–	+	c	0	0	1
6	+	–	+	ac	1	0	1
7	–	+	+	bc	0	1	1
8	+	+	+	abc	1	1	1

There are seven degrees of freedom beween the eight treatment combinations in the 2³ design. Three degrees of freedom are associated with the main effects of A, B, and C. Four degrees of freedom are associated with interactions; one each with AB, AC, and BC and one with ABC.

Consider estimating the main effects. First, consider estimating the main effect A. The effect of A when B and C are at the low level is $[a - (1)]/n$. Similarly, the effect of A when B is at the high level and C is at the low level is $[ab - b]/n$. The effect of A when C is at the high level and B is at the low level is $[ac - c]/n$. Finally, the effect of A when both B and C are at the high

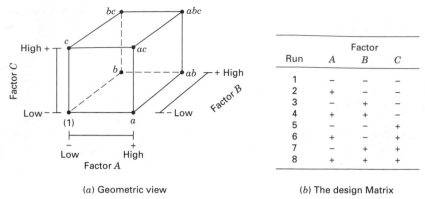

Run	Factor		
	A	B	C
1	−	−	−
2	+	−	−
3	−	+	−
4	+	+	−
5	−	−	+
6	+	−	+
7	−	+	+
8	+	+	+

(a) Geometric view (b) The design Matrix

Figure 7–4. The 2^3 factorial design.

level is $[abc - bc]/n$. Thus, the average effect of A is just the average of these four, or

$$A = \frac{1}{4n}[a - (1) + ab - b + ac - c + abc - bc] \qquad (7\text{–}11)$$

This equation can also be developed as a contrast between the four treatment combinations in the right face of the cube in Figure 7–5a (where A is at the high level) and the four in the left face (where A is at the low level). That is, the A effect is just the average of the four runs where A is at the high level $(\bar{y}_{A^+})$ minus the average of the four runs where A is at the low level $(\bar{y}_{A^-})$, or

$$A = \bar{y}_{A^+} - \bar{y}_{A^-}$$
$$= \frac{a + ab + ac + abc}{4n} - \frac{(1) + b + c + bc}{4n}$$

This equation can be rearranged as

$$A = \frac{1}{4n}[a + ab + ac + abc - (1) - b - c - bc]$$

which is identical to Equation 7–11.

In a similar manner, the effect of B is the difference in averages between the four treatment combinations in the front face of the cube and the four in the back. This yields

$$B = \bar{y}_{B^+} - \bar{y}_{B^-}$$
$$= \frac{1}{4n}[b + ab + bc + abc - (1) - a - c - ac] \qquad (7\text{–}12)$$

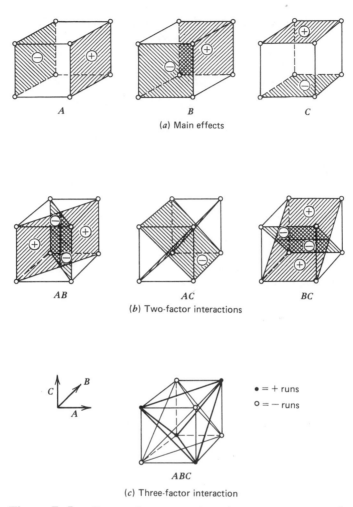

(a) Main effects

(b) Two-factor interactions

(c) Three-factor interaction

• = + runs
○ = − runs

Figure 7–5. Geometric presentation of contrasts corresponding to the main effects and interactions in the 2³ design.

The effect of C is the difference in averages between the four treatment combinations in the top face of the cube and the four in the bottom, that is,

$$C = \bar{y}_{C^+} - \bar{y}_{C^-}$$
$$= \frac{1}{4n}[c + ac + bc + abc - (1) - a - b - ab] \tag{7–13}$$

The two-factor interaction effects may be computed easily. A measure of the AB interaction is the difference between the average A effects at the two levels of B. By convention, one-half of this difference is called the AB interaction. Symbolically,

$\underline{B}$	**Average A Effect**
High (+)	$\dfrac{[(abc - bc) + (ab - b)]}{2n}$
Low (−)	$\dfrac{\{(ac - c) + [a - (1)]\}}{2n}$
Difference	$\dfrac{[abc - bc + ab - b - ac + c - a + (1)]}{2n}$

Since the AB interaction is one-half of this difference,

$$AB = \frac{[abc - bc + ab - b - ac + c - a + (1)]}{4n} \tag{7–14}$$

We could write Equation 7–14 as follows:

$$AB = \frac{abc + ab + c + (1)}{4n} - \frac{bc + b + ac + a}{4n}$$

In this form, the AB interaction is easily seen to be the difference in averages between runs on two diagonal planes in the cube in Figure 7–5b. Using similar logic and referring to Figure 7–5b , the AC and BC interactions are

$$AC = \frac{1}{4n}[(1) - a + b - ab - c + ac - bc + abc] \tag{7–15}$$

and

$$BC = \frac{1}{4n}[(1) + a - b - ab - c - ac + bc + abc] \tag{7–16}$$

The ABC interaction is defined as the average difference between the AB interaction for the two different levels of C. Thus,

$$\begin{aligned} ABC &= \frac{1}{4n}\{[abc - bc] - [ac - c] - [ab - b] + [a - (1)]\} \\ &= \frac{1}{4n}[abc - bc - ac + c - ab + b + a - (1)] \end{aligned} \tag{7–17}$$

As before, we can think of the ABC interaction as the difference in two averages. If the runs in the two averages are isolated, they define the vertices of the two tetrahedra that comprise the cube in Figure 7–5c.

In Equations 7–11 through 7–17, the quantities in brackets are **contrasts** in the treatment combinations. A table of plus and minus signs can be developed from the contrasts and is shown in Table 7–3. Signs for the main effects are

Table 7-3 Algebraic Signs for Calculating Effects in the 2^3 Design

Treatment Combination	Factorial Effect							
	I	A	B	AB	C	AC	BC	ABC
(1)	+	−	−	+	−	+	+	−
a	+	+	−	−	−	−	+	+
b	+	−	+	−	−	+	−	+
ab	+	+	+	+	−	−	−	−
c	+	−	−	+	+	−	−	+
ac	+	+	−	−	+	+	−	−
bc	+	−	+	−	+	−	+	−
abc	+	+	+	+	+	+	+	+

determined by associating a plus with the high level and a minus with the low level. Once the signs for the main effects have been established, the signs for the remaining columns can be obtained by multiplying the appropriate preceding columns, row by row. For example, the signs in the AB column are the product of the A and B column signs in each row. The contrast for any effect can be obtained easily from this table.

Table 7-3 has several interesting properties: (1) Except for column I, every column has an equal number of plus and minus signs. (2) The sum of the products of the signs in any two columns is zero. (3) Column I multiplied times any column leaves that column unchanged. That is, I is an **identity element.** (4) The product of any two columns yields a column in the table. For example, $A \times B = AB$, and

$$AB \times B = AB^2 = A$$

We see that the exponents in the products are formed by using **modulus 2** arithmetic. (That is, the exponent can only be zero or one; if it is greater than one, it is reduced by multiples of two until it is either zero or one.) All of these properties are implied by the orthogonality of the contrasts used to estimate the effects.

Sums of squares for the effects are easily computed, since each effect has a corresponding single-degree-of-freedom contrast. In the 2^3 design with n replicates, the sum of squares for any effect is

$$SS = \frac{(Contrast)^2}{8n} \tag{7-18}$$

Example 7-1

Recall Example 6-3, which presented a study on the effect of percent carbonation, operating pressure, and line speed on the fill height of a carbonated beverage. Suppose that only two levels of carbonation are used so that the experiment is a 2^3 factorial

Table 7-4　Data for the Fill Height Problem, Example 7–1

	Operating Pressure (B)			
	25 psi		30 psi	
	Line Speed (C)		Line Speed (C)	
Percent Carbonation (A)	200	250	200	250
10	-3 -1	-1 0	-1 0	1 1
	$-4 = (1)$	$-1 = c$	$-1 = b$	$2 = bc$
12	0 1	2 1	2 3	6 5
	$1 = a$	$3 = ac$	$5 = ab$	$11 = abc$

design with two replicates. The data, deviations from the target fill height, are shown in Table 7–4, and the design is shown geometrically in Figure 7–6.

Using the totals under the treatment combinations shown in Table 7–4, we may estimate the factor effects as follows:

$$A = \frac{1}{4n} [a - (1) + ab - b + ac - c + abc - bc]$$

$$= \frac{1}{8} [1 - (-4) + 5 - (-1) + 3 - (-1) + 11 - 2]$$

$$= \frac{1}{8} [24] = 3.00$$

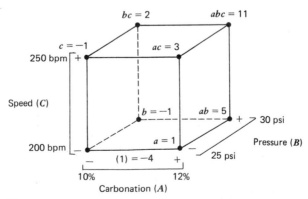

Figure 7–6.　The 2^3 design for the fill height deviation experiment for Example 7–1.

$$B = \frac{1}{4n} [b + ab + bc + abc - (1) - a - c - ac]$$

$$= \frac{1}{8} [-1 + 5 + 2 + 11 - (-4) - 1 - (-1) - 3]$$

$$= \frac{1}{8} [18] = 2.25$$

$$C = \frac{1}{4n} [c + ac + bc + abc - (1) - a - b - ab]$$

$$= \frac{1}{8} [-1 + 3 + 2 + 11 - (-4) - 1 - (-1) - 5]$$

$$= \frac{1}{8} [14] = 1.75$$

$$AB = \frac{1}{4n} [ab - a - b + (1) + abc - bc - ac + c]$$

$$= \frac{1}{8} [5 - 1 - (-1) + (-4) + 11 - 2 - 3 + (-1)]$$

$$= \frac{1}{8} [6] = 0.75$$

$$AC = \frac{1}{4n} [(1) - a + b - ab - c + ac - bc + abc]$$

$$= \frac{1}{8} [-4 - 1 + (-1) - 5 - (-1) + 3 - 2 + 11]$$

$$= \frac{1}{8} [2] = 0.25$$

$$BC = \frac{1}{4n} [(1) + a - b - ab - c - ac + bc + abc]$$

$$= \frac{1}{8} [-4 + 1 - (-1) - 5 - (-1) - 3 + 2 + 11]$$

$$= \frac{1}{8} [4] = 0.50$$

and

$$ABC = \frac{1}{4n} [abc - bc - ac + c - ab + b + a - (1)]$$

$$= \frac{1}{8} [11 - 2 - 3 + (-1) - 5 + (-1) + 1 - (-4)]$$

$$= \frac{1}{8} [4] = 0.50$$

The largest effects are for carbonation ($A = 3.00$), pressure ($B = 2.25$), speed ($C = 1.75$) and the carbonation–pressure interaction ($AB = 0.75$), although the interaction effect does not appear to have as large an impact on fill height deviation as the main effects.

Table 7–5   Analysis of Variance for the Fill Height Data

Source of Variation	Sum of Squares	Degrees of Freedom	Mean Square	F_0	P-Value
Percent carbonation (A)	36.00	1	36.00	57.60	<0.0001
Pressure (B)	20.25	1	20.25	32.40	0.0005
Line speed (C)	12.25	1	12.25	19.60	0.0022
AB	2.25	1	2.25	3.60	0.0943
AC	0.25	1	0.25	0.40	0.5447
BC	1.00	1	1.00	1.60	0.2415
ABC	1.00	1	1.00	1.60	0.2415
Error	5.00	8	0.625		
Total	78.00	15			

The analysis of variance may be used to confirm the magnitude of these effects. From Equation 7–18, the sums of squares are

$$SS_A = \frac{(24)^2}{16} = 36.00$$

$$SS_B = \frac{(18)^2}{16} = 20.25$$

$$SS_C = \frac{(14)^2}{16} = 12.25$$

$$SS_{AB} = \frac{(6)^2}{16} = 2.25$$

$$SS_{AC} = \frac{(2)^2}{16} = 0.25$$

$$SS_{BC} = \frac{(4)^2}{16} = 1.00$$

and

$$SS_{ABC} = \frac{(4)^2}{16} = 1.00$$

The total sum of squares is $SS_T = 78.00$, and by subtraction, $SS_E = 5.00$. The analysis of variance is summarized in Table 7–5, and it confirms the significance of the main effects. The AB interaction is significant at about the 10 percent level; thus there is some mild interaction between carbonation and pressure.

The reader may wish to refer to Example 6–3 for the practical interpretation of this experiment. The process developers decided to run the process at low pressure and high line speed and to reduce the variability in carbonation by controlling the temperature more precisely. This resulted in a substantial reduction in the deviation of fill height from the target value.

■

The Regression Model and Response Surface ▪ The regression model for predicting fill height deviation is

$$\hat{y} = \hat{\beta}_0 + \hat{\beta}_1 x_1 + \hat{\beta}_2 x_2 + \hat{\beta}_3 x_3 + \hat{\beta}_{12} x_1 x_2$$

$$= 1.00 + \left(\frac{3.00}{2}\right) x_1 + \left(\frac{2.25}{2}\right) x_2 + \left(\frac{1.75}{2}\right) x_3 + \left(\frac{0.75}{2}\right) x_1 x_2,$$

where the coded variables x_1, x_2, and x_3 represent A, B, and C, respectively. The $x_1 x_2$ term is the AB interaction. Residuals can be obtained as the difference between observed and predicted fill height deviations. We leave the analysis of these residuals as an exercise for the reader.

Figure 7–7 (page 310) presents the response surface and contour plot for fill height deviation obtained from the regression model, assuming that line speed is at the high level ($x_3 = 1$). Notice that since the model contains interaction, the contour lines of constant fill height deviation are curved (or the response surface is a "twisted" plane). It is desirable to operate this filling process so that fill deviation is as close to zero as possible. The contour plot shows that if line speed is at the high level, then there are several combinations of pressure and carbonation level that will satisfy this objective. However, it will be necessary to control both of these variables very precisely.

Computer Solution ▪ There are many statistics software packages that will set up and analyze two-level factorial designs. The output from one of these computer programs, *Design-Ease,* is shown in Table 7–6 on pages 311–312. In the upper part of the table, before the model is selected, the program displays the effect estimates, the regression coefficients, and the sum of squares for each main effect and interaction. These are in agreement with the manual calculations given previously. Once the full model is selected, an analysis of variance output is presented. The format of this presentation is somewhat different from the results given in Table 7–5. Notice that the analysis of variance is an overall summary for the full model (all main effects and interactions), and the model sum of squares is

$$SS_{\text{Model}} = SS_A + SS_B + SS_C + SS_{AB} + SS_{AC} + SS_{BC} + SS_{ABC}$$

$$= 73.0$$

Thus the statistic

$$F_0 = \frac{MS_{\text{Model}}}{MS_E} = \frac{10.42857}{0.625} = 16.69$$

is testing the hypotheses

$$H_0: \beta_1 = \beta_2 = \beta_3 = \beta_{12} = \beta_{13} = \beta_{23} = \beta_{123} = 0$$
$$H_1: \text{at least one } \beta \neq 0$$

Since F_0 is large, we would conclude that at least one variable has a nonzero effect.

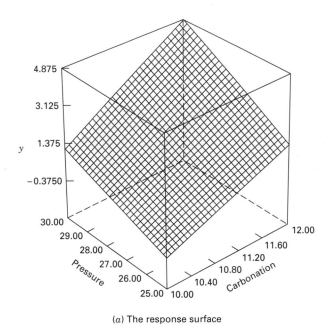

(a) The response surface

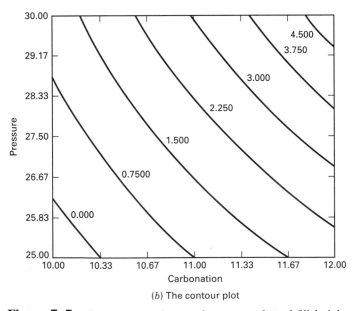

(b) The contour plot

Figure 7-7.   Response surface and contour plot of fill height deviation with speed at the high level (250 bpm), Example 7–1.

Table 7–6 *Design-Ease* Computer Output for Example 7–1

VAR	VARIABLE	UNITS	−1 LEVEL	+1 LEVEL
A	carbonation	percent	10.000	12.000
B	pressure	psi	25.000	30.000
C	speed	bpm	200.000	250.000

VARIABLE	COEFFICIENT	STANDARDIZED EFFECT	SUM OF SQUARES
OVERALL AVERAGE	1.00000		
A	1.50000	3.00000	36.00000
B	1.12500	2.25000	20.25000
C	0.87500	1.75000	12.25000
AB	0.37500	0.75000	2.25000
AC	0.12500	0.25000	0.25000
BC	0.25000	0.50000	1.00000
ABC	0.25000	0.50000	1.00000

Computations done for Factorial

Model selected for Factorial: Full Model

Results of Factorial Model Fitting

ANOVA for Selected Model

SOURCE	SUM OF SQUARES	DF	MEAN SQUARE	F VALUE	PROB > F
MODEL	73.00000	7	10.42857	16.69	0.0003
RESIDUAL	5.00000	8	0.62500		
PURE ERROR	5.00000	8	0.62500		
COR TOTAL	78.00000	15			

ROOT MSE	0.790569		R-SQUARED	0.9359
DEP MEAN	1.000000			
C.V.	79.06%			

VARIABLE	COEFFICIENT ESTIMATE	DF	STANDARD ERROR	t FOR H0 COEFFICIENT=0	PROB > |t|
INTERCEPT	1.000000	1	0.197642		
A	1.500000	1	0.197642	7.589	0.0001
B	1.125000	1	0.197642	5.692	0.0005
C	0.875000	1	0.197642	4.427	0.0022
AB	0.375000	1	0.197642	1.897	0.0943
AC	0.125000	1	0.197642	0.6325	0.5447
BC	0.250000	1	0.197642	1.265	0.2415
ABC	0.250000	1	0.197642	1.265	0.2415

Final Equation in Terms of Coded Variables:

```
fill_dev =
                    1.000000
        +           1.500000 * A
        +           1.125000 * B
        +           0.875000 * C
        +           0.375000 * A * B
        +           0.125000 * A * C
        +           0.250000 * B * C
        +           0.250000 * A * B * C
```

Table 7-6   (continued)

Final Equation in Terms of Uncoded Variables:

```
        fill_dev =
                        -225.500000
                +          21.000000 * carbonation
                +           7.800000 * pressure
                +           1.080000 * speed
                -           0.750000 * carbonation * pressure
                -           0.105000 * carbonation * speed
                -           0.040000 * pressure * speed
                +           0.004000 * carbonation * pressure * speed
```

Model selected for Factorial: A, B, C, AB

Results of Factorial Model Fitting

ANOVA for Selected Model

SOURCE	SUM OF SQUARES	DF	MEAN SQUARE	F VALUE	PROB > F
MODEL	70.75000	4	17.68750	26.84	0.0001
RESIDUAL	7.25000	11	0.65909		
LACK OF FIT	2.25000	3	0.75000	1.200	0.3700
PURE ERROR	5.00000	8	0.62500		
COR TOTAL	78.00000	15			

ROOT MSE	0.811844		R-SQUARED	0.9071
DEP MEAN	1.000000			
C.V.	81.18%			

VARIABLE	COEFFICIENT ESTIMATE	DF	STANDARD ERROR	t FOR H0 COEFFICIENT=0	PROB > |t|
INTERCEPT	1.000000	1	0.202961		
A	1.500000	1	0.202961	7.391	0.0001
B	1.125000	1	0.202961	5.543	0.0002
C	0.875000	1	0.202961	4.311	0.0012
AB	0.375000	1	0.202961	1.848	0.0917

Final Equation in Terms of Coded Variables:

```
        fill_dev =
                        1.000000
                +       1.500000 * A
                +       1.125000 * B
                +       0.875000 * C
                +       0.375000 * A * B
```

Final Equation in Terms of Uncoded Variables:

```
        fill_dev =
                        9.625000
                -       2.625000 * carbonation
                -       1.200000 * pressure
                +       0.035000 * speed
                +       0.150000 * carbonation * pressure
```

In Table 7–5, we constructed F tests for each individual effect. *Design-Ease* uses the t test instead. In Table 7–6, each t statistic is computed from

$$t_0 = \frac{\hat{\beta}}{se(\hat{\beta})} \qquad (7\text{-}19)$$

where $\hat{\beta}$ is the regression coefficient for each factor or interaction, and $se(\hat{\beta})$ is the **standard error** of the regression coefficient, computed from

$$se(\hat{\beta}) = \sqrt{\frac{MS_E}{n2^k}} = \sqrt{\frac{MS_E}{2(2^3)}} = \sqrt{\frac{0.625}{16}} = 0.197642 \qquad (7\text{-}20)$$

These t tests are equivalent to the F tests in Table 7–5. (The square of a t random variable with ν degrees of freedom is an F random variable with 1 numerator and ν denominator degrees of freedom. Note that the square of each t_0 value in Table 7–6 for the full model is approximately equal to the corresponding F_0 value in Table 7–5.) The full model in terms of both the coded variables and the natural variables is also presented.

The last part of the display in Table 7–6 illustrates the output following removal of the nonsignificant interaction terms. The model now contains only the main effects A, B, and C, and the AB interaction. The **error** or **residual** sum of squares is now composed of a **pure error** component arising from the replication of the eight corners of the cube, and a **lack of fit** component consisting of the sums of squares for the interactions that were dropped from the model (BC, AC, and ABC). Once again, the regression model representation of the experimental results is given in terms of both coded and natural variables. The proportion of total variability in fill height deviation that is explained by this model is

$$R^2 = \frac{SS_{\text{Model}}}{SS_{\text{Total}}} = \frac{70.75}{78.00} = 0.9071$$

This computer program will also generate the residuals and construct all of the residual plots that we have previously discussed.

Other Methods for Judging the Significance of Effects ▪ The analysis of variance is a formal way to determine which factor effects are nonzero. There are two other methods that are useful. In the first method, we calculate the **standard error of the effects** and use these standard errors to construct **confidence intervals** on the effects. The second method, which we will illustrate in Section 7–5, uses **normal probability plots** to assess the importance of the effects.

The standard error of an effect is easy to find. If we assume that there are n replicates at each of the 2^k runs in the design, and if $y_{i1}, y_{i2}, \ldots, y_{in}$ are the observations at the ith run, then

$$S_i^2 = \frac{1}{n-1} \sum_{j=1}^{n} (y_{ij} - \bar{y}_i)^2 \qquad i = 1, 2, \ldots, 2^k$$

is an estimate of the variance at the ith run. The 2^k variance estimates can be combined to give an overall variance estimate

$$S^2 = \frac{1}{2^k(n-1)} \sum_{i=1}^{2^k} \sum_{j=1}^{n} (y_{ij} - \bar{y}_i)^2 \qquad (7\text{-}21)$$

This is also the variance estimate given by the error mean square in the analysis of variance. The *variance* of each effect estimate is

$$V(\text{Effect}) = V\left(\frac{\text{Contrast}}{n2^{k-1}}\right)$$

$$= \frac{1}{(n2^{k-1})^2} V(\text{Contrast})$$

Each contrast is a linear combination of 2^k treatment totals, and each total consists of n observations. Therefore,

$$V(\text{Contrast}) = n2^k\sigma^2$$

and the variance of an effect is

$$V(\text{Effect}) = \frac{1}{(n2^{k-1})^2} n2^k\sigma^2$$

$$= \frac{1}{n2^{k-2}} \sigma^2 \qquad (7\text{-}22)$$

The estimated standard error would be found by replacing σ^2 by its estimate S^2 and taking the square root of Equation 7–22.

To illustrate this method, consider the fill height deviation experiment in Example 7–1. The mean square error is $MS_E = 0.625$. Therefore, the standard error of each effect is (using $S^2 = MS_E$)

$$se(\text{Effect}) = \sqrt{\frac{1}{n2^{k-2}} S^2}$$

$$= \sqrt{\frac{1}{2(2^{3-2})} 0.625}$$

$$= 0.40$$

Two standard error limits on the effect estimates are then

$$
\begin{aligned}
A: &\quad 3.00 \pm 0.80 \\
B: &\quad 2.25 \pm 0.80 \\
C: &\quad 1.75 \pm 0.80 \\
AB: &\quad 0.75 \pm 0.80 \\
AC: &\quad 0.25 \pm 0.80 \\
BC: &\quad 0.50 \pm 0.80 \\
ABC: &\quad 0.50 \pm 0.80
\end{aligned}
$$

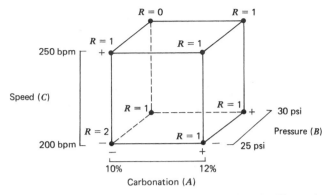

Figure 7–8. Ranges of fill height deviation for Example 7–1.

These intervals are approximate 95 percent confidence intervals on the factor effects. This analysis indicates that *A*, *B*, and *C* are important factors, as they are the only factor effect estimates for which the approximate 95 percent confidence intervals do not include zero.

Dispersion Effects ▪ The process engineer working on the filling process was also interested in **dispersion effects;** that is, do any of the factors affect variability in fill height deviation from run to run? One way to answer the question is to look at the **range** of fill height deviations for each of the eight runs in the 2^3 design. These ranges are plotted on the cube in Figure 7–8. Notice that the ranges are about the same for all eight runs in the design. Consequently, there is no strong evidence indicating that some of the process variables directly affect the variability in fill height deviation in the process.

7–4 THE GENERAL 2^k DESIGN

The methods of analysis that we have presented thus far may be generalized to the case of a **2^k factorial design,** that is, a design with *k* factors each at two levels. The statistical model for a 2^k design would include *k* main effects, $\binom{k}{2}$ two-factor interactions, $\binom{k}{3}$ three-factor interactions, . . . , and one *k*-factor interaction. That is, for a 2^k design the complete model would contain $2^k - 1$ effects. The notation introduced earlier for treatment combinations is also used here. For example, in a 2^5 design *abd* denotes the treatment combination with factors *A*, *B*, and *D* at the high level and factors *C* and *E* at the low level. The treatment combinations may be written in **standard order** by introducing the factors one at a time, with each new factor being successively combined with those that precede it. For example, the standard order for a 2^4 design is (1), *a*, *b*, *ab*, *c*, *ac*, *bc*, *abc*, *d*, *ad*, *bd*, *abd*, *cd*, *acd*, *bcd*, and *abcd*.

The general approach to the statistical analysis of the 2^k design is summarized in Table 7–7. As we have indicated previously, a computer software package is usually employed in this analysis process.

The sequence of steps in Table 7–7 should, by now, be familiar. The first step is to estimate factor effects and examine their signs and magnitudes. This gives the experimenter preliminary information regarding which factors and interactions may be important, and in which directions these factors should be adjusted to improve the response. In forming the initial model for the experiment, we usually choose the **full model,** that is, all main effects and interactions, provided that at least one of the design points has been replicated (in the next section, we discuss a modification to this step). Then in step 3, we use the analysis of variance to formally test for significance of main effects and interaction. Table 7–8 shows the general form of an analysis of variance for a 2^k factorial design with n replicates. Step 4, refine the model, usually consists of removing any nonsignificant variables from the full model. Step 5 is the usual residual analysis to check for model adequacy and to check assumptions. Sometimes model refinement will occur after residual analysis, if we find that the model is inadequate or assumptions are badly violated. The final step usually consists of graphical analysis—either main effect or interaction plots, or response surface and contour plots.

Although the calculations described above are almost always done with a computer, occasionally it is necessary to manually calculate an effect estimate or sum of squares for an effect. To estimate an effect or to compute the sum of squares for an effect, we must first determine the contrast associated with that effect. This can always be done by using a table of plus and minus signs, such as Table 7–2 or Table 7–3. However, for large values of k this is awkward, and we can use an alternate method. In general, we determine the contrast for effect $AB \cdot \cdot \cdot K$ by expanding the right-hand side of

$$\text{Contrast}_{AB \cdots K} = (a \pm 1)(b \pm 1) \cdot \cdot \cdot (k \pm 1) \qquad (7\text{–}23)$$

In expanding Equation 7–23, ordinary algebra is used with "1" being replaced by (1) in the final expression. The sign in each set of parentheses is negative if the factor is included in the effect and positive if the factor is not included.

Table 7–7   Analysis
Procedure for a 2^k Design
_____

1. Estimate factor effects
2. Form initial model
3. Perform statistical testing
4. Refine model
5. Analyze residuals
6. Interpret results
_____

Table 7–8   Analysis of Variance for a 2^k Design

Source of Variation	Sum of Squares	Degrees of Freedom
k main effects		
A	SS_A	1
B	SS_B	1
$\vdots$	$\vdots$	$\vdots$
K	SS_K	1
$\binom{k}{2}$ two-factor interactions		
AB	SS_{AB}	1
AC	SS_{AC}	1
$\vdots$	$\vdots$	$\vdots$
JK	SS_{JK}	1
$\binom{k}{3}$ three-factor interactions		
ABC	SS_{ABC}	1
ABD	SS_{ABD}	1
$\vdots$	$\vdots$	$\vdots$
IJK	SS_{IJK}	1
$\vdots$	$\vdots$	$\vdots$
$\binom{k}{k} = 1$ k-factor interaction		
$ABC \cdots K$	$SS_{ABC \cdots K}$	1
Error	SS_E	$2^k(n-1)$
Total	SS_T	$n2^k - 1$

To illustrate the use of Equation 7–23, consider a 2^3 factorial design. The contrast for AB would be

$$\text{Contrast}_{AB} = (a-1)(b-1)(c+1)$$
$$= abc + ab + c + (1) - ac - bc - a - b$$

As a further example, in a 2^5 design, the contrast for $ABCD$ would be

$$\text{Contrast}_{ABCD} = (a-1)(b-1)(c-1)(d-1)(e+1)$$
$$= abcde + cde + bde + ade + bce$$
$$+ ace + abe + e + abcd + cd + bd$$
$$+ ad + bc + ac + ab + (1) - a - b - c$$
$$- abc - d - abd - acd - bcd - ae$$
$$- be - ce - abce - de - abde - acde - bcde$$

Once the contrasts for the effects have been computed, we may estimate the effects and compute the sums of squares according to

$$AB \cdot \cdot \cdot K = \frac{2}{n2^k} (\text{Contrast}_{AB\cdots K}) \qquad (7\text{--}24)$$

and

$$SS_{AB\cdots K} = \frac{1}{n2^k} (\text{Contrast}_{AB\cdots K})^2 \qquad (7\text{--}25)$$

respectively, where n denotes the number of replicates. We will present another method for estimating the effects in the 2^k design in Section 7–7.

7–5   A SINGLE REPLICATE OF THE 2^k DESIGN

For even a moderate number of factors, the total number of treatment combinations in a 2^k factorial design is large. For example, a 2^5 design has 32 treatment combinations, a 2^6 design has 64 treatment combinations, and so on. Since resources are usually limited, the number of replicates that the experimenter can employ may be restricted. Frequently, available resources only allow a **single replicate** of the design to be run, unless the experimenter is willing to omit some of the original factors. Usually, the experimenter will consider this strategy when he or she is comfortable with the assumption that the random error (or noise) in the process being studied is moderately small.

A single replicate of a 2^k design is sometimes called an **unreplicated factorial.** With only one replicate, there is no internal estimate of error (or "pure error"). One approach to the analysis of an unreplicated factorial is to assume that certain high-order interactions are negligible and combine their mean squares to estimate the error. This is an appeal to the **sparsity of effects principle;** that is, most systems are dominated by some of the main effects and low-order interactions, and most high-order interactions are negligible.

When analyzing data from unreplicated factorial designs, occasionally real high-order interactions occur. The use of an error mean square obtained by pooling high-order interactions is inappropriate in these cases. A method of analysis attributed to Daniel (1959) provides a simple way to overcome this problem. Daniel suggests examining a **normal probability plot** of the estimates of the effects. The effects that are negligible are normally distributed, with mean zero and variance σ^2 and will tend to fall along a straight line on this plot, whereas significant effects will have nonzero means and will not lie along the straight line. Thus the preliminary model will be specified to contain those effects that are apparently nonzero, based on the normal probability plot. The apparently negligible effects are combined as an estimate of error.

Example 7-2

A Single Replicate of the 2^4 Design

A chemical product is produced in a pressure vessel. A factorial experiment is carried out in the pilot plant to study the factors thought to influence the filtration rate of this product. The four factors are temperature (A), pressure (B), concentration of formaldehyde (C), and stirring rate (D). Each factor is present at two levels. The design matrix and the response data obtained from a single replicate of the 2^4 experiment are shown in Table 7-9 and Figure 7-9. The 16 runs are made in random order. The process engineer is interested in maximizing the filtration rate. Current process conditions give filtration rates of around 75 gal/h. The process also currently uses the concentration of formaldehyde, factor C, at the high level. The engineer would like to reduce the formaldehyde concentration as much as possible but has been unable to do so because it always results in lower filtration rates.

We will begin the analysis of this data by constructing a normal probability plot of the effect estimates. The table of plus and minus signs for the contrast constants for the 2^4 design are shown in Table 7-10 on page 321. From these contrasts, we may estimate the 15 factorial effects that follow:

$$
\begin{array}{llll}
A = 21.625 & AB = & 0.125 & ABC = & 1.875 \\
B = & 3.125 & AC = -18.125 & ABD = & 4.125 \\
C = & 9.875 & AD = & 16.625 & ACD = -1.625 \\
D = 14.625 & BC = & 2.375 & BCD = -2.625 \\
& BD = & -0.375 & ABCD = & 1.375 \\
& CD = & -1.125 &
\end{array}
$$

Table 7-9   Pilot Plant Filtration Rate Experiment

Run Number	Factor A	B	C	D	Run Label	Filtration Rate (gal/h)
1	−	−	−	−	(1)	45
2	+	−	−	−	a	71
3	−	+	−	−	b	48
4	+	+	−	−	ab	65
5	−	−	+	−	c	68
6	+	−	+	−	ac	60
7	−	+	+	−	bc	80
8	+	+	+	−	abc	65
9	−	−	−	+	d	43
10	+	−	−	+	ad	100
11	−	+	−	+	bd	45
12	+	+	−	+	abd	104
13	−	−	+	+	cd	75
14	+	−	+	+	acd	86
15	−	+	+	+	bcd	70
16	+	+	+	+	abcd	96

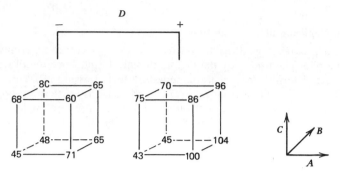

Figure 7–9.   Data from the pilot plant filtration rate experiment for Example 7–2.

The normal probability plot of these effects is shown in Figure 7–10. All of the effects that lie along the line are negligible, whereas the large effects are far from the line. The important effects that emerge from this analysis are the main effects of A, C, and D and the AC and AD interactions.

The main effects of A, C, and D are plotted in Figure 7–11a. All three effects

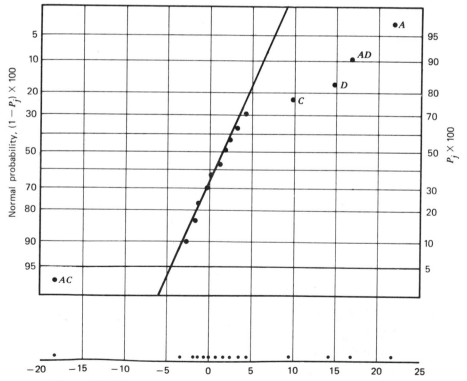

Figure 7–10.   Ordered effects for the 2^4 factorial in Example 7–2.

Table 7-10 Contrast Constants for the 2^4 Design

	A	B	AB	C	AC	BC	ABC	D	AD	BD	ABD	CD	ACD	BCD	ABCD
(1)	−	−	+	−	+	+	−	−	+	+	−	+	−	−	+
a	+	−	−	−	−	+	+	−	−	+	+	+	+	−	−
b	−	+	−	−	+	−	+	−	+	−	+	+	−	+	−
ab	+	+	+	−	−	−	−	−	−	−	−	+	+	+	+
c	−	−	+	+	−	−	+	−	+	+	−	−	+	+	−
ac	+	−	−	+	+	−	−	−	−	+	+	−	−	+	+
bc	−	+	−	+	−	+	−	−	+	−	+	−	+	−	+
abc	+	+	+	+	+	+	+	−	−	−	−	−	−	−	−
d	−	−	+	−	+	+	−	+	−	−	+	−	+	+	−
ad	+	−	−	−	−	+	+	+	+	−	−	−	−	+	+
bd	−	+	−	−	+	−	+	+	−	+	−	−	+	−	+
abd	+	+	+	−	−	−	−	+	+	+	+	−	−	−	−
cd	−	−	+	+	−	−	+	+	−	−	+	+	−	−	+
acd	+	−	−	+	+	−	−	+	+	−	−	+	+	−	−
bcd	−	+	−	+	−	+	−	+	−	+	−	+	−	+	−
abcd	+	+	+	+	+	+	+	+	+	+	+	+	+	+	+

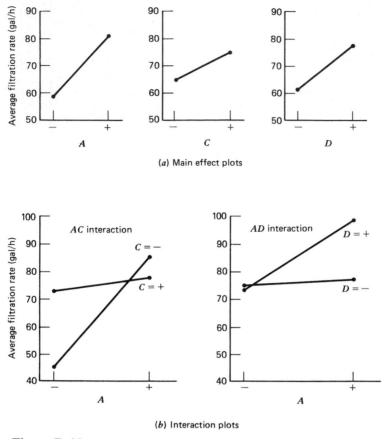

(a) Main effect plots

(b) Interaction plots

Figure 7–11.   Main effect and interaction plots for Example 7–2.

are positive, and if we considered only these main effects, we would run all three factors at the high level to maximize the filtration rate. However, it is always necessary to examine any interactions that are important. Remember that main effects do not have much meaning when they are involved in significant interactions.

The AC and AD interactions are plotted in Figure 7–11b. These interactions are the key to solving the problem. Note from the AC interaction that the temperature effect is very small when the concentration is at the high level and very large when the concentration is at the low level, with the best results obtained with low concentration and high temperature. The AD interaction indicates that stirring rate D has little effect at low temperature but a large positive effect at high temperature. Therefore, the best filtration rates would appear to be obtained when A and D are at the high level and C is at the low level. This would allow the reduction of the formaldehyde concentration to a lower level, another objective of the experimenter.

■

Design Projection ▪ Another interpretation of the effects in Figure 7–10 is possible. Since B (pressure) is not significant and all interactions involving B are negligible, we may discard B from the experiment so that the design becomes a 2^3 factorial in A, C, and D with two replicates. This is easily seen from examining only columns A, C, and D in the design matrix shown in Table 7–9 and noting that those columns form two replicates of a 2^3 design. The analysis of variance for the data using this simplifying assumption is summarized in Table 7–11. The conclusions that we would draw from this analysis are essentially unchanged from those of Example 7–2. Note that by projecting the single replicate of the 2^4 into a replicated 2^3, we now have both an estimate of the ACD interaction and an estimate of error based on replication.

The concept of projecting an unreplicated factorial into a replicated factorial in fewer factors is very useful. In general, if we have a single replicate of a 2^k design and if h ($h < k$) factors are negligible and can be dropped, then the original data correspond to a full two-level factorial in the remaining $k - h$ factors with 2^h replicates.

Diagnostic Checking ▪ The usual diagnostic checks should be applied to the residuals of a 2^k design. Our analysis indicates that the only significant effects are $A = 21.625$, $C = 9.875$, $D = 14.625$, $AC = -18.125$, and $AD = 16.625$. If this is true, then the estimated filtration rates at the vertices of the design are given by

$$\hat{y} = 70.06 + \left(\frac{21.625}{2}\right) x_1 + \left(\frac{9.875}{2}\right) x_3 + \left(\frac{14.625}{2}\right) x_4 - \left(\frac{18.125}{2}\right) x_1 x_3$$

$$+ \left(\frac{16.625}{2}\right) x_1 x_4$$

Table 7–11   Analysis of Variance for the Pilot Plant Filtration Rate Experiment in A, C, and D

Source of Variation	Sum of Squares	Degrees of Freedom	Mean Square	F_0	P-Value
A	1870.56	1	1870.56	83.36	<0.0001
C	390.06	1	390.06	17.38	<0.0001
D	855.56	1	855.56	38.13	<0.0001
AC	1314.06	1	1314.06	58.56	<0.0001
AD	1105.56	1	1105.56	49.27	<0.0001
CD	5.06	1	5.06	<1	
ACD	10.56	1	10.56	<1	
Error	179.52	8	22.44		
Total	5730.94	15			

where 70.06 is the average response and the coded variables x_1, x_3, x_4 take on the values $+1$ or -1. The predicted filtration rate at run (1) is

$$\hat{y} = 70.06 + \left(\frac{21.625}{2}\right)(-1) + \left(\frac{9.875}{2}\right)(-1) + \left(\frac{14.625}{2}\right)(-1)$$
$$- \left(\frac{18.125}{2}\right)(-1)(-1) + \left(\frac{16.625}{2}\right)(-1)(-1)$$
$$= 46.22$$

Since the observed value is 45, the residual is $e = y - \hat{y} = 45 - 46.22 = -1.22$. The values of y, $\hat{y}$, and $e = y - \hat{y}$ for all 16 observations follow.

	y	$\hat{y}$	$e = y - \hat{y}$
(1)	45	46.22	-1.22
a	71	69.39	1.61
b	48	46.22	1.78
ab	65	69.39	-4.39
c	68	74.23	-6.23
ac	60	61.14	-1.14
bc	80	74.23	5.77
abc	65	61.14	3.86
d	43	44.22	-1.22
ad	100	100.65	-0.65
bd	45	44.22	0.78
abd	104	100.65	3.35
cd	75	72.23	2.77
acd	86	92.40	-6.40
bcd	70	72.23	-2.23
abcd	96	92.40	3.60

A normal probability plot of the residuals is shown in Figure 7–12. The points on this plot lie reasonably close to a straight line, lending support to our conclusion that A, C, D, AC, and AD are the only significant effects and that the underlying assumptions of the analysis are satisfied.

The Response Surface ▪ We used the interaction plots in Figure 7–11 to provide a practical interpretation of the results of this experiment. Sometimes we find it helpful to use the response surface for this purpose. The response surface is generated by the regression model

$$\hat{y} = 70.06 + \left(\frac{21.625}{2}\right)x_1 + \left(\frac{9.875}{2}\right)x_3 + \left(\frac{14.625}{2}\right)x_4$$
$$- \left(\frac{18.125}{2}\right)x_1x_3 + \left(\frac{16.625}{2}\right)x_1x_4$$

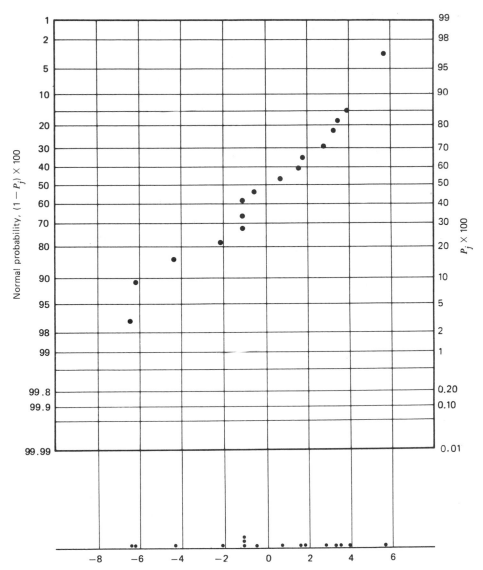

Figure 7-12. Normal probability plot of residuals for Example 7-2.

Figure 7-13a shows the response surface contour plot when stirring rate is at the high level (i.e., $x_4 = 1$). The contours are generated from the above model with $x_4 = 1$, or

$$\hat{y} = 77.3725 + \left(\frac{38.25}{2}\right)x_1 + \left(\frac{9.875}{2}\right)x_3 - \left(\frac{18.125}{2}\right)x_1x_3$$

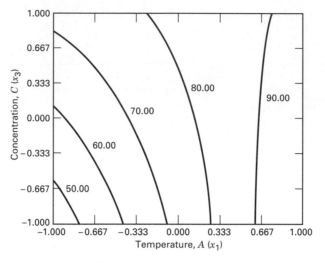

(a) Contour plot with stirring rate (D), $x_4 = 1$

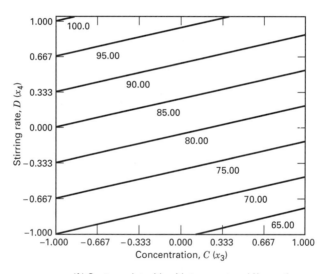

(b) Contour plot with with temperature (A), $x_1 = 1$

Figure 7–13. Contour plots of filtration rate, Example 7–2.

Notice that the contours are curved lines because the model contains an interaction term.

Figure 7–13b is the response surface contour plot when temperature is at the high level (i.e., $x_1 = 1$). When we put $x_1 = 1$ in the regression model we obtain

$$\hat{y} = 80.8725 - \left(\frac{8.25}{2}\right) x_3 + \left(\frac{31.25}{2}\right) x_4$$

These contours are parallel straight lines because the model contains only the main effects of factors C (x_3) and D (x_4).

Both contour plots indicate that if we want to maximize the filtration rate, variables A (x_1) and D (x_4) should be at the high level, and that the process is relatively robust to concentration C. We obtained similar conclusions from the interaction graphs.

Example 7–3

Data Transformation in a Factorial Design

Daniel (1976) describes a 2^4 factorial design used to study the advance rate of a drill as a function of four factors: drill load (A), flow rate (B), rotational speed (C), and the type of drilling mud used (D). The data from the experiment are shown in Figure 7–14.

The normal probability plot of the effect estimates from this experiment is shown in Figure 7–15. Based on this plot, factors B, C, and D along with the BC and BD interactions require interpretation. Figure 7–16 is the normal probability plot of the residuals and Figure 7–17 is the plot of the residuals versus the predicted advance rate from the model containing the identified factors. There are clearly problems with normality and equality of variance. A data transformation is often used to deal with such problems. As the response variable is a rate, the log transformation seems a reasonable candidate. (The methods described in Chapter 3 could be used to select an appropriate transformation, if desired.)

Figure 7–18 presents a normal probability plot of the effect estimates following the transformation $y^* = \ln y$. Notice that a much simpler interpretation now seems possible, as only factors B, C, and D are active. That is, expressing the data in the correct metric has simplified its structure to the point that the two interactions are no longer required in the explanatory model.

Figures 7–19 and 7–20 present, respectively, a normal probability plot of the residuals and a plot of the residuals versus the predicted advance rate for the model in the log scale containing B, C, and D. These plots are now satisfactory. We conclude

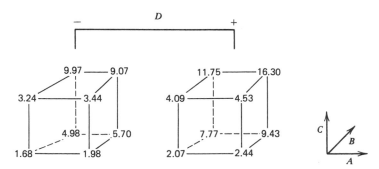

Figure 7–14.   Data from the drilling experiment of Example 7–3.

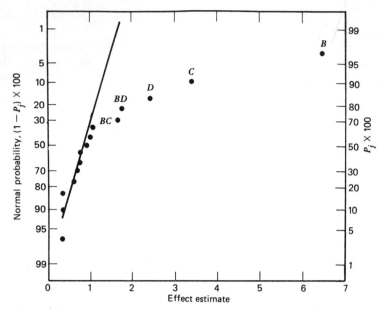

Figure 7–15.   Normal probability plot of effects for Example 7–3.

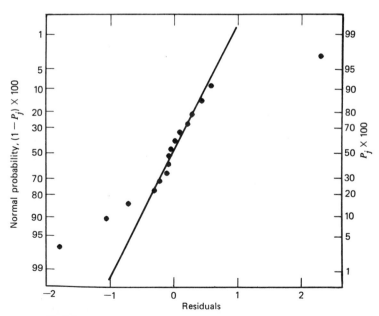

Figure 7–16.   Normal probability plot of residuals for Example 7–3.

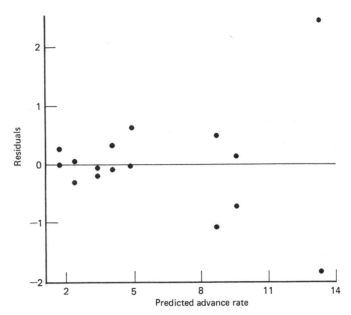

Figure 7–17.   Plot of residuals versus predicted rate for Example 7–3.

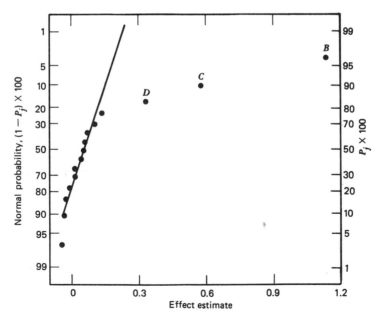

Figure 7–18.   Normal probability plot of effects for Example 7–3 following log transformation.

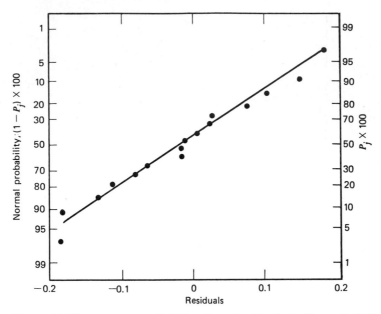

Figure 7–19.   Normal probability plot of residuals for Example 7–3 following log transformation.

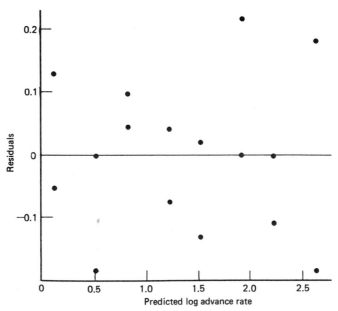

Figure 7–20.   Plot of residuals versus predicted rate for Example 7–3 following log transformation.

Table 7–12 Analysis of Variance for Example 7–3 Following the Log Transformation

Source of Variation	Sum of Squares	Degrees of Freedom	Mean Square	F_0	P-Value
B (Flow)	5.345	1	5.345	381.79	<0.0001
C (Speed)	1.339	1	1.339	95.64	<0.0001
D (Mud)	0.431	1	0.431	30.79	<0.0001
Error	0.173	12	0.014		
Total	7.288	15			

that the model for $y^* = \ln y$ requires only factors B, C, and D for adequate interpretation. The analysis of variance for this model is summarized in Table 7–12. The model sum of squares is

$$SS_{Model} = SS_B + SS_C + SS_D$$
$$= 5.345 + 1.339 + 0.431$$
$$= 7.115$$

and $R^2 = SS_{Model}/SS_T = 7.115/7.288 = 0.98$, so the model explains about 98 percent of the variability in the drill advance rate.

■

Example 7–4

Location and Dispersion Effects in an Unreplicated Factorial

A 2^4 design was run in a manufacturing process producing interior side-wall and window panels for commercial aircraft. The panels are formed in a press, and under present conditions the average number of defects per panel in a press load is much too high. (The current process average is 5.5 defects per panel.) Four factors are investigated using a single replicate of a 2^4 design, with each replicate corresponding to a single press load. The factors are temperature (A), clamp time (B), resin flow (C), and press closing time (D). The data for this experiment are shown in Figure 7–21.

A normal probability plot of the factor effects is shown in Figure 7–22. Clearly the two largest effects are $A = 5.75$ and $C = -4.25$. No other factor effects appear to be large, and A and C explain about 77 percent of the total variability, so we conclude that lower temperature (A) and higher resin flow (C) would reduce the incidence of panel defects.

Careful residual analysis is an important aspect of any experiment. A normal probability plot of the residuals showed no anomalies, but when the experimenter plotted the residuals versus each of the factors A through D, the plot of residuals versus B (clamp time) presented the pattern in Figure 7–23. This factor, which is unimportant insofar as the average number of defects per panel is concerned, is very important in its effect on process variability, with the lower clamp time resulting in less variability in the average number of defects per panel in a press load.

The dispersion effect of clamp time is also very evident from the **cube plot** in Figure 7–24, which plots the average number of defects per panel and the range of the number of defects at each point in the cube defined by factors A, B, and C. The

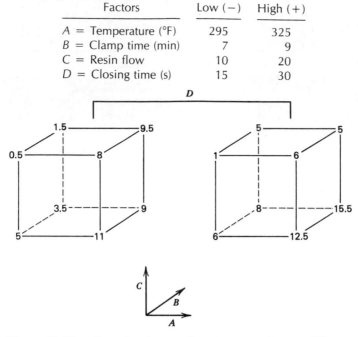

Factors	Low (−)	High (+)
A = Temperature (°F)	295	325
B = Clamp time (min)	7	9
C = Resin flow	10	20
D = Closing time (s)	15	30

Figure 7–21. Data for the panel process experiment of Example 7–4.

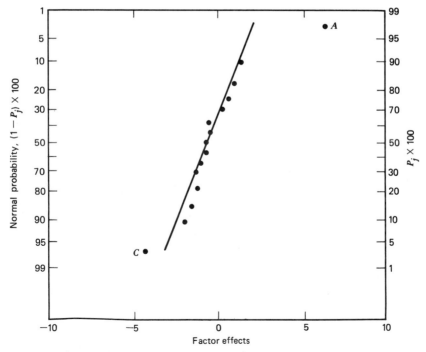

Figure 7–22. Normal probability plot of the factor effects for the panel process experiment of Example 7–4.

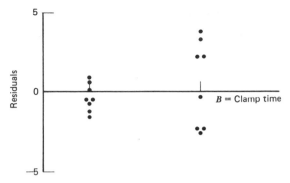

Figure 7-23. Plot of residuals versus clamp time for Example 7-4.

average range when B is at the high level (the back face of the cube in Figure 7-24) is $\overline{R}_{B^+} = 4.75$, and when B is at the low level it is $\overline{R}_{B^-} = 1.25$.

As a result of this experiment, the engineer decided to run the process at low temperature and high resin flow to reduce the average number of defects, at low clamp time to reduce the variability in the number of defects per panel, and at low press closing time (which had no effect on either location or dispersion). The new set of operating conditions resulted in a new process average of less than one defect per panel.

∎

The residuals from a 2^k design provide much information about the problem under study. Since residuals can be thought of as observed values of the noise or error, they often give insight into process variability. We can systematically examine the residuals from an unreplicated 2^k design to provide information about process variability.

Consider the residual plot in Figure 7-23. The standard deviation of the eight residuals where B is at the low level is $S(B^-) = 0.83$, and the standard

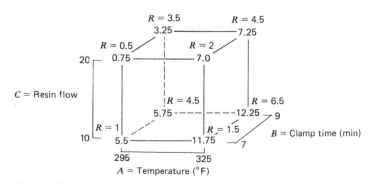

Figure 7-24. Cube plot of temperature, clamp time, and resin flow for Example 7-4.

Table 7-13 Calculation of Dispersion Effects for Example 7-4

Run	A	B	AB	C	AC	BC	ABC	D	AD	BD	ABD	CD	ACD	BCD	ABCD	Residual
1	−	−	+	−	+	+	−	−	+	+	−	+	−	−	+	−0.94
2	+	−	−	−	−	+	+	−	−	+	+	+	+	−	−	−0.69
3	−	+	−	−	+	−	+	−	+	−	+	+	−	+	−	−2.44
4	+	+	+	−	−	−	−	−	−	−	−	+	+	+	+	−2.69
5	−	−	+	+	−	−	+	−	+	+	−	−	+	+	−	−1.19
6	+	−	−	+	+	−	−	−	−	+	+	−	−	+	+	0.56
7	−	+	−	+	−	+	−	−	+	−	+	−	+	−	+	−0.19
8	+	+	+	+	+	+	+	−	−	−	−	−	−	−	−	2.06
9	−	−	+	−	+	+	−	+	−	−	+	−	+	+	−	0.06
10	+	−	−	−	−	+	+	+	+	−	−	−	−	+	+	0.81
11	−	+	−	−	+	−	+	+	−	+	−	−	+	−	+	2.06
12	+	+	+	−	−	−	−	+	+	+	+	−	−	−	−	3.81
13	−	−	+	+	−	−	+	+	−	−	+	+	−	−	+	−0.69
14	+	−	−	+	+	−	−	+	+	−	−	+	+	−	−	−1.44
15	−	+	−	+	−	+	−	+	−	+	−	+	−	+	−	3.31
16	+	+	+	+	+	+	+	+	+	+	+	+	+	+	+	−2.44
$S(i^+)$	2.25	2.72	2.21	1.91	1.81	1.80	1.80	2.24	2.05	2.28	1.97	1.93	1.52	2.09	1.61	
$S(i^-)$	1.85	0.83	1.86	2.20	2.24	2.26	2.24	1.55	1.93	1.61	2.11	1.58	2.16	1.89	2.33	
F_i^*	0.39	2.37	0.34	−0.28	−0.43	−0.46	−0.44	0.74	0.12	0.70	−0.14	0.40	−0.70	0.28	−0.74	

deviation of the eight residuals where B is at the high level is $S(B^+) = 2.72$. The statistic

$$F_B^* = \ln \frac{S^2(B^+)}{S^2(B^-)} \tag{7-26}$$

has an approximate normal distribution if the two variances $\sigma^2(B^+)$ and $\sigma^2(B^-)$ are equal. To illustrate the calculations, the value of F_B^* is

$$F_B^* = \ln \frac{S^2(B^+)}{S^2(B^-)}$$

$$= \ln \frac{(2.72)^2}{(0.83)^2}$$

$$= 2.37$$

Table 7–13 presents the complete set of contrasts for the 2^4 design along with the residuals for each run from the panel process experiment in Example 7–4. Each column in this table contains an equal number of plus and minus signs, and we can calculate the standard deviation of the residuals for each group of signs in each column, say $S(i^+)$ and $S(i^-)$, $i = 1, 2, \ldots, 15$. Then

$$F_i^* = \ln \frac{S^2(i^+)}{S^2(i^-)} \qquad i = 1, 2, \ldots, 15 \tag{7-27}$$

is a statistic that can be used to assess the magnitude of the **dispersion effects** in the experiment. If the variance of the residuals for the runs where factor i is

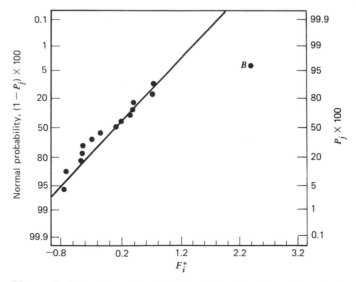

Figure 7–25.   Normal probability plot of the dispersion effects F_i^* for Example 7–4.

positive equals the variance of the residuals for the runs where factor i is negative, then F_i^* has an approximate normal distribution. The values of F_i^* are shown below each column in Table 7–13.

Figure 7–25 is a normal probability plot of the dispersion effects F_i^*. Clearly, B is an important factor with respect to process dispersion. For more discussion of this procedure, see Box and Meyer (1986).

7–6   THE ADDITION OF CENTER POINTS TO THE 2^k DESIGN

A potential concern in the use of two-level factorial designs is the assumption of **linearity** in the factor effects. Of course, perfect linearity is unnecessary, and the 2^k system will work quite well even when the linearity assumption holds only very approximately. In fact, we have noted that if **interaction terms** are added to a main effects or first-order model, resulting in

$$y = \beta_0 + \sum_{j=1}^{k} \beta_j x_j + \sum\sum_{i<j} \beta_{ij} x_i x_j + \epsilon \qquad (7\text{--}28)$$

then we have a model capable of representing some curvature in the response function. This curvature, of course, results from the twisting of the plane induced by the interaction terms $\beta_{ij} x_i x_j$.

There are going to be situations where the curvature in the response function is not adequately modeled by Equation 7–28. In such cases, a logical model to consider is

$$y = \beta_0 + \sum_{j=1}^{k} \beta_j x_j + \sum\sum_{i<j} \beta_{ij} x_i x_j + \sum_{j=1}^{k} \beta_{jj} x_j^2 + \epsilon \qquad (7\text{--}29)$$

where the β_{jj} represent pure second-order or **quadratic effects.** Equation 7–29 is called a **second-order response surface model.**

In running a two-level factorial experiment, we usually anticipate fitting the first-order model in Equation 7–28, but we should be alert to the possibility that the second-order model in Equation 7–29 is really more appropriate. There is a method of replicating certain points in a 2^k factorial that will provide protection against curvature from second-order effects as well as allow an independent estimate of error to be obtained. The method consists of adding **center points** to the 2^k design. These consist of n replicates run at the points $x_i = 0$ ($i = 1, 2, \ldots, k$). One important reason for adding the replicate runs at the design center is that center points do not impact the usual effects estimates in a 2^k design. When we add center points, we assume that the k factors are **quantitative.**

To illustrate the approach, consider a 2^2 design with one observation at each of the factorial points $(-, -)$, $(+, -)$, $(-, +)$ and $(+, +)$ and n_C observations at the center point $(0, 0)$. Figure 7–26 illustrates the situation. Let $\bar{y}_F$ be the average of the four runs at the four factorial points and let $\bar{y}_C$ be the average of

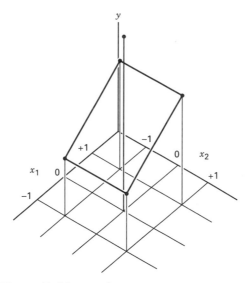

Figure 7–26.   A 2^2 design with center points.

the n_C runs at the center point. If the difference $\bar{y}_F - \bar{y}_C$ is small, then the center points lie on or near the plane passing through the factorial points, and there is no quadratic curvature. On the other hand, if $\bar{y}_F - \bar{y}_C$ is large, then quadratic curvature is present. A single-degree-of-freedom **sum of squares for pure quadratic curvature** is given by

$$SS_{\text{Pure quadratic}} = \frac{n_F n_C (\bar{y}_F - \bar{y}_C)^2}{n_F + n_C} \tag{7–30}$$

where, in general, n_F is the number of factorial design points. This quantity may be compared to the error mean square to test for pure quadratic curvature. More specifically, when points are added to the center of the 2^k design, then the test for curvature (using Equation 7–30) actually tests the hypotheses

$$H_0: \sum_{j=1}^{k} \beta_{jj} = 0$$

$$H_1: \sum_{j=1}^{k} \beta_{jj} \neq 0$$

Furthermore, if the factorial points in the design are unreplicated, one may use the n_C center points to construct an estimate of error with $n_C - 1$ degrees of freedom.

Example 7–5

A chemical engineer is studying the yield of a process. There are two variables of interest, reaction time and reaction temperature. Because she is uncertain about the assumption of linearity over the region of exploration, the engineer decides to conduct

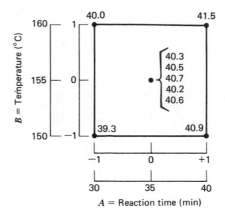

Figure 7–27.   The 2^2 design with five center points for Example 7–5.

a 2^2 design (with a single replicate of each factorial run) augmented with five center points. The design and the yield data are shown in Figure 7–27.

Table 7–14 summarizes the analysis of variance for this experiment. The mean square error is calculated from the center points as follows:

$$MS_E = \frac{SS_E}{n_C - 1}$$

$$= \frac{\sum\limits_{Center\ points} (y_i - \bar{y})^2}{n_C - 1}$$

$$= \frac{\sum\limits_{i=1}^{5} (y_i - 40.46)^2}{4}$$

$$= \frac{0.1720}{4}$$

$$= 0.0430$$

The average of the points in the factorial portion of the design is $\bar{y}_F = 40.425$, and the average of the points at the center is $\bar{y}_C = 40.46$. The difference $\bar{y}_F - \bar{y}_C =$

Table 7–14   Analysis of Variance for Example 7–5

Source of Variation	Sum of Squares	Degrees of Freedom	Mean Square	F_0	P-Value
A (Time)	2.4025	1	2.4025	55.87	0.0017
B (Temperature)	0.4225	1	0.4225	9.83	0.0350
AB	0.0025	1	0.0025	0.06	0.8185
Pure quadratic	0.0027	1	0.0027	0.06	0.8185
Error	0.1720	4	0.0430		
Total	3.0022	8			

$40.425 - 40.46 = -0.035$ appears to be small. The pure quadratic curvature sum of squares in the analysis of variance table is computed from Equation 7–30 as follows:

$$SS_{\text{Pure quadratic}} = \frac{n_F n_C (\bar{y}_F - \bar{y}_C)^2}{n_F + n_C}$$

$$= \frac{(4)(5)(-0.035)^2}{4 + 5}$$

$$= 0.0027$$

The analysis of variance indicates that both factors exhibit significant main effects, that there is no interaction, and that there is no evidence of second-order curvature in the response over the region of exploration. That is, the null hypothesis H_0: $\beta_{11} + \beta_{22} = 0$ cannot be rejected.

■

In Example 7–5, we concluded that there was no indication of quadratic effects; that is, a first-order model

$$y = \beta_0 + \beta_1 x_1 + \beta_2 x_2 + \beta_{12} x_1 x_2 + \epsilon$$

is appropriate (although we probably don't need the interaction term). There will be situations where the quadratic terms will be required. That is, we will now have to assume a second-order model such as

$$y = \beta_0 + \beta_1 x_1 + \beta_2 x_2 + \beta_{12} x_1 x_2 + \beta_{11} x_1^2 + \beta_{22} x_2^2 + \epsilon$$

Unfortunately, we cannot estimate the unknown parameters (the β's) in this model because there are six parameters to estimate and the 2^2 design plus center points in Figure 7–27 only has five independent runs.

A simple and highly effective solution to this problem is to augment the 2^2 design with four **axial runs,** as shown in Figure 7–28a. The resulting design, called a **central composite design,** can now be used to fit the second-order model. Figure 7–28b shows a central composite design for $k = 3$ factors. This design has $14 + n_C$ runs (usually $3 \le n_C \le 5$), and is a very efficient design for fitting the ten-parameter second-order model in $k = 3$ factors.

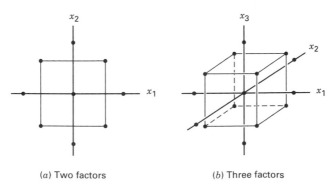

(a) Two factors   (b) Three factors

Figure 7–28. Central composite designs.

Central composite designs are used extensively in building second-order response surface models. These designs will be discussed in more detail in Chapter 14.

7-7 YATES' ALGORITHM FOR THE 2^k DESIGN

While we typically use a computer program for the statistical analysis, there is a very simple technique devised by Yates (1937) for estimating the effects and determining the sums of squares in a 2^k factorial design. The procedure is occasionally useful for manual calculations, and is best learned through the study of a numerical example.

Consider the data for the 2^3 design in Example 7-1. These data have been entered in Table 7-15. The treatment combinations are always written down in standard order, and the column labeled "Response" contains the corresponding observation (or total of all observations) at that treatment combination. The first half of column (1) is obtained by adding the responses in adjacent pairs. The second half of column (1) is obtained by changing the sign of the first entry in each of the pairs in the Response column and adding the adjacent pairs. For example, in column (1) we obtain for the fifth entry $5 = -(-4) + 1$, for the sixth entry $6 = -(-1) + 5$, and so on.

Column (2) is obtained from column (1) just as column (1) is obtained from the Response column. Column (3) is obtained from column (2) similarly. In general, for a 2^k design we would construct k columns of this type. Column (3) [in general, column (k)] is the contrast for the effect designated at the beginning of the row. To obtain the estimate of the effect, we divide the entries in column (3) by $n2^{k-1}$ (in our example, $n2^{k-1} = 8$). Finally, the sums of squares for the effects are obtained by squaring the entries in column (3) and dividing by $n2^k$ (in our example, $n2^k = 16$).

The estimates of the effects and sums of squares obtained by Yates' algorithm

Table 7-15   Yates' Algorithm for the Data in Example 7-1

Treatment Combination	Response	(1)	(2)	(3)	Effect	Estimate of Effect $(3) \div n2^{k-1}$	Sum of Squares $(3)^2 \div n2^k$
(1)	-4	-3	1	16	I	—	—
a	1	4	15	24	A	3.00	36.00
b	-1	2	11	18	B	2.25	20.25
ab	5	13	13	6	AB	0.75	2.25
c	-1	5	7	14	C	1.75	12.25
ac	3	6	11	2	AC	0.25	0.25
bc	2	4	1	4	BC	0.50	1.00
abc	11	9	5	4	ABC	0.50	1.00

for the data in Example 7–1 are in agreement with the results found there by the usual methods. Note that the entry in column (3) [in general, column (k)] for the row corresponding to (1) is always equal to the grand total of the observations.

In spite of its apparent simplicity, it is notoriously easy to make numerical errors in Yates' algorithm, and we should be extremely careful in executing the procedure. As a partial check on the computations, we may use the fact that the sum of the squares of the elements in the jth column is 2^j times the sum of the squares of the elements in the response column. Note, however, that this check is subject to errors in sign in column j. See Davies (1956), Good (1955, 1958), Kempthorne (1952), and Rayner (1967) for other error-checking techniques.

7–8 PROBLEMS

7–1 An engineer is interested in the effects of cutting speed (A), tool geometry (B), and cutting angle (C) on the life (in hours) of a machine tool. Two levels of each factor are chosen, and three replicates of a 2^3 factorial design are run. The results follow:

				Replicate		
A	B	C	Treatment Combination	I	II	III
−	−	−	(1)	22	31	25
+	−	−	a	32	43	29
−	+	−	b	35	34	50
+	+	−	ab	55	47	46
−	−	+	c	44	45	38
+	−	+	ac	40	37	36
−	+	+	bc	60	50	54
+	+	+	abc	39	41	47

(a) Estimate the factor effects. Which effects appear to be large?

(b) Use the analysis of variance to confirm your conclusions for part (a).

(c) Write down a regression model for predicting tool life (in hours) based on the results of this experiment.

(d) Analyze the residuals. Are there any obvious problems?

(e) Based on an analysis of main effect and interaction plots, what levels of A, B, and C would you recommend using?

7–2 Reconsider part (c) of Problem 7–1. Use the regression model to generate response surface and contour plots of the tool life response. Interpret these plots. Do they provide insight regarding the desirable operating conditions for this process?

7-3   Find the standard error of the factor effects and approximate 95 percent confidence limits for the factor effects in Problem 7–1. Do the results of this analysis agree with the conclusions from the analysis of variance?

7-4   Plot the factor effects from Problem 7–1 on a graph relative to an appropriately scaled t distribution. Does this graphical display adequately identify the important factors? Compare the conclusions from this plot with the results from the analysis of variance.

7-5   A router is used to cut locating notches on a printed circuit board. The vibration level at the surface of the board as it is cut is considered to be a major source of dimensional variation in the notches. Two factors are thought to influence vibration: bit size (A) and cutting speed (B). Two bit sizes ($\frac{1}{16}$ and $\frac{1}{8}$ inch) and two speeds (40 and 90 rpm) are selected, and four boards are cut at each set of conditions shown below. The response variable is vibration measured as the resultant vector of three accelerometers (x, y, and z) on each test circuit board.

		Treatment	Replicate			
A	B	Combination	I	II	III	IV
−	−	(1)	18.2	18.9	12.9	14.4
+	−	a	27.2	24.0	22.4	22.5
−	+	b	15.9	14.5	15.1	14.2
+	+	ab	41.0	43.9	36.3	39.9

(a) Analyze the data from this experiment.

(b) Construct a normal probability plot of the residuals, and plot the residuals versus the predicted vibration level. Interpret these plots.

(c) Draw the AB interaction plot. Interpret this plot. What levels of bit size and speed would you recommend for routine operation?

7-6   Reconsider the experiment described in Problem 7–1. Suppose that the experimenter only performed the eight trials from replicate I. In addition, he ran four center points and obtained the following response values: 36, 40, 43, 45.

(a) Estimate the factor effects. Which effects are large?

(b) Perform an analysis of variance, including a check for pure quadratic curvature. What are your conclusions?

(c) Write down an appropriate model for predicting tool life, based on the results of this experiment. Does this model differ in any substantial way from the model in Problem 7–1, part (c)?

(d) Analyze the residuals.

(e) What conclusions would you draw about the appropriate operating conditions for this process?

7-7 An experiment was performed to improve the yield of a chemical process. Four factors were selected, and two replicates of a completely randomized experiment were run. The results are shown in the following table:

Treatment Combination	Replicate		Treatment Combination	Replicate	
	I	II		I	II
(1)	90	93	d	98	95
a	74	78	ad	72	76
b	81	85	bd	87	83
ab	83	80	abd	85	86
c	77	78	cd	99	90
ac	81	80	acd	79	75
bc	88	82	bcd	87	84
abc	73	70	abcd	80	80

(a) Estimate the factor effects.

(b) Prepare an analysis of variance table, and determine which factors are important in explaining yield.

(c) Write down a regression model for predicting yield, assuming that all four factors were varied over the range from -1 to $+1$ (in coded units).

(d) Plot the residuals versus the predicted yield and on a normal probability scale. Does the residual analysis appear satisfactory?

(e) Two three-factor interactions, ABC and ABD, apparently have large effects. Draw a cube plot in the factors A, B, and C with the average yields shown at each corner. Repeat using the factors A, B, and D. Do these two plots aid in data interpretation? Where would you recommend that the process be run with respect to the four variables?

7-8 A bacteriologist is interested in the effects of two different culture media and two different times on the growth of a particular virus. She performs six replicates of a 2^2 design, making the runs in random order. Analyze the bacterial growth data that follow and draw appropriate conclusions. Analyze the residuals and comment on the model's adequacy.

Time	Culture Medium			
	1		2	
12 hr	21	22	25	26
	23	28	24	25
	20	26	29	27
18 hr	37	39	31	34
	38	38	29	33
	35	36	30	35

7-9   An industrial engineer employed by a beverage bottler is interested in the effects of two different types of 32-ounce bottles on the time to deliver 12-bottle cases of the product. The two bottle types are glass and plastic. Two workers are used to perform a task consisting of moving 40 cases of the product 50 feet on a standard type of hand truck and stacking the cases in a display. Four replicates of a 2^2 factorial design are performed, and the times observed are listed in the following table. Analyze the data and draw appropriate conclusions. Analyze the residuals and comment on the model's adequacy.

Bottle Type	Worker			
	1		2	
Glass	5.12	4.89	6.65	6.24
	4.98	5.00	5.49	5.55
Plastic	4.95	4.43	5.28	4.91
	4.27	4.25	4.75	4.71

7-10   In Problem 7–9, the engineer was also interested in potential fatigue differences resulting from the two types of bottles. As a measure of the amount of effort required, he measured the elevation of the heart rate (pulse) induced by the task. The results follow. Analyze the data and draw conclusions. Analyze the residuals and comment on the model's adequacy.

Bottle Type	Worker			
	1		2	
Glass	39	45	20	13
	58	35	16	11
Plastic	44	35	13	10
	42	21	16	15

7-11   Calculate approximate 95 percent confidence limits for the factor effects in Problem 7–10. Do the results of this analysis agree with the analysis of variance performed in Problem 7–10?

7-12   An article in the *AT&T Technical Journal* (March/April 1986, Vol. 65, pp. 39–50) describes the application of two-level factorial designs to integrated circuit manufacturing. A basic processing step is to grow an epitaxial layer on polished silicon wafers. The wafers mounted on a susceptor are positioned inside a bell jar, and chemical vapors are introduced. The susceptor is rotated and heat is applied until the epitaxial layer is thick enough. An experiment was run using two factors: arsenic flow rate (A) and deposition time (B). Four replicates were run, and the epitaxial layer thickness was measured (in μm). The data are shown below:

A	B	I	II	III	IV		Low (−)	High (+)
			Replicate				Factor Levels	
−	−	14.037	16.165	13.972	13.907	A	55%	59%
+	−	13.880	13.860	14.032	13.914			
−	+	14.821	14.757	14.843	14.878	B	Short	Long
+	+	14.888	14.921	14.415	14.932		(10 min)	(15 min)

(a) Estimate the factor effects.

(b) Conduct an analysis of variance. Which factors are important?

(c) Write down a regression equation that could be used to predict epitaxial layer thickness over the region of arsenic flow rate and deposition time used in this experiment.

(d) Analyze the residuals. Are there any residuals that should cause concern?

(e) Discuss how you might deal with the potential outlier found in part (d).

7–13 *Continuation of Problem 7–12.* Use the regression model in part (c) of Problem 7–12 to generate a response surface contour plot for epitaxial layer thickness. Suppose it is critically important to obtain layer thickness of 14.5 μm. What settings of arsenic flow rate and deposition time would you recommend?

7–14 *Continuation of Problem 7–13.* How would your answer to Problem 7–13 change if arsenic flow rate was more difficult to control in the process than the deposition time?

7–15 A nickel–titanium alloy is used to make components for jet turbine aircraft engines. Cracking is a potentially serious problem in the final part, as it can lead to nonrecoverable failure. A test is run at the parts producer to determine the effect of four factors on cracks. The four factors are pouring temperature (A), titanium content (B), heat treatment method (C), and amount of grain refiner used (D). Two replicates of a 2^4 design are run, and the length of crack (in mm) induced in a sample coupon subjected to a standard test is measured. The data are shown below:

A	B	C	D	Treatment Combination	I	II
					Replicate	
−	−	−	−	(1)	1.71	2.01
+	−	−	−	a	1.42	1.58
−	+	−	−	b	1.35	1.63
+	+	−	−	ab	1.67	1.65
−	−	+	−	c	1.23	1.48

(continued)

(continued)

A	B	C	D	Treatment Combination	Replicate I	Replicate II
+	−	+	−	ac	1.25	1.36
−	+	+	−	bc	1.46	1.52
+	+	+	−	abc	1.29	1.37
−	−	−	+	d	2.04	2.29
+	−	−	+	ad	1.86	1.95
−	+	−	+	bd	1.79	2.05
+	+	−	+	abd	1.42	1.69
−	−	+	+	cd	1.81	2.02
+	−	+	+	acd	1.34	1.39
−	+	+	+	bcd	1.46	1.63
+	+	+	+	abcd	0.85	1.00

(a) Estimate the factor effects. Which factor effects appear to be large?

(b) Conduct an analysis of variance. Do any of the factors affect cracking? Use $\alpha = 0.05$.

(c) Write down a regression model that can be used to predict crack length as a function of the significant main effects and interactions you have identified in part (b).

(d) Analyze the residuals from this experiment.

(e) Is there an indication that any of the factors affect the variability in cracking?

(f) What recommendations would you make regarding process operations? Use interaction and/or main effect plots to assist in drawing conclusions.

7–16 *Continuation of Problem 7–15.* One of the variables in the experiment described in Problem 7–15, heat treatment method (C), is a categorical variable. Assume that the remaining factors are continuous.

(a) Write two regression models for predicting crack length, one for each level of the heat treatment method variable. What differences, if any, do you notice in these two equations?

(b) Generate appropriate response surface contour plots for the two regression models in part (a).

(c) What set of conditions would you recommend for the factors A, B, and D if you use heat treatment method $C = +$?

(d) Repeat part (c) assuming that you wish to use heat treatment method $C = -$.

7–17 An experimenter has run a single replicate of a 2^4 design. The following effect estimates have been calculated:

$$A = 76.95 \qquad AB = -51.32 \qquad ABC = -2.82$$
$$B = -67.52 \qquad AC = 11.69 \qquad ABD = -6.50$$
$$C = -7.84 \qquad AD = 9.78 \qquad ACD = 10.20$$

$$D = -18.73 \qquad BC = 20.78 \qquad BCD = -7.98$$
$$BD = 14.74 \qquad ABCD = -6.25$$
$$CD = 1.27$$

(a) Construct a normal probability plot of these effects.

(b) Identify a tentative model, based on the plot of the effects in part (a).

7-18 An article in *Solid State Technology* ("Orthogonal Design for Process Optimization and Its Application in Plasma Etching," May 1987, pp. 127–132) describes the application of factorial designs in developing a nitride etch process on a single-wafer plasma etcher. The process uses C_2F_6 as the reactant gas. Four factors are of interest: anode–cathode gap (A), pressure in the reactor chamber (B), C_2F_6 gas flow (C), and power applied to the cathode (D). The response variable of interest is the etch rate for silicon nitride. A single replicate of a 2^4 design is run, and the data are shown below:

Run Number	Actual Run Order	A	B	C	D	Etch Rate (Å/min)	Factor Levels	Low (−)	High (+)
1	13	−	−	−	−	550	A (cm)	0.80	1.20
2	8	+	−	−	−	669	B (mTorr)	4.50	550
3	12	−	+	−	−	604	C (SCCM)	125	200
4	9	+	+	−	−	650	D (W)	275	325
5	4	−	−	+	−	633			
6	15	+	−	+	−	642			
7	16	−	+	+	−	601			
8	3	+	+	+	−	635			
9	1	−	−	−	+	1037			
10	14	+	−	−	+	749			
11	5	−	+	−	+	1052			
12	10	+	+	−	+	868			
13	11	−	−	+	+	1075			
14	2	+	−	+	+	860			
15	7	−	+	+	+	1063			
16	6	+	+	+	+	729			

(a) Estimate the factor effects. Construct a normal probability plot of the factor effects. Which effects appear large?

(b) Conduct an analysis of variance to confirm your findings for part (a).

(c) What is the regression model relating etch rate to the significant process variables?

(d) Analyze the residuals from this experiment. Comment on the model's adequacy.

(e) If not all the factors are important, project the 2^4 design into a 2^k design with $k < 4$ and conduct the analysis of variance.

(f) Draw graphs to interpret any significant interactions.

(g) Plot the residuals versus the actual run order. What problems might be revealed by this plot?

7-19 *Continuation of Problem 7-18.* Consider the regression model obtained in part (c) of Problem 7-18.

(a) Construct contour plots of the etch rate using this model.

(b) Suppose that it was necessary to operate this process at an etch rate of 800 Å/min. What settings of the process variables would you recommend?

7-20 Consider the single replicate of the 2^4 design in Example 7-2. Suppose we had arbitrarily decided to analyze the data assuming that all three- and four-factor interactions were negligible. Conduct this analysis and compare your results with those obtained in the example. Do you think that it is a good idea to arbitrarily assume interactions to be negligible even if they are relatively high-order ones?

7-21 An experiment was run in a semiconductor fabrication plant in an effort to increase yield. Five factors, each at two levels, were studied. The factors (and levels) were A = aperture setting (small, large), B = exposure time (20% below nominal, 20% above nominal), C = development time (30 s, 45 s), D = mask dimension (small, large), and E = etch time (14.5 min, 15.5 min). The unreplicated 2^5 design shown below was run.

(1) = 7	d = 8	e = 8	de = 6
a = 9	ad = 10	ae = 12	ade = 10
b = 34	bd = 32	be = 35	bde = 30
ab = 55	abd = 50	abe = 52	abde = 53
c = 16	cd = 18	ce = 15	cde = 15
ac = 20	acd = 21	ace = 22	acde = 20
bc = 40	bcd = 44	bce = 45	bcde = 41
abc = 60	abcd = 61	abce = 65	abcde = 63

(a) Construct a normal probability plot of the effect estimates. Which effects appear to be large?

(b) Conduct an analysis of variance to confirm your findings for part (a).

(c) Write down the regression model relating yield to the significant process variables.

(d) Plot the residuals on normal probability paper. Is the plot satisfactory?

(e) Plot the residuals versus the predicted yields and versus each of the five factors. Comment on the plots.

(f) Interpret any significant interactions.

(g) What are your recommendations regarding process operating conditions?

(h) Project the 2^5 design in this problem into a 2^k design in the important factors. Sketch the design and show the average and range of yields at each run. Does this sketch aid in interpreting the results of this experiment?

7-22 *Continuation of Problem 7-21.* Suppose that the experimenter had run four center points in addition to the 32 trials in the original experiment. The yields obtained at the center point runs were 68, 74, 76, and 70.

(a) Reanalyze the experiment, including a test for pure quadratic curvature.

(b) Discuss what your next step would be.

7-23 In a process development study on yield, four factors were studied, each at two levels: time (A), concentration (B), pressure (C), and temperature (D). A single replicate of a 2^4 design was run, and the resulting data are shown in the following table:

Run Number	Actual Run Order	A	B	C	D	Yield (lbs)	Factor Levels		
								Low ($-$)	High ($+$)
1	5	$-$	$-$	$-$	$-$	12	A (h)	2.5	3
2	9	$+$	$-$	$-$	$-$	18	B (%)	14	18
3	8	$-$	$+$	$-$	$-$	13	C (psi)	60	80
4	13	$+$	$+$	$-$	$-$	16	D (°C)	225	250
5	3	$-$	$-$	$+$	$-$	17			
6	7	$+$	$-$	$+$	$-$	15			
7	14	$-$	$+$	$+$	$-$	20			
8	1	$+$	$+$	$+$	$-$	15			
9	6	$-$	$-$	$-$	$+$	10			
10	11	$+$	$-$	$-$	$+$	25			
11	2	$-$	$+$	$-$	$+$	13			
12	15	$+$	$+$	$-$	$+$	24			
13	4	$-$	$-$	$+$	$+$	19			
14	16	$+$	$-$	$+$	$+$	21			
15	10	$-$	$+$	$+$	$+$	17			
16	12	$+$	$+$	$+$	$+$	23			

(a) Construct a normal probability plot of the effect estimates. Which factors appear to have large effects?

(b) Conduct an analysis of variance using the normal probability plot in part (a) for guidance in forming an error term. What are your conclusions?

(c) Write down a regression model relating yield to the important process variables.

(d) Analyze the residuals from this experiment. Does your analysis indicate any potential problems?

(e) Can this design be collapsed into a 2^3 design with two replicates? If so, sketch the design with the average and range of yield shown at each point in the cube. Interpret the results.

7-24   *Continuation of Problem 7-23.* Use the regression model in part (c) of Problem 7-23 to generate a response surface contour plot of yield. Discuss the practical value of this response surface plot.

7-25   *The scrumptious brownie experiment.* The author is an engineer by training and a firm believer in learning by doing. I have taught experimental design for many years to a wide variety of audiences and have always assigned the planning, conduct, and analysis of an actual experiment to the class participants. The participants seem to enjoy this practical experience and always learn a great deal from it. This problem uses the results of an experiment performed by Gretchen Krueger at Arizona State University.

There are many different ways to bake brownies. The purpose of this experiment was to determine how the pan material, the brand of brownie mix, and the stirring method affect the scrumptiousness of brownies. The factor levels were

Factor	Low (−)	High (+)
A = pan material	Glass	Aluminum
B = stirring method	Spoon	Mixer
C = brand of mix	Expensive	Cheap

The response variable was scrumptiousness, a subjective measure derived from a questionnaire given to the subjects who sampled each batch of brownies. (The questionnaire dealt with such issues as taste, appearance, consistency, aroma, and so forth). An eight-person test panel sampled each batch and filled out the questionnaire. The design matrix and the response data are shown below:

Brownie Batch	A	B	C	1	2	3	4	5	6	7	8
				\multicolumn{8}{c}{Test Panel Results}							
1	−	−	−	11	9	10	10	11	10	8	9
2	+	−	−	15	10	16	14	12	9	6	15
3	−	+	−	9	12	11	11	11	11	11	12
4	+	+	−	16	17	15	12	13	13	11	11
5	−	−	+	10	11	15	8	6	8	9	14
6	+	−	+	12	13	14	13	9	13	14	9
7	−	+	+	10	12	13	10	7	7	17	13
8	+	+	+	15	12	15	13	12	12	9	14

(a)   Analyze the data from this experiment as if there were eight replicates of a 2^3 design. Comment on the results.

(b)   Is the analysis in part (a) the correct approach? There are only eight batches; do we really have eight replicates of a 2^3 factorial design?

(c)   Analyze the average and standard deviation of the scrumptiousness ratings. Comment on the results. Is this analysis more appropriate than the one in part (a)? Why or why not?

7–26 An experiment was conducted on a chemical process that produces a polymer. The four factors studied were temperature (A), catalyst concentration (B), time (C), and pressure (D). Two responses, molecular weight and viscosity, were observed. The design matrix and response data are shown below:

Run Number	Actual Run Order	A	B	C	D	Molecular Weight	Viscosity	Factor Levels		
									Low (−)	High (+)
1	18	−	−	−	−	2400	1400	A (°C)	100	120
2	9	+	−	−	−	2410	1500	B (%)	4	8
3	13	−	+	−	−	2315	1520	C (min)	20	30
4	8	+	+	−	−	2510	1630	D (psi)	60	75
5	3	−	−	+	−	2615	1380			
6	11	+	−	+	−	2625	1525			
7	14	−	+	+	−	2400	1500			
8	17	+	+	+	−	2750	1620			
9	6	−	−	−	+	2400	1400			
10	7	+	−	−	+	2390	1525			
11	2	−	+	−	+	2300	1500			
12	10	+	+	−	+	2520	1500			
13	4	−	−	+	+	2625	1420			
14	19	+	−	+	+	2630	1490			
15	15	−	+	+	+	2500	1500			
16	20	+	+	+	+	2710	1600			
17	1	0	0	0	0	2515	1500			
18	5	0	0	0	0	2500	1460			
19	16	0	0	0	0	2400	1525			
20	12	0	0	0	0	2475	1500			

(a) Consider only the molecular weight response. Plot the effect estimates on a normal probability scale. What effects appear important?

(b) Use an analysis of variance to confirm the results from part (a). Is there indication of curvature?

(c) Write down a regression model to predict molecular weight as a function of the important variables.

(d) Analyze the residuals and comment on model adequacy.

(e) Repeat parts (a)–(d) using the viscosity response.

7–27 *Continuation of Problem 7–26.* Use the regression models for molecular weight and viscosity to answer the following questions.

(a) Construct a response surface contour plot for molecular weight. In what direction would you adjust the process variables to increase molecular weight?

(b) Construct a response surface contour plot for viscosity. In what direction would you adjust the process variables to decrease viscosity?

(c) What operating conditions would you recommend if it was necessary to produce a product with molecular weight between 2400 and 2500, and the lowest possible viscosity?

7-28   Analyze the data in Example 7–2 using Yates' algorithm.

7-29   Consider the single replicate of the 2^4 design in Example 7–2. Suppose that we ran five points at the center (0, 0, 0, 0) and observed the following responses: 73, 75, 71, 69, and 76. Test for curvature in this experiment. Interpret the results.

7-30   *A missing value in a 2^k factorial.* It is not unusual to find that one of the observations in a 2^k design is missing due to faulty measuring equipment, a spoiled test, or some other reason. If the design is replicated n times ($n > 1$) some of the techniques discussed in Chapter 6 can be employed. However, for an unreplicated factorial ($n = 1$) some other method must be used. One logical approach is to estimate the missing value with a number that makes the highest-order interaction contrast zero. Apply this technique to the experiment in Example 7–2 assuming that run ab is missing. Compare the results with the results of Example 7–2.

7-31   An engineer has performed an experiment to study the effect of four factors on the surface roughness of a machined part. The factors (and their levels) are $A =$ tool angle (12 degrees, 15 degrees), $B =$ cutting fluid viscosity (300, 400), $C =$ feed rate (10 in/min, 15 in/min), and $D =$ cutting fluid cooler used (no, yes). The data from this experiment (with the factors coded to the usual $-1, +1$ levels) are shown below.

Run	A	B	C	D	Surface Roughness
1	−	−	−	−	0.00340
2	+	−	−	−	0.00362
3	−	+	−	−	0.00301
4	+	+	−	−	0.00182
5	−	−	+	−	0.00280
6	+	−	+	−	0.00290
7	−	+	+	−	0.00252
8	+	+	+	−	0.00160
9	−	−	−	+	0.00336
10	+	−	−	+	0.00344
11	−	+	−	+	0.00308
12	+	+	−	+	0.00184
13	−	−	+	+	0.00269
14	+	−	+	+	0.00284
15	−	+	+	+	0.00253
16	+	+	+	+	0.00163

(a) Estimate the factor effects. Plot the effect estimates on a normal probability plot and select a tentative model.

(b) Fit the model identified in part (a) and analyze the residuals. Is there any indication of model inadequacy?

(c) Repeat the analysis from parts (a) and (b) using $1/y$ as the response variable. Is there an indication that the transformation has been useful?

(d) Fit a model in terms of the coded variables that can be used to predict the surface roughness. Convert this prediction equation into a model in the natural variables.

7-32   Resistivity on a silicon wafer is influenced by several factors. The results of a 2^4 factorial experiment performed during a critical processing step is shown below.

Run	A	B	C	D	Resistivity
1	−	−	−	−	1.92
2	+	−	−	−	11.28
3	−	+	−	−	1.09
4	+	+	−	−	5.75
5	−	−	+	−	2.13
6	+	−	+	−	9.53
7	−	+	+	−	1.03
8	+	+	+	−	5.35
9	−	−	−	+	1.60
10	+	−	−	+	11.73
11	−	+	−	+	1.16
12	+	+	−	+	4.68
13	−	−	+	+	2.16
14	+	−	+	+	9.11
15	−	+	+	+	1.07
16	+	+	+	+	5.30

(a) Estimate the factor effects. Plot the effect estimates on a normal probability plot and select a tentative model.

(b) Fit the model identified in part (a) and analyze the residuals. Is there any indication of model inadequacy?

(c) Repeat the analysis from parts (a) and (b) using $\ln(y)$ as the response variable. Is there an indication that the transformation has been useful?

(d) Fit a model in terms of the coded variables that can be used to predict the resistivity.

Chapter 8
Blocking and Confounding in the 2^k Factorial Design

8-1 INTRODUCTION

There are many situations in which it is impossible to perform all of the runs in a 2^k factorial experiment under homogeneous conditions. For example, a single batch of raw material might not be large enough to make all of the required runs. In other cases, it might be desirable to deliberately vary the experimental conditions to ensure that the treatments are equally effective (that is, are robust) across many situations that are likely to be encountered in practice. For example, a chemical engineer may run a pilot plant experiment with several batches of raw material because he knows that different raw material batches of different quality grades are likely to be used in the actual full-scale process.

The design technique used in these situations is **blocking.** In this chapter, we will concentrate on some special techniques for blocking in the 2^k factorial design.

8-2 BLOCKING A REPLICATED 2^k FACTORIAL DESIGN

Suppose that the 2^k factorial design has been replicated n times. This is identical to the situation discussed in Chapter 6, where we showed how to run a general factorial design in blocks. If there are n replicates, then each set of nonhomogeneous conditions defines a block, and each replicate is run in one of the blocks. The runs in each block (or replicate) would be made in random order. The analysis of the design is similar to that of any blocked factorial experiment; for example, see the discussion in Section 6-6.

Table 8-1   Chemical Process Experiment in Three Blocks

	Block 1	Block 2	Block 3
	$(1) = 28$	$(1) = 25$	$(1) = 27$
	$a = 36$	$a = 32$	$a = 32$
	$b = 18$	$b = 19$	$b = 23$
	$ab = 31$	$ab = 30$	$ab = 29$
Block totals:	$B_1 = 113$	$B_2 = 106$	$B_3 = 111$

Example 8-1

Consider the chemical process experiment first described in Section 7-2. Suppose that only four experimental trials can be made from a single batch of raw material. Therefore, three batches of raw material will be required to run all three replicates of this design. Table 8-1 shows the design, where each batch of raw material corresponds to a block.

The analysis of variance for this blocked design is shown in Table 8-2. All of the sums of squares are calculated exactly as in a standard, unblocked 2^k design. The sum of squares for blocks is calculated from the block totals. Let B_1, B_2, and B_3 represent the block totals (see Table 8-1). Then

$$SS_{\text{Blocks}} = \sum_{i=1}^{3} \frac{B_i^2}{4} - \frac{y_{\ldots}^2}{12}$$

$$= \frac{(113)^2 + (106)^2 + (111)^2}{4} - \frac{(330)^2}{12}$$

$$= 6.50$$

There are two degrees of freedom among the three blocks. Table 8-2 indicates that the conclusions from this analysis, had the design been run in blocks, are identical to those in Section 7-2 and that the block effect is relatively small.

■

Table 8-2   Analysis of Variance for the Chemical Process Experiment in Three Blocks

Source of Variation	Sum of Squares	Degrees of Freedom	Mean Square	F_0	P-Value
Blocks	6.50	2	3.25		
A (concentration)	208.33	1	208.33	50.32	0.0004
B (catalyst)	75.00	1	75.00	18.12	0.0053
AB	8.33	1	8.33	2.01	0.2060
Error	24.84	6	4.14		
Total	323.00	11			

8-3   CONFOUNDING IN THE 2^k FACTORIAL DESIGN

There are many problems for which it is impossible to perform a complete replicate of a factorial design in one block. **Confounding** is a design technique for arranging a complete factorial experiment in blocks, where the block size is smaller than the number of treatment combinations in one replicate. The technique causes information about certain treatment effects (usually high-order interactions) to be **indistinguishable from,** or **confounded with blocks.** In this chapter we concentrate on confounding systems for the 2^k factorial design. Note that even though the designs presented are incomplete block designs because each block does not contain all the treatments or treatment combinations, the special structure of the 2^k factorial system allows a simplified method of analysis.

We consider the construction and analysis of the 2^k factorial design in 2^p incomplete blocks, where $p < k$. Consequently, these designs can be run in two blocks, four blocks, eight blocks, and so on.

8-4   CONFOUNDING THE 2^k FACTORIAL DESIGN IN TWO BLOCKS

Suppose that we wish to run a single replicate of the 2^2 design. Each of the $2^2 = 4$ treatment combinations requires a quantity of raw material, for example, and each batch of raw material is only large enough for two treatment combinations to be tested. Thus, two batches of raw material are required. If batches of raw material are considered as blocks, then we must assign two of the four treatment combinations to each block.

Figure 8–1 shows one possible design for this problem. The geometric view, Figure 8–1a, indicates that treatment combinations on opposing diagonals are assigned to different blocks. Notice from Figure 8–1b that block 1 contains the treatment combinations (1) and ab and that block 2 contains a and b. Of course, the *order* in which the treatment combinations are run within a block is randomly determined. We would also randomly decide which block to run first. Suppose we estimate the main effects of A and B just as if no blocking had occurred. From Equations 7–1 and 7–2, we obtain

$$A = \tfrac{1}{2}[ab + a - b - (1)]$$
$$B = \tfrac{1}{2}[ab + b - a - (1)]$$

Note that both A and B are unaffected by blocking since in each estimate there is one plus and one minus treatment combination from each block. That is, any difference between block 1 and block 2 will cancel out.

Now consider the AB interaction

$$AB = \tfrac{1}{2}[ab + (1) - a - b]$$

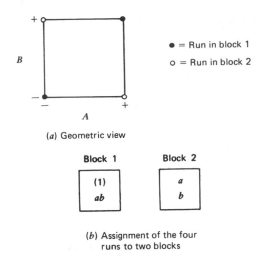

(a) Geometric view

Block 1   **Block 2**

(1)   *a*

ab   *b*

(b) Assignment of the four
runs to two blocks

Figure 8–1.   A 2^2 design in two blocks.

Since the two treatment combinations with the plus sign [*ab* and (1)] are in block 1 and the two with the minus sign (*a* and *b*) are in block 2, the block effect and the *AB* interaction are identical. That is, *AB* is *confounded* with blocks.

The reason for this is apparent from the table of plus and minus signs for the 2^2 design. This was originally given as Table 7–2, but for convenience it is reproduced as Table 8–3. From this table, we see that all treatment combinations that have a plus sign on *AB* are assigned to block 1, whereas all treatment combinations that have a minus sign on *AB* are assigned to block 2. This approach can be used to confound any effect (*A*, *B*, or *AB*) with blocks. For example, if (1) and *b* had been assigned to block 1 and *a* and *ab* to block 2, then the main effect *A* would have been confounded with blocks. The usual practice is to confound the highest-order interaction with blocks.

This scheme can be used to confound any 2^k design in two blocks. As a second example, consider a 2^3 design run in two blocks. Suppose we wish to confound the three-factor interaction *ABC* with blocks. From the table of plus and minus signs shown in Table 8–4, we assign the treatment combinations that are minus on *ABC* to block 1 and those that are plus on *ABC* to block 2. The

Table 8–3   Table of Plus and Minus Signs for the 2^2 Design

Treatment Combination	Factorial Effect			
	I	*A*	*B*	*AB*
(1)	+	−	−	+
a	+	+	−	−
b	+	−	+	−
ab	+	+	+	+

Table 8–4   Table of Plus and Minus Signs for the 2^3 Design

Treatment Combination	Factorial Effect							
	I	A	B	AB	C	AC	BC	ABC
(1)	+	−	−	+	−	+	+	−
a	+	+	−	−	−	−	+	+
b	+	−	+	−	−	+	−	+
ab	+	+	+	+	−	−	−	−
c	+	−	−	+	+	−	−	+
ac	+	+	−	−	+	+	−	−
bc	+	−	+	−	+	−	+	−
abc	+	+	+	+	+	+	+	+

resulting design is shown in Figure 8–2. Once again, we emphasize that the treatment combinations *within* a block are run in random order.

Other Methods for Constructing the Blocks ▪ There is another method for constructing these designs. The method uses the linear combination

$$L = \alpha_1 x_1 + \alpha_2 x_2 + \cdots + \alpha_k x_k \tag{8-1}$$

where x_i is the level of the ith factor appearing in a particular treatment combination and α_i is the exponent appearing on the ith factor in the effect to be confounded. For the 2^k system, we have $\alpha_i = 0$ or 1 and $x_i = 0$ (low level) or

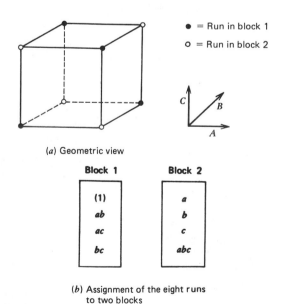

(a) Geometric view

Block 1	Block 2
(1)	a
ab	b
ac	c
bc	abc

(b) Assignment of the eight runs to two blocks

Figure 8–2.   The 2^3 design in two blocks with ABC confounded.

$x_i = 1$ (high level). Equation 8–1 is called a **defining contrast.** Treatment combinations that produce the same value of L (mod 2) will be placed in the same block. Since the only possible values of L (mod 2) are 0 and 1, this will assign the 2^k treatment combinations to exactly two blocks.

To illustrate the approach, consider a 2^3 design with ABC confounded with blocks. Here x_1 corresponds to A, x_2 to B, x_3 to C, and $\alpha_1 = \alpha_2 = \alpha_3 = 1$. Thus, the defining contrast corresponding to ABC is

$$L = x_1 + x_2 + x_3$$

The treatment combination (1) is written 000 in the (0, 1) notation; therefore,

$$L = 1(0) + 1(0) + 1(0) = 0 = 0 \ (\text{mod } 2)$$

Similarly, the treatment combination a is 100, yielding

$$L = 1(1) + 1(0) + 1(0) = 1 = 1 \ (\text{mod } 2)$$

Thus, (1) and a would be run in different blocks. For the remaining treatment combinations, we have

$$
\begin{aligned}
b: &\quad L = 1(0) + 1(1) + 1(0) = 1 = 1 \ (\text{mod } 2)\\
ab: &\quad L = 1(1) + 1(1) + 1(0) = 2 = 0 \ (\text{mod } 2)\\
c: &\quad L = 1(0) + 1(0) + 1(1) = 1 = 1 \ (\text{mod } 2)\\
ac: &\quad L = 1(1) + 1(0) + 1(1) = 2 = 0 \ (\text{mod } 2)\\
bc: &\quad L = 1(0) + 1(1) + 1(1) = 2 = 0 \ (\text{mod } 2)\\
abc: &\quad L = 1(1) + 1(1) + 1(1) = 3 = 1 \ (\text{mod } 2)
\end{aligned}
$$

Thus (1), ab, ac, and bc are run in block 1 and a, b, c, and abc are run in block 2. This is the same design shown in Figure 8–2, which was generated from the table of plus and minus signs.

Another method may be used to construct these designs. The block containing the treatment combination (1) is called the **principal block.** The treatment combinations in this block have a useful group-theoretic property; namely, they form a group with respect to multiplication modulus 2. This implies that any element [except (1)] in the principal block may be generated by multiplying two other elements in the principal block modulus 2. For example, consider the principal block of the 2^3 design with ABC confounded, as shown in Figure 8–2. Note that

$$
\begin{aligned}
ab \cdot ac &= a^2bc = bc\\
ab \cdot bc &= ab^2c = ac\\
ac \cdot bc &= abc^2 = ab
\end{aligned}
$$

Treatment combinations in the other block (or blocks) may be generated by multiplying one element in the new block by each element in the principal block modulus 2. For the 2^3 with ABC confounded, since the principal block is (1), ab,

Replicate I		Replicate II		Replicate III		Replicate IV	
Block 1	**Block 2**	**Block 1**	**Block 2**	**Block 1**	**Block 2**	**Block 1**	**Block 2**
(1)	*abc*	(1)	*abc*	(1)	*abc*	(1)	*abc*
ac	*a*	*ac*	*a*	*ac*	*a*	*ac*	*a*
ab	*b*	*ab*	*b*	*ab*	*b*	*ab*	*b*
bc	*c*	*bc*	*c*	*bc*	*c*	*bc*	*c*

Figure 8–3.   Four replicates of the 2^3 design with ABC confounded.

ac, and bc, we know that b is in the other block. Thus, the elements of this second block are

$$b \cdot (1) \quad\quad = b$$
$$b \cdot ab = ab^2 = a$$
$$b \cdot ac \quad\quad = abc$$
$$b \cdot bc = b^2c = c$$

This agrees with the results obtained previously.

Estimation of Error ▪ When the number of variables is small, say $k = 2$ or 3, it is usually necessary to replicate the experiment to obtain an estimate of error. For example, suppose that a 2^3 factorial must be run in two blocks with ABC confounded, and the experimenter decides to replicate the design four times. The resulting design might appear as in Figure 8–3. Note that ABC is confounded in each replicate.

The analysis of variance for this design is shown in Table 8–5. There are 32

Table 8–5   Analysis of Variance for Four Replicates of a 2^3 Design with ABC Confounded

Source of Variation	Degrees of Freedom
Replicates	3
Blocks (ABC)	1
Error for ABC (replicates × blocks)	3
A	1
B	1
C	1
AB	1
AC	1
BC	1
Error (or replicates × effects)	18
Total	31

observations and 31 total degrees of freedom. Furthermore, since there are eight blocks, seven degrees of freedom must be associated with these blocks. One breakdown of those seven degrees of freedom is shown in Table 8–5. The error sum of squares actually consists of the two-factor interactions between replicates and each of the effects (A, B, C, AB, AC, BC). It is usually safe to consider the interactions to be zero and to treat the resulting mean square as an estimate of error. Main effects and two-factor interactions are tested against the mean square error. Cochran and Cox (1957) state that the block or ABC mean square could be compared to the error for the ABC mean square, which is really replicates × blocks. This test usually is very insensitive.

If resources are sufficient to allow the replication of confounded designs, it is generally better to use a slightly different method of designing the blocks in each replicate. This approach consists of confounding a different effect in each replicate so that some information on all effects is obtained. Such a procedure is called **partial confounding** and is discussed in Section 8–7. If k is moderately large, say $k \geq 4$, we frequently can afford only a single replicate. The experimenter usually assumes higher-order interactions to be negligible and combines their sums of squares as error. The normal probability plot of factor effects can be very helpful in this regard.

Example 8–2

Consider the situation described in Example 7–2. Recall that four factors—temperature (A), pressure (B), concentration of formaldehyde (C), and stirring rate (D)—are studied in a pilot plant to determine their effect on product filtration rate. We will use this experiment to illustrate the ideas of blocking and confounding in an unreplicated design. We will make two modifications to the original experiment. First, suppose that the $2^4 = 16$ treatment combinations cannot all be run using one batch of raw material. The experimenter can run eight treatment combinations from a single batch of material, so a 2^4 design confounded in two blocks seems appropriate. It is logical to confound the highest-order interaction $ABCD$ with blocks. The defining contrast is

$$L = x_1 + x_2 + x_3 + x_4$$

and it is easy to verify that the design is as shown in Figure 8–4 on page 362. Alternatively, one may examine Table 7–10 and observe that the treatment combinations that are $+$ in the $ABCD$ column are assigned to block 1 and those that are $-$ in the $ABCD$ column are in block 2.

The second modification that we will make is to introduce a block effect, so that the utility of blocking can be demonstrated. Suppose that when we select the two batches of raw material required to run the experiment, one of them is of much poorer quality and, as a result, all responses will be 20 units lower in this material batch than in the other. The poor quality batch becomes block 1 and the good quality batch becomes block 2 (it doesn't matter which batch is called block 1 or which batch is called block 2). Now all the tests in block 1 are performed first (the eight runs in the block are, of course, performed in random order), but the responses are 20 units lower than they would have been if good quality material had been used. Figure 8–4(b) shows the resulting responses—note that these have been found by subtracting

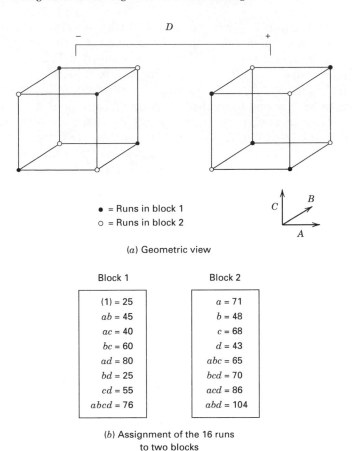

• = Runs in block 1
○ = Runs in block 2

(*a*) Geometric view

Block 1

| (1) = 25 |
| ab = 45 |
| ac = 40 |
| bc = 60 |
| ad = 80 |
| bd = 25 |
| cd = 55 |
| abcd = 76 |

Block 2

| a = 71 |
| b = 48 |
| c = 68 |
| d = 43 |
| abc = 65 |
| bcd = 70 |
| acd = 86 |
| abd = 104 |

(*b*) Assignment of the 16 runs
to two blocks

Figure 8–4.   The 2^4 design in two blocks for Example 8–2.

the block effect from the original observations given in Example 7–2. That is, the original response for treatment combination (1) was 45, and in Figure 8–4(b) it is reported as (1) = 25 (= 45 − 20). The other responses in this block are obtained similarly. After the tests in block 1 are performed, the eight tests in block 2 follow. There is no problem with the raw material in this batch, so the responses are exactly as they were originally in Example 7–2.

Now suppose we calculate the estimates of the factor effects. It is easy to verify that the estimates of the 4 main effects, the 6 two-factor interactions, and the 4 three-factor interactions are identical to the effect estimates obtained in Example 7–2 where there was *no block effect*. When a normal probability of these effect estimates is constructed, factors *A, C, D* and the *AC* and *AD* interactions emerge as the important effects, just as in the original experiment. (The reader should verify this!)

What about the *ABCD* interaction effect? The estimate of this effect in the original experiment (Example 7–2) was *ABCD* = 1.375. In the present example, we would estimate the *ABCD* interaction effect as *ABCD* = −18.625. Since *ABCD* is confounded with blocks, the *ABCD* interaction estimates the *original interaction*

Table 8–6   Analysis of Variance for Example 8–2

Source of Variation	Sum of Squares	Degrees of Freedom	Mean Square	F_0	P-Value
Blocks (ABCD)	1387.5625	1	—		
A	1870.5625	1	1870.5625	89.76	<0.0001
C	390.0625	1	390.0625	18.72	0.0019
D	855.5625	1	855.5625	41.05	0.0001
AC	1314.0625	1	1314.0625	63.05	<0.0001
AD	1105.5625	1	1105.5625	53.05	<0.0001
Error	187.5625	9	20.8403		
Total	7111.4375	15			

effect (1.375) plus the *block effect* (−20), so $ABCD = 1.375 + (-20) = -18.625$. (Do you see why the block effect is −20?) The block effect may also be calculated directly as the difference in average response between the two blocks, or

$$\text{Block Effect} = \bar{y}_{\text{Block 1}} - \bar{y}_{\text{Block 2}}$$

$$= \frac{406}{8} - \frac{555}{8}$$

$$= \frac{-149}{8}$$

$$= -18.625$$

Of course, this effect really estimates Blocks + $ABCD$.

Table 8–6 summarizes the analysis of variance for this experiment. The effects with large estimates are included in the model, and the block sum of squares is

$$SS_{\text{Blocks}} = \frac{(406)^2 + (555)^2}{8} - \frac{(961)^2}{16} = 1387.5625$$

The conclusions from this experiment exactly match those from Example 7–2, where no block effect was present. Notice that if the experiment had not been run in blocks, and if an effect of magnitude −20 had affected the first 8 trials (which would have been selected in a random fashion, since the 16 trials would be run in random order in an unblocked design), the results could have been very different.   ■

8–5   CONFOUNDING THE 2^k FACTORIAL DESIGN IN FOUR BLOCKS

It is possible to construct 2^k factorial designs confounded in four blocks of 2^{k-2} observations each. These designs are particularly useful in situations where the number of factors is moderately large, say $k \geq 4$, and block sizes are relatively small.

As an example, consider the 2^5 design. If each block will hold only eight runs, then four blocks must be used. The construction of this design is relatively straightforward. Select *two* effects to be confounded with blocks, say ADE and BCE. These effects have the two defining contrasts

$$L_1 = x_1 + x_4 + x_5$$
$$L_2 = x_2 + x_3 + x_5$$

associated with them. Now every treatment combination will yield a particular pair of values of L_1 (mod 2) and L_2 (mod 2); that is, either $(L_1, L_2) = (0, 0)$, $(0, 1)$, $(1, 0)$, or $(1, 1)$. Treatment combinations yielding the same values of (L_1, L_2) are assigned to the same block. In our example we find

$$L_1 = 0, L_2 = 0 \quad \text{for} \quad (1), ad, bc, abcd, abe, ace, cde, bde$$
$$L_1 = 1, L_2 = 0 \quad \text{for} \quad a, d, abc, bcd, be, abde, ce, acde$$
$$L_1 = 0, L_2 = 1 \quad \text{for} \quad b, abd, c, acd, ae, de, abce, bcde$$
$$L_1 = 1, L_2 = 1 \quad \text{for} \quad e, ade, bce, abcde, ab, bd, ac, cd$$

These treatment combinations would be assigned to different blocks. The complete design is as shown in Figure 8–5.

With a little reflection we realize that another effect in addition to ADE and BCE must be confounded with blocks. Since there are four blocks with three degrees of freedom between them, and since ADE and BCE have only one degree of freedom each, clearly an additional effect with one degree of freedom must be confounded. This effect is the **generalized interaction** of ADE and BCE, which is defined as the product of ADE and BCE modulus 2. Thus, in our example the generalized interaction $(ADE)(BCE) = ABCDE^2 = ABCD$ is also confounded with blocks. It is easy to verify this by referring to a table of plus and minus signs for the 2^5 design, such as in Davies (1956). Inspection of such a table reveals that the treatment combinations are assigned to the blocks as follows:

Treatment Combinations in	Sign on ADE	Sign on BCE	Sign on $ABCD$
Block 1	−	−	+
Block 2	+	−	−
Block 3	−	+	−
Block 4	+	+	+

Notice that the product of signs of any two effects for a particular block (e.g., ADE and BCE) yields the sign of the other effect for that block (in this case, $ABCD$). Thus, ADE, BCE, and $ABCD$ are all confounded with blocks.

The group-theoretic properties of the principal block mentioned in Section

Block 1		Block 2		Block 3		Block 4	
$L_1 = 0$		$L_1 = 1$		$L_1 = 0$		$L_1 = 1$	
$L_2 = 0$		$L_2 = 0$		$L_2 = 1$		$L_2 = 1$	
(1)	abe	a	be	b	abce	e	abcde
ad	ace	d	abde	abd	ae	ade	bd
bc	cde	abc	ce	c	bcde	bce	ac
abcd	bde	bcd	acde	acd	de	ab	cd

Figure 8–5. The 2^5 design in four blocks with ADE, BCE, and $ABCD$ confounded.

8–4 still hold. For example, we see that the product of two treatment combinations in the principal block yields another element of the principal block. That is,

$$ad \cdot bc = abcd \quad \text{and} \quad abe \cdot bde = ab^2de^2 = ad$$

and so forth. To construct another block, select a treatment combination that is not in the principal block (e.g., b) and multiply b by all the treatment combinations in the principal block. This yields

$$b \cdot (1) = b \quad b \cdot ad = abd \quad b \cdot bc = c \quad b \cdot abcd = ab^2cd = acd$$

and so forth, which will produce the eight treatment combinations in block 3. In practice, the principal block can be obtained from the defining contrasts and the group-theoretic property, and the remaining blocks can be determined from these treatment combinations by the method shown above.

The general procedure for constructing a 2^k design confounded in four blocks is to choose two effects to generate the blocks, automatically confounding a third effect that is the generalized interaction of the first two. Then, the design is constructed by using the two defining contrasts (L_1, L_2) and the group-theoretic properties of the principal block. In selecting effects to be confounded with blocks, care must be exercised to obtain a design that does not confound effects that may be of interest. For example, in a 2^5 design we might choose to confound $ABCDE$ and ABD, which automatically confounds CE, an effect that is probably of interest. A better choice is to confound ADE and BCE, which automatically confounds $ABCD$. It is preferable to sacrifice information on the three-factor interactions ADE and BCE instead of the two-factor interaction CE.

8–6 CONFOUNDING THE 2^k FACTORIAL DESIGN IN 2^p BLOCKS

The methods described above may be extended to the construction of a 2^k factorial design confounded in 2^p blocks ($p < k$), where each block contains exactly 2^{k-p} runs. We select p independent effects to be confounded, where by

Table 8-7 Suggested Blocking Arrangements for the 2^k Factorial Design

Number of Factors, k	Number of Blocks, 2^p	Block Size, 2^{k-p}	Effects Chosen to Generate the Blocks	Interactions Confounded with Blocks
3	2	4	ABC	ABC
	4	2	AB, AC	AB, AC, BC
4	2	8	$ABCD$	$ABCD$
	4	4	ABC, ACD	ABC, ACD, BD
	8	2	AB, BC, CD	$AB, BC, CD, AC, BD, AD, ABCD$
5	2	16	$ABCDE$	$ABCDE$
	4	8	ABC, CDE	$ABC, CDE, ABDE$
	8	4	ABE, BCE, CDE	$ABE, BCE, CDE, AC, ABCD, BD, ADE$
	16	2	AB, AC, CD, DE	All 2-factor and 4-factor interactions (15 effects)
6	2	32	$ABCDEF$	$ABCDEF$
	4	16	$ABCF, CDEF$	$ABCF, CDEF, ABDE$
	8	8	$ABEF, ABCD, ACE$	$ABEF, ABCD, ACE, BCF, BDE, CDEF, ADF$
	16	4	ABF, ACF, BDF, DEF	$ABE, ACF, BDF, DEF, BC, ABCD, ABDE, AD, ACDE,$ $CE, BDF, BCDEF, ABCEF, AEF, BE$
	32	2	AB, BC, CD, DE, EF	All 2-factor, 4-factor, and 6-factor interactions (31 effects)
7	2	64	$ABCDEFG$	$ABCDEFG$
	4	32	$ABCFG, CDEFG$	$ABCFG, CDEFG, ABDE$
	8	16	ABC, DEF, AFG	$ABC, DEF, AFG, ABCDEF, DCFG, ADEG, BCDEG$
	16	8	$ABCD, EFG, CDE, ADG$	$ABCD, EFG, CDE, ADG, ABCDEFG, ABE, BCG,$ $CDFG, ADEF, ACEG, ABFG, BCEF, BDEG, ACF, BDF$
	32	4	ABG, BCG, CDG, DEG, EFG	$ABG, BCG, CDG, DEG, EFG, AC, BD, CE, DF, AE, BE,$ $ABCD, ABDE, ABEF, BCDE, BCEF, CDEF, ABCDEFG,$ $ADG, ACDEG, ACEFG, ABDFG, ABCEG, BEG,$ $BDEFG, CFG, ADEF, ACDF, ABCF, AFG$
	64	2	AB, BC, CD, DE, EF, FG	All 2-factor, 4-factor, and 6-factor interactions (63 effects)

"independent" we mean that no effect chosen is the generalized interaction of the others. The blocks may be generated by use of the p defining contrasts L_1, $L_2, \ldots, L_p$ associated with these effects. In addition, exactly $2^p - p - 1$ other effects will be confounded with blocks, these being the generalized interactions of those p independent effects initially chosen. Care should be exercised in selecting effects to be confounded so that information on effects that may be of potential interest is not sacrificed.

The statistical analysis of these designs is straightforward. Sums of squares for all the effects are computed as if no blocking had occurred. Then, the block sum of squares is found by adding the sums of squares for all the effects confounded with blocks.

Obviously, the choice of the p effects used to generate the block is critical since the confounding structure of the design directly depends on them. Table 8–7 presents a list of useful designs. To illustrate the use of this table, suppose we wish to construct a 2^6 design confounded in $2^3 = 8$ blocks of $2^3 = 8$ runs each. Table 8–7 indicates that we would choose $ABEF$, $ABCD$, and ACE as the $p = 3$ independent effects to generate the blocks. The remaining $2^p - p - 1 = 2^3 - 3 - 1 = 4$ effects that are confounded are the generalized interactions of these three; that is,

$$(ABEF)(ABCD) = A^2B^2CDEF = CDEF$$
$$(ABEF)(ACE) = A^2BCE^2F = BCF$$
$$(ABCD)(ACE) = A^2BC^2ED = BDE$$
$$(ABEF)(ABCD)(ACE) = A^3B^2C^2DE^2F = ADF$$

The reader is asked to generate the eight blocks for this design in Problem 8–11.

8–7 PARTIAL CONFOUNDING

We remarked in Section 8–4 that, unless experimenters have a prior estimate of error or are willing to assume certain interactions to be negligible, they must replicate the design to obtain an estimate of error. Figure 8–3 shows a 2^3 factorial in two blocks with ABC confounded, replicated four times. From the analysis of variance for this design, shown in Table 8–5, we note that information on the ABC interaction cannot be retrieved since ABC is confounded with blocks *in each replicate*. This design is said to be **completely confounded.**

Consider the alternative shown in Figure 8–6. Once again, there are four replicates of the 2^3 design, but a *different* interaction has been confounded in each replicate. That is, ABC is confounded in replicate I, AB is confounded in replicate II, BC is confounded in replicate III, and AC is confounded in replicate IV. As a result, information on ABC can be obtained from the data in replicates II, III, and IV; information on AB can be obtained from replicates I, III, and IV; information on AC can be obtained from replicates I, II, and III; and informa-

Replicate I *ABC* Confounded		Replicate II *AB* Confounded		Replicate III *BC* Confounded		Replicate IV *AC* Confounded	
(1)	*a*	(1)	*a*	(1)	*b*	(1)	*a*
ab	*b*	*c*	*b*	*a*	*c*	*b*	*c*
ac	*c*	*ab*	*ac*	*bc*	*ab*	*ac*	*ab*
bc	*abc*	*abc*	*bc*	*abc*	*ac*	*abc*	*bc*

Figure 8–6.   Partial confounding in the 2^3 design.

tion on BC can be obtained from replicates I, II, and IV. We say that three-quarters information can be obtained on the interactions because they are unconfounded in only three replicates. Yates (1937) calls the ratio 3/4 the **relative information for the confounded effects.** This design is said to be **partially confounded.**

The analysis of variance for this design is shown in Table 8–8. In calculating the interaction sums of squares, only data from the replicates in which an interaction is unconfounded are used. The error sum of squares consists of replicates × main effect sums of squares plus replicates × interaction sums of squares for each replicate in which that interaction is unconfounded (e.g., replicates × ABC for replicates II, III, and IV). Furthermore, there are seven degrees of freedom among the eight blocks. This is usually partitioned into three degrees of freedom for replicates and four degrees of freedom for blocks within replicates. The composition of the sum of squares for blocks is shown in Table 8–8 and follows directly from the choice of the effect confounded in each replicate.

Table 8–8   Analysis of Variance for a Partially Confounded 2^3 Design

Source of Variation	Degrees of Freedom
Replicates	3
Blocks within replicates [or ABC (rep. I) + AB (rep. II) + BC (rep. III) + AC (rep. IV)]	4
A	1
B	1
C	1
AB (from replicates I, III, and IV)	1
AC (from replicates I, II, and III)	1
BC (from replicates I, II, and IV)	1
ABC (from replicates II, III, and IV)	1
Error	17
Total	31

Example 8-3

A 2^3 Design with Partial Confounding

Consider Example 7-1, in which a study was performed to determine the effect of percent carbonation (A), operating pressure (B), and line speed (C) on the fill height of a carbonated beverage. Suppose that each batch of syrup is only large enough to test four treatment combinations. Thus, each replicate of the 2^3 design must be run in two blocks. Two replicates are run, with ABC confounded in replicate I and AB confounded in replicate II. The data are as follows:

	Replicate I ABC Confounded				Replicate II AB Confounded	
$(1) = -3$		$a = 0$		$(1) = -1$		$a = 1$
$ab = 2$		$b = -1$		$c = 0$		$b = 0$
$ac = 2$		$c = -1$		$ab = 3$		$ac = 1$
$bc = 1$		$abc = 6$		$abc = 5$		$bc = 1$

The sums of squares for A, B, C, AC, and BC may be calculated in the usual manner. Yates' algorithm may also be employed, as shown in Table 8-9. Note that in this table, the calculations of the sums of squares for AB and ABC are not completed. We must find SS_{ABC} using only the data in replicate II and SS_{AB} using only the data in replicate I as follows:

$$SS_{ABC} = \frac{[a + b + c + abc - ab - ac - bc - (1)]^2}{n2^k}$$

$$= \frac{[1 + 0 + 0 + 5 - 3 - 1 - 1 - (-1)]^2}{(1)(8)} = 0.50$$

$$S_{AB} = \frac{[(1) + abc - ac + c - a - b + ab - bc]^2}{n2^k}$$

$$= \frac{[-3 + 6 - 2 + (-1) - 0 - (-1) + 2 - 1]^2}{(1)(8)} = 0.50$$

Table 8-9   Yates' Algorithm for the Data in Example 8-3

Treatment Combination	Response	(1)	(2)	(3)	Effect	Sum of Squares
(1)	-4	-3	1	16	I	—
a	1	4	15	24	A	36.00
b	-1	2	11	18	B	20.25
ab	5	13	13	6	AB	Not direct
c	-1	5	7	14	C	12.25
ac	3	6	11	2	AC	0.25
bc	2	4	1	4	BC	1.00
abc	11	9	5	4	ABC	Not direct

Table 8–10   Analysis of Variance for Example 8–3

Source of Variation	Sum of Squares	Degrees of Freedom	Mean Square	F_0	P-Value
Replicates	1.00	1	1.00	—	
Blocks within replicates	2.50	2	1.25	—	
A	36.00	1	36.00	48.00	0.0001
B	20.25	1	20.25	27.00	0.0035
C	12.25	1	12.25	16.33	0.0099
AB (rep. I only)	0.50	1	0.50	0.67	0.4503
AC	0.25	1	0.25	0.33	0.5905
BC	1.00	1	1.00	1.33	0.3009
ABC (rep. II only)	0.50	1	0.50	0.67	0.4503
Error	3.75	5	0.75	—	
Total	78.00	15	—	—	

The sum of squares for the replicates is, in general,

$$SS_{\text{Rep}} = \sum_{h=1}^{n} \frac{R_h^2}{2^k} - \frac{y_{...}^2}{N}$$

$$= \frac{(6)^2 + (10)^2}{8} - \frac{(16)^2}{16} = 1.00$$

where R_h is the total of the observations in the hth replicate. The block sum of squares is the sum of SS_{ABC} from replicate I and SS_{AB} from replicate II, or $SS_{\text{Blocks}} = 2.50$.

The analysis of variance is summarized in Table 8–10. All three main effects are important.

∎

8–8   PROBLEMS

8–1   Consider the experiment described in Problem 7–1. Analyze this experiment assuming that each replicate represents a block of a single production shift.

8–2   Consider the experiment described in Problem 7–5. Analyze this experiment assuming that each one of the four replicates represents a block.

8–3   Consider the alloy cracking experiment described in Problem 7–15. Suppose that only 16 runs could be made on a single day, so each replicate was treated as a block. Analyze the experiment and draw conclusions.

8–4   Consider the data from the first replicate of Problem 7–1. Suppose that these observations could not all be run using the same bar stock. Set up a design to run these observations in two blocks of four observations each with ABC confounded. Analyze the data.

8-5 Consider the data from the first replicate of Problem 7-7. Construct a design with two blocks of eight observations each with $ABCD$ confounded. Analyze the data.

8-6 Repeat Problem 8-5 assuming that four blocks are required. Confound ABD and ABC (and consequently CD) with blocks.

8-7 Using the data from the 2^5 design in Problem 7-21, construct and analyze a design in two blocks with $ABCDE$ confounded with blocks.

8-8 Repeat Problem 8-7 assuming that four blocks are necessary. Suggest a reasonable confounding scheme.

8-9 Consider the data from the 2^5 design in Problem 7-21. Suppose that it was necessary to run this design in four blocks with $ACDE$ and BCD (and consequently ABE) confounded. Analyze the data from this design.

8-10 Design an experiment for confounding a 2^6 factorial in four blocks. Suggest an appropriate confounding scheme, different from the one shown in Table 8-7.

8-11 Consider the 2^6 design in eight blocks of eight runs each with $ABCD$, ACE, and $ABEF$ as the independent effects chosen to be confounded with blocks. Generate the design. Find the other effects confounded with blocks.

8-12 Consider the 2^2 design in two blocks with AB confounded. Prove algebraically that $SS_{AB} = SS_{Blocks}$.

8-13 Consider the data in Example 8-2. Suppose that all the observations in block 2 are increased by 20. Analyze the data that would result. Estimate the block effect. Can you explain its magnitude? Do blocks now appear to be an important factor? Are any other effect estimates impacted by the change you made to the data?

8-14 Suppose that in Problem 7-1 we had confounded ABC in replicate I, AB in replicate II, and BC in replicate III. Construct the analysis of variance table.

8-15 Repeat Problem 7-1 assuming that ABC was confounded with blocks in each replicate.

8-16 Suppose that in Problem 7-7 $ABCD$ was confounded in replicate I and ABC was confounded in replicate II. Perform the statistical analysis of this design.

8-17 Construct a 2^3 design with ABC confounded in the first two replicates and BC confounded in the third. Outline the analysis of variance and comment on the information obtained.

Chapter 9
Two-Level Fractional Factorial Designs

9-1 INTRODUCTION

As the number of factors in a 2^k factorial design increases, the number of runs required for a complete replicate of the design rapidly outgrows the resources of most experimenters. For example, a complete replicate of the 2^6 design requires 64 runs. In this design only 6 of the 63 degrees of freedom correspond to main effects, and only 15 degrees of freedom correspond to two-factor interactions. The remaining 42 degrees of freedom are associated with three-factor and higher interactions.

If the experimenter can reasonably assume that certain high-order interactions are negligible, then information on the main effects and low-order interactions may be obtained by running only a fraction of the complete factorial experiment. These **fractional factorial designs** are among the most widely used types of designs for product and process design and for process improvement.

A major use of fractional factorials is in **screening experiments.** These are experiments in which many factors are considered with the purpose of identifying those factors (if any) that have large effects. Screening experiments are usually performed in the early stages of a project when it is likely that many of the factors initially considered have little or no effect on the response. The factors that are identified as important are then investigated more thoroughly in subsequent experiments.

The successful use of fractional factorial designs is based on three key ideas:

1. *The sparsity of effects principle.* When there are several variables, the system or process is likely to be driven primarily by some of the main effects and low-order interactions.

2. *The projection property.* Fractional factorial designs can be projected into stronger (larger) designs in the subset of significant factors.

3. *Sequential experimentation.* It is possible to combine the runs of two (or more) fractional factorials to assemble sequentially a larger design to estimate the factor effects and interactions of interest.

We will focus on these principles in this chapter and illustrate them with several examples.

9–2   THE ONE-HALF FRACTION OF THE 2^k DESIGN

Consider a situation in which three factors, each at two levels, are of interest, but the experimenters cannot afford to run all $2^3 = 8$ treatment combinations. They can, however, afford four runs. This suggests a one-half fraction of a 2^3 design. Because the design contains $2^{3-1} = 4$ treatment combinations, a one-half fraction of the 2^3 design is often called a **2^{3-1} design.**

The table of plus and minus signs for the 2^3 design is shown in Table 9–1. Suppose we select the four treatment combinations a, b, c, and abc as our one-half fraction. These runs are shown in the top half of Table 9–1 and in Figure 9–1a.

Notice that the 2^{3-1} design is formed by selecting only those treatment combinations that have a plus in the ABC column. Thus, ABC is called the **generator** of this particular fraction. Sometimes we will refer to a generator such as ABC as a **word.** Furthermore, the identity column I is also always plus, so we call

$$I = ABC$$

the **defining relation** for our design. In general, the defining relation for a fractional factorial will always be the set of all columns that are equal to the identity column I.

The treatment combinations in the 2^{3-1} design yield three degrees of freedom that we may use to estimate the main effects. Referring to Table 9–1, we note

Table 9–1   Plus and Minus Signs for the 2^3 Factorial Design

Treatment Combination	Factorial Effect							
	I	A	B	C	AB	AC	BC	ABC
a	$+$	$+$	$-$	$-$	$-$	$-$	$+$	$+$
b	$+$	$-$	$+$	$-$	$-$	$+$	$-$	$+$
c	$+$	$-$	$-$	$+$	$+$	$-$	$-$	$+$
abc	$+$	$+$	$+$	$+$	$+$	$+$	$+$	$+$
ab	$+$	$+$	$+$	$-$	$+$	$-$	$-$	$-$
ac	$+$	$+$	$-$	$+$	$-$	$+$	$-$	$-$
bc	$+$	$-$	$+$	$+$	$-$	$-$	$+$	$-$
(1)	$+$	$-$	$-$	$-$	$+$	$+$	$+$	$-$

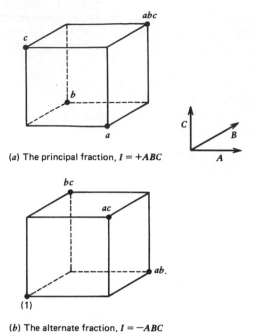

(a) The principal fraction, $I = +ABC$

(b) The alternate fraction, $I = -ABC$

Figure 9–1.   The two one-half fractions of the 2^3 design.

that the linear combinations of the observations used to estimate the main effects of A, B, and C are

$$\ell_A = \tfrac{1}{2}(a - b - c + abc)$$
$$\ell_B = \tfrac{1}{2}(-a + b - c + abc)$$
$$\ell_C = \tfrac{1}{2}(-a - b + c + abc)$$

It is also easy to verify that the linear combinations of the observations used to estimate the two-factor interactions are

$$\ell_{BC} = \tfrac{1}{2}(a - b - c + abc)$$
$$\ell_{AC} = \tfrac{1}{2}(-a + b - c + abc)$$
$$\ell_{AB} = \tfrac{1}{2}(-a - b + c + abc)$$

Thus, $\ell_A = \ell_{BC}$, $\ell_B = \ell_{AC}$, and $\ell_C = \ell_{AB}$; consequently, it is impossible to differentiate between A and BC, B and AC, and C and AB. In fact, when we estimate A, B, and C we are *really* estimating $A + BC$, $B + AC$, and $C + AB$. Two or more effects that have this property are called **aliases.** In our example A and BC are aliases, B and AC are aliases, and C and AB are aliases. We indicate this by the notation $\ell_A \rightarrow A + BC$, $\ell_B \rightarrow B + AC$, and $\ell_C \rightarrow C + AB$.

The alias structure for this design may be easily determined by using the defining relation $I = ABC$. Multiplying any column by the defining relation yields the aliases for that effect. In our example, this yields as the alias of A

$$A \cdot I = A \cdot ABC = A^2BC$$

or, since the square of any column is just the identity I,

$$A = BC$$

Similarly, we find the aliases of B and C as

$$B \cdot I = B \cdot ABC$$
$$B = AB^2C = AC$$

and

$$C \cdot I = C \cdot ABC$$
$$C = ABC^2 = AB$$

This one-half fraction, with $I = +ABC$, is usually called the **principal fraction.**

Now suppose that we had chosen the *other* one-half fraction, that is, the treatment combinations in Table 9–1 associated with minus in the ABC column. This **alternate** or **complementary** one-half fraction (consisting of the runs (1), *ab, ac,* and *bc*) is shown in Figure 9–1*b*. The defining relation for this design is

$$I = -ABC$$

The linear combination of the observations, say ℓ'_A, ℓ'_B, and ℓ'_C, from the alternate fraction gives us

$$\ell'_A \rightarrow A - BC$$
$$\ell'_B \rightarrow B - AC$$
$$\ell'_C \rightarrow C - AB$$

Thus, when we estimate A, B, and C with this particular fraction, we are really estimating $A - BC$, $B - AC$, and $C - AB$.

In practice, it does not matter which fraction is actually used. Both fractions belong to the same **family;** that is, the two one-half fractions form a complete 2^3 design. This is easily seen by reference to parts *a* and *b* of Figure 9–1.

Suppose that after running one of the one-half fractions of the 2^3 design, the other one was also run. Thus, all eight runs associated with the full 2^3 are now available. We may now obtain de-aliased estimates of all the effects by analyzing the eight runs as a full 2^3 design in two blocks of four runs each. This could also be done by adding and subtracting the linear combination of effects from the two individual fractions. For example, consider $\ell_A \rightarrow A + BC$ and $\ell'_A \rightarrow A - BC$. This implies that

$$\tfrac{1}{2}(\ell_A + \ell'_A) = \tfrac{1}{2}(A + BC + A - BC) \rightarrow A$$

and that

$$\tfrac{1}{2}(\ell_A - \ell'_A) = \tfrac{1}{2}(A + BC - A + BC) \rightarrow BC$$

Thus, for all three pairs of linear combinations, we would obtain the following:

i	From $\tfrac{1}{2}(\ell_i + \ell'_i)$	From $\tfrac{1}{2}(\ell_i - \ell'_i)$
A	A	BC
B	B	AC
C	C	AB

Design Resolution ▪ The preceding 2^{3-1} design is called a **resolution III design.** In such a design, main effects are aliased with two-factor interactions. A design is of resolution R if no p-factor effect is aliased with another effect containing less than $R - p$ factors. We usually employ a Roman numeral subscript to denote design resolution; thus, the one-half fraction of the 2^3 design with the defining relation $I = ABC$ (or $I = -ABC$) is a 2^{3-1}_{III} design.

Designs of resolution III, IV, and V are particularly important. The definitions of these designs and an example of each follow.

1. *Resolution III designs.* These are designs in which no main effects are aliased with any other main effect, but main effects are aliased with two-factor interactions and two-factor interactions may be aliased with each other. The 2^{3-1} design in Table 9–1 is of resolution III (2^{3-1}_{III}).

2. *Resolution IV designs.* These are designs in which no main effect is aliased with any other main effect *or* with any two-factor interaction, but two-factor interactions are aliased with each other. A 2^{4-1} design with $I = ABCD$ is a resolution IV design (2^{4-1}_{IV}).

3. *Resolution V designs.* These are designs in which no main effect or two-factor interaction is aliased with any other main effect or two-factor interaction, but two-factor interactions are aliased with three-factor interactions. A 2^{5-1} design with $I = ABCDE$ is a resolution V design (2^{5-1}_{V}).

In general, the resolution of a two-level fractional factorial design is equal to the smallest number of letters in any word in the defining relation. Consequently, we could call the preceding design types three-letter, four-letter, and five-letter designs, respectively. We usually like to employ fractional designs that have the highest possible resolution consistent with the degree of fractionation required. The higher the resolution, the less restrictive the assumptions that are required regarding which interactions are negligible in order to obtain a unique interpretation of the data.

Constructing One-Half Fractions ▪ A one-half fraction of the 2^k design of the highest resolution may be constructed by writing down a **basic design** consisting of the runs for a *full* 2^{k-1} factorial and then adding the kth factor by identifying its plus and minus levels with the plus and minus signs of the highest-order interaction $ABC \cdot \cdot \cdot (K - 1)$. Therefore, the 2_{III}^{3-1} fractional factorial is obtained by writing down the full 2^2 factorial as the basic design and then equating factor C to the AB interaction. The alternate fraction would be obtained by equating factor C to the $-AB$ interaction. This approach is illustrated in Table 9-2. Notice that the basic design always has the right number of runs (rows), but it is missing one column. The generator $I = ABC \cdot \cdot \cdot K$ is then solved for the missing column (K) so that $K = ABC \cdot \cdot \cdot (K - 1)$ defines the product of plus and minus signs to use in each row to produce the levels for the kth factor.

Note that *any* interaction effect could be used to generate the column for the kth factor. However, using any effect other than $ABC \cdot \cdot \cdot (K - 1)$ will not produce a design of the highest possible resolution.

Another way to view the construction of a one-half fraction is to partition the runs into two blocks with the highest-order interaction $ABC \cdot \cdot \cdot K$ confounded. Each block is a 2^{k-1} fractional factorial design of the highest resolution.

Projection of Fractions into Factorials ▪ Any fractional factorial design of resolution R contains complete factorial designs (possibly replicated factorials) in any subset of $R - 1$ factors. This is an important and useful concept. For example, if an experimenter has several factors of potential interest but believes that only $R - 1$ of them have important effects, then a fractional factorial design of resolution R is the appropriate choice of design. If the experimenter is correct, then the fractional factorial design of resolution R will project into a full factorial in the $R - 1$ significant factors. This process is illustrated in Figure 9-2 for the 2_{III}^{3-1} design, which projects into a 2^2 design in every subset of two factors.

Since the maximum possible resolution of a one-half fraction of the 2^k design is $R = k$, every 2^{k-1} design will project into a full factorial in any $(k - 1)$ of the original k factors. Furthermore, a 2^{k-1} design may be projected into two replicates

Table 9-2 The Two One-Half Fractions of the 2^3 Design

Run	Full 2^2 Factorial (Basic Design)		$2_{III}^{3-1}, I = ABC$			$2_{III}^{3-1}, I = -ABC$		
	A	B	A	B	$C = AB$	A	B	$C = -AB$
1	−	−	−	−	+	−	−	−
2	+	−	+	−	−	+	−	+
3	−	+	−	+	−	−	+	+
4	+	+	+	+	+	+	+	−

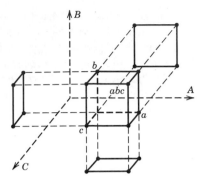

Figure 9–2.   Projection of a 2_{III}^{3-1} design into three 2^2 designs.

of a full factorial in any subset of $k - 2$ factors, four replicates of a full factorial in any subset of $k - 3$ factors, and so on.

Example 9–1

Consider the filtration rate experiment in Example 7–2. The original design, shown in Table 7–9, is a single replicate of the 2^4 design. In that example, we found that the main effects A, C, and D and the interactions AC and AD were different from zero. We will now return to this experiment and simulate what would have happened if a half-fraction of the 2^4 design had been run instead of the full factorial.

We will use the 2^{4-1} design with $I = ABCD$, as this choice of generator will result in a design of the highest possible resolution (IV). To construct the design, we first write down the basic design, which is a 2^3 design, as shown in the first three columns of Table 9–3. This basic design has the necessary number of runs (eight), but only three columns (factors). To find the fourth factor levels, solve $I = ABCD$ for D, or $D = ABC$. Thus, the level of D in each run is the product of the plus and minus signs in columns A, B, and C. The process is illustrated in Table 9–3. Since the generator $ABCD$ is positive, this 2_{IV}^{4-1} design is the principal fraction. The design is shown graphically in Figure 9–3.

Table 9–3   The 2_{IV}^{4-1} Design with the Defining Relation $I = ABCD$

Run	Basic Design			$D = ABC$	Treatment Combination	Filtration Rate
	A	B	C			
1	−	−	−	−	(1)	45
2	+	−	−	+	ad	100
3	−	+	−	+	bd	45
4	+	+	−	−	ab	65
5	−	−	+	+	cd	75
6	+	−	+	−	ac	60
7	−	+	+	−	bc	80
8	+	+	+	+	$abcd$	96

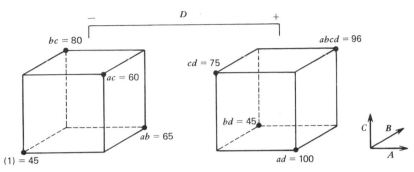

Figure 9–3. The 2_{IV}^{4-1} design for the filtration rate experiment of Example 9–1.

Using the defining relation, we note that each main effect is aliased with a three-factor interaction; that is, $A = A^2BCD = BCD$, $B = AB^2CD = ACD$, $C = ABC^2D = ABD$, and $D = ABCD^2 = ABC$. Furthermore, every two-factor interaction is aliased with another two-factor interaction. These alias relationships are $AB = CD$, $AC = BD$, and $BC = AD$. The four main effects plus the three two-factor interaction alias pairs account for the seven degrees of freedom for the design.

At this point, we would normally randomize the eight runs and perform the experiment. Since we have already run the full 2^4 design, we will simply select the eight observed filtration rates from Example 7–2 that correspond to the runs in the 2_{IV}^{4-1} design. These observations are shown in the last column of Table 9–3 and are also shown in Figure 9–3.

The estimates of the effects obtained from this 2_{IV}^{4-1} design are shown in Table 9–4. To illustrate the calculations, the linear combination of observations associated with the A effect is

$$\ell_A = \tfrac{1}{4}(-45 + 100 - 45 + 65 - 75 + 60 - 80 + 96) = 19.00 \rightarrow A + BCD$$

whereas for the AB effect, we would obtain

$$\ell_{AB} = \tfrac{1}{4}(45 - 100 - 45 + 65 + 75 - 60 - 80 + 96) = -1.00 \rightarrow AB + CD$$

Table 9–4 Estimates of Effects and Aliases from Example 9–1[a]

Estimate	Alias Structure
$\ell_A = 19.00$	$\ell_A \rightarrow A + BCD$
$\ell_B = 1.50$	$\ell_B \rightarrow B + ACD$
$\ell_C = 14.00$	$\ell_C \rightarrow C + ABD$
$\ell_D = 16.50$	$\ell_D \rightarrow D + ABC$
$\ell_{AB} = -1.00$	$\ell_{AB} \rightarrow AB + CD$
$\ell_{AC} = -18.50$	$\ell_{AC} \rightarrow AC + BD$
$\ell_{AD} = 19.00$	$\ell_{AD} \rightarrow AD + BC$

[a] Significant effects are shown in boldface type.

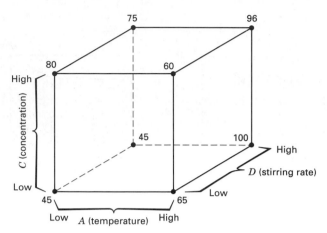

Figure 9–4.   Projection of the 2_{IV}^{4-1} design into a 2^3 design in A, C, and D for Example 9–1.

From inspection of the information in Table 9–4, it is not unreasonable to conclude that the main effects A, C, and D are large and that the AC and AD interactions are also significant. This agrees with the conclusions from the analysis of the complete 2^4 design in Example 7–2.

Since factor B is not significant, we may drop it from consideration. Consequently, we may project this 2_{IV}^{4-1} design into a single replicate of the 2^3 design in factors A, C, and D, as shown in Figure 9–4. Visual examination of this cube plot makes us more comfortable with the conclusions reached above. Notice that if the temperature (A) is at the low level, the concentration (C) has a large positive effect, whereas if the temperature is at the high level, the concentration has a very small effect. This is likely due to an AC interaction. Furthermore, if the temperature is at the low level, the effect of the stirring rate (D) is negligible, whereas if the temperature is at the high level, the stirring rate has a large positive effect. This is likely due to the AD interaction tentatively identified previously.

Based on the above analysis, we now can obtain a model to predict filtration rate over the experimental region. This model is

$$\hat{y} = \hat{\beta}_0 + \hat{\beta}_1 x_1 + \hat{\beta}_3 x_3 + \hat{\beta}_4 x_4 + \hat{\beta}_{13} x_1 x_3 + \hat{\beta}_{14} x_1 x_4$$

where x_1, x_3, and x_4 are coded variables $(-1 \leq x_i \leq +1)$ that represent A, C, and D, and the $\hat{\beta}$'s are regression coefficients that can be obtained from the effect estimates as we did previously. Therefore, the prediction equation is

$$\hat{y} = 70.75 + \left(\frac{19.00}{2}\right) x_1 + \left(\frac{14.00}{2}\right) x_3 + \left(\frac{16.50}{2}\right) x_4 + \left(\frac{-18.50}{2}\right) x_1 x_3 + \left(\frac{19.00}{2}\right) x_1 x_4$$

Remember that the intercept $\hat{\beta}_0$ is the average of all responses at the eight runs in the design. This model is very similar to the one that resulted from the full 2^k factorial design in Example 7–2.

■

Example 9–2

A 2^{5-1} Design Used for Process Improvement

Five factors in a manufacturing process for an integrated circuit were investigated in a 2^{5-1} design with the objective of improving the process yield. The five factors were A = aperture setting (small, large), B = exposure time (20 percent below nominal, 20 percent above nominal), C = development time (30 s, 45 s), D = mask dimension (small, large), and E = etch time (14.5 min, 15.5 min). The construction of the 2^{5-1} design is shown in Table 9–5. Notice that the design was constructed by writing down the basic design having 16 runs (a 2^4 design in A, B, C, and D), selecting $ABCDE$ as the generator, and then setting the levels of the fifth factor $E = ABCD$. Figure 9–5 on page 382 gives a pictorial representation of the design.

The defining relation for the design is $I = ABCDE$. Consequently, every main effect is aliased with a four-factor interaction (for example, $\ell_A \rightarrow A + BCDE$), and every two-factor interaction is aliased with a three-factor interaction (for example, $\ell_{AB} \rightarrow AB + CDE$). Thus, the design is of resolution V. We would expect this 2^{5-1} design to provide excellent information concerning the main effects and two-factor interactions.

Table 9–6 on page 383 contains the effect estimates, sums of squares, and model regression coefficients for the 15 effects from this experiment. Figure 9–6 presents a normal probability plot of the effect estimates from this experiment. The main effects of A, B, and C and the AB interaction are large. Remember that, because of aliasing, these effects are really $A + BCDE$, $B + ACDE$, $C + ABDE$, and $AB + CDE$. However, since it seems plausible that three-factor and higher interactions are negligible, we feel safe in concluding that only A, B, C, and AB are important effects.

Table 9–5 A 2^{5-1} Design for Example 9–2

Run	Basic Design				$E = ABCD$	Treatment Combination	Yield
	A	B	C	D			
1	−	−	−	−	+	e	8
2	+	−	−	−	−	a	9
3	−	+	−	−	−	b	34
4	+	+	−	−	+	abe	52
5	−	−	+	−	−	c	16
6	+	−	+	−	+	ace	22
7	−	+	+	−	+	bce	45
8	+	+	+	−	−	abc	60
9	−	−	−	+	−	d	6
10	+	−	−	+	+	ade	10
11	−	+	−	+	+	bde	30
12	+	+	−	+	−	abd	50
13	−	−	+	+	+	cde	15
14	+	−	+	+	−	acd	21
15	−	+	+	+	−	bcd	44
16	+	+	+	+	+	$abcde$	63

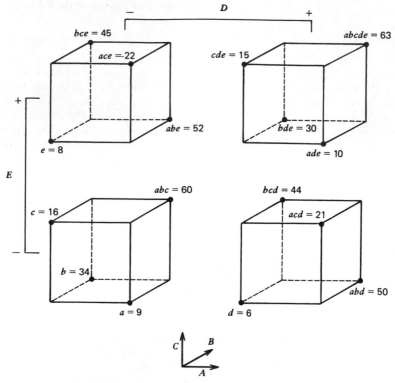

Figure 9–5. The 2_V^{5-1} design for Example 9–2.

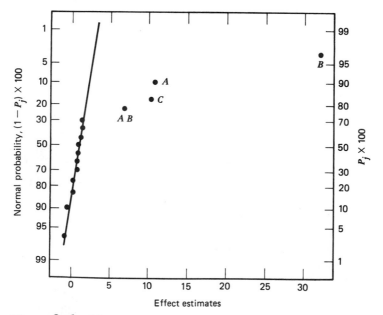

Figure 9–6. Normal probability plot of effects for Example 9–2.

Table 9-6   Effects, Regression Coefficients, and Sums of Squares for Example 9-2

Variable	Name	−1 Level	+1 Level
A	Aperture	−1.000	1.000
B	Development time	−1.000	1.000
C	Exposure time	−1.000	1.000
D	Mask dimension	−1.000	1.000
E	Etch time	−1.000	1.000

Variable	Regression Coefficient	Estimated Effect	Sum of Squares
Overall Average	30.3125		
A	5.5625	11.1250	495.062
B	16.9375	33.8750	4590.062
C	5.4375	10.8750	473.062
D	−0.4375	−0.8750	3.063
E	0.3125	0.6250	1.563
AB	3.4375	6.8750	189.063
AC	0.1875	0.3750	0.563
AD	0.5625	1.1250	5.063
AE	0.5625	1.1250	5.063
BC	0.3125	0.6250	1.563
BD	−0.0625	−0.1250	0.063
BE	−0.0625	−0.1250	0.063
CD	0.4375	0.8750	3.063
CE	0.1875	0.3750	0.563
DE	−0.6875	−1.3750	7.563

Table 9-7 summarizes the analysis of variance for this experiment. The model sum of squares is $SS_{\text{Model}} = SS_A + SS_B + SS_C + SS_{AB} = 5747.25$, and this accounts for over 99 percent of the total variability in yield. Figure 9-7 presents a normal probability plot of the residuals, and Figure 9-8 is a plot of the residuals versus the predicted values. Both plots are satisfactory.

The three factors A, B, and C have large positive effects. The AB or aperture–exposure time interaction is plotted in Figure 9-9 (see page 385). This plot confirms that the yields are higher when both A and B are at the high level.

Table 9-7   Analysis of Variance for Example 9-2

Source of Variation	Sum of Squares	Degrees of Freedom	Mean Square	F_0	P-Value
A (Aperture)	495.0625	1	495.0625	193.20	<0.0001
B (Exposure time)	4590.0625	1	4590.0625	1791.24	<0.0001
C (Development time)	473.0625	1	473.0625	184.61	<0.0001
AB	189.0625	1	189.0625	73.78	<0.0001
Error	28.1875	11	2.5625		
Total	5775.4375	15			

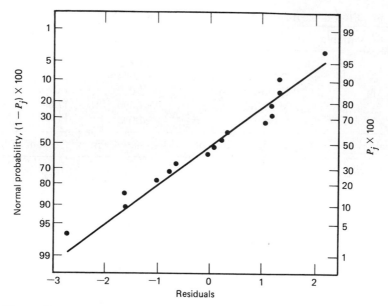

Figure 9-7.   Normal probability plot of the residuals for Example 9–2.

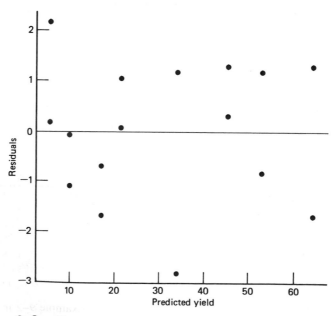

Figure 9-8.   Plot of residuals versus predicted yield for Example 9–2.

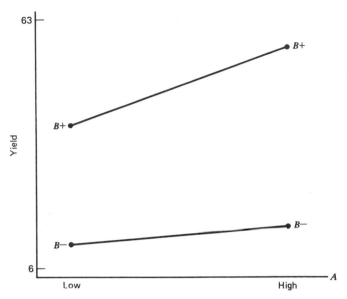

Figure 9–9.   Aperture–exposure time interaction for Example 9–2.

The 2^{5-1} design will collapse into two replicates of a 2^3 design in any three of the original five factors. (Looking at Figure 9–5 will help you visualize this.) Figure 9–10 is a cube plot in the factors A, B, and C with the average yields superimposed on the eight corners. It is clear from inspection of the cube plot that highest yields are achieved with A, B, and C all at the high level. Factors D and E have little effect on average process yield and may be set to values that optimize other objectives (such as cost).

■

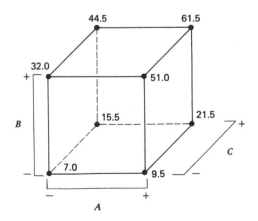

Figure 9–10.   Projection of the 2_V^{5-1} design in Example 9–2 into two replicates of a 2^3 design in the factors A, B, and C.

Sequences of Fractional Factorials ▪ Using fractional factorial designs often leads to great economy and efficiency in experimentation, particularly if the runs can be made **sequentially.** For example, suppose that we are investigating $k = 4$ factors ($2^4 = 16$ runs). It is almost always preferable to run a 2_{IV}^{4-1} fractional design (eight runs), analyze the results, and then decide on the best set of runs to perform next. If it is necessary to resolve ambiguities, we can always run the alternate fraction and complete the 2^4 design. When this method is used to complete the design, both one-half fractions represent **blocks** of the complete design with the highest-order interaction confounded with blocks (here $ABCD$ would be confounded). Thus, sequential experimentation has the result of losing information only on the highest-order interaction. Its advantage is that in many cases we learn enough from the one-half fraction to proceed to the next stage of experimentation, which might involve adding or removing factors, changing responses, or varying some of the factors over new ranges. Some of these possibilities are illustrated graphically in Figure 9–11.

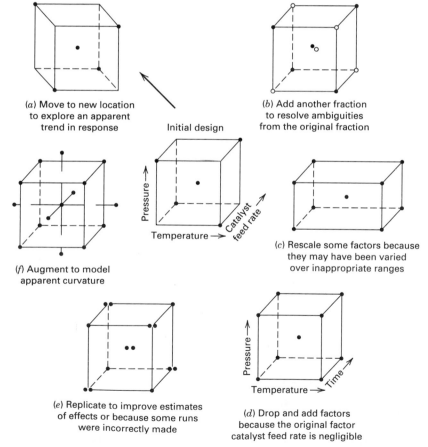

(a) Move to new location to explore an apparent trend in response

Initial design

(b) Add another fraction to resolve ambiguities from the original fraction

(f) Augment to model apparent curvature

(c) Rescale some factors because they may have been varied over inappropriate ranges

(e) Replicate to improve estimates of effects or because some runs were incorrectly made

(d) Drop and add factors because the original factor catalyst feed rate is negligible

Figure 9–11. Possibilities for follow-up experimentation after a fractional factorial experiment (adapted from Box (1992–93), with permission of the publisher).

Example 9-3

Reconsider the experiment in Example 9-2. We have used a 2_{IV}^{4-1} design, and tentatively identified three large main effects—A, C, and D. There are two large effects associated with two-factor interactions, $AC + BD$ and $AD + BC$. In Example 9-2, we used the fact that the main effect of B was negligible to tentatively conclude that the important interactions were AC and AD. Sometimes the experimenter will have process knowledge that can assist in discriminating between interactions likely to be important. However, we can always isolate the significant interaction by running the alternate fraction, given by $I = -ABCD$. It is straightforward to show that the design and the responses are as follows:

Run	A	B	C	$D = -ABC$	Treatment Combination	Filtration Rate
1	−	−	−	+	d	43
2	+	−	−	−	a	71
3	−	+	−	−	b	48
4	+	+	−	+	abd	104
5	−	−	+	−	c	68
6	+	−	+	+	acd	86
7	−	+	+	+	bcd	70
8	+	+	+	−	abc	65

(columns A, B, C grouped under "Basic Design")

The linear combinations of observations obtained from this alternate fraction are

$$\ell_A' = \quad 24.25 \rightarrow A - BCD$$
$$\ell_B' = \quad 4.75 \rightarrow B - ACD$$
$$\ell_C' = \quad 5.75 \rightarrow C - ABD$$
$$\ell_D' = \quad 12.75 \rightarrow D - ABC$$
$$\ell_{AB}' = \quad 1.25 \rightarrow AB - CD$$
$$\ell_{AC}' = -17.75 \rightarrow AC - BD$$
$$\ell_{AD}' = \quad 14.25 \rightarrow AD - BC$$

These estimates may be combined with those obtained from the original one-half fraction to yield the following estimates of the effects:

i	From $\frac{1}{2}(\ell_i + \ell_i')$	From $\frac{1}{2}(\ell_i - \ell_i')$
A	$21.63 \rightarrow A$	$-2.63 \rightarrow BCD$
B	$3.13 \rightarrow B$	$-1.63 \rightarrow ACD$
C	$9.88 \rightarrow C$	$4.13 \rightarrow ABD$
D	$14.63 \rightarrow D$	$1.88 \rightarrow ABC$
AB	$0.13 \rightarrow AB$	$-1.13 \rightarrow CD$
AC	$-18.13 \rightarrow AC$	$-0.38 \rightarrow BD$
AD	$16.63 \rightarrow AD$	$2.38 \rightarrow BC$

These estimates agree exactly with those from the original analysis of the data as a single replicate of a 2^4 factorial design, as reported in Example 7–2. Clearly, it is the AC and AD interactions that are large.

■

Adding the alternate fraction to the principal fraction may be thought of as a type of **confirmation experiment,** in that it provides information that will allow us to strengthen our initial conclusions about the two-factor interaction effects. We will investigate some other aspects of combining fractional factorials to isolate interactions in Section 9–5. Sometimes a confirmation experiment is not this elaborate. For example, one could use the model equation to predict the response at a point of interest within the design space (not one of the points in the current design), then actually run that trial (perhaps several times) and use the comparison between predicted and observed response to confirm the results.

Yates' Algorithm for Fractional Factorials ■ We may use Yates' algorithm for the analysis of a 2^{k-1} fractional factorial design by initially considering the data as having been obtained from a full factorial in $k - 1$ variables. The treatment combinations for this full factorial are listed in standard order, and then an additional letter (or letters) is added in parentheses to these treatment combinations to produce the actual treatment combinations run. Yates' algorithm then proceeds as usual. The actual effects estimated are identified by multiplying the effects associated with the treatment combinations in the full 2^{k-1} design by the defining relation to the 2^{k-1} fractional factorial.

The procedure is demonstrated in Table 9–8 using the data from Example 9–1. The data are arranged as a full 2^3 design in the factors A, B, and C. Then the letter d is added in parentheses to yield the actual treatment combinations. The effect estimated by, say, the second row in this table is $A + BCD$ since A and BCD are aliases.

Table 9–8   Yates' Algorithm for the 2^{4-1}_{IV} Fractional Factorial in Example 9–1

Treatment Combination	Response	(1)	(2)	(3)	Effect	Estimate of Effect $2 \times (3)/N$
(1)	45	145	255	566	—	—
$a(d)$	100	110	311	76	$A + BCD$	19.00
$b(d)$	45	135	75	6	$B + ACD$	1.50
ab	65	176	1	−4	$AB + CD$	−1.00
$c(d)$	75	55	−35	56	$C + ABD$	14.00
ac	60	20	41	−74	$AC + BD$	−18.50
bc	80	−15	−35	76	$BC + AD$	19.00
$abc(d)$	96	16	31	66	$ABC + D$	16.50

9–3 THE ONE-QUARTER FRACTION OF THE 2^k DESIGN

For a moderately large number of factors, smaller fractions of the 2^k design are frequently useful. Consider a one-quarter fraction of the 2^k design. This design contains 2^{k-2} runs and is usually called a 2^{k-2} **fractional factorial.**

The 2^{k-2} design may be constructed by first writing down a **basic design** consisting of the runs associated with a full factorial in $k - 2$ factors and then associating the two additional columns with appropriately chosen interactions involving the first $k - 2$ factors. Thus, a one-quarter fraction of the 2^k design has two generators. If P and Q represent the generators chosen, then $I = P$ and $I = Q$ are called the **generating relations** for the design. The signs of P and Q (either $+$ or $-$) determine which one of the one-quarter fractions is produced. All four fractions associated with the choice of **generators** $\pm P$ and $\pm Q$ are members of the same **family.** The fraction for which both P and Q are positive is the principal fraction.

The **complete defining relation** for the design consists of all the columns that are equal to the identity column I. These will consist of P, Q, and their **generalized interaction** PQ; that is, the defining relation is $I = P = Q = PQ$. We call the elements P, Q, and PQ in the defining relation **words.** The aliases of any effect are produced by the multiplication of the column for that effect by each word in the defining relation. Clearly, each effect has *three* aliases. The experimenter should be careful in choosing the generators so that potentially important effects are not aliased with each other.

As an example, consider the 2^{6-2} design. Suppose we choose $I = ABCE$ and $I = BCDF$ as the design generators. Now the generalized interaction of the generators $ABCE$ and $BCDF$ is $ADEF$; therefore, the complete defining relation for this design is

$$I = ABCE = BCDF = ADEF$$

Consequently, this is a resolution IV design. To find the aliases of any effect (e.g., A), multiply that effect by each word in the defining relation. For A, this produces

$$A = BCE = ABCDF = DEF$$

It is easy to verify that every main effect is aliased by three-factor and five-factor interactions, whereas two-factor interactions are aliased with each other and with higher-order interactions. Thus, when we estimate A, for example, we are really estimating $A + BCE + DEF + ABCDF$. The complete alias structure of this design is shown in Table 9–9. If three-factor and higher interactions are negligible, this design gives clear estimates of the main effects.

To construct the design, first write down the *basic design,* which consists of the 16 runs for a full $2^{6-2} = 2^4$ design in A, B, C, and D. Then the two factors E and F are added by associating their plus and minus levels with the plus and minus signs of the interactions ABC and BCD, respectively. This procedure is shown in Table 9–10.

Table 9–9   Alias Structure for the 2_{IV}^{6-2} Design with $I = ABCE = BCDF = ADEF$

$A = BCE = DEF = ABCDF$	$AB = CE = ACDF = BDEF$
$B = ACE = CDF = ABDEF$	$AC = BE = ABDF = CDEF$
$C = ABE = BDF = ACDEF$	$AD = EF = BCDE = ABCF$
$D = BCF = AEF = ABCDE$	$AE = BC = DF = ABCDEF$
$E = ABC = ADF = BCDEF$	$AF = DE = BCEF = ABCD$
$F = BCD = ADE = ABCEF$	$BD = CF = ACDE = ABEF$
	$BF = CD = ACEF = ABDE$
$ABD = CDE = ACF = BEF$	
$ACD = BDE = ABF = CEF$	

Another way to construct this design is to derive the four blocks of the 2^6 design with $ABCE$ and $BCDF$ confounded and then choose the block with treatment combinations that are positive on $ABCE$ and $BCDF$. This would be a 2^{6-2} fractional factorial with generating relations $I = ABCE$ and $I = BCDF$, and since both generators $ABCE$ and $BCDF$ are positive, this is the principal fraction.

There are, of course, three **alternate fractions** of this particular 2_{IV}^{6-2} design. They are the fractions with generating relationships $I = ABCE$ and $I = -BCDF$; $I = -ABCE$ and $I = BCDF$; and $I = -ABCE$ and $I = -BCDF$. These fractions may be easily constructed by the method shown in Table 9–10. For example, if we wish to find the fraction for which $I = ABCE$ and $I = -BCDF$, then in the

Table 9–10   Construction of the 2_{IV}^{6-2} Design with the Generators $I = ABCE$ and $I = BCDF$

Run	Basic Design				$E = ABC$	$F = BCD$
	A	B	C	D		
1	−	−	−	−	−	−
2	+	−	−	−	+	−
3	−	+	−	−	+	+
4	+	+	−	−	−	+
5	−	−	+	−	+	+
6	+	−	+	−	−	+
7	−	+	+	−	−	−
8	+	+	+	−	+	−
9	−	−	−	+	−	+
10	+	−	−	+	+	+
11	−	+	−	+	+	−
12	+	+	−	+	−	−
13	−	−	+	+	+	−
14	+	−	+	+	−	−
15	−	+	+	+	−	+
16	+	+	+	+	+	+

last column of Table 9–10 we set $F = -BCD$, and the column of levels for factor F becomes

$$+ + - - - - + + - - + + + + - -$$

The complete defining relation for this alternate fraction is $I = ABCE = -BCDF = -ADEF$. Certain signs in the alias structure in Table 9–9 are now changed; for instance, the aliases of A are $A = BCE = -DEF = -ABCDF$. Thus, the linear combination of the observations ℓ_A actually estimates $A + BCE - DEF - ABCDF$.

Finally, note that the 2_{IV}^{6-2} fractional factorial will project into a single replicate of a 2^4 design in any subset of four factors that is not a word in the defining relation. It also collapses to a replicated one-half fraction of a 2^4 in any subset of four factors that is a word in the defining relation. Thus, the design in Table 9–10 becomes two replicates of a 2^{4-1} in the factors $ABCE$, $BCDF$, and $ADEF$, since these are the words in the defining relation. There are 12 other combinations of the six factors, such as $ABCD$, $ABCF$, and so on, for which the design projects to a single replicate of the 2^4. This design also collapses to two replicates of a 2^3 in *any* subset of three of the six factors or four replicates of a 2^2 in any subset of two factors.

In general, any 2^{k-2} fractional factorial design can be collapsed into either a full factorial or a fractional factorial in some subset of $r \leq k - 2$ of the original factors. Those subsets of variables that form full factorials are not words in the complete defining relation.

Example 9–4

Parts manufactured in an injection molding process are showing excessive shrinkage. This is causing problems in assembly operations downstream from the injection molding area. A quality improvement team has decided to use a designed experiment to study the injection molding process so that shrinkage can be reduced. The team decides to investigate six factors—mold temperature (A), screw speed (B), holding time (C), cycle time (D), gate size (E), and holding pressure (F)—each at two levels, with the objective of learning how each factor affects shrinkage and also, something about how the factors interact.

The team decides to use the 16-run two-level fractional factorial design in Table 9–10. The design is shown again in Table 9–11, along with the observed shrinkage $(\times 10)$ for the test part produced at each of the 16 runs in the design. Table 9–12 shows the effect estimates, sums of squares, and the regression coefficients for this experiment.

A normal probability plot of the effect estimates from this experiment is shown in Figure 9–12 (page 393). The only large effects are A (mold temperature), B (screw speed), and the AB interaction. In light of the alias relationships in Table 9–9, it seems reasonable to adopt these conclusions tentatively. The plot of the AB interaction in Figure 9–13 (page 393) shows that the process is very insensitive to temperature if the screw speed is at the low level but very sensitive to temperature if the screw speed is at the high level. With the screw speed at the low level, the process should produce an average shrinkage of around 10 percent regardless of the temperature level chosen.

Table 9–11 A 2_{IV}^{6-2} Design for the Injection Molding Experiment in Example 9–4

| | Basic Design | | | | | | Observed Shrinkage |
Run	A	B	C	D	E = ABC	F = BCD	(× 10)
1	−	−	−	−	−	−	6
2	+	−	−	−	+	−	10
3	−	+	−	−	+	+	32
4	+	+	−	−	−	+	60
5	−	−	+	−	+	+	4
6	+	−	+	−	−	+	15
7	−	+	+	−	−	−	26
8	+	+	+	−	+	−	60
9	−	−	−	+	−	+	8
10	+	−	−	+	+	+	12
11	−	+	−	+	+	−	34
12	+	+	−	+	−	−	60
13	−	−	+	+	+	−	16
14	+	−	+	+	−	−	5
15	−	+	+	+	−	+	37
16	+	+	+	+	+	+	52

Table 9–12 Effects, Sums of Squares, and Regression Coefficients for Example 9–4

Variable	Name	−1 Level	+1 Level
A	mold_temp	−1.000	1.000
B	screw_spd	−1.000	1.000
C	hold_time	−1.000	1.000
D	cycle_time	−1.000	1.000
E	gate_size	−1.000	1.000
F	hold_press	−1.000	1.000

Variable[a]	Regression Coefficient	Estimated Effect	Sum of Squares
Overall Average	27.3125		
A	6.9375	13.8750	770.062
B	17.8125	35.6250	5076.562
C	−0.4375	−0.8750	3.063
D	0.6875	1.3750	7.563
E	0.1875	0.3750	0.563
F	0.1875	0.3750	0.563
AB + CE	5.9375	11.8750	564.063
AC + BE	−0.8125	−1.6250	10.562
AD + EF	−2.6875	−5.3750	115.562
AE + BC + DF	−0.9375	−1.8750	14.063
AF + DE	0.3125	0.6250	1.563
BD + CF	−0.0625	−0.1250	0.063
BF + CD	−0.0625	−0.1250	0.063
ABD	0.0625	0.1250	0.063
ABF	−2.4375	−4.8750	95.063

[a] Only main effects and two-factor interactions.

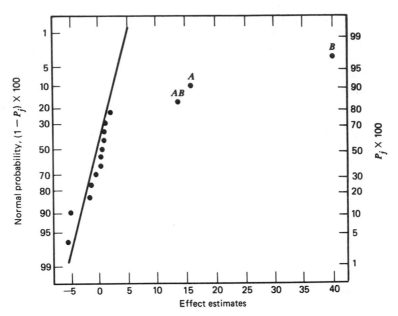

Figure 9–12.   Normal probability plot of effects for Example 9–4.

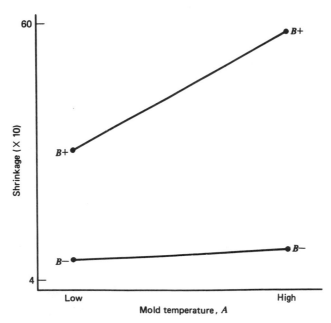

Figure 9–13.   Plot of AB (mold temperature–screw speed) interaction for Example 9–4.

Based on this initial analysis, the team decides to set both the mold temperature and the screw speed at the low level. This set of conditions will reduce the *mean* shrinkage of parts to around 10 percent. However, the variability in shrinkage from part to part is still a potential problem. In effect, the mean shrinkage can be adequately reduced by the above modifications; however, the part-to-part variability in shrinkage over a production run could still cause problems in assembly. One way to address this issue is to see if any of the process factors affect the *variability* in parts shrinkage.

Figure 9–14 presents the normal probability plot of the residuals. This plot appears satisfactory. The plots of residuals versus each factor were then constructed. One of these plots, that for residuals versus factor C (holding time), is shown in Figure 9–15. The plot reveals that there is much less scatter in the residuals at the low holding time than at the high holding time. These residuals were obtained in the usual way from a model for predicted shrinkage

$$\hat{y} = \hat{\beta}_0 + \hat{\beta}_1 x_1 + \hat{\beta}_2 x_2 + \hat{\beta}_{12} x_1 x_2$$
$$= 27.3125 + 6.9375 x_1 + 17.8125 x_2 + 5.9375 x_1 x_2$$

where x_1, x_2, and $x_1 x_2$ are coded variables that correspond to the factors A and B and the AB interaction. The residuals are then

$$e = y - \hat{y}$$

The regression model used to produce the residuals essentially removes the **location effects** of A, B, and AB from the data; the residuals therefore contain information about unexplained variability. Figure 9–15 indicates that there is a *pattern* in the variability and that the variability in the shrinkage of parts may be smaller when the holding time is at the low level.

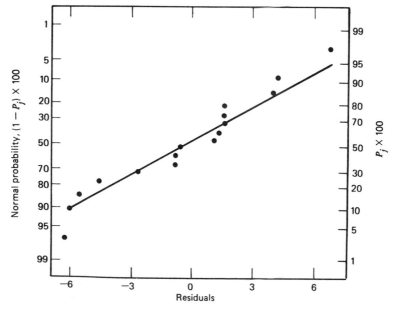

Figure 9–14.   Normal probability plot of residuals for Example 9–4.

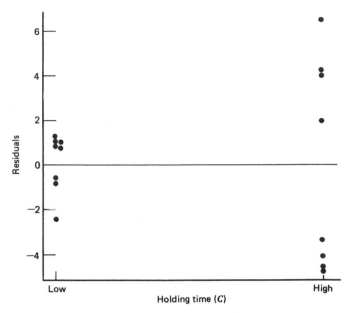

Figure 9–15.   Residuals versus holding time (C) for Example 9–4.

This is further amplified by the analysis of residuals shown in Table 9–13. In this table, the residuals are arranged at the low $(-)$ and high $(+)$ levels of each factor, and the standard deviation of the residuals at the low and high levels of each factor have been calculated. Note that the standard deviation of the residuals with C at the low level $[S(C^-) = 1.63]$ is considerably smaller than the standard deviation of the residuals with C at the high level $[S(C^+) = 5.70]$.

The bottom line of Table 9–13 presents the statistic

$$F_i^* = \ln \frac{S^2(i^+)}{S^2(i^-)}$$

Recall that if the variances of the residuals at the high $(+)$ and low $(-)$ levels of factor i are equal, then this ratio is approximately normally distributed with mean zero and it can be used to judge the difference in the response variability at the two levels of factor i. Since the ratio F_C^* is relatively large, we would conclude that the apparent **dispersion** or **variability effect** observed in Figure 9–15 is real. Thus, setting the holding time at its low level would contribute to reducing the variability in shrinkage from part to part during a production run. Figure 9–16 presents a normal probability plot of the F_i^* values in Table 9–13; this also indicates that factor C has a large dispersion effect.

Figure 9–17 shows the data from this experiment projected onto a cube in the factors A, B, and C. The average observed shrinkage and the range of observed shrinkage are shown at each corner of the cube. From inspection of this figure, we see that running the process with the screw speed (B) at the low level is the key to reducing average parts shrinkage. If B is low, virtually any combination of temperature

Table 9–13 Calculation of Dispersion Effects for Example 9–4

Run	A	B	AB = CE	C	AC = BE	AE = BC	E	D	AD = EF	BD = CE	ABD	BF = CD	ACD	F	AF = DE	Residual
1	−	−	+	−	+	+	−	−	+	+	−	+	−	−	+	−2.50
2	+	−	−	−	−	+	+	−	−	+	+	+	+	−	−	−0.50
3	−	+	−	−	+	−	+	−	+	−	+	+	−	+	−	−0.25
4	+	+	+	−	−	−	−	−	−	−	−	+	+	+	+	2.00
5	−	−	+	+	−	−	+	−	+	+	−	−	+	+	−	−4.50
6	+	−	−	+	+	−	−	−	−	+	+	−	−	+	+	4.50
7	−	+	−	+	−	+	−	−	+	−	+	−	+	−	+	−6.25
8	+	+	+	+	+	+	+	−	−	−	−	−	−	−	−	2.00
9	−	−	+	−	+	+	−	+	−	−	+	−	+	+	−	−0.50
10	+	−	−	−	−	+	+	+	+	−	−	−	−	+	+	1.50
11	−	+	−	−	+	−	+	+	−	+	−	−	+	−	+	1.75
12	+	+	+	−	−	−	−	+	+	+	+	−	−	−	−	2.00
13	−	−	+	+	−	−	+	+	−	−	+	+	−	−	+	7.50
14	+	−	−	+	+	−	−	+	+	−	−	+	+	−	−	−5.50
15	−	+	−	+	−	+	−	+	−	+	−	+	−	+	−	4.75
16	+	+	+	+	+	+	+	+	+	+	+	+	+	+	+	−6.00
$S(i^+)$	3.80	4.01	4.33	5.70	3.68	3.85	4.17	4.64	3.39	4.01	4.72	4.71	3.50	3.88	4.87	
$S(i^-)$	4.60	4.41	4.10	1.63	4.53	4.33	4.25	3.59	2.75	4.41	3.51	3.65	3.12	4.52	3.40	
F_i^*	−0.38	−0.19	0.11	2.50	−0.42	−0.23	−0.04	0.51	0.42	−0.19	0.59	0.51	0.23	−0.31	0.72	

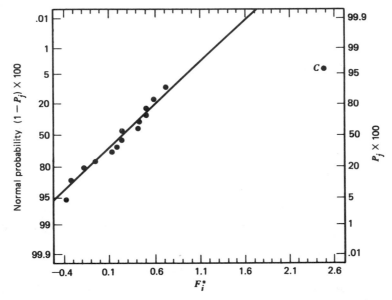

Figure 9–16.   Normal probability plot of the dispersion effects F_i^* for Example 9–4.

(A) and holding time (C) will result in low values of average parts shrinkage. However, from examining the ranges of the shrinkage values at each corner of the cube, it is immediately clear that setting the holding time (C) at the low level is the only reasonable choice if we wish to keep the part-to-part variability in shrinkage low during a production run.

◼

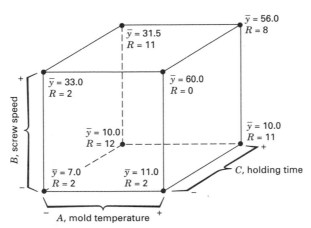

Figure 9–17.   Average shrinkage and range of shrinkage in factors A, B, and C for Example 9–4.

9-4 THE GENERAL 2^{k-p} FRACTIONAL FACTORIAL DESIGN

A 2^k fractional factorial design containing 2^{k-p} runs is called a $1/2^p$ fraction of the 2^k design or, more simply, a **2^{k-p} fractional factorial design.** These designs require the selection of p independent generators. The defining relation for the design consists of the p generators initially chosen and their $2^p - p - 1$ generalized interactions. In this section we discuss the construction and analysis of these designs.

The alias structure may be found by multiplying each effect column by the defining relation. Care should be exercised in choosing the generators so that effects of potential interest are not aliased with each other. Each effect has $2^p - 1$ aliases. For moderately large values of k, we usually assume higher-order interactions (say, third- or fourth-order and higher) to be negligible, and this greatly simplifies the alias structure.

It is important to select the p generators for a 2^{k-p} fractional factorial design in such a way that we obtain the best possible alias relationships. A reasonable criterion is to select the generators such that the resulting 2^{k-p} design has the highest possible design resolution. Table 9–14 presents a selection of 2^{k-p} fractional factorial designs for $k \leq 15$ factors and up to $n \leq 128$ runs. The suggested generators in this table will result in a design of the highest possible resolution.

The alias relationships for all of the designs in Table 9–14 for which $n \leq 64$ are given in Appendix Table XII(a–w). The alias relationships presented in this

Table 9–14   Selected 2^{k-p} Fractional Factorial Designs

Number of Factors, k	Fraction	Number of Runs	Design Generators
3	2_{III}^{3-1}	4	$C = \pm AB$
4	2_{IV}^{4-1}	8	$D = \pm ABC$
5	2_{V}^{5-1}	16	$E = \pm ABCD$
	2_{III}^{5-2}	8	$D = \pm AB$
			$E = \pm AC$
6	2_{VI}^{6-1}	32	$F = \pm ABCDE$
	2_{IV}^{6-2}	16	$E = \pm ABC$
			$F = \pm BCD$
	2_{III}^{6-3}	8	$D = \pm AB$
			$E = \pm AC$
			$F = \pm BC$
7	2_{VII}^{7-1}	64	$G = \pm ABCDEF$
	2_{IV}^{7-2}	32	$F = \pm ABCD$
			$G = \pm ABDE$
	2_{IV}^{7-3}	16	$E = \pm ABC$
			$F = \pm BCD$
			$G = \pm ACD$

Table 9–14   (continued)

Number of Factors, k	Fraction	Number of Runs	Design Generators
	2^{7-4}_{III}	8	$D = \pm AB$
			$E = \pm AC$
			$F = \pm BC$
			$G = \pm ABC$
8	2^{8-2}_{V}	64	$G = \pm ABCD$
			$H = \pm ABEF$
	2^{8-3}_{IV}	32	$F = \pm ABC$
			$G = \pm ABD$
			$H = \pm BCDE$
	2^{8-4}_{IV}	16	$E = \pm BCD$
			$F = \pm ACD$
			$G = \pm ABC$
			$H = \pm ABD$
9	2^{9-2}_{VI}	128	$H = \pm ACDFG$
			$J = \pm BCEFG$
	2^{9-3}_{IV}	64	$G = \pm ABCD$
			$H = \pm ACEF$
			$J = \pm CDEF$
	2^{9-4}_{IV}	32	$F = \pm BCDE$
			$G = \pm ACDE$
			$H = \pm ABDE$
			$J = \pm ABCE$
	2^{9-5}_{III}	16	$E = \pm ABC$
			$F = \pm BCD$
			$G = \pm ACD$
			$H = \pm ABD$
			$J = \pm ABCD$
10	2^{10-3}_{V}	128	$H = \pm ABCG$
			$J = \pm ACDE$
			$K = \pm ACDF$
	2^{10-4}_{IV}	64	$G = \pm BCDF$
			$H = \pm ACDF$
			$J = \pm ABDE$
			$K = \pm ABCE$
	2^{10-5}_{IV}	32	$F = \pm ABCD$
			$G = \pm ABCE$
			$H = \pm ABDE$
			$J = \pm ACDE$
			$K = \pm BCDE$
	2^{10-6}_{III}	16	$E = \pm ABC$
			$F = \pm BCD$
			$G = \pm ACD$
			$H = \pm ABD$
			$J = \pm ABCD$
			$K = \pm AB$

(continued)

Table 9–14 (continued)

Number of Factors, k	Fraction	Number of Runs	Design Generators
11	2_{IV}^{11-5}	64	$G = \pm CDE$
			$H = \pm ABCD$
			$J = \pm ABF$
			$K = \pm BDEF$
			$L = \pm ADEF$
	2_{IV}^{11-6}	32	$F = \pm ABC$
			$G = \pm BCD$
			$H = \pm CDE$
			$J = \pm ACD$
			$K = \pm ADE$
			$L = \pm BDE$
	2_{III}^{11-7}	16	$E = \pm ABC$
			$F = \pm BCD$
			$G = \pm ACD$
			$H = \pm ABD$
			$J = \pm ABCD$
			$K = \pm AB$
			$L = \pm AC$
12	2_{III}^{12-8}	16	$E = \pm ABC$
			$F = \pm ABD$
			$G = \pm ACD$
			$H = \pm BCD$
			$J = \pm ABCD$
			$K = \pm AB$
			$L = \pm AC$
			$M = \pm AD$
13	2_{III}^{13-9}	16	$E = \pm ABC$
			$F = \pm ABD$
			$G = \pm ACD$
			$H = \pm BCD$
			$J = \pm ABCD$
			$K = \pm AB$
			$L = \pm AC$
			$M = \pm AD$
			$N = \pm BC$
14	2_{III}^{14-10}	16	$E = \pm ABC$
			$F = \pm ABD$
			$G = \pm ACD$
			$H = \pm BCD$
			$J = \pm ABCD$
			$K = \pm AB$
			$L = \pm AC$
			$M = \pm AD$
			$N = \pm BC$
			$O = \pm BD$

Table 9–14 (continued)

Number of Factors, k	Fraction	Number of Runs	Design Generators
15	2_{III}^{15-11}	16	$E = \pm ABC$
			$F = \pm ABD$
			$G = \pm ACD$
			$H = \pm BCD$
			$J = \pm ABCD$
			$K = \pm AB$
			$L = \pm AC$
			$M = \pm AD$
			$N = \pm BC$
			$O = \pm BD$
			$P = \pm CD$

table focus on main effects and two-factor and three-factor interactions. The complete defining relation is given for each design. This appendix table makes it very easy to select a design of sufficient resolution to ensure that any interactions of potential interest can be estimated.

Example 9–5

To illustrate the use of Table 9–14, suppose that we have seven factors and that we are interested in estimating the seven main effects and getting some insight regarding the two-factor interactions. We are willing to assume that three-factor and higher interactions are negligible. This information suggests that a resolution IV design would be appropriate.

Table 9–14 shows that there are two resolution IV fractions available: the 2_{IV}^{7-2} with 32 runs and the 2_{IV}^{7-3} with 16 runs. Appendix Table XII contains the complete alias relationships for these two designs. The aliases for the 2_{IV}^{7-3} 16-run design are in Appendix Table XII(i). Notice that all seven main effects are aliased with three-factor interactions. The two-factor interactions are all aliased in groups of three. Therefore, this design will satisfy our objectives; that is, it will allow the estimation of the main effects, and it will give some insight regarding two-factor interactions. It is not necessary to run the 2_{IV}^{7-2} design, which would require 32 runs. Appendix Table XII(j) shows that this design would allow the estimation of all seven main effects and that 15 of the 21 two-factor interactions could also be uniquely estimated. (Recall that three-factor and higher interactions are negligible.) This is more information about interactions than is necessary. The complete design is shown in Table 9–15. Notice that it was constructed by starting with the 16-run 2^4 design in A, B, C, and D as the basic design and then adding the three columns $E = ABC$, $F = BCD$, and $G = ACD$. The generators are $I = ABCE$, $I = BCDF$, and $I = ACDG$ (Table 9–14). The complete defining relation is $I = ABCE = BCDF = ADEF = ACDG = BDEG = CEFG = ABFG$.

■

Table 9–15   A 2_{IV}^{7-3} Fractional Factorial Design

Run	Basic Design				$E = ABC$	$F = BCD$	$G = ACD$
	A	B	C	D			
1	−	−	−	−	−	−	−
2	+	−	−	−	+	−	+
3	−	+	−	−	+	+	−
4	+	+	−	−	−	+	+
5	−	−	+	−	+	+	+
6	+	−	+	−	−	+	−
7	−	+	+	−	−	−	+
8	+	+	+	−	+	−	−
9	−	−	−	+	−	+	+
10	+	−	−	+	+	+	−
11	−	+	−	+	+	−	+
12	+	+	−	+	−	−	−
13	−	−	+	+	+	−	−
14	+	−	+	+	−	−	+
15	−	+	+	+	−	+	−
16	+	+	+	+	+	+	+

Analysis of 2^{k-p} Fractional Factorials ▪ There are many computer programs that can be used to analyze the 2^{k-p} fractional factorial design. For example, the *Design-Ease* program illustrated in Chapter 7 has this capability.

The design may also be analyzed by resorting to first principles; the ith effect is estimated by

$$\ell_i = \frac{2(Contrast_i)}{N} = \frac{Contrast_i}{(N/2)}$$

where the *Contrast$_i$* is found using the plus and minus signs in column i and $N = 2^{k-p}$ is the total number of observations. The 2^{k-p} design allows only $2^{k-p} - 1$ effects (and their aliases) to be estimated. Alternatively, the contrasts may be generated by Yates' algorithm. This is accomplished by considering the data as from a full 2^r factorial, where $r = k - p$ is a subset of the original k factors. The additional factors are incorporated when identifying the actual effects estimated, as discussed in Section 9–2. Usually, one may select the subset of r factors to be those columns in the basic design.

Projection of the 2^{k-p} Fractional Factorial ▪ The 2^{k-p} design collapses into either a full factorial or a fractional factorial in any subset of $r \leq k - p$ of the original factors. Those subsets of factors providing fractional factorials are subsets appearing as words in the complete defining relation. This is particularly useful in screening experiments when we suspect at the outset of the experiment that most of the original factors will have small effects. The original 2^{k-p} fractional factorial can then be projected into a full factorial, say, in the most interesting factors. Conclusions drawn from designs of this type should be considered tentative and subject to further analysis. It is usually possible to find alternative explanations of the data involving higher-order interactions.

As an example, consider the 2_{IV}^{7-3} design from Example 9–5. This is a 16-run design involving seven factors. It will project into a full factorial in any four of the original seven factors that is not a word in the defining relation. There are 35 subsets of four factors, seven of which appear in the complete defining relation (see Table 9–15). Thus, there are 28 subsets of four factors that would form 2^4 designs. One combination that is obvious upon inspecting Table 9–15 is A, B, C, and D.

To illustrate the usefulness of this projection properly, suppose that we are conducting an experiment to improve the efficiency of a ball mill and the seven factors are as follows.

1. Motor speed
2. Gain
3. Feed mode
4. Feed sizing
5. Material type
6. Screen angle
7. Screen vibration level

We are fairly certain that motor speed, feed mode, feed sizing, and material type will affect efficiency and that these factors may interact. The role of the other three factors is less well-known, but it is likely that they are negligible. A reasonable strategy would be to assign motor speed, feed mode, feed sizing, and material type to columns A, B, C, and D, respectively, in Table 9–15. Gain, screen angle, and screen vibration level would be assigned to columns E, F, and G, respectively. If we are correct and the "minor variables" E, F, and G are negligible, we will be left with a full 2^4 design in the key process variables.

Blocking Fractional Factorials ▪ Occasionally, a fractional factorial design requires so many runs that all of them cannot be made under homogeneous conditions. In these situations, fractional factorials may be confounded in blocks. Appendix Table XII contains recommended blocking arrangements for many of the fractional factorial designs in Table 9–14. The minimum block size for these designs is eight runs.

To illustrate the general procedure, consider the 2_{IV}^{6-2} fractional factorial design with the defining relation $I = ABCE = BCDF = ADEF$ shown in Table 9–10. This fractional design contains 16 treatment combinations. Suppose we wish to run the design in two blocks of eight treatment combinations each. In selecting an interaction to confound with blocks, we note from examining the alias structure in Appendix Table XII(f) that there are two alias sets involving

Block 1	Block 2
(1)	ae
abf	acf
cef	bef
abce	bc
aef	df
bde	abd
acd	cde
bcdf	abcdef

Figure 9–18. The 2_{IV}^{6-2} design in two blocks with ABD confounded.

only three-factor interactions. The table suggests selecting ABD (and its aliases) to be confounded with blocks. This would give the two blocks shown in Figure 9–18. Notice that the principal block contains those treatment combinations that have an even number of letters in common with ABD. These are also the treatment combinations for which $L = x_1 + x_2 + x_4 = 0$ (mod 2).

Example 9–6

A five-axis CNC machine is used to machine an impeller used in a jet turbine engine. The blade profiles are an important quality characteristic. Specifically, the deviation of the blade profile from the profile specified on the engineering drawing is of interest. An experiment is run to determine which machine parameters affect profile deviation. The eight factors selected for the design are as follows.

Factor	Low Level (−)	High Level (+)
A = x-Axis shift (0.001 in)	0	15
B = y-Axis shift (0.001 in)	0	15
C = z-Axis shift (0.001 in)	0	15
D = Tool vendor	1	2
E = a-Axis shift (0.001 deg)	0	30
F = Spindle speed (%)	90	110
G = Fixture height (0.001 in)	0	15
H = Feed rate (%)	90	110

One test blade on each part is selected for inspection. The profile deviation is measured using a coordinate measuring machine, and the standard deviation of the difference between the actual profile and the specified profile is used as the response variable.

The machine has four spindles. Because there may be differences in the spindles, the process engineers feel that the spindles should be treated as blocks.

The engineers feel confident that three-factor and higher interactions are not too important, but they are reluctant to ignore the two-factor interactions. From Table 9–14, two designs initially appear appropriate: the 2_{IV}^{8-4} design with 16 runs and the 2_{IV}^{8-3} design with 32 runs. Appendix Table XII(l) indicates that if the 16-run design is used there will be fairly extensive aliasing of two-factor interactions. Furthermore, this design cannot be run in four blocks without confounding four two-factor interactions with blocks. Therefore, the experimenters decide to use the 2_{IV}^{8-3} design in four blocks. This confounds one three-factor interaction alias chain and one two-factor interaction (EH) and its three-factor interaction aliases with blocks. The EH interaction is the interaction between the a-axis shift and the feed rate, and the

Table 9–16   The 2^{8-3} Design in Four Blocks for Example 9–6

	Basic Design					$F =$	$G =$	$H =$		Actual Run	Standard Deviation
Run	A	B	C	D	E	ABC	ABD	$BCDE$	Block	Order	$(\times 10^3 \text{ in})$
1	$-$	$-$	$-$	$-$	$-$	$-$	$-$	$+$	3	18	2.76
2	$+$	$-$	$-$	$-$	$-$	$+$	$+$	$+$	2	16	6.18
3	$-$	$+$	$-$	$-$	$-$	$+$	$+$	$-$	4	29	2.43
4	$+$	$+$	$-$	$-$	$-$	$-$	$-$	$-$	1	4	4.01
5	$-$	$-$	$+$	$-$	$-$	$+$	$-$	$-$	1	6	2.48
6	$+$	$-$	$+$	$-$	$-$	$-$	$+$	$-$	4	26	5.91
7	$-$	$+$	$+$	$-$	$-$	$-$	$+$	$+$	2	14	2.39
8	$+$	$+$	$+$	$-$	$-$	$+$	$-$	$+$	3	22	3.35
9	$-$	$-$	$-$	$+$	$-$	$-$	$+$	$-$	1	8	4.40
10	$+$	$-$	$-$	$+$	$-$	$+$	$-$	$-$	4	32	4.10
11	$-$	$+$	$-$	$+$	$-$	$+$	$-$	$+$	2	15	3.22
12	$+$	$+$	$-$	$+$	$-$	$-$	$+$	$+$	3	19	3.78
13	$-$	$-$	$+$	$+$	$-$	$+$	$+$	$+$	3	24	5.32
14	$+$	$-$	$+$	$+$	$-$	$-$	$-$	$+$	2	11	3.87
15	$-$	$+$	$+$	$+$	$-$	$-$	$-$	$-$	4	27	3.03
16	$+$	$+$	$+$	$+$	$-$	$+$	$+$	$-$	1	3	2.95
17	$-$	$-$	$-$	$-$	$+$	$-$	$-$	$-$	2	10	2.64
18	$+$	$-$	$-$	$-$	$+$	$+$	$+$	$-$	3	21	5.50
19	$-$	$+$	$-$	$-$	$+$	$+$	$+$	$+$	1	7	2.24
20	$+$	$+$	$-$	$-$	$+$	$-$	$-$	$+$	4	28	4.28
21	$-$	$-$	$+$	$-$	$+$	$+$	$-$	$+$	4	30	2.57
22	$+$	$-$	$+$	$-$	$+$	$-$	$+$	$+$	1	2	5.37
23	$-$	$+$	$+$	$-$	$+$	$-$	$+$	$-$	3	17	2.11
24	$+$	$+$	$+$	$-$	$+$	$+$	$-$	$-$	2	13	4.18
25	$-$	$-$	$-$	$+$	$+$	$-$	$+$	$+$	4	25	3.96
26	$+$	$-$	$-$	$+$	$+$	$+$	$-$	$+$	1	1	3.27
27	$-$	$+$	$-$	$+$	$+$	$+$	$-$	$-$	3	23	3.41
28	$+$	$+$	$-$	$+$	$+$	$-$	$+$	$-$	2	12	4.30
29	$-$	$-$	$+$	$+$	$+$	$+$	$+$	$-$	2	9	4.44
30	$+$	$-$	$+$	$+$	$+$	$-$	$-$	$-$	3	20	3.65
31	$-$	$+$	$+$	$+$	$+$	$-$	$-$	$+$	1	5	4.41
32	$+$	$+$	$+$	$+$	$+$	$+$	$+$	$+$	4	31	3.40

Table 9–17   Effect Estimates, Regression Coefficients, and Sums of Squares for Example 9–6

Variable	Name	−1 Level	+1 Level
A	x-Axis shift	0	15
B	y-Axis shift	0	15
C	z-Axis shift	0	15
D	Tool vendor	1	2
E	a-Axis shift	0	30
F	Spindle speed	90	110
G	Fixture height	0	15
H	Feed rate	90	110

Variable[a]	Regression Coefficient	Estimated Effect	Sum of Squares
Overall Average	1.28007		
A	0.14513	0.29026	0.674020
B	−0.10027	−0.20054	0.321729
C	−0.01288	−0.02576	0.005310
D	0.05407	0.10813	0.093540
E	−2.531E-04	−5.063E-04	2.050E-06
F	−0.01936	−0.03871	0.011988
G	0.05804	0.11608	0.107799
H	0.00708	0.01417	0.001606
AB + CF + DG	−0.00294	−0.00588	2.767E-04
AC + BF	−0.03103	−0.06206	0.030815
AD + BG	−0.18706	−0.37412	1.119705
AE	0.00402	0.00804	5.170E-04
AF + BC	−0.02251	−0.04502	0.016214
AG + BD	0.02644	0.05288	0.022370
AH	−0.02521	−0.05042	0.020339
BE	0.04925	0.09851	0.077627
BH	0.00654	0.01309	0.001371
CD + FG	0.01726	0.03452	0.009535
CE	0.01991	0.03982	0.012685
CG + DF	−0.00733	−0.01467	0.001721
CH	0.03040	0.06080	0.029568
DE	0.00854	0.01708	0.002334
DH	0.00784	0.01569	0.001969
EF	−0.00904	−0.01808	0.002616
EG	−0.02685	−0.05371	0.023078
EH	−0.01767	−0.03534	0.009993
FH	−0.01404	−0.02808	0.006308
GH	0.00245	0.00489	1.914E-04
ABE	0.01665	0.03331	0.008874
ABH	−0.00631	−0.01261	0.001273
ACD	−0.02717	−0.05433	0.023617

[a] Only main effects and two-factor interactions.

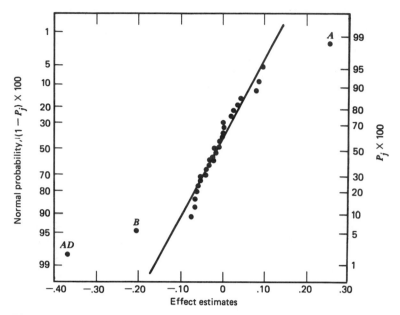

Figure 9–19.   Normal probability plot of the effect estimates for Example 9–6.

engineers consider an interaction between these two variables to be fairly unlikely. (One could always assign other factors that are unlikely to interact to E and H in the event that the original assignment is unsatisfactory.)

Table 9–16 (page 405) contains the design and the resulting responses in standard deviation $\times 10^3$ in. Since the response variable is a standard deviation, it is often best to perform the analysis following a log transformation. The effect estimates are shown in Table 9–17. Figure 9–19 is a normal probability plot of the effect estimates, using ln (standard deviation $\times 10^3$) as the response variable. The only large effects are $A = x$-axis shift, $B = y$-axis shift, and the AD (x-axis shift–tool vendor) interaction. Since the AD interaction is large, we will include the main effect of D in the model to preserve the principal of hierarchy. Table 9–18 is the analysis of variance for this

Table 9–18   Analysis of Variance for Example 9–6

Source of Variation	Sum of Squares	Degrees of Freedom	Mean Square	F_0	P-Value
A	0.6740	1	0.6740	39.42	<0.0001
B	0.3217	1	0.3217	18.81	0.0002
D	0.0935	1	0.0935	5.47	0.0280
AD	1.1197	1	1.1197	65.48	<0.0001
Blocks	0.0201	3	0.0067		
Error	0.4099	24	0.0171		
Total	2.6389	31			

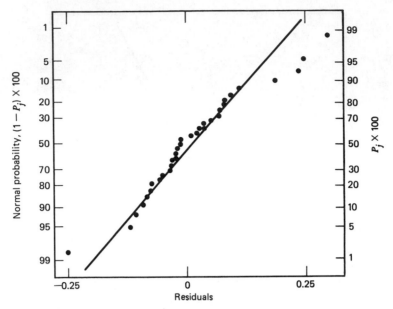

Figure 9–20.   Normal probability plot of the residuals for Example 9–6.

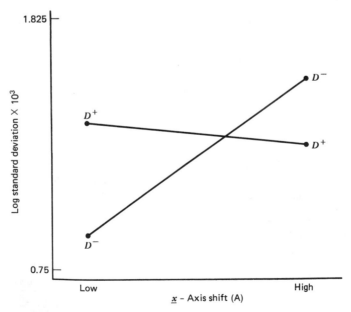

Figure 9–21.   Plot of AD interaction for Example 9–6.

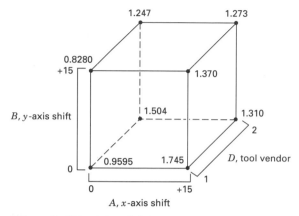

Figure 9–22. The 2_{IV}^{8-3} design in Example 9–6 projected into four replicates of a 2^3 design in factors A, B, and D.

model. Notice that the block effect appears small, suggesting that the machine spindles are not different.

Figure 9–20 is a normal probability plot of the residuals from this experiment. This plot is suggestive of slightly heavier than normal tails, so possibly other transformations should be considered. The AD interaction plot is in Figure 9–21. Notice that tool vendor (D) and the magnitude of the x-axis shift (A) have a profound impact on the variability of the blade profile from design specifications. Running A at the low level (0 offset) and buying tools from vendor 1 gives the best results. Figure 9–22 shows the projection of this 2_{IV}^{8-3} design into four replicates of a 2^3 design in factors A, B, and D. The best combination of operating conditions is A at the low level (0 offset), B at the high level (0.015 in offset), and D at the low level (tool vendor 1).

■

9–5 RESOLUTION III DESIGNS

As indicated earlier, the sequential use of fractional factorial designs is very useful, often leading to great economy and efficiency in experimentation. We now illustrate these ideas using the class of resolution III designs.

It is possible to construct resolution III designs for investigating up to $k = N - 1$ factors in only N runs, where N is a multiple of 4. These designs are frequently useful in industrial experimentation. Designs in which N is a power of 2 can be constructed by the methods presented earlier in this chapter, and these are presented first. Of particular importance are designs requiring 4 runs for up to 3 factors, 8 runs for up to 7 factors and 16 runs for up to 15 factors. If $k = N - 1$, the fractional factorial design is said to be **saturated.**

A design for analyzing up to three factors in four runs is the 2_{III}^{3-1} design, presented in Section 9–2. Another very useful saturated fractional factorial is a design for studying seven factors in eight runs; that is, the 2_{III}^{7-4} design. This design is a one-sixteenth fraction of the 2^7. It may be constructed by first writing down as the basic design the plus and minus levels for a full 2^3 design in A, B, and C and then associating the levels of four additional factors with the interactions of the original three as follows: $D = AB$, $E = AC$, $F = BC$, and $G = ABC$. Thus, the generators for this design are $I = ABD$, $I = ACE$, $I = BCF$, and $I = ABCG$. The design is shown in Table 9–19.

The complete defining relation for this design is obtained by multiplying the four generators ABD, ACE, BCF, and $ABCG$ together two at a time, three at a time, and four at a time, yielding

$$I = ABD = ACE = BCF = ABCG = BCDE = ACDF = CDG$$
$$= ABEF = BEG = AFG = DEF = ADEG = CEFG = BDFG = ABCDEFG$$

To find the aliases of any effect, simply multiply the effect by each word in the defining relation. For example, the aliases of B are

$$B = AD = ABCE = CF = ACG = CDE = ABCDF = BCDG = AEF = EG$$
$$= ABFG = BDEF = ABDEG = BCEFG = DFG = ACDEFG$$

This design is a one-sixteenth fraction, and since the signs chosen for the generators are positive, this is the principal fraction. It is also of resolution III since the smallest number of letters in any word of the defining contrast is three. Any one of the 16 different 2_{III}^{7-4} designs in this family could be constructed by using the generators with one of the 16 possible arrangements of signs in $I = \pm ABD$, $I = \pm ACE$, $I = \pm BCF$, $I = \pm ABCG$.

The seven degrees of freedom in this design may be used to estimate the seven main effects. Each of these effects has 15 aliases; however, if we assume that three-factor and higher interactions are negligible, then considerable simplification in the alias structure results. Making this assumption, each of the linear

Table 9–19   The 2_{III}^{7-4} Design with the Generators $I = ABD$, $I = ACE$, $I = BCF$, and $I = ABCG$

Run	Basic Design							
	A	B	C	$D = AB$	$E = AC$	$F = BC$	$G = ABC$	
1	−	−	−	+	+	+	−	def
2	+	−	−	−	−	+	+	afg
3	−	+	−	−	+	−	+	beg
4	+	+	−	+	−	−	−	abd
5	−	−	+	+	−	−	+	cdg
6	+	−	+	−	+	−	−	ace
7	−	+	+	−	−	+	−	bcf
8	+	+	+	+	+	+	+	abcdefg

combinations associated with the seven main effects in this design actually estimates the main effect and three two-factor interactions:

$$
\begin{aligned}
\ell_A &\rightarrow A + BD + CE + FG \\
\ell_B &\rightarrow B + AD + CF + EG \\
\ell_C &\rightarrow C + AE + BF + DG \\
\ell_D &\rightarrow D + AB + CG + EF \\
\ell_E &\rightarrow E + AC + BG + DF \\
\ell_F &\rightarrow F + BC + AG + DE \\
\ell_G &\rightarrow G + CD + BE + AF
\end{aligned}
\tag{9-1}
$$

These aliases are found in Appendix Table XII(h), ignoring three-factor and higher interactions.

The saturated 2_{III}^{7-4} design in Table 9-19 can be used to obtain resolution III designs for studying fewer than seven factors in eight runs. For example, to generate a design for six factors in eight runs, simply drop any one column in Table 9-19, for example, column G. This produces the design shown in Table 9-20.

It is easy to verify that this design is also of resolution III; in fact, it is a 2_{III}^{6-3} or a one-eighth fraction of the 2^6 design. The defining relation for the 2_{III}^{6-3} design is equal to the defining relation for the original 2_{III}^{7-4} design with any words containing the letter G deleted. Thus, the defining relation for our new design is

$$
I = ABD = ACE = BCF = BCDE = ACDF = ABEF = DEF
$$

In general, when d factors are dropped to produce a new design, the new defining relation is obtained as those words in the original defining relation that do not contain any dropped letters. When constructing designs by this method, care should be exercised to obtain the best arrangement possible. If we drop columns B, D, F, and G from Table 9-19, we obtain a design for three factors in eight runs, yet the treatment combinations correspond to two replicates of a 2^3 design. The experimenter would probably prefer to run a full 2^3 design in A, C, and E.

Table 9-20   A 2_{III}^{6-3} Design with the Generators $I = ABD$, $I = ACE$, and $I = BCF$

	Basic Design						
Run	A	B	C	$D = AB$	$E = AC$	$F = BC$	
1	−	−	−	+	+	+	*def*
2	+	−	−	−	−	+	*af*
3	−	+	−	−	+	−	*be*
4	+	+	−	+	−	−	*abd*
5	−	−	+	+	−	−	*cd*
6	+	−	+	−	+	−	*ace*
7	−	+	+	−	−	+	*bcf*
8	+	+	+	+	+	+	*abcdef*

It is also possible to obtain a resolution III design for studying up to 15 factors in 16 runs. This saturated 2_{III}^{15-11} design can be generated by first writing down the 16 treatment combinations associated with a 2^4 design in A, B, C, and D and then equating 11 new factors with the two-, three-, and four-factor interactions of the original four. In this design, each of the 15 main effects is aliased with seven two-factor interactions. A similar procedure can be used for the 2_{III}^{31-26} design, which allows up to 31 factors to be studied in 32 runs.

Sequential Assembly of Fractions to Separate Effects ▪ By combining fractional factorial designs in which certain signs are switched, we can systematically isolate effects of potential interest. The alias structure for any fraction with the signs for one or more factors reversed is obtained by making changes of sign on the appropriate factors in the alias structure of the original fraction.

Consider the 2_{III}^{7-4} design in Table 9–19. Suppose that along with this principal fraction a second fractional design with the signs reversed in the column for factor D is also run. That is, the column for D in the second fraction is

$$-++--++-$$

The effects that may be estimated from the first fraction are shown in Equation 9–1, and from the second fraction we obtain

$$\ell'_A \rightarrow A - BD + CE + FG$$
$$\ell'_B \rightarrow B - AD + CF + EG$$
$$\ell'_C \rightarrow C + AE + BF + DG$$
$$\ell'_D \rightarrow D - AB - CG - EF$$
$$\text{i.e., } \ell'_{-D} \rightarrow -D + AB + CG + EF \tag{9–2}$$
$$\ell'_E \rightarrow E + AC + BG - DF$$
$$\ell'_F \rightarrow F + BC + AG - DE$$
$$\ell'_G \rightarrow G - CD + BE + AF$$

assuming that three-factor and higher interactions are insignificant. Now from the two linear combinations of effects $\frac{1}{2}(\ell_i + \ell'_i)$ and $\frac{1}{2}(\ell_i - \ell'_i)$ we obtain

i	From $\frac{1}{2}(\ell_i + \ell'_i)$	From $\frac{1}{2}(\ell_i - \ell'_i)$
A	$A + CE + FG$	BD
B	$B + CF + EG$	AD
C	$C + AE + BF$	DG
D	D	$AB + CG + EF$
E	$E + AC + BG$	DF
F	$F + BC + AG$	DE
G	$G + BE + AF$	CD

Thus, we have isolated the main effect of D and all of its two-factor interactions. In general, if we add to a fractional factorial design of resolution III or higher a further fraction with the signs of a *single factor* reversed, then the combined design will provide estimates of the main effect of that factor and its two-factor interactions.

Now suppose we add to a resolution III fractional a second fraction in which the *signs for all the factors are reversed*. This procedure, which is called **folding over** the design, breaks the alias links between main effects and two-factor interactions. That is, we may use the **combined design** to estimate all the main effects clear of any two-factor interactions. The following example illustrates the technique.

Example 9-7

A human performance analyst is conducting an experiment to study eye focus time and has built an apparatus in which several factors can be controlled during the test. The factors he initially regards as important are acuity or sharpness of vision (A), distance from target to eye (B), target shape (C), illumination level (D), target size (E), target density (F), and subject (G). Two levels of each factor are considered. He suspects that only a few of these seven factors are of major importance, and that high-order interactions between the factors can be neglected. On the basis of this assumption, the analyst decides to run a screening experiment to identify the most important factors and then to concentrate further study on those. To screen these seven factors, he runs the treatment combinations from the 2_{III}^{7-4} design in Table 9-19 in random order, obtaining the focus times in milliseconds, as shown in Table 9-21.

Seven main effects and their aliases may be estimated from these data. From Equation 9-1, we see that the effects and their aliases are

$$\ell_A = 20.63 \to A + BD + CE + FG$$
$$\ell_B = 38.38 \to B + AD + CF + EG$$
$$\ell_C = -0.28 \to C + AE + BF + DG$$

Table 9-21   A 2_{III}^{7-4} Design for the Eye Focus Time Experiment

Run	Basic Design A	B	C	$D = AB$	$E = AC$	$F = BC$	$G = ABC$		Time
1	−	−	−	+	+	+	−	def	85.5
2	+	−	−	−	−	+	+	afg	75.1
3	−	+	−	−	+	−	+	beg	93.2
4	+	+	−	+	−	−	−	abd	145.4
5	−	−	+	+	−	−	+	cdg	83.7
6	+	−	+	−	+	−	−	ace	77.6
7	−	+	+	−	−	+	−	bcf	95.0
8	+	+	+	+	+	+	+	abcdefg	141.8

$$\ell_D = 28.88 \rightarrow D + AB + CG + EF$$
$$\ell_E = -0.28 \rightarrow E + AC + BG + DF$$
$$\ell_F = -0.63 \rightarrow F + BC + AG + DE$$
$$\ell_G = -2.43 \rightarrow G + CD + BE + AF$$

For example,

$$\ell_A = \tfrac{1}{4}(-85.5 + 75.1 - 93.2 + 145.4 - 83.7 + 77.6 - 95.0 + 141.8) = 20.63$$

The three largest effects are ℓ_A, ℓ_B, and ℓ_D. The simplest interpretation of the data is that the main effects of A, B, and D are all significant. However, this interpretation is not unique, since one could also logically conclude that A, B, and the AB interaction, or perhaps B, D, and the BD interaction, or perhaps A, D, and the AD interaction are the true effects.

Notice that ABD is a word in the defining relation for this design. Therefore, this 2_{III}^{7-4} design does not project into a full 2^3 factorial in ABD; instead, it projects into two replicates of a 2^{3-1} design, as shown in Figure 9–23. Since the 2^{3-1} design is a resolution III design, A will be aliased with BD, B will be aliased with AD, and D will be aliased with AB, so the interactions cannot be separated from the main effects. The analyst here may have been unlucky. If he had assigned illumination level to C instead of D, the design would have projected into a full 2^3 design.

To separate the main effects and the two-factor interactions, a second fraction is run with all the signs reversed. This fold-over design is shown in Table 9–22 along with the observed responses. Notice that when we fold over a resolution III design in this manner, we (in effect) change the signs on the generators that have an odd number of letters. The effects estimated by this fraction are

$$\ell_A' = -17.68 \rightarrow A - BD - CE - FG$$
$$\ell_B' = 37.73 \rightarrow B - AD - CF - EG$$
$$\ell_C' = -3.33 \rightarrow C - AE - BF - DG$$
$$\ell_D' = 29.88 \rightarrow D - AB - CG - EF$$
$$\ell_E' = 0.53 \rightarrow E - AC - BG - DF$$
$$\ell_F' = 1.63 \rightarrow F - BC - AG - DE$$
$$\ell_G' = 2.68 \rightarrow G - CD - BE - AF$$

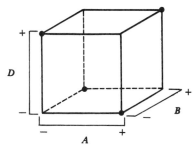

Figure 9–23. The 2_{III}^{7-4} design projected into two replicates of a 2_{III}^{3-1} design in A, B, and D.

Table 9–22   A Second 2_{III}^{7-4} Design for the Eye Focus Time Experiment

Run	Basic Design A	B	C	$D = -AB$	$E = -AC$	$F = -BC$	$G = ABC$		Time
1	+	+	+	−	−	−	+	abcg	91.3
2	−	+	+	+	+	−	−	bcde	136.7
3	+	−	+	+	−	+	−	acdf	82.4
4	−	−	+	−	+	+	+	cefg	73.4
5	+	+	−	−	+	+	−	abef	94.1
6	−	+	−	+	−	+	+	bdfg	143.8
7	+	−	−	+	+	−	+	adeg	87.3
8	−	−	−	−	−	−	−	(1)	71.9

By combining this second fraction with the original one, we obtain the following estimates of the effects:

i	From $\frac{1}{2}(\ell_i + \ell_i')$	From $\frac{1}{2}(\ell_i - \ell_i')$
A	$A =$ 1.48	$BD + CE + FG =$ 19.15
B	$B =$ 38.05	$AD + CF + EG =$ 0.33
C	$C =$ −1.80	$AE + BF + DG =$ 1.53
D	$D =$ 29.38	$AB + CG + EF =$ −0.50
E	$E =$ 0.13	$AC + BG + DF =$ −0.40
F	$F =$ 0.50	$BC + AG + DE =$ −1.53
G	$G =$ 0.13	$CD + BE + AF =$ −2.55

The two largest effects are B and D. Furthermore, the third largest effect is $BD + CE + FG$, so it seems reasonable to attribute this to the BD interaction. The analyst used the two factors distance (B) and illumination level (D) in subsequent experiments with the other factors A, C, E, and F at standard settings and verified the results obtained here. He decided to use subjects as blocks in these new experiments rather than ignore a potential subject effect, since several different subjects had to be used to complete the experiment. ■

The Defining Relation for a Fold-Over Design ▪ Combining fractional factorial designs via fold over as demonstrated in Example 9–7 is a very useful technique. It is often of interest to know the defining relation for the combined design. It can be easily determined. Each separate fraction will have $L + U$ words used as generators: L words of like sign and U words of unlike sign. The combined design will have $L + U - 1$ words used as generators. These will be the L words of like sign and the $U - 1$ words consisting of independent even products of the words of unlike sign. (**Even products** are words taken two at a time, four at a time, and so forth.)

To illustrate this procedure, consider the design in Example 9–7. For the first fraction, the generators are

$$I = ABD, \quad I = ACE, \quad I = BCF, \quad \text{and} \quad I = ABCG$$

and for the second fraction, they are

$$I = -ABD, \quad I = -ACE, \quad I = -BCF, \quad \text{and} \quad I = ABCG$$

Notice that in the second fraction we have switched the signs on the generators with an odd number of letters. Also, notice that $L + U = 1 + 3 = 4$. The combined design will have $I = ABCG$ (the like sign word) as a generator, and two words that are independent even products of the words of unlike sign. For example, take $I = ABD$ and $I = ACE$; then $I = (ABD)(ACE) = BCDE$ is a generator of the combined design. Also, take $I = ABD$ and $I = BCF$; then $I = (ABD)(BCF) = ACDF$ is a generator of the combined design. The complete defining relation for the combined design is

$$I = ABCG = BCDE = ACDF = ADEG = BDFG = ABEF = CEFG$$

Plackett–Burman Designs ▪ These designs, attributed to Plackett and Burman (1946), are two-level fractional factorial designs for studying $k = N - 1$ variables in N runs, where N is a multiple of 4. If N is a power of 2, these designs are identical to those presented earlier in this section. However, for $N = 12, 20, 24, 28,$ and 36, the Plackett–Burman designs are sometimes of interest. Since these designs cannot be represented as cubes, they are sometimes called **nongeometric designs.**

The upper half of Table 9–23 presents rows of plus and minus signs that are used to construct the Plackett–Burman designs for $N = 12, 20, 24,$ and 36, whereas the lower half of the table presents blocks of plus and minus signs for constructing the design for $N = 28$. The designs for $N = 12, 20, 24,$ and 36 are obtained by writing the appropriate row in Table 9–23 as a column (or row). A second column (or row) is then generated from this first one by moving the elements of the column (or row) down (or to the right) one position and placing the last element in the first position. A third column (or row) is produced from the second similarly, and the process is continued until column (or row) k is generated. A row of minus signs is then added, completing the design. For $N = 28$, the three blocks X, Y, and Z are written down in the order

$$X \quad Y \quad Z$$
$$Z \quad X \quad Y$$
$$Y \quad Z \quad X$$

and a row of minus signs is added to these 27 rows. The design for $N = 12$ runs and $k = 11$ factors is shown in Table 9–24.

The nongeometric Plackett–Burman designs for $N = 12, 20, 24, 28,$ and 36

Table 9–23 Plus and Minus Signs for the Plackett–Burman Designs

$k = 11, N = 12$ ++−+++−−−+−
$k = 19, N = 20$ ++−−++++−+−+−−−−++−
$k = 23, N = 24$ +++++−+−++−−++−−+−+−−−−
$k = 35, N = 36$ −+−++++−−−+++++−+++−−+−−−−+−+−++−−+−

$k = 27, N = 28$		
+−++++−−−	−+−−−+−−+	++−+−++−+
++−+++−−−	−−++−−+−−	−++++−++−
−+++++−−−	+−−−+−−+−	+−+−++−++
−−−+−++++	−−+−+−−−+	+−+++−+−+
−−−++−+++	+−−−−++−−	++−−++++−
−−−−+++++	−+−+−−−+−	−+++−+−++
+++−−−+−+	−−+−−+−+−	+−++−+++−
+++−−−++−	+−−+−−−−+	++−++−−++
+++−−−−++	−+−−+−+−−	−++−+++−+

have very messy alias structures. For example, in the 12-run design every main effect is **partially aliased** with every two-factor interaction not involving itself. For example, the AB interaction is aliased with the nine main effects C, D, . . . , K. Furthermore, each main effect is *partially aliased* with 45 two-factor interactions. In the larger designs, the situation is even more complex. We advise the experimenter to use these designs *very carefully*.

The projective properties of the nongeometric Plackett–Burman designs are not terribly attractive. For example, consider the 12-run design in Table 9–24. This design will project into three replicates of a full 2^2 design in any two of the original 11 factors. However, in three factors, the projected design is a full 2^3 factorial plus a 2^{3-1}_{III} fractional factorial (see Figure 9–24a). The four-dimensional

Table 9–24 Plackett–Burman Design for $N = 12, k = 11$

Run	A	B	C	D	E	F	G	H	I	J	K
1	+	−	+	−	−	−	+	+	+	−	+
2	+	+	−	+	−	−	−	+	+	+	−
3	−	+	+	−	+	−	−	−	+	+	+
4	+	−	+	+	−	+	−	−	−	+	+
5	+	+	−	+	+	−	+	−	−	−	+
6	+	+	+	−	+	+	−	+	−	−	−
7	−	+	+	+	−	+	+	−	+	−	−
8	−	−	+	+	+	−	+	+	−	+	−
9	−	−	−	+	+	+	−	+	+	−	+
10	+	−	−	−	+	+	+	−	+	+	−
11	−	+	−	−	−	+	+	+	−	+	+
12	−	−	−	−	−	−	−	−	−	−	−

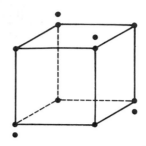

(*a*) Projection into three factors

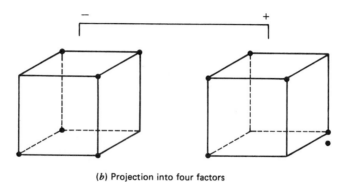

(*b*) Projection into four factors

Figure 9–24.   Projection of the 12-run Plackett–Burman design into three- and four-factor designs.

projections are shown in Figure 9–24*b*. Notice that these three- and four-factor projections are not balanced designs. This would complicate their interpretation.

Example 9–8

We will illustrate some of the potential difficulties associated with the Plackett–Burman designs by using the 11-variable, 12-run design and a set of simulated data. We will assume that the process has three significant main effects (A, B, D) and two significant two-factor interactions (AB and AD). The model is

$$y = 200 + 8x_1 + 10x_2 + 12x_4 - 12x_1x_2 + 9x_1x_4 + \epsilon$$

where each x_i is a coded variable defined on the -1, $+1$ interval and ϵ is an NID(0, 9) random error term. Therefore, three of the $k = 11$ factors are large, and there are two large interactions; not an unreasonable situation.

Table 9–25 presents the 12-run Plackett–Burman design and the simulated responses. This design looks different from the 12-run design in Table 9–24 because it was constructed using the row of signs for $k = 11$, $N = 12$ in Table 9–23 as a row. The effect estimates are shown in Table 9–26. Notice that there are seven large effects; A, B, C, D, E, J, and K (and, of course, their aliases). It is not immediately

Table 9–25 Plackett–Burman Design for Example 9–8

Run	A	B	C	D	E	F	G	H	J	K	L	Response
1	+	+	−	+	+	+	−	−	−	+	−	231
2	−	+	+	−	+	+	+	−	−	−	+	207
3	+	−	+	+	−	+	+	+	−	−	−	230
4	−	+	−	+	+	−	+	+	+	−	−	217
5	−	−	+	−	+	+	−	+	+	+	−	175
6	−	−	−	+	−	+	+	−	+	+	+	176
7	+	−	−	−	+	−	+	+	−	+	+	183
8	+	+	−	−	−	+	−	+	+	−	+	185
9	+	+	+	−	−	−	+	−	+	+	−	181
10	−	+	+	+	−	−	−	+	−	+	+	220
11	+	−	+	+	+	−	−	−	+	−	+	229
12	−	−	−	−	−	−	−	−	−	−	−	168

obvious that some of these effects could be interactions. Some of this ambiguity could be resolved by folding over the design. This will generally resolve the main effects, but it often still leaves the experimenter uncertain about interaction effects. Refer to Problem 9–30.

∎

The difficulty in interpretation in a Plackett–Burman design, illustrated in the above example, occurs fairly often in practice. If your choice is between a geometric 2_{III}^{11-7} design with 16 runs or a 12-run Plackett–Burman design that may have to be folded over (thereby requiring 24 runs) the geometric design may turn out to be a better choice. Under some conditions, we can untangle the

Table 9–26 Effect Estimates, Regression Coefficients, and Sums of Squares for Example 9–8

Variable[a]	Regression Coefficient	Estimated Effect	Sum of Squares
Overall Average	200.167		
A	6.333	12.667	481.333
B	6.667	13.333	533.333
C	6.833	13.667	560.333
D	17.000	34.000	3468.000
E	6.833	13.667	560.333
F	0.500	1.000	3.000
G	−1.167	−2.333	16.333
H	1.500	3.000	27.000
J	−6.333	−12.667	481.333
K	−5.833	−11.667	408.333
L	−0.167	−0.333	0.333

[a] Every main effect is partially aliased with 45 two-factor interactions.

aliases in a nongeometric Plackett–Burman design by using regression model-building techniques. This is discussed by Hamada and Wu (1992).

9–6   RESOLUTION IV AND
V DESIGNS

A 2^{k-p} fractional factorial design is of resolution IV if the main effects are clear of two-factor interactions and some two-factor interactions are aliased with each other. Thus, if three-factor and higher interactions are suppressed, the main effects may be estimated directly in a 2_{IV}^{k-p} design. An example is the 2_{IV}^{6-2} design in Table 9–10. Furthermore, the two combined fractions of the 2_{III}^{7-4} design in Example 9–7 yields a 2_{IV}^{7-3} design.

Any 2_{IV}^{k-p} design must contain at least $2k$ runs. Resolution IV designs that contain exactly $2k$ runs are called **minimal designs.** Resolution IV designs may be obtained from resolution III designs by the process of **fold over.** Recall that to fold over a 2_{III}^{k-p} design, simply add to the original fraction a second fraction with all the signs reversed. Then the plus signs in the identity column I in the first fraction could be switched in the second fraction, and a $(k + 1)$st factor could be associated with this column. The result is a 2_{IV}^{k+1-p} fractional factorial design. The process is demonstrated in Table 9–27 for the 2_{III}^{3-1} design. It is easy to verify that the resulting design is a 2_{IV}^{4-1} design with defining relation $I = ABCD$.

It is also possible to fold over resolution IV designs to separate two-factor interactions that are aliased with each other. To fold over a resolution IV design, run a second fraction in which the sign is reversed on every design generator that has an even number of letters. To illustrate, consider the 2_{IV}^{6-2} design used for the injection molding experiment in Example 9–4. The generators for the

Table 9–27   A 2_{IV}^{4-1} Design
Obtained by Fold Over

$\begin{array}{c}D\\I\end{array}$	A	B	C
\multicolumn{4}{c}{Original 2_{III}^{3-1} $I = ABC$}
+	−	−	+
+	+	−	−
+	−	+	−
+	+	+	+
\multicolumn{4}{c}{Second 2_{III}^{3-1} with Signs Switched}
−	+	+	−
−	−	+	+
−	+	−	+
−	−	−	−

design in Table 9–11 are $I = ABCE$ and $I = BCDF$. The second fraction would use the generators $I = -ABCE$ and $I = -BCDF$, and the single generator for the combined design would be $I = ADEF$. Thus, the combined design is still a resolution IV fractional factorial design. However, the alias relationships will be much simpler than in the original 2_{IV}^{6-2} design. In fact, the only two-factor interactions that will be aliased are $AD = EF$, $AE = DF$, and $AF = DE$. All the other two-factor interactions can be estimated from the combined design.

As another example, consider the 2_{IV}^{8-3} design with 32 runs. Table 9–14 indicates that the best set of generators for this design is $I = ABCF$, $I = ABDG$, and $I = BCDEH$. Appendix Table XII(m) shows the aliases for this design. Notice that there are six pairs of two-factor interactions and one group of three two-factor interactions that are aliases. If we fold over this design, the second fraction would have the generators $I = -ABCF$, $I = -ABDG$, and $I = BCDEH$. The combined design has the generators $I = CDFG$ and $I = BCDEH$, and the complete defining relation is

$$I = CDFG = BCDEH = BEFGH$$

The combined design is of resolution IV, but the only two-factor interactions that are still aliased are $CD = FG$, $CF = DG$, and $CG = DF$. This is a considerable simplification of the aliasing in the original fraction.

Notice that when we start with a resolution III design, the fold-over procedure guarantees that the combined design will be of resolution IV, thereby ensuring that all the main effects can be separated from their two-factor interaction aliases. When folding over a resolution IV design, we will not necessarily separate *all* the two-factor interactions. In fact, if the *original* fraction has an alias structure with more than two two-factor interactions in any alias chain, folding over will not completely separate all the two-factor interactions. Both of the foregoing examples, the 2_{IV}^{6-2} and the 2_{IV}^{8-3}, have at least one such two-factor interaction alias chain.

Resolution V designs are fractional factorials in which the main effects and the two-factor interactions do not have other main effects and two-factor interactions as their aliases. These are very powerful designs, allowing the unique estimation of all the main effects and two-factor interactions, provided that all the three-factor and higher interactions are negligible. The smallest word in the defining relation of such a design must have five letters. The 2^{5-1} design with the generating relation $I = ABCDE$ is of resolution V. Another example is the 2_V^{8-2} design with the generating relations $I = ABCDG$ and $I = ABEFH$. Further examples of these designs are given by Box and Hunter (1961b).

9–7 SUMMARY

This chapter has introduced the 2^{k-p} fractional factorial design. We have emphasized the use of these designs in screening experiments to identify quickly and

Table 9–28   Useful Factorial and Fractional Factorial
Designs from the 2^{k-p} System. The Numbers in the Cells are
the Number of Factors in the Experiment

	Number of Runs			
Design Type	4	8	16	32
Full factorial	2	3	4	5
Half-fraction	3	4	5	6
Resolution IV fraction	—	4	6–8	7–16
Resolution III fraction	3	5–7	9–15	17–31

efficiently the subset of factors that are active and to provide some information on interaction. The projective property of these designs makes it possible in many cases to examine the active factors in more detail. Sequential assembly of these designs via fold over is a very effective way to gain additional information about interactions that an initial experiment may identify as possibly important.

In practice, 2^{k-p} fractional factorial designs with $N = 4, 8, 16$, and 32 runs are highly useful. Table 9–28 summarizes these designs, identifying how many factors can be used with each design to obtain various types of screening experiments. For example, the 16-run design is a full factorial for 4 factors, a one-half fraction for 5 factors, a resolution IV fraction for 6 to 8 factors, and a resolution III fraction for 9 to 15 factors. All of these designs may be constructed using the methods discussed in this chapter, and many of their alias structures are shown in Appendix Table XII.

9–8   PROBLEMS

9–1   Suppose that in the chemical process development experiment described in Problem 7–7, it was only possible to run a one-half fraction of the 2^4 design. Construct the design and perform the statistical analysis, using the data from replicate I.

9–2   Suppose that in Problem 7–15, only a one-half fraction of the 2^4 design could be run. Construct the design and perform the analysis, using the data from replicate I.

9–3   Consider the plasma etch experiment described in Problem 7–18. Suppose that only a one-half fraction of the design could be run. Set up the design and analyze the data.

9–4   Problem 7–21 describes a process improvement study in the manufacturing process of an integrated circuit. Suppose that only eight runs could be made in this process. Set up an appropriate 2^{5-2} design and find the alias structure. Use the appropriate observations from Problem 7–21 as the observations in this design and estimate the factor effects. What conclusions can you draw?

9–5 ***Continuation of Problem 9–4.*** Suppose you have made the eight runs in the 2^{5-2} design in Problem 9–4. What additional runs would be required to identify the factor effects that are of interest? What are the alias relationships in the combined design?

9–6 R. D. Snee ("Experimenting with a Large Number of Variables," in *Experiments in Industry: Design, Analysis and Interpretation of Results,* by R. D. Snee, L. B. Hare, and J. B. Trout, Editors, ASQC, 1985) describes an experiment in which a 2^{5-1} design with $I = ABCDE$ was used to investigate the effects of five factors on the color of a chemical product. The factors are A = solvent/reactant, B = catalyst/reactant, C = temperature, D = reactant purity, and E = reactant pH. The results obtained were as follows:

$e =$	-0.63	$d =$	6.79
$a =$	2.51	$ade =$	5.47
$b =$	-2.68	$bde =$	3.45
$abe =$	1.66	$abd =$	5.68
$c =$	2.06	$cde =$	5.22
$ace =$	1.22	$acd =$	4.38
$bce =$	-2.09	$bcd =$	4.30
$abc =$	1.93	$abcde =$	4.05

(a) Prepare a normal probability plot of the effects. Which effects seem active?

(b) Calculate the residuals. Construct a normal probability plot of the residuals and plot the residuals versus the fitted values. Comment on the plots.

(c) If any factors are negligible, collapse the 2^{5-1} design into a full factorial in the active factors. Comment on the resulting design, and interpret the results.

9–7 An article by J. J. Pignatiello, Jr. and J. S. Ramberg in the *Journal of Quality Technology,* (Vol. 17, 1985, pp. 198–206) describes the use of a replicated fractional factorial to investigate the effect of five factors on the free height of leaf springs used in an automotive application. The factors are A = furnace temperature, B = heating time, C = transfer time, D = hold down time, and E = quench oil temperature. The data are shown below:

A	B	C	D	E	Free Height		
$-$	$-$	$-$	$-$	$-$	7.78	7.78	7.81
$+$	$-$	$-$	$+$	$-$	8.15	8.18	7.88
$-$	$+$	$-$	$+$	$-$	7.50	7.56	7.50
$+$	$+$	$-$	$-$	$-$	7.59	7.56	7.75
$-$	$-$	$+$	$+$	$-$	7.54	8.00	7.88
$+$	$-$	$+$	$-$	$-$	7.69	8.09	8.06
$-$	$+$	$+$	$-$	$-$	7.56	7.52	7.44
$+$	$+$	$+$	$+$	$-$	7.56	7.81	7.69
$-$	$-$	$-$	$-$	$+$	7.50	7.25	7.12
$+$	$-$	$-$	$+$	$+$	7.88	7.88	7.44

(continued)

(continued)

A	B	C	D	E		Free Height	
−	+	−	+	+	7.50	7.56	7.50
+	+	−	−	+	7.63	7.75	7.56
−	−	+	+	+	7.32	7.44	7.44
+	−	+	−	+	7.56	7.69	7.62
−	+	+	−	+	7.18	7.18	7.25
+	+	+	+	+	7.81	7.50	7.59

(a) Write out the alias structure for this design. What is the resolution of this design?

(b) Analyze the data. What factors influence the mean free height?

(c) Calculate the range and standard deviation of the free height for each run. Is there any indication that any of these factors affects variability in the free height?

(d) Analyze the residuals from this experiment, and comment on your findings.

(e) Is this the best possible design for five factors in 16 runs? Specifically, can you find a fractional design for five factors in 16 runs with a higher resolution than this one?

9-8 An article in *Industrial and Engineering Chemistry* ("More on Planning Experiments to Increase Research Efficiency," 1970, pp. 60–65) uses a 2^{5-2} design to investigate the effect of A = condensation temperature, B = amount of material 1, C = solvent volume, D = condensation time, and E = amount of material 2 on yield. The results obtained are as follows:

$$e = 23.2 \qquad ad = 16.9 \qquad cd = 23.8 \qquad bde = 16.8$$
$$ab = 15.5 \qquad bc = 16.2 \qquad ace = 23.4 \qquad abcde = 18.1$$

(a) Verify that the design generators used were $I = ACE$ and $I = BDE$.

(b) Write down the complete defining relation and the aliases for this design.

(c) Estimate the main effects.

(d) Prepare an analysis of variance table. Verify that the AB and AD interactions are available to use as error.

(e) Plot the residuals versus the fitted values. Also construct a normal probability plot of the residuals. Comment on the results.

9-9 Consider the leaf spring experiment in Problem 9–7. Suppose that factor E (quench oil temperature) is very difficult to control during manufacturing. Where would you set factors A, B, C, and D to reduce variability in the free height as much as possible regardless of the quench oil temperature used?

9-10 Construct a 2^{7-2} design by choosing two four-factor interactions as the independent generators. Write down the complete alias structure for this design. Outline the analysis of variance table. What is the resolution of this design?

9–11 Consider the 2^5 design in Problem 7–21. Suppose that only a one-half fraction could be run. Furthermore, two days were required to take the 16 observations, and it was necessary to confound the 2^{5-1} design in two blocks. Construct the design and analyze the data.

9–12 Analyze the data in Problem 7–23 as if it came from a 2_{IV}^{4-1} design with $I = ABCD$. Project the design into a full factorial in the subset of the original four factors that appear to be significant.

9–13 Repeat Problem 9–12 using $I = -ABCD$. Does use of the alternate fraction change your interpretation of the data?

9–14 Project the 2_{IV}^{4-1} design in Example 9–1 into two replicates of a 2^2 design in the factors A and B. Analyze the data and draw conclusions.

9–15 Construct a 2_{III}^{6-3} design. Determine the effects that may be estimated if a second fraction of this design is run with all signs reversed.

9–16 Consider the 2_{III}^{6-3} design in Problem 9–15. Determine the effects that may be estimated if a second fraction of this design is run with the signs for factor A reversed.

9–17 Fold over the 2_{III}^{7-4} design in Table 9–19 to produce an eight-factor design. Verify that the resulting design is a 2_{IV}^{8-4} design. Is this a minimal design?

9–18 Fold over a 2_{III}^{5-2} design to produce a six-factor design. Verify that the resulting design is a 2_{IV}^{6-2} design. Compare this design to the 2_{IV}^{6-2} design in Table 9–10.

9–19 An industrial engineer is conducting an experiment using a Monte Carlo simulation model of an inventory system. The independent variables in her model are the order quantity (A), the reorder point (B), the setup cost (C), the backorder cost (D), and the carrying cost rate (E). The response variable is average annual cost. To conserve computer time, she decides to investigate these factors using a 2_{III}^{5-2} design with $I = ABD$ and $I = BCE$. The results she obtains are $de = 95$, $ae = 134$, $b = 158$, $abd = 190$, $cd = 92$, $ac = 187$, $bce = 155$, and $abcde = 185$.

(a) Verify that the treatment combinations given are correct. Estimate the effects, assuming three-factor and higher interactions are negligible.

(b) Suppose that a second fraction is added to the first, for example, $ade = 136$, $e = 93$, $ab = 187$, $bd = 153$, $acd = 139$, $c = 99$, $abce = 191$, and $bcde = 150$. How was this second fraction obtained? Add this data to the original fraction, and estimate the effects.

(c) Suppose that the fraction $abc = 189$, $ce = 96$, $bcd = 154$, $acde = 135$, $abe = 193$, $bde = 152$, $ad = 137$, and $(1) = 98$ was run. How was this fraction obtained? Add this data to the original fraction and estimate the effects.

9–20 Construct a 2^{5-1} design. Show how the design may be run in two blocks of eight observations each. Are any main effects or two-factor interactions confounded with blocks?

9–21 Construct a 2^{7-2} design. Show how the design may be run in four blocks of eight observations each. Are any main effects or two-factor interactions confounded with blocks?

9-22   *Irregular fractions of the 2ᵏ [John (1971)].* Consider a 2^4 design. We must estimate the four main effects and the six two-factor interactions, but the full 2^4 factorial cannot be run. The largest possible block size contains 12 runs. These 12 runs can be obtained from the four one-quarter replicates defined by $I = \pm AB = \pm ACD = \pm BCD$ by omitting the principal fraction. Show how the remaining three 2^{4-2} fractions can be combined to estimate the required effects, assuming three-factor and higher interactions are negligible. This design could be thought of as a three-quarter fraction.

9-23   Carbon anodes used in a smelting process are baked in a ring furnace. An experiment is run in the furnace to determine which factors influence the weight of packing material that is stuck to the anodes after baking. Six variables are of interest, each at two levels: A = pitch/fines ratio (0.45, 0.55); B = packing material type (1, 2); C = packing material temperature (ambient, 325°C); D = flue location (inside, outside); E = pit temperature (ambient, 195°C); and F = delay time before packing (zero, 24 hours). A 2^{6-3} design is run, and three replicates are obtained at each of the design points. The weight of packing material stuck to the anodes is measured in grams. The data in run order are as follows: abd = (984, 826, 936); $abcdef$ = (1275, 976, 1457); be = (1217, 1201, 890); af = (1474, 1164, 1541); def = (1320, 1156, 913); cd = (765, 705, 821); ace = (1338, 1254, 1294); and bcf = (1325, 1299, 1253). We wish to minimize the amount of stuck packing material.

(a) Verify that the eight runs correspond to a 2_{III}^{6-3} design. What is the alias structure?

(b) Use the average weight as a response. What factors appear to be influential?

(c) Use the range of the weights as a response. What factors appear to be influential?

(d) What recommendations would you make to the process engineers?

9-24   A 16-run experiment was performed in a semiconductor manufacturing plant to study the effects of six factors on the curvature or camber of the substrate devices produced. The six variables and their levels are shown below:

Run	Lamination Temperature (°C)	Lamination Time (s)	Lamination Pressure (tn)	Firing Temperature (°C)	Firing Cycle Time (h)	Firing Dew Point (°C)
1	55	10	5	1580	17.5	20
2	75	10	5	1580	29	26
3	55	25	5	1580	29	20
4	75	25	5	1580	17.5	26
5	55	10	10	1580	29	26
6	75	10	10	1580	17.5	20
7	55	25	10	1580	17.5	26
8	75	25	10	1580	29	20
9	55	10	5	1620	17.5	26
10	75	10	5	1620	29	20

(continued)

Run	Lamination Temperature (°C)	Lamination Time (s)	Lamination Pressure (tn)	Firing Temperature (°C)	Firing Cycle Time (h)	Firing Dew Point (°C)
11	55	25	5	1620	29	26
12	75	25	5	1620	17.5	20
13	55	10	10	1620	29	20
14	75	10	10	1620	17.5	26
15	55	25	10	1620	17.5	20
16	75	25	10	1620	29	26

Each run was replicated four times, and a camber measurement was taken on the substrate. The data are shown below:

Run	Camber for Replicate (in/in) 1	2	3	4	Total (10^{-4} in/in)	Mean (10^{-4} in/in)	Standard Deviation
1	0.0167	0.0128	0.0149	0.0185	629	157.25	24.418
2	0.0062	0.0066	0.0044	0.0020	192	48.00	20.976
3	0.0041	0.0043	0.0042	0.0050	176	44.00	4.083
4	0.0073	0.0081	0.0039	0.0030	223	55.75	25.025
5	0.0047	0.0047	0.0040	0.0089	223	55.75	22.410
6	0.0219	0.0258	0.0147	0.0296	920	230.00	63.639
7	0.0121	0.0090	0.0092	0.0086	389	97.25	16.029
8	0.0255	0.0250	0.0226	0.0169	900	225.00	39.42
9	0.0032	0.0023	0.0077	0.0069	201	50.25	26.725
10	0.0078	0.0158	0.0060	0.0045	341	85.25	50.341
11	0.0043	0.0027	0.0028	0.0028	126	31.50	7.681
12	0.0186	0.0137	0.0158	0.0159	640	160.00	20.083
13	0.0110	0.0086	0.0101	0.0158	455	113.75	31.12
14	0.0065	0.0109	0.0126	0.0071	371	92.75	29.51
15	0.0155	0.0158	0.0145	0.0145	603	150.75	6.75
16	0.0093	0.0124	0.0110	0.0133	460	115.00	17.45

(a) What type of design did the experimenters use?

(b) What are the alias relationships in this design?

(c) Do any of the process variables affect average camber?

(d) Do any of the process variables affect the variability in camber measurements?

(e) If it is important to reduce camber as much as possible, what recommendations would you make?

9-25 A spin coater is used to apply photoresist to a bare silicon wafer. This operation usually occurs early in the semiconductor manufacturing process, and the average coating thickness and the variability in the coating thickness has an important impact on downstream manufacturing steps. Six variables are used in the experiment. The variables and their high and low levels are as follows:

Factor	Low Level	High Level
Final spin speed	7350 rpm	6650 rpm
Acceleration rate	5	20
Volume of resist applied	3 cc	5 cc
Time of spin	14 s	6 s
Resist batch variation	Batch 1	Batch 2
Exhaust pressure	Cover off	Cover on

The experimenter decides to use a 2^{6-1} design and to make three readings on resist thickness on each test wafer. The data are shown in Table 9–29.

(a) Verify that this is a 2^{6-1} design. Discuss the alias relationships in this design.

(b) What factors appear to affect average resist thickness?

(c) Since the volume of resist applied has little effect on average thickness, does this have any important practical implications for the process engineers?

(d) Project this design into a smaller design involving only the significant factors. Graphically display the results. Does this aid in interpretation?

(e) Use the range of resist thickness as a response variable. Is there any indication that any of these factors affect the variability in resist thickness?

(f) Where would you recommend that the process engineers run the process?

9–26   Harry and Judy Peterson-Nedry (two friends of the author) own a vineyard in Oregon. They grow several varieties of grapes and manufacture wine. Harry and Judy have used factorial designs for process and product development in the winemaking segment of their business. This problem describes the experiment conducted for their 1985 Pinot Noir. Eight variables, shown below, were originally studied in this experiment:

Variable	Low Level ($-$)	High Level ($+$)
A = Pinot Noir clone	Pommard	Wadenswil
B = Oak type	Allier	Troncais
C = Age of barrel	Old	New
D = Yeast/skin contact	Champagne	Montrachet
E = Stems	None	All
F = Barrel toast	Light	Medium
G = Whole cluster	None	10%
H = Fermentation temperature	Low (75°F max)	High (92°F max)

Table 9–29 Data for Problem 9–25

	A	B	C	D	E	F	Resist Thickness				
Run	Volume	Batch	Time	Speed	Acc.	Cover	Left	Center	Right	Avg.	Range
1	5	Batch 2	14 sec	7350	5	Off	4531	4531	4515	4525.7	16
2	5	Batch 1	6 sec	7350	5	Off	4446	4464	4428	4446	36
3	3	Batch 1	6 sec	6650	5	Off	4452	4490	4452	4464.7	38
4	3	Batch 2	14 sec	7350	20	Off	4316	4328	4308	4317.3	20
5	3	Batch 1	14 sec	7350	5	Off	4307	4295	4289	4297	18
6	5	Batch 1	6 sec	6650	20	Off	4470	4492	4495	4485.7	25
7	3	Batch 1	6 sec	7350	5	On	4496	4502	4482	4493.3	20
8	5	Batch 2	14 sec	6650	20	Off	4542	4547	4538	4542.3	9
9	5	Batch 1	14 sec	6650	5	Off	4621	4643	4613	4625.7	30
10	3	Batch 1	14 sec	6650	5	On	4653	4670	4645	4656	25
11	3	Batch 2	14 sec	6650	20	On	4480	4486	4470	4478.7	16
12	3	Batch 1	6 sec	7350	20	Off	4221	4233	4217	4223.7	16
13	5	Batch 1	6 sec	6650	5	On	4620	4641	4619	4626.7	22
14	3	Batch 1	6 sec	6650	20	On	4455	4480	4466	4467	25
15	5	Batch 2	14 sec	7350	20	On	4255	4288	4243	4262	45
16	5	Batch 2	6 sec	7350	5	On	4490	4534	4523	4515.7	44
17	3	Batch 2	14 sec	7350	5	On	4514	4551	4540	4535	37
18	3	Batch 1	14 sec	6650	20	Off	4494	4503	4496	4497.7	9
19	5	Batch 2	6 sec	7350	20	Off	4293	4306	4302	4300.3	13
20	3	Batch 2	6 sec	7350	5	Off	4534	4545	4512	4530.3	33
21	5	Batch 1	14 sec	6650	20	On	4460	4457	4436	4451	24
22	3	Batch 2	6 sec	6650	5	On	4650	4688	4656	4664.7	38
23	5	Batch 1	14 sec	7350	20	Off	4231	4244	4230	4235	14
24	3	Batch 2	6 sec	7350	20	On	4225	4228	4208	4220.3	20
25	5	Batch 1	14 sec	7350	5	On	4381	4391	4376	4382.7	15
26	3	Batch 2	6 sec	6650	20	Off	4533	4521	4511	4521.7	22
27	3	Batch 1	14 sec	7350	20	On	4194	4230	4172	4198.7	58
28	5	Batch 2	6 sec	6650	5	Off	4666	4695	4672	4677.7	29
29	5	Batch 1	6 sec	7350	20	On	4180	4213	4197	4196.7	33
30	5	Batch 2	6 sec	6650	20	On	4465	4496	4463	4474.7	33
31	5	Batch 2	14 sec	6650	5	On	4653	4685	4665	4667.7	32
32	3	Batch 2	14 sec	6650	5	Off	4683	4712	4677	4690.7	35

Harry and Judy decided to use a 2_{IV}^{8-4} design with 16 runs. The wine was taste-tested by a panel of experts on 8 March 1986. Each expert ranked the 16 samples of wine tasted, with rank 1 being the best. The design and the taste-test panel results are shown in Table 9–30.

(a) What are the alias relationships in the design selected by Harry and Judy?

(b) Use the average ranks ($\bar{y}$) as a response variable. Analyze the data and draw conclusions. You will find it helpful to examine a normal probability plot of the effect estimates.

Table 9-30 Design and Results for Wine Tasting Experiment

Run	Variable								Panel Rankings					Summary	
	A	B	C	D	E	F	G	H	HPN	JPN	CAL	DCM	RGB	$\bar{y}$	s
1	−	−	−	−	−	−	−	−	12	6	13	10	7	9.6	3.05
2	+	−	−	−	+	+	−	+	10	7	14	14	9	10.8	3.11
3	−	+	−	−	+	+	+	+	14	13	10	11	15	12.6	2.07
4	+	+	−	−	+	+	−	−	9	9	7	9	12	9.2	1.79
5	−	−	+	−	+	−	+	−	8	8	11	8	10	9.0	1.41
6	+	−	+	−	+	+	−	+	16	12	15	16	16	15.0	1.73
7	−	+	+	−	−	−	−	+	6	5	6	5	3	5.0	1.22
8	+	+	+	−	−	+	+	−	15	16	16	15	14	15.2	0.84
9	−	−	−	+	+	+	−	+	1	2	3	3	2	2.2	0.84
10	+	−	−	+	−	−	+	−	7	11	4	7	6	7.0	2.55
11	−	+	−	+	−	+	+	−	13	3	8	12	8	8.8	3.96
12	+	+	−	+	−	−	−	+	3	1	5	1	4	2.8	1.79
13	−	−	+	+	−	+	+	+	2	10	2	4	5	9.6	3.29
14	+	−	+	+	+	+	−	−	4	4	1	2	1	2.4	1.52
15	−	+	+	+	+	−	−	−	5	15	9	6	11	9.2	4.02
16	+	+	+	+	+	+	+	+	11	14	12	13	13	12.6	1.14

(c) Use the standard deviation of the ranks (or some appropriate transformation such as log s) as a response variable. What conclusions can you draw about the effects of the eight variables on variability in wine quality?

(d) After looking at the results, Harry and Judy decide that one of the panel members (DCM) knows more about beer than he does about wine, so they decide to delete his ranking. What affect would this have on the results and conclusions from parts (b) and (c)?

(e) Suppose that just before the start of the experiment, Harry and Judy discovered that the eight new barrels they ordered from France for use in the experiment would not arrive in time, and all 16 runs would have to be made with old barrels. If Harry and Judy just drop column C from their design, what does this do to the alias relationships? Do they need to start over and construct a new design?

(f) Harry and Judy know from experience that some treatment combinations are unlikely to produce good results. For example, the run with all eight variables at the high level generally results in a poorly rated wine. This was confirmed in the 8 March 1986 taste test. They want to set up a new design for their 1986 Pinot Noir using these same eight variables, but they do not want to make the run with all eight factors at the high level. What design would you suggest?

9–27 In an article in *Quality Engineering* ("An Application of Fractional Factorial Experimental Designs," 1988, Vol. 1, pp. 19–23) M. B. Kilgo describes an experiment to determine the effect of CO_2 pressure (A), CO_2 temperature (B), peanut moisture (C), CO_2 flow rate (D), and peanut particle size (E) on the total yield of oil per batch of peanuts (y). The levels that she used for these factors are as follows:

Coded Level	A, Pressure (bar)	B, Temp. (°C)	C, Moisture (% by weight)	D, Flow (liters/min)	E, Part. Size (mm)
−1	415	25	5	40	1.28
1	550	95	15	60	4.05

She conducted the 16-run fractional factorial experiment shown below.

A	B	C	D	E	y
415	25	5	40	1.28	63
550	25	5	40	4.05	21
415	95	5	40	4.05	36
550	95	5	40	1.28	99
415	25	15	40	4.05	24
550	25	15	40	1.28	66

(continued)

(continued)

A	B	C	D	E	y
415	95	15	40	1.28	71
550	95	15	40	4.05	54
415	25	5	60	4.05	23
550	25	5	60	1.28	74
415	95	5	60	1.28	80
550	95	5	60	4.05	33
415	25	15	60	1.28	63
550	25	15	60	4.05	21
415	95	15	60	4.05	44
550	95	15	60	1.28	96

(a) What type of design has been used? Identify the defining relation and the alias relationships.

(b) Estimate the factor effects and use a normal probability plot to tentatively identify the important factors.

(c) Perform an appropriate statistical analysis to test the hypotheses that the factors identified in part (b) above have a significant effect on the yield of peanut oil.

(d) Fit a model that could be used to predict peanut oil yield in terms of the factors that you have identified as important.

(e) Analyze the residuals from this experiment and comment on model adequacy.

9–28 A 16-run fractional factorial experiment in 10 factors on sand-casting of engine manifolds was conducted by engineers at the Essex Aluminum Plant of the Ford Motor Company and described in the article "Evaporative Cast Process 3.0 Liter Intake Manifold Poor Sandfill Study," by D. Becknell (*Fourth Symposium on Taguchi Methods,* American Supplier Institute, Dearborn, MI, 1986, pp. 120–130). The purpose was to determine which of 10 factors has an effect on the proportion of defective castings. The design and the resulting proportion of nondefective castings $\hat{p}$ observed on each run are shown below. This is a resolution III fraction with generators $E = CD$, $F = BD$, $G = BC$, $H = AC$, $J = AB$, and $K = ABC$. Assume that the number of castings made at each run in the design is 1000.

Run	A	B	C	D	E	F	G	H	J	K	$\hat{p}$	arcsin$\sqrt{\hat{p}}$	F&T's Modification
1	−	−	−	−	+	+	+	+	+	−	0.958	1.364	1.363
2	+	−	−	−	+	+	+	−	−	+	1.000	1.571	1.555
3	−	+	−	−	+	−	−	+	−	+	0.977	1.419	1.417
4	+	+	−	−	+	−	−	−	+	−	0.775	1.077	1.076
5	−	−	+	−	−	+	−	−	+	+	0.958	1.364	1.363
6	+	−	+	−	−	+	−	+	−	−	0.958	1.364	1.363
7	−	+	+	−	−	−	+	−	−	−	0.813	1.124	1.123
8	+	+	+	−	−	−	+	+	+	+	0.906	1.259	1.259
9	−	−	−	+	−	−	+	+	+	−	0.679	0.969	0.968

(continued)

Run	A	B	C	D	E	F	G	H	J	K	$\hat{p}$	arcsin$\sqrt{\hat{p}}$	F&T's Modification
10	+	−	−	+	−	−	+	−	−	+	0.781	1.081	1.083
11	−	+	−	+	−	+	−	+	−	+	1.000	1.571	1.556
12	+	+	−	+	−	+	−	−	+	−	0.896	1.241	1.242
13	−	−	+	+	+	−	−	−	+	+	0.958	1.364	1.363
14	+	−	+	+	+	−	−	+	−	−	0.818	1.130	1.130
15	−	+	+	+	+	+	+	−	−	−	0.841	1.161	1.160
16	+	+	+	+	+	+	+	+	+	+	0.955	1.357	1.356

(a) Find the defining relation and the alias relationships in this design.

(b) Estimate the factor effects and use a normal probability plot to tentatively identify the important factors.

(c) Fit an appropriate model using the factors identified in part (b) above.

(d) Plot the residuals from this model versus the predicted proportion of nondefective castings. Also prepare a normal probability plot of the residuals. Comment on the adequacy of these plots.

(e) In part (d) you should have noticed an indication that the variance of the response is not constant (considering that the response is a proportion, you should have expected this). The previous table also shows a transformation on $\hat{p}$, the arcsin square root, that is a widely-used *variance stabilizing transformation* for proportion data (refer to the discussion of variance stabilizing transformations in Chapter 3). Repeat parts (a) through (d) above using the transformed response and comment on your results. Specifically, are the residual plots improved?

(f) There is a modification to the arcsin square root transformation, proposed by Freeman and Tukey ("Transformations Related to the Angular and the Square Root," *Annals of Mathematical Statistics,* Vol. 21, 1950, pp. 607–611) that improves its performance in the tails. F&T's modification is:

$$[\arcsin\sqrt{n\hat{p}/(n+1)} + \arcsin\sqrt{(n\hat{p}+1)/(n+1)}]/2$$

Rework parts (a) through (d) using this transformation and comment on the results. (For an interesting discussion and analysis of this experiment, refer to "Analysis of Factorial Experiments with Defects or Defectives as the Response," by S. Bisgaard and H. T. Fuller, *Quality Engineering,* Vol. 7, 1994–95, pp. 429–443.)

9-29 A 16-run fractional factorial experiment in 9 factors was conducted by Chrysler Motors Engineering and described in the article "Sheet Molded Compound Process Improvement," by P. I. Hsieh and D. E. Goodwin (*Fourth Symposium on Taguchi Methods,* American Supplier Institute, Dearborn, MI, 1986, pp. 13–21). The purpose was to reduce the number of defects in the finish of sheet-molded grill opening panels. The design, and the resulting number of defects, c, observed on each run, is shown below. This is a resolution III fraction with generators $E = BD$, $F = BCD$, $G = AC$, $H = ACD$, and $J = AB$.

Run	A	B	C	D	E	F	G	H	J	c	$\sqrt{c}$	F&T's Modification
1	−	−	−	−	+	−	+	−	+	56	7.48	7.52
2	+	−	−	−	+	−	−	+	−	17	4.12	4.18
3	−	+	−	−	−	+	+	−	−	2	1.41	1.57
4	+	+	−	−	−	+	−	+	+	4	2.00	2.12
5	−	−	+	−	+	+	−	+	+	3	1.73	1.87
6	+	−	+	−	+	+	+	−	−	4	2.00	2.12
7	−	+	+	−	−	−	−	+	−	50	7.07	7.12
8	+	+	+	−	−	−	+	−	+	2	1.41	1.57
9	−	−	−	+	−	+	+	+	+	1	1.00	1.21
10	+	−	−	+	−	+	−	−	−	0	0.00	0.50
11	−	+	−	+	+	−	+	+	−	3	1.73	1.87
12	+	+	−	+	+	−	−	−	+	12	3.46	3.54
13	−	−	+	+	−	−	−	−	+	3	1.73	1.87
14	+	−	+	+	−	−	+	+	−	4	2.00	2.12
15	−	+	+	+	+	+	−	−	−	0	0.00	0.50
16	+	+	+	+	+	+	+	+	+	0	0.00	0.50

(a) Find the defining relation and the alias relationships in this design.

(b) Estimate the factor effects and use a normal probability plot to tentatively identify the important factors.

(c) Fit an appropriate model using the factors identified in part (b) above.

(d) Plot the residuals from this model versus the predicted number of defects. Also, prepare a normal probability plot of the residuals. Comment on the adequacy of these plots.

(e) In part (d) you should have noticed an indication that the variance of the response is not constant (considering that the response is a count, you should have expected this). The previous table also shows a transformation on c, the square root, that is a widely-used *variance stabilizing transformation* for count data (refer to the discussion of variance stabilizing transformations in Chapter 3). Repeat parts (a) through (d) using the transformed response and comment on your results. Specifically, are the residual plots improved?

(f) There is a modification to the square root transformation, proposed by Freeman and Tukey ("Transformations Related to the Angular and the Square Root," *Annals of Mathematical Statistics,* Vol. 21, 1950, pp. 607–611) that improves its performance. F&T's modification to the square root transformation is:

$$[\sqrt{c} + \sqrt{(c + 1)}]/2$$

Rework parts (a) through (d) using this transformation and comment on the results. (For an interesting discussion and analysis of this experiment, refer to "Analysis of Factorial Experiments with Defects or Defectives as the Response," by S. Bisgaard and H. T. Fuller, *Quality Engineering,* Vol. 7, 1994–95, pp. 429–443.)

9-30 In Example 9–8 you saw an illustration of a 12-run Plackett-Burman design in 11 factors. In that example, a set of data was artificially created to simulate how a Plackett-Burman design would perform in the presence of a true underlying model with interaction. The analysis of the resulting data was unrewarding in that many effect estimates were large, and the interpretation of the experiment was very inconclusive. Suppose that the experimenter decided to fold over the original 12 runs and to analyze the combined design in an effort to resolve the important effects. The 12 responses obtained in the fold-over design are as follows: 169, 232, 167, 184, 213, 187, 216, 171, 175, 193, 211, and 277 (these observations were generated from the same true model as in Example 9–8).

(a) Analyze the data from the combined design. Specifically, estimate the effects of all 11 main effects and the 11 two-factor interaction alias chains and use a normal probability plot to identify the significant factors. Did folding over the original design resolve the main effects and interactions that you know are important?

(b) What lessons did you learn about the 12-run Plackett-Burman design from this problem and Example 9–8?

Chapter 10
Three-Level and Mixed-Level Factorial and Fractional Factorial Designs

The two-level series of factorial and fractional factorial designs discussed in Chapters 7, 8, and 9 are widely used in industrial research and development. There are some extensions and variations of these designs that are occasionally useful, such as the designs for cases where all the factors are present at three levels. These 3^k designs will be discussed in this chapter. We will also consider cases where some factors have two levels and other factors have either three or four levels.

10–1   THE 3^k FACTORIAL DESIGN

10–1.1   Notation and Motivation for the 3^k Design

We now discuss the 3^k factorial design; that is, a factorial arrangement with k factors each at three levels. Factors and interactions will be denoted by capital letters. We will refer to the three levels of the factors as low, intermediate, and high. There are several different notations used to represent these factor levels; one possibility is to represent the factor levels by the digits 0 (low), 1 (intermediate), and 2 (high). Each treatment combination in the 3^k design will be denoted by k digits, where the first digit indicates the level of factor A, the second digit indicates the level of factor B, . . . , and the kth digit indicates the level of factor K. For example, in a 3^2 design, 00 denotes the treatment combination corresponding to A and B both at the low level, and 01 denotes the treatment combination corresponding to A at the low level and B at the intermediate level. Figures 10–1 and 10–2 show the geometry of the 3^2 and the 3^3 design, respectively, using this notation.

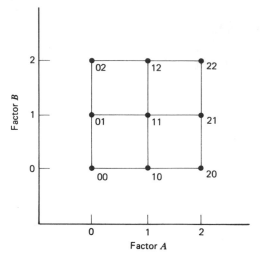

Figure 10–1. Treatment combinations in a 3^2 design.

This system of notation could have been used for the 2^k designs presented previously, with 0 and 1 used in place of the minus and plus ones, respectively. In the 2^k design, we prefer the ± 1 notation because it facilitates the geometric view of the design and because it is directly applicable to regression modeling, blocking, and the construction of fractional factorials.

In the 3^k system of designs, when the factors are **quantitative,** we often

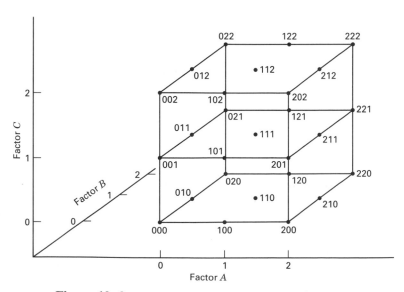

Figure 10–2. Treatment combinations in a 3^3 design.

denote the low, intermediate, and high levels by -1, 0, and $+1$, respectively. This facilitates fitting a **regression model** relating the response to the factor levels. For example, consider the 3^2 design in Figure 10–1, and let x_1 represent factor A and x_2 represent factor B. A regression model relating the response y to x_1 and x_2 that is supported by this design is

$$y = \beta_0 + \beta_1 x_1 + \beta_2 x_2 + \beta_{12} x_1 x_2 + \beta_{11} x_1^2 + \beta_{22} x_2^2 + \epsilon \qquad (10\text{--}1)$$

Notice that the addition of a third factor level allows the relationship between the response and the design factors to be modeled as a quadratic.

The 3^k design is certainly a possible choice by an experimenter who is concerned about curvature in the response function. However, two points need to be considered:

1. The 3^k design is not the most efficient way to model a quadratic relationship; the **response surface designs** discussed in Chapter 14 are superior alternatives.

2. The 2^k design augmented with center points, as discussed in Chapter 7, is an excellent way to obtain an indication of curvature. It allows one to keep the size and complexity of the design low and simultaneously obtain some protection against curvature. Then, if curvature is important, the two-level design can be augmented with **axial runs** to obtain a **central composite design,** as shown in Figure 7–29. This **sequential strategy** of experimentation is far more efficient than running a 3^k factorial design with quantitative factors.

10–1.2 The 3^2 Design

The simplest design in the 3^k system is the 3^2 design, which has two factors, each at three levels. The treatment combinations for this design were shown in Figure 10–1. Since there are $3^2 = 9$ treatment combinations, there are eight degrees of freedom between these treatment combinations. The main effects of A and B each have two degrees of freedom, and the AB interaction has four degrees of freedom. If there are n replicates, there will be $n3^2 - 1$ total degrees of freedom and $3^2(n - 1)$ degrees of freedom for error.

The sums of squares for A, B, and AB may be computed by the usual methods for factorial designs discussed in Chapter 6. Alternatively, in Section 10–1.5 we give an extension of Yates' algorithm to the 3^k series. The sum of squares for any main effect may be partitioned into a linear and a quadratic component, each with a single degree of freedom, using the **orthogonal contrast constants,** as demonstrated in Chapter 6. Of course, this is only meaningful if the factor is quantitative and if the three levels are equally spaced.

As an illustration, consider the tool life problem in Example 6–5. The objective of this experiment was to investigate the effect of cutting speed and tool angle on tool life in a numerically controlled machine. Three cutting speeds and three tool angles were selected for study. This is a 3^2 design with $n = 2$ replicates. The analysis of variance is repeated for convenience in Table 10–1. For details of the sum of squares calculations and the partitioning of the main effects into linear and quadratic components, refer to Chapter 6, Section 6–5.

The two-factor interaction AB may be partitioned in two ways. The first method consists of subdividing AB into the four single-degree-of-freedom components corresponding to $AB_{L \times L}$, $AB_{L \times Q}$, $AB_{Q \times L}$, and $AB_{Q \times Q}$. This is done using orthogonal contrast constants, as demonstrated in Example 6–5. For the tool life data, this yields $SS_{AB_{L \times L}} = 8.00$, $SS_{AB_{L \times Q}} = 42.67$, $SS_{AB_{Q \times L}} = 2.67$, and $SS_{AB_{Q \times Q}} = 8.00$. Since this is an orthogonal partitioning of AB, note that $SS_{AB} = SS_{AB_{L \times L}} + SS_{AB_{L \times Q}} + SS_{AB_{Q \times L}} + SS_{AB_{Q \times Q}}$.

The second method is based on **orthogonal Latin squares.** Consider the totals of the treatment combinations for the data in Example 6–5. These totals are shown in Figure 10–3 as the circled numbers in the squares. The two factors A and B correspond to the rows and columns, respectively, of a 3×3 Latin square. In Figure 10–3, two particular 3×3 Latin squares are shown superimposed on the cell totals. See next page.

These two Latin squares are **orthogonal;** that is, if one square is superimposed on the other, each letter in the first square will appear exactly once with each letter in the second square. The totals for the letters in the (a) square are $Q = 18$, $R = -2$, and $S = 8$, and the sum of squares between these totals is $[18^2 + (-2)^2 + 8^2]/(3)(2) - [24^2/(9)(2)] = 33.34$, with two degrees of freedom. Similarly, the letter totals in the (b) square are $Q = 0$, $R = 6$, and $S = 18$, and the sum of squares between these totals is $[0^2 + 6^2 + 18^2]/(3)(2) - [24^2/(9)(2)] = 28.00$, with two degrees of freedom. Note that the sum of these two components is

$$33.34 + 28.00 = 61.34 = SS_{AB}$$

with $2 + 2 = 4$ degrees of freedom.

In general, the sum of squares computed from square (a) is called the **AB component of interaction,** and the sum of squares computed from square (b) is

Table 10–1 Analysis of Variance for Example 6–5

Source of Variation	Sum of Squares	Degrees of Freedom		Mean Square	F_0	P-Value
Tool angle (A)	24.33	2		12.17	8.45	0.0086
(A_L)	(8.33)		1	(8.33)	5.78	0.0396
(A_Q)	(16.00)		1	(16.00)	11.11	0.0088
Cutting speed (B)	25.33	2		12.67	8.80	0.0076
(B_L)	(21.33)		1	(21.33)	14.81	0.0039
(B_Q)	(4.00)		1	(4.00)	2.77	0.1304
AB interaction	61.34	4		6.82	4.75	0.0245
Error	13.00	9		1.44		
Total	124.00					

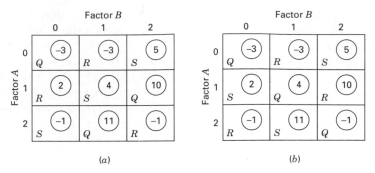

Figure 10–3. Treatment combination totals from Example 6–5 with two orthogonal Latin squares superimposed.

called the **AB^2 component of interaction.** The components AB and AB^2 each have two degrees of freedom. This terminology is used because if we denote the levels $(0, 1, 2)$ for A and B by x_1 and x_2, respectively, then we find that the letters occupy cells according to the following pattern:

Square (*a*)	Square (*b*)
$Q: x_1 + x_2 = 0 \pmod 3$	$Q: x_1 + 2x_2 = 0 \pmod 3$
$R: x_1 + x_2 = 1 \pmod 3$	$S: x_1 + 2x_2 = 1 \pmod 3$
$S: x_1 + x_2 = 2 \pmod 3$	$R: x_1 + 2x_2 = 2 \pmod 3$

For example, in square (b), note that the middle cell corresponds to $x_1 = 1$ and $x_2 = 1$; thus, $x_1 + 2x_2 = 1 + (2)(1) = 3 = 0 \pmod 3$, and Q would occupy the middle cell. When considering expressions of the form $A^p B^q$, we establish the convention that the only exponent allowed on the first letter is one. If the first letter is not one, the entire expression is squared and the exponents are reduced modulus 3. For example, $A^2 B$ is the same as AB^2 since

$$A^2 B = (A^2 B)^2 = A^4 B^2 = AB^2$$

The AB and AB^2 components of the AB interaction have no actual meaning and are usually not displayed in the analysis of variance table. However, this rather arbitrary partitioning of the AB interaction into two orthogonal two-degree-of-freedom components is very useful in constructing more complex designs. Also, there is no connection between the AB and AB^2 components of interaction and the sums of squares for $AB_{L \times L}$, $AB_{L \times Q}$, $AB_{Q \times L}$, and $AB_{Q \times Q}$.

The AB and AB^2 components of interaction may be computed another way. Consider the treatment combination totals in either square in Figure 10–3. If we add the data by diagonals downward from left to right, we obtain the totals $-3 + 4 - 1 = 0$, $-3 + 10 - 1 = 6$, and $5 + 11 + 2 = 18$. The sum of squares between these totals is $28.00 \ (AB^2)$. Similarly, the diagonal totals downward from right to left are $5 + 4 - 1 = 8$, $-3 + 2 - 1 = -2$, and $-3 + 11 + 10 = 18$.

The sum of squares between these totals is 33.34 (AB). Yates called these components of interaction the **I and J components of interaction,** respectively. We use both notations interchangeably; that is,

$$I(AB) = AB^2$$
$$J(AB) = AB$$

10–1.3 The 3^3 Design

Now suppose there are three factors (A, B, and C) under study, and each factor is at three levels arranged in a factorial experiment. This is a 3^3 factorial design, and the experimental layout and treatment combination notation were shown previously in Figure 10–2. The 27 treatment combinations have 26 degrees of freedom. Each main effect has 2 degrees of freedom, each two-factor interaction has 4 degrees of freedom, and the three-factor interaction has 8 degrees of freedom. If there are n replicates, there are $n3^3 - 1$ total degrees of freedom and $3^3(n - 1)$ degrees of freedom for error.

The sums of squares may be calculated using the standard methods for factorial designs. In addition, if the factors are quantitative and equally spaced, the main effects may be partitioned into linear and quadratic components, each with a single degree of freedom. The two-factor interactions may be decomposed into linear $\times$ linear, linear $\times$ quadratic, quadratic $\times$ linear, and quadratic $\times$ quadratic effects. Finally, the three-factor interaction ABC can be partitioned into eight single-degree-of-freedom components corresponding to linear $\times$ linear $\times$ linear, linear $\times$ linear $\times$ quadratic, and so on. Such a breakdown for the three-factor interaction is generally not very useful.

It is also possible to partition the two-factor interactions into their I and J components. These would be designated AB, AB^2, AC, AC^2, BC, and BC^2, and each component would have two degrees of freedom. As in the 3^2 design, these components have no physical significance.

The three-factor interaction ABC may be partitioned into four orthogonal two-degrees-of-freedom components, which are usually called the W, X, Y, and Z components of the interaction. They are also referred to as the AB^2C^2, AB^2C, ABC^2, and ABC components of the ABC interaction, respectively. The two notations are used interchangeably; that is,

$$W(ABC) = AB^2C^2$$
$$X(ABC) = AB^2C$$
$$Y(ABC) = ABC^2$$
$$Z(ABC) = ABC$$

Note that no first letter can have an exponent other than 1. Like the I and J components, the W, X, Y, and Z components have no practical interpretation. They are, however, useful in constructing more complex designs.

Table 10-2 Syrup Loss Data for Example 10-1 (units are cubic centimeters − 70)

Pressure (in psi) (C)	Nozzle Type (A)								
	1			2			3		
	Speed (in RPM) (B)								
	100	120	140	100	120	140	100	120	140
10	−35	−45	−40	17	−65	20	−39	−55	15
	−25	−60	15	24	−58	4	−35	−67	−30
15	110	−10	80	55	−55	110	90	−28	110
	75	30	54	120	−44	44	113	−26	135
20	4	−40	31	−23	−64	−20	−30	−61	54
	5	−30	36	−5	−62	−31	−55	−52	4

Example 10-1

A machine is used to fill 5-gallon metal containers with soft drink syrup. The variable of interest is the amount of syrup loss due to frothing. Three factors are thought to influence frothing: the nozzle design (A), the filling speed (B), and the operating pressure (C). Three nozzles, three filling speeds, and three pressures are chosen and two replicates of a 3^3 factorial experiment are run. The coded data are shown in Table 10-2.

The analysis of variance for the syrup loss data is shown in Table 10-3. The sums of squares have been computed by the usual methods. We see that the filling speed and operating pressure are statistically significant. All three two-factor interactions are also significant. The two-factor interactions are analyzed graphically in Figure 10-4. The middle level of speed gives the best performance, nozzle types 2 and 3, and either the low (10 psi) or high (20 psi) pressure seem most effective in reducing syrup loss.

∎

Table 10-3 Analysis of Variance for Syrup Loss Data

Source of Variation	Sum of Squares	Degrees of Freedom	Mean Square	F_0	P-Value
A, nozzle	993.77	2	496.89	1.17	0.3256
B, speed	61,190.33	2	30,595.17	71.74	<0.0001
C, pressure	69,105.33	2	34,552.67	81.01	<0.0001
AB	6,300.90	4	1,575.22	3.69	0.0383
AC	7,513.90	4	1,878.47	4.40	0.0222
BC	12,854.34	4	3,213.58	7.53	0.0025
ABC	4,628.76	8	578.60	1.36	0.2737
Error	11,515.50	27	426.50		
Total	174,102.83	53			

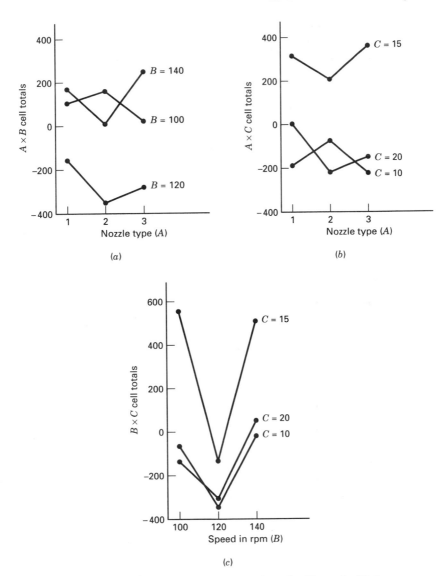

Figure 10-4.   Two-factor interactions for Example 10-1.

Example 10-1 illustrates a situation where the three-level design often finds some application; one or more of the factors is **qualitative,** naturally taking on three levels, and the remaining factors are **quantitative.** In this example, suppose that there are only three nozzle designs that are of interest. This is clearly, then, a qualitative factor that requires three levels. The filling speed and the operating pressure are quantitative factors. Therefore, we could fit a quadratic model such as Equation 10-1 in the two factors speed and pressure at each level of the nozzle factor.

Table 10–4   Regression Models for Example 10–1

Nozzle Type	x_1 = Speed (S), x_2 = Pressure (P) in coded units
1	$\hat{y} = 22.1 + 3.5x_1 + 16.3x_2 + 51.7x_1^2 - 71.8x_2^2 + 2.9x_1x_2$
	$\hat{y} = 1217.3 - 31.256S + 86.017P + 0.12917S^2 - 2.8733P^2 + 0.02875SP$
2	$\hat{y} = 25.6 - 22.8x_1 - 12.3x_2 + 14.1x_1^2 - 56.9x_2^2 - 0.7x_1x_2$
	$\hat{y} = 180.1 - 9.475S + 66.75P + 0.035S^2 - 2.2767P^2 - 0.0075SP$
3	$\hat{y} = 15.1 + 20.3x_1 + 5.9x_2 + 75.8x_1^2 - 94.9x_2^2 + 10.5x_1x_2$
	$\hat{y} = 1940.1 - 40.058S + 102.48P + 0.18958S^2 - 3.7967P^2 + 0.105SP$

Table 10–4 shows these quadratic regression models. The β's in these models were estimated using a standard linear regression computer program. (We will discuss least squares regression in more detail in Chapter 13.) In these models, the variables x_1 and x_2 are coded to the levels -1, 0, $+1$ as discussed previously, and we assumed the following natural levels for pressure and speed:

Coded Level	Speed (psi)	Pressure (RPM)
-1	100	10
0	120	15
$+1$	140	20

Table 10–4 presents models both in terms of these coded variables and in terms of the natural levels of speed and pressure.

Figure 10–5 shows the response surface contour plots of constant syrup loss as a function of speed and pressure for each nozzle type. These plots reveal considerable useful information about the performance of this filling system. Since the objective is to minimize syrup loss, nozzle type 3 would be preferred as the smallest observed contours (-60) appear only on this plot. Filling speed near the middle level of 120 RPM and either the low or high pressure levels should be used.

When constructing contour plots for an experiment that has a mixture of quantitative and qualitative factors, it is not unusual to find that the shapes of the surfaces in the quantitative factors are very different at each level of the qualitative factors. This is noticeable to some degree in Figure 10–5, where the shape of the surface for nozzle type 2 is considerably elongated in comparison to the surfaces for nozzle types 1 and 3. When this occurs, it implies that the optimum operating conditions (and other important conclusions) in terms of the quantitative factors are very different at each level of the qualitative factors.

We can easily show the numerical partitioning of the ABC interaction into its four orthogonal two-degree-of-freedom components using the data in Example 10–1. The general procedure has been described by Cochran and Cox (1957), Davies (1956), and Hicks (1973). First, select any two of the three factors, say

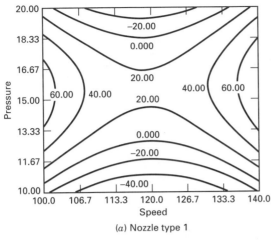

(a) Nozzle type 1

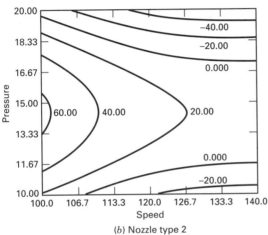

(b) Nozzle type 2

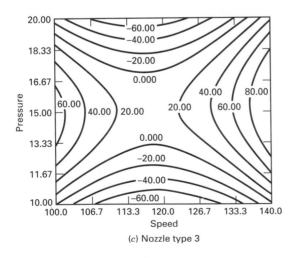

(c) Nozzle type 3

Figure 10-5. Contours of constant syrup loss (units: cc − 70) as a function of speed and pressure for nozzle types 1, 2, and 3, Example 10-1.

AB, and compute the I and J totals of the AB interaction at each level of the third factor C. These calculations follow:

C	B	A 1	A 2	A 3	Totals I	Totals J
	100	−60	41	−74	−198	−222
10	120	−105	−123	−122	−106	−79
	140	−25	24	−15	−155	−158
	100	185	175	203	331	238
15	120	20	−99	−54	255	440
	140	134	154	245	377	285
	100	9	−28	−85	−59	−144
20	120	−70	−126	−113	−74	−40
	140	67	−51	58	−206	−155

The $I(AB)$ and $J(AB)$ totals are now arranged in a two-way table with factor C, and the I and J diagonal totals of this new display are computed:

C	I(AB)			Totals I	Totals J	C	J(AB)			Totals I	Totals J
10	−198	−106	−155	−149	41	10	−222	−79	−158	63	138
15	331	255	377	212	19	15	238	440	285	62	4
20	−59	−74	−206	102	105	20	−144	−40	−155	40	23

The I and J diagonal totals computed above are actually the totals representing the quantities $I[I(AB) \times C] = AB^2C^2$, $J[I(AB) \times C] = AB^2C$, $I[J(AB) \times C] = ABC^2$, and $J[J(AB) \times C] = ABC$, or the W, X, Y, and Z components of ABC. The sums of squares are found in the usual way; that is,

$$I[I(AB) \times C] = AB^2C^2 = W(ABC)$$
$$= \frac{(-149)^2 + (212)^2 + (102)^2}{18} - \frac{(165)^2}{54} = 3804.11$$

$$J[I(AB) \times C] = AB^2C = X(ABC)$$
$$= \frac{(41)^2 + (19)^2 + (105)^2}{18} - \frac{(165)^2}{54} = 221.77$$

$$I[J(AB) \times C] = ABC^2 = Y(ABC)$$
$$= \frac{(63)^2 + (62)^2 + (40)^2}{18} - \frac{(165)^2}{54} = 18.77$$

$$J[J(AB) \times C] = ABC = Z(ABC)$$
$$= \frac{(138)^2 + (4)^2 + (23)^2}{18} - \frac{(165)^2}{54} = 584.11$$

Although this is an orthogonal partitioning of SS_{ABC}, we point out again that it is not customarily displayed in the analysis of variance table. In subsequent sections, we discuss the occasional need for the computation of one or more of these components.

10-1.4 The General 3^k Design

The concepts utilized in the 3^2 and 3^3 designs can be readily extended to the case of k factors, each at three levels, that is, to a 3^k factorial design. The usual digital notation is employed for the treatment combinations, so 0120 represents a treatment combination in a 3^4 design with A and D at the low levels, B at the intermediate level, and C at the high level. There are 3^k treatment combinations, with $3^k - 1$ degrees of freedom between them. These treatment combinations allow sums of squares to be determined for k main effects, each with two degrees of freedom; $\binom{k}{2}$ two-factor interactions, each with four degrees of freedom; . . . ; and one k-factor interaction with 2^k degrees of freedom. In general, an h-factor interaction has 2^h degrees of freedom. If there are n replicates, there are $n3^k - 1$ total degrees of freedom and $3^k(n - 1)$ degrees of freedom for error.

Sums of squares for effects and interactions are computed by the usual methods for factorial designs. Typically, three-factor and higher interactions are not broken down any further. However, any h-factor interaction has 2^{h-1} orthogonal two-degrees-of-freedom components. For example, the four-factor interaction $ABCD$ has $2^{4-1} = 8$ orthogonal two-degrees-of-freedom components, denoted by $ABCD^2$, ABC^2D, AB^2CD, $ABCD$, ABC^2D^2, AB^2C^2D, AB^2CD^2, and $AB^2C^2D^2$. In writing these components, note that the only exponent allowed on the first letter is 1. If the exponent on the first letter is not 1, then the entire expression must be squared and the exponents reduced modulus 3. To demonstrate, consider

$$A^2BCD = (A^2BCD)^2 = A^4B^2C^2D^2 = AB^2C^2D^2$$

These interaction components have no physical interpretation, but they are useful in constructing more complex designs.

The size of the design increases rapidly with k. For example, a 3^3 design has 27 treatment combinations per replication, a 3^4 design has 81, a 3^5 design has 243, and so on. Therefore, frequently only a single replicate of the 3^k design is considered, and higher-order interactions are combined to provide an estimate of error. As an illustration, if three-factor and higher interactions are negligible, then a single replicate of the 3^3 design provides 8 degrees of freedom for error, and a single replicate of the 3^4 design provides 48 degrees of freedom for error. These are still large designs for $k \geqslant 3$ factors, and, consequently, not too useful.

10–1.5   Yates' Algorithm for the 3^k Design

Yates' algorithm can be modified for use in the 3^k factorial design. We illustrate the procedure using the data in Example 6–1. The data for this example are given in Table 6–3. This is a 3^2 design used to investigate the effect of material type (A) and temperature (B) on the life of a battery. There are $n = 4$ replicates.

The procedure is displayed in Table 10–5. The treatment combinations are written down in standard order; that is, the factors are introduced one at a time, each level being combined successively with every set of factor levels above it in the table. (The standard order for a 3^3 design would be 000, 100, 200, 010, 110, 210, 020, 120, 220, 001, ...) The Response column contains the total of all observations taken under the corresponding treatment combination. The entries in column (1) are computed as follows. The first third of the column consists of the sums of each of the three sets of three values in the Response column. The second third of the column is the third minus the first observation in the same set of three. This operation computes the linear component of the effect. The last third of the column is obtained by taking the sum of the first and third value minus twice the second in each set of three observations. This computes the quadratic component. For example, in column (1), the second, fifth, and eighth entries are $229 + 479 + 583 = 1291$, $-229 + 583 = 354$, and $229 - (2)(479) + 583 = -146$, respectively. Column (2) is obtained similarly from column (1). In general, k columns must be constructed.

The Effect column is determined by converting the treatment combinations at the left of the row into corresponding effects. That is, 10 represents the linear effect of A, A_L, and 11 represents the $AB_{L \times L}$ component of the AB interaction. The entries in the Divisor column are found from

$$2^r 3^t n$$

where r is the number of factors in the effect considered, t is the number of factors in the experiment minus the number of linear terms in this effect, and n is the number of replicates. For example, B_L has the divisor $2^1 \times 3^1 \times 4 = 24$.

The sums of squares are obtained by squaring the element in column (2)

Table 10–5   Yates' Algorithm for the 3^2 Design in Example 6–1

Treatment Combination	Response	(1)	(2)	Effect	Divisor	Sum of Squares
00	539	1738	3799	—	—	—
10	623	1291	503	A_L	$2^1 \times 3^1 \times 4$	10,542.04
20	576	770	-101	A_Q	$2^1 \times 3^2 \times 4$	141.68
01	229	37	-968	B_L	$2^1 \times 3^1 \times 4$	39,042.66
11	479	354	75	$AB_{L \times L}$	$2^2 \times 3^0 \times 4$	351.56
21	583	112	307	$AB_{Q \times L}$	$2^2 \times 3^1 \times 4$	1,963.52
02	230	-131	-74	B_Q	$2^1 \times 3^2 \times 4$	76.06
12	198	-146	-559	$AB_{L \times Q}$	$2^2 \times 3^1 \times 4$	6,510.02
22	342	176	337	$AB_{Q \times Q}$	$2^2 \times 3^2 \times 4$	788.67

Table 10-6   Analysis of Variance for the 3^2 Design in Example 6-1

Source of Variation	Sum of Squares	Degrees of Freedom	Mean Square	F_0	P-Value
$A = $ "A_L" $+ $ "A_Q"	10,683.72	2	5,341.86	7.91	0.0020
B, temperature	39,118.72	2	19,558.36	28.97	<0.0001
(B_L)	(39,042.67)	1	39,042.67	57.82	<0.0001
(B_Q)	(76.05)	1	76.05	0.12	0.7314
AB	9,613.78	4	2,403.44	3.56	0.0186
$(A \times B_L = $ "$AB_{L\times L}$" $+ $ "$AB_{Q\times L}$")	(2,315.08)	2	1,157.54	1.71	0.1999
$(A \times B_Q = $ "$AB_{L\times Q}$" $+ $ "$AB_{Q\times Q}$")	(7,298.70)	2	3,649.75	5.41	0.0106
Error	18,230.75	27	675.21		
Total	77,646.97	35			

and dividing by the corresponding entry in the Divisor column. The Sum of Squares column now contains all of the required quantities to construct an analysis of variance table if both factors A and B are quantitative. However, in this example, factor A (material type) is qualitative; thus, the linear and quadratic partitioning of A is not appropriate. Individual observations are required to compute the total sum of squares, and the error sum of squares is obtained by subtraction.

The analysis of variance is summarized in Table 10-6. A comparison of this table with Table 6-18 shows that substantially the same results were obtained by conventional analysis of variance methods. Examples 6-1 and 6-4 provide an interpretation of this experiment.

10-2   CONFOUNDING IN THE 3^k FACTORIAL DESIGN

Even when a single replicate of the 3^k design is considered, the design requires so many runs that it is unlikely that all 3^k runs can be made under uniform conditions. Thus, confounding in blocks is often necessary. The 3^k design may be confounded in 3^p incomplete blocks, where $p < k$. Thus, these designs may be confounded in three blocks, nine blocks, and so on.

10-2.1   The 3^k Factorial Design in Three Blocks

Suppose that we wish to confound the 3^k design in three incomplete blocks. These three blocks have two degrees of freedom among them; thus, there must be two degrees of freedom confounded with blocks. Recall that in the 3^k factorial series each main effect has two degrees of freedom. Furthermore, every two-factor interaction has four degrees of freedom and can be decomposed into two components of interaction (e.g., AB and AB^2), each with two degrees of freedom; every three-factor interaction has eight degrees of freedom and can be decomposed into four components of interaction (e.g., ABC, ABC^2, AB^2C, and AB^2C^2), each with two degrees of freedom; and so on. Therefore, it is convenient to confound a component of interaction with blocks.

The general procedure is to construct a **defining contrast**

$$L = \alpha_1 x_1 + \alpha_2 x_2 + \cdots + \alpha_k x_k \tag{10-2}$$

where α_i represents the exponent on the ith factor in the effect to be confounded and x_i is the level of the ith factor in a particular treatment combination. For the 3^k series, we have $\alpha_i = 0$, 1, or 2 with the first nonzero α_i being unity, and $x_i = 0$ (low level), 1 (intermediate level), or 2 (high level). The treatment combinations in the 3^k design are assigned to blocks based on the value of L (mod 3). Since L (mod 3) can take on only the values 0, 1, or 2, three blocks are uniquely defined. The treatment combinations satisfying $L = 0$ (mod 3) constitute the **principal block.** This block will always contain the treatment combination $00 \ldots 0$.

For example, suppose we wish to construct a 3^2 factorial design in three blocks. Either component of the AB interaction, AB or AB^2, may be confounded with blocks. Arbitrarily choosing AB^2, we obtain the defining contrast

$$L = x_1 + 2x_2$$

The value of L (mod 3) of each treatment combination may be found as follows.

00:   $L = 1(0) + 2(0) = 0 = 0$ (mod 3)   11: $L = 1(1) + 2(1) = 3 = 0$ (mod 3)
01:   $L = 1(0) + 2(1) = 2 = 2$ (mod 3)   21: $L = 1(2) + 2(1) = 4 = 1$ (mod 3)
02:   $L = 1(0) + 2(2) = 4 = 1$ (mod 3)   12: $L = 1(1) + 2(2) = 5 = 2$ (mod 3)
10:   $L = 1(1) + 2(0) = 1 = 1$ (mod 3)   22: $L = 1(2) + 2(2) = 6 = 0$ (mod 3)
20:   $L = 1(2) + 2(0) = 2 = 2$ (mod 3)

The blocks are shown in Figure 10–6.

The elements in the principal block form a group with respect to addition modulus 3. Referring to Figure 10–6, we see that $11 + 11 = 22$, and $11 + 22 = 00$. Treatment combinations in the other two blocks may be generated by adding, modulus 3, any element in the new block to the elements of the principal block. Thus, for block 2 we use 10 and obtain

$$10 + 00 = 10 \qquad 10 + 11 = 21 \qquad \text{and} \qquad 10 + 22 = 02$$

To generate block 3, using 01 we find

$$01 + 00 = 01 \qquad 01 + 11 = 12 \qquad \text{and} \qquad 01 + 22 = 20$$

Example 10–2

We illustrate the statistical analysis of the 3^2 design confounded in three blocks by using the following data, which come from the single replicate of the 3^2 design shown

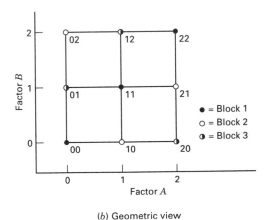

(a) Assignment of the treatment
combinations to blocks

(b) Geometric view

Figure 10-6.   The 3^2 design in three blocks with AB^2 confounded.

in Figure 10-6. Using conventional methods for the analysis of factorials, we find
that $SS_A = 131.56$ and $SS_B = 0.22$. We also find that

$$SS_{\text{Blocks}} = \frac{(0)^2 + (7)^2 + (0)^2}{3} - \frac{(7)^2}{9} = 10.89$$

Block 1	Block 2	Block 3
00 = 4	10 = −2	01 = 5
11 = −4	21 = 1	12 = −5
22 = 0	02 = 8	20 = 0

Block Totals =   0   7   0

However, SS_{Blocks} is exactly equal to the AB^2 component of interaction. To see this,
write the observations as follows:

		Factor B		
		0	1	2
	0	4	5	8
Factor A	1	−2	−4	−5
	2	0	1	0

Table 10–7 Analysis of Variance for Data in Example 10–2

Source of Variation	Sum of Squares	Degrees of Freedom
Blocks (AB^2)	10.89	2
A	131.56	2
B	0.22	2
AB	2.89	2
Total	145.56	8

Recall from Section 10–1.2 that the I or AB^2 component of the AB interaction may be found by computing the sum of squares between the left-to-right diagonal totals in the above layout. This yields

$$SS_{AB^2} = \frac{(0)^2 + (0)^2 + (7)^2}{3} - \frac{(7)^2}{9} = 10.89$$

which is identical to SS_{Blocks}.

The analysis of variance is shown in Table 10–7. Because there is only one replicate, no formal tests can be performed. It is not a good idea to use the AB component of interaction as an estimate of error.

∎

We now look at a slightly more complicated design—a 3^3 factorial confounded in three blocks of nine runs each. The AB^2C^2 component of the three-factor interaction will be confounded with blocks. The defining contrast is

$$L = x_1 + 2x_2 + 2x_3$$

It is easy to verify that the treatment combinations 000, 012, and 101 belong in the principal block. The remaining runs in the principal block are generated as follows:

(1) 000   (4) 101 + 101 = 202   (7) 101 + 021 = 122
(2) 012   (5) 012 + 012 = 021   (8) 012 + 202 = 211
(3) 101   (6) 101 + 012 = 110   (9) 021 + 202 = 220

To find the runs in another block, note that the treatment combination 200 is not in the principal block. Thus, the elements of block 2 are

(1) 200 + 000 = 200   (4) 200 + 202 = 102   (7) 200 + 122 = 022
(2) 200 + 012 = 212   (5) 200 + 021 = 221   (8) 200 + 211 = 111
(3) 200 + 101 = 001   (6) 200 + 110 = 010   (9) 200 + 220 = 120

Notice that these runs all satisfy $L = 2 \pmod 3$. The final block is found by observing that 100 does not belong in block 1 or 2. Using 100 as above yields

(1) $100 + 000 = 100$ (4) $100 + 202 = 002$ (7) $100 + 122 = 222$

(2) $100 + 012 = 112$ (5) $100 + 021 = 121$ (8) $100 + 211 = 011$

(3) $100 + 101 = 201$ (6) $100 + 110 = 210$ (9) $100 + 220 = 020$

The blocks are shown in Figure 10–7.

The analysis of variance for this design is shown in Table 10–8. Using this confounding scheme, information on all the main effects and two-factor interactions is available. The remaining components of the three-factor interaction

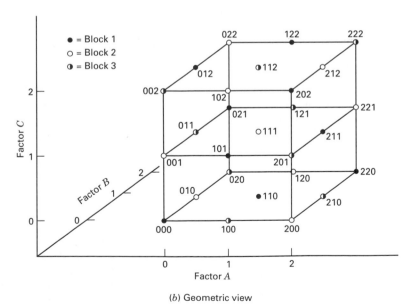

(a) Assignment of the treatment combinations to blocks

(b) Geometric view

Figure 10–7. The 3^3 design in three blocks with AB^2C^2 confounded.

Table 10–8   Analysis of Variance for a
3^3 Design with AB^2C^2 Confounded

Source of Variation	Degrees of Freedom
Blocks (AB^2C^2)	2
A	2
B	2
C	2
AB	4
AC	4
BC	4
Error $(ABC + AB^2C + ABC^2)$	6
Total	26

$(ABC, AB^2C,$ and $ABC^2)$ are combined as an estimate of error. The sum of squares for those three components could be obtained by subtraction. In general, for the 3^k design in three blocks, we would always select a component of the highest-order interaction to confound with blocks. The remaining unconfounded components of this interaction could be obtained by computing the k-factor interaction in the usual way and subtracting from this quantity the sum of squares for blocks.

10–2.2   The 3^k Factorial Design in Nine Blocks

In some experimental situations it may be necessary to confound the 3^k design in nine blocks. Thus, eight degrees of freedom will be confounded with blocks. To construct these designs, we choose *two* components of interaction and, as a result, two more will be confounded automatically, yielding the required eight degrees of freedom. These two are the generalized interactions of the two effects originally chosen. In the 3^k system, the **generalized interactions** of two effects (e.g., P and Q) are defined as PQ and PQ^2 (or P^2Q).

The two components of interaction initially chosen yield *two* defining contrasts

$$L_1 = \alpha_1 x_1 + \alpha_2 x_2 + \cdots + \alpha_k x_k = u \pmod{3} \qquad u = 0, 1, 2$$
$$L_2 = \beta_1 x_1 + \beta_2 x_2 + \cdots + \beta_k x_k = h \pmod{3} \qquad h = 0, 1, 2 \qquad (10\text{–}3)$$

where $\{\alpha_i\}$ and $\{\beta_j\}$ are the exponents in the first and second generalized interactions, respectively, with the convention that the first nonzero α_i and β_j are unity. The defining contrasts in Equation 10–3 imply nine simultaneous equations specified by the pair of values for L_1 and L_2. Treatment combinations having the same pair of values for (L_1, L_2) are assigned to the same block.

The principal block consists of treatment combinations satisfying $L_1 = L_2 = 0 \pmod{3}$. The elements of this block form a group with respect to addition

Block 1	Block 2	Block 3	Block 4	Block 5	Block 6	Block 7	Block 8	Block 9
0000	0001	2000	0200	0020	0010	1000	0100	0002
0122	0120	2122	0022	0112	0102	1122	0222	0121
0211	0212	2211	0111	0201	0221	1211	0011	0210
1021	1022	0021	0221	1011	1001	2021	1121	1020
1110	1111	0110	1010	1100	1120	2110	1210	1112
1201	1200	0202	1102	1222	1212	2202	1002	1201
2012	2010	1012	2212	2002	2022	0012	2112	2011
2101	2012	1101	2001	2121	2111	0101	2201	2100
2220	2221	1220	2120	2210	2200	0220	2020	2222

$(L_1, L_2) =$ (0,0)   (0,1)   (2,2)   (2,0)   (2,1)   (1,2)   (1,1)   (1,0)   (0,2)

Figure 10-8. The 3^4 design in nine blocks with ABC, AB^2D^2, AC^2D, and BC^2D^2 confounded.

modulus 3; thus, the scheme given in Section 10–2.1 can be used to generate the blocks.

As an example, consider the 3^4 factorial design confounded in nine blocks of nine runs each. Suppose we choose to confound ABC and AB^2D^2. Their generalized interactions

$$(ABC)(AB^2D^2) = A^2B^3CD^2 = (A^2B^3CD^2)^2 = AC^2D$$
$$(ABC)(AB^2D^2)^2 = A^3B^5CD^4 = B^2CD = (B^2CD)^2 = BC^2D^2$$

are also confounded with blocks. The defining contrasts for ABC and AB^2D^2 are

$$L_1 = x_1 + x_2 + x_3$$
$$L_2 = x_1 + 2x_2 + 2x_4$$

$$(10\text{--}4)$$

The nine blocks may be constructed by using the defining contrasts (Equation 10–4) and the group-theoretic property of the principal block. The design is shown in Figure 10–8.

For the 3^k design in nine blocks, there will be four components of interaction confounded. The remaining unconfounded components of these interactions can be determined by subtracting the sum of squares for the confounded component from the sum of squares for the entire interaction. The method described in Section 10–1.3 may be useful in computing the components of interaction.

10-2.3   The 3^k Factorial Design in 3^p Blocks

The 3^k factorial design may be confounded in 3^p blocks of 3^{k-p} observations each, where $p < k$. The procedure is to select p independent effects to be confounded with blocks. As a result, exactly $(3^p - 2p - 1)/2$ other effects are automatically confounded. These effects are the generalized interactions of those effects originally chosen.

As an illustration, consider a 3^7 design to be confounded in 27 blocks. Since $p = 3$, we would select three independent components of interaction and automatically confound $[3^3 - 2(3) - 1]/2 = 10$ others. Suppose we choose ABC^2DG, BCE^2F^2G, and $BDEFG$. Three defining contrasts can be constructed from these effects and the 27 blocks can be generated by the methods previously described. The other 10 effects confounded with blocks are

$$(ABC^2DG)(BCE^2F^2G) = AB^2DE^2F^2G^2$$
$$(ABC^2DG)(BCE^2F^2G)^2 = AB^3C^4DE^4F^4G^3 = ACDEF$$
$$(ABC^2DG)(BDEFG) = AB^2C^2D^2EFG^2$$
$$(ABC^2DG)(BDEFG)^2 = AB^3C^2D^3E^2F^2G^3 = AC^2E^2F^2$$
$$(BCE^2F^2G)(BDEFG) = B^2CDE^3F^3G^2 = BC^2D^2G$$
$$(BCE^2F^2G)(BDEFG)^2 = B^3CD^2E^4F^4G^3 = CD^2EF$$
$$(ABC^2DG)(BCE^2F^2G)(BDEFG) = AB^3C^3D^2E^3F^3G^3 = AD^2$$
$$(ABC^2DG)^2(BCE^2F^2G)(BDEFG) = A^2B^4C^5D^3G^4 = AB^2CG^2$$
$$(ABC^2DG)(BCE^2F^2G)^2(BDEFG) = ABCD^2E^2F^2G$$
$$(ABC^2DG)(BCE^2F^2G)(BDEFG)^2 = ABC^3D^3E^4F^4G^4 = ABEFG$$

This is a huge design requiring $3^7 = 2187$ observations arranged in 27 blocks of 81 observations each.

10–3   FRACTIONAL REPLICATION OF THE 3^k FACTORIAL DESIGN

The concept of fractional replication can be extended to the 3^k factorial designs. Because a complete replicate of the 3^k design can require a rather large number of runs even for moderate values of k, fractional replication of these designs is of interest. As we shall see, however, some of these designs have unattractive alias structures.

10–3.1   The One-Third Fraction of the 3^k Factorial Design

The largest fraction of the 3^k design is a one-third fraction containing 3^{k-1} runs. Consequently, we refer to this as a 3^{k-1} fractional factorial design. To construct a 3^{k-1} fractional factorial design select a two-degrees-of-freedom component of interaction (generally, the highest-order interaction) and partition the *full* 3^k design into three blocks. Each of the three resulting blocks is a 3^{k-1} fractional design, and any one of the blocks may be selected for use. If $AB^{\alpha_2}C^{\alpha_3} \cdots K^{\alpha_k}$ is the component of interaction used to define the blocks, then $I = AB^{\alpha_2}C^{\alpha_3} \cdots K^{\alpha_k}$ is called the **defining relation** of the fractional factorial design. Each main effect or component of interaction estimated from the 3^{k-1} design has two aliases, which may be found by multiplying the effect by *both* I and I^2 modulus 3.

As an example, consider a one-third fraction of the 3^3 design. We may select any component of the ABC interaction to construct the design, that is, ABC, AB^2C, ABC^2, or AB^2C^2. Thus, there are actually 12 *different* one-third fractions of the 3^3 design defined by

$$x_1 + \alpha_2 x_2 + \alpha_3 x_3 = u \text{ (mod 3)}$$

where $\alpha = 1$ or 2 and $u = 0$, 1, or 2. Suppose we select the component AB^2C^2. Each fraction of the resulting 3^{3-1} design will contain exactly $3^2 = 9$ treatment combinations that must satisfy

$$x_1 + 2x_2 + 2x_3 = u \text{ (mod 3)}$$

where $u = 0$, 1, or 2. It is easy to verify that the three one-third fractions are as shown in Figure 10–9.

Design 1 $u = 0$	Design 1 $u = 1$	Design 1 $u = 2$
000	100	200
012	112	212
101	201	001
202	002	102
021	121	221
110	210	010
122	222	022
211	011	111
220	020	120

(*a*) Treatment combinations

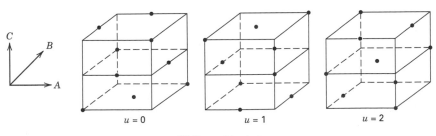

(*b*) Geometric view

Figure 10–9.   The three one-third fractions of the 3^3 design with defining relation $I = AB^2C^2$.

If any one of the 3^{3-1} designs in Figure 10–9 is run, the resulting alias structure is

$$A = A(AB^2C^2) = A^2B^2C^2 = ABC$$
$$A = A(AB^2C^2)^2 = A^3B^4C^4 = BC$$
$$B = B(AB^2C^2) = AB^3C^2 = AC^2$$
$$B = B(AB^2C^2)^2 = A^2B^5C^4 = ABC^2$$
$$C = C(AB^2C^2) = AB^2C^3 = AB^2$$
$$C = C(AB^2C^2)^2 = A^2B^4C^5 = AB^2C$$
$$AB = AB(AB^2C^2) = A^2B^3C^2 = AC$$
$$AB = AB(AB^2C^2)^2 = A^3B^5C^4 = BC^2$$

Consequently, the four effects that are actually estimated from the eight degrees of freedom in the design are $A + BC + ABC$, $B + AC^2 + ABC^2$, $C + AB^2 + AB^2C$, and $AB + AC + BC^2$. This design would be of practical value only if all the interactions were small relative to the main effects. Since the main effects are aliased with two-factor interactions, this is a resolution III design. Notice how complex the alias relationships are in this design. Each main effect is aliased with a *component* of interaction. If, for example, the two-factor interaction BC is large, this will potentially distort the estimate of the main effect of A and make the $AB + AC + BC^2$ effect very difficult to interpret. It is very difficult to see how this design could be useful unless we assume that all interactions are negligible.

Before leaving the 3^{3-1}_{III} design, note that for the design with $u = 0$ (see Figure 10–9) if we let A denote the row and B denote the column, then the design can be written as

000	012	021
101	110	122
202	211	220

which is a 3×3 Latin square. The assumption of negligible interactions required for unique interpretations of the 3^{3-1}_{III} design is paralleled in the Latin square design. However, the two designs arise from different motives, one as a consequence of fractional replication and the other from randomization restrictions. From Table 5–13 we observe that there are only twelve 3×3 Latin squares and that each one corresponds to one of the twelve different 3^{3-1} fractional factorial designs.

The treatment combinations in a 3^{k-1} design with the defining relation $I = AB^{\alpha_2}C^{\alpha_3} \cdot \cdot \cdot K^{\alpha_k}$ can be constructed using a method similar to that employed in the 2^{k-p} series. First, write down the 3^{k-1} runs for a **full** three-level factorial design in $k - 1$ factors, with the usual 0, 1, 2 notation. This is the **basic design** in the terminology of Chapter 9. Then introduce the kth factor by equating its levels x_k to the appropriate component of the highest-order interaction, say $AB^{\alpha_2}C^{\alpha_3}\cdots(K-1)^{\alpha_{k-1}}$, through the relationship

$$x_k = \beta_1 x_1 + \beta_2 x_2 + \cdots + \beta_{k-1} x_{k-1} \tag{10-5}$$

where $\beta_i = (3 - \alpha_k)\alpha_i \pmod{3}$ for $1 \leqslant i \leqslant k - 1$. This yields a design of the highest possible resolution.

As an illustration, we use this method to generate the 3_{IV}^{4-1} design with the defining relation $I = AB^2CD$ shown in Table 10-9. It is easy to verify that the first three digits of each treatment combination in this table are the 27 runs of a full 3^3 design. This is the basic design. For AB^2CD, we have $\alpha_1 = \alpha_3 = \alpha_4 = 1$ and $\alpha_2 = 2$. This implies that $\beta_1 = (3 - 1)\alpha_1 \pmod{3} = (3-1)(1) = 2$, $\beta_2 = (3 - 1)\alpha_2 \pmod{3} = (3 - 1)(2) = 4 = 1 \pmod{3}$, and $\beta_3 = (3 - 1)\alpha_3 \pmod{3} = (3 - 1)(1) = 2$. Thus, Equation 10-5 becomes

$$x_4 = 2x_1 + x_2 + 2x_3 \tag{10-6}$$

The levels of the fourth factor satisfy Equation 10-6. For example, we have $2(0) + 1(0) + 2(0) = 0$, $2(0) + 1(1) + 2(0) = 1$, $2(1) + 1(1) + 2(0) = 3 = 0$, and so on.

The resulting 3_{IV}^{4-1} design has 26 degrees of freedom that may be used to compute the sums of squares for the 13 main effects and components of interactions (and their aliases). The aliases of any effect are found in the usual manner; for example, the aliases of A are $A(AB^2CD) = ABC^2D^2$ and $A(AB^2CD)^2 = BC^2D^2$. One may verify that the four main effects are clear of any two-factor interaction components, but that some two-factor interaction components are aliased with each other. Once again, we notice the complexity of the alias structure. If any two-factor interactions are large, it will likely be very difficult to isolate them with this design.

The statistical analysis of a 3^{k-1} design is accomplished by the usual analysis of variance procedures for factorial experiments. The sums of squares for the components of interaction may be computed as in Section 10-1. Remember when interpreting results that the components of interactions have no practical interpretation.

Table 10-9 A 3_{IV}^{4-1} Design with $I = AB^2CD$

0000	0012	2221
0101	0110	0021
1100	0211	0122
1002	1011	0220
0202	1112	1020
1201	1210	1121
2001	2010	1222
2102	2111	2022
2200	2212	2120

10–3.2   Other 3^{k-p} Fractional Factorial Designs

For moderate to large values of k, even further fractionation of the 3^k design is potentially desirable. In general, we may construct a $(\frac{1}{3})^p$ fraction of the 3^k design for $p < k$, where the fraction contains 3^{k-p} runs. Such a design is called a 3^{k-p} fractional factorial design. Thus, a 3^{k-2} design is a one-ninth fraction, a 3^{k-3} design is a one–twenty-seventh fraction, and so on.

The procedure for constructing a 3^{k-p} fractional factorial design is to select p components of interaction and use these effects to partition the 3^k treatment combinations into 3^p blocks. Each block is then a 3^{k-p} fractional factorial design. The defining relation I of any fraction consists of the p effects initially chosen and their $(3^p - 2p - 1)/2$ generalized interactions. The alias of any main effect or component of interaction is produced by multiplication modulus 3 of the effect by I and I^2.

We may also generate the runs defining a 3^{k-p} fractional factorial design by first writing down the treatment combinations of a full 3^{k-p} factorial design and then introducing the additional p factors by equating them to components of interaction, as we did in Section 10–3.1.

We illustrate the procedure by constructing a 3^{4-2} design, that is, a one-ninth fraction of the 3^4 design. Let AB^2C and BCD be the two components of interaction chosen to construct the design. Their generalized interactions are (AB^2C) $(BCD) = AC^2D$ and $(AB^2C)(BCD)^2 = ABD^2$. Thus, the defining relation for this design is $I = AB^2C = BCD = AC^2D = ABD^2$, and the design is of resolution III. The nine treatment combinations in the design are found by writing down a 3^2 design in the factors A and B, and then adding two new factors by setting

$$x_3 = 2x_1 + x_2$$
$$x_4 = 2x_2 + 2x_3$$

This is equivalent to using AB^2C and BCD to partition the full 3^4 design into nine blocks and then selecting one of these blocks as the desired fraction. The complete design is shown in Table 10–10.

This design has eight degrees of freedom that may be used to estimate four main effects and their aliases. The aliases of any effect may be found by multiplying the effect modulus 3 by AB^2C, BCD, AC^2D, ABD^2, and their squares. The complete alias structure for the design is given in Table 10–11.

From the alias structure, we see that this design is useful only in the absence

Table 10–10 A 3^{4-2}_{III} Design with $I = AB^2C$ and $I = BCD$

0000	0111	0222
1021	1102	1210
2012	2120	2201

Table 10-11   Alias Structure for the 3^{4-2}_{III} Design in Table 10-10

Effect	Aliases							
			I				I^2	
A	ABC^2	$ABCD$	ACD^2	AB^2D	BC^2	$AB^2C^2D^2$	CD^2	BD^2
B	AC	BC^2D^2	ABC^2D	AB^2D^2	ABC	CD	AB^2C^2D	AD^2
$C\cdot$	AB^2C^2	BC^2D	AD	$ABCD^2$	AB^2	BD	ACD	ABC^2D^2
D	AB^2CD	BCD^2	AC^2D^2	AB	AB^2CD^2	BC	AC^2	ABD

of interaction. Furthermore, if A denotes the rows and B denotes the columns, then from examining Table 10-10 we see that the 3^{4-2}_{III} design is also a Graeco-Latin square.

The publication by Connor and Zelen (1959) contains an extensive selection of designs for $4 \leq k \leq 10$. This pamphlet was prepared for the National Bureau of Standards and is the most complete table of fractional 3^{k-p} plans available.

In this section, we have noted several times the complexity of the alias relationships in 3^{k-p} fractional factorial designs. In general, if k is moderately large, say $k \geq 4$ or 5, the size of the 3^k design will drive most experimenters to consider fairly small fractions. Unfortunately, these designs have alias relationships that involve the **partial aliasing** of two-degrees-of-freedom components of interaction. This, in turn, results in a design that will be difficult if not impossible to interpret if interactions are not negligible. Furthermore, there are no simple augmentation schemes (such as fold over) that can be used to combine two or more fractions to isolate significant interactions. The 3^k design is often suggested as appropriate when curvature is present. However, there are more efficient alternatives (see Chapter 14). For these reasons, I feel that the 3^{k-p} fractional factorial designs are solutions looking for a problem—they are not generally good designs.

10-4 FACTORIALS WITH MIXED LEVELS

We have emphasized factorial and fractional factorial designs in which all the factors have the same number of levels. The two-level system discussed in Chapters 7, 8, and 9 is particularly useful. The three-level system presented earlier in this chapter is much less useful because the designs are relatively large even for a modest number of factors, and most of the small fractions have complex alias relationships that would require very restrictive assumptions regarding interactions to be useful.

It is our belief that the two-level factorial and fractional factorial designs should be the cornerstone of industrial experimentation for product and process development, troubleshooting, and improvement. There are, however, some situations in which it is necessary to include a factor (or a few factors) that have more than two levels. This often occurs when there are both quantitative and qualitative factors in the experiment, and the qualitative factor has (say) three levels. In this section we show how some three-level and four-level factors can be accommodated in a 2^k design.

10–4.1   Factors at Two and Three Levels

Designs in which some factors have two levels and other factors have three levels can be derived from the table of plus and minus signs for the usual 2^k design. The general procedure is best illustrated with an example. Suppose we have two variables, with A at two levels and X at three levels. Consider a table of plus and minus signs for the usual eight-run 2^3 design. The signs in columns B and C have the pattern shown on the left side of Table 10–12. Let the levels of X be represented by x_1, x_2, and x_3. The right side of Table 10–12 shows how the sign patterns for B and C are combined to form the levels of the three-level factor.

Now factor X has two degrees of freedom and can generally be thought of as consisting of a linear and a quadratic component, each component having one degree of freedom. Table 10–13 shows a 2^3 design with the columns labeled to show the actual effects that they estimate, with X_L and X_Q denoting the linear and quadratic effects of X, respectively. Note that the linear effect of X is the sum of the two effect estimates computed from the columns usually associated with B and C, and that the effect of A can only be computed from the runs where X is at either the low or high levels; namely runs 1, 2, 7, and 8. Similarly, the $A \times X_L$ effect is the sum of the two effects that would be computed from the columns usually labeled AB and AC. Furthermore, note that runs 3 and 5 are replicates. Therefore, a one-degree-of-freedom estimate of error can be made using these two runs. Similarly, runs 4 and 6 are replicates, and this would lead to a second one-degree-of-freedom estimate of error. The average variance at

Table 10–12   Use of Two-Level Factors to Form a Three-Level Factor

Two-Level Factors		Three-Level Factor
B	C	X
−	−	x_1
+	−	x_2
−	+	x_2
+	+	x_3

Table 10-13   One Two-Level and One Three-Level Factor in a 2^3 Design

	A	X_L	X_L	$A \times X_L$	$A \times X_L$	X_Q	$A \times X_Q$	Actual Treatment Combinations	
Run	A	B	C	AB	AC	BC	ABC	A	X
1	−	−	−	+	+	+	−	Low	Low
2	+	−	−	−	−	+	+	High	Low
3	−	+	−	−	+	−	+	Low	Med
4	+	+	−	+	−	−	−	High	Med
5	−	−	+	+	−	−	+	Low	Med
6	+	−	+	−	+	−	−	High	Med
7	−	+	+	−	−	+	−	Low	High
8	+	+	+	+	+	+	+	High	High

these two pairs of runs could be used as a mean square for error with two degrees of freedom. The complete analysis of variance is summarized in Table 10-14.

If we are willing to assume that the two-factor and higher interactions are negligible, we can convert the design in Table 10-13 into a resolution III fraction with up to four two-level factors and a single three-level factor. This would be accomplished by associating the two-level factors with columns A, AB, AC, and ABC. Column BC cannot be used for a two-level factor because it contains the quadratic effect of the three-level factor X.

This same procedure can be applied to the 16-run, 32-run, and 64-run 2^k designs. For 16 runs, it is possible to construct resolution V fractional factorials with two two-level factors and either two or three factors at three levels. A 16-run resolution V fraction can also be obtained with three two-level factors and one three-level factor. If we include four two-level factors and a single three-level factor in 16 runs, the design will be of resolution III. The 32- and 64-run designs allow similar arrangements. For additional discussion of some of these designs, see Addleman (1962).

Table 10-14   Analysis of Variance for the Design in Table 10-13

Source of Variation	Sum of Squares	Degrees of Freedom	Mean Square
A	SS_A	1	MS_A
X $(X_L + X_Q)$	SS_X	2	MS_B
AX $(A \times X_L + A \times X_Q)$	SS_{AX}	2	MS_{AX}
Error (from runs 3 and 5 and runs 4 and 6)	SS_E	2	MS_E
Total	SS_T	7	

10–4.2   Factors at Two and Four Levels

It is very easy to accommodate a four-level factor in a 2^k design. The procedure for doing this involves using two two-level factors to represent the four-level factor. For example, suppose that A is a four-level factor with levels a_1, a_2, a_3, and a_4. Consider two columns of the usual table of plus and minus signs, say columns P and Q. The pattern of signs in these two columns is as shown on the left side of Table 10–15. The right side of this table shows how these four sign patterns would correspond to the four levels of factor A. The effects represented by columns P and Q and the PQ interaction are mutually orthogonal and correspond to the three-degrees-of-freedom A effect.

To illustrate this idea more completely, suppose that we have one four-level factor and two two-level factors and that we need to estimate all the main effects and interactions involving these factors. This can be done with a 16-run design. Table 10–16 shows the usual table of plus and minus signs for the 16-run 2^4 design, with columns A and B used to form the four-level factor, say X, with levels x_1, x_2, x_3, and x_4. Sums of squares would be calculated for each column $A, B, \ldots, ABCD$ just as in the usual 2^k system. Then the sums of squares for all factors X, C, D, and their interactions are formed as follows:

$$SS_X = SS_A + SS_B + SS_{AB} \qquad \text{(3 degrees of freedom)}$$

$$SS_C = SS_C \qquad \text{(1 degree of freedom)}$$

$$SS_D = SS_D \qquad \text{(1 degree of freedom)}$$

$$SS_{CD} = SS_{CD} \qquad \text{(1 degree of freedom)}$$

$$SS_{XC} = SS_{AC} + SS_{BC} + SS_{ABC} \qquad \text{(3 degrees of freedom)}$$

$$SS_{XD} = SS_{AD} + SS_{BD} + SS_{ABD} \qquad \text{(3 degrees of freedom)}$$

$$SS_{XCD} = SS_{ACD} + SS_{BCD} + SS_{ABCD} \qquad \text{(3 degrees of freedom)}$$

This could be called a 4×2^2 design. If we are willing to ignore two-factor interactions, up to nine additional two-level factors can be associated with the two-factor interaction (except AB), three-factor interaction, and four-factor interaction columns.

Table 10–15   Four-Level Factor A Expressed as Two Two-Level Factors

	Two-Level Factors		Four-Level Factor
Run	P	Q	A
1	$-$	$-$	a_1
2	$+$	$-$	a_2
3	$-$	$+$	a_3
4	$+$	$+$	a_4

Table 10–16 A Single Four-Level Factor and Two Two-Level Factors in 16 Runs

Run	(A	B)	= X	C	D	AB	AC	BC	ABC	AD	BD	ABD	CD	ACD	BCD	ABCD
1	−	−	x_1	−	−	+	+	+	−	+	+	−	+	−	−	+
2	+	−	x_2	−	−	−	−	+	+	−	+	+	+	+	−	−
3	−	+	x_3	−	−	−	+	−	+	+	−	+	+	−	+	−
4	+	+	x_4	−	−	+	−	−	−	−	−	−	+	+	+	+
5	−	−	x_1	+	−	+	−	−	+	+	+	−	−	+	+	−
6	+	−	x_2	+	−	−	+	−	−	−	+	+	−	−	+	+
7	−	+	x_3	+	−	−	−	+	−	+	−	+	−	+	−	+
8	+	+	x_4	+	−	+	+	+	+	−	−	−	−	−	−	−
9	−	−	x_1	−	+	+	+	+	−	−	−	+	−	+	+	−
10	+	−	x_2	−	+	−	−	+	+	+	−	−	−	−	+	+
11	−	+	x_3	−	+	−	+	−	+	−	+	−	−	+	−	+
12	+	+	x_4	−	+	+	−	−	−	+	+	+	−	−	−	−
13	−	−	x_1	+	+	+	−	−	+	−	−	+	+	−	−	+
14	+	−	x_2	+	+	−	+	−	−	+	−	−	+	+	−	−
15	−	+	x_3	+	+	−	−	+	−	−	+	−	+	−	+	−
16	+	+	x_4	+	+	+	+	+	+	+	+	+	+	+	+	+

10-5 PROBLEMS

10-1 The effects of developer strength (A) and development time (B) on the density of photographic plate film are being studied. Three strengths and three times are used, and four replicates of a 3^2 factorial experiment are run. The data from this experiment follow. Analyze the data using the standard methods for factorial experiments.

Developer Strength	Development Time (minutes)					
	10		14		18	
1	0	2	1	3	2	5
	5	4	4	2	4	6
2	4	6	6	8	9	10
	7	5	7	7	8	5
3	7	10	10	10	12	10
	8	7	8	7	9	8

10-2 Analyze the data in Problem 10-1 using Yates' algorithm.

10-3 Compute the I and J components of the two-factor interaction in Problem 10-1.

10-4 An experiment was performed to study the effect of three different types of 32-ounce bottles (A) and three different shelf types (B)—smooth permanent shelves, end-aisle displays with grilled shelves, and beverage coolers—on the time it takes to stock ten 12-bottle cases on the shelves. Three workers (factor C) were employed in the experiment, and two replicates of a 3^3 factorial design were run. The observed time data are shown in the following table. Analyze the data and draw conclusions.

Worker	Bottle Type	Replicate I			Replicate II		
		Permanent	End Aisle	Cooler	Permanent	End Aisle	Cooler
	Plastic	3.45	4.14	5.80	3.36	4.19	5.23
1	28-mm glass	4.07	4.38	5.48	3.52	4.26	4.85
	38-mm glass	4.20	4.26	5.67	3.68	4.37	5.58
	Plastic	4.80	5.22	6.21	4.40	4.70	5.88
2	28-mm glass	4.52	5.15	6.25	4.44	4.65	6.20
	38-mm glass	4.96	5.17	6.03	4.39	4.75	6.38
	Plastic	4.08	3.94	5.14	3.65	4.08	4.49
3	28-mm glass	4.30	4.53	4.99	4.04	4.08	4.59
	38-mm glass	4.17	4.86	4.85	3.88	4.48	4.90

10-5 A medical researcher is studying the effect of lidocaine on the enzyme level in the heart muscle of beagle dogs. Three different commercial brands of lidocaine

(A), three dosage levels (B), and three dogs (C) are used in the experiment, and two replicates of a 3^3 factorial design are run. The observed enzyme levels follow. Analyze the data from this experiment.

Lidocaine Brand	Dosage Strength	Replicate I			Replicate II		
		Dog			Dog		
		1	2	3	1	2	3
1	1	86	84	85	84	85	86
	2	94	99	98	95	97	90
	3	101	106	98	105	104	103
2	1	85	84	86	80	82	84
	2	95	98	97	93	99	95
	3	108	114	109	110	102	100
3	1	84	83	81	83	80	79
	2	95	97	93	92	96	93
	3	105	100	106	102	111	108

10-6 Compute the *I* and *J* components of the two-factor interactions for Example 10–1.

10-7 An experiment is run in a chemical process using a 3^2 factorial design. The design factors are temperature and pressure, and the response variable is yield. The data that result from this experiment are shown below.

Temperature, °C	Pressure, psig		
	100	120	140
80	47.58, 48.77	64.97, 69.22	80.92, 72.60
90	51.86, 82.43	88.47, 84.23	93.95, 88.54
100	71.18, 92.77	96.57, 88.72	76.58, 83.04

(a) Analyze the data from this experiment by conducting an analysis of variance. What conclusions can you draw?

(b) Graphically analyze the residuals. Are there any concerns about underlying assumptions or model adequacy?

(c) Verify that if we let the low, medium, and high levels of both factors in this design take on the levels −1, 0, and +1, then a least squares fit to a second-order model for yield is

$$\hat{y} = 86.81 + 10.4x_1 + 8.42x_2 - 7.17x_1^2 - 7.86x_2^2 - 7.69x_1x_2$$

(d) Confirm that the model in part (c) can be written in terms of the natural variables temperature (T) and pressure (P) as

$$\hat{y} = -1335.63 + 18.56T + 8.59P - 0.072T^2 - 0.0196P^2 - 0.0384TP$$

    (e) Construct a contour plot for yield as a function of pressure and temperature. Based on examination of this plot, where would you recommend running this process?

10–8   (a) Confound a 3^3 design in three blocks using the ABC^2 component of the three-factor interaction. Compare your results with the design in Figure 10–7.

    (b) Confound a 3^3 design in three blocks using the AB^2C component of the three-factor interaction. Compare your results with the design in Figure 10–7.

    (c) Confound a 3^3 design in three blocks using the ABC component of the three-factor interaction. Compare your results with the design in Figure 10–7.

    (d) After looking at the designs in parts (a), (b), and (c) and Figure 10–7, what conclusions can you draw?

10–9   Confound a 3^4 design in three blocks using the AB^2CD component of the four-factor interaction.

10–10   Consider the data from the first replicate of Problem 10–4. Assuming that all 27 observations could not be run on the same day, set up a design for conducting the experiment over three days with AB^2C confounded with blocks. Analyze the data.

10–11   Outline the analysis of variance table for the 3^4 design in nine blocks. Is this a practical design?

10–12   Consider the data in Problem 10–4. If ABC is confounded in replicate I and ABC^2 is confounded in replicate II, perform the analysis of variance.

10–13   Consider the data from replicate I of Problem 10–4. Suppose that only a one-third fraction of this design with $I = ABC$ is run. Construct the design, determine the alias structure, and analyze the data.

10–14   From examining Figure 10–9, what type of design would remain if after completing the first 9 runs, one of the three factors could be dropped?

10–15   Construct a 3_{IV}^{4-1} design with $I = ABCD$. Write out the alias structure for this design.

10–16   Verify that the design in Problem 10–15 is a resolution IV design.

10–17   Construct a 3^{5-2} design with $I = ABC$ and $I = CDE$. Write out the alias structure for this design. What is the resolution of this design?

10–18   Construct a 3^{9-6} design, and verify that it is a resolution III design.

10–19   Construct a 4×2^3 design confounded in two blocks of 16 observations each. How could Yates' algorithm be employed to analyze this design?

10–20   Outline the analysis of variance table for a $2^2 3^2$ factorial design. Discuss how this design may be confounded in blocks.

10–21   Starting with a 16-run 2^4 design, show how two three-level factors can be incorporated in this experiment. How many two-level factors can be included if we want some information on two-factor interactions?

10–22   Starting with a 16-run 2^4 design, show how one three-level factor and three two-level factors can be accommodated and still allow the estimation of two-factor interactions.

10-23 In Problem 9–26, you met Harry and Judy Peterson-Nedry, two friends of the author who have a winery and vineyard in Newberg, Oregon. That problem described the application of two-level fractional factorial designs to their 1985 Pinot Noir product. In 1987, they wanted to conduct another Pinot Noir experiment. The variables for this experiment were

Variable	Levels
Clone of Pinot Noir	Wadenswil, Pommard
Berry size	Small, large
Fermentation temperature	80°F, 85°F, 90/80°F, 90°F
Whole berry	None, 10%
Maceration time	10 days, 21 days
Yeast type	Assmanhau, Champagne
Oak type	Tronçais, Allier

Harry and Judy decided to use a 16-run two-level fractional factorial design, treating the four levels of fermentation temperature as two two-level variables. As in Problem 9–26, they used the rankings from a taste-test panel as the response variable. The design and the resulting average ranks are shown below:

Run	Clone	Berry Size	Ferm. Temp.	Whole Berry	Macer. Time	Yeast Type	Oak Type	Average Rank
1	−	−	−	−	−	−	−	−	4
2	+	−	−	−	−	+	+	+	10
3	−	+	−	−	+	−	+	+	6
4	+	+	−	−	+	+	−	−	9
5	−	−	+	−	+	+	+	−	11
6	+	−	+	−	+	−	−	+	1
7	−	+	+	−	−	+	−	+	15
8	+	+	+	−	−	−	+	−	5
9	−	−	−	+	+	+	−	+	12
10	+	−	−	+	+	−	+	−	2
11	−	+	−	+	−	+	+	−	16
12	+	+	−	+	−	−	−	+	3
13	−	−	+	+	−	−	+	+	8
14	+	−	+	+	−	+	−	−	14
15	−	+	+	+	+	−	−	−	7
16	+	+	+	+	+	+	+	+	13

(a) Describe the aliasing in this design.

(b) Analyze the data and draw conclusions.

(c) What comparisons can you make between this experiment and the 1985 Pinot Noir experiment from Problem 9–26?

Chapter 11
Factorial Experiments with Random Factors

Throughout most of this book we have assumed that the factors in an experiment were **fixed factors;** that is, the levels of the factors used by the experimenter were the specific levels of interest. The implication of this, of course, is that the statistical inferences made about these factors is confined to the specific levels studied. That is, if three material types are investigated as in the battery life experiment of Example 6–1, then our conclusions are valid only about those specific material types. A variation of this occurs when the factor or factors are **quantitative.** In these situations, we often use a regression model relating the response to the factors to predict the response over the region spanned by the factor levels used in the experimental design. Several examples of this were presented in Chapters 6 through 10. In general, with a fixed effect, we say that the **inference space** of the experiment is the specific set of factor levels investigated.

In some experimental situations, the factor levels are chosen at random from a larger population of possible levels, and the experimenter wishes to draw conclusions about the entire population of levels, not just those that were used in the experimental design. In this situation the factor is said to be a **random factor.** We briefly introduced the concept of a random factor and the **random effects** or **components of variance** model for the analysis of variance in Chapter 3. There we focused on experiments with a single factor; however, random factors occur regularly in factorial experiments and other types of experiments as well. In this chapter we will introduce methods for the design and analysis of factorial experiments with random factors. In Chapter 12 we will present nested and split-plot designs, two situations where random factors are frequently encountered in practice.

470

11–1   THE TWO-FACTOR FACTORIAL WITH RANDOM FACTORS

Suppose that we have two factors, A and B, and that both factors have a large number of levels that are of interest (as in Chapter 3, we will assume that the number of levels is infinite). We will choose at random a levels of factor A and b levels of factor B and arrange these factor levels in a factorial experimental design. If the experiment is replicated n times, then we may represent the observations by the linear model

$$y_{ijk} = \mu + \tau_i + \beta_j + (\tau\beta)_{ij} + \epsilon_{ijk} \qquad \begin{cases} i = 1, 2, \ldots, a \\ j = 1, 2, \ldots, b \\ k = 1, 2, \ldots, n \end{cases} \qquad (11\text{--}1)$$

where the model parameters τ_i, β_j, $(\tau\beta)_{ij}$, and ϵ_{ijk} are all independent random variables. We are also going to assume that the random variables τ_i, β_j, $(\tau\beta)_{ij}$, and ϵ_{ijk} are normally distributed with mean zero and variances given by $V(\tau_i) = \sigma_\tau^2$, $V(\beta_j) = \sigma_\beta^2$, $V[(\tau\beta)_{ij}] = \sigma_{\tau\beta}^2$, and $V(\epsilon_{ijk}) = \sigma^2$. Therefore the variance of any observation is

$$V(y_{ijk}) = \sigma_\tau^2 + \sigma_\beta^2 + \sigma_{\tau\beta}^2 + \sigma^2 \qquad (11\text{--}2)$$

and σ_τ^2, σ_β^2, $\sigma_{\tau\beta}^2$, and σ^2 are called **variance components.** The hypotheses that we are interested in testing are $H_0: \sigma_\tau^2 = 0$, $H_0: \sigma_\beta^2 = 0$, and $H_0: \sigma_{\tau\beta}^2 = 0$. Notice the similarity to the single-factor random effects model.

The numerical calculations in the analysis of variance remains unchanged; that is, SS_A, SS_B, SS_{AB}, SS_T, and SS_E are all calculated as in the fixed effects case. However, to form the test statistics, we must examine the **expected mean squares.** It may be shown that

$$E(MS_A) = \sigma^2 + n\sigma_{\tau\beta}^2 + bn\sigma_\tau^2$$
$$E(MS_B) = \sigma^2 + n\sigma_{\tau\beta}^2 + an\sigma_\beta^2 \qquad (11\text{--}3)$$
$$E(MS_{AB}) = \sigma^2 + n\sigma_{\tau\beta}^2$$

and

$$E(MS_E) = \sigma^2$$

From the expected mean squares, we see that the appropriate statistic for testing the no-interaction hypothesis $H_0: \sigma_{\tau\beta}^2 = 0$ is

$$F_0 = \frac{MS_{AB}}{MS_E} \qquad (11\text{--}4)$$

because under H_0 both numerator and denominator of F_0 have expectation σ^2, and only if H_0 is false is $E(MS_{AB})$ greater than $E(MS_E)$. The ratio F_0 is distributed as $F_{(a-1)(b-1),ab(n-1)}$. Similarly, for testing $H_0: \sigma_\tau^2 = 0$ we would use

$$F_0 = \frac{MS_A}{MS_{AB}} \tag{11-5}$$

which is distributed as $F_{a-1,(a-1)(b-1)}$, and for testing $H_0: \sigma_\beta^2 = 0$ the statistic is

$$F_0 = \frac{MS_B}{MS_{AB}} \tag{11-6}$$

which is distributed as $F_{b-1,(a-1)(b-1)}$. These are all upper-tail, one-tail tests. Notice that these test statistics are not the same as those used if both factors A and B are fixed. The expected mean squares are always used as a guide to test statistic construction.

In many experiments involving random factors, interest centers at least as much on estimating the variance components as on hypothesis testing. The variance components may be estimated by the **analysis of variance method,** that is, by equating the observed mean squares in the lines of the analysis of variance table to their expected values and solving for the variance components. This yields

$$\hat{\sigma}^2 = MS_E$$
$$\hat{\sigma}_{\tau\beta}^2 = \frac{MS_{AB} - MS_E}{n}$$
$$\hat{\sigma}_\beta^2 = \frac{MS_B - MS_{AB}}{an} \tag{11-7}$$
$$\hat{\sigma}_\tau^2 = \frac{MS_A - MS_{AB}}{bn}$$

as the point estimates of the variance components in the two-factor random effects model. We will discuss other methods for obtaining point estimates of the variance components and procedures for constructing confidence intervals in Section 11–6.

Example 11–1

A Measurement Systems Capability Study

Statistically designed experiments are frequently used to investigate the sources of variability that impact a system. A common industrial application is to use a designed experiment to study the components of variability in a measurement system. These studies are often called **gauge capability studies** or **gauge repeatability and reproducibility (R&R) studies,** because these are the components of variability that are of interest.

A typical gauge R&R experiment [from Montgomery (1991)] is shown in Table 11–1. An instrument or gauge is used to measure a critical dimension on a part. Twenty parts have been selected from the production process, and three randomly selected operators measure each part twice with this gauge. The order in which the measurements are made is completely randomized, so this is a two-factor factorial

Table 11-1 The Repeatability and Reproducibility Experiment in Example 11-1

Part Number	Operator 1		Operator 2		Operator 3	
1	21	20	20	20	19	21
2	24	23	24	24	23	24
3	20	21	19	21	20	22
4	27	27	28	26	27	28
5	19	18	19	18	18	21
6	23	21	24	21	23	22
7	22	21	22	24	22	20
8	19	17	18	20	19	18
9	24	23	25	23	24	24
10	25	23	26	25	24	25
11	21	20	20	20	21	20
12	18	19	17	19	18	19
13	23	25	25	25	25	25
14	24	24	23	25	24	25
15	29	30	30	28	31	30
16	26	26	25	26	25	27
17	20	20	19	20	20	20
18	19	21	19	19	21	23
19	25	26	25	24	25	25
20	19	19	18	17	19	17

experiment with design factors parts and operators, with two replications. Both parts and operators are random factors. The variance component identity in Equation 11-1 applies; namely,

$$\sigma_y^2 = \sigma_\tau^2 + \sigma_\beta^2 + \sigma_{\tau\beta}^2 + \sigma^2$$

where σ_y^2 is the total variability (including variability due to the different parts, variability due to the different operators, and variability due to the gauge), σ_τ^2 is the variance component for parts, σ_β^2 is the variance component for operators, $\sigma_{\tau\beta}^2$ is the variance component that represents interaction between parts and operators, and σ^2 is the random experimental error. Typically, the variance component σ^2 is called the gauge repeatability, because σ^2 can be thought of as reflecting the variation observed when the same part is measured by the same operator, and

$$\sigma_\beta^2 + \sigma_{\tau\beta}^2$$

is usually called the reproducibility of the gauge, because it reflects the additional variability in the measurement system resulting from use of the instrument by the operator. These experiments are usually performed with the objective of estimating the variance components.

Table 11-2 shows the analysis of variance for this experiment. The computations were performed using Statgraphics. Based on the P-values, we conclude that the effect of parts is large, operators may have a small effect, and that there is no

Table 11–2   Analysis of Variance (from Statgraphics) for the Two-Factor Random Model in Example 11–1

Source of Variation	Sum of Squares	d.f.	Mean Square	F-Ratio	Sig. Level
MAIN EFFECTS					
A:part	1185.4250	19	62.390789	87.64(1)	.0000
B:operator	2.6167	2	1.308333	1.83(1)	.1730
INTERACTIONS					
AB	27.050000	38	.7118421	.71(0)	.8614
RESIDUAL	59.500000	60	.9916667		
TOTAL (CORRECTED)	1274.5917	119			

0 missing values have been excluded.
F-ratios are based on the following mean squares:
  (0)RESIDUAL
  (1)AB

significant part–operator interaction. We may use Equation 11–7 to estimate the variance components as follows:

$$\hat{\sigma}_\tau^2 = \frac{62.39 - 0.71}{(3)(2)} = 10.28$$

$$\hat{\sigma}_\beta^2 = \frac{1.31 - 0.71}{(20)(2)} = 0.015$$

$$\hat{\sigma}_{\tau\beta}^2 = \frac{0.71 - 0.99}{2} = -0.14$$

and

$$\hat{\sigma}^2 = 0.99$$

Notice that the estimate of one of the variance components, $\hat{\sigma}_{\tau\beta}^2$, is negative. This is certainly not reasonable, since by definition variances are nonnegative. Unfortunately, negative estimates of variance components can result when we use the analysis of variance method of estimation (this is considered one of its drawbacks). There are a variety of ways to deal with this. One possibility is to assume that the negative estimate means that the variance component is really zero and just set it to zero, leaving the other nonnegative estimates unchanged. Another approach is to estimate the variance components with a method that assures nonnegative estimates (we will discuss this briefly in Section 11–6). Finally, we could note that the P-value for the interaction term in Table 11–2 is very large, take this as evidence that $\sigma_{\tau\beta}^2$ really is zero, that there is no interaction effect, and fit a **reduced model** of the form

$$y_{ijk} = \mu + \tau_i + \beta_j + \epsilon_{ijk}$$

that does not include the interaction term. This is a relatively easy approach and one that often works nearly as well as more sophisticated methods.

Table 11–3 shows the analysis of variance for the reduced model. Since there is no interaction term in the model, both main effects are tested against the error

Table 11–3 Analysis of Variance (from Statgraphics) for the Reduced Model, Example 11–1

Source of Variation	Sum of Squares	d.f.	Mean Square	F-Ratio	Sig. Level
MAIN EFFECTS					
A:part	1185.4250	19	62.390789	70.645	.0000
B:operator	2.6167	2	1.308333	1.481	.2324
RESIDUAL	86.550000	98	.8831633		
TOTAL (CORRECTED)	1274.5917	119			

0 missing values have been excluded.
All F-ratios are based on the residual mean square error.

term, and the estimates of the variance components are

$$\hat{\sigma}_\tau^2 = \frac{62.39 - 0.88}{(3)(2)} = 10.25$$

$$\hat{\sigma}_\beta^2 = \frac{1.31 - 0.88}{(20)(2)} = 0.0108$$

$$\hat{\sigma}^2 = 0.88$$

Finally, we could estimate the variance of the gauge as the sum of the variance component estimates $\hat{\sigma}^2$ and $\hat{\sigma}_\beta^2$ as

$$\hat{\sigma}_{\text{gauge}}^2 = \hat{\sigma}^2 + \hat{\sigma}_\beta^2$$

$$= 0.88 + 0.0108$$

$$= 0.8908$$

The variability in the gauge appears small relative to the variability in the product. This is generally a desirable situation, implying that the gauge is capable of distinguishing among different grades of product. ∎

11–2 THE TWO-FACTOR MIXED MODEL

We now consider the situation where one of the factors A is fixed and the other B is random. This is called the **mixed model** analysis of variance. The linear statistical model is

$$y_{ijk} = \mu + \tau_i + \beta_j + (\tau\beta)_{ij} + \epsilon_{ijk} \qquad \begin{cases} i = 1, 2, \ldots, a \\ j = 1, 2, \ldots, b \\ k = 1, 2, \ldots, n \end{cases} \qquad (11-8)$$

Here τ_i is a fixed effect, β_j is a random effect, the interaction $(\tau\beta)_{ij}$ is assumed to be a random effect, and ϵ_{ijk} is a random error. We also assume that the $\{\tau_i\}$

are fixed effects such that $\sum_{i=1}^{a} \tau_i = 0$ and β_j is a NID$(0, \sigma_\beta^2)$ random variable. The interaction effect, $(\tau\beta)_{ij}$, is a normal random variable with mean 0 and variance $[(a-1)/a]\sigma_{\tau\beta}^2$; however, summing the interaction component over the fixed factor equals zero. That is,

$$\sum_{i=1}^{a} (\tau\beta)_{ij} = (\tau\beta)_{.j} = 0 \qquad j = 1, 2, \ldots, b$$

This restriction implies that certain interaction elements at different levels of the fixed factor are not independent. In fact, we may show (see Problem 11–16) that

$$\text{Cov}[(\tau\beta)_{ij}, (\tau\beta)_{i'j}] = -\frac{1}{a}\sigma_{\tau\beta}^2 \qquad i \neq i'$$

The covariance between $(\tau\beta)_{ij}$ and $(\tau\beta)_{ij'}$ for $j \neq j'$ is zero, and the random error ϵ_{ijk} is NID$(0, \sigma^2)$. Because the sum of the interaction effects over the levels of the fixed factor equals zero, this version of the mixed model is often called the **restricted model.**

In this model the variance of $(\tau\beta)_{ij}$ is defined as $[(a-1)/a]\sigma_{\tau\beta}^2$ rather than $\sigma_{\tau\beta}^2$ in order to simplify the expected mean squares. The assumption $(\tau\beta)_{.j} = 0$ also has an effect on the expected mean squares, which we may show are

$$E(MS_A) = \sigma^2 + n\sigma_{\tau\beta}^2 + \frac{bn\sum_{i=1}^{a}\tau_i^2}{a-1}$$

$$E(MS_B) = \sigma^2 + an\sigma_\beta^2 \qquad\qquad (11\text{--}9)$$

$$E(MS_{AB}) = \sigma^2 + n\sigma_{\tau\beta}^2$$

and

$$E(MS_E) = \sigma^2$$

Therefore, the appropriate test statistic for testing that the means of the fixed factor effects are equal, or $H_0: \tau_i = 0$, is

$$F_0 = \frac{MS_A}{MS_{AB}}$$

for which the reference distribution is $F_{a-1,(a-1)(b-1)}$. For testing $H_0: \sigma_\beta^2 = 0$, the test statistic is

$$F_0 = \frac{MS_B}{MS_E}$$

with reference distribution $F_{b-1,ab(n-1)}$. Finally, for testing the interaction hypothesis $H_0: \sigma_{\tau\beta}^2 = 0$, we would use

$$F_0 = \frac{MS_{AB}}{MS_E}$$

which has reference distribution $F_{(a-1)(b-1),ab(n-1)}$.

In the mixed model, it is possible to estimate the fixed factor effects as

$$\hat{\mu} = \bar{y}...$$
$$\hat{\tau}_i = \bar{y}_{i..} - \bar{y}... \qquad i = 1, 2, \ldots, a \tag{11-10}$$

The variance components σ_β^2, $\sigma_{\tau\beta}^2$, and σ^2 may be estimated using the analysis of variance method. Eliminating the first equation from Equations 11–9 leaves three equations in three unknowns, whose solutions are

$$\hat{\sigma}_\beta^2 = \frac{MS_B - MS_E}{an}$$

$$\hat{\sigma}_{\tau\beta}^2 = \frac{MS_{AB} - MS_E}{n} \tag{11-11}$$

and

$$\hat{\sigma}^2 = MS_E$$

This general approach can be used to estimate the variance components in *any* mixed model. After eliminating the mean squares containing fixed factors, there will always be a set of equations remaining that can be solved for the variance components.

In mixed models the experimenter may be interested in testing hypotheses or constructing confidence intervals about individual treatment means for the fixed factor. In using such procedures, care must be exercised to use the proper standard error of the treatment mean. The standard error of the fixed effect treatment mean is

$$\left[\frac{\text{Mean square for testing the fixed effect}}{\text{Number of observations in each treatment mean}} \right]^{1/2} = \sqrt{\frac{MS_{AB}}{bn}}$$

Example 11–2

The Measurement Systems Capability Experiment Revisited

Reconsider the gauge R&R experiment described in Example 11–1. Suppose now that there are only three operators that use this gauge, so the operators are a fixed factor. However, since the parts are chosen at random, the experiment now involves a mixed model.

The analysis of variance for the mixed model is shown in Table 11–4. The computations were performed using Statgraphics. The conclusions are similar to Example 11–1. The variance components may be estimated from Equation (11–11) as

$$\hat{\sigma}_{\text{Parts}}^2 = \frac{MS_{\text{Parts}} - MS_E}{an} = \frac{62.39 - 0.99}{(3)(2)} = 10.23$$

$$\hat{\sigma}_{\text{Parts} \times \text{operators}}^2 = \frac{MS_{\text{Parts} \times \text{operators}} - MS_E}{n} = \frac{0.71 - 0.99}{2} = -0.14$$

$$\hat{\sigma}^2 = MS_E = 0.99$$

Once again, a negative estimate of the interaction variance component results. An appropriate course of action would be to fit a reduced model, as we did in Example

Table 11–4 Analysis of Variance (from Statgraphics) for the Mixed Model in Example 11–2

Source of Variation	Sum of Squares	d.f.	Mean Square	F-Ratio	Sig. Level
MAIN EFFECTS					
A:part	1185.4250	19	62.390789	87.64(1)	.0000
B:operator	2.6167	2	1.308333	1.31(0)	.2750
INTERACTIONS					
AB	27.050000	38	.7118421	.71(0)	.8614
RESIDUAL	59.500000	60	.9916667		
TOTAL (CORRECTED)	1274.5917	119			

```
0 missing values have been excluded.
F-ratios are based on the following mean squares:
  (0)RESIDUAL
  (1)AB
```

11–1. In the case of a mixed model with two factors, this leads to the same results as in Example 11–1.

■

Alternate Mixed Models ▪ Several different versions of the mixed model have been proposed. These models differ from the restricted version of the mixed model discussed above in the assumptions made about the random components. One of these alternate models is now briefly discussed.

Consider the model

$$y_{ijk} = \mu + \alpha_i + \gamma_j + (\alpha\gamma)_{ij} + \epsilon_{ijk}$$

where the α_i $(i = 1, 2, \ldots, a)$ are fixed effects such that $\Sigma_{i=1}^{a} \alpha_i = 0$, and γ_j, $(\alpha\gamma)_{ij}$, and ϵ_{ijk} are uncorrelated random variables having zero means and variances $V(\gamma_j) = \sigma_\gamma^2$, $V[(\alpha\gamma)_{ij}] = \sigma_{\alpha\gamma}^2$, and $V(\epsilon_{ijk}) = \sigma^2$. Note that the restriction imposed previously on the interaction effect is not used here; consequently, this version of the mixed model is often called the **unrestricted mixed model.**

The expected mean squares for this model are

$$E(MS_A) = \sigma^2 + n\sigma_{\alpha\gamma}^2 + \frac{bn \sum_{i=1}^{a} \alpha_i^2}{a - 1}$$

$$E(MS_B) = \sigma^2 + n\sigma_{\alpha\gamma}^2 + an\sigma_\gamma^2 \qquad (11\text{–}12)$$

$$E(MS_{AB}) = \sigma^2 + n\sigma_{\alpha\gamma}^2$$

and

$$E(MS_E) = \sigma^2$$

Comparing these expected mean squares with those in Equation 11–9, we note that the only obvious difference is the presence of the variance component

$\sigma_{\alpha\gamma}^2$ in the expected mean square for the random effect. (Actually, there are other differences because of the different definitions of the variance of the interaction effect in the two models.) Consequently, we would test the hypothesis that the variance component for the random effect equals zero ($H_0: \sigma_\gamma^2 = 0$) using the statistic

$$F_0 = \frac{MS_B}{MS_{AB}}$$

as contrasted with testing $H_0: \sigma_\beta^2 = 0$ with $F_0 = MS_B/MS_E$ in the restricted model. The test should be more conservative using this model since MS_{AB} will generally be larger than MS_E.

The parameters in the two models are closely related. In fact, we may show that

$$\beta_j = \gamma_j + \overline{(\alpha\gamma)}_{.j}$$
$$(\tau\beta)_{ij} = (\alpha\gamma)_{ij} + \overline{(\alpha\gamma)}_{.j}$$
$$\sigma_\gamma^2 = \sigma_\beta^2 + \frac{1}{a}\sigma_{\alpha\gamma}^2$$

and

$$\sigma_{\tau\beta}^2 = \sigma_{\alpha\gamma}^2$$

The analysis of variance method may be used to estimate the variance components. Referring to the expected mean squares, we find that the only change from Equations 11–11 is that

$$\hat{\sigma}_\gamma^2 = \frac{MS_B - MS_{AB}}{an}$$

Both of these models are special cases of the mixed model proposed by Scheffé (1956a, 1959). This model assumes that the observations may be represented by

$$y_{ijk} = m_{ij} + \epsilon_{ijk} \qquad \begin{cases} i = 1, 2, \ldots, a \\ j = 1, 2, \ldots, b \\ k = 1, 2, \ldots, n \end{cases}$$

where m_{ij} and ϵ_{ijk} are independent random variables. The structure of m_{ij} is

$$m_{ij} = \mu + \tau_i + b_j + c_{ij}$$
$$E(m_{ij}) = \mu + \tau_i$$
$$\sum_{i=1}^{a} \tau_i = 0$$

and

$$c_{.j} = 0 \qquad j = 1, 2, \ldots, b$$

The variances and covariances of b_j and c_{ij} are expressed through the covariances of the m_{ij}. Furthermore, the random effects parameters in other formulations of the mixed model can be related to b_j and c_{ij}. The statistical analysis of Scheffé's model is identical to that for our restricted model, except that in general the statistic MS_A/MS_{AB} is not always distributed as F when $H_0 : \tau_i = 0$ is true.

In light of this multiplicity of mixed models, a logical question is which model should one use? Most statisticians tend to prefer the restricted model, and it is the most widely encountered in the literature. The restricted model is actually slightly more general than the unrestricted model, because in the restricted model the covariance between two observations from the same level of the random factor can be either positive or negative, while this covariance can only be positive in the unrestricted model. If the correlative structure of the random components is not large, then either mixed model is appropriate, and there are only minor differences between these models. When we subsequently refer to mixed models, we assume the restricted model structure. However, if there are large correlations in the data, then Scheffé's model may have to be employed. The choice of model should always be dictated by the data. The article by Hocking (1973) is a clear summary of various mixed models.

11-3   USE OF OPERATING CHARACTERISTIC CURVES IN MODELS WITH RANDOM FACTORS

The operating characteristic curves in the Appendix may be used for the analysis of variance of the two-factor random effects model and the mixed model. Appendix Chart VI is used for the random effects model. The parameter λ, numerator degrees of freedom, and denominator degrees of freedom are shown in the top half of Table 11–5. For the mixed model, both Charts V and VI in the Appendix must be used. The appropriate values for Φ^2 and λ are shown in the bottom half of Table 11–5.

11-4   RULES FOR EXPECTED MEAN SQUARES

An important part of any experimental design problem is conducting the analysis of variance. This involves determining the sum of squares for each component in the model and the number of degrees of freedom associated with each sum of squares. Then, to construct appropriate test statistics, the expected mean squares must be determined. In complex design situations, particularly those involving random or mixed models, it is frequently helpful to have a formal procedure for this process.

We will present a set of rules for writing down the expected mean squares

Table 11–5 Operating Characteristic Curve Parameters for Tables V and VI of the Appendix for the Two-Factor Random Effects and Mixed Models

	The Random Effects Model		
Factor	λ	Numerator Degrees of Freedom	Denominator Degrees of Freedom
A	$\sqrt{1 + \dfrac{bn\sigma_\tau^2}{\sigma^2 + n\sigma_{\tau\beta}^2}}$	$a - 1$	$(a - 1)(b - 1)$
B	$\sqrt{1 + \dfrac{an\sigma_\beta^2}{\sigma^2 + n\sigma_{\tau\beta}^2}}$	$b - 1$	$(a - 1)(b - 1)$
AB	$\sqrt{1 + \dfrac{n\sigma_{\tau\beta}^2}{\sigma^2}}$	$(a - 1)(b - 1)$	$ab(n - 1)$

	The Mixed Model			
Factor	Parameter	Numerator Degrees of Freedom	Denominator Degrees of Freedom	Appendix Chart
A (Fixed)	$\Phi^2 = \dfrac{bn\sum\limits_{i=1}^{a}\tau_i^2}{a[\sigma^2 + n\sigma_{\tau\beta}^2]}$	$a - 1$	$(a - 1)(b - 1)$	V
B (Random)	$\lambda = \sqrt{1 + \dfrac{an\sigma_\beta^2}{\sigma^2}}$	$b - 1$	$ab(n - 1)$	VI
AB	$\lambda = \sqrt{1 + \dfrac{n\sigma_{\tau\beta}^2}{\sigma^2}}$	$(a - 1)(b - 1)$	$ab(n - 1)$	VI

for any balanced factorial, nested,[1] or nested factorial experiment. (Note that partially balanced arrangements, such as Latin squares and incomplete block designs, are specifically excluded.) These rules are discussed by several authors, including Scheffé (1959), Bennett and Franklin (1954), Hicks (1973), and Searle (1971a, 1971b). By examining the expected mean squares, one may develop the appropriate statistic for testing hypotheses about any model parameter. The test statistic is a ratio of mean squares that is chosen such that the expected value of the **numerator** mean square differs from the expected value of the **denominator** mean square only by the variance component or the fixed factor in which we are interested.

It is always possible to determine the expected mean squares in any model as we did in Chapter 3; that is, by the direct application of the expectation operator. This "brute force" method, as it is often called, can be very tedious. The rules that follow always produce the expected mean squares without resorting to the brute force approach, and, with practice, they become relatively simple

---

[1] Nested designs are discussed in Chapter 12.

to use. When applied to a mixed model, these rules produce expected mean squares that are consistent with the assumptions for the restricted mixed model of Section 11–2. We illustrate the rules using the two-factor fixed effects factorial model.

RULE 1. The error term in the model, $\epsilon_{ij\ldots m}$, is written as $\epsilon_{(ij\ldots)m}$, where the subscript m denotes the replication subscript. For the two-factor model, this rule implies that ϵ_{ijk} becomes $\epsilon_{(ij)k}$.

RULE 2. In addition to an overall mean (μ) and an error term $[\epsilon_{(ij\ldots)m}]$, the model contains all the main effects and any interactions that the experimenter assumes exist. If all possible interactions between k factors exist, then there are $\binom{k}{2}$ two-factor interactions, $\binom{k}{3}$ three-factor interactions, . . . , 1 k-factor interaction. If one of the factors in a term appears in parentheses, then there is no interaction between that factor and the other factors in that term.

RULE 3. For each term in the model, divide the subscripts into three classes: (a) live—those subscripts that are present in the term and are not in parentheses; (b) dead—those subscripts that are present in the term and are in parentheses; and (c) absent—those subscripts that are present in the model but not in that particular term.

   Thus, in $(\tau\beta)_{ij}$, i and j are live and k is absent, and in $\epsilon_{(ij)k}$, k is live and i and j are dead.

RULE 4. Degrees of freedom. The number of degrees of freedom for any term in the model is the product of the number of levels associated with each dead subscript and the number of levels minus one associated with each live subscript.

   For example, the number of degrees of freedom associated with $(\tau\beta)_{ij}$ is $(a - 1)(b - 1)$, and the number of degrees of freedom associated with $\epsilon_{(ij)k}$ is $ab(n - 1)$.

RULE 5. Each term in the model has either a variance component (random effect) or a fixed factor (fixed effect) associated with it. If an interaction contains at least one random effect, the entire interaction is considered as random. A variance component has Greek letters as subscripts to identify the particular random effect. Thus, in a two-factor mixed model with factor A fixed and factor B random, the variance component for B is σ_β^2, and the variance component for AB is $\sigma_{\tau\beta}^2$. A fixed effect is always represented by the sum of squares of the model components associated with that factor divided by its degrees of freedom. In our example, the effect of A is

$$\frac{\sum\limits_{i=1}^{a} \tau_i^2}{a - 1}$$

RULE 6. Expected mean squares. To obtain the expected mean squares, prepare the following table. There is a row for each model component (mean square) and a column for each subscript. Over each subscript, write the number of levels of the factor associated with that subscript and whether the factor is fixed (F) or random (R). Replicates are always considered to be random.

(a) In each row, write 1 if one of the dead subscripts in the row component matches the subscript in the column:

Factor	F a i	F b j	R n k
τ_i			
β_j			
$(\tau\beta)_{ij}$			
$\epsilon_{(ij)k}$	1	1	

(b) In each row, if any of the subscripts on the row component match the subscript in the column, write 0 if the column is headed by a fixed factor and 1 if the column is headed by a random factor:

Factor	F a i	F b j	R n k
τ_i	0		
β_j		0	
$(\tau\beta)_{ij}$	0	0	
$\epsilon_{(ij)k}$	1	1	1

(c) In the remaining empty row positions, write the number of levels shown above the column heading:

Factor	F a i	F b j	R n k
τ_i	0	b	n
β_j	a	0	n
$(\tau\beta)_{ij}$	0	0	n
$\epsilon_{(ij)k}$	1	1	1

Table 11–6   Expected Mean Square Derivation, Two-Factor Fixed Effects Model

Factor	F a i	F b j	R n k	Expected Mean Square
τ_i	0	b	n	$\sigma^2 + \dfrac{bn\Sigma\tau_i^2}{a-1}$
β_j	a	0	n	$\sigma^2 + \dfrac{an\Sigma\beta_j^2}{b-1}$
$(\tau\beta)_{ij}$	0	0	n	$\sigma^2 + \dfrac{n\Sigma\Sigma(\tau\beta)_{ij}^2}{(a-1)(b-1)}$
$\epsilon_{(ij)k}$	1	1	1	σ^2

(d) To obtain the expected mean square for any model component, first cover all columns headed by live subscripts on that component. Then, in each row that contains *at least* the *same* subscripts as those on the component being considered, take the product of the visible numbers and multiply by the appropriate fixed or random factor from rule 1. The sum of these quantities is the expected mean square of the model component being considered. To find $E(MS_A)$, for example, cover column i. The product of the visible numbers in the rows that contain at least subscript i are bn (row 1), 0 (row 3), and 1 (row 4). Note that i is missing in row 2. Therefore, the expected mean square is

$$E(MS_A) = \sigma^2 + \frac{bn \sum_{i=1}^{a} \tau_i^2}{a - 1}$$

The complete table of expected mean squares for this design is shown in Table 11–6. Tables 11–7 and 11–8 display the expected mean square derivations for the two-factor random and mixed models, respectively. A three-factor factorial design is treated in the following example.

Table 11–7   Expected Mean Square Derivation, Two-Factor Random Effects Model

Factor	R a i	R b j	R n k	Expected Mean Square
τ_i	1	b	n	$\sigma^2 + n\sigma_{\tau\beta}^2 + bn\sigma_\tau^2$
β_j	a	1	n	$\sigma^2 + n\sigma_{\tau\beta}^2 + an\sigma_\beta^2$
$(\tau\beta)_{ij}$	1	1	n	$\sigma^2 + n\sigma_{\tau\beta}^2$
$\epsilon_{(ij)k}$	1	1	1	σ^2

Table 11-8 Expected Mean Square Derivation, Two-Factor Mixed Model

Factor	F a i	R b j	R n k	Expected Mean Square
τ_i	0	b	n	$\sigma^2 + n\sigma_{\tau\beta}^2 + \dfrac{bn\Sigma\tau_i^2}{a-1}$
β_j	a	1	n	$\sigma^2 + an\sigma_\beta^2$
$(\tau\beta)_{ij}$	0	1	n	$\sigma^2 + n\sigma_{\tau\beta}^2$
$\epsilon_{(ij)k}$	1	1	1	σ^2

Example 11-3

Consider a three-factor factorial experiment with a levels of factor A, b levels of factor B, c levels of factor C, and n replicates. The analysis of this design, assuming that all the factors are fixed effects, is given in Section 6-4. We now determine the expected mean squares assuming that all the factors are *random*. The appropriate statistical model is

$$y_{ijkl} = \mu + \tau_i + \beta_j + \gamma_k + (\tau\beta)_{ij} + (\tau\gamma)_{ik} + (\beta\gamma)_{jk} + (\tau\beta\gamma)_{ijk} + \epsilon_{ijkl}$$

Using the rules previously described, the expected mean squares are derived in Table 11-9.

We notice, by examining the expected mean squares in Table 11-9, that if A, B, and C are all random factors, then no exact test exists for the main effects. That is, if we wish to test the hypothesis $\sigma_\tau^2 = 0$, we cannot form a ratio of two expected mean squares such that the only term in the numerator that is not in the denominator is $bcn\sigma_\tau^2$. The same phenomenon occurs for the main effects of B and C. Notice that proper tests do exist for the two-factor and three-factor interactions. However, it is likely that tests on the main effects are of central importance to the experimenter. Therefore, how should the main effects be tested? This problem is considered in the next section.

∎

Table 11-9 Expected Mean Square Derivation, Three-Factor Random Effects Model

Factor	R a i	R b j	R c k	R n l	Expected Mean Squares
τ_i	1	b	c	n	$\sigma^2 + cn\sigma_{\tau\beta}^2 + bn\sigma_{\tau\gamma}^2 + n\sigma_{\tau\beta\gamma}^2 + bcn\sigma_\tau^2$
β_j	a	1	c	n	$\sigma^2 + cn\sigma_{\tau\beta}^2 + an\sigma_{\beta\gamma}^2 + n\sigma_{\tau\beta\gamma}^2 + acn\sigma_\beta^2$
γ_k	a	b	1	n	$\sigma^2 + bn\sigma_{\tau\gamma}^2 + an\sigma_{\beta\gamma}^2 + n\sigma_{\tau\beta\gamma}^2 + abn\sigma_\gamma^2$
$(\tau\beta)_{ij}$	1	1	c	n	$\sigma^2 + n\sigma_{\tau\beta\gamma}^2 + cn\sigma_{\tau\beta}^2$
$(\tau\gamma)_{ik}$	1	b	1	n	$\sigma^2 + n\sigma_{\tau\beta\gamma}^2 + bn\sigma_{\tau\gamma}^2$
$(\beta\gamma)_{jk}$	a	1	1	n	$\sigma^2 + n\sigma_{\tau\beta\gamma}^2 + an\sigma_{\beta\gamma}^2$
$(\tau\beta\gamma)_{ijk}$	1	1	1	n	$\sigma^2 + n\sigma_{\tau\beta\gamma}^2$
ϵ_{ijkl}	1	1	1	1	σ^2

11–5   APPROXIMATE F TESTS

In factorial experiments with three or more factors involving a random or mixed model and certain other, more complex designs, there are frequently no exact test statistics for certain effects in the models. One possible solution to this dilemma is to assume that certain interactions are negligible. To illustrate, if we could reasonably assume that all the two-factor interactions in Example 11–3 are negligible, then we could put $\sigma_{\tau\beta}^2 = \sigma_{\tau\gamma}^2 = \sigma_{\beta\gamma}^2 = 0$, and tests for main effects could be conducted.

Although this seems to be an attractive possibility, we must point out that there must be something in the nature of the process—or some strong prior knowledge—in order for us to assume that one or more of the interactions are negligible. In general, this assumption is not easily made, nor should it be taken lightly. We should not eliminate certain interactions from the model without conclusive evidence that it is appropriate to do so. A procedure advocated by some experimenters is to test the interactions first, then set at zero those interactions found to be insignificant, and then assume that these interactions are zero when testing other effects in the same experiment. Although sometimes done in practice, this procedure can be dangerous because any decision regarding an interaction is subject to both type I and type II errors.

A variation of this idea is to **pool** certain mean squares in the analysis of variance to obtain an estimate of error with more degrees of freedom. For instance, suppose that in Example 11–3 the test statistic $F_0 = MS_{ABC}/MS_E$ was not significant. Thus, $H_0: \sigma_{\tau\beta\gamma}^2 = 0$ is not rejected, and *both* MS_{ABC} and MS_E estimate the error variance σ^2. The experimenter might consider pooling or combining MS_{ABC} and MS_E according to

$$MS_{E'} = \frac{abc(n-1)MS_E + (a-1)(b-1)(c-1)MS_{ABC}}{abc(n-1) + (a-1)(b-1)(c-1)}$$

so that $E(MS_{E'}) = \sigma^2$. Note that $MS_{E'}$ has $abc(n-1) + (a-1)(b-1)(c-1)$ degrees of freedom, compared to $abc(n-1)$ degrees of freedom for the original MS_E.

The danger of pooling is that one may make a type II error and combine the mean square for a factor that really *is* significant with error, thus obtaining a new residual mean square $(MS_{E'})$ that is too large. This will make other significant effects more difficult to detect. On the other hand, if the original error mean square has a very small number of degrees of freedom (e.g., less than 6), then the experimenter may have much to gain by pooling because it could potentially increase the precision of further tests considerably. A reasonably practical procedure is as follows. If the original error mean square has 6 or more degrees of freedom, do not pool. If the original error mean square has fewer than 6 degrees of freedom, pool only if the F statistic for the mean square to be pooled is not significant at a large value of α, such as $\alpha = 0.25$.

If we cannot assume that certain interactions are negligible and we still need to make inferences about those effects for which exact tests do not exist, a procedure attributed to Satterthwaite (1946) can be employed. Satterthwaite's method uses **linear combinations of mean squares,** for example,

$$MS' = MS_r + \cdots + MS_s \tag{11-13}$$

and

$$MS'' = MS_u + \cdots + MS_v \tag{11-14}$$

where the mean squares in Equations 11–13 and 11–14 are chosen so that $E(MS') - E(MS'')$ is equal to a multiple of the effect (the model parameter or variance component) considered in the null hypothesis. Then the test statistic would be

$$F = \frac{MS'}{MS''} \tag{11-15}$$

which is distributed approximately as $F_{p,q}$, where

$$p = \frac{(MS_r + \cdots + MS_s)^2}{MS_r^2/f_r + \cdots + MS_s^2/f_s} \tag{11-16}$$

and

$$q = \frac{(MS_u + \cdots + MS_v)^2}{MS_u^2/f_u + \cdots + MS_v^2/f_v} \tag{11-17}$$

In p and q, f_i is the number of degrees of freedom associated with the mean square MS_i. There is no assurance that p and q will be integers, so it will be necessary to interpolate in the tables of the F distribution. For example, in the three-factor random effects model (Table 11–9), it is relatively easy to see that an appropriate test statistic for $H_0: \sigma_\tau^2 = 0$ would be $F = MS'/MS''$, with

$$MS' = MS_A + MS_{ABC}$$

and

$$MS'' = MS_{AB} + MS_{AC}$$

The degrees of freedom for F would be computed from Equations 11–16 and 11–17.

The theory underlying this test is that both the numerator and the denominator of the test statistic (Equation 11–15) are distributed approximately as multiples of chi-square random variables, and because no mean square appears in both the numerator or denominator of Equation 11–15, the numerator and denominator are independent. Thus F in Equation 11–15 is distributed approximately as $F_{p,q}$. Satterthwaite remarks that caution should be used in applying the procedure when some of the mean squares in MS' and MS'' are involved negatively. Gaylor and Hopper (1969) report that if $MS' = MS_1 - MS_2$, then

Satterthwaite's approximation holds reasonably well if

$$\frac{MS_1}{MS_2} > F_{0.025, f_2, f_1} \times F_{0.50, f_2, f_2}$$

if $f_1 \leq 100$ and $f_2 \geq f_1/2$.

Example 11–4

The pressure drop measured across an expansion valve in a turbine is being studied. The design engineer considers the important variables that influence pressure drop reading to be gas temperature on the inlet side (A), operator (B), and the specific pressure gauge used by the operator (C). These three factors are arranged in a factorial design, with gas temperature fixed, and operator and pressure gauge random. The coded data for two replicates are shown in Table 11–10. The linear model for this design is

$$y_{ijkl} = \mu + \tau_i + \beta_j + \gamma_k + (\tau\beta)_{ij} + (\tau\gamma)_{ik} + (\beta\gamma)_{jk} + (\tau\beta\gamma)_{ijk} + \epsilon_{ijkl}$$

where τ_i is the effect of the gas temperature (A), β_j is the operator effect (B), and γ_k is the effect of the pressure gauge (C).

The analysis of variance is shown in Table 11–11. A column entitled "Expected Mean Squares" has been added to this table, and the entries in this column are derived by the methods discussed in Section 11–4. From the Expected Mean Squares column, we observe that exact tests exist for all effects except the main effect A. Results for these tests are shown in Table 11–11. To test the gas temperature effect, or $H_0: \tau_i = 0$, we could use the statistic

$$F = \frac{MS'}{MS''}$$

where

$$MS' = MS_A + MS_{ABC}$$

Table 11–10   Coded Pressure Drop Data for the Turbine Experiment

Pressure Gauge (C)	Gas Temperature (A)											
	60°F				75°F				90°F			
	Operator (B)				Operator (B)				Operator (B)			
	1	2	3	4	1	2	3	4	1	2	3	4
1	−2	0	−1	4	14	6	1	−7	−8	−2	−1	−2
	−3	−9	−8	4	14	0	2	6	−8	20	−2	1
2	−6	−5	−8	−3	22	8	6	−5	−8	1	−9	−8
	4	−1	−2	−7	24	6	2	2	3	−7	−8	3
3	−1	−4	0	−2	20	2	3	−5	−2	−1	−4	1
	−2	−8	−7	4	16	0	0	−1	−1	−2	−7	3

Table 11-11 Analysis of Variance for the Pressure Drop Data

Source of Variation	Sum of Squares	Degrees of Freedom	Expected Mean Squares	Mean Square	F_0	P-Value
Temperature, A	616.78	2	$\sigma^2 + bn\sigma_{\tau\gamma}^2 + cn\sigma_{\tau\beta}^2 + n\sigma_{\tau\beta\gamma}^2 + \dfrac{bcn\sum\tau_i^2}{a-1}$	308.39	1.82	0.24
Operator, B	175.56	3	$\sigma^2 + an\sigma_{\beta\gamma}^2 + acn\sigma_{\beta}^2$	58.52	1.45	0.32
Pressure gauge, C	5.03	2	$\sigma^2 + an\sigma_{\beta\gamma}^2 + abn\sigma_{\gamma}^2$	2.52	0.06	0.94
AB	809.44	6	$\sigma^2 + n\sigma_{\tau\beta\gamma}^2 + cn\sigma_{\tau\beta}^2$	134.91	7.00	2.21×10^{-3}
AC	179.06	4	$\sigma^2 + n\sigma_{\tau\beta\gamma}^2 + bn\sigma_{\tau\gamma}^2$	44.77	2.32	0.12
BC	242.19	6	$\sigma^2 + an\sigma_{\beta\gamma}^2$	40.37	1.16	0.35
ABC	231.07	12	$\sigma^2 + n\sigma_{\tau\beta\gamma}^2$	19.26	0.56	0.86
Error	1248.00	36	σ^2	34.67		
Total	3507.11	71				

and

$$MS'' = MS_{AB} + MS_{AC}$$

since

$$E(MS') - E(MS'') = \frac{bcn \sum \tau_i^2}{a - 1}$$

To determine the test statistic for $H_0: \tau_i = 0$, we compute

$$MS' = MS_A + MS_{ABC}$$
$$= 308.39 + 19.26 = 327.65$$
$$MS'' = MS_{AB} + MS_{AC}$$
$$= 134.91 + 44.77 = 179.68$$

and

$$F = \frac{MS'}{MS''} = \frac{327.65}{179.68} = 1.82$$

The degrees of freedom for this statistic are found from Equations 11–16 and 11–17 as follows:

$$p = \frac{(MS_A + MS_{ABC})^2}{MS_A^2/2 + MS_{ABC}^2/12}$$

$$= \frac{(327.65)^2}{(308.39)^2/2 + (19.26)^2/12} = 2.26 \simeq 2$$

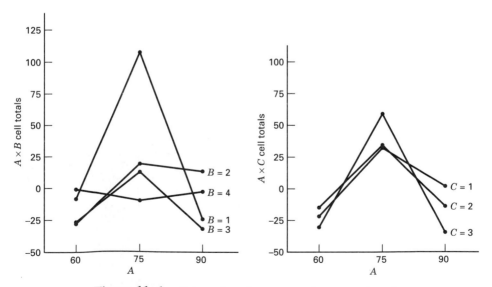

Figure 11–1.   Interactions in pressure drop experiment.

and

$$q = \frac{(MS_{AB} + MS_{AC})^2}{MS_{AB}^2/6 + MS_{AC}^2/4}$$

$$= \frac{(179.68)^2}{(134.91)^2/6 + (44.77)^2/4} = 6.34 \simeq 6$$

Comparing $F = 1.82$ to $F_{0.05,2,6} = 5.14$, we cannot reject H_0. The P-value is approximately $P = 0.24$.

The AB, or temperature–operator, interaction is large, and there is some indication of an AC, or temperature–gauge, interaction. The graphical analysis of the AB and AC interactions, shown in Figure 11–1, indicates that the effect of temperature may be large when operator 1 and gauge 3 are used. Thus, it seems possible that the main effects of temperature and operator are masked by the large AB interaction. ∎

11–6   SOME ADDITIONAL TOPICS ON ESTIMATION OF VARIANCE COMPONENTS

As we have previously observed, estimating the variance components in a random or mixed model is frequently a subject of considerable importance to the experimenter. In this section, we present some further results and techniques useful in estimating variance components. We concentrate on procedures for finding confidence intervals on variance components, and we also illustrate how to find maximum likelihood estimates of variance components. The maximum likelihood method may be a useful alternative when the analysis of variance method produces negative estimates.

11–6.1   Approximate Confidence Intervals on Variance Components

When the random effects model was first introduced in Chapter 3, we presented exact $100(1 - \alpha)$ percent confidence intervals for σ^2 and for other functions of the variance components in that simple experimental design. It is always possible to find an exact confidence interval on any function of the variance components that is the expected value of one of the mean squares in the analysis of variance. For example, consider the error mean square. Since $E(MS_E) = \sigma^2$, we can always find an exact confidence interval on σ^2 because the quantity

$$f_E MS_E / \sigma^2 = f_E \hat{\sigma}^2 / \sigma^2$$

has a chi-square distribution with f_E degrees of freedom. The exact $100(1 - \alpha)$ percent confidence interval is

$$\frac{f_E MS_E}{\chi^2_{\alpha/2, f_E}} \leq \sigma^2 \leq \frac{f_E MS_E}{\chi^2_{1-\alpha/2, f_E}} \tag{11-18}$$

Unfortunately, in more complex experiments involving several design factors, it is generally not possible to find exact confidence intervals on the variance components of interest, because these variances are not the expected value of a single mean square from the analysis of variance. However, the concepts underlying Satterthwaite's approximate "pseudo" F tests, introduced in Section 11-5, can be employed to construct approximate confidence intervals on variance components for which no exact confidence interval is available.

Recall that Satterthwaite's method uses two linear combinations of mean squares

$$MS' = MS_r + \cdots + MS_s$$

and

$$MS'' = MS_u + \cdots + MS_v$$

with the test statistic

$$F = \frac{MS'}{MS''}$$

having an approximate F distribution. Using appropriate degrees of freedom for MS' and MS'', defined in Equations 11–16 and 11–17, this F statistic can be used in an approximate test of the significance of the parameter or variance component of interest.

For testing the significance of a variance component, say σ_0^2, the two linear combinations, MS' and MS'', are chosen such that the difference in their expected values is equal to a multiple of the component, say

$$E(MS') - E(MS'') = k\sigma_0^2$$

or

$$\sigma_0^2 = \frac{E(MS') - E(MS'')}{k}. \tag{11-19}$$

Equation 11–19 provides a basis for a point estimate of σ_0^2:

$$\begin{aligned}
\hat{\sigma}_0^2 &= \frac{MS' - MS''}{k} \\
&= \frac{1}{k} MS' - \frac{1}{k} MS'' \tag{11-20} \\
&= \frac{1}{k} MS_r + \cdots + \frac{1}{k} MS_s - \frac{1}{k} MS_u - \cdots - \frac{1}{k} MS_v
\end{aligned}$$

The mean squares (MS_i) in Equation 11–20 are independent with $f_i MS_i/\sigma_i^2 = SS_i/\sigma_i^2$ having chi-square distributions with f_i degrees of freedom. The estimate of the variance component, $\hat{\sigma}_0^2$, is a linear combination of multiples of the mean squares and $r\hat{\sigma}_0^2/\sigma_0^2$ has an approximate chi-square distribution with r degrees of freedom, where

$$
\begin{aligned}
r &= \frac{(\hat{\sigma}_0^2)^2}{\displaystyle\sum_{i=1}^{m} \frac{1}{k^2} \frac{MS_i^2}{f_i}} \\[2em]
&= \frac{(MS_r + \cdots + MS_s - MS_u - \cdots - MS_v)^2}{\dfrac{MS_r^2}{f_r} + \cdots + \dfrac{MS_s^2}{f_s} + \dfrac{MS_u^2}{f_u} + \cdots + \dfrac{MS_v^2}{f_v}}
\end{aligned}
\tag{11–21}
$$

This result can only be used if $\hat{\sigma}_0^2 > 0$. As r will usually not be an integer, interpolation from the chi-square tables will generally be required. Graybill (1961) derives a general result for r.

Now since $r\hat{\sigma}_0^2/\sigma^2$ has an approximate chi-square distribution with r degrees of freedom, then

$$
P\left\{\chi_{1-\alpha/2,r}^2 \leq \frac{r\hat{\sigma}_0^2}{\sigma_0^2} \leq \chi_{\alpha/2,r}^2\right\} = 1 - \alpha
$$

and

$$
P\left\{\frac{r\hat{\sigma}_0^2}{\chi_{\alpha/2,r}^2} \leq \sigma_0^2 \leq \frac{r\hat{\sigma}_0^2}{\chi_{1-\alpha/2,r}^2}\right\} = 1 - \alpha
$$

Therefore an approximate $100(1 - \alpha)$ percent confidence interval on σ_0^2 is

$$
\frac{r\hat{\sigma}_0^2}{\chi_{\alpha/2,r}^2} \leq \sigma_0^2 \leq \frac{r\hat{\sigma}_0^2}{\chi_{1-\alpha/2,r}^2}
\tag{11–22}
$$

Example 11–5

To illustrate this procedure, reconsider the experiment in Example 11–4, where a three-factor mixed model is used on a study of the pressure drop across an expansion valve of a turbine. The model is

$$
y_{ijkl} = \mu + \tau_i + \beta_j + \gamma_k + (\tau\beta)_{ij} + (\tau\gamma)_{ik} + (\beta\gamma)_{jk} + (\tau\beta\gamma)_{ijk} + \epsilon_{ijkl}
$$

where τ_i is a fixed effect and all other effects are random. We will find an approximate confidence interval on $\sigma_{\tau\beta}^2$. Using the expected mean squares in Table 11–11, we note that the difference in the expected values of the mean squares for the two-way interaction effect AB and the three-way interaction effect ABC is a multiple of the variance component of interest, $\sigma_{\tau\beta}^2$:

$$
\begin{aligned}
E(MS_{AB}) - E(MS_{ABC}) &= \sigma^2 + n\sigma_{\tau\beta\gamma}^2 + cn\sigma_{\tau\beta}^2 - (\sigma^2 + n\sigma_{\tau\beta\gamma}^2) \\
&= cn\sigma_{\tau\beta}^2
\end{aligned}
$$

Therefore, the point estimate of $\sigma_{\tau\beta}^2$ is

$$\hat{\sigma}_{\tau\beta}^2 = \frac{MS_{AB} - MS_{ABC}}{cn} = \frac{134.91 - 19.26}{(3)(2)} = 19.28$$

and

$$r = \frac{(MS_{AB} - MS_{ABC})^2}{\dfrac{MS_{AB}^2}{(a-1)(b-1)} + \dfrac{MS_{ABC}^2}{(a-1)(b-1)(c-1)}} = \frac{(134.91 - 19.26)^2}{\dfrac{(134.91)^2}{(2)(3)} + \dfrac{(19.26)^2}{(2)(3)(2)}} = 4.36$$

The approximate 95 percent confidence interval on $\sigma_{\tau\beta}^2$ is then found from Equation 11–22 as follows:

$$\frac{r\hat{\sigma}_{\tau\beta}^2}{\chi_{0.025,r}^2} \le \sigma_{\tau\beta}^2 \le \frac{r\hat{\sigma}_{\tau\beta}^2}{\chi_{0.975,r}^2}$$

$$\frac{(4.36)(19.28)}{11.58} \le \sigma_{\tau\beta}^2 \le \frac{(4.36)(19.28)}{0.61}$$

$$7.26 \le \sigma_{\tau\beta}^2 \le 137.81$$

This result is consistent with the results of the exact F test on $\sigma_{\tau\beta}^2$, in that there is strong evidence that this variance component is not zero.

■

11–6.2   The Modified Large-Sample Method

The Satterthwaite method in the previous section is a relatively simple way to find an approximate confidence interval on a variance component that can be expressed as a linear combination of mean squares, say

$$\hat{\sigma}_0^2 = \sum_{i=1}^{Q} c_i MS_i \tag{11–23}$$

The Satterthwaite method works well when the degrees of freedom on each mean square MS_i are all relatively large, and when the constants c_i in Equation 11–23 are all positive. However, sometimes some of the c_i are negative. Graybill and Wang (1980) proposed a procedure called the **modified large-sample method,** which can be a very useful alternative to Satterthwaite's method. If all of the constants c_i in Equation 11–23 are positive, then the approximate $100(1 - \alpha)$ percent modified large-sample confidence interval on σ_0^2 is

$$\hat{\sigma}_0^2 - \sqrt{\sum_{i=1}^{Q} G_i^2 c_i^2 MS_i^2} \le \sigma_0^2 \le \hat{\sigma}_0^2 + \sqrt{\sum_{i=1}^{Q} H_i^2 c_i^2 MS_i^2} \tag{11–24}$$

where

$$G_i = 1 - \frac{1}{F_{\alpha, f_i, \infty}} \quad \text{and} \quad H_i = \frac{1}{F_{1-\alpha, f_i, \infty}} - 1$$

Note that an F random variable with an infinite number of denominator degrees of freedom is equivalent to a chi-square random variable divided by its degrees of freedom.

Now consider the more general case of Equation 11–23, where the constants c_i are unrestricted in sign. This may be written as

$$\hat{\sigma}_0^2 = \sum_{i=1}^{P} c_i MS_i - \sum_{j=P+1}^{Q} c_j MS_j, \qquad c_i, c_j \geq 0 \tag{11-25}$$

Ting et al. (1990) give an approximate $100(1 - \alpha)$ percent lower confidence limit on σ_0^2 as

$$L = \hat{\sigma}_0^2 - \sqrt{V_L} \tag{11-26}$$

where

$$V_L = \sum_{i=1}^{P} G_i^2 c_i^2 MS_i^2 + \sum_{j=P+1}^{Q} H_j^2 c_j^2 MS_j^2 + \sum_{i=1}^{P} \sum_{j=P+1}^{Q} G_{ij}^2 c_i c_j MS_i MS_j$$

$$+ \sum_{i=1}^{P-1} \sum_{t>i}^{P} G_{it}^* c_i c_t MS_i MS_t$$

$$G_i = 1 - \frac{1}{F_{\alpha, f_i, \infty}}$$

$$H_i = \frac{1}{F_{1-\alpha, f_i, \infty}} - 1$$

$$G_{ij} = \frac{(F_{\alpha, f_i, f_j} - 1)^2 - G_i^2 F_{\alpha, f_i, f_j}^2 - H_j^2}{F_{\alpha, f_i, f_j}}$$

$$G_{ir}^* = \left[\left(1 - \frac{1}{F_{\alpha, f_i + f_t, \infty}}\right)^2 \frac{(f_i + f_t)^2}{f_i f_t} - \frac{G_i^2 f_i}{f_t} - \frac{G_t^2 f_t}{f_i} \right] (P - 1),$$

$$\text{if } P > 1 \quad \text{and} \quad G_{ir}^* = 0 \quad \text{if } P = 1$$

These results can also be extended to include approximate confidence intervals on ratios of variance components. For a complete account of these methods, refer to the excellent book by Burdick and Graybill (1990).

Example 11–6

To illustrate the modified large-sample method, reconsider the three-factor mixed model in Example 11–4. We will find an approximate 95 percent lower confidence interval on $\sigma^2_{\tau\beta}$. Recall that the point estimate of $\sigma^2_{\tau\beta}$ is

$$\hat{\sigma}^2_{\tau\beta} = \frac{MS_{AB} - MS_{ABC}}{cn} = \frac{134.91 - 19.26}{(3)(2)} = 19.28$$

Therefore, in the notation of Equation 11–25, $c_1 = c_2 = \frac{1}{6}$, and

$$G_1 = 1 - \frac{1}{F_{0.05,6,\infty}} = 1 - \frac{1}{2.1} = 0.524$$

$$H_1 = \frac{1}{F_{0.95,12,\infty}} - 1 = \frac{1}{0.435} - 1 = 1.30$$

$$G_{11} = \frac{(F_{0.05,6,12} - 1)^2 - (G_1)^2 F^2_{0.05,6,12} - (H_1)^2}{F_{0.05,6,12}} = \frac{(3.00 - 1)^2 - (0.524)^2(3.00)^2 - (1.3)^2}{3.00} = -0.054$$

$$G^*_{1t} = 0$$

From Equation 11–26

$$\begin{aligned} V_L &= G^2_1 c^2_1 MS^2_{AB} + H^2_1 c^2_2 MS^2_{ABC} + G_{11} c_1 c_2 MS_{AB} MS_{ABC} \\ &= (0.524)^2(1/6)^2(134.91)^2 + (1.3)^2(1/6)^2(19.26)^2 + (-0.054)(1/6)(1/6)(134.91)(19.26) \\ &= 152.36 \end{aligned}$$

So an approximate 95 percent lower confidence limit on $\sigma^2_{\tau\beta}$ is:

$$L = \hat{\sigma}^2_{\tau\beta} - \sqrt{V_L} = 19.28 - \sqrt{152.36} = 6.94$$

This result is consistent with the results of the exact F test for this effect.

■

11–6.3   Maximum Likelihood Estimation of Variance Components

This chapter has emphasized the analysis of variance method of variance component estimation because it is relatively straightforward and makes use of familiar quantities—the mean squares in the analysis of variance table. However, the method has some disadvantages, including an embarrassing tendency to produce negative estimates on occasion. Furthermore, the analysis of variance method is really a **method of moments estimator,** a technique that mathematical statisticians generally do not prefer to use for parameter estimation because it often results in parameter estimates that do not have good statistical properties. The preferred parameter estimation technique is called the **method of maximum likelihood.** The implementation of this method can be somewhat involved, particularly for an experimental design model, but in a sense, the method of maximum likelihood selects parameter estimates that, for a specified model and error distribution, maximize the probability of occurrence of the sample results. A very nice overview of the method of maximum likelihood applied to experimental design models is given by Milliken and Johnson (1984).

A complete presentation of the method of maximum likelihood is beyond the scope of this book, but the general idea can be illustrated very easily. Suppose that x is a random variable with probability distribution $f(x; \theta)$, where θ is an unknown parameter. Let $x_1, x_2, \ldots, x_n$ be a random sample of n observations. Then the likelihood function of the sample is

$$L(\theta) = f(x_1; \theta) \cdot f(x_2; \theta) \cdot \ldots \cdot f(x_n; \theta)$$

Note that the likelihood function is now a function of only the unknown parameter θ. The maximum likelihood estimator of θ is the value of θ that maximizes the likelihood function $L(\theta)$.

To illustrate how this applies to an experimental design model with random effects, consider a two-factor model with $a = b = n = 2$. The model is

$$y_{ijk} = \mu + \tau_i + \beta_j + (\tau\beta)_{ij} + \epsilon_{ijk}$$

with $i = 1, 2$, $j = 1, 2$, and $k = 1, 2$. The variance of any observation is

$$V(y_{ijk}) = \sigma_y^2 = \sigma_\tau^2 + \sigma_\beta^2 + \sigma_{\tau\beta}^2 + \sigma^2$$

and the covariances are

$$
\begin{aligned}
\mathrm{Cov}(y_{ijk}, y_{i'j'k'}) &= \sigma_\tau^2 + \sigma_\beta^2 + \sigma_{\tau\beta}^2 & i = i', j = j', k \neq k' \\
&= \sigma_\tau^2 & i = i', j \neq j' \\
&= \sigma_\beta^2 & i \neq i', j = j' \\
&= 0 & i \neq i', j \neq j'
\end{aligned}
\qquad (11\text{-}27)
$$

It is convenient to think of the observations as an 8×1 vector, say

$$
\mathbf{y} = \begin{bmatrix} y_{111} \\ y_{112} \\ y_{211} \\ y_{212} \\ y_{121} \\ y_{122} \\ y_{221} \\ y_{222} \end{bmatrix}
$$

and the variances and covariances can be expressed as an 8×8 **covariance matrix**

$$
\Sigma = \begin{bmatrix} \Sigma_{11} & \Sigma_{12} \\ \Sigma_{21} & \Sigma_{22} \end{bmatrix}
$$

where Σ_{11}, Σ_{22}, Σ_{12} and $\Sigma_{21} = \Sigma'_{12}$ are 4×4 matrices defined as follows:

$$\Sigma_{11} = \Sigma_{22} = \begin{bmatrix} \sigma_y^2 & \sigma_\tau^2 + \sigma_\beta^2 + \sigma_{\tau\beta}^2 & \sigma_\tau^2 & \sigma_\tau^2 \\ \sigma_\tau^2 + \sigma_\beta^2 + \sigma_{\tau\beta}^2 & \sigma_y^2 & \sigma_\tau^2 & \sigma_\tau^2 \\ \sigma_\tau^2 & \sigma_\tau^2 & \sigma_y^2 & \sigma_\tau^2 + \sigma_\beta^2 + \sigma_{\tau\beta}^2 \\ \sigma_\tau^2 & \sigma_\tau^2 & \sigma_\tau^2 + \sigma_\beta^2 + \sigma_{\tau\beta}^2 & \sigma_y^2 \end{bmatrix}$$

$$\Sigma_{12} = \begin{bmatrix} \sigma_\beta^2 & \sigma_\beta^2 & 0 & 0 \\ \sigma_\beta^2 & \sigma_\beta^2 & 0 & 0 \\ 0 & 0 & \sigma_\beta^2 & \sigma_\beta^2 \\ 0 & 0 & \sigma_\beta^2 & \sigma_\beta^2 \end{bmatrix}$$

and Σ_{21} is just the transpose of Σ_{12}. Now each observation is normally distributed with variance σ_y^2, and if we assume that all $N = abn$ observations have a joint normal distribution, then the likelihood function for the random model becomes

$$L(\mu, \sigma_\tau^2, \sigma_\beta^2, \sigma_{\tau\beta}^2, \sigma^2) = \frac{1}{(2\pi)^{N/2} |\Sigma|^{1/2}} \exp\left[-\frac{1}{2} (\mathbf{y} - \mathbf{j}_N\mu)' \Sigma^{-1} (\mathbf{y} - \mathbf{j}_N\mu) \right]$$

where $\mathbf{j}_N$ is an $N \times 1$ vector of ones. The maximum likelihood estimates of μ, $\sigma_\tau^2, \sigma_\beta^2, \sigma_{\tau\beta}^2$, and σ^2 are those values of these parameters that maximize the likelihood function. It would also be desirable to restrict the variance component estimates to nonnegative values. So in practice, we would maximize the likelihood function subject to this constraint.

Estimating variance components by maximum likelihood requires specialized computer software. Some general statistical software packages have this capability. The SAS system computes maximum likelihood estimates of variance components in random or mixed models through SAS PROC MIXED. We will illustrate the use of PROC MIXED by applying it to the two-factor factorial model introduced in Examples 11–1 and 11–2.

First consider Example 11–1. The model is a two-factor factorial, random effects model. The analysis of variance method has produced a negative estimate of the interaction variance component. Negative estimates of variance components can be avoided in PROC MIXED by specifying that a **restricted maximum likelihood (REML)** method be used. REML essentially constrains the variance component estimates to nonnegative values.

SAS PROC MIXED requires as input the covariance matrix of the model parameters. The structure for a random model where all the random variables are mutually independent is

$$\mathbf{G} = \begin{bmatrix} \sigma_\tau^2\mathbf{I} & 0 & 0 \\ 0 & \sigma_\beta^2\mathbf{I} & 0 \\ 0 & 0 & \sigma_{\tau\beta}^2\mathbf{I} \end{bmatrix} \tag{11-28}$$

where the **I**'s are identity matrices. (The covariance structure of a model can be specified in PROC MIXED by the TYPE=*structure* option in the RANDOM statement. The covariance structure for the model in Example 11–1 is specified by TYPE=SIM (the default in PROC MIXED), which specifies the simple covariance structure for the model parameters given in Equation 11–28.)

Table 11–12 (page 500) presents the output from SAS PROC MIXED for the experiment in Example 11–1. We specified the REML method of variance component estimation. The output has been annotated with numbers to facilitate the description shown below:

1. Variance component estimates and related output.

2. Covariance parameter. This identifies the model parameters: σ_τ^2, σ_β^2, $\sigma_{\tau\beta}^2$, and σ^2.

3. Ratio of estimated variance of effect to estimated residual error variance.

$$\hat{\sigma}_i^2/\hat{\sigma}^2$$

4. Parameter estimates. These are the REML estimates of the variance components $\hat{\sigma}_\tau^2$, $\hat{\sigma}_\beta^2$, $\hat{\sigma}_{\tau\beta}^2$, and $\hat{\sigma}^2$. Notice that the REML estimate of $\sigma_{\tau\beta}^2$ is zero.

5. Standard error of estimate. This is the large-sample standard error of the parameter estimate; $se(\hat{\sigma}_i^2) = \sqrt{V(\hat{\sigma}_i^2)}$.

6. Z statistic associated with the estimated variance.

$$Z = \hat{\sigma}_i^2/se(\hat{\sigma}_i^2).$$

7. *P*-value of the computed Z statistic.

8. Alpha level used to compute confidence interval.

9. Lower and upper bounds of a $100(1 - \alpha)$ percent large-sample normal theory confidence interval on the variance components

$$L = \hat{\sigma}_i^2 - Z_{\alpha/2}se(\hat{\sigma}_i^2), \qquad U = \hat{\sigma}_i^2 + Z_{\alpha/2}se(\hat{\sigma}_i^2)$$

10. Asymptotic covariance matrix of estimates. This is the large-sample covariance matrix of the variance components estimates.

11. Model fitting measures for comparing the fit of alternative models.

Notice that the results from SAS PROC MIXED fairly closely match those reported in Example 11–1 when the reduced model (without the interaction term) was fit to the data.

In Example 11–2 we considered the same experiment, but assumed that the operators were a fixed factor, leading to a mixed model. SAS PROC MIXED can be employed to estimate the variance components for this situation. The covariance structure for the observations, assuming that all random variables

Table 11-12 SAS PROC MIXED Output for Analysis of Gauge Repeatability and Reproducibility Study in Example 11-1 Using Restricted Maximum Likelihood (REML) Estimation of Variance Components

The MIXED Procedure

Class Level Information

Class	Levels	Values
PART	20	1 2 3 4 5 6 7 8 9 10 11 12 13 14 15 16 17 18 19 20
OPERATOR	3	1 2 3
REPLICAT	2	1 2

Covariance Parameter Estimates (REML)

Cov Parm	Ratio	Estimate	Std Error	Z	Pr > |Z|	Alpha	Lower	Upper
OPERATOR	0.01203539	0.01062922	0.03286000	0.32	0.7463	0.05	-0.0538	0.0750
PART	11.60743820	10.25126446	3.37376878	3.04	0.0024	0.05	3.6388	16.8637
PART*OPERATOR	-0.00000000	-0.00000000	.	.	.	.	.	.
Residual	1.00000000	0.88316339	0.12616620	7.00	0.0000	0.05	0.6359	1.1304

Asymptotic Covariance Matrix of Estimates

Cov Parm	OPERATOR	PART	PART*OPERATOR	Residual
OPERATOR	0.00107978	0.00006632	0.00000000	-0.00039795
PART	0.00006632	11.38231579	0.00000000	-0.00265287
PART*OPERATOR	0.00000000	0.00000000	0.00000000	-0.00000000
Residual	-0.00039795	-0.00265287	-0.00000000	0.01591791

Model Fitting Information for VALUE

Description	Value
Observations	120.0000
Variance Estimate	0.8832
Standard Deviation Estimate	0.9398
REML Log Likelihood	-204.696
Akaike's Information Criterion	-208.696
Schwarz's Bayesian Criterion	-214.254
-2 REML Log Likelihood	409.3913

Table 11-13 SAS PROC MIXED Output for Analysis of Gauge Repeatability and Reproducibility Study with the Operator as a Fixed Effect Using Restricted Maximum Likelihood (REML) Estimation of Variance Components

The MIXED Procedure

Covariance Parameter Estimates (REML)

Cov Parm	Ratio	Estimate	Std Error	Z	Pr > |Z|	Alpha	Lower	Upper
PART	11.60743876	10.25126472	3.37376895	3.04	0.0024	0.05	3.6388	16.8637
PART*OPERATOR	0.00000000	0.00000000	.	.	.	.	.	.
Residual	1.00000000	0.88316337	0.12616620	7.00	0.0000	0.05	0.6359	1.1304

Asymptotic Covariance Matrix of Estimates

Cov Parm	PART	PART*OPERATOR	Residual
PART	11.38231693	0.00000000	-0.00265287
PART*OPERATOR	0.00000000	0.00000000	-0.00000000
Residual	-0.00265287	0.00000000	0.01591791

Model Fitting Information for VALUE

Description	Value
Observations	120.0000
Variance Estimate	0.8832
Standard Deviation Estimate	0.9398
REML Log Likelihood	-204.729
Akaike's Information Criterion	-207.729
Schwarz's Bayesian Criterion	-211.872
-2 REML Log Likelihood	409.4572

Tests of Fixed Effects

Source	NDF	DDF	Type III F	Pr > F
OPERATOR	2	38	1.48	0.2401

are mutually independent (that is, the unrestricted mixed model), is

$$
\begin{aligned}
\mathrm{Cov}(y_{ijk}, y_{i'j'k'}) &= \sigma_\beta^2 + \sigma_{\tau\beta}^2 + \sigma^2 && i = i', j = j', k = k' \\
&= \sigma_\beta^2 + \sigma_{\tau\beta}^2 && i = i', j = j', k \neq k' \\
&= \sigma_\beta^2 && i \neq i', j = j' \\
&= 0 && j \neq j'
\end{aligned}
\tag{11–29}
$$

The covariance matrix of the model parameters is

$$
\mathbf{G} = \begin{bmatrix} \sigma_\beta^2 \mathbf{I} & 0 \\ 0 & \sigma_{\tau\beta}^2 \mathbf{I} \end{bmatrix}
\tag{11–30}
$$

(This is specified by the TYPE=SIM statement in the input to PROC MIXED.) The SAS PROC MIXED output for the unrestricted form of the mixed model of Example 11–2 is shown in Table 11–13 (page 501). Once again, the REML method was selected. The estimate for the variance component due to part is very similar to the estimate obtained using the random model. The estimate for the residual error variance is also similar. In addition, the output includes an F test for the fixed effect.

11–7 PROBLEMS

11–1 An experiment was performed to investigate the capability of a measurement system. Ten parts were randomly selected, and two randomly selected operators measured each part three times. The tests were made in random order, and the data below resulted.

Part Number	Operator 1 Measurements			Operator 2 Measurements		
	1	2	3	1	2	3
1	50	49	50	50	48	51
2	52	52	51	51	51	51
3	53	50	50	54	52	51
4	49	51	50	48	50	51
5	48	49	48	48	49	48
6	52	50	50	52	50	50
7	51	51	51	51	50	50
8	52	50	49	53	48	50
9	50	51	50	51	48	49
10	47	46	49	46	47	48

(a) Analyze the data from this experiment.

(b) Find point estimates of the variance components using the analysis of variance method.

11-2   Reconsider the data in Problem 6-6. Suppose that both factors, machines and operators, are chosen at random.

(a) Analyze the data from this experiment.

(b) Find point estimates of the variance components using the analysis of variance method.

11-3   Reconsider the data in Problem 6-13. Suppose that both factors are random.

(a) Analyze the data from this experiment.

(b) Estimate the variance components.

11-4   Suppose that in Problem 6-11 the furnace positions were randomly selected, resulting in a mixed model experiment. Reanalyze the data from this experiment under this new assumption. Estimate the appropriate model components.

11-5   Reanalyze the measurement systems experiment in Problem 11-1, assuming that operators are a fixed factor. Estimate the appropriate model components.

11-6   In Problem 6-6, suppose that there are only four machines of interest, but the operators were selected at random.

(a) What type of model is appropriate?

(b) Perform the analysis and estimate the model components.

11-7   By application of the expectation operator, develop the expected mean squares for the two-factor factorial, mixed model. Check your results with the expected mean squares given in Table 11-8 to see that they agree.

11-8   Consider the three-factor factorial design in Example 11-3. Propose appropriate test statistics for all main effects and interactions. Repeat for the case where A and B are fixed and C is random.

11-9   Consider the experiment in Example 11-4. Analyze the data for the case where A, B, and C are random.

11-10   Derive the expected mean squares shown in Table 11-11.

11-11   Consider a four-factor factorial experiment where factor A is at a levels, factor B is at b levels, factor C is at c levels, factor D is at d levels, and there are n replicates. Write down the sums of squares, the degrees of freedom, and the expected mean squares for the following cases.

(a) A, B, C, and D are fixed factors.

(b) A, B, C, and D are random factors.

(c) A is fixed and B, C, and D are random.

(d) A and B are fixed and C and D are random.

(e) A, B, and C are fixed and D is random.

Do exact tests exist for all effects? If not, propose test statistics for those effects that cannot be directly tested.

11–12   In Problem 6–17, assume that the three operators were selected at random. Analyze the data under these conditions and draw conclusions. Estimate the variance components.

11–13   Consider the three-factor factorial model

$$y_{ijk} = \mu + \tau_i + \beta_j + \gamma_k + (\tau\beta)_{ij} + (\beta\gamma)_{jk} + \epsilon_{ijk} \qquad \begin{cases} i = 1, 2, \ldots, a \\ j = 1, 2, \ldots, b \\ k = 1, 2, \ldots, c \end{cases}$$

Assuming that all the factors are random, develop the analysis of variance table, including the expected mean squares. Propose appropriate test statistics for all effects.

11–14   The three-factor factorial model for a single replicate is

$$y_{ijk} = \mu + \tau_i + \beta_j + \gamma_k + (\tau\beta)_{ij} + (\beta\gamma)_{jk} + (\tau\gamma)_{ik} + (\tau\beta\gamma)_{ijk} + \epsilon_{ijk}$$

If all the factors are random, can any effects be tested? If the three-factor interaction and the $(\tau\beta)_{ij}$ interaction do not exist, can all the remaining effects be tested?

11–15   In Problem 6–6, assume that both machines and operators were chosen randomly. Determine the power of the test for detecting a machine effect such that $\sigma_\beta^2 = \sigma^2$, where σ_β^2 is the variance component for the machine factor. Are two replicates sufficient?

11–16   In the two-factor mixed model analysis of variance, show that $\text{Cov}[(\tau\beta)_{ij}, (\tau\beta)_{i'j}] = -(1/a)\sigma_{\tau\beta}^2$ for $i \neq i'$.

11–17   Show that the method of analysis of variance always produces unbiased point estimates of the variance components in any random or mixed model.

11–18   Invoking the usual normality assumptions, find an expression for the probability that a negative estimate of a variance component will be obtained by the analysis of variance method. Using this result, write a statement giving the probability that $\hat\sigma_\tau^2 < 0$ in a one-factor analysis of variance. Comment on the usefulness of this probability statement.

11–19   Analyze the data in Problem 11–1, assuming that operators are fixed, using both of the mixed models discussed in Section 11–2. Compare the results obtained from the two models.

11–20   Consider the two-factor mixed model. Show that the standard error of the fixed factor mean (e.g., A) is $[MS_{AB}/bn]^{1/2}$.

11–21   Consider the variance components in the random model from Problem 11–1.

(a) Find an exact 95 percent confidence interval on σ^2.

(b) Find approximate 95 percent confidence intervals on the other variance components using the Satterthwaite method.

11–22   Use the experiment described in Problem 6–6 and assume that both factors are random. Find an exact 95 percent confidence interval on σ^2. Construct approximate 95 percent confidence intervals on the other variance components using the Satterthwaite method.

11-23 Consider the three-factor experiment in Problem 6–17 and assume that operators were selected at random. Find an approximate 95 percent confidence interval on the operator variance component.

11-24 Rework Problem 11–21 using the modified large-sample approach described in Section 11–6.2. Compare the two sets of confidence intervals obtained and discuss.

11-25 Rework Problem 11–23 using the modified large-sample method described in Section 11–6.2. Compare this confidence interval with the one obtained previously and discuss.

Chapter 12
Nested and Split-Plot Designs

This chapter introduces two important types of experimental designs, the nested design and the split-plot design. Both of these designs find reasonably widespread application in the industrial use of designed experiments. They also frequently involve one or more random factors, and so some of the concepts introduced in Chapter 11 will find application here.

12–1 THE TWO-STAGE NESTED DESIGN

In certain multifactor experiments the levels of one factor (e.g., factor B) are similar but not identical for different levels of another factor (e.g., A). Such an arrangement is called a **nested** or **hierarchical design,** with the levels of factor B nested under the levels of factor A. For example, consider a company that purchases its raw material from three different suppliers. The company wishes to determine if the purity of the raw material is the same from each supplier. There are four batches of raw material available from each supplier, and three determinations of purity are to be taken from each batch. The situation is depicted in Figure 12–1.

This is a **two-stage nested design,** with batches nested under suppliers. At first glance, you may ask why this is not a factorial experiment. If this were a factorial, then batch 1 would always refer to the same batch, batch 2 would always refer to the same batch, and so on. This is clearly not the case since the batches from each supplier are unique for that particular supplier. That is, batch 1 from supplier 1 has no connection with batch 1 from any other supplier, batch 2 from supplier 1 has no connection with batch 2 from any other supplier, and so forth. To emphasize the fact that the batches from each supplier are different batches, we may renumber the batches as 1, 2, 3, and 4 from supplier 1; 5, 6, 7,

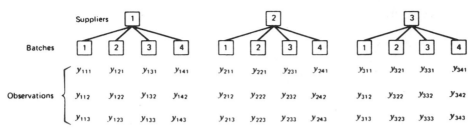

Figure 12–1. A two-stage nested design.

and 8 from supplier 2; and 9, 10, 11, and 12 from supplier 3, as shown in Figure 12–2.

Sometimes we may be uncertain as to whether a factor is crossed in a factorial arrangement or nested. If the levels of the factor can be renumbered arbitrarily as in Figure 12–2, then the factor is nested.

12–1.1 Statistical Analysis

The linear statistical model for the two-stage nested design is

$$y_{ijk} = \mu + \tau_i + \beta_{j(i)} + \epsilon_{(ij)k} \qquad \begin{cases} i = 1, 2, \ldots, a \\ j = 1, 2, \ldots, b \\ k = 1, 2, \ldots, n \end{cases} \qquad (12\text{–}1)$$

That is, there are a levels of factor A, b levels of factor B nested under each level of A, and n replicates. The subscript $j(i)$ indicates that the jth level of factor B is nested under the ith level of factor A. It is convenient to think of the replicates as being nested within the combination of levels of A and B; thus, the subscript $(ij)k$ is used for the error term. This is a **balanced nested design** since there are an equal number of levels of B within each level of A and an equal number of replicates. Since every level of factor B does not appear with every level of factor A, there can be no interaction between A and B.

We may write the total corrected sum of squares as

$$\sum_{i=1}^{a}\sum_{j=1}^{b}\sum_{k=1}^{n}(y_{ijk} - \bar{y}_{...})^2 = \sum_{i=1}^{a}\sum_{j=1}^{b}\sum_{k=1}^{n}[(\bar{y}_{i..} - \bar{y}_{...}) + (\bar{y}_{ij.} - \bar{y}_{i..}) + (y_{ijk} - \bar{y}_{ij.})]^2 \quad (12\text{–}2)$$

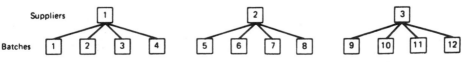

Figure 12–2. Alternate layout for the two-stage nested design.

Expanding the right-hand side of Equation 12–2 yields

$$\sum_{i=1}^{a}\sum_{j=1}^{b}\sum_{k=1}^{n}(y_{ijk}-\bar{y}...)^2 = bn\sum_{i=1}^{a}(\bar{y}_{i..}-\bar{y}...)^2 + n\sum_{i=1}^{a}\sum_{j=1}^{b}(\bar{y}_{ij.}-\bar{y}_{i..})^2$$
$$+ \sum_{i=1}^{a}\sum_{j=1}^{b}\sum_{k=1}^{n}(y_{ijk}-\bar{y}_{ij.})^2 \tag{12-3}$$

since the three cross-product terms are zero. Equation 12–3 indicates that the total sum of squares can be partitioned into a sum of squares due to factor A, a sum of squares due to factor B under the levels of A, and a sum of squares due to error. Symbolically, we may write Equation 12–3 as

$$SS_T = SS_A + SS_{B(A)} + SS_E \tag{12-4}$$

There are $abn - 1$ degrees of freedom for SS_T, $a - 1$ degrees of freedom for SS_A, $a(b - 1)$ degrees of freedom for $SS_{B(A)}$, and $ab(n - 1)$ degrees of freedom for error. Note that $abn - 1 = (a - 1) + a(b - 1) + ab(n - 1)$. If the errors are NID$(0, \sigma^2)$, we may divide each sum of squares on the right of Equation 12–4 by its degrees of freedom to obtain independently distributed mean squares such that the ratio of any two mean squares is distributed as F.

The appropriate statistics for testing the effects of factors A and B depend on whether A and B are fixed or random. If factors A and B are fixed, we assume that $\sum_{i=1}^{a} \tau_i = 0$ and $\sum_{j=1}^{b} \beta_{j(i)} = 0$ $(i = 1, 2, \ldots, a)$. That is, the A treatment effects sum to zero, and the B treatment effects sum to zero within each level of A. Alternatively, if A and B are random, then we assume that τ_i is NID$(0, \sigma_\tau^2)$ and $\beta_{j(i)}$ is NID$(0, \sigma_\beta^2)$. Mixed models with A fixed and B random are also widely encountered. The expected mean squares can be determined by a straightforward application of the rules in Chapter 11. Table 12–1 gives the expected mean squares for these situations.

Table 12–1 indicates that if the levels of A and B are fixed, then $H_0: \tau_i = 0$ is tested by MS_A/MS_E and $H_0: \beta_{j(i)} = 0$ is tested by $MS_{B(A)}/MS_E$. If A is a fixed factor and B is random, then $H_0: \tau_i = 0$ is tested by $MS_A/MS_{B(A)}$ and $H_0: \sigma_\beta^2 = 0$

Table 12–1   Expected Mean Squares in the Two-Stage Nested Design

$E(MS)$	A Fixed B Fixed	A Fixed B Random	A Random B Random
$E(MS_A)$	$\sigma^2 + \dfrac{bn\sum\tau_i^2}{a-1}$	$\sigma^2 + n\sigma_\beta^2 + \dfrac{bn\sum\tau_i^2}{a-1}$	$\sigma^2 + n\sigma_\beta^2 + bn\sigma_\tau^2$
$E(MS_{B(A)})$	$\sigma^2 + \dfrac{n\sum\sum\beta_{j(i)}^2}{a(b-1)}$	$\sigma^2 + n\sigma_\beta^2$	$\sigma^2 + n\sigma_\beta^2$
$E(MS_E)$	σ^2	σ^2	σ^2

Table 12–2 Analysis of Variance Table for the Two-Stage
Nested Design

Source of Variation	Sum of Squares	Degrees of Freedom	Mean Square
A	$bn \sum (\bar{y}_{i..} - \bar{y}_{...})^2$	$a - 1$	MS_A
B within A	$n \sum \sum (\bar{y}_{ij.} - \bar{y}_{i..})^2$	$a(b - 1)$	$MS_{B(A)}$
Error	$\sum \sum \sum (y_{ijk} - \bar{y}_{ij.})^2$	$ab(n - 1)$	MS_E
Total	$\sum \sum \sum (y_{ijk} - \bar{y}_{...})^2$	$abn - 1$	

is tested by $MS_{B(A)}/MS_E$. Finally, if both A and B are random factors, we test $H_0: \sigma_\tau^2 = 0$ by $MS_A/MS_{B(A)}$ and $H_0: \sigma_\beta^2 = 0$ by $MS_{B(A)}/MS_E$. The test procedure is summarized in an analysis of variance table as shown in Table 12–2. Computing formulas for the sums of squares may be obtained by expanding the quantities in Equation 12–3 and simplifying. They are

$$SS_A = \frac{1}{bn} \sum_{i=1}^{a} y_{i..}^2 - \frac{y_{...}^2}{abn} \qquad (12\text{–}5)$$

$$SS_{B(A)} = \frac{1}{n} \sum_{i=1}^{a} \sum_{j=1}^{b} y_{ij.}^2 - \frac{1}{bn} \sum_{i=1}^{a} y_{i..}^2 \qquad (12\text{–}6)$$

$$SS_E = \sum_{i=1}^{a} \sum_{j=1}^{b} \sum_{k=1}^{n} y_{ijk}^2 - \frac{1}{n} \sum_{i=1}^{a} \sum_{j=1}^{b} y_{ij.}^2 \qquad (12\text{–}7)$$

$$SS_T = \sum_{i=1}^{a} \sum_{j=1}^{b} \sum_{k=1}^{n} y_{ijk}^2 - \frac{y_{...}^2}{abn} \qquad (12\text{–}8)$$

We see that Equation 12–6 for $SS_{B(A)}$ can be written as

$$SS_{B(A)} = \sum_{i=1}^{a} \left[\frac{1}{n} \sum_{j=1}^{b} y_{ij.}^2 - \frac{y_{i..}^2}{bn} \right]$$

This expresses the idea that $SS_{B(A)}$ is the sum of squares between levels of B for each level of A, summed over all the levels of A.

Example 12–1

Consider a company that buys raw material in batches from three different suppliers. The purity of this raw material varies considerably, which causes problems in manufacturing the finished product. We wish to determine if the variability in purity is attributable to differences between the suppliers. Four batches of raw material are selected at random from each supplier, and three determinations of purity are made on each batch. This is, of course, a two-stage nested design. The data, after coding

Table 12–3   Coded Purity Data for Example 12–1 (Code: y_{ijk} = purity − 93)

		Supplier 1				Supplier 2				Supplier 3			
Batches		1	2	3	4	1	2	3	4	1	2	3	4
		1	−2	−2	1	1	0	−1	0	2	−2	1	3
		−1	−3	0	4	−2	4	0	3	4	0	−1	2
		0	−4	1	0	−3	2	−2	2	0	2	2	1
Batch totals	$y_{ij.}$	0	−9	−1	5	−4	6	−3	5	6	0	2	6
Supplier totals	$y_{i..}$		−5				4				14		

by subtracting 93, are shown in Table 12–3. The sums of squares are computed as follows:

$$SS_T = \sum_{i=1}^{a} \sum_{j=1}^{b} \sum_{k=1}^{n} y_{ijk}^2 - \frac{y_{...}^2}{abn}$$

$$= 153.00 - \frac{(13)^2}{36} = 148.31$$

$$SS_A = \frac{1}{bn} \sum_{i=1}^{a} y_{i..}^2 - \frac{y_{...}^2}{abn}$$

$$= \frac{1}{(4)(3)} [(-5)^2 + (4)^2 + (14)^2] - \frac{(13)^2}{36}$$

$$= 19.75 - 4.69 = 15.06$$

$$SS_{B(A)} = \frac{1}{n} \sum_{i=1}^{a} \sum_{j=1}^{b} y_{ij.}^2 - \frac{1}{bn} \sum_{i=1}^{a} y_{i..}^2$$

$$= \frac{1}{3} [(0)^2 + (-9)^2 + (-1)^2 + \cdots + (2)^2 + (6)^2] - 19.75$$

$$= 89.67 - 19.75 = 69.92$$

and

$$SS_E = \sum_{i=1}^{a} \sum_{j=1}^{b} \sum_{k=1}^{n} y_{ijk}^2 - \frac{1}{n} \sum_{i=1}^{a} \sum_{j=1}^{b} y_{ij.}^2$$

$$= 153.00 - 89.67 = 63.33$$

The analysis of variance is summarized in Table 12–4. Suppliers are fixed and batches are random, so the expected mean squares are obtained from the middle column of Table 12–1. They are repeated for convenience in Table 12–4. From examining the P-values, we would conclude that there is no significant effect on purity due to

Table 12-4   Analysis of Variance for the Data in Example 12-1

Source of Variation	Sum of Squares	Degrees of Freedom	Mean Square	Expected Mean Square	F_0	P-Value
Suppliers	15.06	2	7.53	$\sigma^2 + 3\sigma_\beta^2 + 6\sum \tau_i^2$	0.97	0.42
Batches (within suppliers)	69.92	9	7.77	$\sigma^2 + 3\sigma_\beta^2$	2.94	0.02
Error	63.33	24	2.64	σ^2		
Total	148.31	35				

suppliers but the purity of batches of raw material from the same supplier does differ significantly. ∎

The practical implications of this experiment and the analysis are very important. The objective of the experimenter is to find the source of the variability in raw material purity. If it results from differences among suppliers, then we may be able to solve the problem by selecting the "best" supplier. However, that solution is not applicable here because the major source of variability is the batch-to-batch purity variation *within* suppliers. Therefore, we must attack the problem by working with the suppliers to reduce their batch-to-batch variability. This may involve modifications to the suppliers' production processes or their internal quality assurance system.

Notice what would have happened if we had incorrectly analyzed this design as a two-factor factorial experiment. If batches are considered to be crossed with suppliers, we obtain batch totals of 2, −3, −2, and 16, with each batch × suppliers cell containing three replicates. Thus, a sum of squares due to batches and an interaction sum of squares can be computed. The complete factorial analysis of variance is shown in Table 12-5, assuming the mixed model.

This analysis indicates that batches differ significantly and that there is a

Table 12-5   Incorrect Analysis of the Two-Stage Nested Design in Example 12-1 as a Factorial (Suppliers Fixed, Batches Random)

Source of Variation	Sum of Squares	Degrees of Freedom	Mean Square	F_0	P-Value
Suppliers (S)	15.06	2	7.53	1.02	0.42
Batches (B)	25.64	3	8.55	3.24	0.04
S × B interaction	44.28	6	7.38	2.80	0.03
Error	63.33	24	2.64		
Total	148.31	35			

significant interaction between batches and suppliers. However, it is difficult to give a practical interpretation of the batches × suppliers interaction. For example, does this significant interaction mean that the supplier effect is not constant from batch to batch? Furthermore, the significant interaction coupled with the nonsignificant supplier effect could lead the analyst to conclude that suppliers really differ but their effect is masked by the significant interaction.

Comparing Tables 12–4 and 12–5, notice that

$$SS_B + SS_{S \times B} = 25.64 + 44.28 = 69.92 \equiv SS_{B(S)}$$

That is, the sum of squares for batches within suppliers consists of the sum of squares of the batches and the sum of squares for the batches × suppliers interaction. The degrees of freedom have a similar property; that is,

$$\frac{\text{Batches}}{3} + \frac{\text{Batches} \times \text{Suppliers}}{6} = \frac{\text{Batches within Suppliers}}{9}$$

Therefore, a computer program for analyzing factorial designs could also be used for the analysis of nested designs by pooling the "main effect" of the nested factor and interactions of that factor with the factor under which it is nested. This is useful information, since a computer program specifically written for the analysis of nested designs may not be readily available, whereas programs for analyzing factorial designs are widely available.

12–1.2   Diagnostic Checking

The major tool used in diagnostic checking is **residual analysis.** For the two-stage nested design, the residuals are

$$e_{ijk} = y_{ijk} - \hat{y}_{ijk}$$

The fitted value is

$$\hat{y}_{ijk} = \hat{\mu} + \hat{\tau}_i + \hat{\beta}_{j(i)}$$

and if we make the usual restrictions on the model parameters ($\Sigma_i \hat{\tau}_i = 0$ and $\Sigma_j \hat{\beta}_{j(i)} = 0, i = 1, 2, \ldots, a$), then $\hat{\mu} = \bar{y}_{...}$, $\hat{\tau}_i = \bar{y}_{i..} - \bar{y}_{...}$, and $\hat{\beta}_{j(i)} = \bar{y}_{ij.} - \bar{y}_{i..}$. Consequently, the fitted value is

$$\hat{y}_{ijk} = \bar{y}_{...} + (\bar{y}_{i..} - \bar{y}_{...}) + (\bar{y}_{ij.} - \bar{y}_{i..})$$
$$= \bar{y}_{ij.}$$

Thus, the residuals from the two-stage nested design are

$$e_{ijk} = y_{ijk} - \bar{y}_{ij.}. \tag{12–9}$$

where $\bar{y}_{ij.}$ are the individual batch averages.

The observations, fitted values, and residuals for the purity data in Example 12–1 follow:

Observed Value y_{ijk}	Fitted Value $\hat{y}_{ijk} = \bar{y}_{ij.}$	$e_{ijk} = y_{ijk} - \bar{y}_{ij.}$
1	0.00	1.00
−1	0.00	−1.00
0	0.00	0.00
−2	−3.00	1.00
−3	−3.00	0.00
−4	−3.00	−1.00
−2	−0.33	−1.67
0	−0.33	0.33
1	−0.33	1.33
1	1.67	−0.67
4	1.67	2.33
0	1.67	−1.67
1	−1.33	2.33
−2	−1.33	−0.67
−3	−1.33	−1.67
0	2.00	−2.00
4	2.00	2.00
2	2.00	0.00
−1	−1.00	0.00
0	−1.00	1.00
−2	−1.00	−1.00
0	1.67	−1.67
3	1.67	1.33
2	1.67	0.33
2	2.00	0.00
4	2.00	2.00
0	2.00	−2.00
−2	0.00	−2.00
0	0.00	0.00
2	0.00	2.00
1	0.67	0.33
−1	0.67	−1.67
2	0.67	1.33
3	2.00	1.00
2	2.00	0.00
1	2.00	−1.00

The usual diagnostic checks—including normal probability plots, checking for outliers, and plotting the residuals versus fitted values—may now be performed. As an illustration, the residuals are plotted versus the fitted values and against the levels of the supplier factor in Figure 12–3.

In a problem situation such as that described in Example 12–1, the residual plots are particularly useful because of the additional diagnostic information they contain. For instance, the analysis of variance has indicated that the mean

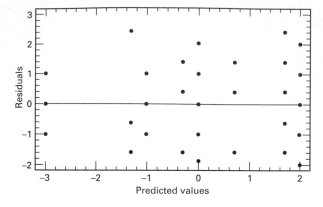

(a) Plot of residuals versus the predicted values

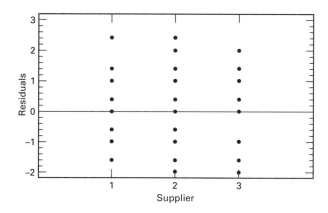

(b) Plot of residuals versus supplier

Figure 12–3.   Residual plots for Example 12–1.

purity of all three suppliers does not differ, but that there is statistically significant batch-to-batch variability (that is, $\sigma_\beta^2 > 0$). But is the variability within batches the same for all suppliers? In effect we have assumed this to be the case, and if it's not true we would certainly like to know it, because it has considerable practical impact on our interpretation of the results of the experiment. The plot of residuals versus suppliers in Figure 12–3b is a simple but effective way to check this assumption. Since the spread of the residuals is about the same for all three suppliers, we would conclude that the batch-to-batch variability in purity is about the same for all three suppliers.

12–1.3   Variance Components

For the random effects case, the analysis of variance method can be used to estimate the variance components σ^2, σ_β^2, and σ_τ^2. From the expected mean squares in the last column of Table 12–1, we obtain

$$\hat{\sigma}^2 = MS_E \tag{12-10}$$

$$\hat{\sigma}_\beta^2 = \frac{MS_{B(A)} - MS_E}{n} \tag{12-11}$$

and

$$\hat{\sigma}_\tau^2 = \frac{MS_A - MS_{B(A)}}{bn} \tag{12-12}$$

Many applications of nested designs involve a mixed model, with the main factor (A) fixed and the nested factor (B) random. This is the case for the problem described in Example 12–1; suppliers (factor A) are fixed, and batches of raw material (factor B) are random. The effects of the suppliers may be estimated by

$$\hat{\tau}_1 = \bar{y}_{1..} - \bar{y}_{...} = \frac{-5}{12} - \frac{13}{36} = \frac{-28}{36}$$

$$\hat{\tau}_2 = \bar{y}_{2..} - \bar{y}_{...} = \frac{4}{12} - \frac{13}{36} = \frac{-1}{36}$$

$$\hat{\tau}_3 = \bar{y}_{3..} - \bar{y}_{...} = \frac{14}{12} - \frac{13}{36} = \frac{29}{36}$$

To estimate the variance components σ^2 and σ_β^2, we eliminate the line in the analysis of variance table pertaining to suppliers and apply the analysis of variance estimation method to the next two lines. This yields

$$\hat{\sigma}^2 = MS_E = 2.64$$

and

$$\hat{\sigma}_\beta^2 = \frac{MS_{B(A)} - MS_E}{n} = \frac{7.77 - 2.64}{3} = 1.71$$

From the analysis in Example 12–1, we know that the τ_i do not differ significantly from zero, whereas the variance component σ_β^2 is greater than zero.

12-1.4   Staggered Nested Designs

A potential problem in the application of nested designs is that sometimes to get a reasonable number of degrees of freedom at the highest level, we can end up with many degrees of freedom (perhaps too many) at lower stages. To illustrate, suppose that we are investigating potential differences in chemical analysis among different lots of material. We plan to take five samples per lot, and each sample will be measured twice. If we want to estimate a variance component for lots, then 10 lots would not be an unreasonable choice. This results in 9 degrees of

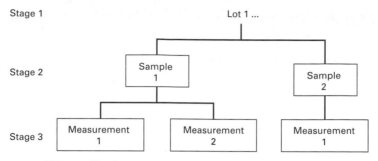

Figure 12–4. A three-stage staggered nested design.

freedom for lots, 40 degrees of freedom for samples, and 50 degrees of freedom for measurements.

One way to avoid this is to use a particular type of unbalanced nested design called a **staggered nested design.** A three-stage staggered nested design is shown in Figure 12–4. Notice that only two samples are taken from each lot; one of the samples is measured twice, while the other sample is measured once. If there are a lots, then there will be $a - 1$ degrees of freedom for lots (or, in general, the upper stage), and all lower stages will have exactly a degrees of freedom. For more information on the use and analysis of these designs, see Bainbridge (1965), Smith and Beverly (1981), and Nelson (1983, 1995a, 1995b).

12–2 THE GENERAL m-STAGE NESTED DESIGN

The results of Section 12–1 can be easily extended to the case of m completely nested factors. Such a design would be called an ***m*-stage nested design.** As an example, suppose a foundry wishes to investigate the hardness of two different formulations of a metal alloy. Three heats of each alloy formulation are prepared, two ingots are selected at random from each heat for testing, and two hardness measurements are made on each ingot. The situation is illustrated in Figure 12–5.

In this experiment, heats are nested under the levels of the factor alloy formulation, and ingots are nested under the levels of the factor heats. Thus, this is a three-stage nested design with two replicates.

The **model** for the general three-stage nested design is

$$y_{ijkl} = \mu + \tau_i + \beta_{j(i)} + \gamma_{k(ij)} + \epsilon_{(ijk)l} \qquad \begin{cases} i = 1, 2, \ldots, a \\ j = 1, 2, \ldots, b \\ k = 1, 2, \ldots, c \\ l = 1, 2, \ldots, n \end{cases} \qquad (12\text{–}13)$$

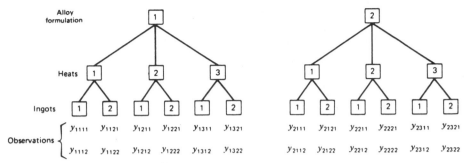

Figure 12-5. A three-stage nested design.

For our example, τ_i is the effect of the *i*th alloy formulation, $\beta_{j(i)}$ is the effect of the *j*th heat within the *i*th alloy, $\gamma_{k(ij)}$ is the effect of the *k*th ingot within the *j*th heat and *i*th alloy, and $\epsilon_{(ijk)l}$ is the usual NID$(0, \sigma^2)$ error term. Extension of this model to *m* factors is straightforward.

Notice that in the above example the overall variability in hardness consists of three components: one that results from alloy formulations, one that results from heats, and one that results from analytical test error. These components of the variability in overall hardness are illustrated in Figure 12-6.

This example demonstrates how the nested design is often used in analyzing processes to identify the major sources of variability in the output. For instance, if the alloy formulation variance component is large, then this implies that overall hardness variability could be reduced by using only one alloy formulation.

The calculation of the sums of squares and the analysis of variance for the *m*-stage nested design are similar to the analysis presented in Section 12-1. For example, the analysis of variance for the three-stage nested design is summarized in Table 12-6. Definitions of the sums of squares are also shown in this table.

Table 12-6 Analysis of Variance for the Three-Stage Nested Design

Source of Variation	Sum of Squares	Degrees of Freedom	Mean Square
A	$bcn \sum_i (\bar{y}_{i...} - \bar{y}_{....})^2$	$a - 1$	MS_A
B (within A)	$cn \sum_i \sum_j (\bar{y}_{ij..} - \bar{y}_{....})^2$	$a(b - 1)$	$MS_{B(A)}$
C (within B)	$n \sum_i \sum_j \sum_k (\bar{y}_{ijk.} - \bar{y}_{....})^2$	$ab(c - 1)$	$MS_{C(B)}$
Error	$\sum_i \sum_j \sum_k \sum_l (y_{ijkl} - \bar{y}_{ijk.})^2$	$abc(n - 1)$	MS_E
Total	$\sum_i \sum_j \sum_k \sum_l (y_{ijkl} - \bar{y}_{....})^2$	$abcn - 1$	

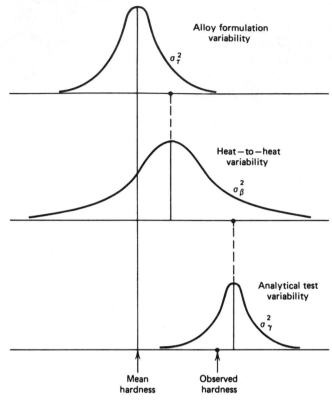

Figure 12–6.   Sources of variation in the three-stage nested design example.

Table 12–7   Expected Mean Square Derivation for a Three-Stage Nested Design with A and B Fixed and C Random

Factor	$\begin{matrix}F\\a\\i\end{matrix}$	$\begin{matrix}F\\b\\j\end{matrix}$	$\begin{matrix}R\\c\\k\end{matrix}$	$\begin{matrix}R\\n\\l\end{matrix}$	Expected Mean Square
τ_i	0	b	c	n	$\sigma^2 + n\sigma_\gamma^2 + \dfrac{bcn\sum \tau_i^2}{a-1}$
$\beta_{j(i)}$	1	0	c	n	$\sigma^2 + n\sigma_\gamma^2 + \dfrac{cn\sum\sum \beta_{j(i)}^2}{a(b-1)}$
$\gamma_{k(ij)}$	1	1	1	n	$\sigma^2 + n\sigma_\gamma^2$
$\epsilon_{l(ijk)}$	1	1	1	1	σ^2

Notice that they are a simple extension of the formulas for the two-stage nested design.

To determine the proper test statistics we must find the expected mean squares using the methods of Chapter 11. For example, if factors A and B are fixed and factor C is random, then we may derive the expected mean squares as shown in Table 12–7. This table indicates the proper test statistics for this situation.

12–3 DESIGNS WITH BOTH NESTED AND FACTORIAL FACTORS

Occasionally in a multifactor experiment, some factors will be arranged in a factorial layout and other factors will be nested. We sometimes call these designs **nested-factorial designs.** The statistical analysis of one such design with three factors is illustrated in the following example.

Example 12–2

An industrial engineer is studying the hand insertion of electronic components on printed circuit boards to improve the speed of the assembly operation. He has designed three assembly fixtures and two workplace layouts that seem promising. Operators are required to perform the assembly, and it is decided to randomly select four operators for each fixture–layout combination. However, because the workplaces are in different locations within the plant, it is difficult to use the *same* four operators for each layout. Therefore, the four operators chosen for layout 1 are different individuals from the four operators chosen for layout 2. The treatment combinations in this design are run in random order and two replicates are obtained. The assembly times are measured in seconds and are shown in Table 12–8.

In this experiment, operators are nested within the levels of layouts, whereas fixtures and layouts are arranged in a factorial. Thus, this design has both nested and factorial factors. The **linear model** for this design is

$$y_{ijkl} = \mu + \tau_i + \beta_j + \gamma_{k(j)} + (\tau\beta)_{ij} + (\tau\gamma)_{ik(j)} + \epsilon_{(ijk)l} \qquad \begin{cases} i = 1, 2, 3 \\ j = 1, 2 \\ k = 1, 2, 3, 4 \\ l = 1, 2 \end{cases} \qquad (12\text{–}14)$$

where τ_i is the effect of the ith fixture, β_j is the effect of the jth layout, $\gamma_{k(j)}$ is the effect of the kth operator within the jth level of layout, $(\tau\beta)_{ij}$ is the fixture $\times$ layout interaction, $(\tau\gamma)_{ik(j)}$ is the fixture $\times$ operators within layout interaction, and $\epsilon_{(ijk)l}$ is the usual error term. Notice that no layout $\times$ operator interaction can exist because all the operators do not use all the layouts. Similarly, there can be no three-way fixture $\times$ layout $\times$ operator interaction. Furthermore, fixtures and layouts are fixed factors, whereas the operator is a random factor. The expected mean squares are derived in Table 12–9. The proper test statistic for any effect or interaction can be found from inspection of this table.

Table 12–8   Assembly Time Data For Example 12–2

Operator	Layout 1				Layout 2				$y_{i...}$
	1	2	3	4	1	2	3	4	
Fixture 1	22	23	28	25	26	27	28	24	404
	24	24	29	23	28	25	25	23	
Fixture 2	30	29	30	27	29	30	24	28	447
	27	28	32	25	28	27	23	30	
Fixture 3	25	24	27	26	27	26	24	28	401
	21	22	25	23	25	24	27	27	
Operator totals, $y_{.jk.}$	149	150	171	149	163	159	151	160	
Layout totals, $y_{.j..}$			619				633		$1252 = y_{....}$

The complete analysis of variance is shown in Table 12–10. The SAS software system was used to perform the calculation. We see that assembly fixtures are significant and that operators within layouts also differ significantly. There is also a significant interaction between fixtures and operators within layouts, indicating that the effects of the different fixtures are not the same for all operators. The workplace layouts seem to have little effect on the assembly time. Therefore, to minimize assembly time, we should concentrate on fixture types 1 and 3. (Note that the fixture totals in Table 12–8 are smaller for fixture types 1 and 3 than for type 2. This difference in fixture type means could be formally tested using multiple comparisons.) Furthermore, the interaction between operators and fixtures implies that some operators are more effective than others using the same fixtures. Perhaps these operator–fixture effects could be isolated and the less-effective operators' performance improved by retraining them.

■

If specialized computer software such as SAS is not available, then a program for the analysis of factorial experiments may also be used to analyze designs

Table 12–9   Expected Mean Square Derivation for Example 12–2

Factor	F 3 i	F 2 j	R 4 k	R 2 l	Expected Mean Square
τ_i	0	2	4	2	$\sigma^2 + 2\sigma_{\tau\gamma}^2 + 8\sum \tau_i^2$
β_j	3	0	4	2	$\sigma^2 + 6\sigma_\gamma^2 + 24\sum \beta_j^2$
$\gamma_{k(j)}$	3	1	1	2	$\sigma^2 + 6\sigma_\gamma^2$
$(\tau\beta)_{ij}$	0	0	4	2	$\sigma^2 + 2\sigma_{\tau\gamma}^2 + 4\sum\sum (\tau\beta)_{ij}^2$
$(\tau\gamma)_{ik(j)}$	0	1	1	2	$\sigma^2 + 2\sigma_{\tau\gamma}^2$
$\epsilon_{(ijk)l}$	1	1	1	1	σ^2

Table 12–10 Analysis of Variance for Example 12–2

Source of Variation	Sum of Squares	Degrees of Freedom	Mean Square	F_0	P-Value
Fixtures (F)	82.80	2	41.40	7.54	0.01
Layouts (L)	4.08	1	4.09	0.34	0.58
Operators (within layouts), $O(L)$	71.91	6	11.99	5.15	<0.01
FL	19.04	2	9.52	1.73	0.22
$FO(L)$	65.84	12	5.49	2.36	0.04
Error	56.00	24	2.33		
Total	299.67	47			

with nested and factorial factors. For instance, the experiment in Example 12–2 could be considered as a three-factor factorial, with fixtures (F), operators (O), and layouts (L) as the factors. Then certain sums of squares and degrees of freedom from the factorial analysis would be pooled to form the appropriate quantities required for the design with nested and factorial factors as follows:

Factorial Analysis		Nested-Factorial Analysis	
Sum of Squares	Degrees of Freedom	Sum of Squares	Degrees of Freedom
SS_F	2	SS_F	2
SS_L	1	SS_L	1
SS_{FL}	2	SS_{FL}	2
SS_O	3	$SS_{O(L)} = SS_O + SS_{LO}$	6
SS_{LO}	3		
SS_{FO}	6	$SS_{FO(L)} = SS_{FO} + SS_{FOL}$	12
SS_{FOL}	6		
SS_E	24	SS_E	24
SS_T	47	SS_T	47

12–4 THE SPLIT-PLOT DESIGN

In some multifactor designs involving randomized blocks, we may be unable to completely randomize the order of the runs within the block. This often results in a generalization of the randomized block design called a **split-plot design.**

As an example, consider a paper manufacturer who is interested in three different pulp preparation methods and four different cooking temperatures for the pulp and who wishes to study the effect of these two factors on the tensile strength of the paper. Each replicate of a factorial experiment requires 12 observations, and the experimenter has decided to run three replicates. However, the pilot plant is only capable of making 12 runs per day, so the experimenter decides to run one replicate on each of the three days and to consider the days or replicates as blocks. On any day, he conducts the experiment as follows. A batch of pulp is produced by one of the three methods under study. Then this batch is divided into four samples, and each sample is cooked at one of the four temperatures. Then a second batch of pulp is made up using another of the three methods. This second batch is also divided into four samples that are tested at the four temperatures. The process is then repeated, using a batch of pulp produced by the third method. The data are shown in Table 12–11.

Initially, we might consider this to be a factorial experiment with three levels of preparation method (factor A) and four levels of temperature (factor B) in a randomized block. If this is the case, then the order of experimentation within the block should be completely randomized. That is, within a block, we should randomly select a treatment combination (a preparation method and a temperature) and obtain an observation, then we should randomly select another treatment combination and obtain a second observation, and so on, until the 12 observations in the block have been taken. However, the experimenter did not collect the data this way. He made up a batch of pulp and obtained observations for all four temperatures from that batch. Because of the economics of preparing the batches and the size of the batches, this is the only feasible way to run this experiment.

Each block in the design is divided into three parts called **whole plots,** and the preparation methods are called the **whole plot** or **main treatments.** Each whole plot is divided into four parts called **subplots** (or **split-plots**), and one temperature is assigned to each. Temperature is called the **subplot treatment.** Note that if there are other uncontrolled or undesigned factors present, and if these uncontrolled factors vary as the pulp preparation methods are changed, then any effect of the undesigned factors on the response will be completely confounded with the effect of the pulp preparation methods. Since the whole plot treatments in a split-plot design are confounded with the whole plots and

Table 12–11   The Experiment on the Tensile Strength of Paper

Pulp Preparation Method	Block 1			Block 2			Block 3		
	1	2	3	1	2	3	1	2	3
Temperature (°F)									
200	30	34	29	28	31	31	31	35	32
225	35	41	26	32	36	30	37	40	34
250	37	38	33	40	42	32	41	39	39
275	36	42	36	41	40	40	40	44	45

the subplot treatments are not confounded, it is best to assign the factor we are most interested in to the subplots, if possible.

The **linear model** for the split-plot design is

$$y_{ijk} = \mu + \tau_i + \beta_j + (\tau\beta)_{ij} + \gamma_k + (\tau\gamma)_{ik}$$

$$+ (\beta\gamma)_{jk} + (\tau\beta\gamma)_{ijk} + \epsilon_{ijk} \quad \begin{cases} i = 1, 2, \ldots, a \\ j = 1, 2, \ldots, b \\ k = 1, 2, \ldots, c \end{cases} \quad (12\text{--}15)$$

where τ_i, β_j, and $(\tau\beta)_{ij}$ represent the whole plot and correspond respectively to blocks (factor A), main treatments (factor B), and **whole plot error** (AB); and γ_k, $(\tau\gamma)_{ik}$, $(\beta\gamma)_{jk}$, and $(\tau\beta\gamma)_{ijk}$ represent the subplot and correspond respectively to the subplot treatment (factor C), the AC and BC interactions, and the **subplot error** (ABC). Note that the whole plot error is the AB interaction and the subplot error is the three-factor interaction ABC. The sums of squares for these factors are computed as in the three-way analysis of variance without replication.

The expected mean squares for the split-plot design, with blocks random and main treatments and subplot treatments fixed, are derived in Table 12–12. Note that the main factor (B) in the whole plot is tested against the whole plot error, whereas the subtreatment (C) is tested against the block $\times$ subtreatment (AC) interaction. The BC interaction is tested against the subplot error. Notice

Table 12–12   Expected Mean Square Derivation for Split-Plot Design

	Factor	a R i	b F j	c F k	1 R h	Expected Mean Square
	τ_i	1	b	c	1	$\sigma^2 + bc\sigma_\tau^2$
Whole plot	β_j	a	0	c	1	$\sigma^2 + c\sigma_{\tau\beta}^2 + \dfrac{ac\sum\beta_j^2}{b-1}$
	$(\tau\beta)_{ij}$	1	0	c	1	$\sigma^2 + c\sigma_{\tau\beta}^2$
	γ_k	a	b	0	1	$\sigma^2 + b\sigma_{\tau\gamma}^2 + \dfrac{ab\sum\gamma_k^2}{(c-1)}$
	$(\tau\gamma)_{ik}$	1	b	0	1	$\sigma^2 + b\sigma_{\tau\gamma}^2$
Subplot	$(\beta\gamma)_{jk}$	a	0	0	1	$\sigma^2 + \sigma_{\tau\beta\gamma}^2 + \dfrac{a\sum\sum(\beta\gamma)_{jk}^2}{(b-1)(c-1)}$
	$(\tau\beta\gamma)_{ijk}$	1	0	0	1	$\sigma^2 + \sigma_{\tau\beta\gamma}^2$
	$\epsilon_{(ijk)h}$	1	1	1	1	σ^2 (not estimable)

that there are no tests for the block effect (A) or the block $\times$ subtreatment (AC) interaction. Some authors believe that the subplot error should be found by pooling the ABC and AC interactions and then testing both the subtreatment and the BC interaction by this pooled estimate. If the experimenter is reasonably confortable with the assumption that blocks do not interact with the various treatments, this is appropriate. On the other hand, if one cannot be reasonably sure that the interactions are negligible, then pooling is inadvisable.

The analysis of variance for the tensile strength data in Table 12–11 is summarized in Table 12–13. Since both preparation methods and temperatures are fixed and blocks are random, the expected mean squares in Table 12–12 apply. The mean square for preparation methods is compared to the whole plot error mean square, and the mean square for temperatures is compared to the block $\times$ temperature AC mean square. Finally, the preparation method $\times$ temperature mean square is tested against the subplot error. Both preparation methods and temperature have a significant effect on strength.

Note from Table 12–13 that the subplot error (4.24) is less than the whole plot error (9.07). This is the usual case in split-plot designs since the subplots are generally more homogeneous than the whole plots. This results in *two different error structures* for the experiment. Because the subplot treatments are compared with greater precision, it is preferable to assign the treatment we are most interested in to the subplots, if possible.

The split-plot design has an agricultural heritage, with the whole plots usually being large areas of land and the subplots being smaller areas of land within the large areas. For example, several varieties of a crop could be planted in different fields (whole plots), one variety to a field. Then each field could be divided into, say, four subplots, and each subplot could be treated with a different type of fertilizer. Here the crop varieties are the main treatments and the different fertilizers are the subtreatments.

Table 12–13   Analysis of Variance for the Split-Plot Design Using Tensile Strength Data from Table 12–11

Source of Variation	Sum of Squares	Degrees of Freedom	Mean Square	F_0	P-Value
Blocks (A)	77.55	2	38.78		
Preparation method (B)	128.39	2	64.20	7.08	0.05
AB (whole plot error)	36.28	4	9.07		
Temperature (C)	434.08	3	144.69	41.94	<0.01
AC	20.67	6	3.45		
BC	75.17	6	12.53	2.96	0.05
ABC (subplot error)	50.83	12	4.24		
Total	822.97	35			

Despite its agricultural basis, the split-plot design is useful in many scientific and industrial experiments. In these experimental settings, it is not unusual to find that some factors require large experimental units whereas other factors require small experimental units, such as in the tensile strength problem described above. Alternatively, we sometimes find that complete randomization is not feasible because it is more difficult to change the levels of some factors than others. The hard-to-vary factors form the whole plots whereas the easy-to-vary factors are run in the subplots.

In principle, we must carefully consider how the experiment must be conducted and incorporate all restrictions on randomization into the analysis. We illustrate this point using a modification of the eye focus time experiment in Chapter 9. Suppose there are only two factors, visual acuity (A) and illumination level (B). A factorial experiment with a levels of acuity, b levels of illumination, and n replicates would require that all abn observations be taken in random order. However, in the test apparatus, it is fairly difficult to adjust these two factors to different levels, so the experimenter decides to obtain the n replicates by adjusting the device to one of the a acuities and one of the b illumination levels and running all n observations at once. In the factorial design, the error actually represents the scatter or noise in the system plus the ability of the subject to reproduce the same focus time. The model for the factorial design could be written as

$$y_{ijk} = \mu + \tau_i + \beta_j + (\tau\beta)_{ij} + \phi_{ijk} + \theta_{ijk} \quad \begin{cases} i = 1, 2, \ldots, a \\ j = 1, 2, \ldots, b \\ k = 1, 2, \ldots, n \end{cases} \quad (12\text{--}16)$$

where ϕ_{ijk} represents the scatter or noise in the system that results from "experimental error" (that is, our failure to duplicate exactly the same levels of acuity and illumination on different runs, variability in environmental conditions, and the like), and θ_{ijk} represents the "reproducibility error" of the subject. Usually, we combine these components into one overall error term, say $\epsilon_{ijk} = \phi_{ijk} + \theta_{ijk}$. Assume that $V(\epsilon_{ijk}) = \sigma^2 = \sigma_\phi^2 + \sigma_\theta^2$. Now, in the factorial design, the error mean square has expectation $\sigma^2 = \sigma_\phi^2 + \sigma_\theta^2$, with $ab(n-1)$ degrees of freedom.

If we restrict the randomization as in the second design above, then the "error" mean square in the analysis of variance provides an estimate of the "reproducibility error" σ_θ^2 with $ab(n-1)$ degrees of freedom, but it yields no information on the "experimental error" σ_ϕ^2. Thus, the mean square for error in this second design is too small; consequently, we will wrongly reject the null hypothesis very frequently. As pointed out by John (1971), this design is similar to a split-plot design with ab whole plots, each divided into n subplots, and no subtreatment. The situation is also similar to **subsampling,** as described by Ostle (1963). Assuming that A and B are fixed, the expected mean squares in this case are

$$E(MS_A) = \sigma_\theta^2 + n\sigma_\phi^2 + \frac{bn \sum \tau_i^2}{a - 1}$$

$$E(MS_B) = \sigma_\theta^2 + n\sigma_\phi^2 + \frac{an \sum \beta_j^2}{b - 1} \qquad (12\text{–}17)$$

$$E(MS_{AB}) = \sigma_\theta^2 + n\sigma_\phi^2 + \frac{n \sum\sum (\tau\beta)_{ij}^2}{(a - 1)(b - 1)}$$

$$E(MS_E) = \sigma_\theta^2$$

Thus, there are no tests on the main effects unless interaction is negligible. The situation is exactly that of a two-way analysis of variance with one observation per cell. If both factors are random, then the main effects may be tested against the AB interaction. If only one factor is random, then the fixed factor can be tested against the AB interaction.

In general, if one analyzes a factorial design and all the main effects and interactions are significant, then one should examine carefully *how the experiment was actually conducted.* There may be randomization restrictions in the model not accounted for in the analysis, and consequently, the data should not be analyzed as a factorial.

12–5   THE SPLIT–SPLIT-PLOT DESIGN

The concept of split-plot designs can be extended to situations in which randomization restrictions may occur at any number of levels within a block. If there are two levels of randomization restrictions within a block, then the layout is called a **split–split-plot design.** The following example illustrates such a design.

Example 12–3

A researcher is studying the absorption times of a particular type of antibiotic capsule. There are three technicians, three dosage strengths, and four capsule wall thicknesses of interest to the researcher. Each replicate of a factorial experiment would require 36 observations. The experimenter has decided on four replicates, and it is necessary to run each replicate on a different day. Thus, the days are blocks. Within a block (day), the experiment is performed by assigning a unit of antibiotic to a technician who conducts the experiment on the three dosage strengths and the four wall thicknesses. Once a particular dosage strength is formulated, all four wall thicknesses are tested at that strength. Then another dosage strength is selected and all four wall thicknesses are tested. Finally, the third dosage strength and the four wall thicknesses are tested. Meanwhile, two other laboratory technicians also follow this plan, each starting with a unit of antibiotic.

Note that there are *two* randomization restrictions within a block: technician and dosage strength. The whole plots correspond to the technician. The order in which the technicians are assigned the units of antibiotic is randomly determined. The dosage strengths form three subplots. Dosage strength may be randomly assigned to a subplot. Finally, within a particular dosage strength, the four capsule wall thicknesses are tested in random order, forming four sub-subplots. The wall thicknesses are usually called sub-subtreatments. Because there are *two* randomization restrictions within a block (some authors say two "splits" in the design), the design is called a split–split-plot design. Figure 12–7 illustrates the randomization restrictions and experimental layout in this design. See page 528.

■

The linear statistical model for the split–split-plot design is

$$
\begin{aligned}
y_{ijkh} = {} & \mu + \tau_i + \beta_j + (\tau\beta)_{ij} + \gamma_k + (\tau\gamma)_{ik} + (\beta\gamma)_{jk} + (\tau\beta\gamma)_{ijk} \\
& + \delta_h + (\tau\delta)_{ih} + (\beta\delta)_{jh} \\
& + (\tau\beta\delta)_{ijh} + (\gamma\delta)_{kh} + (\tau\gamma\delta)_{ikh} + (\beta\gamma\delta)_{jkh} \\
& + (\tau\beta\gamma\delta)_{ijkh} + \epsilon_{ijkh}
\end{aligned}
\qquad
\begin{cases}
i = 1, 2, \ldots, a \\
j = 1, 2, \ldots, b \\
k = 1, 2, \ldots, c \\
h = 1, 2, \ldots, d
\end{cases}
\qquad (12\text{–}18)
$$

where τ_i, β_j, and $(\tau\beta)_{ij}$ represent the whole plot and correspond to blocks (factor A), main treatments (factor B), and whole plot error (AB), respectively; and γ_k, $(\tau\gamma)_{jk}$, $(\beta\gamma)_{jk}$, and $(\tau\beta\gamma)_{ijk}$ represent the subplot and correspond to the subplot treatment (factor C), the AC and BC interactions, and the subplot error, respectively; and δ_h and the remaining parameters correspond to the sub-subplot and represent respectively the sub-subplot treatment (factor D) and the remaining interactions. The four-factor interaction $(\tau\beta\gamma\delta)_{ijkh}$ is called the sub-subplot error.

Assuming that blocks are random and that the other factors are fixed, we may derive the expected mean squares as shown in Table 12–14. See page 529. Tests on the main treatments, subtreatments, sub-subtreatments and their interactions are obvious from inspection of this table. Note that no tests on blocks or interactions involving the blocks exist.

The statistical analysis of a split–split-plot design is like that of a single replicate of a four-factor factorial. The number of degrees of freedom for each test are determined in the usual manner. To illustrate, in Example 12–3, where we had four blocks, three technicians, three dosage strengths, and four wall thicknesses, we would have only $(a - 1)(b - 1) = (4 - 1)(3 - 1) = 6$ whole plot error degrees of freedom for testing technicians. This is a relatively small number of degrees of freedom, and the experimenter might consider using additional blocks to increase the precision of the test. If there are a blocks, then we will have $2(a - 1)$ degrees of freedom for whole plot error. Thus, five blocks

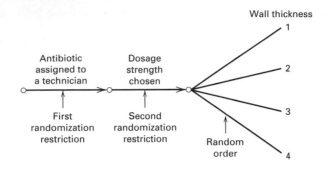

		Technician								
		1			2			3		
Blocks	Dosage strength	1	2	3	1	2	3	1	2	3
1	Wall thicknesses	1	1	1	1	1	1	1	1	1
		2	2	2	2	2	2	2	2	2
		3	3	3	3	3	3	3	3	3
		4	4	4	4	4	4	4	4	4
2	Wall thicknesses	1	1	1	1	1	1	1	1	1
		2	2	2	2	2	2	2	2	2
		3	3	3	3	3	3	3	3	3
		4	4	4	4	4	4	4	4	4
3	Wall thicknesses	1	1	1	1	1	1	1	1	1
		2	2	2	2	2	2	2	2	2
		3	3	3	3	3	3	3	3	3
		4	4	4	4	4	4	4	4	4
4	Wall thicknesses	1	1	1	1	1	1	1	1	1
		2	2	2	2	2	2	2	2	2
		3	3	3	3	3	3	3	3	3
		4	4	4	4	4	4	4	4	4

Figure 12–7. A split–split-plot design.

yield $2(5 - 1) = 8$ error degrees of freedom, six blocks yield $2(6 - 1) = 10$ error degrees of freedom, seven blocks yield $2(7 - 1) = 12$ error degrees of freedom, and so on. Consequently, we would probably not want to run fewer than four blocks since this would yield only 4 error degrees of freedom. Each additional block allows us to gain 2 degrees of freedom for error. If we could afford to run five blocks, we could increase the precision of the test by one-third (from 6 to 8 degrees of freedom). Also, in going from five to six blocks, there is an additional 25 percent gain in precision. If resources permit, the experimenter should run five or six blocks.

Table 12–14 Expected Mean Square Derivation for the Split–Split-Plot Design

	Factor	a R i	b F j	c F k	d F h	1 R l	Expected Mean Square
	τ_i	1	b	c	d	1	$\sigma^2 + bcd\sigma_\tau^2$
Whole plot	β_j	a	0	c	d	1	$\sigma^2 + cd\sigma_{\tau\beta}^2 + \dfrac{acd\sum \beta_j^2}{(b-1)}$
	$(\tau\beta)_{ij}$	1	0	c	d	1	$\sigma^2 + cd\sigma_{\tau\beta}^2$
	γ_k	a	b	0	d	1	$\sigma^2 + bd\sigma_{\tau\gamma}^2 + \dfrac{abd\sum \gamma_k^2}{(c-1)}.$
	$(\tau\gamma)_{ik}$	1	b	0	d	1	$\sigma^2 + bd\sigma_{\tau\gamma}^2$
Subplot	$(\beta\gamma)_{jk}$	a	0	0	d	1	$\sigma^2 + d\sigma_{\tau\beta\gamma}^2 + \dfrac{ad\sum\sum (\beta\gamma)_{jh}^2}{(b-1)(c-1)}$
	$(\tau\beta\gamma)_{ijk}$	1	0	0	d	1	$\sigma^2 + d\sigma_{\tau\beta\gamma}^2$
	δ_h	a	b	c	0	1	$\sigma^2 + bc\sigma_{\tau\delta}^2 + \dfrac{abc\sum \gamma_k^2}{(c-1)}$
	$(\tau\delta)_{ih}$	1	b	c	0	1	$\sigma^2 + bc\sigma_{\tau\delta}^2$
	$(\beta\delta)_{jh}$	a	0	c	0	1	$\sigma^2 + c\sigma_{\tau\beta\delta}^2 + \dfrac{ac\sum\sum (\beta\delta)_{jh}^2}{(b-1)(d-1)}$
	$(\tau\beta\delta)_{ijh}$	1	0	c	0	1	$\sigma^2 + c\sigma_{\tau\beta\delta}^2$
Sub-subplot	$(\gamma\delta)_{kh}$	a	b	0	0	1	$\sigma^2 + b\sigma_{\tau\gamma\delta}^2 + \dfrac{ab\sum\sum (\gamma\delta)_{kh}^2}{(c-1)(d-1)}$
	$(\tau\gamma\delta)_{ikh}$	1	b	0	0	1	$\sigma^2 + b\sigma_{\tau\gamma\delta}^2$
	$(\beta\gamma\delta)_{jkh}$	a	0	0	0	1	$\sigma^2 + \sigma_{\tau\beta\gamma\delta}^2 + \dfrac{a\sum\sum\sum (\beta\gamma\delta)_{ijk}^2}{(b-1)(c-1)(d-1)}$
	$(\tau\beta\gamma\delta)_{ijkh}$	1	0	0	0	1	$\sigma^2 + \sigma_{\tau\beta\gamma\delta}^2$
	$\epsilon_{l(ijkh)}$	1	1	1	1	1	σ^2 (not estimable)

12–6 PROBLEMS

12–1 An ammunition manufacturer is studying the burning rate of powder from three production processes. Four batches of powder are randomly selected from the output of each process and three determinations of burning rate are made on each batch. The results follow. Analyze the data and draw conclusions.

	Process 1				Process 2				Process 3			
Batch	1	2	3	4	1	2	3	4	1	2	3	4
	25	19	15	15	19	23	18	35	14	35	38	25
	30	28	17	16	17	24	21	27	15	21	54	29
	26	20	14	13	14	21	17	25	20	24	50	33

12–2 The surface finish of metal parts made on four machines is being studied. An experiment is conducted in which each machine is run by three different operators and two specimens from each operator are collected and tested. Because of the location of the machines, different operators are used on each machine, and the operators are chosen at random. The data are shown in the following table. Analyze the data and draw conclusions.

	Machine 1			Machine 2			Machine 3			Machine 4		
Operator	1	2	3	1	2	3	1	2	3	1	2	3
	79	94	46	92	85	76	88	53	46	36	40	62
	62	74	57	99	79	68	75	56	57	53	56	47

12–3 A manufacturing engineer is studying the dimensional variability of a particular component that is produced on three machines. Each machine has two spindles, and four components are randomly selected from each spindle. The results follow. Analyze the data, assuming that machines and spindles are fixed factors.

	Machine 1		Machine 2		Machine 3	
Spindle	1	2	1	2	1	2
	12	8	14	12	14	16
	9	9	15	10	10	15
	11	10	13	11	12	15
	12	8	14	13	11	14

12–4 To simplify production scheduling, an industrial engineer is studying the possibility of assigning one time standard to a particular class of jobs, believing that differences between jobs are negligible. To see if this simplification is possible, six jobs are randomly selected. Each job is given to a different group of three operators. Each operator completes the job twice at different times during the week, and the following results are obtained. What are your conclusions about the use of a common time standard for all jobs in this class? What value would you use for the standard?

Job	Operator 1		Operator 2		Operator 3	
1	158.3	159.4	159.2	159.6	158.9	157.8
2	154.6	154.9	157.7	156.8	154.8	156.3
3	162.5	162.6	161.0	158.9	160.5	159.5
4	160.0	158.7	157.5	158.9	161.1	158.5
5	156.3	158.1	158.3	156.9	157.7	156.9
6	163.7	161.0	162.3	160.3	162.6	161.8

12–5 Consider the three-stage nested design shown in Figure 12–5 to investigate alloy hardness. Using the data that follow, analyze the design, assuming that alloy chemistry and heats are fixed factors and ingots are random.

Alloy Chemistry											
	1						2				
Heats	1		2		3		1		2		3	
Ingots	1	2	1	2	1	2	1	2	1	2	1	2
	40	27	95	69	65	78	22	23	83	75	61	35
	63	30	67	47	54	45	10	39	62	64	77	42

12–6 Derive the expected mean squares for a balanced three-stage nested design, assuming that A is fixed and that B and C are random. Obtain formulas for estimating the variance components.

12–7 Derive the expected mean squares for a balanced three-stage nested design if all three factors are random. Obtain formulas for estimating the variance components.

12–8 Verify the expected mean squares given in Table 12–1.

12–9 *Unbalanced nested designs.* Consider an unbalanced two-stage nested design with b_j levels of B under the ith level of A and n_{ij} replicates in the ijth cell.

(a) Write down the least squares normal equations for this situation. Solve the normal equations.

(b) Construct the analysis of variance table for the unbalanced two-stage nested design.

(c) Analyze the following data, using the results in part (b).

Factor A	1		2		
Factor B	1	2	1	2	3
	6	-3	5	2	1
	4	1	7	4	0
	8		9	3	-3
			6		

12–10 ***Variance components in the unbalanced two-stage nested design.*** Consider the
model

$$y_{ijk} = \mu + \tau_i + \beta_{j(i)} + \epsilon_{k(ij)} \qquad \begin{cases} i = 1, 2, \ldots, a \\ j = 1, 2, \ldots, b_i \\ k = 1, 2, \ldots, n_{ij} \end{cases}$$

where A and B are random factors. Show that

$$E(MS_A) = \sigma^2 + c_1\sigma_\beta^2 + c_2\sigma_\tau^2$$
$$E(MS_{B(A)}) = \sigma^2 + c_0\sigma_\beta^2$$
$$E(MS_E) = \sigma^2$$

where

$$c_0 = \frac{N - \sum\limits_{i=1}^{a}\left(\sum\limits_{j=1}^{b_i} n_{ij}^2/n_{i.}\right)}{b - a}$$

$$c_1 = \frac{\sum\limits_{i=1}^{a}\left(\sum\limits_{j=1}^{b_i} n_{ij}^2/n_{i.}\right) - \sum\limits_{i=1}^{a}\sum\limits_{j=1}^{b_i} n_{ij}^2/N}{a - 1}$$

$$c_2 = \frac{N - \dfrac{\sum\limits_{i=1}^{a} n_{i.}^2}{N}}{a - 1}$$

12–11 A process engineer is testing the yield of a product manufactured on three
machines. Each machine can be operated at two power settings. Furthermore,
a machine has three stations on which the product is formed. An experiment is
conducted in which each machine is tested at both power settings, and three
observations on yield are taken from each station. The runs are made in random
order, and the results follow. Analyze this experiment, assuming that all three
factors are fixed.

Station	Machine 1			Machine 2			Machine 3		
	1	2	3	1	2	3	1	2	3
	34.1	33.7	36.2	32.1	33.1	32.8	32.9	33.8	33.6
Power setting 1	30.3	34.9	36.8	33.5	34.7	35.1	33.0	33.4	32.8
	31.6	35.0	37.1	34.0	33.9	34.3	33.1	32.8	31.7
	24.3	28.1	25.7	24.1	24.1	26.0	24.2	23.2	24.7
Power setting 2	26.3	29.3	26.1	25.0	25.1	27.1	26.1	27.4	22.0
	27.1	28.6	24.9	26.3	27.9	23.9	25.3	28.0	24.8

12-12 Suppose that in Problem 12-11 a large number of power settings could have been used and that the two selected for the experiment were chosen randomly. Obtain the expected mean squares for this situation and modify the previous analysis appropriately.

12-13 A structural engineer is studying the strength of aluminum alloy purchased from three vendors. Each vendor submits the alloy in standard-sized bars of 1.0, 1.5, or 2.0 inches. The processing of different sizes of bar stock from a common ingot involves different forging techniques, and so this factor may be important. Furthermore, the bar stock is forged from ingots made in different heats. Each vendor submits two test specimens of each size bar stock from three heats. The resulting strength data follow. Analyze the data, assuming that vendors and bar size are fixed and heats are random.

Heat	Vendor 1			Vendor 2			Vendor 3		
	1	2	3	1	2	3	1	2	3
Bar size: 1 inch	1.230	1.346	1.235	1.301	1.346	1.315	1.247	1.275	1.324
	1.259	1.400	1.206	1.263	1.392	1.320	1.296	1.268	1.315
$1\frac{1}{2}$ inch	1.316	1.329	1.250	1.274	1.384	1.346	1.273	1.260	1.392
	1.300	1.362	1.239	1.268	1.375	1.357	1.264	1.265	1.364
2 inch	1.287	1.346	1.273	1.247	1.362	1.336	1.301	1.280	1.319
	1.292	1.382	1.215	1.215	1.328	1.342	1.262	1.271	1.323

12-14 Suppose that in Problem 12-13 the bar stock may be purchased in many sizes and that the three sizes actually used in the experiment were selected randomly. Obtain the expected mean squares for this situation and modify the previous analysis appropriately.

12-15 Steel is normalized by heating above the critical temperature, soaking, and then air cooling. This process increases the strength of the steel, refines the grain, and homogenizes the structure. An experiment is performed to determine the effect of temperature and heat treatment time on the strength of normalized steel. Two temperatures and three times are selected. The experiment is performed by heating the oven to a randomly selected temperature and inserting three specimens. After 10 minutes one specimen is removed, after 20 minutes the second is removed, and after 30 minutes the final specimen is removed. Then the temperature is changed to the other level and the process is repeated. Four shifts are required to collect the data, which are shown below. Analyze the data and draw conclusions, assuming both factors are fixed.

		Temperature (°F)	
Shift	Time (minutes)	1500	1600
	10	63	89
1	20	54	91
	30	61	62
	10	50	80
2	20	52	72
	30	59	69
	10	48	73
3	20	74	81
	30	71	69
	10	54	88
4	20	48	92
	30	59	64

12–16   Repeat Problem 12–15, assuming that the factor levels for temperature and time are randomly selected.

12–17   An experiment is designed to study pigment dispersion in paint. Four different mixes of a particular pigment are studied. The procedure consists of preparing a particular mix and then applying that mix to a panel by three application methods (brushing, spraying, and rolling). The response measured is the percentage reflectance of pigment. Three days are required to run the experiment, and the data obtained follow. Analyze the data and draw conclusions, assuming that mixes and application methods are fixed.

	Application	Mix			
Day	Method	1	2	3	4
	1	64.5	66.3	74.1	66.5
1	2	68.3	69.5	73.8	70.0
	3	70.3	73.1	78.0	72.3
	1	65.2	65.0	73.8	64.8
2	2	69.2	70.3	74.5	68.3
	3	71.2	72.8	79.1	71.5
	1	66.2	66.5	72.3	67.7
3	2	69.0	69.0	75.4	68.6
	3	70.8	74.2	80.1	72.4

12-18 Repeat Problem 12-17, assuming that the mixes are random and the application methods are fixed.

12-19 Consider the split–split-plot design described in Example 12-3. Suppose that this experiment is conducted as described and that the data shown below are obtained. Analyze the data and draw conclusions.

| | | Technician | | | | | | | | |
| | | 1 | | | 2 | | | 3 | | |
Blocks	Dosage Strengths	1	2	3	1	2	3	1	2	3
	Wall Thickness									
	1	95	71	108	96	70	108	95	70	100
1	2	104	82	115	99	84	100	102	81	106
	3	101	85	117	95	83	105	105	84	113
	4	108	85	116	97	85	109	107	87	115
	1	95	78	110	100	72	104	92	69	101
2	2	106	84	109	101	79	102	100	76	104
	3	103	86	116	99	80	108	101	80	109
	4	109	84	110	112	86	109	108	86	113
	1	96	70	107	94	66	100	90	73	98
3	2	105	81	106	100	84	101	97	75	100
	3	106	88	112	104	87	109	100	82	104
	4	113	90	117	121	90	117	110	91	112
	1	90	68	109	98	68	106	98	72	101
4	2	100	84	112	102	81	103	102	78	105
	3	102	85	115	100	85	110	105	80	110
	4	114	88	118	118	85	116	110	95	120

12-20 Rework Problem 12-19, assuming that the dosage strengths are chosen at random.

12-21 Suppose that in Problem 12-19 four technicians had been used. Assuming that all the factors are fixed, how many blocks should be run to obtain an adequate number of degrees of freedom on the test for differences among technicians?

12-22 Consider the experiment described in Example 12-3. Demonstrate how the order in which the treatment combinations are run would be determined if this experiment were run as (a) a split–split-plot, (b) a split-plot, (c) a factorial design in a randomized block, and (d) a completely randomized factorial design.

Chapter 13
Fitting Regression Models

13–1 INTRODUCTION

In many problems there are two or more variables that are related, and it is of interest to model and explore this relationship. For example, in a chemical process the yield of product is related to the operating temperature. The chemical engineer may want to build a model relating yield to temperature and then use the model for prediction, process optimization, or process control.

In general, suppose that there is a single **dependent variable** or **response** y that depends on k **independent** or **regressor variables,** for example, $x_1, x_2, \ldots, x_k$. The relationship between these variables is characterized by a mathematical model called a **regression model.** The regression model is fit to a set of sample data. In some instances, the experimenter knows the exact form of the true functional relationship between y and $x_1, x_2, \ldots, x_k$, say $y = \phi(x_1, x_2, \ldots, x_k)$. However, in most cases, the true functional relationship is unknown, and the experimenter chooses an appropriate function to approximate ϕ. Low-order polynomial models are widely used as approximating functions.

There is a strong interplay between design of experiments and regression analysis. Throughout this book we have emphasized the importance of expressing the results of an experiment quantitatively, in terms of an **empirical model,** to facilitate understanding, interpretation, and implementation. Regression models are the basis for this. On numerous occasions we have shown the regression model that represented the results of an experiment. In this chapter, we present some aspects of fitting these models. More complete presentations of regression are available in Montgomery and Peck (1992) and Myers (1990).

Regression methods are frequently used to analyze data from **unplanned experiments,** such as might arise from observation of uncontrolled phenomena or historical records. Regression analysis is also highly useful in designed experiments where something has "gone wrong." We will illustrate some of these situations in this chapter.

13–2 LINEAR REGRESSION MODELS

We will focus on fitting linear regression models. To illustrate, suppose that we wish to develop an empirical model relating the viscosity of a polymer to the temperature and the catalyst feed rate. A model that might describe this relationship is

$$y = \beta_0 + \beta_1 x_1 + \beta_2 x_2 + \epsilon \tag{13–1}$$

where y represents the viscosity, x_1 represents the temperature, and x_2 represents the catalyst feed rate. This is a **multiple linear regression model** with two independent variables. We often call the independent variables **predictor variables** or **regressors.** The term *linear* is used because Equation 13–1 is a linear function of the unknown parameters β_0, β_1, and β_2. The model describes a plane in the two-dimensional x_1, x_2 space. The parameter β_0 defines the intercept of the plane. We sometimes call β_1 and β_2 *partial regression coefficients,* because β_1 measures the expected change in y per unit change in x_1 when x_2 is held constant, and β_2 measures the expected change in y per unit change in x_2 when x_1 is held constant.

In general, the response variable y may be related to k regressor variables. The model

$$y = \beta_0 + \beta_1 x_1 + \beta_2 x_2 + \cdots + \beta_k x_k + \epsilon \tag{13–2}$$

is called a *multiple linear regression model* with k regressor variables. The parameters $\beta_j, j = 0, 1, \ldots, k$, are called the *regression coefficients.* This model describes a hyperplane in the k-dimensional space of the regressor variables $\{x_j\}$. The parameter β_j represents the expected change in response y per unit change in x_j when all the remaining independent variables x_i $(i \neq j)$ are held constant.

Models that are more complex in appearance than Equation 13–2 may often still be analyzed by multiple linear regression techniques. For example, consider adding an interaction term to the first-order model in two variables, say

$$y = \beta_0 + \beta_1 x_1 + \beta_2 x_2 + \beta_{12} x_1 x_2 + \epsilon \tag{13–3}$$

If we let $x_3 = x_1 x_2$ and $\beta_3 = \beta_{12}$, then Equation 13–3 can be written as

$$y = \beta_0 + \beta_1 x_1 + \beta_2 x_2 + \beta_3 x_3 + \epsilon \tag{13–4}$$

which is a standard multiple linear regression model with three regressors. Recall that we presented empirical models like Equations 13–2 and 13–4 in several examples in Chapters 7, 8, and 9 to quantitatively express the results of a two-level factorial design. As another example, consider the second-order **response surface model** in two variables:

$$y = \beta_0 + \beta_1 x_1 + \beta_2 x_2 + \beta_{11} x_1^2 + \beta_{22} x_2^2 + \beta_{12} x_1 x_2 + \epsilon \tag{13–5}$$

If we let $x_3 = x_1^2$, $x_4 = x_2^2$, $x_5 = x_1 x_2$, $\beta_3 = \beta_{11}$, $\beta_4 = \beta_{22}$, and $\beta_5 = \beta_{12}$, then this becomes

$$y = \beta_0 + \beta_1 x_1 + \beta_2 x_2 + \beta_3 x_3 + \beta_4 x_4 + \beta_5 x_5 + \epsilon \tag{13–6}$$

which is a linear regression model. We have also seen this model in examples earlier in the text. In general, any regression model that is linear in the parameters (the β values) is a linear regression model, regardless of the shape of the response surface that it generates.

In this chapter we will summarize methods for estimating the parameters in multiple linear regression models. This is often called **model fitting.** We will also discuss methods for testing hypotheses and constructing confidence intervals for these models, as well as for checking the adequacy of the model fit. Our focus is primarily on those aspects of regression analysis useful in designed experiments. For more complete presentations of regression, refer to Montgomery and Peck (1992) and Myers (1990).

13–3   ESTIMATION OF THE PARAMETERS IN LINEAR REGRESSION MODELS

The method of least squares is typically used to estimate the regression coefficients in a multiple linear regression model. Suppose that $n > k$ observations on the response variable are available, say $y_1, y_2, \ldots, y_n$. Along with each observed response y_i, we will have an observation on each regressor variable and let x_{ij} denote the ith observation or level of variable x_j. The data will appear as in Table 13–1. We assume that the error term ϵ in the model has $E(\epsilon) = 0$ and $V(\epsilon) = \sigma^2$ and that the $\{\epsilon_i\}$ are uncorrelated random variables.

We may write the model equation (Equation 13–2) in terms of the observations in Table 13–1 as

$$
\begin{aligned}
y_i &= \beta_0 + \beta_1 x_{i1} + \beta_2 x_{i2} + \cdots + \beta_k x_{ik} + \epsilon_i \\
&= \beta_0 + \sum_{j=1}^{k} \beta_j x_{ij} + \epsilon_i \qquad i = 1, 2, \ldots, n
\end{aligned}
\tag{13-7}
$$

The method of least squares chooses the β's in Equation 13–7 so that the sum of the squares of the errors, ϵ_i, is minimized. The least squares function is

$$
\begin{aligned}
L &= \sum_{i=1}^{n} \epsilon_i^2 \\
&= \sum_{i=1}^{n} \left(y_i - \beta_0 - \sum_{j=1}^{k} \beta_j x_{ij} \right)^2
\end{aligned}
\tag{13-8}
$$

The function L is to be minimized with respect to $\beta_0, \beta_1, \ldots, \beta_k$. The least squares estimators, say $\hat{\beta}_0, \hat{\beta}_1, \ldots, \hat{\beta}_k$, must satisfy

$$
\left. \frac{\partial L}{\partial \beta_0} \right|_{\hat{\beta}_0, \hat{\beta}_1, \ldots, \hat{\beta}_k} = -2 \sum_{i=1}^{n} \left(y_i - \hat{\beta}_0 - \sum_{j=1}^{k} \hat{\beta}_j x_{ij} \right) = 0
\tag{13-9a}
$$

Table 13-1   Data for Multiple
Linear Regression

y	x_1	x_2	$\cdots$	x_k
y_1	x_{11}	x_{12}	$\cdots$	x_{1k}
y_2	x_{21}	x_{22}	$\cdots$	x_{2k}
$\vdots$	$\vdots$	$\vdots$		$\vdots$
y_n	x_{n1}	x_{n2}	$\cdots$	x_{nk}

and

$$\left.\frac{\partial L}{\partial \beta_j}\right|_{\hat{\beta}_0,\hat{\beta}_1,\dots,\hat{\beta}_k} = -2 \sum_{i=1}^{n} \left(y_i - \hat{\beta}_0 - \sum_{j=1}^{k} \hat{\beta}_j x_{ij} \right) x_{ij} = 0 \qquad j = 1, 2, \dots, k \quad (13\text{-}9\text{b})$$

Simplifying Equation 13-9, we obtain

$$n\hat{\beta}_0 + \hat{\beta}_1 \sum_{i=1}^{n} x_{i1} + \hat{\beta}_2 \sum_{i=1}^{n} x_{i2} + \cdots + \hat{\beta}_k \sum_{i=1}^{n} x_{ik} = \sum_{i=1}^{n} y_i$$

$$\hat{\beta}_0 \sum_{i=1}^{n} x_{i1} + \hat{\beta}_1 \sum_{i=1}^{n} x_{i1}^2 + \hat{\beta}_2 \sum_{i=1}^{n} x_{i1}x_{i2} + \cdots + \hat{\beta}_k \sum_{i=1}^{n} x_{i1}x_{ik} = \sum_{i=1}^{n} x_{i1}y_i \qquad (13\text{-}10)$$

$$\hat{\beta}_0 \sum_{i=1}^{n} x_{ik} + \hat{\beta}_i \sum_{i=1}^{n} x_{ik}x_{i1} + \hat{\beta}_2 \sum_{i=1}^{n} x_{ik}x_{i2} + \cdots + \hat{\beta}_k \sum_{i=1}^{n} x_{ik}^2 = \sum_{i=1}^{n} x_{ik}y_i$$

These equations are called the **least squares normal equations.** Note that there are $p = k + 1$ normal equations, one for each of the unknown regression coefficients. The solution to the normal equations will be the least squares estimators of the regression coefficients $\hat{\beta}_0, \hat{\beta}_1, \dots, \hat{\beta}_k$.

It is simpler to solve the normal equations if they are expressed in matrix notation. We now give a matrix development of the normal equations that parallels the development of Equation 13-10. The model in terms of the observations, Equation 13-7, may be written in matrix notation as

$$\mathbf{y} = \mathbf{X}\boldsymbol{\beta} + \boldsymbol{\epsilon}$$

where

$$\mathbf{y} = \begin{bmatrix} y_1 \\ y_2 \\ \vdots \\ y_n \end{bmatrix}, \qquad \mathbf{X} = \begin{bmatrix} 1 & x_{11} & x_{12} & \cdots & x_{1k} \\ 1 & x_{21} & x_{22} & \cdots & x_{2k} \\ \vdots & \vdots & \vdots & & \vdots \\ 1 & x_{n1} & x_{n2} & \cdots & x_{nk} \end{bmatrix},$$

$$\boldsymbol{\beta} = \begin{bmatrix} \beta_0 \\ \beta_1 \\ \vdots \\ \beta_k \end{bmatrix}, \quad \text{and} \quad \boldsymbol{\epsilon} = \begin{bmatrix} \epsilon_1 \\ \epsilon_2 \\ \vdots \\ \epsilon_n \end{bmatrix}$$

In general, $\mathbf{y}$ is an $(n \times 1)$ vector of the observations, $\mathbf{X}$ is an $(n \times p)$ matrix of the levels of the independent variables, $\boldsymbol{\beta}$ is a $(p \times 1)$ vector of the regression coefficients, and $\boldsymbol{\epsilon}$ is an $(n \times 1)$ vector of random errors.

We wish to find the vector of least squares estimators, $\hat{\boldsymbol{\beta}}$, that minimizes

$$L = \sum_{i=1}^{n} \epsilon_i^2 = \boldsymbol{\epsilon}'\boldsymbol{\epsilon} = (\mathbf{y} - \mathbf{X}\boldsymbol{\beta})'(\mathbf{y} - \mathbf{X}\boldsymbol{\beta})$$

Note that L may be expressed as

$$\begin{aligned} L &= \mathbf{y}'\mathbf{y} - \boldsymbol{\beta}'\mathbf{X}'\mathbf{y} - \mathbf{y}'\mathbf{X}\boldsymbol{\beta} + \boldsymbol{\beta}'\mathbf{X}'\mathbf{X}\boldsymbol{\beta} \\ &= \mathbf{y}'\mathbf{y} - 2\boldsymbol{\beta}'\mathbf{X}'\mathbf{y} + \boldsymbol{\beta}'\mathbf{X}'\mathbf{X}\boldsymbol{\beta} \end{aligned} \tag{13–11}$$

since $\boldsymbol{\beta}'\mathbf{X}'\mathbf{y}$ is a (1×1) matrix, or a scalar, and its transpose $(\boldsymbol{\beta}'\mathbf{X}'\mathbf{y})' = \mathbf{y}'\mathbf{X}\boldsymbol{\beta}$ is the same scalar. The least squares estimators must satisfy

$$\left. \frac{\partial L}{\partial \boldsymbol{\beta}} \right|_{\hat{\boldsymbol{\beta}}} = -2\mathbf{X}'\mathbf{y} + 2\mathbf{X}'\mathbf{X}\hat{\boldsymbol{\beta}} = \mathbf{0}$$

which simplifies to

$$\mathbf{X}'\mathbf{X}\hat{\boldsymbol{\beta}} = \mathbf{X}'\mathbf{y} \tag{13–12}$$

Equation 13–12 is the matrix form of the least squares normal equations. It is identical to Equation 13–10. To solve the normal equations, multiply both sides of Equation 13–12 by the inverse of $\mathbf{X}'\mathbf{X}$. Thus, the least squares estimator of $\boldsymbol{\beta}$ is

$$\hat{\boldsymbol{\beta}} = (\mathbf{X}'\mathbf{X})^{-1}\mathbf{X}'\mathbf{y} \tag{13–13}$$

It is easy to see that the matrix form of the normal equations is identical to the scalar form. Writing out Equation 13–12 in detail, we obtain

$$\begin{bmatrix} n & \sum_{i=1}^{n} x_{i1} & \sum_{i=1}^{n} x_{i2} & \cdots & \sum_{i=1}^{n} x_{ik} \\ \sum_{i=1}^{n} x_{i1} & \sum_{i=1}^{n} x_{i1}^2 & \sum_{i=1}^{n} x_{i1}x_{i2} & \cdots & \sum_{i=1}^{n} x_{i1}x_{ik} \\ \vdots & \vdots & \vdots & & \vdots \\ \sum_{i=1}^{n} x_{ik} & \sum_{i=1}^{n} x_{ik}x_{i1} & \sum_{i=1}^{n} x_{ik}x_{i2} & \cdots & \sum_{i=1}^{n} x_{ik}^2 \end{bmatrix} \begin{bmatrix} \hat{\beta}_0 \\ \hat{\beta}_1 \\ \vdots \\ \hat{\beta}_k \end{bmatrix} = \begin{bmatrix} \sum_{i=1}^{n} y_i \\ \sum_{i=1}^{n} x_{i1}y_i \\ \vdots \\ \sum_{i=1}^{n} x_{ik}y_i \end{bmatrix}$$

If the indicated matrix multiplication is performed, the scalar form of the normal equations (i.e., Equation 13–10) will result. In this form it is easy to see that

X'X is a $(p \times p)$ symmetric matrix and **X'y** is a $(p \times 1)$ column vector. Note the special structure of the **X'X** matrix. The diagonal elements of **X'X** are the sums of squares of the elements in the columns of **X**, and the off-diagonal elements are the sums of cross-products of the elements in the columns of **X**. Furthermore, note that the elements of **X'y** are the sums of cross-products of the columns of **X** and the observations $\{y_i\}$.

The fitted regression model is

$$\hat{\mathbf{y}} = \mathbf{X}\hat{\boldsymbol{\beta}} \tag{13-14}$$

In scalar notation, the fitted model is

$$\hat{y}_i = \hat{\beta}_0 + \sum_{i=1}^{k} \hat{\beta}_j x_{ij} \qquad i = 1, 2, \ldots, n$$

The difference between the actual observation y_i and the corresponding fitted value $\hat{y}_i$ is the **residual,** say $e_i = y_i - \hat{y}_i$. The $(n \times 1)$ vector of residuals is denoted by

$$\mathbf{e} = \mathbf{y} - \hat{\mathbf{y}} \tag{13-15}$$

Estimating σ^2 ▪ It is also usually necessary to estimate σ^2. To develop an estimator of this parameter, consider the sum of squares of the residuals, say

$$SS_E = \sum_{i=1}^{n} (y_i - \hat{y}_i)^2$$

$$= \sum_{i=1}^{n} e_i^2$$

$$= \mathbf{e}'\mathbf{e}$$

Substituting $\mathbf{e} = \mathbf{y} - \hat{\mathbf{y}} = \mathbf{y} - \mathbf{X}\hat{\boldsymbol{\beta}}$, we have

$$SS_E = (\mathbf{y} - \mathbf{X}\hat{\boldsymbol{\beta}})'(\mathbf{y} - \mathbf{X}\hat{\boldsymbol{\beta}})$$
$$= \mathbf{y}'\mathbf{y} - \hat{\boldsymbol{\beta}}'\mathbf{X}'\mathbf{y} - \mathbf{y}'\mathbf{X}\hat{\boldsymbol{\beta}} + \hat{\boldsymbol{\beta}}'\mathbf{X}'\mathbf{X}\hat{\boldsymbol{\beta}}$$
$$= \mathbf{y}'\mathbf{y} - 2\hat{\boldsymbol{\beta}}'\mathbf{X}'\mathbf{y} + \hat{\boldsymbol{\beta}}'\mathbf{X}'\mathbf{X}\hat{\boldsymbol{\beta}}$$

Because $\mathbf{X}'\mathbf{X}\hat{\boldsymbol{\beta}} = \mathbf{X}'\mathbf{y}$, this last equation becomes

$$SS_E = \mathbf{y}'\mathbf{y} - \hat{\boldsymbol{\beta}}'\mathbf{X}'\mathbf{y} \tag{13-16}$$

Equation 13-16 is called the **error** or **residual sum of squares,** and it has $n - p$ degrees of freedom associated with it. It can be shown that

$$E(SS_E) = \sigma^2(n - p)$$

so an unbiased estimator of σ^2 is given by

$$\hat{\sigma}^2 = \frac{SS_E}{n - p} \tag{13-17}$$

Properties of the Estimators ▪ The method of least squares produces an unbiased estimator of the parameter $\boldsymbol{\beta}$ in the linear regression model. This may be easily demonstrated by taking the expected value of $\hat{\boldsymbol{\beta}}$ as follows:

$$
\begin{aligned}
E(\hat{\boldsymbol{\beta}}) &= E[(\mathbf{X'X})^{-1}\mathbf{X'y}] \\
&= E[(\mathbf{X'X})^{-1}\mathbf{X'}(\mathbf{X}\boldsymbol{\beta} + \boldsymbol{\epsilon})] \\
&= E[(\mathbf{X'X})^{-1}\mathbf{X'X}\boldsymbol{\beta} + (\mathbf{X'X})^{-1}\mathbf{X'}\boldsymbol{\epsilon}] \\
&= \boldsymbol{\beta}
\end{aligned}
$$

since $E(\boldsymbol{\epsilon}) = \mathbf{0}$ and $(\mathbf{X'X})^{-1}\mathbf{X'X} = \mathbf{I}$. Thus, $\hat{\boldsymbol{\beta}}$ is an unbiased estimator of $\boldsymbol{\beta}$.

The variance property of $\hat{\boldsymbol{\beta}}$ is expressed in the **covariance matrix:**

$$
\mathrm{Cov}(\hat{\boldsymbol{\beta}}) \equiv E\{[\hat{\boldsymbol{\beta}} - E(\hat{\boldsymbol{\beta}})][\hat{\boldsymbol{\beta}} - E(\hat{\boldsymbol{\beta}})]'\} \tag{13–18}
$$

which is just a symmetric matrix whose ith main diagonal element is the variance of the individual regression coefficient $\hat{\beta}_i$ and whose (ij)th element is the covariance between $\hat{\beta}_i$ and $\hat{\beta}_j$. The covariance matrix of $\hat{\boldsymbol{\beta}}$ is

$$
\mathrm{Cov}(\hat{\boldsymbol{\beta}}) = \sigma^2(\mathbf{X'X})^{-1} \tag{13–19}
$$

Example 13–1

Sixteen observations on the viscosity of a polymer (y) and two process variables—reaction temperature (x_1) and catalyst feed rate (x_2)—are shown in Table 13–2. We will fit a multiple linear regression model

$$
y = \beta_0 + \beta_1 x_1 + \beta_2 x_2 + \epsilon
$$

to these data. The $\mathbf{X}$ matrix and $\mathbf{y}$ vector are

$$
\mathbf{X} =
\begin{bmatrix}
1 & 80 & 8 \\
1 & 93 & 9 \\
1 & 100 & 10 \\
1 & 82 & 12 \\
1 & 90 & 11 \\
1 & 99 & 8 \\
1 & 81 & 8 \\
1 & 96 & 10 \\
1 & 94 & 12 \\
1 & 93 & 11 \\
1 & 97 & 13 \\
1 & 95 & 11 \\
1 & 100 & 8 \\
1 & 85 & 12 \\
1 & 86 & 9 \\
1 & 87 & 12
\end{bmatrix}
, \quad
\mathbf{y} =
\begin{bmatrix}
2256 \\
2340 \\
2426 \\
2293 \\
2330 \\
2368 \\
2250 \\
2409 \\
2364 \\
2379 \\
2440 \\
2364 \\
2404 \\
2317 \\
2309 \\
2328
\end{bmatrix}
$$

Table 13–2 Viscosity Data for Example 13–1 (viscosity in centistokes @ 100°C)

Observation	Temperature $(x_1, °C)$	Catalyst Feed Rate $(x_2, lb/hr)$	Viscosity
1	80	8	2256
2	93	9	2340
3	100	10	2426
4	82	12	2293
5	90	11	2330
6	99	8	2368
7	81	8	2250
8	96	10	2409
9	94	12	2364
10	93	11	2379
11	97	13	2440
12	95	11	2364
13	100	8	2404
14	85	12	2317
15	86	9	2309
16	87	12	2328

The $\mathbf{X'X}$ matrix is

$$\mathbf{X'X} = \begin{bmatrix} 1 & 1 & \cdots & 1 \\ 80 & 93 & \cdots & 87 \\ 8 & 9 & \cdots & 12 \end{bmatrix} \begin{bmatrix} 1 & 80 & 8 \\ 1 & 93 & 9 \\ \vdots & \vdots & \vdots \\ 1 & 87 & 12 \end{bmatrix}$$

$$= \begin{bmatrix} 16 & 1458 & 164 \\ 1458 & 133,560 & 14,946 \\ 164 & 14,946 & 1,726 \end{bmatrix}$$

and the $\mathbf{X'y}$ vector is

$$\mathbf{X'y} = \begin{bmatrix} 1 & 1 & \cdots & 1 \\ 80 & 93 & \cdots & 87 \\ 8 & 9 & \cdots & 12 \end{bmatrix} \begin{bmatrix} 2256 \\ 2340 \\ \vdots \\ 2328 \end{bmatrix}$$

$$= \begin{bmatrix} 37,577 \\ 3,429,550 \\ 385,562 \end{bmatrix}$$

The least squares estimate of $\boldsymbol{\beta}$ is

$$\hat{\boldsymbol{\beta}} = (\mathbf{X'X})^{-1}\mathbf{X'y}$$

Table 13–3   Predicted Values, Residuals, and Other Diagnostics from Example 13–1

Observation i	y_i	Predicted Value $\hat{y}_i$	Residual e_i	h_{ii}	Studentized Residual	D_i	R-Student
1	2256	2244.5	11.5	0.350	0.87	0.137	0.87
2	2340	2352.1	−12.1	0.102	−0.78	0.023	−0.77
3	2426	2414.1	11.9	0.177	0.80	0.046	0.79
4	2293	2294.0	−1.0	0.251	−0.07	0.001	−0.07
5	2330	2346.4	−16.4	0.077	−1.05	0.030	−1.05
6	2368	2389.3	−21.3	0.265	−1.52	0.277	−1.61
7	2250	2252.1	−2.1	0.319	−0.15	0.004	−0.15
8	2409	2383.6	25.4	0.098	1.64	0.097	1.76
9	2364	2385.5	−21.5	0.142	−1.42	0.111	−1.48
10	2379	2369.3	9.7	0.080	0.62	0.011	0.60
11	2440	2416.9	23.1	0.278	1.66	0.354	1.80
12	2364	2384.5	−20.5	0.096	−1.32	0.062	−1.36
13	2404	2396.9	7.1	0.289	0.52	0.036	0.50
14	2317	2316.9	0.1	0.185	0.01	0.000	<0.01
15	2309	2298.8	10.2	0.134	0.67	0.023	0.66
16	2328	2332.1	−4.1	0.156	−0.28	0.005	−0.27

or

$$\hat{\beta} = \begin{bmatrix} 14.176004 & -0.129746 & -0.223453 \\ -0.129746 & 1.429184 \times 10^{-3} & -4.763947 \times 10^{-5} \\ -0.223453 & -4.763947 \times 10^{-5} & 2.222381 \times 10^{-2} \end{bmatrix} \begin{bmatrix} 37{,}577 \\ 3{,}429{,}550 \\ 385{,}562 \end{bmatrix}$$

$$= \begin{bmatrix} 1566.07777 \\ 7.62129 \\ 8.58485 \end{bmatrix}$$

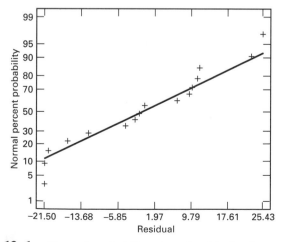

Figure 13–1.   Normal probability plot of residuals, Example 13–1.

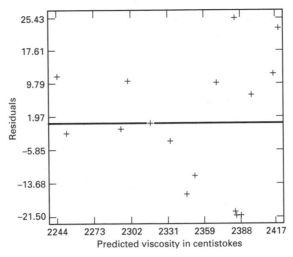

Figure 13–2. Plot of residuals versus predicted viscosity, Example 13–1.

The least squares fit, with the regression coefficients reported to two decimal places, is

$$\hat{y} = 1566.08 + 7.62x_1 + 8.58x_2$$

The first three columns of Table 13–3 present the actual observations y_i, the predicted or fitted values $\hat{y}_i$, and the residuals. Figure 13–1 is a normal probability plot of the residuals. Plots of the residuals versus the predicted values $\hat{y}_i$ and versus the two variables x_1 and x_2 are shown in Figures 13–2, 13–3, and 13–4, respectively.

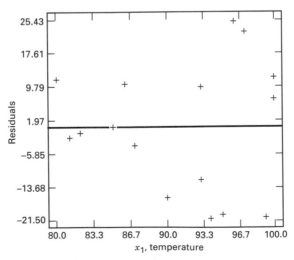

Figure 13–3. Plot of residuals versus x_1 (temperature), Example 13–1.

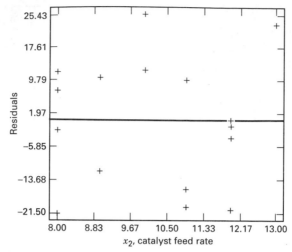

Figure 13–4. Plot of residuals versus x_2 (feed rate), Example 13–1.

Just as in designed experiments, residual plotting is an integral part of regression model building. These plots indicate that there is a tendency for the variance of the observed viscosity to increase with the magnitude of viscosity. Figure 13–3 suggests that the variability in viscosity is increasing as temperature increases.

■

Using the Computer ▪ Regression model fitting is almost always done using a statistical software package. Table 13–4 shows some of the output obtained when Statgraphics is used to fit the viscosity regression model in Example 13–1. Many of the quantities in this output should be familiar, as they have similar meanings to the quantities in the output displays for computer analysis of data from

Table 13–4   Statgraphics Output for the Viscosity Regression Model, Example 13–1

Source	Sum of Squares	DF	Mean Square	F Ratio	P-Value
Model	44157.1	2	22078.5	82.5046	.0000
Error	3478.85	13	267.604		
Total (Corr.)	47635.9	15			

R-squared = 0.92697
R-squared (Adj) = 0.915735

Independent Variable	Coefficient	Std. Error	t Value	Sig. Level
CONSTANT	1566.077771	61.591836	25.4267	0.0000
REGRESS.temp	7.62129	0.61843	12.3236	0.0000
REGRESS.feed_rate	8.584846	2.438684	3.5203	0.0038

designed experiments. We have seen many such computer outputs previously in the book. In subsequent sections, we will discuss the analysis of variance and t test information in Table 13–4 in detail, and show exactly how these quantities were computed.

Fitting Regression Models in Designed Experiments ▪ We have often used a regression model to present the results of a designed experiment in a quantitative form. We now give a complete example showing how this is done. This is followed by three other brief examples that illustrate other useful applications of regression analysis in designed experiments.

Example 13–2

Regression Analysis of a 2^3 Factorial Design

A chemical engineer is investigating the yield of a process. Three process variables are of interest: temperature, pressure, and catalyst concentration. Each variable can be run at a low and a high level, and the engineer decides to run a 2^3 design with four center points. The design and the resulting yields are shown in Figure 13–5, where we have shown both the natural levels of the design factors and the $+1$, -1 coded variable notation normally employed in 2^k factorial designs to represent the factor levels.

Suppose that the engineer decides to fit a main effects only model, say

$$y = \beta_0 + \beta_1 x_1 + \beta_2 x_2 + \beta_3 x_3 + \epsilon$$

For this model the $\mathbf{X}$ matrix and $\mathbf{y}$ vector are

$$\mathbf{X} = \begin{bmatrix} 1 & -1 & -1 & -1 \\ 1 & 1 & -1 & -1 \\ 1 & -1 & 1 & -1 \\ 1 & 1 & 1 & -1 \\ 1 & -1 & -1 & 1 \\ 1 & 1 & -1 & 1 \\ 1 & -1 & 1 & 1 \\ 1 & 1 & 1 & 1 \\ 1 & 0 & 0 & 0 \\ 1 & 0 & 0 & 0 \\ 1 & 0 & 0 & 0 \\ 1 & 0 & 0 & 0 \end{bmatrix}, \quad \mathbf{y} = \begin{bmatrix} 32 \\ 46 \\ 57 \\ 65 \\ 36 \\ 48 \\ 57 \\ 68 \\ 50 \\ 44 \\ 53 \\ 56 \end{bmatrix}$$

It is easy to show that

$$\mathbf{X'X} = \begin{bmatrix} 12 & 0 & 0 & 0 \\ 0 & 8 & 0 & 0 \\ 0 & 0 & 8 & 0 \\ 0 & 0 & 0 & 8 \end{bmatrix}, \quad \mathbf{X'y} = \begin{bmatrix} 612 \\ 45 \\ 85 \\ 9 \end{bmatrix}$$

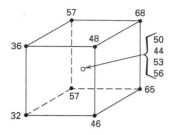

Process Variables			Coded Variables			Yield, y
Run	Temp (°C)	Pressure (psig)	Conc (g/l)	x_1	x_2	x_3	
1	120	40	15	−1	−1	−1	32
2	160	40	15	1	−1	−1	46
3	120	80	15	−1	1	−1	57
4	160	80	15	1	1	−1	65
5	120	80	30	−1	−1	1	36
6	160	40	30	1	−1	1	48
7	120	80	30	−1	1	1	57
8	160	80	30	1	1	1	68
9	140	60	22.5	0	0	0	50
10	140	60	22.5	0	0	0	44
11	140	60	22.5	0	0	0	53
12	140	60	22.5	0	0	0	56

$$x_1 = \frac{\text{Temp} - 140}{20}, \quad x_2 = \frac{\text{Pressure} - 60}{20}, \quad x_3 = \frac{\text{Conc} - 22.5}{7.5}$$

Figure 13–5. Experimental design for Example 13–2.

Since $\mathbf{X'X}$ is *diagonal,* the required inverse is also diagonal, and the least squares estimates of the regression coefficients is

$$\hat{\beta} = (\mathbf{X'X})^{-1}\mathbf{X'y} = \begin{bmatrix} 1/12 & 0 & 0 & 0 \\ 0 & 1/8 & 0 & 0 \\ 0 & 0 & 1/8 & 0 \\ 0 & 0 & 0 & 1/8 \end{bmatrix} \begin{bmatrix} 612 \\ 45 \\ 85 \\ 9 \end{bmatrix} = \begin{bmatrix} 51.000 \\ 5.625 \\ 10.625 \\ 1.125 \end{bmatrix}$$

The fitted regression model is

$$\hat{y} = 51.000 + 5.625x_1 + 10.625x_2 + 1.125x_3$$

As we have made use of on many occasions, the regression coefficients are closely connected to the effect estimates that would be obtained from the usual analysis of a 2^3 design. For example, the effect of temperature is (refer to Figure 13–5)

$$T = \bar{y}_{T^+} - \bar{y}_{T^-}$$
$$= 56.75 - 45.50$$
$$= 11.25$$

Notice that the regression coefficient for x_1 is

$$(11.25)/2 = 5.625$$

That is, the regression coefficient is exactly one-half the usual effect estimate. This will always be true for a 2^k design. As noted above, we used this result in Chapters 7 through 9 to produce regression models, fitted values, and residuals for several two-level experiments. This example demonstrates that the effect estimates from a 2^k design are least squares estimates.

∎

In Example 13–2, the inverse matrix is easy to obtain because $\mathbf{X'X}$ is diagonal. Intuitively, this seems to be advantageous, not only because of the computational simplicity but also because the estimators of all the regression coefficients are uncorrelated; that is, $\text{Cov}(\hat{\beta}_i, \hat{\beta}_j) = 0$. If we can choose the levels of the x variables before the data are collected, we might wish to design the experiment so that a diagonal $\mathbf{X'X}$ will result.

In practice, it can be relatively easy to do this. We know that the off-diagonal elements in $\mathbf{X'X}$ are the sums of cross-products of the columns in $\mathbf{X}$. Therefore, we must make the inner product of the columns of $\mathbf{X}$ equal to zero; that is, these columns must be **orthogonal.** Experimental designs that have this property for fitting a regression model are called **orthogonal designs.** In general, the 2^k factorial design is an orthogonal design for fitting the multiple linear regression model.

Regression methods are extremely useful when something "goes wrong" in a designed experiment. This is illustrated in the next two examples.

Example 13–3

A 2^3 Factorial Design with a Missing Observation

Consider the 2^3 factorial design with four center points from Example 13–2. Suppose that when this experiment was performed, the run with all variables at the high level (run 8 in Figure 13–5) was missing. This can happen for a variety of reasons; the measurement system can produce a faulty reading, the combination of factor levels may prove infeasible, the experimental unit may be damaged, and so forth.

We will fit the main effects model

$$y = \beta_0 + \beta_1 x_1 + \beta_2 x_2 + \beta_3 x_3 + \epsilon$$

using the 11 remaining observations. The $\mathbf{X}$ matrix and $\mathbf{y}$ vector are

$$\mathbf{X} = \begin{bmatrix} 1 & -1 & -1 & -1 \\ 1 & 1 & -1 & -1 \\ 1 & -1 & 1 & -1 \\ 1 & 1 & 1 & -1 \\ 1 & -1 & -1 & 1 \\ 1 & 1 & -1 & 1 \\ 1 & -1 & 1 & 1 \\ 1 & 0 & 0 & 0 \\ 1 & 0 & 0 & 0 \\ 1 & 0 & 0 & 0 \\ 1 & 0 & 0 & 0 \end{bmatrix}, \quad \mathbf{y} = \begin{bmatrix} 32 \\ 46 \\ 57 \\ 65 \\ 36 \\ 48 \\ 57 \\ 50 \\ 44 \\ 53 \\ 56 \end{bmatrix}$$

To estimate the model parameters, we form

$$\mathbf{X'X} = \begin{bmatrix} 11 & -1 & -1 & -1 \\ -1 & 7 & -1 & -1 \\ -1 & -1 & 7 & -1 \\ -1 & -1 & -1 & 7 \end{bmatrix}, \quad \mathbf{X'y} = \begin{bmatrix} 544 \\ -23 \\ 17 \\ -59 \end{bmatrix}$$

and then

$$\hat{\boldsymbol{\beta}} = (\mathbf{X'X})^{-1}\mathbf{X'y}$$

$$= \begin{bmatrix} 9.61538 \times 10^{-2} & 1.92307 \times 10^{-2} & 1.92307 \times 10^{-2} & 1.92307 \times 10^{-2} \\ 1.92307 \times 10^{-2} & 0.15385 & 2.88462 \times 10^{-2} & 2.88462 \times 10^{-2} \\ 1.92307 \times 10^{-2} & 2.88462 \times 10^{-2} & 0.15385 & 2.88462 \times 10^{-2} \\ 1.92307 \times 10^{-2} & 2.88462 \times 10^{-2} & 2.88462 \times 10^{-2} & 0.15385 \end{bmatrix} \begin{bmatrix} 544 \\ -23 \\ 17 \\ -59 \end{bmatrix}$$

$$= \begin{bmatrix} 51.25 \\ 5.75 \\ 10.75 \\ 1.25 \end{bmatrix}$$

Therefore, the fitted model is

$$\hat{y} = 51.25 + 5.75x_1 + 10.75x_2 + 1.25x_3$$

Compare this model to the one obtained in Example 13–2, where all 12 observations were used. The regression coefficients are very similar. Since the regression coefficients are closely related to the factor effects, our conclusions would not be seriously impacted by the missing observation. However, notice that the effect estimates are no longer orthogonal, because $(\mathbf{X'X})$ and its inverse are no longer diagonal.

■

Example 13–4

Inaccurate Levels in Design Factors

When running a designed experiment, it is sometimes difficult to reach and hold the precise factor levels required by the design. Small discrepancies are not important, but large ones are potentially of more concern. Regression methods are useful in the analysis of a designed experiment where the experimenter has been unable to obtain the required factor levels.

To illustrate, the experiment in Table 13–5 shows a variation of the 2^3 design from Example 13–2, where many of the test combinations are not exactly the ones specified in the design. Most of the difficulty seems to have occurred with the temperature variable.

We will fit the main effects model

$$y = \beta_0 + \beta_1 x_1 + \beta_2 x_2 + \beta_3 x_3 + \epsilon$$

Table 13-5   Experimental Design for Example 13-4

Run	Temp (°C)	Pressure (psig)	Conc (g/l)	x_1	x_2	x_3	Yield y
1	125	41	14	−0.75	−0.95	−1.133	32
2	158	40	15	0.90	−1	−1	46
3	121	82	15	−0.95	1.1	−1	57
4	160	80	15	1	1	−1	65
5	118	39	33	−1.10	−1.05	1.4	36
6	163	40	30	1.15	−1	1	48
7	122	80	30	−0.90	1	1	57
8	165	83	30	1.25	1.15	1	68
9	140	60	22.5	0	0	0	50
10	140	60	22.5	0	0	0	44
11	140	60	22.5	0	0	0	53
12	140	60	22.5	0	0	0	56

The **X** matrix and **y** vector are

$$
\mathbf{X} = \begin{bmatrix}
1 & -0.75 & -0.95 & -1.133 \\
1 & 0.90 & -1 & -1 \\
1 & -0.95 & 1.1 & -1 \\
1 & 1 & 1 & -1 \\
1 & -1.10 & -1.05 & 1.4 \\
1 & 1.15 & -1 & 1 \\
1 & -0.90 & 1 & 1 \\
1 & 1.25 & 1.15 & 1 \\
1 & 0 & 0 & 0 \\
1 & 0 & 0 & 0 \\
1 & 0 & 0 & 0 \\
1 & 0 & 0 & 0
\end{bmatrix}, \quad
\mathbf{y} = \begin{bmatrix}
32 \\
46 \\
57 \\
65 \\
36 \\
48 \\
57 \\
68 \\
50 \\
44 \\
53 \\
56
\end{bmatrix}
$$

To estimate the model parameters, we need

$$
\mathbf{X'X} = \begin{bmatrix}
12 & 0.60 & 0.25 & 0.2670 \\
0.60 & 8.18 & 0.31 & -0.1403 \\
0.25 & 0.31 & 8.5375 & -0.3437 \\
0.2670 & -0.1403 & -0.3437 & 9.2437
\end{bmatrix}, \quad
\mathbf{X'y} = \begin{bmatrix}
612 \\
77.55 \\
161.50 \\
19.144
\end{bmatrix}
$$

Then

$$\hat{\beta} = (\mathbf{X}'\mathbf{X})^{-1}\mathbf{X}'\mathbf{y}$$

$$
= \begin{bmatrix}
8.37447 \times 10^{-2} & -6.09871 \times 10^{-3} & -2.33542 \times 10^{-3} & -2.59833 \times 10^{-3} \\
-6.09871 \times 10^{-3} & 0.12289 & -4.20766 \times 10^{-3} & 1.88490 \times 10^{-3} \\
-2.33542 \times 10^{-3} & -4.20766 \times 10^{-3} & 0.11753 & 4.37851 \times 10^{-3} \\
-2.59833 \times 10^{-3} & 1.88490 \times 10^{-3} & 4.37351 \times 10^{-3} & 0.10845
\end{bmatrix}
\begin{bmatrix}
612 \\
77.55 \\
161.50 \\
19.144
\end{bmatrix}
$$

$$
= \begin{bmatrix}
50.36496 \\
5.41932 \\
10.16672 \\
1.07653
\end{bmatrix}
$$

The fitted regression model, with the coefficients reported to two decimal places, is

$$\hat{y} = 50.36 + 5.42x_1 + 10.17x_2 + 1.08x_3$$

Comparing this to the original model in Example 13–2, where the factor levels were exactly those specified by the design, we note very little difference. The practical interpretation of the results of this experiment would not be seriously affected by the inability of the experimenter to achieve the desired factor levels exactly.

■

Example 13–5

Dealiasing Interactions in a Fractional Factorial

We observed in Chapter 9 that it is possible to dealias interactions in a fractional factorial design by a process called fold over. For a resolution III design, a full fold over is constructed by running a second fraction in which the signs are reversed from those in the original fraction. Then the combined design can be used to dealias all main effects from the two-factor interactions.

A difficulty with fold over is that it requires a second group of runs of identical size as the original design. It is usually possible to dealias certain interactions of interest by augmenting the original design with fewer runs than required in a full fold over. Regression methods are an easy way to formulate this problem and see how to solve it.

To illustrate, suppose that we have run a 2_{IV}^{4-1} design. Table 9–3 shows the principal fraction of this design, in which $I = ABCD$. Suppose that after the data from the first eight trials were observed, the largest effects were A, B, C, D (we ignore the three-factor interactions that are aliased with these main effects), and the $AB + CD$ alias chain. The other two alias chains can be ignored, but clearly either AB, or CD, or both two-factor interactions are large. To find out which interactions are important we could, of course, run the alternate fraction, which would require another eight trials. Then all 16 runs could be used to estimate the main effects and the two-factor interactions.

It is possible to dealias AB and CD in fewer than eight additional trials. Suppose that we wish to fit the model

$$y = \beta_0 + \beta_1 x_1 + \beta_2 x_2 + \beta_3 x_3 + \beta_4 x_4 + \beta_{12} x_1 x_2 + \beta_{34} x_3 x_4 + \epsilon$$

where x_1, x_2, x_3, and x_4 are the coded variables representing A, B, C, and D. Using the design in Table 9–3, the **X** matrix for this model is

$$
\mathbf{X} =
\begin{array}{ccccccc}
x_1 & x_2 & x_3 & x_4 & x_1 x_2 & x_3 x_4 \\
\end{array}
\left[
\begin{array}{rrrrrrr}
1 & -1 & -1 & -1 & -1 & 1 & 1 \\
1 & 1 & -1 & -1 & 1 & -1 & -1 \\
1 & -1 & 1 & -1 & 1 & -1 & -1 \\
1 & 1 & 1 & -1 & -1 & 1 & 1 \\
1 & -1 & -1 & 1 & 1 & 1 & 1 \\
1 & 1 & -1 & 1 & -1 & -1 & -1 \\
1 & -1 & 1 & 1 & -1 & -1 & -1 \\
1 & 1 & 1 & 1 & 1 & 1 & 1 \\
\end{array}
\right]
$$

where we have written the variables above the columns to facilitate understanding. Notice that the $x_1 x_2$ column is identical to the $x_3 x_4$ column (as anticipated, since AB or $x_1 x_2$ is aliased with CD or $x_3 x_4$), implying a linear dependency in the columns of **X**. Therefore we cannot estimate both β_{12} and β_{34} in the model. However, suppose that we add a single run $x_1 = -1$, $x_2 = -1$, $x_3 = -1$, and $x_4 = 1$ from the alternate fraction to the original eight runs. The **X** matrix for the model now becomes

$$
\mathbf{X} =
\begin{array}{ccccccc}
x_1 & x_2 & x_3 & x_4 & x_1 x_2 & x_3 x_4 \\
\end{array}
\left[
\begin{array}{rrrrrrr}
1 & -1 & -1 & -1 & -1 & 1 & 1 \\
1 & 1 & -1 & -1 & 1 & -1 & -1 \\
1 & -1 & 1 & -1 & 1 & -1 & -1 \\
1 & 1 & 1 & -1 & -1 & 1 & 1 \\
1 & -1 & -1 & 1 & 1 & 1 & 1 \\
1 & 1 & -1 & 1 & -1 & -1 & -1 \\
1 & -1 & 1 & 1 & -1 & -1 & -1 \\
1 & 1 & 1 & 1 & 1 & 1 & 1 \\
1 & -1 & -1 & -1 & 1 & 1 & -1 \\
\end{array}
\right]
$$

Notice that the columns $x_1 x_2$ and $x_3 x_4$ are now no longer identical, and we can fit the model including both the $x_1 x_2$ (AB) and $x_3 x_4$ (CD) interactions. The magnitudes of the regression coefficients will give insight regarding which interactions are important.

While adding a single run will dealias the AB and CD interactions, this approach does have a disadvantage. Suppose that there is a time effect (or a block effect) between the first eight runs and the last run added above. Add a column to the **X** matrix for blocks, and you obtain the following:

$$
\mathbf{X} =
\begin{array}{c}
 \\

\end{array}
\begin{array}{cccccccc}
x_1 & x_2 & x_3 & x_4 & x_1x_2 & x_3x_4 & \text{blocks} \\
\end{array}
$$

$$
\mathbf{X} =
\begin{bmatrix}
1 & -1 & -1 & -1 & -1 & 1 & 1 & -1 \\
1 & 1 & -1 & -1 & 1 & -1 & -1 & -1 \\
1 & -1 & 1 & -1 & 1 & -1 & -1 & -1 \\
1 & 1 & 1 & -1 & -1 & 1 & 1 & -1 \\
1 & -1 & -1 & 1 & 1 & 1 & 1 & -1 \\
1 & 1 & -1 & 1 & -1 & -1 & -1 & -1 \\
1 & -1 & 1 & 1 & -1 & -1 & -1 & -1 \\
1 & 1 & 1 & 1 & 1 & 1 & 1 & -1 \\
1 & -1 & -1 & -1 & 1 & 1 & -1 & 1 \\
\end{bmatrix}
$$

We have assumed the block factor was at the low or "−" level during the first eight runs, and at the high or "+" level during the ninth run. It is easy to see that the sum of the cross-products of every column with the block column does not sum to zero, meaning that blocks are no longer orthogonal to treatments, or that the block effect now impacts the estimates of the model regression coefficients. To block orthogonally, you must add an even number of runs. For example, the four runs

x_1	x_2	x_3	x_4
−1	−1	−1	1
1	−1	−1	−1
−1	1	1	1
1	1	1	−1

will dealias AB from CD and allow orthogonal blocking (you can see this by writing out the $\mathbf{X}$ matrix as we did previously).

In general, it is usually straightforward to examine the $\mathbf{X}$ matrix for the reduced model obtained from a fractional factorial and determine which runs to augment the original design with to dealias interactions of potential interest. Furthermore, the impact of specific augmentation strategies can be evaluated using the general results for regression models given later in this chapter.

■

13–4 HYPOTHESIS TESTING IN MULTIPLE REGRESSION

In multiple linear regression problems, certain tests of hypotheses about the model parameters are helpful in measuring the usefulness of the model. In this section, we describe several important hypothesis-testing procedures. These procedures require that the errors ϵ_i in the model be normally and independently distributed with mean zero and variance σ^2, abbreviated $\epsilon \sim NID(0, \sigma^2)$. As a result of this assumption, the observations y_i are normally and independently distributed with mean $\beta_0 + \Sigma_{j=1}^{k} \beta_j x_{ij}$ and variance σ^2.

13-4.1 Test for Significance of Regression

The test for significance of regression is a test to determine if there is a linear relationship between the response variable y and a subset of the regressor variables $x_1, x_2, \ldots, x_k$. The appropriate hypotheses are

$$H_0: \beta_1 = \beta_2 = \cdots = \beta_k = 0$$
$$H_1: \beta_j \neq 0 \quad \text{for at least one } j \tag{13-20}$$

Rejection of H_0 in Equation 13–20 implies that at least one of the regressor variables $x_1, x_2, \ldots, x_k$ contributes significantly to the model. The test procedure involves an analysis of variance partitioning of the total sum of squares SS_T into a sum of squares due to the model (or to regression) and a sum of squares due to residual (or error), say

$$SS_T = SS_R + SS_E \tag{13-21}$$

Now if the null hypothesis $H_0: \beta_1 = \beta_2 = \cdots = \beta_k = 0$ is true, then SS_R/σ^2 is distributed as χ_k^2, where the number of degrees of freedom for χ^2 are equal to the number of regressor variables in the model. Also, we can show that SS_E/σ^2 is distributed as χ_{n-k-1}^2 and that SS_E and SS_R are independent. The test procedure for $H_0: \beta_1 = \beta_2 = \cdots = \beta_k = 0$ is to compute

$$F_0 = \frac{SS_R/k}{SS_E/(n-k-1)} = \frac{MS_R}{MS_E} \tag{13-22}$$

and to reject H_0 if F_0 exceeds $F_{\alpha,k,n-k-1}$. Alternatively, we could use the P-value approach to hypothesis testing and, thus, reject H_0 if the P-value for the statistic F_0 is less than α. The test is usually summarized in an analysis of variance table such as Table 13–6.

A computational formula for SS_R may be found easily. We have derived a computational formula for SS_E in Equation 13–16—that is,

$$SS_E = \mathbf{y'y} - \hat{\boldsymbol{\beta}}'\mathbf{X'y}$$

Now, because $SS_T = \Sigma_{i=1}^{n} y_i^2 - (\Sigma_{i=1}^{n} y_i)^2/n = \mathbf{y'y} - (\Sigma_{i=1}^{n} y_i)^2/n$, we may rewrite the foregoing equation as

$$SS_E = \mathbf{y'y} - \frac{\left(\sum_{i=1}^{n} y_i\right)^2}{n} - \left[\hat{\boldsymbol{\beta}}'\mathbf{X'y} - \frac{\left(\sum_{i=1}^{n} y_i\right)^2}{n}\right]$$

Table 13–6   Analysis of Variance for Significance of Regression in Multiple Regression

Source of Variation	Sum of Squares	Degrees of Freedom	Mean Square	F_0
Regression	SS_R	k	MS_R	MS_R/MS_E
Error or residual	SS_E	$n - k - 1$	MS_E	
Total	SS_T	$n - 1$		

or

$$SS_E = SS_T - SS_R$$

Therefore, the regression sum of squares is

$$SS_R = \hat{\boldsymbol{\beta}}'\mathbf{X}'\mathbf{y} - \frac{\left(\sum_{i=1}^{n} y_i\right)^2}{n} \tag{13-23}$$

and the error sum of squares is

$$SS_E = \mathbf{y}'\mathbf{y} - \hat{\boldsymbol{\beta}}'\mathbf{X}'\mathbf{y} \tag{13-24}$$

and the total sum of squares is

$$SS_T = \mathbf{y}'\mathbf{y} - \frac{\left(\sum_{i=1}^{n} y_i\right)^2}{n} \tag{13-25}$$

These computations are almost always performed with regression software. For instance, Table 13–4 shows some of the output from Statgraphics for the viscosity regression model in Example 13–1. The upper portion in this display is the analysis of variance for the model. The test of significance of regression in this example involves the hypotheses

$$H_0: \beta_1 = \beta_2 = 0$$
$$H_1: \beta_j \neq 0 \quad \text{for at least one } j$$

The P-value in Table 13–4 for the F statistic (Equation 13–22) is very small, so we would conclude that at least one of the two variables—temperature (x_1) and feed rate (x_2)—has a nonzero regression coefficient.

Table 13–4 also reports the coefficient of multiple determination R^2, where

$$R^2 = \frac{SS_R}{SS_T} = 1 - \frac{SS_E}{SS_T} \tag{13-26}$$

Just as in designed experiments, R^2 is a measure of the amount of reduction in the variability of y obtained by using the regressor variables $x_1, x_2, \ldots, x_k$ in the model. However, as we have noted previously, a large value of R^2 does not necessarily imply that the regression model is a good one. Adding a variable to the model will always increase R^2, regardless of whether the additional variable is statistically significant or not. Thus it is possible for models that have large values of R^2 to yield poor predictions of new observations or estimates of the mean response.

Because R^2 always increases as we add terms to the model, some regression model builders prefer to use an **adjusted R^2 statistic** defined as

$$R_{adj}^2 = 1 - \frac{SS_E/(n-p)}{SS_T/(n-1)} = 1 - \left(\frac{n-1}{n-p}\right)(1-R^2) \qquad (13\text{-}27)$$

In general, the adjusted R^2 statistic will not always increase as variables are added to the model. In fact, if unnecessary terms are added, the value of R_{adj}^2 will often decrease.

For example, consider the viscosity regression model. The adjusted R^2 for the model is shown in Table 13–4. It is computed as

$$R_{adj}^2 = 1 - \left(\frac{n-1}{n-p}\right)(1-R^2)$$

$$= 1 - \left(\frac{15}{13}\right)(1 - 0.92697)$$

$$= 0.915735$$

which is very close to the ordinary R^2. When R^2 and R_{adj}^2 differ dramatically there is a good chance that nonsignificant terms have been included in the model.

13–4.2 Tests on Individual Regression Coefficients and Groups of Coefficients

We are frequently interested in testing hypotheses on the individual regression coefficients. Such tests would be useful in determining the value of each of the regressor variables in the regression model. For example, the model might be more effective with the inclusion of additional variables, or perhaps with the deletion of one or more of the variables already in the model.

Adding a variable to the regression model always causes the sum of squares for regression to increase and the error sum of squares to decrease. We must decide whether the increase in the regression sum of squares is sufficient to warrant using the additional variable in the model. Furthermore, adding an unimportant variable to the model can actually increase the mean square error, thereby decreasing the usefulness of the model.

The hypotheses for testing the significance of any individual regression coefficient, say β_j, are

$$H_0: \beta_j = 0$$
$$H_1: \beta_j \neq 0$$

If $H_0: \beta_j = 0$ is not rejected, then this indicates that x_j can be deleted from the model. The test statistic for this hypothesis is

$$t_0 = \frac{\hat{\beta}_j}{\sqrt{\hat{\sigma}^2 C_{jj}}} \qquad (13\text{--}28)$$

where C_{jj} is the diagonal element of $(\mathbf{X'X})^{-1}$ corresponding to $\hat{\beta}_j$. The null hypothesis $H_0: \beta_j = 0$ is rejected if $|t_0| > t_{\alpha/2, n-k-1}$. Note that this is really a partial or marginal test, because the regression coefficient $\hat{\beta}_j$ depends on all the other regressor variables x_i $(i \neq j)$ that are in the model.

The denominator of Equation 13–28, $\sqrt{\hat{\sigma}^2 C_{jj}}$, is often called the **standard error** of the regression coefficient $\hat{\beta}_j$. That is,

$$se(\hat{\beta}_j) = \sqrt{\hat{\sigma}^2 C_{jj}} \qquad (13\text{--}29)$$

Therefore, an equivalent way to write the test statistic in Equation (13–28) is

$$t_0 = \frac{\hat{\beta}_j}{se(\hat{\beta}_j)} \qquad (13\text{--}30)$$

Most regression computer programs provide the t test for each model parameter. For example, consider Table 13–4, which contains the Statgraphics output for Example 13–1. The lower portion of this table gives the least squares estimate of each parameter, the standard error, the t statistic, and the corresponding P-value. We would conclude that both variables, temperature and feed rate, contribute significantly to the model.

We may also directly examine the contribution to the regression sum of squares for a particular variable, say x_j, given that other variables x_i $(i \neq j)$ are included in the model. The procedure for doing this is the general regression significance test or, as it is often called, the **extra sum of squares method.** This procedure can also be used to investigate the contribution of a *subset* of the regressor variables to the model. Consider the regression model with k regressor variables:

$$\mathbf{y} = \mathbf{X}\boldsymbol{\beta} + \boldsymbol{\epsilon}$$

where $\mathbf{y}$ is $(n \times 1)$, $\mathbf{X}$ is $(n \times p)$, $\boldsymbol{\beta}$ is $(p \times 1)$, $\boldsymbol{\epsilon}$ is $(n \times 1)$, and $p = k + 1$. We would like to determine if the subset of regressor variables $x_1, x_2, \ldots, x_r$ $(r < k)$ contribute significantly to the regression model. Let the vector of regression coefficients be partitioned as follows:

$$\boldsymbol{\beta} = \begin{bmatrix} \boldsymbol{\beta}_1 \\ \boldsymbol{\beta}_2 \end{bmatrix}$$

where $\boldsymbol{\beta}_1$ is $(r \times 1)$ and $\boldsymbol{\beta}_2$ is $[(p - r) \times 1]$. We wish to test the hypotheses

$$\begin{aligned} H_0&: \boldsymbol{\beta}_1 = \mathbf{0} \\ H_1&: \boldsymbol{\beta}_1 \neq \mathbf{0} \end{aligned} \qquad (13\text{--}31)$$

The model may be written as

$$\mathbf{y} = \mathbf{X}\boldsymbol{\beta} + \boldsymbol{\epsilon} = \mathbf{X}_1\boldsymbol{\beta}_1 + \mathbf{X}_2\boldsymbol{\beta}_2 + \boldsymbol{\epsilon} \qquad (13\text{-}32)$$

where $\mathbf{X}_1$ represents the columns of $\mathbf{X}$ associated with $\boldsymbol{\beta}_1$ and $\mathbf{X}_2$ represents the columns of $\mathbf{X}$ associated with $\boldsymbol{\beta}_2$.

For the **full model** (including both $\boldsymbol{\beta}_1$ and $\boldsymbol{\beta}_2$), we know that $\hat{\boldsymbol{\beta}} = (\mathbf{X}'\mathbf{X})^{-1}\mathbf{X}'\mathbf{y}$. Also, the regression sum of squares for all variables including the intercept is

$$SS_R(\boldsymbol{\beta}) = \hat{\boldsymbol{\beta}}'\mathbf{X}'\mathbf{y} \qquad (p \text{ degrees of freedom})$$

and

$$MS_E = \frac{\mathbf{y}'\mathbf{y} - \hat{\boldsymbol{\beta}}\mathbf{X}'\mathbf{y}}{n - p}$$

$SS_R(\boldsymbol{\beta})$ is called the regression sum of squares due to $\boldsymbol{\beta}$. To find the contribution of the terms in $\boldsymbol{\beta}_1$ to the regression, we fit the model assuming the null hypothesis $H_0: \boldsymbol{\beta}_1 = \mathbf{0}$ to be true. The **reduced model** is found from Equation 13–32 with $\boldsymbol{\beta}_1 = \mathbf{0}$:

$$\mathbf{y} = \mathbf{X}_2\boldsymbol{\beta}_2 + \boldsymbol{\epsilon} \qquad (13\text{-}33)$$

The least squares estimator of $\boldsymbol{\beta}_2$ is $\hat{\boldsymbol{\beta}}_2 = (\mathbf{X}_2'\mathbf{X}_2)^{-1}\mathbf{X}_2'\mathbf{y}$, and

$$SS_R(\boldsymbol{\beta}_2) = \hat{\boldsymbol{\beta}}_2'\mathbf{X}_2'\mathbf{y} \qquad (p - r \text{ degrees of freedom}) \qquad (13\text{-}34)$$

The regression sum of squares due to $\boldsymbol{\beta}_1$ given that $\boldsymbol{\beta}_2$ is already in the model is

$$SS_R(\boldsymbol{\beta}_1|\boldsymbol{\beta}_2) = SS_R(\boldsymbol{\beta}) - SS_R(\boldsymbol{\beta}_2) \qquad (13\text{-}35)$$

This sum of squares has r degrees of freedom. It is the "extra sum of squares" due to $\boldsymbol{\beta}_1$. Note that $SS_R(\boldsymbol{\beta}_1|\boldsymbol{\beta}_2)$ is the increase in the regression sum of squares due to including the variables $x_1, x_2, \ldots, x_r$ in the model.

Now, $SS_R(\boldsymbol{\beta}_1|\boldsymbol{\beta}_2)$ is independent of MS_E, and the null hypothesis $\boldsymbol{\beta}_1 = \mathbf{0}$ may be tested by the statistic

$$F_0 = \frac{SS_R(\boldsymbol{\beta}_1|\boldsymbol{\beta}_2)/r}{MS_E} \qquad (13\text{-}36)$$

If $F_0 > F_{\alpha,r,n-p}$, we reject H_0, concluding that at least one of the parameters in $\boldsymbol{\beta}_1$ is not zero and, consequently, at least one of the variables $x_1, x_2, \ldots, x_r$ in $\mathbf{X}_1$ contributes significantly to the regression model. Some authors call the test in Equation 13–36 a **partial F test**.

The partial F test is very useful. We can use it to measure the contribution of x_j as if it were the last variable added to the model by computing

$$SS_R(\beta_j|\beta_0, \beta_1, \ldots, \beta_{j-1}, \beta_{j+1}, \ldots, \beta_k)$$

This is the increase in the regression sum of squares due to adding x_j to a model that already includes $x_1, \ldots, x_{j-1}, x_{j+1}, \ldots, x_k$. Note that the partial F test on a single variable x_j is equivalent to the t test in Equation 13–28. However, the partial F test is a more general procedure in that we can measure the effect of sets of variables.

Example 13–6

Consider the viscosity data in Example 13–1. Suppose that we wish to investigate the contribution of the variable x_2 (feed rate) to the model. That is, the hypotheses we wish to test are

$$H_0: \beta_2 = 0$$
$$H_1: \beta_2 \neq 0$$

This will require the extra sum of squares due to β_2, or

$$SS_R(\beta_2|\beta_1, \beta_0) = SS_R(\beta_0, \beta_1, \beta_2) - SS_R(\beta_0, \beta_1)$$
$$= SS_R(\beta_1, \beta_2|\beta_0) - SS_R(\beta_1|\beta_0)$$

Now from Table 13–4, where we tested for significance of regression, we have

$$SS_R(\beta_1, \beta_2|\beta_0) = 44{,}157.1$$

which was called the model sum of squares in the table. This sum of squares has 2 degrees of freedom.

The reduced model is

$$y = \beta_0 + \beta_1 x_1 + \epsilon$$

The least squares fit for this model is

$$\hat{y} = 1652.3955 + 7.6397 x_1$$

and the regression sum of squares for this model (with 1 degree of freedom) is

$$SS_R(\beta_1|\beta_0) = 40{,}840.8$$

Therefore,

$$SS_R(\beta_2|\beta_0, \beta_1) = 44{,}157.1 - 40{,}840.8$$
$$= 3316.3$$

with $2 - 1 = 1$ degree of freedom. This is the increase in the regression sum of squares that results from adding x_2 to a model already containing x_1. To test $H_0: \beta_2 = 0$, from the test statistic we obtain

$$F_0 = \frac{SS_R(\beta_2|\beta_0, \beta_1)/1}{MS_E} = \frac{3316.3/1}{267.604} = 12.3926$$

Note that MS_E from the full model (Table 13–4) is used in the denominator of F_0. Now, because $F_{0.05,1,13} = 1.67$, we would reject $H_0: \beta_2 = 0$ and conclude that x_2 (feed rate) contributes significantly to the model.

Because this partial F test involves only a single regressor, it is equivalent to the t test because the square of a t random variable with ν degrees of freedom is an F random variable with 1 and ν degrees of freedom. To see this, note from Table 13–4 that the t statistic for $H_0: \beta_2 = 0$ resulted in $t_0 = 3.5203$ and that $t_0^2 = (3.5203)^2 = 12.3925 \simeq F_0$.

■

13-5 CONFIDENCE INTERVALS IN MULTIPLE REGRESSION

It is often necessary to construct confidence interval estimates for the regression coefficients $\{\beta_j\}$ and for other quantities of interest from the regression model. The development of a procedure for obtaining these confidence intervals requires that we assume the errors $\{\epsilon_i\}$ to be normally and independently distributed with mean zero and variance σ^2, the same assumption made in the section on hypothesis testing in Section 13-4.

13-5.1 Confidence Intervals on the Individual Regression Coefficients

Because the least squares estimator $\hat{\boldsymbol{\beta}}$ is a linear combination of the observations, it follows that $\hat{\boldsymbol{\beta}}$ is normally distributed with mean vector $\boldsymbol{\beta}$ and covariance matrix $\sigma^2(\mathbf{X}'\mathbf{X})^{-1}$. Then each of the statistics

$$\frac{\hat{\beta}_j - \beta_j}{\sqrt{\hat{\sigma}^2 C_{jj}}} \qquad j = 0, 1, \ldots, k \qquad (13-37)$$

is distributed as t with $n - p$ degrees of freedom, where C_{jj} is the (jj)th element of the $(\mathbf{X}'\mathbf{X})^{-1}$ matrix, and $\hat{\sigma}^2$ is the estimate of the error variance, obtained from Equation 13-17. Therefore, a $100(1 - \alpha)$ percent confidence interval for the regression coefficient β_j, $j = 0, 1, \ldots, k$, is

$$\hat{\beta}_j - t_{\alpha/2, n-p}\sqrt{\hat{\sigma}^2 C_{jj}} \leq \beta_j \leq \hat{\beta}_j + t_{\alpha/2, n-p}\sqrt{\hat{\sigma}^2 C_{jj}} \qquad (13-38)$$

Note that this confidence interval could also be written as

$$\hat{\beta}_j - t_{\alpha/2, n-p}se(\hat{\beta}_j) \leq \beta_j \leq \hat{\beta}_j + t_{\alpha/2, n-p}se(\hat{\beta}_j)$$

because $se(\hat{\beta}_j) = \sqrt{\hat{\sigma}^2 C_{jj}}$.

Example 13-7

We will construct a 95 percent confidence interval for the parameter β_1 in Example 13-1. Now $\hat{\beta}_1 = 7.62129$, and because $\hat{\sigma}^2 = 267.604$ and $C_{11} = 1.429184 \times 10^{-3}$, we find that

$$\hat{\beta}_1 - t_{0.025,13}\sqrt{\hat{\sigma}^2 C_{11}} \leq \beta_1 \leq \hat{\beta}_1 + t_{0.025,13}\sqrt{\hat{\sigma}^2 C_{11}}$$

$$7.62129 - 2.16\sqrt{(267.604)(1.429184 \times 10^{-3})} \leq \beta_1$$

$$\leq 7.62129 + 2.16\sqrt{(267.604)(1.429184 \times 10^{-3})}$$

$$7.62129 - 2.16(0.6184) \leq \beta_1 \leq 7.62129 + 2.16(0.6184)$$

and the 95 percent confidence interval on β_1 is

$$6.2855 \leq \beta_1 \leq 8.9570$$

∎

13–5.2   Confidence Interval on the Mean Response

We may also obtain a confidence interval on the mean response at a particular point, say, $x_{01}, x_{02}, \ldots, x_{0k}$. We first define the vector

$$\mathbf{x}_0 = \begin{bmatrix} 1 \\ x_{01} \\ x_{02} \\ \vdots \\ x_{0k} \end{bmatrix}$$

The mean response at this point is

$$\mu_{y|\mathbf{x}_0} = \beta_0 + \beta_1 x_{01} + \beta_2 x_{02} + \cdots + \beta_k x_{0k} = \mathbf{x}_0' \boldsymbol{\beta}$$

The estimated mean response at this point is

$$\hat{y}(\mathbf{x}_0) = \mathbf{x}_0' \hat{\boldsymbol{\beta}} \tag{13–39}$$

This estimator is unbiased, because $E[\hat{y}(\mathbf{x}_0)] = E(\mathbf{x}_0' \hat{\boldsymbol{\beta}}) = \mathbf{x}_0' \boldsymbol{\beta} = \mu_{y|\mathbf{x}_0}$, and the variance of $\hat{y}(\mathbf{x}_0)$ is

$$V[\hat{y}(\mathbf{x}_0)] = \sigma^2 \mathbf{x}_0' (\mathbf{X}'\mathbf{X})^{-1} \mathbf{x}_0 \tag{13–40}$$

Therefore, a $100(1 - \alpha)$ percent confidence interval on the mean response at the point $x_{01}, x_{02}, \ldots, x_{0k}$ is

$$\begin{aligned} \hat{y}(\mathbf{x}_0) - t_{\alpha/2, n-p} &\sqrt{\hat{\sigma}^2 \mathbf{x}_0' (\mathbf{X}'\mathbf{X})^{-1} \mathbf{x}_0} \\ &\leq \mu_{y|\mathbf{x}_0} \leq \hat{y}(\mathbf{x}_0) + t_{\alpha/2, n-p} \sqrt{\hat{\sigma}^2 \mathbf{x}_0' (\mathbf{X}'\mathbf{X})^{-1} \mathbf{x}_0} \end{aligned} \tag{13–41}$$

13–6   PREDICTION OF NEW RESPONSE OBSERVATIONS

A regression model can be used to predict future observations on the response y corresponding to particular values of the regressor variables, say $x_{01}, x_{02}, \ldots, x_{0k}$. If $\mathbf{x}_0' = [1, x_{01}, x_{02}, \ldots, x_{0k}]$, then a point estimate for the future observation y_0 at the point $x_{01}, x_{02}, \ldots, x_{0k}$ is computed from Equation 13–39:

$$\hat{y}(\mathbf{x}_0) = \mathbf{x}_0' \hat{\boldsymbol{\beta}}$$

A $100(1 - \alpha)$ percent **prediction interval** for this future observation is

$$\begin{aligned} \hat{y}(\mathbf{x}_0) - t_{\alpha/2, n-p} &\sqrt{\hat{\sigma}^2 (1 + \mathbf{x}_0' (\mathbf{X}'\mathbf{X})^{-1} \mathbf{x}_0)} \\ &\leq y_0 \leq \hat{y}(\mathbf{x}_0) + t_{\alpha/2, n-p} \sqrt{\hat{\sigma}^2 (1 + \mathbf{x}_0' (\mathbf{X}'\mathbf{X})^{-1} \mathbf{x}_0)} \end{aligned} \tag{13–42}$$

In predicting new observations and in estimating the mean response at a given point $x_{01}, x_{02}, \ldots, x_{0k}$, we must be careful about extrapolating beyond the region containing the original observations. It is very possible that a model that fits well in the region of the original data will no longer fit well outside of that region.

13–7 REGRESSION MODEL DIAGNOSTICS

As we emphasized in designed experiments, **model adequacy checking** is an important part of the data analysis procedure. This is equally important in building regression models, and as we illustrated in Example 13–1, the **residual plots** that we used with designed experiments should always be examined for a regression model. In general, it is always necessary to (a) examine the fitted model to ensure that it provides an adequate approximation to the true system and (b) verify that none of the least squares regression assumptions are violated. The regression model will likely give poor or misleading results unless it is an adequate fit.

In addition to residual plots, there are other model diagnostics that are frequently useful in regression. This section briefly summarizes some of these procedures. For more complete presentations, see Montgomery and Peck (1992) and Myers (1990).

13–7.1 Scaled Residuals and PRESS

Standardized and Studentized Residuals ▪ Many model builders prefer to work with **scaled residuals** in contrast to the ordinary least squares residuals. These scaled residuals often convey more information than do the ordinary residuals.

One type of scaled residual is the **standardized residual:**

$$d_i = \frac{e_i}{\hat{\sigma}} \qquad i = 1, 2, \ldots, n \qquad (13\text{–}43)$$

where we generally use $\hat{\sigma} = \sqrt{MS_E}$ in the computation. These standardized residuals have mean zero and approximately unit variance; consequently, they are useful in looking for **outliers.** Most of the standardized residuals should lie in the interval $-3 \le d_i \le 3$, and any observation with a standardized residual outside of this interval is potentially unusual with respect to its observed response. These outliers should be carefully examined, because they may represent something as simple as a data-recording error or something of more serious concern, such as a region of the regressor variable space where the fitted model is a poor approximation to the true response surface.

The standardizing process in Equation (13–43) scales the residuals by dividing them by their average standard deviation. In some data sets, residuals may have standard deviations that differ greatly. We now present a scaling that takes this into account.

The vector of fitted values $\hat{y}_i$ corresponding to the observed values y_i is

$$\hat{\mathbf{y}} = \mathbf{X}\hat{\boldsymbol{\beta}}$$
$$= \mathbf{X}(\mathbf{X'X})^{-1}\mathbf{X'y} \qquad (13\text{–}44)$$
$$= \mathbf{Hy}$$

The $n \times n$ matrix $\mathbf{H} = \mathbf{X}(\mathbf{X'X})^{-1}\mathbf{X'}$ is usually called the "hat" matrix because it maps the vector of observed values into a vector of fitted values. The hat matrix and its properties play a central role in regression analysis.

The residuals from the fitted model may be conveniently written in matrix notation as

$$\mathbf{e} = \mathbf{y} - \hat{\mathbf{y}}$$

and it turns out that the covariance matrix of the residuals is

$$\text{Cov}(\mathbf{e}) = \sigma^2(\mathbf{I} - \mathbf{H}) \qquad (13\text{–}45)$$

The matrix $\mathbf{I} - \mathbf{H}$ is generally not diagonal, so the residuals have different variances and they are correlated.

Thus, the variance of the ith residual is

$$V(e_i) = \sigma^2(1 - h_{ii}) \qquad (13\text{–}46)$$

where h_{ii} is the ith diagonal element of $\mathbf{H}$. Because $0 \leqslant h_{ii} \leqslant 1$, using the residual mean square MS_E to estimate the variance of the residuals actually overestimates $V(e_i)$. Furthermore, because h_{ii} is a measure of the location of the ith point in x space, the variance of e_i depends on where the point x_i lies. Generally, residuals near the center of the x space have larger variance than do residuals at more remote locations. Violations of model assumptions are more likely at remote points, and these violations may be hard to detect from inspection of e_i (or d_i) because their residuals will usually be smaller.

We recommend taking this inequality of variance into account when scaling the residuals. We suggest plotting the **studentized residuals:**

$$r_i = \frac{e_i}{\sqrt{\hat{\sigma}^2(1 - h_{ii})}} \qquad i = 1, 2, \ldots, n \qquad (13\text{–}47)$$

with $\hat{\sigma}^2 = MS_E$ instead of e_i (or d_i). The studentized residuals have constant variance $V(r_i) = 1$ regardless of the location of $\mathbf{x}_i$ when the form of the model is correct. In many situations the variance of the residuals stabilizes, particularly for large data sets. In these cases there may be little difference between the standardized and studentized residuals. Thus standardized and studentized residuals often convey equivalent information. However, because any point with a large residual and a large h_{ii} is potentially highly influential on the least squares fit, examination of the studentized residuals is generally recommended. Table

13-3 displays the hat diagonals h_{ii} and the studentized residuals for the viscosity regression model in Example 13-1.

PRESS Residuals ▪ The prediction error sum of squares (PRESS) provides a useful residual scaling. To calculate PRESS, we select an observation—for example, i. We fit the regression model to the remaining $n - 1$ observations and use this equation to predict the withheld observation y_i. Denoting this predicted value $\hat{y}_{(i)}$, we may find the prediction error for point i as $e_{(i)} = y_i - \hat{y}_{(i)}$. The prediction error is often called the ith PRESS residual. This procedure is repeated for each observation $i = 1, 2, \ldots, n$, producing a set of n PRESS residuals $e_{(1)}$, $e_{(2)}, \ldots, e_{(n)}$. Then the PRESS statistic is defined as the sum of squares of the n PRESS residuals as in

$$\text{PRESS} = \sum_{i=1}^{n} e_{(i)}^2 = \sum_{i=1}^{n} [y_i - \hat{y}_{(i)}]^2 \qquad (13\text{-}48)$$

Thus PRESS uses each possible subset of $n - 1$ observations as an estimation data set, and every observation in turn is used to form a prediction data set.

It would initially seem that calculating PRESS requires fitting n different regressions. However, it is possible to calculate PRESS from the results of a single least squares fit to all n observations. It turns out that the ith PRESS residual is

$$e_{(i)} = \frac{e_i}{1 - h_{ii}} \qquad (13\text{-}49)$$

Thus because PRESS is just the sum of the squares of the PRESS residuals, a simple computing formula is

$$\text{PRESS} = \sum_{i=1}^{n} \left(\frac{e_i}{1 - h_{ii}} \right)^2 \qquad (13\text{-}50)$$

From Equation 13-49 it is easy to see that the PRESS residual is just the ordinary residual weighted according to the diagonal elements of the hat matrix h_{ii}. Data points for which h_{ii} are large will have large PRESS residuals. These observations will generally be **high influence** points. Generally, a large difference between the ordinary residual and the PRESS residuals will indicate a point where the model fits the data well, but a model built without that point predicts poorly. In the next section we will discuss some other measures of influence.

Finally, we note that PRESS can be used to compute an approximate R^2 for prediction, say

$$R^2_{\text{Prediction}} = 1 - \frac{\text{PRESS}}{S_{yy}} \qquad (13\text{-}51)$$

This statistic gives some indication of the predictive capability of the regression model. For the viscosity regression model from Example 13-1, we can compute the PRESS residuals using the ordinary residuals and the values of h_{ii} found in Table 13–3. The corresponding value of the PRESS statistic is PRESS $=$ 5207.7. Then

$$R^2_{\text{Prediction}} = 1 - \frac{\text{PRESS}}{S_{yy}}$$

$$= 1 - \frac{5207.7}{47,635.9}$$

$$= 0.8907$$

Therefore we could expect this model to "explain" about 89 percent of the variability in predicting new observations, as compared to the approximately 93 percent of the variability in the original data explained by the least squares fit. The overall predictive capability of the model based on this criterion seems very satisfactory.

R- Student ▪ The studentized residual r_i discussed above is often considered an outlier diagnostic. It is customary to use MS_E as an estimate of σ^2 in computing r_i. This is referred to as internal scaling of the residual because MS_E is an internally generated estimate of σ^2 obtained from fitting the model to all n observations. Another approach would be to use an estimate of σ^2 based on a data set with the ith observation removed. We denote the estimate of σ^2 so obtained by $S^2_{(i)}$. We can show that

$$S^2_{(i)} = \frac{(n-p)MS_E - e_i^2/(1 - h_{ii})}{n - p - 1} \tag{13–52}$$

The estimate of σ^2 in Equation 13–52 is used instead of MS_E to produce an externally studentized residual, usually called R-student, given by

$$t_i = \frac{e_i}{\sqrt{S^2_{(i)}(1 - h_{ii})}} \qquad i = 1, 2, \ldots, n \tag{13–53}$$

In many situations, t_i will differ little from the studentized residual r_i. However, if the ith observation is influential, then $S^2_{(i)}$ can differ significantly from MS_E, and thus the R-student will be more sensitive to this point. Furthermore, under the standard assumptions, t_i has a t_{n-p-1} distribution. Thus R-student offers a more formal procedure for outlier detection via hypothesis testing. Table 13–3 displays the values of R-student for the viscosity regression model in Example 13–1. None of those values is unusually large.

13–7.2   Influence Diagnostics

We occasionally find that a small subset of the data exerts a disproportionate influence on the fitted regression model. That is, parameter estimates or predictions may depend more on the influential subset than on the majority of the data. We would like to locate these influential points and assess their impact on the model. If these influential points are "bad" values, then they should be eliminated. On the other hand, there may be nothing wrong with these points. But if they control key model properties, we would like to know it, because it could affect the use of the model. In this section we describe and illustrate some useful measures of influence.

Leverage Points ▪ The disposition of points in x space is important in determining model properties. In particular, remote observations potentially have dispropȯrtionate leverage on the parameter estimates, predicted values, and the usual summary statistics.

The hat matrix $\mathbf{H} = \mathbf{X}(\mathbf{X'X})^{-1}\mathbf{X'}$ is very useful in identifying influential observations. As noted earlier, $\mathbf{H}$ determines the variances and covariances of $\hat{\mathbf{y}}$ and $\mathbf{e}$, because $V(\hat{\mathbf{y}}) = \sigma^2\mathbf{H}$ and $V(\mathbf{e}) = \sigma^2(\mathbf{I} - \mathbf{H})$. The elements h_{ij} of $\mathbf{H}$ may be interpreted as the amount of leverage exerted by y_j on $\hat{y}_i$. Thus, inspection of the elements of $\mathbf{H}$ can reveal points that are potentially influential by virtue of their location in x space. Attention is usually focused on the diagonal elements h_{ii}. Because $\Sigma_{i=1}^{n} h_{ii} = \text{rank}(\mathbf{H}) = \text{rank}(\mathbf{X}) = p$, the average size of the diagonal element of the $\mathbf{H}$ matrix is p/n. As a rough guideline, then, if a diagonal element h_{ii} is greater than $2p/n$, observation i is a high-leverage point. To apply this to the viscosity model in Example 13–1, note that $2p/n = 2(3)/16 = 0.375$. Table 13–3 gives the hat diagonals h_{ii} for the first-order model; because none of the h_{ii} exceeds 0.375, we would conclude that there are no leverage points in these data.

Influence on Regression Coefficients ▪ The hat diagonals will identify points that are potentially influential due to their location in x space. It is desirable to consider both the location of the point and the response variable in measuring influence. Cook (1977, 1979) has suggested using a measure of the squared distance between the least squares estimate based on all n points $\hat{\boldsymbol{\beta}}$ and the estimate obtained by deleting the ith point, say $\hat{\boldsymbol{\beta}}_{(i)}$. This distance measure can be expressed as

$$D_i = \frac{(\hat{\boldsymbol{\beta}}_{(i)} - \hat{\boldsymbol{\beta}})'\mathbf{X'X}(\hat{\boldsymbol{\beta}}_{(i)} - \hat{\boldsymbol{\beta}})}{pMS_E} \qquad i = 1, 2, \ldots, n \qquad (13\text{–}54)$$

A reasonable cutoff for D_i is unity. That is, we usually consider observations for which $D_i > 1$ to be influential.

The D_i statistic is actually calculated from

$$D_i = \frac{r_i^2}{p} \frac{V[\hat{y}(x_i)]}{V(e_i)} = \frac{r_i^2}{p} \frac{h_{ii}}{(1 - h_{ii})} \qquad i = 1, 2, \ldots, n \qquad (13\text{–}55)$$

Note that, apart from the constant p, D_i is the product of the square of the ith studentized residual and $h_{ii}/(1 - h_{ii})$. This ratio can be shown to be the distance from the vector $\mathbf{x}_i$ to the centroid of the remaining data. Thus, D_i is made up of a component that reflects how well the model fits the ith observation y_i and a component that measures how far that point is from the rest of the data. Either component (or both) may contribute to a larger value of D_i.

Table 13–3 presents the values of D_i for the regression model fit to the viscosity data in Example 13–1. None of these values of D_i exceeds 1, so there is no strong evidence of influential observations in these data.

13–8   TESTING FOR LACK OF FIT

In Section 7–6 we showed how adding center points to a 2^k factorial design allows the experimenter to obtain an estimate of pure experimental error. This allows the partitioning of the residual sum of squares SS_E into two components; that is,

$$SS_E = SS_{PE} + SS_{LOF}$$

where SS_{PE} is the sum of squares due to pure error and SS_{LOF} is the sum of squares due to lack of fit.

We may give a general development of this partitioning in the context of a regression model. Suppose that we have n_i observations on the response at the ith level of the regressors $\mathbf{x}_i$, $i = 1, 2, \ldots, m$. Let y_{ij} denote the jth observation on the response at $\mathbf{x}_i$, $i = 1, 2, \ldots, m$ and $j = 1, 2, \ldots, n_i$. There are $n = \sum_{i=1}^{m} n_i$ total observations. We may write the (ij)th residual as

$$y_{ij} - \hat{y}_i = (y_{ij} - \bar{y}_i) + (\bar{y}_i - \hat{y}_i) \tag{13–56}$$

where $\bar{y}_i$ is the average of the n_i observations at $\mathbf{x}_i$. Squaring both sides of Equation 13–56 and summing over i and j yields

$$\sum_{i=1}^{m} \sum_{j=1}^{n_i} (y_{ij} - \hat{y}_i)^2 = \sum_{i=1}^{m} \sum_{j=1}^{n_i} (y_{ij} - \bar{y}_i)^2 + \sum_{i=1}^{m} n_i(\bar{y}_i - \hat{y}_i)^2 \tag{13–57}$$

The left-hand side of Equation 13–57 is the usual residual sum of squares. The two components on the right-hand side measure pure error and lack of fit. We see that the pure error sum of squares

$$SS_{PE} = \sum_{i=1}^{m} \sum_{j=1}^{n_i} (y_{ij} - \bar{y}_i)^2 \tag{13–58}$$

is obtained by computing the corrected sum of squares of the repeat observations at each level of $\mathbf{x}$ and then pooling over the m levels of $\mathbf{x}$. If the assumption of constant variance is satisfied, this is a **model-independent** measure of pure error because only the variability of the y's at each $\mathbf{x}_i$ level is used to compute SS_{PE}. Because there are $n_i - 1$ degrees of freedom for pure error at each level $\mathbf{x}_i$, the total number of degrees of freedom associated with the pure error sum of squares is

$$\sum_{i=1}^{m} (n_i - 1) = n - m \tag{13–59}$$

The sum of squares for lack of fit

$$SS_{LOF} = \sum_{i=1}^{m} n_i(\bar{y}_i - \hat{y}_i)^2 \tag{13-60}$$

is a weighted sum of squared deviations between the mean response $\bar{y}_i$ at each x_i level and the corresponding fitted value. If the fitted values $\hat{y}_i$ are close to the corresponding average responses $\bar{y}_i$, then there is a strong indication that the regression function is linear. If the $\hat{y}_i$ deviate greatly from the $\bar{y}_i$, then it is likely that the regression function is not linear. There are $m - p$ degrees of freedom associated with SS_{LOF} because there are m levels of $\mathbf{x}$, and p degrees of freedom are lost because p parameters must be estimated for the model. Computationally we usually obtain SS_{LOF} by subtracting SS_{PE} from SS_E.

The test statistic for lack of fit is

$$F_0 = \frac{SS_{LOF}/(m-p)}{SS_{PE}/(n-m)} = \frac{MS_{LOF}}{MS_{PE}} \tag{13-61}$$

The expected value of MS_{PE} is σ^2, and the expected value of MS_{LOF} is

$$E(MS_{LOF}) = \sigma^2 + \frac{\sum_{i=1}^{m} n_i \left[E(y_i) - \beta_0 - \sum_{j=1}^{k} \beta_j x_{ij} \right]^2}{m-2} \tag{13-62}$$

If the true regression function is linear, then $E(y_i) = \beta_0 + \sum_{j=1}^{k} \beta_j x_{ij}$, and the second term of Equation 13-62 is zero, resulting in $E(MS_{LOF}) = \sigma^2$. However, if the true regression function is not linear, then $E(y_i) \neq \beta_0 + \sum_{j=1}^{k} \beta_j x_{ij}$, and $E(MS_{LOF}) > \sigma^2$. Furthermore, if the true regression function is linear, then the statistic F_0 follows the $F_{m-p,n-m}$ distribution. Therefore, to test for lack of fit, we would compute the test statistic F_0 and conclude that the regression function is not linear if $F_0 > F_{\alpha,m-p,n-m}$.

This test procedure may be easily incorporated into the analysis of variance. If we conclude that the regression function is not linear, then the tentative model must be abandoned and attempts made to find a more appropriate equation. Alternatively, if F_0 does not exceed $F_{\alpha,m-p,n-m}$, there is no strong evidence of lack of fit and MS_{PE} and MS_{LOF} are often combined to estimate σ^2. Example 7-5 is a very complete illustration of this procedure, where the replicate runs are center points in a 2^2 factorial design.

13-9 PROBLEMS

13-1 The tensile strength of a paper product is related to the amount of hardwood in the pulp. Ten samples are produced in the pilot plant, and the data obtained are shown in the following table.

Strength	Percent Hardwood	Strength	Percent Hardwood
160	10	181	20
171	15	188	25
175	15	193	25
182	20	195	28
184	20	200	30

(a) Fit a linear regression model relating strength to percent hardwood.

(b) Test the model in part (a) for significance of regression.

(c) Find a 95 percent confidence interval on the parameter β_1.

13–2 A plant distills liquid air to produce oxygen, nitrogen, and argon. The percentage of impurity in the oxygen is thought to be linearly related to the amount of impurities in the air as measured by the "pollution count" in parts per million (ppm). A sample of plant operating data is shown below.

Purity (%)	93.3	92.0	92.4	91.7	94.0	94.6	93.6
Pollution count (ppm)	1.10	1.45	1.36	1.59	1.08	0.75	1.20

	93.1	93.2	92.9	92.2	91.3	90.1	91.6	91.9
	0.99	0.83	1.22	1.47	1.81	2.03	1.75	1.68

(a) Fit a linear regression model to the data.

(b) Test for significance of regression.

(c) Find a 95 percent confidence interval on β_1.

13–3 Plot the residuals from Problem 13–1 and comment on model adequacy.

13–4 Plot the residuals from Problem 13–2 and comment on model adequacy.

13–5 Using the results of Problem 13–1, test the regression model for lack of fit.

13–6 A study was performed on wear of a bearing y and its relationship to $x_1 =$ oil viscosity and $x_2 =$ load. The following data were obtained.

y	x_1	x_2
193	1.6	851
230	15.5	816
172	22.0	1058
91	43.0	1201
113	33.0	1357
125	40.0	1115

(a) Fit a multiple linear regression model to the data.

(b) Test for significance of regression.

(c) Compute t statistics for each model parameter. What conclusions can you draw?

13-7 The brake horsepower developed by an automobile engine on a dynomometer is thought to be a function of the engine speed in revolutions per minute (rpm), the road octane number of the fuel, and the engine compression. An experiment is run in the laboratory and the data that follow are collected.

Brake Horsepower	rpm	Road Octane Number	Compression
225	2000	90	100
212	1800	94	95
229	2400	88	110
222	1900	91	96
219	1600	86	100
278	2500	96	110
246	3000	94	98
237	3200	90	100
233	2800	88	105
224	3400	86	97
223	1800	90	100
230	2500	89	104

(a) Fit a multiple regression model to these data.

(b) Test for significance of regression. What conclusions can you draw?

(c) Based on t tests, do you need all three regressor variables in the model?

13-8 Analyze the residuals from the regression model in Problem 13-7. Comment on model adequacy.

13-9 The yield of a chemical process is related to the concentration of the reactant and the operating temperature. An experiment has been conducted with the following results.

Yield	Concentration	Temperature
81	1.00	150
89	1.00	180
83	2.00	150
91	2.00	180
79	1.00	150
87	1.00	180
84	2.00	150
90	2.00	180

(a) Suppose we wish to fit a main effects model to this data. Set up the $\mathbf{X'X}$ matrix using the data exactly as it appears in the table.

(b) Is the matrix you obtained in part (a) diagonal? Discuss your response.

(c) Suppose we write our model in terms of the "usual" coded variables

$$x_1 = \frac{\text{Conc} - 1.5}{0.5}, \qquad x_2 = \frac{\text{Temp} - 165}{15}$$

Set up the $\mathbf{X'X}$ matrix for the model in terms of these coded variables. Is this matrix diagonal? Discuss your response.

(d) Define a new set of coded variables

$$x_1 = \frac{\text{Conc} - 1.0}{1.0}, \qquad x_2 = \frac{\text{Temp} - 150}{30}$$

Set up the $\mathbf{X'X}$ matrix for the model in terms of this set of coded variables. Is this matrix diagonal? Discuss your response.

(e) Summarize what you have learned from this problem about coding the variables.

13–10 Consider the 2^4 factorial experiment in Example 7–2. Suppose that the last observation is missing. Reanalyze the data and draw conclusions. How do these conclusions compare with those from the original example?

13–11 Consider the 2^4 factorial experiment in Example 7–2. Suppose that the last two observations are missing. Reanalyze the data and draw conclusions. How do these conclusions compare with those from the original example?

13–12 Given the following data, fit the second-order polynomial regression model

$$y = \beta_0 + \beta_1 x_1 + \beta_2 x_2 + \beta_{11} x_1^2 + \beta_{22} x_2^2 + \beta_{12} x_1 x_2 + \epsilon$$

y	x_1	x_2
26	1.0	1.0
24	1.0	1.0
175	1.5	4.0
160	1.5	4.0
163	1.5	4.0
55	0.5	2.0
62	1.5	2.0
100	0.5	3.0
26	1.0	1.5
30	0.5	1.5
70	1.0	2.5
71	0.5	2.5

After you have fit the model, test for significance of regression.

13-13 (a) Consider the quadratic regression model from Problem 13–12. Compute t statistics for each model parameter and comment on the conclusions that follow from these quantities.

(b) Use the extra sum of squares method to evaluate the value of the quadratic terms $x_1^2, x_2^2,$ and x_1x_2 to the model.

13-14 *Relationship between analysis of variance and regression.* Any analysis of variance model can be expressed in terms of the general linear model $\mathbf{y} = \mathbf{X}\boldsymbol{\beta} + \boldsymbol{\epsilon}$, where the $\mathbf{X}$ matrix consists of zeros and ones. Show that the single-factor model $y_{ij} = \mu + \tau_i + \epsilon_{ij}, i = 1, 2, 3, j = 1, 2, 3, 4$ can be written in general linear model form. Then

(a) Write the normal equations $(\mathbf{X}'\mathbf{X})\hat{\boldsymbol{\beta}} = \mathbf{X}'\mathbf{y}$ and compare them with the normal equations found for this model in Chapter 3.

(b) Find the rank of $\mathbf{X}'\mathbf{X}$. Can $(\mathbf{X}'\mathbf{X})^{-1}$ be obtained?

(c) Suppose the first normal equation is deleted and the restriction $\sum_{i=1}^{3} n\hat{\tau}_i = 0$ is added. Can the resulting system of equations be solved? If so, find the solution. Find the regression sum of squares $\hat{\boldsymbol{\beta}}'\mathbf{X}'\mathbf{y}$, and compare it to the treatment sum of squares in the single-factor model.

13-15 Suppose that we are fitting a straight line and we desire to make the variance of $\hat{\beta}_1$ as small as possible. Restricting ourselves to an even number of experimental points, where should we place these points so as to minimize $V(\hat{\beta}_1)$? (Note: Use the design called for in this exercise with *great* caution because, even though it minimizes $V(\hat{\beta}_1)$, it has some undesirable properties; for example, see Myers and Montgomery (1995). Only if you are *very sure* the true functional relationship is linear should you consider using this design.)

13-16 *Weighted least squares.* Suppose that we are fitting the straight line $y = \beta_0 + \beta_1 x + \epsilon$, but the variance of the y's now depends on the level of x; that is,

$$V(y|x_i) = \sigma_i^2 = \frac{\sigma^2}{w_i} \qquad i = 1, 2, \dots, n$$

where the w_i are known constants, often called weights. Show that if we choose estimates of the regression coefficients to minimize the weighted sum of squared errors given by $\sum_{i=1}^{n} w_i(y_i - \beta_0 - \beta_1 x_i)^2$, the resulting least squares normal equations are

$$\hat{\beta}_0 \sum_{i=1}^{n} w_i + \hat{\beta}_1 \sum_{i=1}^{n} w_i x_i = \sum_{i=1}^{n} w_i y_i$$

$$\hat{\beta}_0 \sum_{i=1}^{n} w_i x_i + \hat{\beta}_1 \sum_{i=1}^{n} w_i x_i^2 = \sum_{i=1}^{n} w_i x_i y_i$$

13-17 Consider the 2_{IV}^{4-1} design discussed in Example 13–5.

(a) Suppose you elect to augment the design with the single run selected in that example. Find the variances and covariances of the regression coefficients in the model (ignoring blocks):

$$y = \beta_0 + \beta_1 x_1 + \beta_2 x_2 + \beta_3 x_3 + \beta_4 x_4$$
$$+ \beta_{12} x_1 x_2 + \beta_{34} x_3 x_4 + \epsilon$$

(b) Are there any other runs in the alternate fraction that would dealias AB from CD?

(c) Suppose you augment the design with the four runs suggested in Example 13–5. Find the variances and covariances of the regression coefficients (ignoring blocks) for the model in part (a).

(d) Considering parts (a) and (c), which augmentation strategy would you prefer, and why?

13–18   Consider a 2_{III}^{7-4} design. Suppose after running the experiment, the largest observed effects are $A + BD$, $B + AD$, and $D + AB$. You wish to augment the original design with a group of four runs to dealias these effects.

(a) Which four runs would you make?

(b) Find the variances and covariances of the regression coefficients in the model

$$y = \beta_0 + \beta_1 x_1 + \beta_2 x_2 + \beta_4 x_4 + \beta_{12} x_1 x_2 + \beta_{14} x_1 x_4$$
$$+ \beta_{24} x_2 x_4 + \epsilon.$$

(c) Is it possible to dealias these effects with fewer than four additional runs?

Chapter 14
Response Surface Methods and Other Approaches to Process Optimization

14-1 INTRODUCTION TO RESPONSE SURFACE METHODOLOGY

Response surface methodology, or **RSM,** is a collection of mathematical and statistical techniques that are useful for the modeling and analysis of problems in which a response of interest is influenced by several variables and the objective is to optimize this response. For example, suppose that a chemical engineer wishes to find the levels of temperature (x_1) and pressure (x_2) that maximize the yield (y) of a process. The process yield is a function of the levels of temperature and pressure, say

$$y = f(x_1, x_2) + \epsilon$$

where ϵ represents the noise or error observed in the response y. If we denote the expected response by $E(y) = f(x_1, x_2) = \eta$, then the surface represented by

$$\eta = f(x_1, x_2)$$

is called a **response surface.**

We usually represent the response surface graphically, such as in Figure 14-1, where η is plotted versus the levels of x_1 and x_2. We have seen response surface plots such as this before, particularly in the chapters on factorial designs. To help visualize the shape of a response surface, we often plot the contours of the response surface as shown in Figure 14-2. In the contour plot, lines of constant response are drawn in the x_1, x_2 plane. Each contour corresponds to a particular height of the response surface. We have also previously seen the utility of contour plots.

In most RSM problems, the form of the relationship between the response and the independent variables is unknown. Thus, the first step in RSM is to find

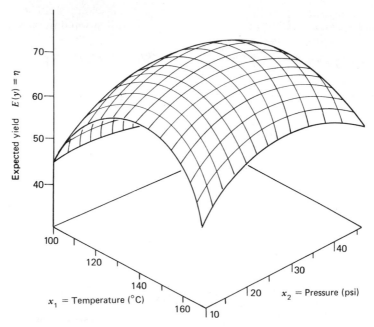

Figure 14–1.   A three-dimensional response surface showing the expected yield (η) as a function of temperature (x_1) and pressure (x_2).

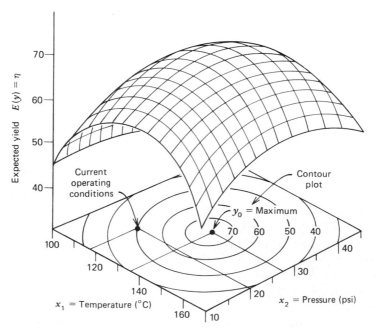

Figure 14–2.   A contour plot of a response surface.

a suitable approximation for the true functional relationship between y and the set of independent variables. Usually, a low-order polynomial in some region of the independent variables is employed. If the response is well modeled by a linear function of the independent variables, then the approximating function is the **first-order model**

$$y = \beta_0 + \beta_1 x_1 + \beta_2 x_2 + \cdots + \beta_k x_k + \epsilon \qquad (14\text{-}1)$$

If there is curvature in the system, then a polynomial of higher degree must be used, such as the **second-order model**

$$y = \beta_0 + \sum_{i=1}^{k} \beta_i x_i + \sum_{i=1}^{k} \beta_{ii} x_i^2 + \sum\sum_{i<j} \beta_{ij} x_i x_j + \epsilon \qquad (14\text{-}2)$$

Almost all RSM problems utilize one or both of these models. Of course, it is unlikely that a polynomial model will be a reasonable approximation of the true functional relationship over the entire space of the independent variables, but for a relatively small region they usually work quite well.

The method of least squares, discussed in Chapter 13, is used to estimate the parameters in the approximating polynomials. The response surface analysis is then done in terms of the fitted surface. If the fitted surface is an adequate approximation of the true response function, then analysis of the fitted surface will be approximately equivalent to analysis of the actual system. The model parameters can be estimated most effectively if proper experimental designs are used to collect the data. Designs for fitting response surfaces are called **response surface designs.** These designs are discussed in Section 14–4.

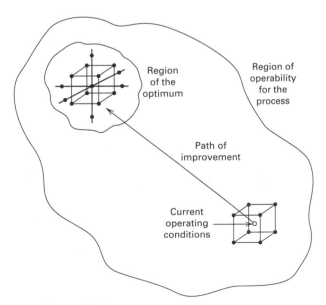

Figure 14–3. The sequential nature of RSM.

RSM is a **sequential procedure.** Often, when we are at a point on the response surface that is remote from the optimum, such as the current operating conditions in Figure 14–3, there is little curvature in the system and the first-order model will be appropriate. Our objective here is to lead the experimenter rapidly and efficiently along a path of improvement toward the general vicinity of the **optimum.** Once the region of the optimum has been found, a more elaborate model, such as the second-order model, may be employed, and an analysis may be performed to locate the optimum. From Figure 14–3, we see that the analysis of a response surface can be thought of as "climbing a hill," where the top of the hill represents the point of maximum response. If the true optimum is a point of minimum response, then we may think of "descending into a valley."

The eventual objective of RSM is to determine the optimum operating conditions for the system or to determine a region of the factor space in which operating requirements are satisfied. More extensive presentations of RSM are in Myers and Montgomery (1995), Khuri and Cornell (1987), and Box and Draper (1987).

14–2   THE METHOD OF STEEPEST ASCENT

Frequently, the initial estimate of the optimum operating conditions for the system will be far from the actual optimum. In such circumstances, the objective of the experimenter is to move rapidly to the general vicinity of the optimum. We wish to use a simple and economically efficient experimental procedure. When we are remote from the optimum, we usually assume that a first-order model is an adequate approximation to the true surface in a small region of the x's.

The **method of steepest ascent** is a procedure for moving sequentially along the path of steepest ascent, that is, in the direction of the maximum increase in the response. Of course, if minimization is desired, then we call this technique the **method of steepest descent.** The fitted first-order model is

$$\hat{y} = \hat{\beta}_0 + \sum_{i=1}^{k} \hat{\beta}_i x_i \tag{14–3}$$

and the first-order response surface, that is, the contours of $\hat{y}$, is a series of parallel lines such as that shown in Figure 14–4. The direction of steepest ascent is the direction in which $\hat{y}$ increases most rapidly. This direction is parallel to the normal to the fitted response surface. We usually take as the **path of steepest ascent** the line through the center of the region of interest and normal to the fitted surface. Thus, the steps along the path are proportional to the regression coefficients $\{\hat{\beta}_i\}$. The actual step size is determined by the experimenter based on process knowledge or other practical considerations.

Experiments are conducted along the path of steepest ascent until no further increase in response is observed. Then a new first-order model may be fit, a new

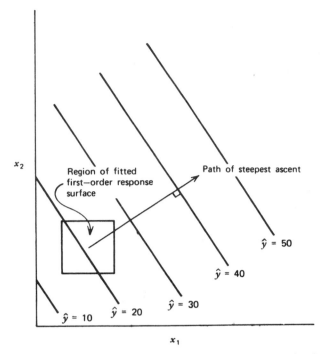

Figure 14-4.   First-order response surface and path of steepest ascent.

path of steepest ascent determined, and the procedure continued. Eventually, the experimenter will arrive in the vicinity of the optimum. This is usually indicated by lack of fit of a first-order model. At that time additional experiments are conducted to obtain a more precise estimate of the optimum.

Example 14-1

A chemical engineer is interested in determining the operating conditions that maximize the yield of a process. Two controllable variables influence process yield: reaction time and reaction temperature. The engineer is currently operating the process with a reaction time of 35 minutes and a temperature of 155°F, which result in yields of around 40 percent. Since it is unlikely that this region contains the optimum, she fits a first-order model and applies the method of steepest ascent.

The engineer decides that the region of exploration for fitting the first-order model should be (30, 40) minutes of reaction time and (150, 160)°F. To simplify the calculations, the independent variables will be coded to the usual $(-1, 1)$ interval. Thus, if ξ_1 denotes the **natural variable** time and ξ_2 denotes the **natural variable** temperature, then the **coded variables** are

$$x_1 = \frac{\xi_1 - 35}{5} \quad \text{and} \quad x_2 = \frac{\xi_2 - 155}{5}$$

Table 14–1   Process Data for Fitting the First-Order Model

Natural Variables		Coded Variables		Response
ξ_1	ξ_2	x_1	x_2	y
30	150	−1	−1	39.3
30	160	−1	1	40.0
40	150	1	−1	40.9
40	160	1	1	41.5
35	155	0	0	40.3
35	155	0	0	40.5
35	155	0	0	40.7
35	155	0	0	40.2
35	155	0	0	40.6

The experimental design is shown in Table 14–1. Note that the design used to collect this data is a 2^2 factorial augmented by five center points. Replicates at the center are used to estimate the experimental error and to allow for checking the adequacy of the first-order model. Also, the design is centered about the current operating conditions for the process.

A first-order model may be fit to these data by least squares. Following the methods of Chapter 13, we obtain the following model in the coded variables:

$$\hat{y} = 40.44 + 0.775x_1 + 0.325x_2$$

Before exploring along the path of steepest ascent, the adequacy of the first-order model should be investigated. The 2^2 design with center points allows the experimenter to

1.   Obtain an estimate of error.
2.   Check for interactions (cross-product terms) in the model.
3.   Check for quadratic effects (curvature).

The replicates at the center can be used to calculate an estimate of error as follows:

$$\hat{\sigma}^2 = \frac{(40.3)^2 + (40.5)^2 + (40.7)^2 + (40.2)^2 + (40.6)^2 - (202.3)^2/5}{4}$$

$$= 0.0430$$

The first-order model assumes that the variables x_1 and x_2 have an **additive effect** on the response. Interaction between the variables would be represented by the coefficient β_{12} of a cross-product term x_1x_2 added to the model. The least squares estimate of this coefficient is just one-half the interaction effect calculated as in an ordinary 2^2 factorial design or

$$\hat{\beta}_{12} = \tfrac{1}{4}[(1 \times 39.3) + (1 \times 41.5) + (-1 \times 40.0) + (-1 \times 40.9)]$$
$$= \tfrac{1}{4}(-0.1)$$
$$= -0.025$$

The single degree of freedom sum of squares for interaction is

$$SS_{\text{Interaction}} = \frac{(-0.1)^2}{4}$$

$$= 0.0025$$

Comparing $SS_{\text{Interaction}}$ to $\hat{\sigma}^2$ gives a lack-of-fit statistic

$$F = \frac{SS_{\text{Interaction}}}{\hat{\sigma}^2}$$

$$= \frac{0.0025}{0.0430}$$

$$= 0.058$$

which is small, indicating that interaction is negligible.

Another check of the adequacy of the straight-line model is obtained by applying the check for quadratic curvature described in Section 7–6. Recall that this consists of comparing the average response at the four points in the factorial portion of the design, say $\bar{y}_f = 40.425$, with the average response at the design center, say $\bar{y}_c = 40.46$. If there is quadratic curvature in the true response function, then $\bar{y}_f - \bar{y}_c$ is a measure of this curvature. If β_{11} and β_{22} are the coefficients of the "pure quadratic" terms x_1^2 and x_2^2, then $\bar{y}_f - \bar{y}_c$ is an estimate of $\beta_{11} + \beta_{22}$. In our example, an estimate of the pure quadratic term is

$$\hat{\beta}_{11} + \hat{\beta}_{22} = \bar{y}_f - \bar{y}_c$$

$$= 40.425 - 40.46$$

$$= -0.035$$

The single-degree-of-freedom sum of squares associated with the null hypothesis, $H_0 : \beta_{11} + \beta_{22} = 0$, is

$$SS_{\text{Pure Quadratic}} = \frac{n_f n_c (\bar{y}_f - \bar{y}_c)^2}{n_f + n_c}$$

$$= \frac{(4)(5)(-0.035)^2}{4 + 5}$$

$$= 0.0027$$

where n_f and n_c are the number of points in the factorial portion and the number of center points, respectively. Since

$$F = \frac{SS_{\text{Pure Quadratic}}}{\hat{\sigma}^2}$$

$$= \frac{0.0027}{0.0430}$$

$$= 0.063$$

is small, there is no indication of a pure quadratic effect.

The analysis of variance for this model is summarized in Table 14–2. Both the interaction and curvature checks are not significant, whereas the F test for the overall regression is significant. Furthermore, the standard error of $\hat{\beta}_1$ and $\hat{\beta}_2$ is

$$se(\hat{\beta}_i) = \sqrt{\frac{MS_E}{4}} = \sqrt{\frac{\hat{\sigma}^2}{4}} = \sqrt{\frac{0.0430}{4}} = 0.10 \qquad i = 1, 2$$

Table 14–2   Analysis of Variance for the First-Order Model

Source of Variation	Sum of Squares	Degrees of Freedom	Mean Square	F_0	P-Value
Regression (β_1, β_2)	2.8250	2	1.4125	47.83	0.0002
Residual	0.1772	6			
(Interaction)	(0.0025)	1	0.0025	0.058	0.8215
(Pure quadratic)	(0.0027)	1	0.0027	0.063	0.8142
(Pure error)	(0.1720)	4	0.0430		
Total	3.0022	8			

Both regression coefficients $\hat{\beta}_1$ and $\hat{\beta}_2$ are large relative to their standard errors. At this point, we have no reason to question the adequacy of the first-order model.

To move away from the design center—the point ($x_1 = 0$, $x_2 = 0$)—along the path of steepest ascent, we would move 0.775 units in the x_1 direction for every 0.325 units in the x_2 direction. Thus, the path of steepest ascent passes through the point ($x_1 = 0$, $x_2 = 0$) and has a slope 0.325/0.775. The engineer decides to use 5 minutes of reaction time as the basic step size. Using the relationship between ξ_1 and x_1, we see that 5 minutes of reaction time is equivalent to a step in the *coded* variable x_1 of $\Delta x_1 = 1$. Therefore, the steps along the path of steepest ascent are $\Delta x_1 = 1.0000$ and $\Delta x_2 = (0.325/0.775)\Delta x_1 = 0.42$.

The engineer computes points along this path and observes the yields at these points until a decrease in response is noted. The results are shown in Table 14–3 in both coded and natural variables. Although the coded variables are easier to manipulate mathematically, the natural variables must be used in running the process. Figure 14–5 plots the yield at each step along the path of steepest ascent. Increases in

Table 14–3   Steepest Ascent Experiment for Example 14–1

Steps	Coded Variables		Natural Variables		Response
	x_1	x_2	ξ_1	ξ_2	y
Origin	0	0	35	155	
Δ	1.00	0.42	5	2	
Origin + Δ	1.00	0.42	40	157	41.0
Origin + 2Δ	2.00	0.84	45	159	42.9
Origin + 3Δ	3.00	1.26	50	161	47.1
Origin + 4Δ	4.00	1.68	55	163	49.7
Origin + 5Δ	5.00	2.10	60	165	53.8
Origin + 6Δ	6.00	2.52	65	167	59.9
Origin + 7Δ	7.00	2.94	70	169	65.0
Origin + 8Δ	8.00	3.36	75	171	70.4
Origin + 9Δ	9.00	3.78	80	173	77.6
Origin + 10Δ	10.00	4.20	85	175	80.3
Origin + 11Δ	11.00	4.62	90	179	76.2
Origin + 12Δ	12.00	5.04	95	181	75.1

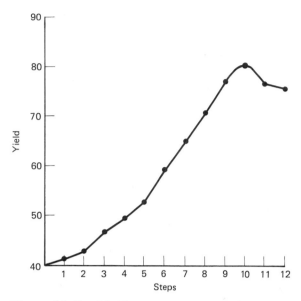

Figure 14–5.   Yield versus steps along the path of steepest ascent for Example 14–1.

response are observed through the tenth step; however, all steps beyond this point result in a decrease in yield. Therefore, another first-order model should be fit in the general vicinity of the point ($\xi_1 = 85$, $\xi_2 = 175$).

A new first-order model is fit around the point ($\xi_1 = 85$, $\xi_2 = 175$). The region of exploration for ξ_1 is [80, 90], and for ξ_2 it is [170, 180]. Thus, the coded variables are

$$x_1 = \frac{\xi_1 - 85}{5} \quad \text{and} \quad x_2 = \frac{\xi_2 - 175}{5}$$

Once again, a 2^2 design with five center points is used. The experimental design is shown in Table 14–4.

Table 14–4   Data for Second First-Order Model

Natural Variables		Coded Variables		Response
ξ_1	ξ_2	x_1	x_2	y
80	170	−1	−1	76.5
80	180	−1	1	77.0
90	170	1	−1	78.0
90	180	1	1	79.5
85	175	0	0	79.9
85	175	0	0	80.3
85	175	0	0	80.0
85	175	0	0	79.7
85	175	0	0	79.8

Table 14–5   Analysis of Variance for the Second First-Order Model

Source of Variation	Sum of Squares	Degrees of Freedom	Mean Square	F_0	P-Value
Regression	5.00	2			
Residual	11.1200	6			
(Interaction)	(0.2500)	1	0.2500	4.72	0.0955
(Pure quadratic)	(10.6580)	1	10.6580	201.09	0.0001
(Pure error)	(0.2120)	4	0.0530		
Total	16.1200	8			

The first-order model fit to the coded variables in Table 14–4 is

$$\hat{y} = 78.97 + 1.00x_1 + 0.50x_2$$

The analysis of variance for this model, including the interaction and pure quadratic term checks, is shown in Table 14–5. The interaction and pure quadratic checks imply that the first-order model is not an adequate approximation. This curvature in the true surface may indicate that we are near the optimum. At this point, additional analysis must be done to locate the optimum more precisely.

∎

From Example 14–1 we notice that the *path of steepest ascent is proportional to the signs and magnitudes of the regression coefficients* in the fitted first-order model

$$\hat{y} = \hat{\beta}_0 + \sum_{i=1}^{k} \hat{\beta}_i x_i$$

It is easy to give a general algorithm for determining the coordinates of a point on the path of steepest ascent. Assume that the point $x_1 = x_2 = \ldots = x_k = 0$ is the base or origin point. Then

1. Choose a step size in one of the process variables, say Δx_j. Usually, we would select the variable we know the most about, or we would select the variable that has the largest absolute regression coefficient $|\hat{\beta}_j|$.
2. The step size in the other variables is

$$\Delta x_i = \frac{\hat{\beta}_i}{\hat{\beta}_j / \Delta x_j} \qquad i = 1, 2, \ldots, k; \quad i \neq j$$

3. Convert the Δx_i from coded variables to the natural variables.

To illustrate, consider the path of steepest ascent computed in Example 14–1. Since x_1 has the largest regression coefficient, we select reaction time as the variable in step 1 of the above procedure. Five minutes of reaction time is the step size (based on process knowledge). In terms of the coded variables, this is $\Delta x_1 = 1.0$. Therefore, from guideline 2, the step size in temperature is

$$\Delta x_2 = \frac{\hat{\beta}_2}{\hat{\beta}_1 / \Delta x_1} = \frac{0.325}{(0.775/1.0)} = 0.42$$

To convert the coded step sizes ($\Delta x_1 = 1.0$ and $\Delta x_2 = 0.42$) to the natural units of time and temperature, we use the relationships

$$\Delta x_1 = \frac{\Delta \xi_1}{5} \quad \text{and} \quad \Delta x_2 = \frac{\Delta \xi_2}{5}$$

which results in

$$\Delta \xi_1 = \Delta x_1(5) = 1.0(5) = 5 \text{ min}$$

and

$$\Delta \xi_2 = \Delta x_2(5) = 0.42(5) = 2°\text{F}$$

14–3   ANALYSIS OF A SECOND-ORDER RESPONSE SURFACE

When the experimenter is relatively close to the optimum, a model that incorporates curvature is usually required to approximate the response. In most cases, the second-order model

$$y = \beta_0 + \sum_{i=1}^{k} \beta_i x_i + \sum_{i=1}^{k} \beta_{ii} x_i^2 + \sum\sum_{i<j} \beta_{ij} x_i x_j + \epsilon \tag{14-4}$$

is adequate. In this section we will show how to use this fitted model to find the optimum set of operating conditions for the x's and to characterize the nature of the response surface.

14–3.1   Location of the Stationary Point

Suppose we wish to find the levels of $x_1, x_2, \ldots, x_k$ that optimize the predicted response. This point, if it exists, will be the set of $x_1, x_2, \ldots, x_k$ for which the partial derivatives $\partial \hat{y}/\partial x_1 = \partial \hat{y}/\partial x_2 = \cdots = \partial \hat{y}/\partial x_k = 0$. This point, say $x_{1,s}, x_{2,s}, \ldots, x_{k,s}$, is called the **stationary point.** The stationary point could represent (1) a point of **maximum response,** (2) a point of **minimum response,** or (3) a **saddle point.** These three possibilities are shown in Figure 14–6.

Contour plots play a very important role in the study of the response surface. By generating contour plots using computer software for response surface analysis, the experimenter can usually characterize the shape of the surface and locate the optimum with reasonable precision.

We may obtain a general mathematical solution for the location of the

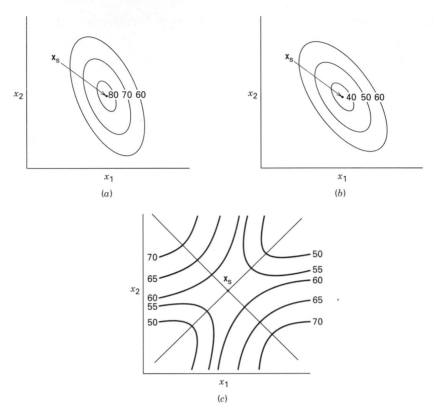

Figure 14–6.   Stationary points in a fitted second-order response surface.
(*a*) Maximum response. (*b*) Minimum response. (*c*) Saddle point.

stationary point. Writing the second-order model in matrix notation, we have

$$\hat{y} = \hat{\beta}_0 + \mathbf{x'b} + \mathbf{x'Bx} \qquad (14\text{–}5)$$

where

$$\mathbf{x} = \begin{bmatrix} x_1 \\ x_2 \\ \vdots \\ x_k \end{bmatrix} \quad \mathbf{b} = \begin{bmatrix} \hat{\beta}_1 \\ \hat{\beta}_2 \\ \vdots \\ \hat{\beta}_k \end{bmatrix} \quad \text{and} \quad \mathbf{B} = \begin{bmatrix} \hat{\beta}_{11}, \hat{\beta}_{12}/2, \ldots, \hat{\beta}_{1k}/2 \\ \hat{\beta}_{22}, \ldots, \hat{\beta}_{2k}/2 \\ \ddots \vdots \\ \text{sym.} \hat{\beta}_{kk} \end{bmatrix}$$

That is, **b** is a $(k \times 1)$ vector of the first-order regression coefficients and **B** is a $(k \times k)$ symmetric matrix whose main diagonal elements are the *pure* quadratic coefficients $(\hat{\beta}_{ii})$ and whose off-diagonal elements are one-half the *mixed* quadratic coefficients $(\hat{\beta}_{ij}, i \neq j)$. The derivative of $\hat{y}$ with respect to the elements of the vector $\mathbf{x}$ equated to $\mathbf{0}$ is

$$\frac{\partial \hat{y}}{\partial \mathbf{x}} = \mathbf{b} + 2\mathbf{Bx} = \mathbf{0} \tag{14-6}$$

The stationary point is the solution to Equation 14–6, or

$$\mathbf{x}_s = -\tfrac{1}{2}\mathbf{B}^{-1}\mathbf{b} \tag{14-7}$$

Furthermore, by substituting Equation 14–7 into Equation 14–5, we can find the predicted response at the stationary point as

$$\hat{y}_s = \hat{\beta}_0 + \tfrac{1}{2}\mathbf{x}_s'\mathbf{b} \tag{14-8}$$

14-3.2   Characterizing the Response Surface

Once we have found the stationary point, it is usually necessary to characterize the response surface in the immediate vicinity of this point. By **characterize,** we mean determine whether the stationary point is a point of maximum or minimum response or a saddle point. We also usually want to study the relative sensitivity of the response to the variables $x_1, x_2, \ldots, x_k$.

As we mentioned previously, the most straightforward way to do this is to examine a **contour plot** of the fitted model. If there are only two or three process variables (the x's), the construction and interpretation of this contour plot is relatively easy. However, even when there are relatively few variables, a more formal analysis, called the **canonical analysis,** can be useful.

It is helpful first to transform the model into a new coordinate system with the origin at the stationary point $\mathbf{x}_s$ and then to rotate the axes of this system until they are parallel to the principal axes of the fitted response surface. This transformation is shown in Figure 14–7 on the following page. We can show that this results in the fitted model

$$\hat{y} = \hat{y}_s + \lambda_1 w_1^2 + \lambda_2 w_2^2 + \cdots + \lambda_k w_k^2 \tag{14-9}$$

where the $\{w_i\}$ are the transformed independent variables and the $\{\lambda_i\}$ are constants. Equation 14–9 is called the **canonical form** of the model. Furthermore, the $\{\lambda_i\}$ are just the **eigenvalues** or **characteristic roots** of the matrix $\mathbf{B}$.

The nature of the response surface can be determined from the stationary point and the *sign* and *magnitude* of the $\{\lambda_i\}$. First, suppose that the stationary point is within the region of exploration for fitting the second-order model. If the $\{\lambda_i\}$ are all positive, then $\mathbf{x}_s$ is a point of minimum response; if the $\{\lambda_i\}$ are all negative, then $\mathbf{x}_s$ is a point of maximum response; and if the $\{\lambda_i\}$ have different signs, $\mathbf{x}_s$ is a saddle point. Furthermore, the surface is steepest in the w_i direction for which $|\lambda_i|$ is the greatest. For example, Figure 14–7 depicts a system for which $\mathbf{x}_s$ is a maximum (λ_1 and λ_2 are negative) with $|\lambda_1| > |\lambda_2|$.

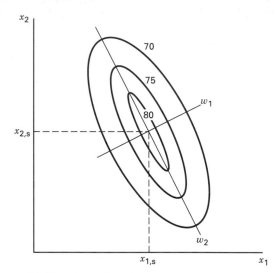

Figure 14–7. Canonical form of the second-order model.

Example 14–2

We will continue the analysis of the chemical process in Example 14–1. A second-order model in the variables x_1 and x_2 cannot be fit using the design in Table 14–4. The experimenter decides to augment this design with enough points to fit a second-order model.[1] She obtains four observations at $(x_1 = 0, x_2 = \pm 1.414)$ and $(x_1 = \pm 1.414, x_2 = 0)$. The complete experiment is shown in Table 14–6, and the design is displayed in Figure 14–8. This design is called a **central composite design** (or a CCD) and will be discussed in more detail in Section 14–4.2. In this second phase of the study, two additional responses were of interest, the viscosity and the molecular weight of the product. The responses are also shown in Table 14–6.

We will focus on fitting a quadratic model to the yield response y_1 (the other responses will be discussed in Section 14–3.4). We generally use computer software to fit a response surface and to construct the contour plots. Table 14–7 contains the output from Design-Expert, a popular PC package for fitting and optimizing response surfaces. (See pages 590–591.) From examining this table we notice that this software package first computes the "sequential or extra sums of squares" for the linear, quadratic, and cubic terms in the model (there is a warning message concerning aliasing in the cubic model because the CCD does not contain enough runs to support a cubic model). Based on the small P-value for the quadratic terms, we decided to fit the second-order model to the yield response. The computer output shows the final model in terms of both the coded variables and the natural or actual factor levels.

Figure 14–9 (page 592) shows the three-dimensional response surface plot and the contour plot for the yield response, in terms of the process variables time and temperature. It is relatively easy to see from examining these figures that the optimum is very near 175°F and 85 minutes of reaction time and that the response is at a

---

[1] The engineer ran the additional four observations at about the same time she ran the original nine observations. If substantial time had elapsed between the two sets of runs, then blocking would have been necessary. Blocking in response surface designs is discussed in Section 14–4.3.

Table 14–6 Central Composite Design for Example 14–2

Natural Variables		Coded Variables		Responses		
ξ_1	ξ_2	x_1	x_2	y_1 (yield)	y_2 (viscosity)	y_3 (molecular weight)
80	170	−1	−1	76.5	62	2940
80	180	−1	1	77.0	60	3470
90	170	1	−1	78.0	66	3680
90	180	1	1	79.5	59	3890
85	175	0	0	79.9	72	3480
85	175	0	0	80.3	69	3200
85	175	0	0	80.0	68	3410
85	175	0	0	79.7	70	3290
85	175	0	0	79.8	71	3500
92.07	175	1.414	0	78.4	68	3360
77.93	175	−1.414	0	75.6	71	3020
85	182.07	0	1.414	78.5	58	3630
85	167.93	0	−1.414	77.0	57	3150

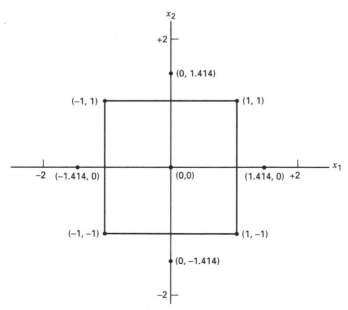

Figure 14–8. Central composite design for Example 14–2.

Table 14–7 Computer Output from Design-Expert for Fitting a Quadratic Model to the Yield Response, Example 14–2

Response: yield; File = No File   Run on 01/16/96 at 14:31:24

FAC	FACTOR	UNITS	−1 LEVEL	+1 LEVEL
A	time	minutes	80.000	90.000
B	temperature	degF	170.000	180.000

***** WARNING: The Cubic Model is Aliased! *****

Sequential Model Sum of Squares

SOURCE	SUM OF SQUARES	DF	MEAN SQUARE	F VALUE	PROB > F
MEAN	80062.16	1	80062.16		
Linear	10.04	2	5.02	2.69	0.117
Quadratic	18.20	3	6.07	85.57	< 0.001
Cubic	0.00	2	0.00	0.01	0.990
RESIDUAL	0.49	5	0.10		
TOTAL	80090.90	13			

Lack of Fit Tests

MODEL	SUM OF SQUARES	DF	MEAN SQUARE	F VALUE	PROB > F
Linear	18.49	6	3.08	58.14	< 0.001
Quadratic	0.28	3	0.09	1.79	0.289
Cubic	0.28	1	0.28	5.33	0.082
PURE ERR	0.21	4	0.05		

Model Summary Statistics

SOURCE	ROOT MSE	R-SQR	ADJ R-SQR	PRED R-SQR	PRESS
Linear	1.367	0.3494	0.2193	−0.0435	29.994
Quadratic	0.266	0.9827	0.9704	0.9181	2.353
Cubic	0.314	0.9828	0.9587	0.3598	18.400

Response: yield; File = No File   Run on 01/16/96 at 14:31:33

FAC	FACTOR	UNITS	−1 LEVEL	+1 LEVEL
A	time	minutes	80.000	90.000
B	temperature	degF	170.000	180.000

ANOVA for Quadratic Model

SOURCE	SUM OF SQUARES	DF	MEAN SQUARE	F VALUE	PROB > F
MODEL	28.25	5	5.649	79.67	< 0.001
RESIDUAL	0.50	7	0.071		
Lack Of Fit	0.28	3	0.095	1.79	0.289
Pure Error	0.21	4	0.053		
COR TOTAL	28.74	12			

ROOT MSE	0.266		R-SQUARED	0.9827
DEP MEAN	78.477		ADJ R-SQUARED	0.9704
C.V.	0.34%		PRED R-SQUARED	0.9181

Predicted Residual Sum of Squares (PRESS) = 2.35

Table 14–7   *continued*

| Response: yield; File = No File | | | | | | Run on 01/16/96 at 14:31:33 | |

FACTOR	COEFFICIENT ESTIMATE	DF	STD ERROR	t FOR H0 COEF=0	PROB $>$ |t|	VIF
Intercept	79.940	1	0.119	671.26		
A-time	0.995	1	0.094	10.57	$<$ 0.001	1.00
B-temperature	0.515	1	0.094	5.47	$<$ 0.001	1.00
A2	−1.376	1	0.101	−13.63	$<$ 0.001	1.02
B2	−1.001	1	0.101	−9.92	$<$ 0.001	1.02
AB	0.250	1	0.133	1.88	0.103	1.00

Final Equation in Terms of Coded Factors:

```
        yield =
                        79.940
                 +       0.995 * A
                 +       0.515 * B
                 −       1.376 * A2
                 −       1.001 * B2
                 +       0.250 * AB
```

Final Equation in Terms of Actual Factors:

```
        yield =
                     −1430.688
                 +       7.8089 * time
                 +      13.272 * temperature
                 −     5.506E-02 * time^2
                 −     4.005E-02 * temperature^2
                 +     1.000E-02 * time * temperature
```

Obs Ord	ACTUAL VALUE	PREDICTED VALUE	RESIDUAL	LEVER	STUDENT RESID	COOK'S DIST	OUTLIER t	Run Ord
1	76.500	76.302	0.198	0.625	1.21	0.410	1.27	3
2	78.000	77.792	0.208	0.625	1.28	0.452	1.35	6
3	77.000	76.832	0.168	0.625	1.03	0.294	1.03	7
4	79.500	79.322	0.178	0.625	1.09	0.330	1.11	1
5	75.600	75.781	−0.181	0.625	−1.11	0.342	−1.13	9
6	78.400	78.595	−0.195	0.625	−1.20	0.397	−1.24	10
7	77.000	77.209	−0.209	0.625	−1.28	0.458	−1.36	11
8	78.500	78.666	−0.166	0.625	−1.02	0.289	−1.02	8
9	79.900	79.940	−0.040	0.200	−0.17	0.001	−0.16	12
10	80.300	79.940	0.360	0.200	1.51	0.095	1.71	13
11	80.000	79.940	0.060	0.200	0.25	0.003	0.23	2
12	79.700	79.940	−0.240	0.200	−1.01	0.042	−1.01	4
13	79.800	79.940	−0.140	0.200	−0.59	0.014	−0.56	5

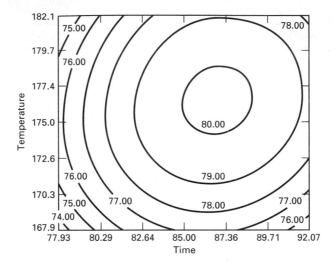

(*a*) The contour plot

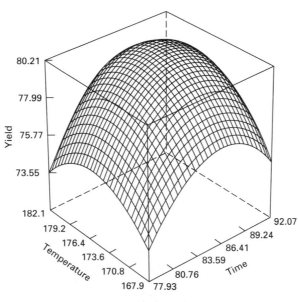

(*b*) The response surface plot

Figure 14–9.   Contour and response surface plots of the
yield response, Example 14–2.

maximum at this point. From examination of the contour plot, we note that the process may be slightly more sensitive to changes in reaction time than to changes in temperature.

We could also find the location of the stationary point using the general solution in Equation 14–7. Note that

$$\mathbf{b} = \begin{bmatrix} 0.995 \\ 0.515 \end{bmatrix} \qquad \mathbf{B} = \begin{bmatrix} -1.376 & 0.1250 \\ 0.1250 & -1.001 \end{bmatrix}$$

and from Equation 14–7 the stationary point is

$$\mathbf{x}_s = -\tfrac{1}{2}\mathbf{B}^{-1}\mathbf{b}$$

$$= -\tfrac{1}{2}\begin{bmatrix} -0.7345 & -0.0917 \\ -0.0917 & -1.0096 \end{bmatrix}\begin{bmatrix} 0.995 \\ 0.515 \end{bmatrix} = \begin{bmatrix} 0.389 \\ 0.306 \end{bmatrix}$$

That is, $x_{1,s} = 0.389$ and $x_{2,s} = 0.306$. In terms of the natural variables, the stationary point is

$$0.389 = \frac{\xi_1 - 85}{5} \qquad 0.306 = \frac{\xi_2 - 175}{5}$$

which yields $\xi_1 = 86.95 \simeq 87$ minutes of reaction time and $\xi_2 = 176.53 \simeq 176.5°F$. This is very close to the stationary point found by visual examination of the contour plot in Figure 14–9. Using Equation 14–8, we may find the predicted response at the stationary point as $\hat{y}_s = 80.21$.

We may also use the canonical analysis described in this section to characterize the response surface. First, it is necessary to express the fitted model in canonical form (Equation 14–9). The eigenvalues λ_1 and λ_2 are the roots of the determinantal equation

$$\mathbf{B} - \lambda\mathbf{I} = 0$$

$$\begin{vmatrix} -1.3770 - \lambda & 0.1250 \\ 0.1250 & -1.0018 - \lambda \end{vmatrix} = 0$$

which reduces to

$$\lambda^2 + 2.3788\lambda + 1.3639 = 0$$

The roots of this quadratic equation are $\lambda_1 = -0.9641$ and $\lambda_2 = -1.4147$. Thus, the canonical form of the fitted model is

$$\hat{y} = 80.21 - 0.9641w_1^2 - 1.4147w_2^2$$

Since both λ_1 and λ_2 are negative and the stationary point is within the region of exploration, we conclude that the stationary point is a maximum.

■

In some RSM problems it may be necessary to find the relationship between the **canonical variables** $\{w_i\}$ and the **design variables** $\{x_i\}$. This is particularly true if it is impossible to operate the process at the stationary point. As an illustration, suppose that in Example 14–2 we could not operate the process at $\xi_1 = 87$ minutes and $\xi_2 = 176.5°F$ because this combination of factors results in excessive cost. We now wish to "back away" from the stationary point to a point of lower cost without incurring large losses in yield. The canonical form of the model indicates that the surface is less sensitive to yield loss in the w_1 direction. Exploration of the canonical form requires converting points in the (w_1, w_2) space to points in the (x_1, x_2) space.

In general, the variables $\mathbf{x}$ are related to the canonical variables $\mathbf{w}$ by

$$\mathbf{w} = \mathbf{M}'(\mathbf{x} - \mathbf{x}_s)$$

where $\mathbf{M}$ is a $(k \times k)$ orthogonal matrix. The columns of $\mathbf{M}$ are the normalized eigenvectors associated with the $\{\lambda_i\}$. That is, if $\mathbf{m}_i$ is the ith column of $\mathbf{M}$, then $\mathbf{m}_i$ is the solution to

$$(\mathbf{B} - \lambda_i\mathbf{I})\mathbf{m}_i = \mathbf{0} \tag{14-10}$$

for which $\Sigma_{j=1}^k m_{ji}^2 = 1$.

We illustrate the procedure using the fitted second-order model in Example 14-2. For $\lambda_1 = -0.9641$, Equation 14-10 becomes

$$\begin{bmatrix} (-1.3770 + 0.9641) & 0.1250 \\ 0.1250 & (-1.0018 + 0.9641) \end{bmatrix} \begin{bmatrix} m_{11} \\ m_{21} \end{bmatrix} = \begin{bmatrix} 0 \\ 0 \end{bmatrix}$$

or

$$-0.4129m_{11} + 0.1250m_{21} = 0$$
$$0.1250m_{11} - 0.0377m_{21} = 0$$

We wish to obtain the normalized solution to these equations, that is, the one for which $m_{11}^2 + m_{21}^2 = 1$. There is no unique solution to these equations, so it is most convenient to assign an arbitrary value to one unknown, solve the system, and normalize the solution. Letting $m_{21}^* = 1$, we find $m_{11}^* = 0.3027$. To normalize this solution, we divide m_{11}^* and m_{21}^* by

$$\sqrt{(m_{11}^*)^2 + (m_{21}^*)^2} = \sqrt{(0.3027)^2 + (1)^2} = 1.0448$$

This yields the normalized solution

$$m_{11} = \frac{m_{11}^*}{1.0448} = \frac{0.3027}{1.0448} = 0.2897$$

and

$$m_{21} = \frac{m_{21}^*}{1.0448} = \frac{1}{1.0448} = 0.9571$$

which is the first column of the $\mathbf{M}$ matrix.

Using $\lambda_2 = -1.4147$, we can repeat the above procedure, obtaining $m_{12} = -0.9574$ and $m_{22} = 0.2888$ as the second column of $\mathbf{M}$. Thus, we have

$$\mathbf{M} = \begin{bmatrix} 0.2897 & -0.9574 \\ 0.9571 & 0.2888 \end{bmatrix}$$

The relationship between the **w** and **x** variables is

$$\begin{bmatrix} w_1 \\ w_2 \end{bmatrix} = \begin{bmatrix} 0.2897 & 0.9571 \\ -0.9574 & 0.2888 \end{bmatrix} \begin{bmatrix} x_1 - 0.3890 \\ x_2 - 0.3056 \end{bmatrix}$$

or

$$w_1 = 0.2897(x_1 - 0.3890) + 0.9571(x_2 - 0.3056)$$
$$w_2 = -0.9574(x_1 - 0.3890) + 0.2888(x_2 - 0.3056)$$

If we wished to explore the response surface in the vicinity of the stationary point, we could determine appropriate points at which to take observations in the (w_1, w_2) space and then use the above relationship to convert these points into the (x_1, x_2) space so that the runs may be made.

14-3.3 Ridge Systems

It is not unusual to encounter variations of the pure maximum, minimum, or saddle point response surfaces discussed in the previous section. Ridge systems, in particular, are fairly common. Consider the canonical form of the second-order model given previously in Equation 14–9:

$$\hat{y} = \hat{y}_s + \lambda_1 w_1^2 + \lambda_2 w_2^2 + \cdots + \lambda_k w_k^2$$

Now suppose that the stationary point $\mathbf{x}_s$ is within the region of experimentation; furthermore, let one or more of the λ_i be very small (e.g., $\lambda_i \approx 0$). The response variable is then very insensitive to the variables w_i multiplied by the small λ_i.

An example is shown in Figure 14–10 for $k = 2$ variables with $\lambda_1 = 0$. (In practice, λ_1 would be close to zero.) The canonical model for this response surface is theoretically

$$\hat{y} = \hat{y}_s + \lambda_2 w_2^2$$

with λ_2 negative. Notice that the severe elongation in the w_1 direction has resulted in a line of centers at $\hat{y} = 70$ and the optimum may be taken anywhere along that line. This type of response surface is called a **stationary ridge system.**

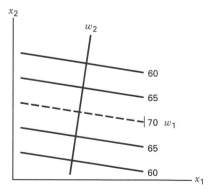

Figure 14–10.   A stationary ridge system.

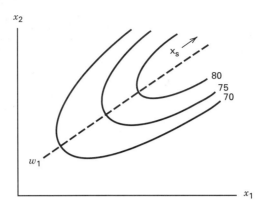

Figure 14–11.   A rising ridge system.

If the stationary point is far outside the region of exploration for fitting the second-order model and one (or more) λ_i is near zero, then the surface may be a **rising ridge.** Figure 14–11 illustrates a rising ridge for $k = 2$ variables with λ_1 near zero and λ_2 negative. In this type of ridge system, we cannot draw inferences about the true surface or the stationary point since $\mathbf{x}_s$ is outside the region where we have fit the model. However, further exploration is warranted in the w_1 direction. If λ_2 had been positive, we would call this system a falling ridge.

14–3.4   Multiple Responses

Many response surface problems involve the analysis of several responses. For instance, in Example 14–2, three responses were measured by the experimenter. In that example, we optimized the process with respect to only the yield response y_1.

Simultaneous consideration of multiple responses involves first building an appropriate response surface model for each response and then trying to find a set of operating conditions that in some sense optimizes all responses, or at least keeps them in desired ranges. An extensive treatment of the multiple response problem is given in Myers and Montgomery (1995).

We may obtain models for the viscosity and molecular weight responses (y_2 and y_3, respectively) in Example 14–2 as follows:

$$\hat{y}_2 = 70.00 - 0.16x_1 - 0.95x_2 - 0.09x_1^2 - 0.69x_2^2 - 1.25x_1x_2$$

$$\hat{y}_3 = 3386.2 + 205.1x_1 + 177.4x_2$$

In terms of the natural levels of time (ξ_1) and temperature (ξ_2), these models are

$$\hat{y}_2 = -9030.74 + 13.393\xi_1 + 97.708\xi_2$$
$$+ 2.75 \times 10^{-2}\xi_1^2 + 0.26757\xi_2^2 + 5 \times 10^{-2}\xi_1\xi_2$$

and

$$\hat{y}_3 = -6308.8 + 41.025\xi_1 + 35.473\xi_2$$

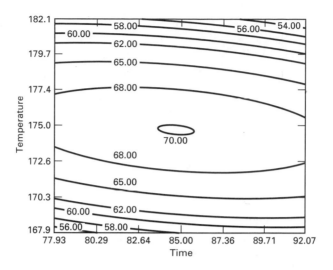

(a) The contour plot

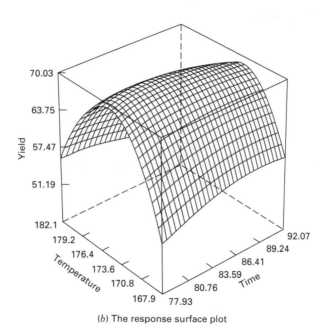

(b) The response surface plot

Figure 14–12.   Contour plot and response surface plot of viscosity, Example 14–2.

Figures 14–12 and 14–13 present the contour and response surface plots for these models.

A relatively straightforward approach to optimizing several responses that works well when there are only a few process variables is to overlay the contour plots for each response. Figure 14–14 shows an overlay plot for the three re-

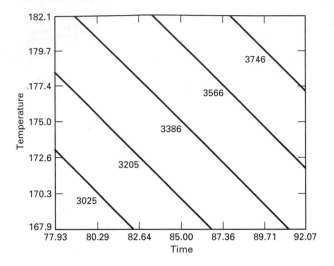

(a) The contour plot

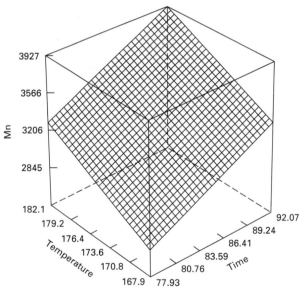

(b) The response surface plot

Figure 14–13.   Contour plot and response surface plot of molecular weight, Example 14–2.

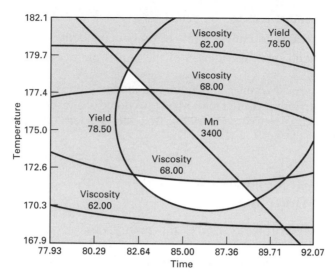

Figure 14–14. Region of the optimum found by overlaying yield, viscosity, and molecular weight response surfaces, Example 14–2.

sponses in Example 14–2, with contours for which y_1 (yield) ≥ 78.5, $62 \leq y_2$ (viscosity) ≤ 68, and y_3 (molecular weight Mn) ≤ 3400. If these boundaries represent important conditions that must be met by the process, then as the unshaded portion of Figure 14–14 shows, there are a number of combinations of time and temperature that will result in a satisfactory process.

14–4 EXPERIMENTAL DESIGNS FOR FITTING RESPONSE SURFACES

Fitting and analyzing response surfaces is greatly facilitated by the proper choice of an experimental design. In this section, we discuss some aspects of selecting appropriate designs for fitting response surfaces.

When selecting a response surface design, some of the features of a desirable design are as follows:

1. Provides a reasonable distribution of data points (and hence information) throughout the region of interest.
2. Allows model adequacy, including lack of fit, to be investigated.
3. Allows experiments to be performed in blocks.
4. Allows designs of higher order to be built up sequentially.
5. Provides an internal estimate of error.

6. Does not require a large number of runs.

7. Does not require too many levels of the independent variables.

8. Ensures simplicity of calculation of the model parameters.

These features are sometimes conflicting, so judgment must often be applied in design selection. For more information on the choice of a response surface design, refer to Myers and Montgomery (1995), Box and Draper (1987), and Khuri and Cornell (1987).

14–4.1 Designs for Fitting the First-Order Model

Suppose we wish to fit the first-order model in k variables

$$y = \beta_0 + \sum_{i=1}^{k} \beta_i x_i + \epsilon \tag{14–11}$$

There is a unique class of designs that minimize the variance of the regression coefficients $\{\hat{\beta}_i\}$. These are the **orthogonal first-order designs.** A first-order design

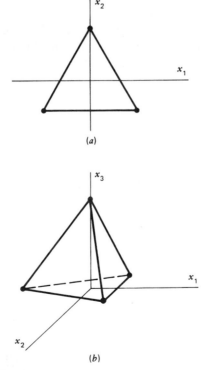

(a)

(b)

Figure 14–15. The simplex design for (a) $k = 2$ variables and (b) $k = 3$ variables.

is orthogonal if the off-diagonal elements of the $(\mathbf{X'X})$ matrix are all zero. This implies that the cross-products of the columns of the $\mathbf{X}$ matrix sum to zero.

The class of orthogonal first-order designs includes the 2^k factorial and fractions of the 2^k series in which main effects are not aliased with each other. In using these designs, we assume that the low and high levels of the k factors are coded to the usual ± 1 levels.

The 2^k design does not afford an estimate of the experimental error unless some runs are replicated. A common method of including replication in the 2^k design is to augment the design with several observations at the center (the point $x_i = 0, i = 1, 2, \ldots, k$). The addition of center points to the 2^k design does not influence the $\{\hat{\beta}_i\}$ for $i \geq 1$, but the estimate of β_0 becomes the grand average of all observations. Furthermore, the addition of center points does not alter the orthogonality property of the design. Example 14-1 illustrates the use of a 2^2 design augmented with five center points to fit a first-order model.

Another orthogonal first-order design is the **simplex.** The simplex is a regularly sided figure with $k + 1$ vertices in k dimensions. Thus, for $k = 2$ the simplex design is an equilateral triangle and for $k = 3$ it is a regular tetrahedron. Simplex designs in two and three dimensions are shown in Figure 14-15.

14-4.2 Designs for Fitting the Second-Order Model

We have informally introduced in Example 14-2 (and even earlier, in Example 7-5) the **central composite design** or **CCD** for fitting a second-order model. This is the most popular class of designs used for fitting these models. Generally, the CCD consists of a 2^k factorial (or fractional factorial of resolution V) with n_f runs, $2k$ axial or star runs, and n_c center runs. Figure 14-16 shows the CCD for $k = 2$ and $k = 3$ factors.

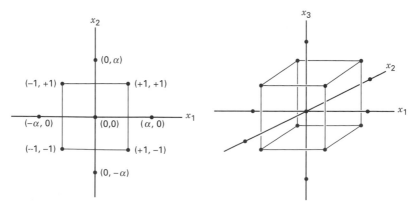

Figure 14-16. Central composite designs for $k = 2$ and $k = 3$.

The practical deployment of a CCD often arises through **sequential experimentation,** as in Examples 14–1 and 14–2. That is, a 2^k has been used to fit a first-order model, this model has exhibited lack of fit, and the axial runs are then added to allow the quadratic terms to be incorporated into the model. The CCD is a very efficient design for fitting the second-order model. There are two parameters in the design that must be specified; the distance α of the axial runs from the design center, and the number of center points n_c. We now discuss the choice of these two parameters.

Rotatability ▪ It is important for the second-order model to provide good predictions throughout the region of interest. One way to define "good" is to require that the model have a reasonably consistent and stable variance of the predicted response at points of interest **x**. Recall from Chapter 13, Equation 13–41, that the variance of the predicted response at some point **x** is

$$V[\hat{y}(\mathbf{x})] = \sigma^2 \mathbf{x}'(\mathbf{X}'\mathbf{X})\mathbf{x}$$

Box and Hunter (1957) suggested that a second-order response surface design should be **rotatable.** This means that the $V[\hat{y}(\mathbf{x})]$ is the same at all points **x** that are the same distance from the design center. That is, the variance of predicted response is constant on spheres.

Figure 14–17 shows contours of constant $\sqrt{V[\hat{y}(\mathbf{x})]}$ for the second-order model fit using the CCD in Example 14–2. Notice that the contours of constant standard deviation of predicted response are concentric circles. A design with this property will leave the variance of $\hat{y}$ unchanged when the design is rotated about the center $(0, 0, \ldots, 0)$; hence, the name *rotatable* design.

Rotatability is a reasonable basis for the selection of a response surface design. Since the purpose of RSM is optimization and the location of the optimum is unknown prior to running the experiment, it makes sense to use a design that provides equal precision of estimation in all directions (it can be shown that any first-order orthogonal design is rotatable).

A central composite design is made rotatable by the choice of α. The value of α for rotatability depends on the number of points in the factorial portion of the design; in fact, $\alpha = (n_f)^{1/4}$ yields a rotatable central composite design where n_f is the number of points used in the factorial portion of the design.

The Spherical CCD ▪ Rotatability is a spherical property; that is, it makes the most sense as a design criterion when the region of interest is a sphere. However, it is not important to have exact rotatability in order to have a good design. In fact, for a spherical region of interest, the best choice of α from a prediction variance viewpoint for the CCD is to set $\alpha = \sqrt{k}$. This design, called a **spherical CCD,** puts all the factorial and axial design points on the surface of a sphere of radius $\sqrt{k}$. For more discussion of this, see Myers and Montgomery (1995).

Center Runs in the CCD ▪ The choice of α in the CCD is dictated primarily by the region of interest. When this region is a sphere, the design must include center

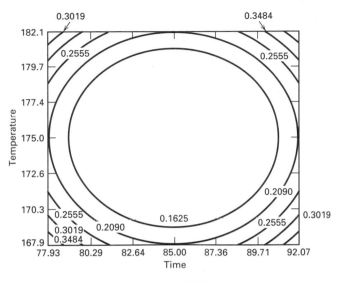

(a) contours of $\sqrt{V[\hat{y}(\mathbf{x})]}$

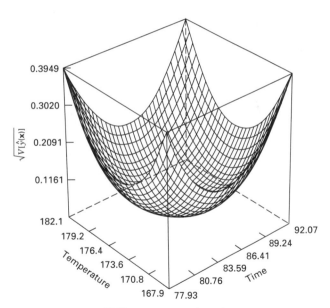

(b) The response surface plot

Figure 14–17. Contours of constant standard deviation of predicted response for the rotatable CCD, Example 14–2.

Table 14–8   A Three-Variable Box–Behnken Design

Run	x_1	x_2	x_3
1	−1	−1	0
2	−1	1	0
3	1	−1	0
4	1	1	0
5	−1	0	−1
6	−1	0	1
7	1	0	−1
8	1	0	1
9	0	−1	−1
10	0	−1	1
11	0	1	−1
12	0	1	1
13	0	0	0
14	0	0	0
15	0	0	0

runs to provide reasonably stable variance of predicted response. Generally, three to five center runs are recommended.

The Box–Behnken Design ▪ Box and Behnken (1960) have proposed some three-level designs for fitting response surfaces. These designs are formed by combining 2^k factorials with incomplete block designs. The resulting designs are usually very efficient in terms of the number of required runs, and they are either rotatable or nearly rotatable.

Table 14–8 shows a three-variable Box–Behnken design. The design is also shown geometrically in Figure 14–18. Notice that the Box–Behnken design is a spherical design, with all points lying on a sphere of radius $\sqrt{2}$. Also, the Box–Behnken design does not contain any points at the vertices of the cubic region created by the upper and lower limits for each variable. This could be advanta-

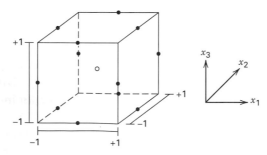

Figure 14–18.   A Box–Behnken design for three factors.

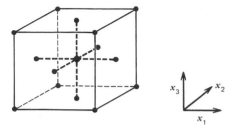

Figure 14–19. A face-centered central composite design for $k = 3$.

geous when the points on the corners of the cube represent factor-level combinations that are prohibitively expensive or impossible to test because of physical process constraints.

Cuboidal Region of Interest ▪ There are many situations where the region of interest is cuboidal rather than spherical. In these cases, a useful variation of the central composite design is the **face-centered central composite design** or the **face-centered cube,** in which $\alpha = 1$. This design locates the star or axial points on the centers of the faces of the cube, as shown in Figure 14–19 for $k = 3$. This variation of the central composite design is also sometimes used because it requires only three levels of each factor, and in practice it is frequently difficult to change factor levels. However, note that face-centered central composite designs are not rotatable.

The face-centered cube does not require as many center points as the spherical CCD. In practice, $n_c = 2$ is sufficient to provide good variance of prediction throughout the experimental region. It should be noted that sometimes more center runs will be employed to give a reasonable estimate of experimental error.

Other Designs ▪ There are many other response surface designs that are occasionally useful in practice. For two variables, we could use designs consisting of points that are equally spaced on a circle and that form regular polygons. Because the design points are equidistant from the origin, these arrangements are often called **equiradial designs.**

For $k = 2$, a rotatable equiradial design is obtained by combining $n_2 \geqslant 5$ points equally spaced on a circle with $n_1 \geqslant 1$ points at the center of the circle. Particularly useful designs for $k = 2$ are the pentagon and hexagon. These designs are shown in Figure 14–20. Other useful designs include the small composite designs, consisting of a fractional factorial in the cube of resolution III* (main effects aliased with two-factor interactions and no two-factor interactions aliased with each other) and the usual axial and center runs, and the class of hybrid designs, which may be of considerable value when it is important to reduce the number of runs as much as possible. For details of these designs, see Myers and Montgomery (1995).

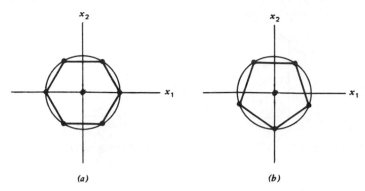

Figure 14-20.   Equiradial designs for two variables. (*a*) Hexagon. (*b*) Pentagon.

14-4.3   Blocking in Response Surface Designs

When using response surface designs, it is often necessary to consider blocking to eliminate nuisance variables. For example, this problem may occur when a second-order design is assembled sequentially from a first-order design, as was illustrated in Examples 14-1 and 14-2. Considerable time may elapse between the running of the first-order design and the running of the supplemental experiments required to build up a second-order design, and during this time test conditions may change, thus necessitating blocking.

A response surface design is said to **block orthogonally** if it is divided into blocks such that block effects do not affect the parameter estimates of the response surface model. If a 2^k or 2^{k-p} design is used as a first-order response surface design, the methods of Chapter 8 may be used to arrange the runs in 2^r blocks. The center points in these designs should be allocated equally among the blocks.

For a second-order design to block orthogonally, two conditions must be satisfied. If there are n_b observations in the bth block, then these conditions are

1.   Each block must be a first-order orthogonal design; that is,

$$\sum_{u=1}^{n_b} x_{iu}x_{ju} = 0 \qquad i \neq j = 0, 1, \ldots, k \quad \text{for all } b$$

where x_{iu} and x_{ju} are the levels of ith and jth variables in the uth run of the experiment with $x_{0u} = 1$ for all u.

2.   The fraction of the total sum of squares for each variable contributed by every block must be equal to the fraction of the total observations that occur

in the block; that is,

$$\frac{\sum\limits_{u=1}^{n_b} x_{iu}^2}{\sum\limits_{u=1}^{N} x_{iu}^2} = \frac{n_b}{N} \qquad i = 1, 2, \ldots, k \quad \text{for all } b$$

where N is the number of runs in the design.

As an example of applying these conditions, consider a rotatable central composite design in $k = 2$ variables with $N = 12$ runs. We may write the levels of x_1 and x_2 for this design in the design matrix

$$\mathbf{D} = \begin{bmatrix} x_1 & x_2 \\ -1 & -1 \\ 1 & -1 \\ -1 & 1 \\ 1 & 1 \\ 0 & 0 \\ 0 & 0 \\ 1.414 & 0 \\ -1.414 & 0 \\ 0 & 1.414 \\ 0 & -1.414 \\ 0 & 0 \\ 0 & 0 \end{bmatrix} \begin{array}{l} \left. \vphantom{\begin{matrix}1\\1\\1\\1\\1\\1\end{matrix}} \right\} \text{Block 1} \\ \left. \vphantom{\begin{matrix}1\\1\\1\\1\\1\\1\end{matrix}} \right\} \text{Block 2} \end{array}$$

Notice that the design has been arranged in two blocks, with the first block consisting of the factorial portion of the design plus two center points and the second block consisting of the axial points plus two additional center points. It is clear that condition 1 is met; that is, both blocks are first-order orthogonal designs. To investigate condition two, consider first block 1 and note that

$$\sum_{u=1}^{n_1} x_{1u}^2 = \sum_{u}^{n_1} x_{2u}^2 = 4$$

$$\sum_{u=1}^{N} x_{1u}^2 = \sum_{u=1}^{N} x_{2u}^2 = 8 \qquad \text{and} \qquad n_1 = 6$$

Therefore,

$$\frac{\sum\limits_{u=1}^{n_1} x_{iu}^2}{\sum\limits_{u=1}^{N} x_{iu}^2} = \frac{n_1}{N}$$

$$\frac{4}{8} = \frac{6}{12}$$

Thus, condition 2 is satisfied in block 1. For block 2, we have

$$\sum_{u=1}^{n_2} x_{1u}^2 = \sum_{u=1}^{n_2} x_{2u}^2 = 4 \qquad \text{and} \qquad n_2 = 6$$

Therefore,

$$\frac{\sum\limits_{u=1}^{n_2} x_{iu}^2}{\sum\limits_{u=1}^{N} x_{iu}^2} = \frac{n_2}{N}$$

$$\frac{4}{8} = \frac{6}{12}$$

Since condition 2 is also satisfied in block 2, this design blocks orthogonally.

In general, the central composite design can always be constructed to block orthogonally in two blocks with the first block consisting of n_f factorial points plus n_{cf} center points and the second block consisting of $n_a = 2k$ axial points plus n_{ca} center points. The first condition for orthogonal blocking will always hold regardless of the value used for α in the design. For the second condition to hold,

$$\frac{\sum\limits_{u}^{n_2} x_{iu}^2}{\sum\limits_{u}^{n_1} x_{iu}^2} = \frac{n_a + n_{ca}}{n_f + n_{cf}} \qquad (14\text{--}12)$$

The left-hand side of Equation 14–12 is $2\alpha^2/n_f$, and after substituting in this quantity, we may solve the equation for the value of α that will result in orthogonal blocking as

$$\alpha = \left[\frac{n_f(n_a + n_{ca})}{2(n_f + n_{cf})} \right]^{1/2} \qquad (14\text{--}13)$$

This value of α does not, in general, result in a rotatable or spherical design. If the design is also required to be rotatable, then $\alpha = (n_f)^{1/4}$ and

$$(n_f)^{1/2} = \frac{n_f(n_a + n_{ca})}{2(n_f + n_{cf})} \tag{14-14}$$

It is not always possible to find a design that exactly satisfies Equation 14–14. For example if $k = 3$ then $n_f = 8$ and $n_a = 6$, and Equation 14–14 reduces to

$$(8)^{1/2} = \frac{8(6 + n_{ca})}{2(8 + n_{cf})}$$

$$2.83 = \frac{48 + 8n_{ca}}{16 + 2n_{cf}}$$

It is impossible to find values of n_{ca} and n_{cf} that exactly satisfy this last equation. However, note that if $n_{cf} = 3$ and $n_{ca} = 2$, then the right-hand side is

$$\frac{48 + 8(2)}{16 + 2(3)} = 2.91$$

so the design nearly blocks orthogonally. In practice, one could relax somewhat the requirement of either rotatability or orthogonal blocking without any major loss of information.

The central composite design is very versatile in its ability to accommodate blocking. If k is large enough, the factorial portion of the design can be divided into two or more blocks. (The number of factorial blocks must be a power of 2, with the axial portion forming a single block.) Table 14–9 presents several useful blocking arrangements for the central composite design.

There are two important points about the analysis of variance when the response surface design has been run in blocks. The first concerns the use of center points to calculate an estimate of pure error. Only center points that are run in the same block can be considered to be replicates, so the pure error term can only be calculated within each block. If the variability is consistent across blocks, then these pure error estimates could be pooled. The second point concerns the block effect. If the design blocks orthogonally in m blocks, the sum of squares for blocks is

$$SS_{\text{Blocks}} = \sum_{b=1}^{m} \frac{B_b^2}{n_b} - \frac{G^2}{N} \tag{14-15}$$

where B_b is the total of the n_b observations in the bth block and G is the grand total of all N observations in all m blocks. When blocks are not exactly orthogonal, the general regression significance test (the "extra sum of squares" method) described in Chapter 13 can be used.

Table 14–9 Some Rotatable and Near-Rotatable Central Composite Designs that Block Orthogonally

k	2	3	4	5	5 ½ rep.	6	6 ½ rep.	7	7 ½ rep.
Factorial Block(s)									
n_f	4	8	16	32	16	64	32	128	64
Number of blocks	1	2	2	4	1	8	2	16	8
Number of points in each block	4	4	8	8	16	8	16	8	8
Number of center points in each block	3	2	2	2	6	1	4	1	1
Total number of points in each block	7	6	10	10	22	9	20	9	9
Axial Block									
n_a	4	6	8	10	10	12	12	14	14
n_{ca}	3	2	2	4	1	6	2	11	4
Total number of points in the axial block	7	8	10	14	11	18	14	25	18
Total number of points N in the design	14	20	30	54	33	90	54	169	80
Values of α									
Orthogonal blocking	1.4142	1.6330	2.0000	2.3664	2.0000	2.8284	2.3664	3.3636	2.8284
Rotatability	1.4142	1.6818	2.0000	2.3784	2.0000	2.8284	2.3784	3.3333	2.8284

14–5   MIXTURE EXPERIMENTS

In previous sections, we have presented response surface designs for those situations in which the levels of each factor are independent of the levels of other factors. In **mixture experiments,** the factors are the components or ingredients of a mixture, and consequently, their levels are not independent. For example, if $x_1, x_2, \ldots, x_p$ denote the proportions of p components of a mixture, then

$$0 \leq x_i \leq 1 \qquad i = 1, 2, \ldots, p$$

and

$$x_1 + x_2 + \cdots + x_p = 1 \qquad \text{(i.e., 100 percent)}$$

These restrictions are illustrated graphically in Figure 14–21 for $p = 2$ and $p = 3$ components. For two components, the factor space for the design includes all values of the two components that lie on the line segment $x_1 + x_2 = 1$, with each component being bounded by 0 and 1. With three components, the mixture space is a triangle with vertices corresponding to formulations that are **pure blends** (mixtures that are 100 percent of a single component).

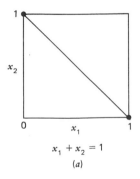

$x_1 + x_2 = 1$

(a)

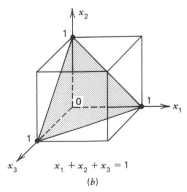

$x_1 + x_2 + x_3 = 1$

(b)

Figure 14–21.   Constrained factor space for mixtures with (a) $p = 2$ components and (b) $p = 3$ components.

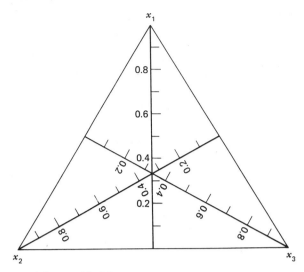

Figure 14–22. Trilinear coordinate system.

When there are three components of the mixture, the constrained experimental region can be conveniently represented on **trilinear coordinate paper** as shown in Figure 14–22. Each of the three sides of the graph in Figure 14–22 represents a mixture that has none of one of the three components (the component labeled on the opposite vertex). The nine grid lines in each direction mark off 10 percent increments in the respective components.

Simplex designs are used to study the effects of mixture components on the response variable. A $\{p, m\}$ **simplex lattice design** for p components consists of points defined by the following coordinate settings: the proportions assumed by each component take the $m + 1$ equally spaced values from 0 to 1,

$$x_i = 0, \frac{1}{m}, \frac{2}{m}, \ldots, 1 \qquad i = 1, 2, \ldots, p \qquad (14\text{–}16)$$

and all possible combinations (mixtures) of the proportions from Equation 14–16 are used. As an example, let $p = 3$ and $m = 2$. Then

$$x_i = 0, \tfrac{1}{2}, 1 \qquad i = 1, 2, 3$$

and the simplex lattice consists of the following six runs:

$$(x_1, x_2, x_3) = (1, 0, 0), (0, 1, 0), (0, 0, 1), (\tfrac{1}{2}, \tfrac{1}{2}, 0), (\tfrac{1}{2}, 0, \tfrac{1}{2}), (0, \tfrac{1}{2}, \tfrac{1}{2})$$

This design is shown in Figure 14–23. The three vertices $(1, 0, 0)$, $(0, 1, 0)$, and $(0, 0, 1)$ are the pure blends, whereas the points $(\tfrac{1}{2}, \tfrac{1}{2}, 0)$, $(\tfrac{1}{2}, 0, \tfrac{1}{2})$, and $(0, \tfrac{1}{2}, \tfrac{1}{2})$ are binary blends or two-component mixtures located at the midpoints of the three sides of the triangle. Figure 14–23 also shows the $\{3, 3\}$, the $\{4, 2\}$, and the $\{4, 3\}$ simplex lattice designs. In general, the number of points in a $\{p, m\}$ simplex lattice design is

$$N = \frac{(p + m - 1)!}{m!(p - 1)!}$$

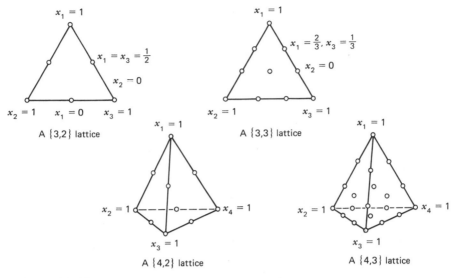

Figure 14–23. Some simplex lattice designs for $p = 3$ and $p = 4$ components.

An alternative to the simplex lattice design is the **simplex centroid design.** In a p-component simplex centroid design, there are $2^p - 1$ points, corresponding to the p permutations of $(1, 0, 0, \ldots, 0)$, the $\binom{p}{2}$ permutations of $(\frac{1}{2}, \frac{1}{2}, 0, \ldots, 0)$, the $\binom{p}{3}$ permutations of $(\frac{1}{3}, \frac{1}{3}, \frac{1}{3}, 0, \ldots, 0), \ldots$, and the overall centroid $(\frac{1}{p}, \frac{1}{p}, \ldots, \frac{1}{p})$. Figure 14–24 shows some simplex centroid designs.

A criticism of the simplex designs described above is that most of the experimental runs occur on the boundary of the region and, consequently, include only $p - 1$ of the p components. It is usually desirable to augment the simplex lattice or simplex centroid with additional points in the interior of the region where

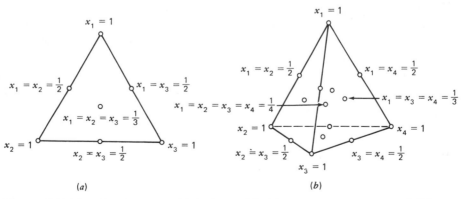

Figure 14–24. Simplex centroid designs with (a) $p = 3$ components and (b) $p = 4$ components.

the blends will consist of all p mixture components. For more discussion, see Cornell (1990) and Myers and Montgomery (1995).

Mixture models differ from the usual polynomials employed in response surface work, because of the constraint $\Sigma x_i = 1$. The standard forms of the mixture models that are in widespread use are

Linear:

$$E(y) = \sum_{i=1}^{p} \beta_i x_i \qquad (14\text{--}17)$$

Quadratic:

$$E(y) = \sum_{i=1}^{p} \beta_i x_i + \sum \sum_{i<j}^{p} \beta_{ij} x_i x_j \qquad (14\text{--}18)$$

Full Cubic:

$$E(y) = \sum_{i=1}^{p} \beta_i x_i + \sum \sum_{i<j}^{p} \beta_{ij} x_i x_j$$

$$+ \sum \sum_{i<j}^{p} \delta_{ij} x_i x_j (x_i - x_j) \qquad (14\text{--}19)$$

$$+ \sum \sum \sum_{i<j<k} \beta_{ijk} x_i x_j x_k$$

Special Cubic:

$$E(y) = \sum_{i=1}^{p} \beta_i x_i + \sum \sum_{i<j}^{p} \beta_{ij} x_i x_j$$

$$+ \sum \sum \sum_{i<j<k} \beta_{ijk} x_i x_j x_k \qquad (14\text{--}20)$$

The terms in these models have relatively simple interpretations. In Equations 14–17 through 14–20, the parameter β_i represents the expected response to the pure blend $x_i = 1$ and $x_j = 0$ when $j \neq i$. The portion $\sum_{i=1}^{p} \beta_i x_i$ is called the **linear blending portion.** When there is curvature arising from nonlinear blending between component pairs, the parameters β_{ij} represent either **synergistic** or **antagonistic blending.** Higher-order terms are frequently necessary in mixture models because (1) the phenomena studied may be complex and (2) the experimental region is frequently the entire operability region and is therefore large, requiring an elaborate model.

Example 14–3

A Three-Component Mixture

Cornell (1990) describes a mixture experiment in which three components—polyethylene (x_1), polystyrene (x_2), and polypropylene (x_3)—were blended to form fiber that will be spun into yarn for draperies. The response variable of interest is yarn elongation in kilograms of force applied. A {3, 2} simplex lattice design is

Table 14-10 The {3, 2} Simplex Lattice Design for the Yarn Elongation Problem

Design Point	Component Proportions			Observed Elongation Values	Average Elongation Value ($\bar{y}$)
	x_1	x_2	x_3		
1	1	0	0	11.0, 12.4	11.7
2	$\frac{1}{2}$	$\frac{1}{2}$	0	15.0, 14.8, 16.1	15.3
3	0	1	0	8.8, 10.0	9.4
4	0	$\frac{1}{2}$	$\frac{1}{2}$	10.0, 9.7, 11.8	10.5
5	0	0	1	16.8, 16.0	16.4
6	$\frac{1}{2}$	0	$\frac{1}{2}$	17.7, 16.4, 16.6	16.9

used to study the product. The design and the observed responses are shown in Table 14-10. Notice that all of the design points involve either **pure** or **binary blends;** that is, at most only two of the three components are used in any formulation of the product. Replicate observations are also run, with two replicates at each of the pure blends and three replicates at each of the binary blends. The error standard deviation can be estimated from these replicate observations as $\hat{\sigma} = 0.85$. Cornell fits the second-order mixture polynomial to the data, resulting in

$$\hat{y} = 11.7x_1 + 9.4x_2 + 16.4x_3 + 19.0x_1x_2 + 11.4x_1x_3 - 9.6x_2x_3$$

This model can be shown to be an adequate representation of the response. Note that since $\hat{\beta}_3 > \hat{\beta}_1 > \hat{\beta}_2$, we would conclude that component 3 (polypropylene) produces yarn with the highest elongation. Furthermore, since $\hat{\beta}_{12}$ and $\hat{\beta}_{13}$ are positive, blending components 1 and 2 or components 1 and 3 produces higher elongation values than would be expected just by averaging the elongations of the pure blends. This is an example of "synergistic" blending effects. Components 2 and 3 have antagonistic blending effects since $\hat{\beta}_{23}$ is negative.

Figure 14-25 plots the contours of elongation, and this may be helpful in inter-

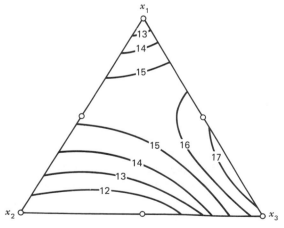

Figure 14-25. Contours of constant estimated yarn elongation from the second-order mixture model for Example 14-3.

preting the results. From examining the figure, we note that if maximum elongation is desired, a blend of components 1 and 3 should be chosen consisting of about 80 percent component 3 and 20 percent component 1.

■

In some mixture problems, constraints on the individual components arise. Lower bound constraints of the form

$$l_i \leq x_i \leq 1 \qquad i = 1, 2, \ldots, p$$

are fairly common. When only lower bound constraints are present, the feasible design region is still a simplex, but it is inscribed inside the original simplex region. This situation may be simplified by the introduction of pseudocomponents, defined as

$$x_i' = \frac{x_i - l_i}{\left(1 - \sum_{j=1}^{p} l_j\right)} \tag{14–21}$$

with $\sum_{j=1}^{p} l_j < 1$. Now

$$x_1' + x_2' + \cdots + x_p' = 1$$

so the use of pseudocomponents allows the use of simplex-type designs when lower bounds are a part of the experimental situation. The formulations specified by the simplex design for the pseudocomponents are transformed into formulations for the original components by reversing the transformation Equation 14–21. That is, if x_i' is the value assigned to the ith pseudocomponent on one of the runs in the experiment, then the ith original mixture component is

$$x_i = l_i + \left(1 - \sum_{j=1}^{p} l_j\right) x_i' \tag{14–22}$$

If the components have both upper and lower bound constraints, then the feasible region is an irregular polytope and simplex-type designs cannot be used. See Cornell (1990) and Myers and Montgomery (1995) for more information on designs for constrained mixture spaces.

14–6 EVOLUTIONARY OPERATION

Response surface methodology is often applied to pilot plant operations by research and development personnel. When it is applied to a full-scale production process, it is usually only done once (or very infrequently) because the experimental procedure is relatively elaborate. However, conditions that were optimum for the pilot plant may not be optimum for the full-scale process. The pilot plant may produce 2 pounds of product per day whereas the full-scale process may produce 2000 pounds per day. This "scale-up" of the pilot plant to the full-scale production process usually results in distortion of the optimum conditions. Even if the full-scale plant begins operation at the optimum, it will eventually "drift" away from that point because of variations in raw materials, environmental changes, and operating personnel.

A method is needed for the continuous monitoring and improvement of a full-scale process with the goal of moving the operating conditions toward the optimum or following a "drift." The method should not require large or sudden changes in operating conditions that might disrupt production. **Evolutionary operation (EVOP)** was proposed by Box (1957) as such an operating procedure. It is designed as a method of routine plant operation that is carried out by manufacturing personnel with minimum assistance from the research and development staff.

EVOP consists of systematically introducing small changes in the levels of the operating variables under consideration. Usually, a 2^k design is employed to do this. The changes in the variables are assumed to be small enough that serious disturbances in yield, quality, or quantity will not occur, yet large enough that potential improvements in process performance will eventually be discovered. Data are collected on the response variables of interest at each point of the 2^k design. When one observation has been taken at each design point, a **cycle** is said to have been completed. The effects and interactions of the process variables may then be computed. Eventually, after several cycles, the effect of one or more process variables or their interactions may appear to have a significant effect on the response. At this point, a decision may be made to change the basic operating conditions to improve the response. When improved conditions have been detected, a **phase** is said to have been completed.

In testing the significance of process variables and interactions, an estimate of experimental error is required. This is calculated from the cycle data. Also, the 2^k design is usually centered about the current best operating conditions. By comparing the response at this point with the 2^k points in the factorial portion, we may check on curvature or **change in mean** (CIM); that is, if the process is really centered at the maximum, say, then the response at the center should be significantly greater than the responses at the 2^k-peripheral points.

In theory, EVOP can be applied to k process variables. In practice, only two or three variables are usually considered. We will give an example of the procedure for two variables. Box and Draper (1969) give a detailed discussion of the three-variable case, including necessary forms and worksheets.

Example 14–4

Consider a chemical process whose yield is a function of temperature (x_1) and pressure (x_2). The current operating conditions are $x_1 = 250°F$ and $x_2 = 145$ psi. The EVOP procedure uses the 2^2 design plus the center point shown in Figure 14–26. The cycle is completed by running each design point in numerical order (1, 2, 3, 4, 5). The yields in the first cycle are also shown in Figure 14–26.

The yields from the first cycle are entered in the EVOP calculation sheet, shown in Table 14–11. At the end of the first cycle, no estimate of the standard deviation

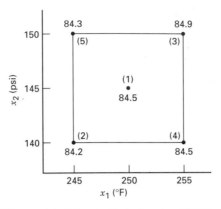

Figure 14–26.   A 2^2 design for EVOP.

Table 14–11   EVOP Calculation Sheet for Example 14–4, $n = 1$

	Cycle: $n = 1$ Response: Yield	Phase: 1 Date: 1/11/96

	Calculation of Averages					Calculation of Standard Deviation
Operating Conditions	(1)	(2)	(3)	(4)	(5)	
(i) Previous cycle sum						Previous sum $S =$
(ii) Previous cycle average						Previous average $S =$
(iii) New observations	84.5	84.2	84.9	84.5	84.3	New $S =$ range $\times$ $f_{5,n} =$
(iv) Differences [(ii) $-$ (iii)]						Range of (iv) $=$
(v) New sums [(i) $+$ (iii)]	84.5	84.2	84.9	84.5	84.3	New sum $S =$
(vi) New averages [$\bar{y}_i = $ (v)/n]	84.5	84.2	84.9	84.5	84.3	New average $S =$ $\dfrac{\text{New sum } S}{n-1}$

Calculation of Effects	Calculation of Error Limits
Temperature effect $= \frac{1}{2}(\bar{y}_3 + \bar{y}_4 - \bar{y}_2 - \bar{y}_5) = 0.45$	For new average $= \dfrac{2}{\sqrt{n}} S =$
Pressure effect $= \frac{1}{2}(\bar{y}_3 + \bar{y}_5 - \bar{y}_2 - \bar{y}_4) = 0.25$	For new effects $\dfrac{2}{\sqrt{n}} S =$
$T \times P$ interaction effect $= \frac{1}{2}(\bar{y}_2 + \bar{y}_3 - \bar{y}_4 - \bar{y}_5) = 0.15$	
Change-in-mean effect $= \frac{1}{5}(\bar{y}_2 + \bar{y}_3 + \bar{y}_4 + \bar{y}_5 - 4\bar{y}_1) = 0.02$	For change in mean $\dfrac{1.78}{\sqrt{n}} S =$

can be made. The effects and interaction for temperature and pressure are calculated in the usual manner for a 2^2 design.

A second cycle is then run and the yield data entered in another EVOP calculation sheet, shown in Table 14–12. At the end of the second cycle, the experimental error can be estimated and the estimates of the effects compared to approximate 95 percent (two standard deviation) limits. Note that the range refers to the range of the differences in row (iv); thus the range is $+1.0 - (-1.0) = 2.0$. Since none of the effects in Table 14–12 exceed their error limits, the true effect is probably zero, and no changes in operating conditions are contemplated.

The results of a third cycle are shown in Table 14–13. The effect of pressure now exceeds its error limit and the temperature effect is equal to the error limit. A change in operating conditions is now probably justified.

Table 14–12   EVOP Calculation Sheet for Example 14–4, $n = 2$

	Cycle: $n = 2$ Response: Yield	Phase: 1 Date: 1/11/96

	Calculation of Averages					Calculation of Standard Deviation
Operating Conditions	(1)	(2)	(3)	(4)	(5)	
(i) Previous cycle sum	84.5	84.2	84.9	84.5	84.3	Previous sum $S =$
(ii) Previous cycle average	84.5	84.2	84.9	84.5	84.3	Previous average $S =$
(iii) New observations	84.9	84.6	85.9	83.5	84.0	New $S =$ range $\times$ $f_{5,n} = 0.60$
(iv) Differences [(ii) − (iii)]	−0.4	−0.4	−1.0	+1.0	0.3	Range of (iv) = 2.0
(v) New sums [(i) + (iii)]	169.4	168.8	170.8	168.0	168.3	New sum $S = 0.60$
(vi) New averages $[\bar{y}_i = (v)/n]$	84.70	84.40	85.40	84.00	84.15	New average $S =$ $\dfrac{\text{New sum } S}{n - 1} =$ 0.60

Calculation of Effects	Calculation of Error Limits
Temperature effect $= \frac{1}{2}(\bar{y}_3 + \bar{y}_4 - \bar{y}_2 - \bar{y}_5) = 0.43$	For new average $= \dfrac{2}{\sqrt{n}} S = 0.85$
Pressure effect $= \frac{1}{2}(\bar{y}_3 + \bar{y}_5 - \bar{y}_2 - \bar{y}_4) = 0.58$	For new effects $\dfrac{2}{\sqrt{n}} S = 0.85$
$T \times P$ interaction effect $= \frac{1}{2}(\bar{y}_2 + \bar{y}_3 - \bar{y}_4 - \bar{y}_5) = 0.83$	
Change-in-mean effect $= \frac{1}{5}(\bar{y}_2 + \bar{y}_3 + \bar{y}_4 + \bar{y}_5 - 4\bar{y}_1) = -0.17$	For change in mean $\dfrac{1.78}{\sqrt{n}} S = 0.76$

Table 14–13   EVOP Calculation Sheet for Example 14–4, $n = 3$

5 ⎡•1⎤ 3 / 2 ⎣ ⎦ 4		Cycle: $n = 3$ Response: Yield		Phase: 1 Date: 1/11/96	

		Calculation of Averages				Calculation of Standard Deviation
Operating Conditions	(1)	(2)	(3)	(4)	(5)	
(i) Previous cycle sum	169.4	168.8	170.8	168.0	168.3	Previous sum $S = 0.60$
(ii) Previous cycle average	84.70	84.40	85.40	84.00	84.15	Previous average $S = 0.60$
(iii) New observations	85.0	84.0	86.6	84.9	85.2	New $S =$ range $\times$ $f_{5,n} = 0.56$
(iv) Differences [(ii) − (iii)]	−0.30	+0.40	−1.20	−0.90	−1.05	Range of (iv) $= 1.60$
(v) New sums [(i) + (iii)]	254.4	252.8	257.4	252.9	253.5	New sum $S = 1.16$
(vi) New averages $[\bar{y}_i = (v)/n]$	84.80	84.27	85.80	84.30	84.50	New average $S =$ $\dfrac{\text{New sum } S}{n-1} =$ 0.58

Calculation of Effects	Calculation of Error Limits
Temperature effect $= \frac{1}{2}(\bar{y}_3 + \bar{y}_4 - \bar{y}_2 - \bar{y}_5) = 0.67$	For new average $\dfrac{2}{\sqrt{n}} S = 0.67$
Pressure effect $= \frac{1}{2}(\bar{y}_3 + \bar{y}_5 - \bar{y}_2 - \bar{y}_4) = 0.87$	For new effects $\dfrac{2}{\sqrt{n}} S = 0.67$
$T \times P$ interaction effect $= \frac{1}{2}(\bar{y}_2 + \bar{y}_3 - \bar{y}_4 - \bar{y}_5) = 0.64$	
Change-in-mean effect $= \frac{1}{5}(\bar{y}_2 + \bar{y}_3 + \bar{y}_4 + \bar{y}_5 - 4\bar{y}_1) = -0.07$	For change in mean $\dfrac{1.78}{\sqrt{n}} S = 0.60$

In light of the results, it seems reasonable to begin a new EVOP phase about point (3). Thus, $x_1 = 225°F$ and $x_2 = 150$ psi would become the center of the 2^2 design in the second phase.

An important aspect of EVOP is feeding the information generated back to the process operators and supervisors. This is accomplished by a prominently displayed EVOP information board. The information board for this example at the end of cycle three is shown in Table 14–14. ∎

Most of the quantities on the EVOP calculation sheet follow directly from the analysis of the 2^k factorial design. For example, the variance of any effect,

Table 14-14   EVOP Information Board, Cycle Three

Response: Percent Yield

Requirement: Maximize

Error Limits for Averages: ±0.67

Effects with	Temperature	0.67	±0.67
95 percent error	Pressure	0.87	±0.67
Limits:	$T \times P$	0.64	±0.67
	Change in mean	0.07	±0.60
Standard deviation:	0.58		

such as $\frac{1}{2}(\bar{y}_3 + \bar{y}_5 - \bar{y}_2 - \bar{y}_4)$, is simply

$$V\left[\frac{1}{2}(\bar{y}_3 + \bar{y}_5 - \bar{y}_2 - \bar{y}_4)\right] = \frac{1}{4}(\sigma^2_{\bar{y}_3} + \sigma^2_{\bar{y}_5} + \sigma^2_{\bar{y}_2} + \sigma^2_{\bar{y}_4})$$

$$= \frac{1}{4}(4\sigma^2_{\bar{y}}) = \frac{\sigma^2}{n}$$

where σ^2 is the variance of the observations (y). Thus, two standard deviation (corresponding to 95 percent) error limits on any effect would be $\pm 2\sigma/\sqrt{n}$. The variance of the change in mean is

$$V(\text{CIM}) = V\left[\frac{1}{5}(\bar{y}_2 + \bar{y}_3 + \bar{y}_4 + \bar{y}_5 - 4\bar{y}_1)\right]$$

$$= \frac{1}{25}(4\sigma^2_{\bar{y}} + 16\sigma^2_{\bar{y}}) = \left(\frac{20}{25}\right)\frac{\sigma^2}{n}$$

Thus, two standard deviation error limits on the CIM are $\pm(2\sqrt{20/25})\sigma/\sqrt{n} = \pm 1.78\sigma/\sqrt{n}$.

The standard deviation σ is estimated by the range method. Let $y_i(n)$ denote the observation at the ith design point in cycle n and $\bar{y}_i(n)$ denote the corresponding average of $y_i(n)$ after n cycles. The quantities in row (iv) of the EVOP calculation sheet are the differences $y_i(n) - \bar{y}_i(n-1)$. The variance of these differences is

$$V[y_i(n) - \bar{y}_i(n-1)] \equiv \sigma^2_D = \sigma^2\left(1 + \frac{1}{(n-1)}\right) = \sigma^2\frac{n}{(n-1)}$$

Table 14-15   Values of $f_{k,n}$

$n =$	2	3	4	5	6	7	8	9	10
$k = 5$	0.30	0.35	0.37	0.38	0.39	0.40	0.40	0.40	0.41
9	0.24	0.27	0.29	0.30	0.31	0.31	0.31	0.32	0.32
10	0.23	0.26	0.28	0.29	0.30	0.30	0.30	0.31	0.31

The range of the differences, say R_D, is related to the estimate of the standard deviation of the differences by $\hat{\sigma}_D = R_D/d_2$. The factor d_2 depends on the number of observations used in computing R_D. Now $R_D/d_2 = \hat{\sigma}\sqrt{n/(n-1)}$, so

$$\hat{\sigma} = \sqrt{\frac{(n-1)}{n}}\frac{R_D}{d_2} = (f_{k,n})R_D \equiv S$$

can be used to estimate the standard deviation of the observations, where k denotes the number of points used in the design. For a 2^2 design with one center point we have $k = 5$, and for a 2^3 design with one center point we have $k = 9$. Values of $f_{k,n}$ are given in Table 14-15.

14-7   TAGUCHI'S CONTRIBUTIONS TO EXPERIMENTAL DESIGN AND QUALITY ENGINEERING

Throughout this book, we have emphasized the importance of using designed experiments for product and process improvement. One important goal of quality improvement is to design quality into every product and the processes that build them. Statistically designed experiments are a major element of this activity.

During the 1960–1980 time period, the principles of experimental design (and statistical methods, in general) were not as widely used in the West as in Japan. Japanese engineers had much greater exposure to these concepts, and consequently, experimental design methods were more of an engineering tool in Japan than they were in the United States. In recent years, this situation has been changing, as many engineers now receive some training in statistical methods, experimental design, and quality engineering tools as part of their formal academic training. Furthermore, these techniques have enjoyed a greatly expanded industrial application in the United States.

In the early 1980s, Professor Genechi Taguchi introduced his approach to using experimental design for

1. Designing products or processes so that they are robust to environmental conditions.

2. Designing/developing products so that they are robust to component variation.

3. Minimizing variation around a target value.

[See Taguchi and Wu (1980) and Taguchi (1986).] By *robust,* we mean that the product or process performs consistently on target and is relatively insensitive to factors that are difficult to control. Taguchi refers to the three goals of his approach as **parameter design.**

The philosophy that Taguchi recommends is sound and should be included in the quality improvement process of any organization. However, he has advocated some novel methods of statistical data analysis and some approaches to the design of experiments that are unnecessarily complicated, inefficient, and sometimes ineffective. In this section, we will briefly overview Taguchi's philosophy regarding quality engineering and experimental design. We will present some examples of his approach to parameter design, and we will use these examples to highlight the problems with his technical methods. As we will see, it is possible to combine his sound engineering concepts with more efficient and effective experimental design and analysis based on response surface methods.

14–7.1   The Taguchi Philosophy

Professor Taguchi advocates a philosophy of quality engineering that is broadly applicable. He considers three stages in a product's (or process's) development: system design, parameter design, and tolerance design. In **system design,** the engineer uses scientific and engineering principles to determine the basic configuration. For example, if we wish to measure an unknown resistance, we may use our knowledge of electrical circuits to determine that the basic system should be configured as a Wheatstone bridge. If we are designing a process to assemble printed circuit boards, we will determine the need for specific types of axial insertion machines, surface-mount placement machines, flow solder machines, and so forth.

In the **parameter design** stage, the specific values for the system parameters are determined. This would involve choosing the nominal resistor and power supply values for the Wheatstone bridge, the number and type of component placement machines for the printed circuit board assembly process, and so forth. Usually, the objective is to specify these nominal parameter values such that the variability transmitted from uncontrollable (or noise) variables is minimized.

Tolerance design is used to determine the best tolerances for the parameters. For example, in the Wheatstone bridge, tolerance design methods would reveal which components in the design were most sensitive and where the tolerances should be set. If a component does not have much effect on the performance of the circuit, it can be specified with a wide tolerance.

Taguchi recommends that statistical experimental design methods be employed to assist in quality improvement, particularly during parameter design and tolerance design. In this section, we will focus on parameter design. Experimental design methods can be used to find a best product or process design, where by "best" we mean a product or process that is robust or insensitive to uncontrollable factors that will influence the product or process once it is in routine operation.

The notion of **robust design** is not new. Engineers have always tried to design products so that they will work well under uncontrollable conditions. For example, commercial transport aircraft fly about as well in a thunderstorm as they do in clear air. Taguchi deserves recognition for realizing that experimental design can be used as a formal part of the **engineering design process** to help accomplish this objective.

A key component of Taguchi's philosophy is the **reduction of variability.** Generally, each product or process performance characteristic will have a target or **nominal** value. The objective is to reduce the variability around this target value. Taguchi models the departures that may occur from this target value with a **loss function.** The loss refers to the cost that is incurred by *society* when the consumer uses a product whose quality characteristics differ from the nominal. The concept of societal loss is a departure from traditional thinking. Taguchi imposes a quadratic loss function of the form

$$L(y) = k(y - T)^2 \tag{14--23}$$

shown in Figure 14–27. Clearly this type of function will penalize even small departures of y from the target T. Again, this is a departure from traditional thinking, which usually attaches penalties only to cases where y is outside of the upper and lower specifications (say $y > USL$ or $y < LSL$ in Figure 14–27). However, the Taguchi philosophy regarding reduction of variability and the emphasis on minimizing costs is entirely consistent with the continuous improvement philosophy of Deming and Juran.

In summary, Taguchi's philosophy involves three central ideas:

1. Products and processes should be designed so that they are robust to external sources of variability.

2. Experimental design methods are an engineering tool to help accomplish this objective.

3. Operation on-target is more important than conformance to specifications.

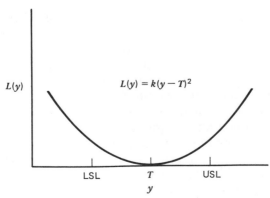

Figure 14–27.   Taguchi's quadratic loss function.

These are sound concepts, and their value should be readily apparent. Furthermore, experimental design methods can play a major role in translating these ideas into practice.

We now turn to a discussion of the specific methods that Professor Taguchi recommends for applying his concepts in practice. As we will see, his approach to experimental design and data analysis can be improved.

14–7.2 The Taguchi Approach to Parameter Design

We will use an example to illustrate the approach to parameter design suggested by Professor Taguchi. The example was published in *Quality Progress* in December 1987 (see "The Taguchi Approach to Parameter Design," by D. M. Byrne and S. Taguchi, *Quality Progress,* December 1987, pp. 19–26).

The Problem ▪ The experiment involves finding a method to assemble an elastomeric connector to a nylon tube that would deliver the required pull-off performance to be suitable for use in an automotive engine application. The specific objective of the experiment is to maximize the pull-off force. Four controllable and three uncontrollable noise factors were identified. These factors are defined in Table 14–16. We want to find the levels of the controllable factors that are the least influenced by the noise factors and that provide the maximum pull-off force. Notice that although the noise factors are not controllable during routine operations, they can be controlled for the purposes of a test. Each controllable factor is tested at three levels, and each noise factor is tested at two levels.

The Experimental Design ▪ In the Taguchi parameter design methodology, one experimental design is selected for the controllable factors and another experi-

Table 14–16   Factors and Levels for the Parameter Design Example

Controllable Factors	Levels		
A = Interference	Low	Medium	High
B = Connector wall thickness	Thin	Medium	Thick
C = Insertion depth	Shallow	Medium	Deep
D = Percent adhesive in connector pre-dip	Low	Medium	High

Uncontrollable Factors	Levels	
E = Conditioning time	24 h	120 h
F = Conditioning temperature	72°F	150°F
G = Conditioning relative humidity	25%	75%

Table 14–17   Designs for the Controllable and Uncontrollable Factors

(a) L_9 Orthogonal Array for the Controllable Factors				(b) L_8 Orthogonal Array for the Uncontrollable Factors							
	Variable							Variable			
Run	A	B	C	D	Run	E	F	$E \times F$	G	$E \times G$	$F \times G$	e
1	1	1	1	1	1	1	1	1	1	1	1	1
2	1	2	2	2	2	1	1	1	2	2	2	2
3	1	3	3	3	3	1	2	2	1	1	2	2
4	2	1	2	3	4	1	2	2	2	2	1	1
5	2	2	3	1	5	2	1	2	1	2	1	2
6	2	3	1	2	6	2	1	2	2	1	2	1
7	3	1	3	2	7	2	2	1	1	2	2	1
8	3	2	1	3	8	2	2	1	2	1	1	2
9	3	3	2	1								

mental design is selected for the noise or uncontrollable factors. These designs are shown in Table 14–17. Panel (a) of Table 14–17 contains an L_9 **orthogonal array**—a table of integers whose column elements (1, 2, and 3) represent the low, medium, and high levels of the column factors. Each row of the orthogonal array represents a run, that is, a specific set of factor levels to be tested. The L_9 orthogonal array will accommodate four factors at three levels each in nine runs. Panel (b) of Table 14–17 contains the L_8 orthogonal array—a design for up to seven factors at two levels each in eight runs. The L_8 array in this example contains only three factors—E, F, and G—so the remaining columns can be used to estimate interactions. The purpose of the noise factor array (L_8) is to create variation so that we can identify the controllable factor levels that are the least sensitive to the noise factors.

The two designs are combined as shown in Table 14–18. In this complete **parameter design layout,** the L_9 array containing the controllable factors is called the **inner array,** and the L_8 array containing the noise factors is called the **outer array.** Literally, each of the 9 runs from the inner array is tested across the 8 runs from the outer array, for a total sample size of 72 runs. This type of design is also called a **crossed array** design. The observed pull-off force is reported in Table 14–18.

Data Analysis and Conclusions ▪ The data from this experiment may now be analyzed. Taguchi recommends analyzing the mean response for each run in the inner array (see Table 14–18), and he also suggests analyzing variation using an appropriately chosen **signal-to-noise ratio (SN).** These signal-to-noise ratios are derived from the quadratic loss function, and three of them are considered to

Table 14–18 Parameter Design with Both Inner and Outer Arrays

	Inner Array (L_9)				Outer Array (L_8)								Responses	
					E	1	1	1	1	2	2	2	2	
					F	1	1	2	2	1	1	2	2	
					G	1	2	1	2	1	2	1	2	
Run	A	B	C	D										$\bar{y}$	SN_L
1	1	1	1	1		15.6	9.5	16.9	19.9	19.6	19.6	20.0	19.1	17.525	24.025
2	1	2	2	2		15.0	16.2	19.4	19.2	19.7	19.8	24.2	21.9	19.475	25.522
3	1	3	3	3		16.3	16.7	19.1	15.6	22.6	18.2	23.3	20.4	19.025	25.335
4	2	1	2	3		18.3	17.4	18.9	18.6	21.0	18.9	23.2	24.7	20.125	25.904
5	2	2	3	1		19.7	18.6	19.4	25.1	25.6	21.4	27.5	25.3	22.825	26.908
6	2	3	1	2		16.2	16.3	20.0	19.8	14.7	19.6	22.5	24.7	19.225	25.326
7	3	1	3	2		16.4	19.1	18.4	23.6	16.8	18.6	24.3	21.6	19.850	25.711
8	3	2	1	3		14.2	15.6	15.1	16.8	17.8	19.6	23.2	24.2	18.338	24.852
9	3	3	2	1		16.1	19.9	19.3	17.3	23.1	22.7	22.6	28.6	21.200	26.152

be "standard" and widely applicable. They are

1. **Nominal the best:**

$$SN_T = 10 \log \left(\frac{\bar{y}^2}{S^2} \right) \tag{14-24}$$

2. **Larger the better:**

$$SN_L = -10 \log \left(\frac{1}{n} \sum_{i=1}^{n} \frac{1}{y_i^2} \right) \tag{14-25}$$

3. **Smaller the better:**

$$SN_S = -10 \log \left(\frac{1}{n} \sum_{i=1}^{n} y_i^2 \right) \tag{14-26}$$

Notice that these SN ratios are expressed on a decibel scale. We would use SN_T if the objective is to reduce variability around a specific target, SN_L if the system is optimized when the response is as large as possible, and SN_S if the system is optimized when the response is as small as possible. Factor levels that maximize the appropriate SN ratio are optimal.

In this problem, we would use SN_L because the objective is to maximize the pull-off force. The last *two* columns of Table 14–18 contain $\bar{y}$ and SN_L values for each of the nine inner-array runs. Taguchi-oriented practitioners often use the analysis of variance to determine the factors that influence $\bar{y}$ and the factors that influence the signal-to-noise ratio. They also employ graphs of the "marginal

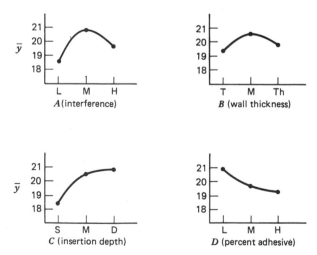

Figure 14–28. The effects of controllable factors on mean response.

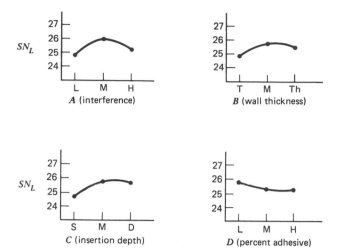

Figure 14–29. The effects of controllable factors on SN_L.

means" of each factor, such as the ones shown in Figures 14–28 and 14–29. The usual approach is to examine the graphs and "pick the winner." In this case, factors A and C have larger effects than do B and D. In terms of maximizing SN_L, we would select A_{Medium}, C_{Deep}, B_{Medium}, and D_{Low}. In terms of maximizing the average pull-off force $\bar{y}$, we would choose A_{Medium}, C_{Medium}, B_{Medium}, and D_{Low}. Notice that there is almost no difference between C_{Medium} and C_{Deep}. The implication is that this choice of levels will maximize the mean pull-off force and reduce variability in the pull-off force.

Taguchi advocates claim that the use of the SN ratio generally eliminates the need for examining specific interactions between the controllable and noise factors, although sometimes looking at these interactions improves process understanding. The authors of this study found that the AG and DE interactions were large. Analysis of these interactions, shown in Figure 14–30, suggests that A_{Medium} is best. (It gives the highest pull-off force and a slope close to zero, indicating that if we choose A_{Medium}, the effect of relative humidity is minimized.) The analysis also suggests that D_{Low} gives the highest pull-off force regardless of the conditioning time.

When cost and other factors were taken into account, the experimenters in this example finally decided to use A_{Medium}, B_{Thin}, C_{Medium}, and D_{Low}. (B_{Thin} was much less expensive than B_{Medium}, and C_{Medium} was felt to give slightly less variability than C_{Deep}.) Since this combination was not a run in the original nine inner array trials, five additional tests were made at this set of conditions as a **confirmation experiment.** For this confirmation experiment, the levels used on the noise variables were E_{Low}, F_{Low}, and G_{Low}. The authors report that good results were obtained from the confirmation test.

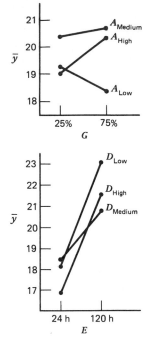

Figure 14–30. *AG* and *DE* interactions.

Critique of the Experimental Design ▪ The advocates of Taguchi's approach to parameter design utilize the orthogonal array designs, two of which (the L_8 and the L_9) were presented in the foregoing example. Figure 14–31 presents several additional orthogonal arrays: the L_4, L_{12}, L_{16}, L_{18}, and L_{27}. These designs were not developed by Taguchi; for example, the L_8 is a 2_{III}^{7-4} fractional factorial, the L_9 is a 3_{III}^{4-2} fractional factorial, the L_{12} is a Plackett–Burman design, the L_{16} is a 2_{III}^{15-11} fractional factorial, and so on. Box, Bisgaard, and Fung (1988) trace the origin of these designs. Some of these designs have very complex alias structures. In particular, the L_{12} and all of the designs that use three-level factors will involve partial aliasing of two-factor interactions with main effects. If any two-factor interactions are large, this may lead to a situation in which the experimenter does not get the correct answer.

Taguchi argues that we do not need to consider two-factor interactions explicitly. He claims that it is possible to eliminate these interactions either by correctly specifying the response and design factors or by using a sliding setting approach to choose factor levels. As an example of the latter approach, consider the two factors pressure and temperature. Varying these factors independently will probably produce an interaction. However, if temperature levels are chosen contingent on the pressure levels, then the interaction effect can be minimized. In practice, these two approaches are usually difficult to implement unless we

$L_4(2^3)$

No. \ Col.	1	2	3
1	1	1	1
2	1	2	2
3	2	1	2
4	2	2	1

$L_{12}(2^{11})$

No. \ Col.	1	2	3	4	5	6	7	8	9	10	11
1	1	1	1	1	1	1	1	1	1	1	1
2	1	1	1	1	1	2	2	2	2	2	2
3	1	1	2	2	2	1	1	1	2	2	2
4	1	2	1	2	2	1	2	2	1	1	2
5	1	2	2	1	2	2	1	2	1	2	1
6	1	2	2	2	1	2	2	1	2	1	1
7	2	1	2	2	1	1	2	2	1	2	1
8	2	1	2	1	2	2	2	1	1	1	2
9	2	1	1	2	2	2	1	2	2	1	1
10	2	2	2	1	1	1	1	2	2	1	2
11	2	2	1	2	1	2	1	1	1	2	2
12	2	2	1	1	2	1	2	1	2	2	1

$L_{16}(2^{15})$

No. \ Col.	1	2	3	4	5	6	7	8	9	10	11	12	13	14	15
1	1	1	1	1	1	1	1	1	1	1	1	1	1	1	1
2	1	1	1	1	1	1	1	2	2	2	2	2	2	2	2
3	1	1	1	2	2	2	2	1	1	1	1	2	2	2	2
4	1	1	1	2	2	2	2	2	2	2	2	1	1	1	1
5	1	2	2	1	1	2	2	1	1	2	2	1	1	2	2
6	1	2	2	1	1	2	2	2	2	1	1	2	2	1	1
7	1	2	2	2	2	1	1	1	1	2	2	2	2	1	1
8	1	2	2	2	2	1	1	2	2	1	1	1	1	2	2
9	2	1	2	1	2	1	2	1	2	1	2	1	2	1	2
10	2	1	2	1	2	1	2	2	1	2	1	2	1	2	1
11	2	1	2	2	1	2	1	1	2	1	2	2	1	2	1
12	2	1	2	2	1	2	1	2	1	2	1	1	2	1	2
13	2	2	1	1	2	2	1	1	2	2	1	1	2	2	1
14	2	2	1	1	2	2	1	2	1	1	2	2	1	1	2
15	2	2	1	2	1	1	2	1	2	2	1	2	1	1	2
16	2	2	1	2	1	1	2	2	1	1	2	1	2	2	1

$L_{27}(3^{13})$

No. \ Col.	1	2	3	4	5	6	7	8	9	10	11	12	13
1	1	1	1	1	1	1	1	1	1	1	1	1	1
2	1	1	1	1	2	2	2	2	2	2	2	2	2
3	1	1	1	1	3	3	3	3	3	3	3	3	3
4	1	2	2	2	1	1	1	2	2	2	3	3	3
5	1	2	2	2	2	2	2	3	3	3	1	1	1
6	1	2	2	2	3	3	3	1	1	1	2	2	2
7	1	3	3	3	1	1	1	3	3	3	2	2	2
8	1	3	3	3	2	2	2	1	1	1	3	3	3
9	1	3	3	3	3	3	3	2	2	2	1	1	1
10	2	1	2	3	1	2	3	1	2	3	1	2	3
11	2	1	2	3	2	3	1	2	3	1	2	3	1
12	2	1	2	3	3	1	2	3	1	2	3	1	2
13	2	2	3	1	1	2	3	2	3	1	3	1	2
14	2	2	3	1	2	3	1	3	1	2	1	2	3
15	2	2	3	1	3	1	2	1	2	3	2	3	1
16	2	3	1	2	1	2	3	3	1	2	2	3	1
17	2	3	1	2	2	3	1	1	2	3	3	1	2
18	2	3	1	2	3	1	2	2	3	1	1	2	3
19	3	1	3	2	1	3	2	1	3	2	1	3	2
20	3	1	3	2	2	1	3	2	1	3	2	1	3
21	3	1	3	2	3	2	1	3	2	1	3	2	1
22	3	2	1	3	1	3	2	2	1	3	3	2	1
23	3	2	1	3	2	1	3	3	2	1	1	3	2
24	3	2	1	3	3	2	1	1	3	2	2	1	3
25	3	3	2	1	1	3	2	3	2	1	2	1	3
26	3	3	2	1	2	1	3	1	3	2	3	2	1
27	3	3	2	1	3	2	1	2	1	3	1	3	2

$L_{18}(2^1 \times 3^7)$

No. \ Col.	1	2	3	4	5	6	7	8
1	1	1	1	1	1	1	1	1
2	1	1	2	2	2	2	2	2
3	1	1	3	3	3	3	3	3
4	1	2	1	1	2	2	3	3
5	1	2	2	2	3	3	1	1
6	1	2	3	3	1	1	2	2
7	1	3	1	2	1	3	2	3
8	1	3	2	3	2	1	3	1
9	1	3	3	1	3	2	1	2
10	2	1	1	3	3	2	2	1
11	2	1	2	1	1	3	3	2
12	2	1	3	2	2	1	1	3
13	2	2	1	2	3	1	3	2
14	2	2	2	3	1	2	1	3
15	2	2	3	1	2	3	2	1
16	2	3	1	3	2	3	1	2
17	2	3	2	1	3	1	2	3
18	2	3	3	2	1	2	2	1

Figure 14–31. Some orthogonal array designs.

have an unusually high level of process knowledge. The lack of provision for adequately dealing with potential interactions between the controllable process factors is a major weakness of the Taguchi approach to parameter design.

Instead of designing the experiment to investigate potential interactions, Taguchi prefers to use three-level factors to estimate curvature. For example, in the inner and outer array design used by Byrne and Taguchi, all four controllable factors were run at three levels. Let x_1, x_2, x_3, and x_4 represent the controllable factors and let z_1, z_2, and z_3 represent the three noise factors. Recall that the noise factors were run at two levels in a complete factorial design. The design they used allows us to fit the following model:

$$y = \beta_0 + \sum_{j=1}^{4} \beta_j x_j + \sum_{j=1}^{4} \beta_{jj} x_j^2 + \sum_{j=1}^{3} \gamma_j z_j$$

$$+ \sum_{i<j} \sum_{j=2}^{3} \gamma_{ij} z_i z_j + \sum_{i=1}^{3} \sum_{j=1}^{4} \delta_{ij} z_i x_j + \epsilon$$

Notice that we can fit the linear and quadratic effects of the controllable factors but not their two-factor interactions (which are aliased with the main effects). We can also fit the linear effects of the noise factors and all the two-factor interactions involving the noise factors. Finally, we can fit the two-factor interactions involving the controllable factors and the noise factors. It may be unwise to ignore potential interactions in the controllable factors. A much safer strategy is to identify potential effects and interactions that may be important and then consider curvature only in the important variables if there is evidence that the curvature is important. This will usually lead to fewer experiments, simpler interpretation of the data, and better overall process understanding.

Another criticism of the Taguchi approach to parameter design is that the crossed array structure usually leads to a very large experiment. For example, in the foregoing application, the authors used 72 tests to investigate only seven factors, and they still could not estimate any of the two-factor interactions among the four controllable factors. There are several alternative experimental designs that would be superior to the inner and outer method used in this example. Suppose that we run all seven factors at two levels in a **combined array** approach. Consider the 2_{IV}^{7-2} fractional factorial design. The alias relationships for this design are shown in the top half of Table 14–19. Notice that this design requires only 32 runs (as compared to 72). In the bottom half of Table 14–19, two different possible schemes for assigning process controllable variables and noise variables to the letters A through G are given. The first assignment scheme allows all the interactions between controllable factors and noise factors to be estimated, and it allows main effect estimates to be made that are clear of two-factor interactions. The second assignment scheme allows all the controllable factor main effects and their two-factor interactions to be estimated; it allows all noise factor main effects to be estimated clear of two-factor interactions; and it aliases only three interactions between controllable factors and noise factors with a two-factor interaction between two noise factors. Both of these arrangements present much cleaner alias relationships than are obtained from the inner and outer array parameter design, which also required over twice as many runs.

In general, the crossed array approach is often unnecessary. A better strategy is to use a **combined design** that incorporates both controllable factors and noise factors. Of course, the design chosen must have a high enough resolution to

Table 14-19   An Alternative Parameter Design

A one-quarter fraction of 7 factors in 32 runs. Resolution IV.
$$I = ABCDF = ABDEG = CEFG.$$
Aliases:

A	$AF = BCD$	$CG = EF$
B	$AG = BDE$	$DE = ABG$
$C = EFG$	$BC = ADF$	$DF = ABC$
D	$BD = ACF = AEG$	$DG = ABE$
$E = CFG$	$BE = ADG$	$ACE = AFG$
$F = CEG$	$BF = ACD$	$ACG = AEF$
$G = CEF$	$BG = ADE$	$BCE = BFG$
$AB = CDF = DEG$	$CD = ABF$	$BCG = BEF$
$AC = BDF$	$CE = FG$	$CDE = DFG$
$AD = BCF = BEG$	$CF = ABD = EG$	$CDG = DEF$
$AF = BDG$		

Factor Assignment Schemes:
1. Controllable factors are assigned to the letters C, E, F, and G. Noise factors are assigned to the letters A, B, and D. All interactions between controllable factors and noise factors can be estimated, and all controllable factor main effects can be estimated clear of two-factor interactions.
2. Controllable factors are assigned to the letters A, B, C, and D. Noise factors are assigned to the letters E, F, and G. All controllable factor main effects and two-factor interactions can be estimated; only the CE, CF, and CG interactions are aliased with interactions of the noise factors.

allow all of the interactions of interest to be estimated. This approach will almost always lead to a dramatic reduction in the size of the experiment, and at the same time, it will produce information that is more likely to improve process understanding. For more discussion of this approach, see Myers and Montgomery (1995).

A final aspect of Taguchi's parameter design is the use of **linear graphs** to assign factors to the columns of the orthogonal array. A set of linear graphs for the L_8 design is shown in Figure 14–32. In these graphs, each number is a column in the design. A line segment on the graph corresponds to an interaction between the nodes it connects. To assign variables to columns in an orthogonal array, assign the variables to nodes first; then when the nodes are used up, assign the variables to the line segments. When you assign variables to the nodes, strike out any line segments that correspond to interactions that might be important. The linear graphs in Figure 14–32 imply that column 3 in the L_8 design contains the interaction between columns 1 and 2, column 5 contains the interaction between columns 1 and 4, and so forth. If we had four factors, we would assign them to columns 1, 2, 4, and 7. This would ensure that each main effect is clear

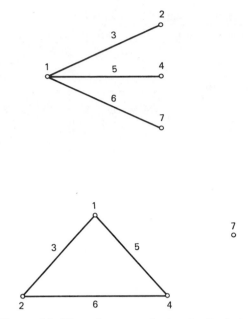

Figure 14–32.   Linear graphs for the L_8 design.

of two-factor interactions. What is *not* clear is the two-factor interaction aliasing. If the main effects are in columns 1, 2, 4, and 7, then column 3 contains the 1–2 *and* the 4–7 interaction, column 5 contains the 1–4 *and* the 2–7 interaction, and column 6 contains the 1–7 *and* the 2–4 interaction. This is clearly the case because four variables in eight runs is a resolution IV plan with all pairs of two-factor interactions aliased. In order to understand fully the two-factor interaction aliasing, Taguchi would refer the experiment designer to a supplementary interaction table.

Taguchi (1986) gives a collection of linear graphs for each of his recommended orthogonal array designs. These linear graphs seem to have been developed heuristically. Unfortunately, their use can lead to inefficient designs. For examples, see his car engine experiment [Taguchi and Wu (1980)] and his cutting tool experiment [Taguchi (1986)]. Both of these are 16-run designs that he sets up as resolution III designs in which main effects are aliased with two-factor interactions. Conventional methods for setting up these designs would have resulted in resolution IV plans in which the main effects are clear of the two-factor interactions. For the experimenter who simply wants to generate a good design, the linear graph approach may not produce the best result. A better approach is to use a simple table that presents the design and its full alias structure such as in Appendix Table XII. These tables are easy to construct and are routinely displayed by several widely available and inexpensive computer programs.

Table 14–20   Data for the "Marginal Means" Plots in Figure 14–33

		A			
		1	2	3	Averages for B
	1	10	10	13	11.00
B	2	8	10	14	9.67
	3	6	9	10	8.33
Averages for A		8.00	9.67	11.67	

Critique of the Data Analysis Methods ▪ Several of Taguchi's data analysis methods are questionable. For example, he recommends some variations of the analysis of variance that are known to produce spurious results, and he also proposes some unique methods for the analysis of attribute and life testing data. For a discussion and critique of these methods, refer to Box, Bisgaard, and Fung (1988), Myers and Montgomery (1995), and the references contained therein. In this section we focus on three aspects of data analysis Taguchi-style—the use of "marginal means" plots to optimize factor settings, the use of signal-to-noise ratios, and some of his uses of the analysis of variance.

Consider the use of "marginal means" plots and the associated "pick the winner" optimization that was demonstrated previously in the pull-off force problem. To keep the situation simple, suppose that we have two factors A and B, each at three levels, as shown in Table 14–20. The "marginal means" plots are shown in Figure 14–33. From looking at these graphs, we would select A_3 and B_1 as the optimum combination, assuming that we wish to maximize y. However, this is the wrong answer. Direct inspection of Table 14–20 or the AB interaction plot in Figure 14–34 shows that the combination of A_3 and B_2 produces the maximum value of y. In general, playing "pick the winner" with marginal

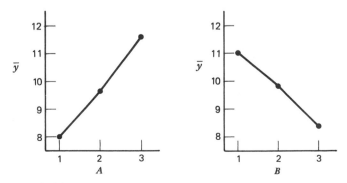

Figure 14–33.   "Marginal means" plots for the data in Table 14–20.

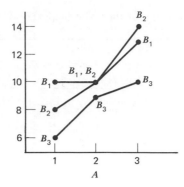

Figure 14–34.   The AB interaction plot for the data in Table 14–20.

averages can never be guaranteed to produce the optimum. The Taguchi advocates recommend that a confirmation experiment be run, although this offers no guarantees either. We might be confirming a response that differs dramatically from the optimum. The best way to find a set of optimum conditions is with response surface methods, as discussed earlier in this chapter.

Taguchi's signal-to-noise ratios SN_T, SN_L, and SN_S (see Equations 14–24, 14–25, and 14–26) are his recommended performance measures in a wide variety of situations. By maximizing the appropriate SN ratio, he claims that variability is minimized.

Consider first the SN_T ratio,

$$SN_T = 10 \log \left(\frac{\bar{y}^2}{S^2} \right)$$

This ratio would be used if we wish to minimize variability around a fixed target value. It has been suggested by Taguchi that it is preferable to work with SN_T instead of the standard deviation because in many cases the process mean and standard deviation are related. (As μ gets larger, σ gets larger, for example.) In such cases, he argues that we cannot directly minimize the standard deviation and then bring the mean on target. Taguchi claims he found empirically that the use of the SN_T ratio coupled with a two-stage optimization procedure would lead to a combination of factor levels where the standard deviation is minimized and the mean is on target. The optimization procedure consists of (1) finding the set of controllable factors that affect SN_T, called the **control factors,** and setting them to levels that maximize SN_T and then (2) finding the set of factors that have significant effects on the mean but do not influence the SN_T ratio, called the **signal factors,** and using these factors to bring the mean on target.

Given that this partitioning of factors is possible, SN_T is an example of a performance measure independent of adjustment (PERMIA) [see Leon et al. (1987)]. The signal factors would be the adjustment factors. The motivation behind the signal-to-noise ratio is to uncouple location and dispersion effects. It can be shown that the use of SN_T is equivalent to an analysis of the standard deviation of the logarithm of the original data. Thus, using SN_T implies that a log transformation will *always* uncouple location and dispersion effects. There is no assurance that this will happen. A much safer approach is to investigate what type of transformation is appropriate.

Note that we can write the SN_T ratio as

$$SN_T = 10 \log \left(\frac{\bar{y}^2}{S^2} \right)$$

$$= 10 \log (\bar{y}^2) - 10 \log (S^2)$$

If the mean is fixed at a target value (estimated by $\bar{y}$), then maximizing the SN_T ratio is equivalent to minimizing $\log (S^2)$. Using $\log (S^2)$ would require fewer calculations, is more intuitively appealing, and would provide a clearer understanding of the factor relationships that influence process variability—in other words, it would provide better process understanding. Furthermore, if we minimize $\log (S^2)$ directly, we eliminate the risk of obtaining wrong answers from the maximization of SN_T if some of the manipulated factors drive the mean $\bar{y}$ upward instead of driving S^2 downward. In general, if the response variable can be expressed in terms of the model

$$y = \mu(x_d, x_a)\epsilon(x_d)$$

where x_d is the subset of factors that drive the dispersion effects and x_a is the subset of adjustment factors that do not affect variability, then maximizing SN_T will be equivalent to minimizing the standard deviation. Considering the other potential problems surrounding SN_T, it is likely to be safer to work directly with the standard deviation (or its logarithm) as a response variable, as suggested by Myers and Montgomery (1995).

The ratios SN_L and SN_S are even more troublesome. Schmidt and Boudot (1989) have conducted a simulation study that addresses the effectiveness of these signal-to-noise ratios in detecting dispersion or variability effects. They show that SN_L and SN_S are completely *ineffective in identifying dispersion effects,* although they may serve to identify **location effects,** that is, factors that drive the mean. The reason for this is relatively easy to see. Consider the SN_S (smaller-the-better) ratio:

$$SN_S = -10 \log \left(\frac{1}{n} \sum_{i=1}^{n} y_i^2 \right)$$

The ratio is motivated by the assumption of a quadratic loss function with y nonnegative. The loss function for such a case would be

$$L = C \frac{1}{n} \sum_{i=1}^{n} y_i^2$$

where C is a constant. Now

$$\log L = \log C + \log \left(\frac{1}{n} \sum_{i=1}^{n} y_i^2 \right)$$

and

$$SN_S = 10 \log C - 10 \log L$$

so maximizing SN_S will minimize L. However, it is easy to show that

$$\frac{1}{n} \sum_{i=1}^{n} y_i^2 = \bar{y}^2 + \frac{1}{n} \left(\sum_{i=1}^{n} y_i^2 - n\bar{y}^2 \right)$$

$$= \bar{y}^2 + \left(\frac{n-1}{n} \right) S^2$$

Therefore, the use of SN_S as a response variable confounds location and dispersion effects.

The confounding of location and dispersion effects was observed in the analysis of the SN_L ratio in the pull-off force example. In Figures 14–28 and 14–29 notice that the plots of $\bar{y}$ and SN_L versus each factor have approximately the same shape, implying that both responses measure location. In fact, it can be shown that *none* of the factors A through D in this example have *any* major effect on process variability. Furthermore, since the SN_S and SN_L ratios involve y^2 and $1/y^2$, they will be very sensitive to outliers or values near zero, and they are not invariant to linear transformation of the original response. We strongly recommend that these signal-to-noise ratios not be used.

A better approach for isolating location and dispersion effects is to develop separate response surface models for $\bar{y}$ and $\log (S)$. If no replication is available to estimate variability at each run in the design, methods for analyzing residuals can be used, or a response surface for the variance can be derived from the response surface for the mean, as shown in Myers and Montgomery (1995). Then standard response surface methods can be used to optimize the mean and variance.

Finally, we turn to some of the applications of the analysis of variance recommended by Taguchi. As an example for discussion, consider the experiment reported by Quinlan (1985) at a symposium on Taguchi methods sponsored by the American Supplier Institute. The experiment concerned the quality improvement of speedometer cables. Specifically, the objective was to reduce the shrinkage in the plastic casing material. (Excessive shrinkage causes the cables to be noisy.) The experiment used an L_{16} orthogonal array (the 2_{III}^{15-11} design). The shrinkage values for four samples taken from 3000-foot lengths of the product manufactured at each set of test conditions are shown in Table 14–21, along with $\bar{y}$ and SN_S.

Quinlan, following the Taguchi approach to data analysis, used SN_S as the response variable in an analysis of variance. The error mean square was formed by pooling the mean squares associated with the seven effects that had the smallest absolute magnitude. This resulted in all eight remaining factors having significant effects (in order of magnitude: E, G, K, A, C, F, D, H). The author did note that E and G were the most important.

Table 14-21 The Speedometer Cable Experiment

Run	H	D	-L	B	-J	F	N	A	-I	-E	M	-C	K	G	-O	y₁	y₂	y₃	y₄	SN_S	ȳ

Variables / Observed Shrinkage / Responses

Run	H	D	-L	B	-J	F	N	A	-I	-E	M	-C	K	G	-O	y_1	y_2	y_3	y_4	SN_S	$\bar{y}$
1	−	−	+	−	+	−	−	−	+	+	−	+	−	−	+	0.49	0.54	0.46	0.45	6.2626	0.4850
2	+	−	−	−	−	−	−	−	−	+	+	+	+	−	−	0.55	0.60	0.57	0.58	4.8024	0.5750
3	−	+	−	−	+	+	+	−	+	−	+	+	+	+	−	0.07	0.09	0.11	0.08	21.0375	0.0875
4	+	+	+	−	+	+	+	−	−	−	−	+	+	+	+	0.16	0.16	0.19	0.19	15.1074	0.1750
5	−	−	+	+	−	+	−	−	+	+	−	−	+	+	−	0.13	0.22	0.20	0.23	14.0285	0.1950
6	+	−	−	+	+	+	−	−	−	+	+	−	−	+	+	0.16	0.17	0.13	0.12	16.6857	0.1450
7	−	+	−	+	−	+	+	−	+	−	+	−	−	−	+	0.24	0.22	0.19	0.25	12.9115	0.2250
8	+	+	+	+	+	−	+	−	−	−	−	−	+	−	−	0.13	0.19	0.19	0.19	15.0446	0.1750
9	−	−	+	−	+	−	+	+	−	+	+	−	+	+	−	0.08	0.10	0.14	0.18	17.6700	0.1250
10	+	−	−	−	−	+	+	+	+	+	−	−	−	+	+	0.07	0.04	0.19	0.18	17.2700	0.1200
11	−	+	−	−	+	+	−	+	−	+	−	+	+	−	+	0.48	0.49	0.44	0.41	6.8183	0.4550
12	+	+	+	−	−	+	−	+	+	+	+	+	−	−	−	0.54	0.53	0.53	0.54	5.4325	0.5350
13	−	−	+	+	+	+	−	+	−	−	+	+	−	−	+	0.13	0.17	0.21	0.17	15.2724	0.1700
14	+	−	−	+	−	+	−	+	+	−	−	+	+	−	−	0.28	0.26	0.26	0.30	11.1976	0.2750
15	−	+	−	+	+	−	+	+	+	+	−	+	−	+	−	0.34	0.32	0.30	0.41	9.2436	0.3425
16	+	+	+	+	+	−	+	+	+	+	+	+	+	+	+	0.58	0.62	0.59	0.54	4.6836	0.5825

Table 14–22   Bias Introduced by Pooling

NID(0, 1) Random Numbers	Mean Squares with One Degree of Freedom	F_0
−0.8607	0.7408	10.19
−0.8820	0.7779	10.70
0.3608*	0.1302	—
0.0227*	0.0005	—
0.1903*	0.0362	—
−0.3071*	0.0943	—
1.2075	1.4581	20.06
0.5641	0.3182	4.38
−0.3936*	0.1549	—
−0.6940	0.4816	6.63
−0.3028*	0.0917	—
0.5832	0.3401	4.68
0.0324*	0.0010	—
1.0202	1.0408	14.32
−0.6347	0.4028	5.54

The pooling of mean squares as in this example is a procedure that has long been known to produce considerable bias in the test results. To illustrate the problem, consider the 15 NID(0, 1) random numbers shown in column 1 of Table 14–22. The square of each of these numbers, shown in column 2 of the table, is a single-degree-of-freedom mean square corresponding to the observed random number. The seven smallest random numbers are marked with an asterisk in column 1 of Table 14–22. The corresponding mean squares are pooled to form a mean square for error with seven degrees of freedom. This quantity is

$$MS_E = \frac{0.5088}{7} = 0.0727$$

Finally, column 3 of Table 14–22 presents the F ratio formed by dividing each of the eight remaining mean squares by MS_E. Now $F_{0.05,1,7} = 5.59$, and this implies that five of the eight effects would be judged significant at the 0.05 level. Recall that since the original data came from a normal distribution with mean zero, *none* of the effects is different from zero.

Analysis methods such as this virtually guarantee erroneous conclusions. The normal probability plotting of effects avoids this invalid pooling of mean squares and provides a simple, easy to interpret method of analysis. Box (1988) provides an alternate analysis of the Quinlan data that correctly reveals factors

E and G to be important along with other interesting results not apparent in the original analysis.

It is important to note that the Taguchi analysis identified negligible factors as significant. This can have profound impact on our use of experimental design to enhance process knowledge. Experimental design methods should make gaining process knowledge easier, not harder.

Some Final Remarks ▪ In this section we have directed some major criticisms toward the specific methods of experimental design and data analysis used in the Taguchi approach to parameter design. Remember that these comments have focused on technical issues, and that the broad *philosophy* recommended by Taguchi, which we discussed in Section 14-7.1, is inherently sound.

On the other hand, many companies have reported success with the use of Taguchi's parameter design methods. If the methods are flawed, why do they produce successful results? Taguchi advocates often refute criticism with the remark that "they work." We must remember that the "best guess" and "one-factor-at-a-time" methods will also work—and occasionally they produce good results. This is no reason to claim that they are good methods. Most of the successful applications of Taguchi's technical methods have been in industries where there was no history of good experimental design practice. Designers and developers were using the best guess and one-factor-at-a-time methods (or other unstructured approaches), and since the Taguchi approach is based on the factorial design concept, it often produced better results than the methods it replaced. In other words, the factorial design is so powerful that, even when it is used inefficiently, it will work better than almost anything else.

As pointed out earlier, the Taguchi approach to parameter design often leads to *large, comprehensive experiments* with 70 or more runs. Many of the successful applications of this approach have been in industries characterized by a high-volume, low-cost manufacturing environment. In such situations, large designs may not be a real problem, as it is really no more difficult to make 72 runs than to make 16 or 32 runs. On the other hand, in industries characterized by low-volume and/or high-cost manufacturing (such as the aerospace industry, chemical and process industries, electronics and semiconductor manufacturing, and so forth), these methodological inefficiencies can be significant.

A final point concerns the learning process. If the Taguchi approach to parameter design works and yields good results, we may still not know what has caused the result because of the aliasing of critical interactions. In other words, we may have solved a problem (a short-term success), but we may not have gained **process knowledge,** which could be invaluable in future problems.

In summary, we should support Taguchi's philosophy of quality engineering. However, we must rely on simpler, more efficient methods that are easier to learn and apply to carry this philosophy into practice. The response surface modeling framework is an ideal approach to optimizing processes and it is fully adaptable to the Taguchi robust design objectives.

14-8   PROBLEMS

14-1   A chemical plant produces oxygen by liquifying air and separating it into its component gases by fractional distillation. The purity of the oxygen is a function of the main condenser temperature and the pressure ratio between the upper and lower columns. Current operating conditions are temperature (ξ_1) = $-220°C$ and pressure ratio (ξ_2) = 1.2. Using the following data, find the path of steepest ascent.

Temperature (ξ_1)	Pressure Ratio (ξ_2)	Purity
-225	1.1	82.8
-225	1.3	83.5
-215	1.1	84.7
-215	1.3	85.0
-220	1.2	84.1
-220	1.2	84.5
-220	1.2	83.9
-220	1.2	84.3

14-2   An industrial engineer has developed a computer simulation model of a two-item inventory system. The decision variables are the order quantity and the reorder point for each item. The response to be minimized is total inventory cost. The simulation model is used to produce the data shown in the following table. Identify the experimental design. Find the path of steepest descent.

Item 1		Item 2		
Order Quantity (ξ_1)	Reorder Point (ξ_2)	Order Quantity (ξ_3)	Reorder Point (ξ_4)	Total Cost
100	25	250	40	625
140	45	250	40	670
140	25	300	40	663
140	25	250	80	654
100	45	300	40	648
100	45	250	80	634
100	25	300	80	692
140	45	300	80	686
120	35	275	60	680
120	35	275	60	674
120	35	275	60	681

14-3   Verify that the following design is a simplex. Fit the first-order model and find the path of steepest ascent.

x_1	x_2	x_3	y
0	$\sqrt{2}$	-1	18.5
$-\sqrt{2}$	0	1	19.8
0	$-\sqrt{2}$	-1	17.4
$\sqrt{2}$	0	1	22.5

14-4 For the first-order model

$$\hat{y} = 60 + 1.5x_1 - 0.8x_2 + 2.0x_3$$

find the path of steepest ascent. The variables are coded as $-1 \leqslant x_i \leqslant 1$.

14-5 The region of experimentation for three factors are time ($40 \leqslant T_1 \leqslant 80$ min), temperature ($200 \leqslant T_2 \leqslant 300°C$), and pressure ($20 \leqslant P \leqslant 50$ psig). A first-order model in coded variables has been fit to yield data from a 2^3 design. The model is

$$\hat{y} = 30 + 5x_1 + 2.5x_2 + 3.5x_3$$

Is the point $T_1 = 85$, $T_2 = 325$, $P = 60$ on the path of steepest ascent?

14-6 The path of steepest ascent is usually computed assuming that the model is truly first-order; that is, there is no interaction. However, even if there is interaction, steepest ascent ignoring the interaction still usually produces good results. To illustrate, suppose that we have fit the model

$$\hat{y} = 20 + 5x_1 - 8x_2 + 3x_1x_2$$

using coded variables ($-1 \leqslant x_i \leqslant +1$).

(a) Draw the path of steepest ascent that you would obtain if the interaction were ignored.

(b) Draw the path of steepest ascent that you would obtain with the interaction included in the model. Compare this with the path found in part (a).

14-7 The data shown in the following table were collected in an experiment to optimize crystal growth as a function of three variables x_1, x_2, and x_3. Large values of y (yield in grams) are desirable. Fit a second-order model and analyze the fitted surface. Under what set of conditions is maximum growth achieved?

x_1	x_2	x_3	y
-1	-1	-1	66
-1	-1	1	70
-1	1	-1	78
-1	1	1	60

(continued)

(continued)

x_1	x_2	x_3	y
1	−1	−1	80
1	−1	1	70
1	1	−1	100
1	1	1	75
−1.682	0	0	100
1.682	0	0	80
0	−1.682	0	68
0	1.682	0	63
0	0	−1.682	65
0	0	1.682	82
0	0	0	113
0	0	0	100
0	0	0	118
0	0	0	88
0	0	0	100
0	0	0	85

14-8   The following data were collected by a chemical engineer. The response y is filtration time, x_1 is temperature, and x_2 is pressure. Fit a second-order model.

x_1	x_2	y
−1	−1	54
−1	1	45
1	−1	32
1	1	47
−1.414	0	50
1.414	0	53
0	−1.414	47
0	1.414	51
0	0	41
0	0	39
0	0	44
0	0	42
0	0	40

(a) What operating conditions would you recommend if the objective is to minimize the filtration time?

(b) What operating conditions would you recommend if the objective is to operate the process at a mean filtration rate very close to 46?

14-9   The hexagon design that follows is used in an experiment that has the objective of fitting a second-order model.

x_1	x_2	y
1	0	68
0.5	$\sqrt{0.75}$	74
−0.5	$\sqrt{0.75}$	65
−1	0	60
−0.5	$-\sqrt{0.75}$	63
0.5	$-\sqrt{0.75}$	70
0	0	58
0	0	60
0	0	57
0	0	55
0	0	69

(a) Fit the second-order model.

(b) Perform the canonical analysis. What type of surface has been found?

(c) What operating conditions on x_1 and x_2 lead to the stationary point?

(d) Where would you run this process if the objective is to obtain a response that is as close to 65 as possible?

14–10   An experimenter has run a Box–Behnken design and has obtained the results below, where the response variable is the viscosity of a polymer.

Level	Temp.	Agitation Rate	Pressure	x_1	x_2	x_3
High	200	10.0	25	+1	+1	+1
Middle	175	7.5	20	0	0	0
Low	150	5.0	15	−1	−1	−1

Run	x_1	x_2	x_3	y_1
1	−1	−1	0	535
2	+1	−1	0	580
3	−1	+1	0	596
4	+1	+1	0	563
5	−1	0	−1	645
6	+1	0	−1	458
7	−1	0	+1	350
8	+1	0	+1	600
9	0	−1	−1	595
10	0	+1	−1	648
11	0	−1	+1	532
12	0	+1	+1	656

(continued)

(continued)

Run	x_1	x_2	x_3	y_1
13	0	0	0	653
14	0	0	0	599
15	0	0	0	620

(a) Fit the second-order model.

(b) Perform the canonical analysis. What type of surface has been found?

(c) What operating conditions on x_1, x_2, and x_3 lead to the stationary point?

(d) What operating conditions would you recommend if it is important to obtain a viscosity that is as close to 600 as possible?

14-11   Consider the three-variable central composite design shown below. Analyze the data and draw conclusions, assuming that we wish to maximize conversion (y_1) with activity (y_2) between 55 and 60.

Run	Time (min)	Temperature (°C)	Catalyst (%)	Conversion (%) y_1	Activity y_2
1	-1.000	-1.000	-1.000	74.00	53.20
2	1.000	-1.000	-1.000	51.00	62.90
3	-1.000	1.000	-1.000	88.00	53.40
4	1.000	1.000	-1.000	70.00	62.60
5	-1.000	-1.000	1.000	71.00	57.30
6	1.000	-1.000	1.000	90.00	67.90
7	-1.000	1.000	1.000	66.00	59.80
8	1.000	1.000	1.000	97.00	67.80
9	0.000	0.000	0.000	81.00	59.20
10	0.000	0.000	0.000	75.00	60.40
11	0.000	0.000	0.000	76.00	59.10
12	0.000	0.000	0.000	83.00	60.60
13	-1.682	0.000	0.000	76.00	59.10
14	1.682	0.000	0.000	79.00	65.90
15	0.000	-1.682	0.000	85.00	60.00
16	0.000	1.682	0.000	97.00	60.70
17	0.000	0.000	-1.682	55.00	57.40
18	0.000	0.000	1.682	81.00	63.20
19	0.000	0.000	0.000	80.00	60.80
20	0.000	0.000	0.000	91.00	58.90

14-12   A manufacturer of cutting tools has developed two empirical equations for tool life in hours (y_1) and for tool cost in dollars (y_2). Both models are linear functions of steel hardness (x_1) and manufacturing time (x_2). The two equations are

$$\hat{y}_1 = 10 + 5x_1 + 2x_2$$
$$\hat{y}_2 = 23 + 3x_1 + 4x_2$$

and both equations are valid over the range $-1.5 \leqslant x_i \leqslant 1.5$. Unit tool cost must be below \$27.50 and life must exceed 12 hours for the product to be competitive. Is there a feasible set of operating conditions for this process? Where would you recommend that the process be run?

14-13 Verify that an orthogonal first-order design is also first-order rotatable.

14-14 Show that augmenting a 2^k design with n_c center points does not affect the estimates of the β_i $(i = 1, 2, \ldots, k)$, but that the estimate of the intercept β_0 is the average of all $2^k + n_c$ observations.

14-15 *The rotatable central composite design.* It can be shown that a second-order design is rotatable if $\sum_{u=1}^{n} x_{iu}^a x_{ju}^b = 0$ if a or b (or both) are odd and if $\sum_{u=1}^{n} x_{iu}^4 = 3\sum_{u=1}^{n} x_{iu}^2 x_{ju}^2$. Show that for the central composite design these conditions lead to $\alpha = (n_f)^{1/4}$ for rotatability, where n_f is the number of points in the factorial portion.

14-16 Verify that the central composite design shown below blocks orthogonally.

Block 1			Block 2			Block 3		
x_1	x_2	x_3	x_1	x_2	x_3	x_1	x_2	x_3
0	0	0	0	0	0	-1.633	0	0
0	0	0	0	0	0	1.633	0	0
1	1	1	1	1	-1	0	-1.633	0
1	-1	-1	1	-1	1	0	1.633	0
-1	-1	1	-1	1	1	0	0	-1.633
-1	1	-1	-1	-1	-1	0	0	1.633
						0	0	0
						0	0	0

14-17 *Blocking in the central composite design.* Consider a central composite design for $k = 4$ variables in two blocks. Can a rotatable design always be found that blocks orthogonally?

14-18 How could a hexagon design be run in two orthogonal blocks?

14-19 Yield during the first four cycles of a chemical process is shown in the following table. The variables are percent concentration (x_1) at levels 30, 31, and 32 and temperature (x_2) at 140, 142, and 144°F. Analyze by EVOP methods.

			Conditions		
Cycle	(1)	(2)	(3)	(4)	(5)
1	60.7	59.8	60.2	64.2	57.5
2	59.1	62.8	62.5	64.6	58.3
3	56.6	59.1	59.0	62.3	61.1
4	60.5	59.8	64.5	61.0	60.1

14-20 Suppose that we approximate a response surface with a model of order d_1, such as $\mathbf{y} = \mathbf{X}_1\boldsymbol{\beta}_1 + \boldsymbol{\epsilon}$, when the true surface is described by a model of order $d_2 > d_1$; that is, $E(\mathbf{y}) = \mathbf{X}_1\boldsymbol{\beta}_1 + \mathbf{X}_1\boldsymbol{\beta}_2$.

(a) Show that the regression coefficients are biased, that is, that $E(\hat{\boldsymbol{\beta}}_1) = \boldsymbol{\beta}_1 + \mathbf{A}\boldsymbol{\beta}_2$, where $\mathbf{A} = (\mathbf{X}_1'\mathbf{X}_1)^{-1}\mathbf{X}_1'\mathbf{X}_2$. $\mathbf{A}$ is usually called the alias matrix.

(b) If $d_1 = 1$ and $d_2 = 2$, and a full 2^k is used to fit the model, use the result in part (a) to determine the alias structure.

(c) If $d_1 = 1$, $d_2 = 2$, and $k = 3$, find the alias structure assuming that a 2^{3-1} design is used to fit the model.

(d) If $d_1 = 1$, $d_2 = 2$, $k = 3$, and the simplex design in Problem 14-3 is used to fit the model, determine the alias structure and compare the results with part (c).

14-21 In an article ("Let's All Beware the Latin Square," *Quality Engineering*, Vol. 1, 1989, pp. 453–465), J. S. Hunter illustrates some of the problems associated with 3^{k-p} fractional factorial designs. Factor A is the amount of ethanol added to a standard fuel and factor B represents the air/fuel ratio. The response variable is carbon monoxide (CO) emission in g/m^3. The design is shown below.

		Design			Observations	
A	B	x_1	x_2		y	
0	0	-1	-1	66	62
1	0	0	-1	78	81
2	0	$+1$	-1	90	94
0	1	-1	0	72	67
1	1	0	0	80	81
2	1	$+1$	0	75	78
0	2	-1	$+1$	68	66
1	2	0	$+1$	66	69
2	2	$+1$	$+1$	60	58

Notice that we have used the notation system of 0, 1, and 2 to represent the low, medium, and high levels for the factors. We have also used a "geometric notation" of -1, 0, and $+1$. Each run in the design is replicated twice.

(a) Verify that the second-order model

$$\hat{y} = 78.5 + 4.5x_1 - 7.0x_2 - 4.5x_1^2 - 4.0x_2^2 - 9.0x_1x_2$$

is a reasonable model for this experiment. Sketch the CO concentration contours in the x_1, x_2 space.

(b) Now suppose that instead of only two factors, we had used *four* factors in a 3^{4-2} fractional factorial design and obtained *exactly* the same data as in part (a). The design would be as follows:

			Design					Observations	
A	B	C	D	x_1	x_2	x_3	x_4		y
0	0	0	0	−1	−1	−1	−1	66	62
1	0	1	1	0	−1	0	0	78	81
2	0	2	2	+1	−1	+1	+1	90	94
0	1	2	1	−1	0	+1	0	72	67
1	1	0	2	0	0	−1	+1	80	81
2	1	1	0	+1	0	0	−1	75	78
0	2	1	2	−1	+1	0	+1	68	66
1	2	2	0	0	+1	+1	−1	66	69
2	2	0	1	+1	+1	−1	0	60	58

Confirm that this design is an L_9 orthogonal array.

(c) Calculate the marginal averages of the CO response at each level of the four factors A, B, C, and D. Construct plots of these marginal averages and interpret the results. Do factors C and D appear to have strong effects? Do these factors *really* have any effect on CO emission? Why is their apparent effect strong?

(d) The design in part (b) allows the model

$$y = \beta_0 + \sum_{i=1}^{4} \beta_i x_i + \sum_{i=1}^{4} \beta_{ii} x_i^2 + \epsilon$$

to be fitted. Suppose that the *true* model is

$$y = \beta_0 + \sum_{i=1}^{4} \beta_i x_i + \sum_{i=1}^{4} \beta_{ii} x_i^2 + \sum_{i<j} \beta_{ij} x_i x_j + \epsilon$$

Show that if $\hat{\beta}_j$ represents the least squares estimates of the coefficients in the fitted model, then

$$E(\hat{\beta}_0) = \beta_0 - \beta_{13} - \beta_{14} - \beta_{34}$$
$$E(\hat{\beta}_1) = \beta_1 - (\beta_{23} + \beta_{24})/2$$
$$E(\hat{\beta}_2) = \beta_2 - (\beta_{13} + \beta_{14} + \beta_{34})/2$$
$$E(\hat{\beta}_3) = \beta_3 - (\beta_{12} + \beta_{24})/2$$

$$E(\hat{\beta}_4) = \beta_4 - (\beta_{12} + \beta_{23})/2$$
$$E(\hat{\beta}_{11}) = \beta_{11} - (\beta_{23} - \beta_{24})/2$$
$$E(\hat{\beta}_{22}) = \beta_{22} + (\beta_{13} + \beta_{14} + \beta_{34})/2$$
$$E(\hat{\beta}_{33}) = \beta_{33} - (\beta_{24} - \beta_{12})/2 + \beta_{14}$$
$$E(\hat{\beta}_{44}) = \beta_{44} - (\beta_{12} - \beta_{23})/2 + \beta_{13}$$

Does this help explain the strong effects for factors C and D observed graphically in part (c)?

14-22   An experiment has been run in a process that applies a coating material to a wafer. Each run in the experiment produced a wafer, and the coating thickness was measured several times at different locations on the wafer. Then the mean y_1 and standard deviation y_2 of the thickness measurement was obtained. The data [adapted from Box and Draper (1987)] are shown in the table below.

Run	Speed	Pressure	Distance	Mean y_1	Std. Dev. y_2
1	−1.000	−1.000	−1.000	24.0	12.5
2	0.000	−1.000	−1.000	120.3	8.4
3	1.000	−1.000	−1.000	213.7	42.8
4	−1.000	0.000	−1.000	86.0	3.5
5	0.000	0.000	−1.000	136.6	80.4
6	1.000	0.000	−1.000	340.7	16.2
7	−1.000	1.000	−1.000	112.3	27.6
8	0.000	1.000	−1.000	256.3	4.6
9	1.000	1.000	−1.000	271.7	23.6
10	−1.000	−1.000	0.000	81.0	0.0
11	0.000	−1.000	0.000	101.7	17.7
12	1.000	−1.000	0.000	357.0	32.9
13	−1.000	0.000	0.000	171.3	15.0
14	0.000	0.000	0.000	372.0	0.0
15	1.000	0.000	0.000	501.7	92.5
16	−1.000	1.000	0.000	264.0	63.5
17	0.000	1.000	0.000	427.0	88.6
18	1.000	1.000	0.000	730.7	21.1
19	−1.000	−1.000	1.000	220.7	133.8
20	0.000	−1.000	1.000	239.7	23.5
21	1.000	−1.000	1.000	422.0	18.5
22	−1.000	0.000	1.000	199.0	29.4
23	0.000	0.000	1.000	485.3	44.7
24	1.000	0.000	1.000	673.7	158.2
25	−1.000	1.000	1.000	176.7	55.5
26	0.000	1.000	1.000	501.0	138.9
27	1.000	1.000	1.000	1010.0	142.4

(a) What type of design did the experimenters use? Is this a good choice of design for fitting a quadratic model?

(b) Build models of both responses.

(c) Find a set of optimum conditions that result in the mean exceeding 900 with the standard deviation less than 60.

Bibliography

1. Addelman, S. (1961). "Irregular Functions of the 2^n Factorial Experiments." *Technometrics,* Vol. 3, pp. 479–496.

2. Addelman, S. (1962). "Orthogonal Main Effect Plans for Asymmetric Factorial Experiments." *Technometrics,* Vol. 4, pp. 21–46.

3. Addelman, S. (1963). "Techniques for Constructing Fractional Replicate Plans." *Journal of the American Statistical Association,* Vol. 58, pp. 45–71.

4. Anderson, V. L. and R. A. McLean (1974). *Design of Experiments: A Realistic Approach.* Dekker, New York.

5. Anscombe, F. J. (1960). "Rejection of Outliers." *Technometrics,* Vol. 2, pp. 123–147.

6. Anscombe, F. J. and J. W. Tukey (1963). "The Examination and Analysis of Residuals." *Technometrics,* Vol. 5, pp. 141–160.

7. Bainbridge, T. R. (1965). "Staggered, Nested Designs for Estimating Variance Components." *Industrial Quality Control,* Vol. 22, pp. 12–20.

8. Bancroft, T. A. (1968). *Topics in Intermediate Statistical Methods.* Iowa State University Press, Ames, Iowa.

9. Barnett, V. and T. Lewis (1978). *Outliers in Statistical Data.* Wiley, New York.

10. Barlett, M. S. (1947). "The Use of Transformations." *Biometrics,* Vol. 3, pp. 39–52.

11. Bennett, C. A. and N. L. Franklin (1954). *Statistical Analysis in Chemistry and the Chemical Industry.* Wiley, New York.

12. Bose, R. C. and T. Shimamoto (1952). "Classification and Analysis of Partially Balanced Incomplete Block Designs with Two Associate Classes." *Journal of the American Statistical Association,* Vol. 47, pp. 151–184.

13. Bose, R. C., W. H. Clatworthy, and S. S. Shrikhande (1954). *Tables of Partially Balanced Designs with Two Associate Classes.* Technical Bulletin No. 107, North Carolina Agricultural Experiment Station.

14. Bowker, A. H. and G. J. Lieberman (1972). *Engineering Statistics.* 2nd edition. Prentice-Hall, Englewood Cliffs, N.J.

15. Box, G. E. P. and K. G. Wilson (1951). "On the Experimental Attainment of Optimum Conditions." *Journal of the Royal Statistical Society,* B, Vol. 13, pp. 1–45.

16. Box, G. E. P. (1954a). "Some Theorems on Quadratic Forms Applied in the Study of Analysis of Variance Problems: I. Effect of Inequality of Variance in the One-Way Classification." *Annals of Mathematical Statistics,* Vol. 25, pp. 290–302.

17. Box, G. E. P. (1954b). "Some Theorems on Quadratic Forms Applied in the Study of Analysis of Variance Problems: II. Effect of Inequality of Variance and of Correlation of Errors in the Two-Way Classification." *Annals of Mathematical Statistics,* Vol. 25, pp. 484–498.

18. Box, G. E. P. (1957). "Evolutionary Operation: A Method for Increasing Industrial Productivity." *Applied Statistics,* Vol. 6, pp. 81–101.

19. Box, G. E. P. and J. S. Hunter (1957). "Multifactor Experimental Designs for Exploring Response Surfaces." *Annals of Mathematical Statistics,* Vol. 28, pp. 195–242.

20. Box, G. E. P. and D. W. Behnken (1960). "Some New Three Level Designs for the Study of Quantitative Variables." *Technometrics,* Vol. 2, pp. 455–476.

21. Box, G. E. P. and J. S. Hunter (1961a). "The 2^{k-p} Fractional Factorial Designs, Part I." *Technometrics,* Vol. 3, pp. 311–352.

22. Box, G. E. P. and J. S. Hunter (1961b). "The 2^{k-p} Fractional Factorial Designs, Part II." *Technometrics,* Vol. 3, pp. 449–458.

23. Box, G. E. P. and D. R. Cox (1964). "An Analysis of Transformations." *Journal of the Royal Statistical Society,* B, Vol. 26, pp. 211–243.

24. Box, G. E. P. and N. R. Draper (1969). *Evolutionary Operation.* Wiley, New York.

25. Box, G. E. P., W. G. Hunter, and J. S. Hunter (1978). *Statistics for Experimenters.* Wiley, New York.

26. Box, G. E. P. and R. D. Meyer (1986). "An Analysis of Unreplicated Fractional Factorials." *Technometrics,* Vol. 28, pp. 11–18.

27. Box, G. E. P. and N. R. Draper (1987). *Empirical Model Building and Response Surfaces.* Wiley, New York.

28. Box, G. E. P. (1988). "Signal-to-Noise Ratios, Performance Criteria, and Transformation." *Technometrics,* Vol. 30, pp. 1–40.

29. Box, G. E. P., S. Bisgaard, and C. A. Fung (1988). "An Explanation and Critique of Taguchi's Contributions to Quality Engineering." *Quality and Reliability Engineering International,* Vol. 4, pp. 123–131.

30. Box, G. E. P. (1992–1993). "Sequential Experimentation and Sequential Assembly of Designs." *Quality Engineering,* Vol. 5, No. 2, pp. 321–330.

31. Box, J. F. (1978). *R. A. Fisher: The Life of a Scientist.* Wiley, New York.

32. Burdick, R. K. and F. A. Graybill (1992). *Confidence Intervals on Variance Components.* Dekker, New York.

33. Carmer, S. G. and M. R. Swanson (1973). "Evaluation of Ten Pairwise Multiple Comparison Procedures by Monte Carlo Methods." *Journal of the American Statistical Association,* Vol. 68, No. 314, pp. 66–74.

34. Cochran, W. G. (1947). "Some Consequences when the Assumptions for the Analysis of Variance Are Not Satisfied." *Biometrics,* Vol. 3, pp. 22–38.

35. Cochran, W. G. (1957). "Analysis of Covariance: Its Nature and Uses." *Biometrics,* Vol. 13, No. 3, pp. 261–281.

36. Cochran, W. G. and G. M. Cox (1957). *Experimental Designs.* 2nd edition. Wiley, New York.

37. Connor, W. S. and M. Zelen (1959). *Fractional Factorial Experimental Designs for Factors at Three Levels.* National Bureau of Standards, Washington, D.C., Applied Mathematics Series, No. 54.

38. Conover, W. J. (1980). *Practical Nonparametric Statistics,* 2nd edition. Wiley, New York.

39. Conover, W. J. and R. L. Iman (1976). "On Some Alternative Procedures Using Ranks for the Analysis of Experimental Designs." *Communications in Statistics,* Vol. A5, pp. 1349–1368.

40. Conover, W. J. and R. L. Iman (1981). "Rank Transformations as a Bridge Between Parametric and Nonparametric Statistics" (with discussion). *The American Statistician,* Vol. 35, pp. 124–133.

41. Cornell, J. A. (1990). *Experiments with Mixtures: Designs, Models, and the Analysis of Mixture Data.* 2nd Edition. Wiley, New York.

42. Daniel, C. (1959). "Use of Half-Normal Plots in Interpreting Factorial Two Level Experiments." *Technometrics,* Vol. 1, pp. 311–342.

43. Daniel, C. (1976). *Applications of Statistics to Industrial Experimentation.* Wiley, New York.

44. Davies, O. L. (1956). *Design and Analysis of Industrial Experiments.* 2nd edition. Hafner Publishing Company, New York.

45. Dolby, J. L. (1963). "A Quick Method for Choosing a Transformation." *Technometrics,* Vol. 5, pp. 317–326.

46. Draper, N. R. and W. G. Hunter (1969). "Transformations: Some Examples Revisited." *Technometrics,* Vol. 11, pp. 23–40.

47. Draper, N. R. and H. Smith (1981). *Applied Regression Analysis.* 2nd edition. Wiley, New York.

48. Duncan, A. J. (1986). *Quality Control and Industrial Statistics.* 5th edition. Richard D. Irwin, Homewood, Ill.

49. Duncan, D. B. (1955). "Multiple Range and Multiple F Tests." *Biometrics,* Vol. 11, pp. 1–42.

50. Dunnett, C. W. (1964). "New Tables for Multiple Comparisons with a Control." *Biometrics,* Vol. 20, pp. 482–491.

51. Eisenhart, C. (1947). "The Assumptions Underlying the Analysis of Variance." *Biometrics,* Vol. 3, pp. 1–21.

52. Federer, W. T. (1957). "Variance and Covariance Analysis for Unbalanced Classifications." *Biometrics,* Vol. 13, No. 3, pp. 333–362.

53. Fisher, R. A. and F. Yates (1953). *Statistical Tables for Biological, Agricultural, and Medical Research.* 4th edition. Oliver and Boyd, Edinburgh.

54. Fisher, R. A. (1958). *Statistical Methods for Research Workers.* 13th edition. Oliver and Boyd, Edinburgh.

55. Fisher, R. A. (1966). *The Design of Experiments.* 8th edition. Hafner Publishing Company, New York.

56. Gaylor, D. W. and T. D. Hartwell (1969). "Expected Mean Squares for Nested Classifications." *Biometrics,* Vol. 25, pp. 427–430.

57. Gaylor, D. W. and F. N. Hopper (1969). "Estimating the Degrees of Freedom for Linear Combinations of Mean Squares by Satterthwaite's Formula." *Technometrics,* Vol. 11, No. 4, pp. 699–706.

58. Good, I. J. (1955). "The Interaction Algorithm and Practical Fourier Analysis." *Journal of the Royal Statistical Society,* B, Vol. 20, pp. 361–372.

59. Good, I. J. (1958). Addendum to "The Interaction Algorithm and Practical Fourier Analysis." *Journal of the Royal Statistical Society,* B, Vol. 22, pp. 372–375.

60. Graybill, F. A. and D. L. Weeks (1959). "Combining Interblock and Intrablock Information in Balanced Incomplete Blocks." *Annals of Mathematical Statistics,* Vol. 30, pp. 799–805.

61. Graybill, F. A. (1961). *An Introduction to Linear Statistical Models.* Vol. 1. McGraw-Hill, New York.

62. Graybill, F. A. and C. M. Wang (1980). "Confidence Intervals on Nonnegative Linear Combinations of Variances." *Journal of the American Statistical Association,* Vol. 75, pp. 869–873.

63. Hamada, M. and C. F. J. Wu (1992). "Analysis of Designed Experiments with Complex Aliasing." *Journal of Quality Technology,* Vol. 24, No. 3, pp. 130–137.

64. Hill, W. G. and W. G. Hunter (1966). "A Review of Response Surface Methodology: A Literature Survey." *Technometrics,* Vol. 8, pp. 571–590.

65. Hines, W. W. and D. C. Montgomery (1990). *Probability and Statistics in Engineering and Management Science.* 3rd edition. Wiley, New York.

66. Hocking, R. R. (1973). "A Discussion of the Two-Way Mixed Model." *The American Statistician,* Vol. 27, No. 4, pp. 148–152.

67. Hocking, R. R. and F. M. Speed (1975). "A Full Rank Analysis of Some Linear Model Problems." *Journal of the American Statistical Association,* Vol. 70, pp. 706–712.

68. Hocking, R. R., O. P. Hackney, and F. M. Speed (1978). "The Analysis of Linear Models with Unbalanced Data." In *Contributions to Survey Sampling and Applied Statistics,* H. A. David (ed.), Academic Press, New York.

69. Hunter, J. S. (1966). "The Inverse Yates Algorithm." *Technometrics,* Vol. 8, pp. 177–183.

70. John, J. A. and P. Prescott (1975). "Critical Values of a Test to Detect Outliers in Factorial Experiments." *Applied Statistics,* Vol. 24, pp. 56–59.

71. John, P. W. M. (1961). "The Three-Quarter Replicates of 2^4 and 2^5 Designs." *Biometrics,* Vol. 17, pp. 319–321.

72. John, P. W. M. (1962). "Three-Quarter Replicates of 2^n Designs." *Biometrics,* Vol. 18, pp. 171–184.

73. John, P. W. M. (1964). "Blocking a $3(2^{n-k})$ Design." *Technometrics,* Vol. 6, pp. 371–376.

74. John, P. W. M. (1971). *Statistical Design and Analysis of Experiments.* Macmillan, New York.

75. Kempthorne, O. (1952). *The Design and Analysis of Experiments.* Wiley, New York.

76. Keuls, M. (1952). "The Use of the Studentized Range in Connection with an Analysis of Variance." *Euphytica,* Vol. 1, pp. 112–122.

77. Khuri, A. I. and J. A. Cornell (1987). *Response Surfaces: Designs and Analyses.* Dekker, New York.

78. Kruskal, W. H. and W. A. Wallis (1952). "Use of Ranks on One Criterion Variance Analysis." *Journal of the American Statistical Association,* Vol. 47, pp. 583–621 (Corrections appear in Vol. 48, pp. 907–911).

79. Leon, R. V., A. C. Shoemaker, and R. N. Kackar (1987). "Performance Measures Independent of Adjustment." *Technometrics,* Vol. 29, pp. 253–265.

80. Margolin, B. H. (1967). "Systematic Methods of Analyzing $2^n 3^m$ Factorial Experiments with Applications." *Technometrics,* Vol. 9, pp. 245–260.

81. Margolin, B. H. (1969). "Results on Factorial Designs of Resolution IV for the 2^n and $2^n 3^m$ Series." *Technometrics,* Vol. 11, pp. 431–444.

82. Miller, R. G. (1966). *Simultaneous Statistical Inference.* McGraw-Hill, New York.

83. Miller, R. G., Jr. (1977). "Developments in Multiple Comparisons, 1966–1976." *Journal of the American Statistical Association,* Vol. 72, pp. 779–788.

84. Milliken, G. A. and D. E. Johnson (1984). *Analysis of Messy Data, Vol. 1.* Van Nostrand Reinhold, New York.

85. Montgomery, D. C. (1996). *Introduction to Statistical Quality Control.* 3rd edition. Wiley, New York.

86. Montgomery, D. C. and E. A. Peck (1992). *Introduction to Linear Regression Analysis.* 2nd edition. Wiley, New York.

87. Montgomery, D. C. and G. C. Runger (1993a). "Gauge Capability Analysis and Designed Experiments. Part I: Basic Methods." *Quality Engineering,* Vol. 6, pp. 115–135.

88. Montgomery, D. C. and G. C. Runger (1993b). "Gauge Capability Analysis and Designed Experiments. Part II: Experimental Design Models and Variance Component Estimation." *Quality Engineering,* Vol. 6, pp. 289–305.

89. Myers, R. H. and D. C. Montgomery (1995). *Response Surface Methodology: Process and Product Optimization Using Designed Experiments.* Wiley, New York.

90. Nelson, L. S. (1995a). "Using Nested Designs I: Estimation of Standard Deviations." *Journal of Quality Technology,* Vol. 27, No. 2, pp. 169–171.

91. Nelson, L. S. (1995b). "Using Nested Designs II: Confidence Limits for Standard Deviations." *Journal of Quality Technology,* Vol. 27, No. 3, pp. 265–267.

92. Nelson, L. S. (1983). "Variance Estimation Using Staggered, Nested Designs." *Journal of Quality Technology,* Vol. 15, pp. 195–198.

93. Nelson, P. R. (1989). "Multiple Comparison of Means Using Simultaneous Confidence Intervals." *Journal of Quality Technology,* Vol. 21, No. 4, pp. 232–241.

94. Neter, J., W. Wasserman, and M. Kunter (1990). *Applied Linear Statistical Models.* 3rd edition. Richard D. Irwin, Homewood, Ill.

95. Newman, D. (1939). "The Distribution of the Range in Samples from a Normal Population, Expressed in Terms of an Independent Estimate of Standard Deviation." *Biometrika,* Vol. 31, pp. 20–30.

96. O'Neill, R. and G. B. Wetherill (1971). "The Present State of Multiple Comparison Methods." *Journal of the Royal Statistical Society,* B, Vol. 33, pp. 218–241.

97. Ostle, B. (1963). *Statistics in Research.* 2nd edition. Iowa State Press, Ames, Iowa.

98. Pearson, E. S. and H. O. Hartley (1966). *Biometrika Tables for Statisticians,* Vol. 1. 3rd edition. Cambridge University Press, Cambridge.

99. Pearson, E. S. and H. O. Hartley (1972). *Biometrika Tables for Statisticians,* Vol. 2. Cambridge University Press, Cambridge.

100. Plackett, R. L. and J. P. Burman (1946). "The Design of Optimum Multifactorial Experiments." *Biometrika,* Vol. 33, pp. 305–325.

101. Quenouille, M. H. (1953). *The Design and Analysis of Experiments.* Charles Griffin and Company, London.

102. Quenouille, M. H. (1955). "Checks on the Calculation of Main Effects and Interactions in a 2^n Factorial Experiment." *Annals of Eugenics,* Vol. 19, pp. 151–152.

103. Quinlan, J. (1985). "Product Improvement by Application of Taguchi Methods." *Third Supplier Symposium on Taguchi Methods,* American Supplier Institute, Inc., Dearborn, Mich.

104. Rayner, A. A. (1967). "The Square Summing Check on the Main Effects and Interactions in a 2^n Experiment as Calculated by Yates' Algorithm." *Biometrics,* Vol. 23, pp. 571–573.

105. Satterthwaite, F. E. (1946). "An Approximate Distribution of Estimates of Variance Components." *Biometrics Bull.,* Vol. 2, pp. 110–112.

106. Scheffé, H. (1953). "A Method for Judging All Contrasts in the Analysis of Variance." *Biometrika,* Vol. 40, pp. 87–104.

107. Scheffé, H. (1956a). "A 'Mixed Model' for the Analysis of Variance." *Annals of Mathematical Statistics,* Vol. 27, pp. 23–36.

108. Scheffé, H. (1956b). "Alternative Models for the Analysis of Variance." *Annals of Mathematical Statistics,* Vol. 27, pp. 251–271.

109. Scheffé, H. (1959). *The Analysis of Variance.* Wiley, New York.

110. Schmidt, S. R. and J. R. Boudot (1989). "A Monte Carlo Simulation Study Comparing Effectiveness of Signal-to-Noise Ratios and Other Methods for Identifying Dispersion Effects." Rocky Mountain Quality Control Conference.

111. Searle, S. R. and R. F. Fawcett (1970). "Expected Mean Squares in Variance Component Models Having Finite Populations." *Biometrics,* Vol. 26, pp. 243–254.

112. Searle, S. R. (1971a). *Linear Models.* Wiley, New York.

113. Searle, S. R. (1971b). "Topics in Variance Component Estimation." *Biometrics,* Vol. 27, pp. 1–76.

114. Searle, S. R., F. M. Speed, and H. V. Henderson (1981). "Some Computational and Model Equivalences in Analyses of Variance of Unequal-Subclass-Numbers Data." *The American Statistician,* Vol. 35, pp. 16–33.

115. Searle, S. R. (1987). *Linear Models for Unbalanced Data.* Wiley, New York.

116. Smith, C. A. B. and H. O. Hartley (1948). "Construction of Youden Squares." *Journal of the Royal Statistical Society,* B, Vol. 10, pp. 262–264.

117. Smith, H. F. (1957). "Interpretations of Adjusted Treatment Means and Regressions in Analysis of Covariance." *Biometrics,* Vol. 13, No. 3, pp. 282–308.

118. Smith, J. R. and J. M. Beverly (1981). "The Use and Analysis of Staggered Nested Factorial Designs." *Journal of Quality Technology,* Vol. 13, pp. 166–173.

119. Speed, F. M. and R. R. Hocking (1976). "The Use of the $R(\ )$-Notation with Unbalanced Data." *The American Statistician,* Vol. 30, pp. 30–33.

120. Speed, F. M., R. R. Hocking, and O. P. Hackney (1978). "Methods of Analysis of Linear Models with Unbalanced Data." *Journal of the American Statistical Association,* Vol. 73, pp. 105–112.

121. Stefansky, W. (1972). "Rejecting Outliers in Factorial Designs." *Technometrics,* Vol. 14, pp. 469–479.

122. Taguchi, G. and Y. Wu (1980). *Introduction to Off-Line Quality Control.* Central Japan Quality Control Association, Nagoya, Japan.

123. Taguchi, G. (1986). *Introduction to Quality Engineering.* Asian Productivity Organization, UNIPUB, White Plains, New York.

124. Ting, N., R. K. Burdick, F. A. Graybill, S. Jeyaratnam, and T.-F. C. Lu (1990). "Confidence Intervals on Linear Combinations of Variance Components That Are Unrestricted in Sign." *Journal of Statistical Computation and Simulation,* Vol. 35, pp. 135–143.

125. Tukey, J. W. (1949a). "One Degree of Freedom for Non-Additivity." *Biometrics,* Vol. 5, pp. 232–242.

126. Tukey, J. W. (1949b). "Comparing Individual Means in the Analysis of Variance." *Biometrics,* Vol. 5, pp. 99–114.

127. Tukey, J. W. (1951). "Quick and Dirty Methods in Statistics, Part II, Simple Analysis for Standard Designs." *Proceedings of the Fifth Annual Convention, American Society for Quality Control,* pp. 189–197.

128. Tukey, J. W. (1953). "The Problem of Multiple Comparisons." Unpublished Notes, Princeton University.

129. Wine, R. L. (1964). *Statistics for Scientists and Engineers.* Prentice-Hall, Englewood Cliffs, N.J.

130. Winer, B. J. (1971). *Statistical Principles in Experimental Design.* 2nd edition. McGraw-Hill, New York.

131. Yates, F. (1934). "The Analysis of Multiple Classifications with Unequal Numbers in the Different Classes." *Journal of the American Statistical Association,* Vol. 29, pp. 52–66.

132. Yates, F. (1937). *Design and Analysis of Factorial Experiments.* Tech. Comm. No. 35, Imperial Bureau of Soil Sciences, London.

133. Yates, F. (1940). "The Recovery of Interblock Information in Balanced Incomplete Block Designs." *Annals of Eugenics,* Vol. 10, pp. 317–325.

Appendix

I. Cumulative Standard Normal Distribution[a]

$$\Phi(z) = \int_{-\infty}^{z} \frac{1}{\sqrt{2\pi}} e^{-u^2/2} \, du$$

z	.00	.01	.02	.03	.04	z
.0	.50000	.50399	.50798	.51197	.51595	.0
.1	.53983	.54379	.54776	.55172	.55567	.1
.2	.57926	.58317	.58706	.59095	.59483	.2
.3	.61791	.62172	.62551	.62930	.63307	.3
.4	.65542	.65910	.66276	.66640	.67003	.4
.5	.69146	.69497	.69847	.70194	.70540	.5
.6	.72575	.72907	.73237	.73565	.73891	.6
.7	.75803	.76115	.76424	.76730	.77035	.7
.8	.78814	.79103	.79389	.79673	.79954	.8
.9	.81594	.81859	.82121	.82381	.82639	.9
1.0	.84134	.84375	.84613	.84849	.85083	1.0
1.1	.86433	.86650	.86864	.87076	.87285	1.1
1.2	.88493	.88686	.88877	.89065	.89251	1.2
1.3	.90320	.90490	.90658	.90824	.90988	1.3
1.4	.91924	.92073	.92219	.92364	.92506	1.4
1.5	.93319	.93448	.93574	.93699	.93822	1.5
1.6	.94520	.94630	.94738	.94845	.94950	1.6
1.7	.95543	.95637	.95728	.95818	.95907	1.7
1.8	.96407	.96485	.96562	.96637	.96711	1.8
1.9	.97128	.97193	.97257	.97320	.97381	1.9
2.0	.97725	.97778	.97831	.97882	.97932	2.0
2.1	.98214	.98257	.98300	.98341	.93882	2.1
2.2	.98610	.98645	.98679	.98713	.98745	2.2
2.3	.98928	.98956	.98983	.99010	.99036	2.3
2.4	.99180	.99202	.99224	.99245	.99266	2.4
2.5	.99379	.99396	.99413	.99430	.99446	2.5
2.6	.99534	.99547	.99560	.99573	.99585	2.6
2.7	.99653	.99664	.99674	.99683	.99693	2.7
2.8	.99744	.99752	.99760	.99767	.99774	2.8
2.9	.99813	.99819	.99825	.99831	.99836	2.9
3.0	.99865	.99869	.99874	.99878	.99882	3.0
3.1	.99903	.99906	.99910	.99913	.99916	3.1
3.2	.99931	.99934	.99936	.99938	.99940	3.2
3.3	.99952	.99953	.99955	.99957	.99958	3.3
3.4	.99966	.99968	.99969	.99970	.99971	3.4
3.5	.99977	.99978	.99978	.99979	.99980	3.5
3.6	.99984	.99985	.99985	.99986	.99986	3.6
3.7	.99989	.99990	.99990	.99990	.99991	3.7
3.8	.99993	.99993	.99993	.99994	.99994	3.8
3.9	.99995	.99995	.99996	.99996	.99996	3.9

[a] Reproduced with permission from *Probability and Statistics in Engineering and Management Science,* 3rd edition, by W. W. Hines and D. C. Montgomery, Wiley, New York, 1990.

I.   Cumulative Standard Normal Distribution (*continued*)

$$\Phi(z) = \int_{-\infty}^{z} \frac{1}{\sqrt{2\pi}} e^{-u^2/2} \, du$$

z	.05	.06	.07	.08	.09	z
.0	.51994	.52392	.52790	.53188	.53586	.0
.1	.55962	.56356	.56749	.57142	.57534	.1
.2	.59871	.60257	.60642	.61026	.61409	.2
.3	.63683	.64058	.64431	.64803	.65173	.3
.4	.67364	.67724	.68082	.68438	.68793	.4
.5	.70884	.71226	.71566	.71904	.72240	.5
.6	.74215	.74537	.74857	.75175	.75490	.6
.7	.77337	.77637	.77935	.78230	.78523	.7
.8	.80234	.80510	.80785	.81057	.81327	.8
.9	.82894	.83147	.83397	.83646	.83891	.9
1.0	.85314	.85543	.85769	.85993	.86214	1.0
1.1	.87493	.87697	.87900	.88100	.88297	1.1
1.2	.89435	.89616	.89796	.89973	.90147	1.2
1.3	.91149	.91308	.91465	.91621	.91773	1.3
1.4	.92647	.92785	.92922	.93056	.93189	1.4
1.5	.93943	.94062	.94179	.94295	.94408	1.5
1.6	.95053	.95154	.95254	.95352	.95448	1.6
1.7	.95994	.96080	.96164	.96246	.96327	1.7
1.8	.96784	.96856	.96926	.96995	.97062	1.8
1.9	.97441	.97500	.97558	.97615	.97670	1.9
2.0	.97982	.98030	.98077	.98124	.98169	2.0
2.1	.98422	.98461	.98500	.98537	.98574	2.1
2.2	.98778	.98809	.98840	.98870	.98899	2.2
2.3	.99061	.99086	.99111	.99134	.99158	2.3
2.4	.99286	.99305	.99324	.99343	.99361	2.4
2.5	.99461	.99477	.99492	.99506	.99520	2.5
2.6	.99598	.99609	.99621	.99632	.99643	2.6
2.7	.99702	.99711	.99720	.99728	.99736	2.7
2.8	.99781	.99788	.99795	.99801	.99807	2.8
2.9	.99841	.99846	.99851	.99856	.99861	2.9
3.0	.99886	.99889	.99893	.99897	.99900	3.0
3.1	.99918	.99921	.99924	.99926	.99929	3.1
3.2	.99942	.99944	.99946	.99948	.99950	3.2
3.3	.99960	.99961	.99962	.99964	.99965	3.3
3.4	.99972	.99973	.99974	.99975	.99976	3.4
3.5	.99981	.99981	.99982	.99983	.99983	3.5
3.6	.99987	.99987	.99988	.99988	.99989	3.6
3.7	.99991	.99992	.99992	.99992	.99992	3.7
3.8	.99994	.99994	.99995	.99995	.99995	3.8
3.9	.99996	.99996	.99996	.99997	.99997	3.9

II. Percentage Points of the t Distribution[a]

ν \ α	.40	.25	.10	.05	.025	.01	.005	.0025	.001	.0005
1	.325	1.000	3.078	6.314	12.706	31.821	63.657	127.32	318.31	636.62
2	.289	.816	1.886	2.920	4.303	6.965	9.925	14.089	23.326	31.598
3	.277	.765	1.638	2.353	3.182	4.541	5.841	7.453	10.213	12.924
4	.271	.741	1.533	2.132	2.776	3.747	4.604	5.598	7.173	8.610
5	.267	.727	1.476	2.015	2.571	3.365	4.032	4.773	5.893	6.869
6	.265	.727	1.440	1.943	2.447	3.143	3.707	4.317	5.208	5.959
7	.263	.711	1.415	1.895	2.365	2.998	3.499	4.019	4.785	5.408
8	.262	.706	1.397	1.860	2.306	2.896	3.355	3.833	4.501	5.041
9	.261	.703	1.383	1.833	2.262	2.821	3.250	3.690	4.297	4.781
10	.260	.700	1.372	1.812	2.228	2.764	3.169	3.581	4.144	4.587
11	.260	.697	1.363	1.796	2.201	2.718	3.106	3.497	4.025	4.437
12	.259	.695	1.356	1.782	2.179	2.681	3.055	3.428	3.930	4.318
13	.259	.694	1.350	1.771	2.160	2.650	3.012	3.372	3.852	4.221
14	.258	.692	1.345	1.761	2.145	2.624	2.977	3.326	3.787	4.140
15	.258	.691	1.341	1.753	2.131	2.602	2.947	3.286	3.733	4.073
16	.258	.690	1.337	1.746	2.120	2.583	2.921	3.252	3.686	4.015
17	.257	.689	1.333	1.740	2.110	2.567	2.898	3.222	3.646	3.965
18	.257	.688	1.330	1.734	2.101	2.552	2.878	3.197	3.610	3.922
19	.257	.688	1.328	1.729	2.093	2.539	2.861	3.174	3.579	3.883
20	.257	.687	1.325	1.725	2.086	2.528	2.845	3.153	3.552	3.850
21	.257	.686	1.323	1.721	2.080	2.518	2.831	3.135	3.527	3.819
22	.256	.686	1.321	1.717	2.074	2.508	2.819	3.119	3.505	3.792
23	.256	.685	1.319	1.714	2.069	2.500	2.807	3.104	3.485	3.767
24	.256	.685	1.318	1.711	2.064	2.492	2.797	3.091	3.467	3.745
25	.256	.684	1.316	1.708	2.060	2.485	2.787	3.078	3.450	3.725
26	.256	.684	1.315	1.706	2.056	2.479	2.779	3.067	3.435	3.707
27	.256	.684	1.314	1.703	2.052	2.473	2.771	3.057	3.421	3.690
28	.256	.683	1.313	1.701	2.048	2.467	2.763	3.047	3.408	3.674
29	.256	.683	1.311	1.699	2.045	2.462	2.756	3.038	3.396	3.659
30	.256	.683	1.310	1.697	2.042	2.457	2.750	3.030	3.385	3.646
40	.255	.681	1.303	1.684	2.021	2.423	2.704	2.971	3.307	3.551
60	.254	.679	1.296	1.671	2.000	2.390	2.660	2.915	3.232	3.460
120	.254	.677	1.289	1.658	1.980	2.358	2.617	2.860	3.160	3.373
∞	.253	.674	1.282	1.645	1.960	2.326	2.576	2.807	3.090	3.291

ν = degrees of freedom.

[a] Adapted with permission from *Biometrika Tables for Statisticians*, Vol. 1, 3rd edition, by E. S. Pearson and H. O. Hartley, Cambridge University Press, Cambridge, 1966.

III.   Percentage Points of the χ^2 Distribution[a]

ν	.995	.990	.975	.950	α .500	.050	.025	.010	.005
1	0.00 +	0.00 +	0.00 +	0.00 +	0.45	3.84	5.02	6.63	7.88
2	0.01	0.02	0.05	0.10	1.39	5.99	7.38	9.21	10.60
3	0.07	0.11	0.22	0.35	2.37	7.81	9.35	11.34	12.84
4	0.21	0.30	0.48	0.71	3.36	9.49	11.14	13.28	14.86
5	0.41	0.55	0.83	1.15	4.35	11.07	12.38	15.09	16.75
6	0.68	0.87	1.24	1.64	5.35	12.59	14.45	16.81	18.55
7	0.99	1.24	1.69	2.17	6.35	14.07	16.01	18.48	20.28
8	1.34	1.65	2.18	2.73	7.34	15.51	17.53	20.09	21.96
9	1.73	2.09	2.70	3.33	8.34	16.92	19.02	21.67	23.59
10	2.16	2.56	3.25	3.94	9.34	18.31	20.48	23.21	25.19
11	2.60	3.05	3.82	4.57	10.34	19.68	21.92	24.72	26.76
12	3.07	3.57	4.40	5.23	11.34	21.03	23.34	26.22	28.30
13	3.57	4.11	5.01	5.89	12.34	22.36	24.74	27.69	29.82
14	4.07	4.66	5.63	6.57	13.34	23.68	26.12	29.14	31.32
15	4.60	5.23	6.27	7.26	14.34	25.00	27.49	30.58	32.80
16	5.14	5.81	6.91	7.96	15.34	26.30	28.85	32.00	34.27
17	5.70	6.41	7.56	8.67	16.34	27.59	30.19	33.41	35.72
18	6.26	7.01	8.23	9.39	17.34	28.87	31.53	34.81	37.16
19	6.84	7.63	8.91	10.12	18.34	30.14	32.85	36.19	38.58
20	7.43	8.26	9.59	10.85	19.34	31.41	34.17	37.57	40.00
25	10.52	11.52	13.12	14.61	24.34	37.65	40.65	44.31	46.93
30	13.79	14.95	16.79	18.49	29.34	43.77	46.98	50.89	53.67
40	20.71	22.16	24.43	26.51	39.34	55.76	59.34	63.69	66.77
50	27.99	29.71	32.36	34.76	49.33	67.50	71.42	76.15	79.49
60	35.53	37.48	40.48	43.19	59.33	79.08	83.30	88.38	91.95
70	43.28	45.44	48.76	51.74	69.33	90.53	95.02	100.42	104.22
80	51.17	53.54	57.15	60.39	79.33	101.88	106.63	112.33	116.32
90	59.20	61.75	65.65	69.13	89.33	113.14	118.14	124.12	128.30
100	67.33	70.06	74.22	77.93	99.33	124.34	129.56	135.81	140.17

ν = degrees of freedom

[a] Adapted with permission from *Biometrika Tables for Statisticians,* Vol. 1, 3rd edition by E. S. Pearson and H. O. Hartley, Cambridge University Press, Cambridge, 1966.

IV. Percentage Points of the F Distribution[a]

$$F_{0.25,\nu_1,\nu_2}$$

ν_2 \ ν_1	1	2	3	4	5	6	7	8	9	10	12	15	20	24	30	40	60	120	∞
1	5.83	7.50	8.20	8.58	8.82	8.98	9.10	9.19	9.26	9.32	9.41	9.49	9.58	9.63	9.67	9.71	9.76	9.80	9.85
2	2.57	3.00	3.15	3.23	3.28	3.31	3.34	3.35	3.37	3.38	3.39	3.41	3.43	3.43	3.44	3.45	3.46	3.47	3.48
3	2.02	2.28	2.36	2.39	2.41	2.42	2.43	2.44	2.44	2.44	2.45	2.46	2.46	2.46	2.47	2.47	2.47	2.47	2.47
4	1.81	2.00	2.05	2.06	2.07	2.08	2.08	2.08	2.08	2.08	2.08	2.08	2.08	2.08	2.08	2.08	2.08	2.08	2.08
5	1.69	1.85	1.88	1.89	1.89	1.89	1.89	1.89	1.89	1.89	1.89	1.89	1.88	1.88	1.88	1.88	1.87	1.87	1.87
6	1.62	1.76	1.78	1.79	1.79	1.78	1.78	1.78	1.77	1.77	1.77	1.76	1.76	1.75	1.75	1.75	1.74	1.74	1.74
7	1.57	1.70	1.72	1.72	1.71	1.71	1.70	1.70	1.70	1.69	1.68	1.68	1.67	1.67	1.66	1.66	1.65	1.65	1.65
8	1.54	1.66	1.67	1.66	1.66	1.65	1.64	1.64	1.63	1.63	1.62	1.62	1.61	1.60	1.60	1.59	1.59	1.58	1.58
9	1.51	1.62	1.63	1.63	1.62	1.61	1.60	1.60	1.59	1.59	1.58	1.57	1.56	1.56	1.55	1.54	1.54	1.53	1.53
10	1.49	1.60	1.60	1.59	1.59	1.58	1.57	1.56	1.56	1.55	1.54	1.53	1.52	1.52	1.51	1.51	1.50	1.49	1.48
11	1.47	1.58	1.58	1.57	1.56	1.55	1.54	1.53	1.53	1.52	1.51	1.50	1.49	1.49	1.48	1.47	1.47	1.46	1.45
12	1.46	1.56	1.56	1.55	1.54	1.53	1.52	1.51	1.51	1.50	1.49	1.48	1.47	1.46	1.45	1.45	1.44	1.43	1.42
13	1.45	1.55	1.55	1.53	1.52	1.51	1.50	1.49	1.49	1.48	1.47	1.46	1.45	1.44	1.43	1.42	1.42	1.41	1.40
14	1.44	1.53	1.53	1.52	1.51	1.50	1.49	1.48	1.47	1.46	1.45	1.44	1.43	1.42	1.41	1.41	1.40	1.39	1.38
15	1.43	1.52	1.52	1.51	1.49	1.48	1.47	1.46	1.46	1.45	1.44	1.43	1.41	1.41	1.40	1.39	1.38	1.37	1.36
16	1.42	1.51	1.51	1.50	1.48	1.47	1.46	1.45	1.44	1.44	1.43	1.41	1.40	1.39	1.38	1.37	1.36	1.35	1.34
17	1.42	1.51	1.50	1.49	1.47	1.46	1.45	1.44	1.43	1.43	1.41	1.40	1.39	1.38	1.37	1.36	1.35	1.34	1.33
18	1.41	1.50	1.49	1.48	1.46	1.45	1.44	1.43	1.42	1.42	1.40	1.39	1.38	1.37	1.36	1.35	1.34	1.33	1.32
19	1.41	1.49	1.49	1.47	1.46	1.44	1.43	1.42	1.41	1.41	1.40	1.38	1.37	1.36	1.35	1.34	1.33	1.32	1.30
20	1.40	1.49	1.48	1.47	1.45	1.44	1.43	1.42	1.41	1.40	1.39	1.37	1.36	1.35	1.34	1.33	1.32	1.31	1.29
21	1.40	1.48	1.48	1.46	1.44	1.43	1.42	1.41	1.40	1.39	1.38	1.37	1.35	1.34	1.33	1.32	1.31	1.30	1.28
22	1.40	1.48	1.47	1.45	1.44	1.42	1.41	1.40	1.39	1.39	1.37	1.36	1.34	1.33	1.32	1.31	1.30	1.29	1.28
23	1.39	1.47	1.47	1.45	1.43	1.42	1.41	1.40	1.39	1.38	1.37	1.35	1.34	1.33	1.32	1.31	1.30	1.28	1.27
24	1.39	1.47	1.46	1.44	1.43	1.41	1.40	1.39	1.38	1.38	1.36	1.35	1.33	1.32	1.31	1.30	1.29	1.28	1.26
25	1.39	1.47	1.46	1.44	1.42	1.41	1.40	1.39	1.38	1.37	1.36	1.34	1.33	1.32	1.31	1.29	1.28	1.27	1.25
26	1.38	1.46	1.45	1.44	1.42	1.41	1.39	1.38	1.37	1.37	1.35	1.34	1.32	1.31	1.30	1.29	1.28	1.26	1.25
27	1.38	1.46	1.45	1.43	1.42	1.40	1.39	1.38	1.37	1.36	1.35	1.33	1.32	1.31	1.30	1.28	1.27	1.26	1.24
28	1.38	1.46	1.45	1.43	1.41	1.40	1.39	1.38	1.37	1.36	1.34	1.33	1.31	1.30	1.29	1.28	1.27	1.25	1.24
29	1.38	1.45	1.45	1.43	1.41	1.40	1.38	1.37	1.36	1.35	1.34	1.32	1.31	1.30	1.29	1.27	1.26	1.25	1.23
30	1.38	1.45	1.44	1.42	1.41	1.39	1.38	1.37	1.36	1.35	1.34	1.32	1.30	1.29	1.28	1.27	1.26	1.24	1.23
40	1.36	1.44	1.42	1.40	1.39	1.37	1.36	1.35	1.34	1.33	1.31	1.30	1.28	1.26	1.25	1.24	1.22	1.21	1.19
60	1.35	1.42	1.41	1.38	1.37	1.35	1.33	1.32	1.31	1.30	1.29	1.27	1.25	1.24	1.22	1.21	1.19	1.17	1.15
120	1.34	1.40	1.39	1.37	1.35	1.33	1.31	1.30	1.29	1.28	1.26	1.24	1.22	1.21	1.19	1.18	1.16	1.13	1.10
∞	1.32	1.39	1.37	1.35	1.33	1.31	1.29	1.28	1.27	1.25	1.24	1.22	1.19	1.18	1.16	1.14	1.12	1.08	1.00

Degrees of Freedom for the Numerator (ν_1)

Degrees of Freedom for the Denominator (ν_2)

ν = degrees of freedom

[a] Adapted with permission from *Biometrika Tables for Statisticians*, Vol. 1, 3rd edition by E. S. Pearson and H. O. Hartley, Cambridge University Press, Cambridge, 1966.

IV. Percentage Points of the F Distribution (continued)

$$F_{0.10, \nu_1, \nu_2}$$

Degrees of Freedom for the Numerator (ν_1)

ν_2	1	2	3	4	5	6	7	8	9	10	12	15	20	24	30	40	60	120	∞
1	39.86	49.50	53.59	55.83	57.24	58.20	58.91	59.44	59.86	60.19	60.71	61.22	61.74	62.00	62.26	62.53	62.79	63.06	63.33
2	8.53	9.00	9.16	9.24	9.29	9.33	9.35	9.37	9.38	9.39	9.41	9.42	9.44	9.45	9.46	9.47	9.47	9.48	9.49
3	5.54	5.46	5.39	5.34	5.31	5.28	5.27	5.25	5.24	5.23	5.22	5.20	5.18	5.18	5.17	5.16	5.15	5.14	5.13
4	4.54	4.32	4.19	4.11	4.05	4.01	3.98	3.95	3.94	3.92	3.90	3.87	3.84	3.83	3.82	3.80	3.79	3.78	3.76
5	4.06	3.78	3.62	3.52	3.45	3.40	3.37	3.34	3.32	3.30	3.27	3.24	3.21	3.19	3.17	3.16	3.14	3.12	3.10
6	3.78	3.46	3.29	3.18	3.11	3.05	3.01	2.98	2.96	2.94	2.90	2.87	2.84	2.82	2.80	2.78	2.76	2.74	2.72
7	3.59	3.26	3.07	2.96	2.88	2.83	2.78	2.75	2.72	2.70	2.67	2.63	2.59	2.58	2.56	2.54	2.51	2.49	2.47
8	3.46	3.11	2.92	2.81	2.73	2.67	2.62	2.59	2.56	2.54	2.50	2.46	2.42	2.40	2.38	2.36	2.34	2.32	2.29
9	3.36	3.01	2.81	2.69	2.61	2.55	2.51	2.47	2.44	2.42	2.38	2.34	2.30	2.28	2.25	2.23	2.21	2.18	2.16
10	3.29	2.92	2.73	2.61	2.52	2.46	2.41	2.38	2.35	2.32	2.28	2.24	2.20	2.18	2.16	2.13	2.11	2.08	2.06
11	3.23	2.86	2.66	2.54	2.45	2.39	2.34	2.30	2.27	2.25	2.21	2.17	2.12	2.10	2.08	2.05	2.03	2.00	1.97
12	3.18	2.81	2.61	2.48	2.39	2.33	2.28	2.24	2.21	2.19	2.15	2.10	2.06	2.04	2.01	1.99	1.96	1.93	1.90
13	3.14	2.76	2.56	2.43	2.35	2.28	2.23	2.20	2.16	2.14	2.10	2.05	2.01	1.98	1.96	1.93	1.90	1.88	1.85
14	3.10	2.73	2.52	2.39	2.31	2.24	2.19	2.15	2.12	2.10	2.05	2.01	1.96	1.94	1.91	1.89	1.86	1.83	1.80
15	3.07	2.70	2.49	2.36	2.27	2.21	2.16	2.12	2.09	2.06	2.02	1.97	1.92	1.90	1.87	1.85	1.82	1.79	1.76
16	3.05	2.67	2.46	2.33	2.24	2.18	2.13	2.09	2.06	2.03	1.99	1.94	1.89	1.87	1.84	1.81	1.78	1.75	1.72
17	3.03	2.64	2.44	2.31	2.22	2.15	2.10	2.06	2.03	2.00	1.96	1.91	1.86	1.84	1.81	1.78	1.75	1.72	1.69
18	3.01	2.62	2.42	2.29	2.20	2.13	2.08	2.04	2.00	1.98	1.93	1.89	1.84	1.81	1.78	1.75	1.72	1.69	1.66
19	2.99	2.61	2.40	2.27	2.18	2.11	2.06	2.02	1.98	1.96	1.91	1.86	1.81	1.79	1.76	1.73	1.70	1.67	1.63
20	2.97	2.59	2.38	2.25	2.16	2.09	2.04	2.00	1.96	1.94	1.89	1.84	1.79	1.77	1.74	1.71	1.68	1.64	1.61
21	2.96	2.57	2.36	2.23	2.14	2.08	2.02	1.98	1.95	1.92	1.87	1.83	1.78	1.75	1.72	1.69	1.66	1.62	1.59
22	2.95	2.56	2.35	2.22	2.13	2.06	2.01	1.97	1.93	1.90	1.86	1.81	1.76	1.73	1.70	1.67	1.64	1.60	1.57
23	2.94	2.55	2.34	2.21	2.11	2.05	1.99	1.96	1.92	1.89	1.84	1.80	1.74	1.72	1.69	1.66	1.62	1.59	1.55
24	2.93	2.54	2.33	2.19	2.10	2.04	1.98	1.94	1.91	1.88	1.83	1.78	1.73	1.70	1.67	1.64	1.61	1.57	1.53
25	2.92	2.53	2.32	2.18	2.09	2.02	1.97	1.93	1.89	1.87	1.82	1.77	1.72	1.69	1.66	1.63	1.59	1.56	1.52
26	2.91	2.52	2.31	2.17	2.08	2.01	1.96	1.92	1.88	1.86	1.81	1.76	1.71	1.68	1.65	1.61	1.58	1.54	1.50
27	2.90	2.51	2.30	2.17	2.07	2.00	1.95	1.91	1.87	1.85	1.80	1.75	1.70	1.67	1.64	1.60	1.57	1.53	1.49
28	2.89	2.50	2.29	2.16	2.06	2.00	1.94	1.90	1.87	1.84	1.79	1.74	1.69	1.66	1.63	1.59	1.56	1.52	1.48
29	2.89	2.50	2.28	2.15	2.06	1.99	1.93	1.89	1.86	1.83	1.78	1.73	1.68	1.65	1.62	1.58	1.55	1.51	1.47
30	2.88	2.49	2.28	2.14	2.03	1.98	1.93	1.88	1.85	1.82	1.77	1.72	1.67	1.64	1.61	1.57	1.54	1.50	1.46
40	2.84	2.44	2.23	2.09	2.00	1.93	1.87	1.83	1.79	1.76	1.71	1.66	1.61	1.57	1.54	1.51	1.47	1.42	1.38
60	2.79	2.39	2.18	2.04	1.95	1.87	1.82	1.77	1.74	1.71	1.66	1.60	1.54	1.51	1.48	1.44	1.40	1.35	1.29
120	2.75	2.35	2.13	1.99	1.90	1.82	1.77	1.72	1.68	1.65	1.60	1.55	1.48	1.45	1.41	1.37	1.32	1.26	1.19
∞	2.71	2.30	2.08	1.94	1.85	1.77	1.72	1.67	1.63	1.60	1.55	1.49	1.42	1.38	1.34	1.30	1.24	1.17	1.00

Degrees of Freedom for the Denominator (ν_2)

(continued)

663

IV. Percentage Points of the F Distribution (*continued*)

$$F_{0.05,\nu_1,\nu_2}$$

ν_2	\multicolumn{19}{c}{Degrees of Freedom for the Numerator (ν_1)}																		
	1	2	3	4	5	6	7	8	9	10	12	15	20	24	30	40	60	120	∞
1	161.4	199.5	215.7	224.6	230.2	234.0	236.8	238.9	240.5	241.9	243.9	245.9	248.0	249.1	250.1	251.1	252.2	253.3	254.3
2	18.51	19.00	19.16	19.25	19.30	19.33	19.35	19.37	19.38	19.40	19.41	19.43	19.45	19.45	19.46	19.47	19.48	19.49	19.50
3	10.13	9.55	9.28	9.12	9.01	8.94	8.89	8.85	8.81	8.79	8.74	8.70	8.66	8.64	8.62	8.59	8.57	8.55	8.53
4	7.71	6.94	6.59	6.39	6.26	6.16	6.09	6.04	6.00	5.96	5.91	5.86	5.80	5.77	5.75	5.72	5.69	5.66	5.63
5	6.61	5.79	5.41	5.19	5.05	4.95	4.88	4.82	4.77	4.74	4.68	4.62	4.56	4.53	4.50	4.46	4.43	4.40	4.36
6	5.99	5.14	4.76	4.53	4.39	4.28	4.21	4.15	4.10	4.06	4.00	3.94	3.87	3.84	3.81	3.77	3.74	3.70	3.67
7	5.59	4.74	4.35	4.12	3.97	3.87	3.79	3.73	3.68	3.64	3.57	3.51	3.44	3.41	3.38	3.34	3.30	3.27	3.23
8	5.32	4.46	4.07	3.84	3.69	3.58	3.50	3.44	3.39	3.35	3.28	3.22	3.15	3.12	3.08	3.04	3.01	2.97	2.93
9	5.12	4.26	3.86	3.63	3.48	3.37	3.29	3.23	3.18	3.14	3.07	3.01	2.94	2.90	2.86	2.83	2.79	2.75	2.71
10	4.96	4.10	3.71	3.48	3.33	3.22	3.14	3.07	3.02	2.98	2.91	2.85	2.77	2.74	2.70	2.66	2.62	2.58	2.54
11	4.84	3.98	3.59	3.36	3.20	3.09	3.01	2.95	2.90	2.85	2.79	2.72	2.65	2.61	2.57	2.53	2.49	2.45	2.40
12	4.75	3.89	3.49	3.26	3.11	3.00	2.91	2.85	2.80	2.75	2.69	2.62	2.54	2.51	2.47	2.43	2.38	2.34	2.30
13	4.67	3.81	3.41	3.18	3.03	2.92	2.83	2.77	2.71	2.67	2.60	2.53	2.46	2.42	2.38	2.34	2.30	2.25	2.21
14	4.60	3.74	3.34	3.11	2.96	2.85	2.76	2.70	2.65	2.60	2.53	2.46	2.39	2.35	2.31	2.27	2.22	2.18	2.13
15	4.54	3.68	3.29	3.06	2.90	2.79	2.71	2.64	2.59	2.54	2.48	2.40	2.33	2.29	2.25	2.20	2.16	2.11	2.07
16	4.49	3.63	3.24	3.01	2.85	2.74	2.66	2.59	2.54	2.49	2.42	2.35	2.28	2.24	2.19	2.15	2.11	2.06	2.01
17	4.45	3.59	3.20	2.96	2.81	2.70	2.61	2.55	2.49	2.45	2.38	2.31	2.23	2.19	2.15	2.10	2.06	2.01	1.96
18	4.41	3.55	3.16	2.93	2.77	2.66	2.58	2.51	2.46	2.41	2.34	2.27	2.19	2.15	2.11	2.06	2.02	1.97	1.92
19	4.38	3.52	3.13	2.90	2.74	2.63	2.54	2.48	2.42	2.38	2.31	2.23	2.16	2.11	2.07	2.03	1.98	1.93	1.88
20	4.35	3.49	3.10	2.87	2.71	2.60	2.51	2.45	2.39	2.35	2.28	2.20	2.12	2.08	2.04	1.99	1.95	1.90	1.84
21	4.32	3.47	3.07	2.84	2.68	2.57	2.49	2.42	2.37	2.32	2.25	2.18	2.10	2.05	2.01	1.96	1.92	1.87	1.81
22	4.30	3.44	3.05	2.82	2.66	2.55	2.46	2.40	2.34	2.30	2.23	2.15	2.07	2.03	1.98	1.94	1.89	1.84	1.78
23	4.28	3.42	3.03	2.80	2.64	2.53	2.44	2.37	2.32	2.27	2.20	2.13	2.05	2.01	1.96	1.91	1.86	1.81	1.76
24	4.26	3.40	3.01	2.78	2.62	2.51	2.42	2.36	2.30	2.25	2.18	2.11	2.03	1.98	1.94	1.89	1.84	1.79	1.73
25	4.24	3.39	2.99	2.76	2.60	2.49	2.40	2.34	2.28	2.24	2.16	2.09	2.01	1.96	1.92	1.87	1.82	1.77	1.71
26	4.23	3.37	2.98	2.74	2.59	2.47	2.39	2.32	2.27	2.22	2.15	2.07	1.99	1.95	1.90	1.85	1.80	1.75	1.69
27	4.21	3.35	2.96	2.73	2.57	2.46	2.37	2.31	2.25	2.20	2.13	2.06	1.97	1.93	1.88	1.84	1.79	1.73	1.67
28	4.20	3.34	2.95	2.71	2.56	2.45	2.36	2.29	2.24	2.19	2.12	2.04	1.96	1.91	1.87	1.82	1.77	1.71	1.65
29	4.18	3.33	2.93	2.70	2.55	2.43	2.35	2.28	2.22	2.18	2.10	2.03	1.94	1.90	1.85	1.81	1.75	1.70	1.64
30	4.17	3.32	2.92	2.69	2.53	2.42	2.33	2.27	2.21	2.16	2.09	2.01	1.93	1.89	1.84	1.79	1.74	1.68	1.62
40	4.08	3.23	2.84	2.61	2.45	2.34	2.25	2.18	2.12	2.08	2.00	1.92	1.84	1.79	1.74	1.69	1.64	1.58	1.51
60	4.00	3.15	2.76	2.53	2.37	2.25	2.17	2.10	2.04	1.99	1.92	1.84	1.75	1.70	1.65	1.59	1.53	1.47	1.39
120	3.92	3.07	2.68	2.45	2.29	2.17	2.09	2.02	1.96	1.91	1.83	1.75	1.66	1.61	1.55	1.55	1.43	1.35	1.25
∞	3.84	3.00	2.60	2.37	2.21	2.10	2.01	1.94	1.88	1.83	1.75	1.67	1.57	1.52	1.46	1.39	1.32	1.22	1.00

Degrees of Freedom for the Denominator (ν_2)

IV. Percentage Points of the F Distribution (continued)

$$F_{0.025,\nu_1,\nu_2}$$

									Degrees of Freedom for the Numerator (ν_1)										
ν_2	1	2	3	4	5	6	7	8	9	10	12	15	20	24	30	40	60	120	∞
1	647.8	799.5	864.2	899.6	921.8	937.1	948.2	956.7	963.3	968.6	976.7	984.9	993.1	997.2	1001	1006	1010	1014	1018
2	38.51	39.00	39.17	39.25	39.30	39.33	39.36	39.37	39.39	39.40	39.41	39.43	39.45	39.46	39.46	39.47	39.48	39.49	39.50
3	17.44	16.04	15.44	15.10	14.88	14.73	14.62	14.54	14.47	14.42	14.34	14.25	14.17	14.12	14.08	14.04	13.99	13.95	13.90
4	12.22	10.65	9.98	9.60	9.36	9.20	9.07	8.98	8.90	8.84	8.75	8.66	8.56	8.51	8.46	8.41	8.36	8.31	8.26
5	10.01	8.43	7.76	7.39	7.15	6.98	6.85	6.76	6.68	6.62	6.52	6.43	6.33	6.28	6.23	6.18	6.12	6.07	6.02
6	8.81	7.26	6.60	6.23	5.99	5.82	5.70	5.60	5.52	5.46	5.37	5.27	5.17	5.12	5.07	5.01	4.96	4.90	4.85
7	8.07	6.54	5.89	5.52	5.29	5.12	4.99	4.90	4.82	4.76	4.67	4.57	4.47	4.42	4.36	4.31	4.25	4.20	4.14
8	7.57	6.06	5.42	5.05	4.82	4.65	4.53	4.43	4.36	4.30	4.20	4.10	4.00	3.95	3.89	3.84	3.78	3.73	3.67
9	7.21	5.71	5.08	4.72	4.48	4.32	4.20	4.10	4.03	3.96	3.87	3.77	3.67	3.61	3.56	3.51	3.45	3.39	3.33
10	6.94	5.46	4.83	4.47	4.24	4.07	3.95	3.85	3.78	3.72	3.62	3.52	3.42	3.37	3.31	3.26	3.20	3.14	3.08
11	6.72	5.26	4.63	4.28	4.04	3.88	3.76	3.66	3.59	3.53	3.43	3.33	3.23	3.17	3.12	3.06	3.00	2.94	2.88
12	6.55	5.10	4.47	4.12	3.89	3.73	3.61	3.51	3.44	3.37	3.28	3.18	3.07	3.02	2.96	2.91	2.85	2.79	2.72
13	6.41	4.97	4.35	4.00	3.77	3.60	3.48	3.39	3.31	3.25	3.15	3.05	2.95	2.89	2.84	2.78	2.72	2.66	2.60
14	6.30	4.86	4.24	3.89	3.66	3.50	3.38	3.29	3.21	3.15	3.05	2.95	2.84	2.79	2.73	2.67	2.61	2.55	2.49
15	6.20	4.77	4.15	3.80	3.58	3.41	3.29	3.20	3.12	3.06	2.96	2.86	2.76	2.70	2.64	2.59	2.52	2.46	2.40
16	6.12	4.69	4.08	3.73	3.50	3.34	3.22	3.12	3.05	2.99	2.89	2.79	2.68	2.63	2.57	2.51	2.45	2.38	2.32
17	6.04	4.62	4.01	3.66	3.44	3.28	3.16	3.06	2.98	2.92	2.82	2.72	2.62	2.56	2.50	2.44	2.38	2.32	2.25
18	5.98	4.56	3.95	3.61	3.38	3.22	3.10	3.01	2.93	2.87	2.77	2.67	2.56	2.50	2.44	2.38	2.32	2.26	2.19
19	5.92	4.51	3.90	3.56	3.33	3.17	3.05	2.96	2.88	2.82	2.72	2.62	2.51	2.45	2.39	2.33	2.27	2.20	2.13
20	5.87	4.46	3.86	3.51	3.29	3.13	3.01	2.91	2.84	2.77	2.68	2.57	2.46	2.41	2.35	2.29	2.22	2.16	2.09
21	5.83	4.42	3.82	3.48	3.25	3.09	2.97	2.87	2.80	2.73	2.64	2.53	2.42	2.37	2.31	2.25	2.18	2.11	2.04
22	5.79	4.38	3.78	3.44	3.22	3.05	2.93	2.84	2.76	2.70	2.60	2.50	2.39	2.33	2.27	2.21	2.14	2.08	2.00
23	5.75	4.35	3.75	3.41	3.18	3.02	2.90	2.81	2.73	2.67	2.57	2.47	2.36	2.30	2.24	2.18	2.11	2.04	1.97
24	5.72	4.32	3.72	3.38	3.15	2.99	2.87	2.78	2.70	2.64	2.54	2.44	2.33	2.27	2.21	2.15	2.08	2.01	1.94
25	5.69	4.29	3.69	3.35	3.13	2.97	2.85	2.75	2.68	2.61	2.51	2.41	2.30	2.24	2.18	2.12	2.05	1.98	1.91
26	5.66	4.27	3.67	3.33	3.10	2.94	2.82	2.73	2.65	2.59	2.49	2.39	2.28	2.22	2.16	2.09	2.03	1.95	1.88
27	5.63	4.24	3.65	3.31	3.08	2.92	2.80	2.71	2.63	2.57	2.47	2.36	2.25	2.19	2.13	2.07	2.00	1.93	1.85
28	5.61	4.22	3.63	3.29	3.06	2.90	2.78	2.69	2.61	2.55	2.45	2.34	2.23	2.17	2.11	2.05	1.98	1.91	1.83
29	5.59	4.20	1.61	3.27	3.04	2.88	2.76	2.67	2.59	2.53	2.43	2.32	2.21	2.15	2.09	2.03	1.96	1.89	1.81
30	5.57	4.18	3.59	3.25	3.03	2.87	2.75	2.65	2.57	2.51	2.41	2.31	2.20	2.14	2.07	2.01	1.94	1.87	1.79
40	5.42	4.05	3.46	3.13	2.90	2.74	2.62	2.53	2.45	2.39	2.29	2.18	2.07	2.01	1.94	1.88	1.80	1.72	1.64
60	5.29	3.93	3.34	3.01	2.79	2.63	2.51	2.41	2.33	2.27	2.17	2.06	1.94	1.88	1.82	1.74	1.67	1.58	1.48
120	5.15	3.80	3.23	2.89	2.67	2.52	2.39	2.30	2.22	2.16	2.05	1.94	1.82	1.76	1.69	1.61	1.53	1.43	1.31
∞	5.02	3.69	3.12	2.79	2.57	2.41	2.29	2.19	2.11	2.05	1.94	1.83	1.71	1.64	1.57	1.48	1.39	1.27	1.00

Degrees of Freedom for the Denominator (ν_2)

(continued)

IV. Percentage Points of the F Distribution (continued)

$$F_{0.01,\nu_1,\nu_2}$$

Degrees of Freedom for the Numerator (ν_1)

ν_2	1	2	3	4	5	6	7	8	9	10	12	15	20	24	30	40	60	120	∞
1	4052	4999.5	5403	5625	5764	5859	5928	5982	6022	6056	6106	6157	6209	6235	6261	6287	6313	6339	6366
2	98.50	99.00	99.17	99.25	99.30	99.33	99.36	99.37	99.39	99.40	99.42	99.43	99.45	99.46	99.47	99.47	99.48	99.49	99.50
3	34.12	30.82	29.46	28.71	28.24	27.91	27.67	27.49	27.35	27.23	27.05	26.87	26.69	26.60	26.50	26.41	26.32	26.22	26.13
4	21.20	18.00	16.69	15.98	15.52	15.21	14.98	14.80	14.66	14.55	14.37	14.20	14.02	13.93	13.84	13.75	13.65	13.56	13.46
5	16.26	13.27	12.06	11.39	10.97	10.67	10.46	10.29	10.16	10.05	9.89	9.72	9.55	9.47	9.38	9.29	9.20	9.11	9.02
6	13.75	10.92	9.78	9.15	8.75	8.47	8.26	8.10	7.98	7.87	7.72	7.56	7.40	7.31	7.23	7.14	7.06	6.97	6.88
7	12.25	9.55	8.45	7.85	7.46	7.19	6.99	6.84	6.72	6.62	6.47	6.31	6.16	6.07	5.99	5.91	5.82	5.74	5.65
8	11.26	8.65	7.59	7.01	6.63	6.37	6.18	6.03	5.91	5.81	5.67	5.52	5.36	5.28	5.20	5.12	5.03	4.95	4.86
9	10.56	8.02	6.99	6.42	6.06	5.80	5.61	5.47	5.35	5.26	5.11	4.96	4.81	4.73	4.65	4.57	4.48	4.40	4.31
10	10.04	7.56	6.55	5.99	5.64	5.39	5.20	5.06	4.94	4.85	4.71	4.56	4.41	4.33	4.25	4.17	4.08	4.00	3.91
11	9.65	7.21	6.22	5.67	5.32	5.07	4.89	4.74	4.63	4.54	4.40	4.25	4.10	4.02	3.94	3.86	3.78	3.69	3.60
12	9.33	6.93	5.95	5.41	5.06	4.82	4.64	4.50	4.39	4.30	4.16	4.01	3.86	3.78	3.70	3.62	3.54	3.45	3.36
13	9.07	6.70	5.74	5.21	4.86	4.62	4.44	4.30	4.19	4.10	3.96	3.82	3.66	3.59	3.51	3.43	3.34	3.25	3.17
14	8.86	6.51	5.56	5.04	4.69	4.46	4.28	4.14	4.03	3.94	3.80	3.66	3.51	3.43	3.35	3.27	3.18	3.09	3.00
15	8.68	6.36	5.42	4.89	4.56	4.32	4.14	4.00	3.89	3.80	3.67	3.52	3.37	3.29	3.21	3.13	3.05	2.96	2.87
16	8.53	6.23	5.29	4.77	4.44	4.20	4.03	3.89	3.78	3.69	3.55	3.41	3.26	3.18	3.10	3.02	2.93	2.84	2.75
17	8.40	6.11	5.18	4.67	4.34	4.10	3.93	3.79	3.68	3.59	3.46	3.31	3.16	3.08	3.00	2.92	2.83	2.75	2.65
18	8.29	6.01	5.09	4.58	4.25	4.01	3.84	3.71	3.60	3.51	3.37	3.23	3.08	3.00	2.92	2.84	2.75	2.66	2.57
19	8.18	5.93	5.01	4.50	4.17	3.94	3.77	3.63	3.52	3.43	3.30	3.15	3.00	2.92	2.84	2.76	2.67	2.58	2.49
20	8.10	5.85	4.94	4.43	4.10	3.87	3.70	3.56	3.46	3.37	3.23	3.09	2.94	2.86	2.78	2.69	2.61	2.52	2.42
21	8.02	5.78	4.87	4.37	4.04	3.81	3.64	3.51	3.40	3.31	3.17	3.03	2.88	2.80	2.72	2.64	2.55	2.46	2.36
22	7.95	5.72	4.82	4.31	3.99	3.76	3.59	3.45	3.35	3.26	3.12	2.98	2.83	2.75	2.67	2.58	2.50	2.40	2.31
23	7.88	5.66	4.76	4.26	3.94	3.71	3.54	3.41	3.30	3.21	3.07	2.93	2.78	2.70	2.62	2.54	2.45	2.35	2.26
24	7.82	5.61	4.72	4.22	3.90	3.67	3.50	3.36	3.26	3.17	3.03	2.89	2.74	2.66	2.58	2.49	2.40	2.31	2.21
25	7.77	5.57	4.68	4.18	3.85	3.63	3.46	3.32	3.22	3.13	2.99	2.85	2.70	2.62	2.54	2.45	2.36	2.27	2.17
26	7.72	5.53	4.64	4.14	3.82	3.59	3.42	3.29	3.18	3.09	2.96	2.81	2.66	2.58	2.50	2.42	2.33	2.23	2.13
27	7.68	5.49	4.60	4.11	3.78	3.56	3.39	3.26	3.15	3.06	2.93	2.78	2.63	2.55	2.47	2.38	2.29	2.20	2.10
28	7.64	5.45	4.57	4.07	3.75	3.53	3.36	3.23	3.12	3.03	2.90	2.75	2.60	2.52	2.44	2.35	2.26	2.17	2.06
29	7.60	5.42	4.54	4.04	3.73	3.50	3.33	3.20	3.09	3.00	2.87	2.73	2.57	2.49	2.41	2.33	2.23	2.14	2.03
30	7.56	5.39	4.51	4.02	3.70	3.47	3.30	3.17	3.07	2.98	2.84	2.70	2.55	2.47	2.39	2.30	2.21	2.11	2.01
40	7.31	5.18	4.31	3.83	3.51	3.29	3.12	2.99	2.89	2.80	2.66	2.52	2.37	2.29	2.20	2.11	2.02	1.92	1.80
60	7.08	4.98	4.13	3.65	3.34	3.12	2.95	2.82	2.72	2.63	2.50	2.35	2.20	2.12	2.03	1.94	1.84	1.73	1.60
120	6.85	4.79	3.95	3.48	3.17	2.96	2.79	2.66	2.56	2.47	2.34	2.19	2.03	1.95	1.86	1.76	1.66	1.53	1.38
∞	6.63	4.61	3.78	3.32	3.02	2.80	2.64	2.51	2.41	2.32	2.18	2.04	1.88	1.79	1.70	1.59	1.47	1.32	1.00

Degrees of Freedom for the Denominator (ν_2)

V. Operating Characteristic Curves for the Fixed Effects Model
Analysis of Variance[a]

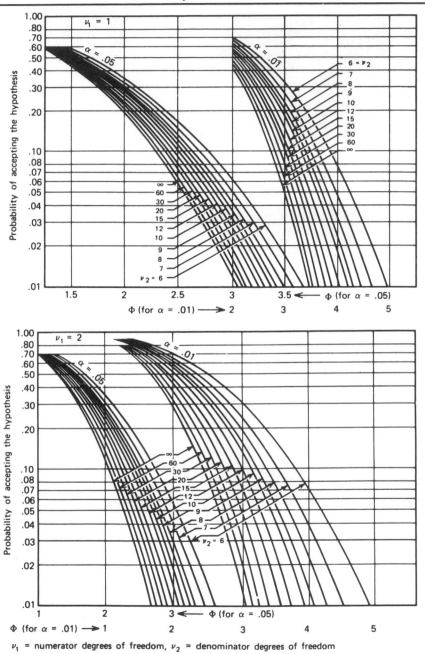

ν_1 = numerator degrees of freedom, ν_2 = denominator degrees of freedom

[a] Adapted with permission from *Biometrika Tables for Statisticians*, Vol. 2, by E. S. Pearson and H. O. Hartley, Cambridge University Press, Cambridge, 1972.

(continued)

V.   Operating Characteristic Curves for the Fixed Effects Model
Analysis of Variance (*continued*)

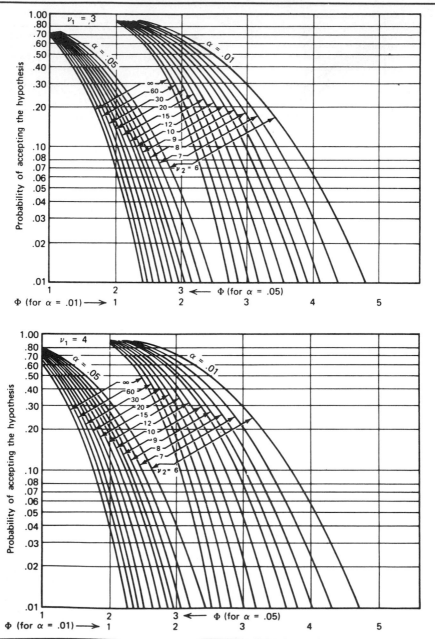

V. Operating Characteristic Curves for the Fixed Effects Model
Analysis of Variance (*continued*)

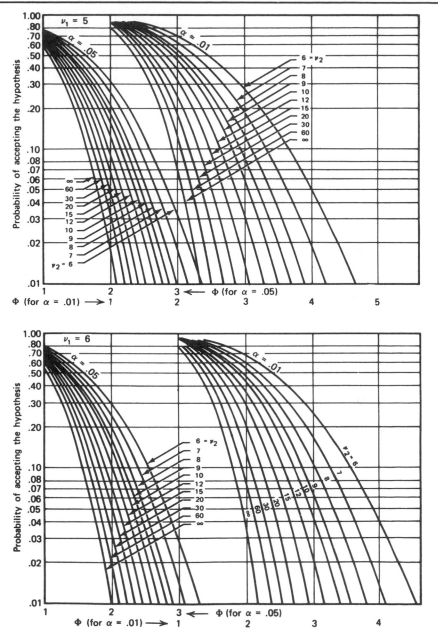

(continued)

V.   Operating Characteristic Curves for the Fixed Effects Model
Analysis of Variance (*continued*)

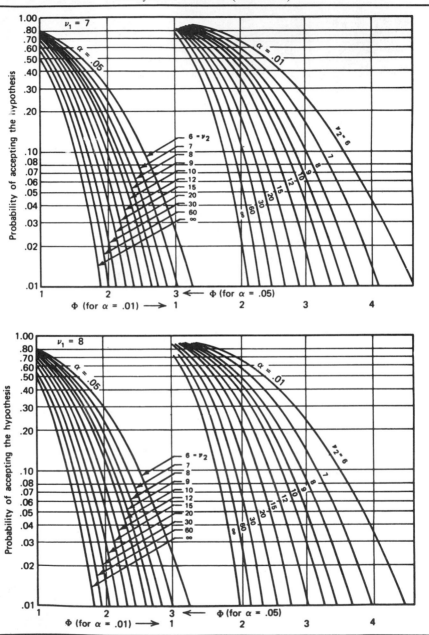

VI. Operating Characteristic Curves for the Random Effects Model
Analysis of Variance[a]

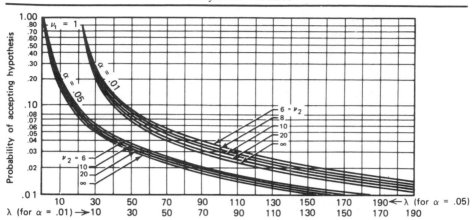

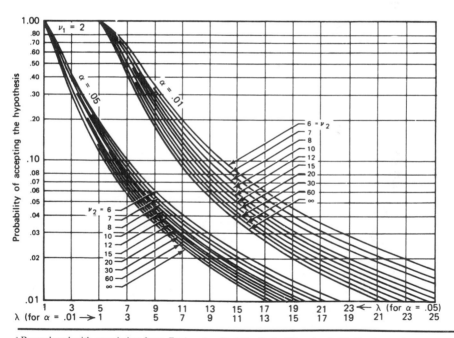

[a] Reproduced with permission from *Engineering Statistics,* 2nd edition, by A. H. Bowker and G. J. Lieberman, Prentice-Hall, Inc., Englewood Cliffs, N.J., 1972.

(continued)

VI.   Operating Characteristic Curves for the Random Effects Model
Analysis of Variance (*continued*)

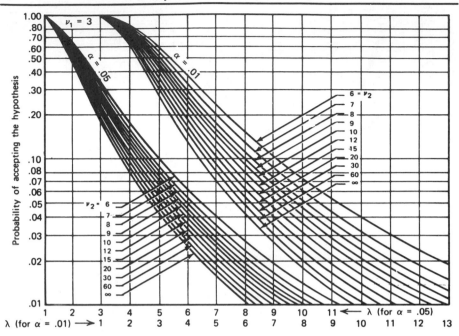

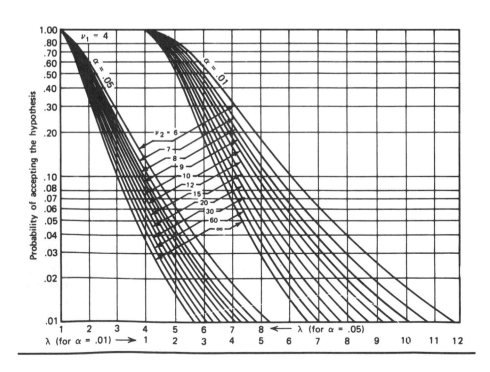

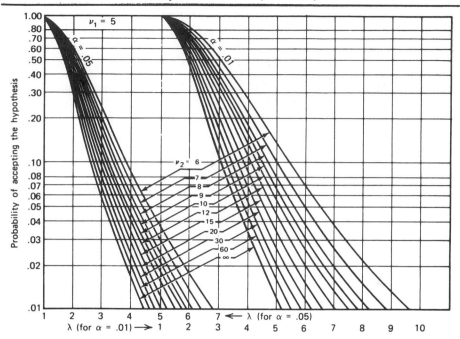

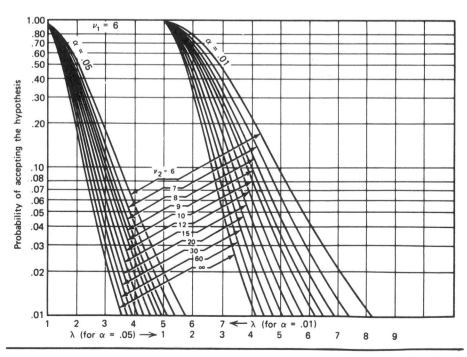

(continued)

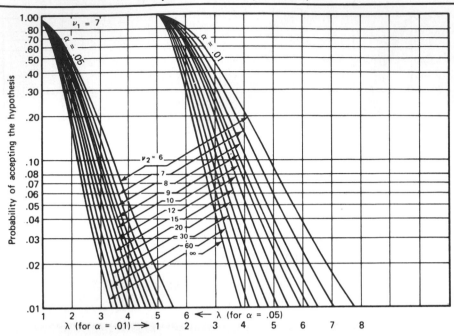

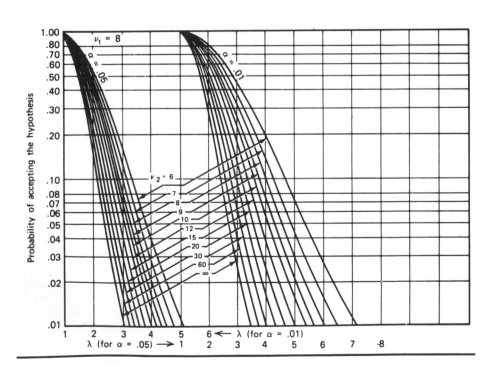

VII. Significant Ranges for Duncan's Multiple Range Test[a]

$$r_{0.01}(p, f)$$

f	\multicolumn{12}{c}{p}
	2	3	4	5	6	7	8	9	10	20	50	100
1	90.0	90.0	90.0	90.0	90.0	90.0	90.0	90.0	90.0	90.0	90.0	90.0
2	14.0	14.0	14.0	14.0	14.0	14.0	14.0	14.0	14.0	14.0	14.0	14.0
3	8.26	8.5	8.6	8.7	8.8	8.9	8.9	9.0	9.0	9.3	9.3	9.3
4	6.51	6.8	6.9	7.0	7.1	7.1	7.2	7.2	7.3	7.5	7.5	7.5
5	5.70	5.96	6.11	6.18	6.26	6.33	6.40	6.44	6.5	6.8	6.8	6.8
6	5.24	5.51	5.65	5.73	5.81	5.88	5.95	6.00	6.0	6.3	6.3	6.3
7	4.95	5.22	5.37	5.45	5.53	5.61	5.69	5.73	5.8	6.0	6.0	6.0
8	4.74	5.00	5.14	5.23	5.32	5.40	5.47	5.51	5.5	5.8	5.8	5.8
9	4.60	4.86	4.99	5.08	5.17	5.25	5.32	5.36	5.4	5.7	5.7	5.7
10	4.48	4.73	4.88	4.96	5.06	5.13	5.20	5.24	5.28	5.55	5.55	5.55
11	4.39	4.63	4.77	4.86	4.94	5.01	5.06	5.12	5.15	5.39	5.39	5.39
12	4.32	4.55	4.68	4.76	4.84	4.92	4.96	5.02	5.07	5.26	5.26	5.26
13	4.26	4.48	4.62	4.69	4.74	4.84	4.88	4.94	4.98	5.15	5.15	5.15
14	4.21	4.42	4.55	4.63	4.70	4.78	4.83	4.87	4.91	5.07	5.07	5.07
15	4.17	4.37	4.50	4.58	4.64	4.72	4.77	4.81	4.84	5.00	5.00	5.00
16	4.13	4.34	4.45	4.54	4.60	4.67	4.72	4.76	4.79	4.94	4.94	4.94
17	4.10	4.30	4.41	4.50	4.56	4.63	4.68	4.73	4.75	4.89	4.89	4.89
18	4.07	4.27	4.38	4.46	4.53	4.59	4.64	4.68	4.71	4.85	4.85	4.85
19	4.05	4.24	4.35	4.43	4.50	4.56	4.61	4.64	4.67	4.82	4.82	4.82
20	4.02	4.22	4.33	4.40	4.47	4.53	4.58	4.61	4.65	4.79	4.79	4.79
30	3.89	4.06	4.16	4.22	4.32	4.36	4.41	4.45	4.48	4.65	4.71	4.71
40	3.82	3.99	4.10	4.17	4.24	4.30	4.34	4.37	4.41	4.59	4.69	4.69
60	3.76	3.92	4.03	4.12	4.17	4.23	4.27	4.31	4.34	4.53	4.66	4.66
100	3.71	3.86	3.98	4.06	4.11	4.17	4.21	4.25	4.29	4.48	4.64	4.65
∞	3.64	3.80	3.90	3.98	4.04	4.09	4.14	4.17	4.20	4.41	4.60	4.68

f = degrees of freedom.
[a] Reproduced with permission from "Multiple Range and Multiple F Tests," by D. B. Duncan, *Biometrics,* Vol. 1, No. 1, pp. 1–42, 1955.

$$r_{0.05}(p, f)$$

f	\multicolumn{12}{c}{p}
	2	3	4	5	6	7	8	9	10	20	50	100
1	18.0	18.0	18.0	18.0	18.0	18.0	18.0	18.0	18.0	18.0	18.0	18.0
2	6.09	6.09	6.09	6.09	6.09	6.09	6.09	6.09	6.09	6.09	6.09	6.09
3	4.50	4.50	4.50	4.50	4.50	4.50	4.50	4.50	4.50	4.50	4.50	4.50
4	3.93	4.01	4.02	4.02	4.02	4.02	4.02	4.02	4.02	4.02	4.02	4.02
5	3.64	3.74	3.79	3.83	3.83	3.83	3.83	3.83	3.83	3.83	3.83	3.83
6	3.46	3.58	3.64	3.68	3.68	3.68	3.68	3.68	3.68	3.68	3.68	3.68
7	3.35	3.47	3.54	3.58	3.60	3.61	3.61	3.61	3.61	3.61	3.61	3.61
8	3.26	3.39	3.47	3.52	3.55	3.56	3.56	3.56	3.56	3.56	3.56	3.56
9	3.20	3.34	3.41	3.47	3.50	3.52	3.52	3.52	3.52	3.52	3.52	3.52
10	3.15	3.30	3.37	3.43	3.46	3.47	3.47	3.47	3.47	3.48	3.48	3.48
11	3.11	3.27	3.35	3.39	3.43	3.44	3.45	3.46	3.46	3.48	3.48	3.48
12	3.08	3.23	3.33	3.36	3.40	3.42	3.44	3.44	3.46	3.48	3.48	3.48
13	3.06	3.21	3.30	3.35	3.38	3.41	3.42	3.44	3.45	3.47	3.47	3.47
14	3.03	3.18	3.27	3.33	3.37	3.39	3.41	3.42	3.44	3.47	3.47	3.47
15	3.01	3.16	3.25	3.31	3.36	3.38	3.40	3.42	3.43	3.47	3.47	3.47
16	3.00	3.15	3.23	3.30	3.34	3.37	3.39	3.41	3.43	3.47	3.47	3.47
17	2.98	3.13	3.22	3.28	3.33	3.36	3.38	3.40	3.42	3.47	3.47	3.47
18	2.97	3.12	3.21	3.27	3.32	3.35	3.37	3.39	3.41	3.47	3.47	3.47
19	2.96	3.11	3.19	3.26	3.31	3.35	3.37	3.39	3.41	3.47	3.47	3.47
20	2.95	3.10	3.18	3.25	3.30	3.34	3.36	3.38	3.40	3.47	3.47	3.47
30	2.89	3.04	3.12	3.20	3.25	3.29	3.32	3.35	3.37	3.47	3.47	3.47
40	2.86	3.01	3.10	3.17	3.22	3.27	3.30	3.33	3.35	3.47	3.47	3.47
60	2.83	2.98	3.08	3.14	3.20	3.24	3.28	3.31	3.33	3.47	3.48	3.48
100	2.80	2.95	3.05	3.12	3.18	3.22	3.26	3.29	3.32	3.47	3.53	3.53
∞	2.77	2.92	3.02	3.09	3.15	3.19	3.23	3.26	3.29	3.47	3.61	3.67

VIII. Percentage Points of the Studentized Range Statistic[a]

$$q_{0.001}(p, f)$$

f	2	3	4	5	6	7	8	9	10	11	12	13	14	15	16	17	18	19	20
1	90.0	135	164	186	202	216	227	237	246	253	260	266	272	272	282	286	290	294	298
2	14.0	19.0	22.3	24.7	26.6	28.2	29.5	30.7	31.7	32.6	33.4	34.1	34.8	35.4	36.0	36.5	37.0	37.5	37.9
3	8.26	10.6	12.2	13.3	14.2	15.0	15.6	16.2	16.7	17.1	17.5	17.9	18.2	18.5	18.8	19.1	19.3	19.5	19.8
4	6.51	8.12	9.17	9.96	10.6	11.1	11.5	11.9	12.3	12.6	12.8	13.1	13.3	13.5	13.7	13.9	14.1	14.2	14.4
5	5.70	6.97	7.80	8.42	8.91	9.32	9.67	9.97	10.24	10.48	10.70	10.89	11.08	11.24	11.40	11.55	11.68	11.81	11.93
6	5.24	6.33	7.03	7.56	7.97	8.32	8.61	8.87	9.10	9.30	9.49	9.65	9.81	9.95	10.08	10.21	10.32	10.43	10.54
7	4.95	5.92	6.54	7.01	7.37	7.68	7.94	8.17	8.37	8.55	8.71	8.86	9.00	9.12	9.24	9.35	9.46	9.55	9.65
8	4.74	5.63	6.20	6.63	6.96	7.24	7.47	7.68	7.87	8.03	8.18	8.31	8.44	8.55	8.66	8.76	8.85	8.94	9.03
9	4.60	5.43	5.96	6.35	6.66	6.91	7.13	7.32	7.49	7.65	7.78	7.91	8.03	8.13	8.23	8.32	8.41	8.49	8.57
10	4.48	5.27	5.77	6.14	6.43	6.67	6.87	7.05	7.21	7.36	7.48	7.60	7.71	7.81	7.91	7.99	8.07	8.15	8.22
11	4.39	5.14	5.62	5.97	6.25	6.48	6.67	6.84	6.99	7.13	7.25	7.36	7.46	7.56	7.65	7.73	7.81	7.88	7.95
12	4.32	5.04	5.50	5.84	6.10	6.32	6.51	6.67	6.81	6.94	7.06	7.17	7.26	7.36	7.44	7.52	7.59	7.66	7.73
13	4.26	4.96	5.40	5.73	5.98	6.19	6.37	6.53	6.67	6.79	6.90	7.01	7.10	7.19	7.27	7.34	7.42	7.48	7.55
14	4.21	4.89	5.32	5.63	5.88	6.08	6.26	6.41	6.54	6.66	6.77	6.87	6.96	7.05	7.12	7.20	7.27	7.33	7.39
15	4.17	4.83	5.25	5.56	5.80	5.99	6.16	6.31	6.44	6.55	6.66	6.76	6.84	6.93	7.00	7.07	7.14	7.20	7.26
16	4.13	4.78	5.19	5.49	5.72	5.92	6.08	6.22	6.35	6.46	6.56	6.66	6.74	6.82	6.90	6.97	7.03	7.09	7.15
17	4.10	4.74	5.14	5.43	5.66	5.85	6.01	6.15	6.27	6.38	6.48	6.57	6.66	6.73	6.80	6.87	6.94	7.00	7.05
18	4.07	4.70	5.09	5.38	5.60	5.79	5.94	6.08	6.20	6.31	6.41	6.50	6.58	6.65	6.72	6.79	6.85	6.91	6.96
19	4.05	4.67	5.05	5.33	5.55	5.73	5.89	6.02	6.14	6.25	6.34	6.43	6.51	6.58	6.65	6.72	6.78	6.84	6.89
20	4.02	4.64	5.02	5.29	5.51	5.69	5.84	5.97	6.09	6.19	6.29	6.37	6.45	6.52	6.59	6.65	6.71	6.76	6.82
24	3.96	4.54	4.91	5.17	5.37	5.54	5.69	5.81	5.92	6.02	6.11	6.19	6.26	6.33	6.39	6.45	6.51	6.56	6.61
30	3.89	4.45	4.80	5.05	5.24	5.40	5.54	5.65	5.76	5.85	5.93	6.01	6.08	6.14	6.20	6.26	6.31	6.36	6.41
40	3.82	4.37	4.70	4.93	5.11	5.27	5.39	5.50	5.60	5.69	5.77	5.84	5.90	5.96	6.02	6.07	6.12	6.17	6.21
60	3.76	4.28	4.60	4.82	4.99	5.13	5.25	5.36	5.45	5.53	5.60	5.67	5.73	5.79	5.84	5.89	5.93	5.98	6.02
120	3.70	4.20	4.50	4.71	4.87	5.01	5.12	5.21	5.30	5.38	5.44	5.51	5.56	5.61	5.66	5.71	5.75	5.79	5.83
∞	3.64	4.12	4.40	4.60	4.76	4.88	4.99	5.08	5.16	5.23	5.29	5.35	5.40	5.45	5.49	5.54	5.57	5.61	5.65

f = degrees of freedom.

[a] From J. M. May, "Extended and Corrected Tables of the Upper Percentage Points of the Studentized Range," Biometrika, Vol. 39, pp. 192–193, 1952. Reproduced by permission of the trustees, of Biometrika.

VIII. Percentage Points of the Studentized Range Statistic (*continued*)

$$q_{0.05}(p, f)$$

f	p 2	3	4	5	6	7	8	9	10	11	12	13	14	15	16	17	18	19	20
1	18.1	26.7	32.8	37.2	40.5	43.1	45.4	47.3	49.1	50.6	51.9	53.2	54.3	55.4	56.3	57.2	58.0	58.8	59.6
2	6.09	8.28	9.80	10.89	11.73	12.43	13.03	13.54	13.99	14.39	14.75	15.08	15.38	15.65	15.91	16.14	16.36	16.57	16.77
3	4.50	5.88	6.83	7.51	8.04	8.47	8.85	9.18	9.46	9.72	9.95	10.16	10.35	10.52	10.69	10.84	10.98	11.12	11.24
4	3.93	5.00	5.76	6.31	6.73	7.06	7.35	7.60	7.83	8.03	8.21	8.37	8.52	8.67	8.80	8.92	9.03	9.14	9.24
5	3.64	4.60	5.22	5.67	6.03	6.33	6.58	6.80	6.99	7.17	7.32	7.47	7.60	7.72	7.83	7.93	8.03	8.12	8.21
6	3.46	4.34	4.90	5.31	5.63	5.89	6.12	6.32	6.49	6.65	6.79	6.92	7.04	7.14	7.24	7.34	7.43	7.51	7.59
7	3.34	4.16	4.68	5.06	5.35	5.59	5.80	5.99	6.15	6.29	6.42	6.54	6.65	6.75	6.84	6.93	7.01	7.08	7.16
8	3.26	4.04	4.53	4.89	5.17	5.40	5.60	5.77	5.92	6.05	6.18	6.29	6.39	6.48	6.57	6.65	6.73	6.80	6.87
9	3.20	3.95	4.42	4.76	5.02	5.24	5.43	5.60	5.74	5.87	5.98	6.09	6.19	6.28	6.36	6.44	6.51	6.58	6.65
10	3.15	3.88	4.33	4.66	4.91	5.12	5.30	5.46	5.60	5.72	5.83	5.93	6.03	6.12	6.20	6.27	6.34	6.41	6.47
11	3.11	3.82	4.26	4.58	4.82	5.03	5.20	5.35	5.49	5.61	5.71	5.81	5.90	5.98	6.06	6.14	6.20	6.27	6.33
12	3.08	3.77	4.20	4.51	4.75	4.95	5.12	5.27	5.40	5.51	5.61	5.71	5.80	5.88	5.95	6.02	6.09	6.15	6.21
13	3.06	3.73	4.15	4.46	4.69	4.88	5.05	5.19	5.32	5.43	5.53	5.63	5.71	5.79	5.86	5.93	6.00	6.06	6.11
14	3.03	3.70	4.11	4.41	4.64	4.83	4.99	5.13	5.25	5.36	5.46	5.56	5.64	5.72	5.79	5.86	5.92	5.98	6.03
15	3.01	3.67	4.08	4.37	4.59	4.78	4.94	5.08	5.20	5.31	5.40	5.49	5.57	5.65	5.72	5.79	5.85	5.91	5.96
16	3.00	3.65	4.05	4.34	4.56	4.74	4.90	5.03	5.15	5.26	5.35	5.44	5.52	5.59	5.66	5.73	5.79	5.84	5.90
17	2.98	3.62	4.02	4.31	4.52	4.70	4.86	4.99	5.11	5.21	5.31	5.39	5.47	5.55	5.61	5.68	5.74	5.79	5.84
18	2.97	3.61	4.00	4.28	4.49	4.67	4.83	4.96	5.07	5.17	5.27	5.35	5.43	5.50	5.57	5.63	5.69	5.74	5.79
19	2.96	3.59	3.98	4.26	4.47	4.64	4.79	4.92	5.04	5.14	5.23	5.32	5.39	5.46	5.53	5.59	5.65	5.70	5.75
20	2.95	3.58	3.96	4.24	4.45	4.62	4.77	4.90	5.01	5.11	5.20	5.28	5.36	5.43	5.50	5.56	5.61	5.66	5.71
24	2.92	3.53	3.90	4.17	4.37	4.54	4.68	4.81	4.92	5.01	5.10	5.18	5.25	5.32	5.38	5.44	5.50	5.55	5.59
30	2.89	3.48	3.84	4.11	4.30	4.46	4.60	4.72	4.83	4.92	5.00	5.08	5.15	5.21	5.27	5.33	5.38	5.43	5.48
40	2.86	3.44	3.79	4.04	4.23	4.39	4.52	4.63	4.74	4.82	4.90	4.98	5.05	5.11	5.17	5.22	5.27	5.32	5.36
60	2.83	3.40	3.74	3.98	4.16	4.31	4.44	4.55	4.65	4.73	4.81	4.88	4.94	5.00	5.06	5.11	5.15	5.20	5.24
120	2.80	3.36	3.69	3.92	4.10	4.24	4.36	4.47	4.56	4.64	4.71	4.78	4.84	4.90	4.95	5.00	5.04	5.09	5.13
∞	2.77	3.32	3.63	3.86	4.03	4.17	4.29	4.39	4.47	4.55	4.62	4.68	4.74	4.80	4.84	4.98	4.93	4.97	5.01

IX. Critical Values for Dunnett's Test for Comparing Treatments with a Control[a]
$$d_{0.05}(a - 1, f)$$
Two-Sided Comparisons

f	$a - 1 =$ Number of Treatment Means (excluding control)								
	1	2	3	4	5	6	7	8	9
5	2.57	3.03	3.29	3.48	3.62	3.73	3.82	3.90	3.97
6	2.45	2.86	3.10	3.26	3.39	3.49	3.57	3.64	3.71
7	2.36	2.75	2.97	3.12	3.24	3.33	3.41	3.47	3.53
8	2.31	2.67	2.88	3.02	3.13	3.22	3.29	3.35	3.41
9	2.26	2.61	2.81	2.95	3.05	3.14	3.20	3.26	3.32
10	2.23	2.57	2.76	2.89	2.99	3.07	3.14	3.19	3.24
11	2.20	2.53	2.72	2.84	2.94	3.02	3.08	3.14	3.19
12	2.18	2.50	2.68	2.81	2.90	2.98	3.04	3.09	3.14
13	2.16	2.48	2.65	2.78	2.87	2.94	3.00	3.06	3.10
14	2.14	2.46	2.63	2.75	2.84	2.91	2.97	3.02	3.07
15	2.13	2.44	2.61	2.73	2.82	2.89	2.95	3.00	3.04
16	2.12	2.42	2.59	2.71	2.80	2.87	2.92	2.97	3.02
17	2.11	2.41	2.58	2.69	2.78	2.85	2.90	2.95	3.00
18	2.10	2.40	2.56	2.68	2.76	2.83	2.89	2.94	2.98
19	2.09	2.39	2.55	2.66	2.75	2.81	2.87	2.92	2.96
20	2.09	2.38	2.54	2.65	2.73	2.80	2.86	2.90	2.95
24	2.06	2.35	2.51	2.61	2.70	2.76	2.81	2.86	2.90
30	2.04	2.32	2.47	2.58	2.66	2.72	2.77	2.82	2.86
40	2.02	2.29	2.44	2.54	2.62	2.68	2.73	2.77	2.81
60	2.00	2.27	2.41	2.51	2.58	2.64	2.69	2.73	2.77
120	1.98	2.24	2.38	2.47	2.55	2.60	2.65	2.69	2.73
∞	1.96	2.21	2.35	2.44	2.51	2.57	2.61	2.65	2.69

f = degrees of freedom.

[a] Reproduced with permission from C. W. Dunnett, "New Tables for Multiple Comparison with a Control," *Biometrics,* Vol. 20, No. 3, 1964, and from C. W. Dunnett, "A Multiple Comparison Procedure for Comparing Several Treatments with a Control," *Journal of the American Statistical Association,* Vol. 50, 1955.

IX.   Critical Values for Dunnett's Test for Comparing Treatments with a Control

$$d_{0.01}(a - 1, f)$$

Two-Sided Comparisons (*continued*)

	\multicolumn{9}{c}{$a - 1$ = Number of Treatment Means (excluding control)}
f	1	2	3	4	5	6	7	8	9
5	4.03	4.63	4.98	5.22	5.41	5.56	5.69	5.80	5.89
6	3.71	4.21	4.51	4.71	4.87	5.00	5.10	5.20	5.28
7	3.50	3.95	4.21	4.39	4.53	4.64	4.74	4.82	4.89
8	3.36	3.77	4.00	4.17	4.29	4.40	4.48	4.56	4.62
9	3.25	3.63	3.85	4.01	4.12	4.22	4.30	4.37	4.43
10	3.17	3.53	3.74	3.88	3.99	4.08	4.16	4.22	4.28
11	3.11	3.45	3.65	3.79	3.89	3.98	4.05	4.11	4.16
12	3.05	3.39	3.58	3.71	3.81	3.89	3.96	4.02	4.07
13	3.01	3.33	3.52	3.65	3.74	3.82	3.89	3.94	3.99
14	2.98	3.29	3.47	3.59	3.69	3.76	3.83	3.88	3.93
15	2.95	3.25	3.43	3.55	3.64	3.71	3.78	3.83	3.88
16	2.92	3.22	3.39	3.51	3.60	3.67	3.73	3.78	3.83
17	2.90	3.19	3.36	3.47	3.56	3.63	3.69	3.74	3.79
18	2.88	3.17	3.33	3.44	3.53	3.60	3.66	3.71	3.75
19	2.86	3.15	3.31	3.42	3.50	3.57	3.63	3.68	3.72
20	2.85	3.13	3.29	3.40	3.48	3.55	3.60	3.65	3.69
24	2.80	3.07	3.22	3.32	3.40	3.47	3.52	3.57	3.61
30	2.75	3.01	3.15	3.25	3.33	3.39	3.44	3.49	3.52
40	2.70	2.95	3.09	3.19	3.26	3.32	3.37	3.41	3.44
60	2.66	2.90	3.03	3.12	3.19	3.25	3.29	3.33	3.37
120	2.62	2.85	2.97	3.06	3.12	3.18	3.22	3.26	3.29
∞	2.58	2.79	2.92	3.00	3.06	3.11	3.15	3.19	3.22

$$d_{0.05}(a - 1, f)$$

One-Sided Comparisons

	\multicolumn{9}{c}{$a - 1$ = Number of Treatment Means (excluding control)}
f	1	2	3	4	5	6	7	8	9
5	2.02	2.44	2.68	2.85	2.98	3.08	3.16	3.24	3.30
6	1.94	2.34	2.56	2.71	2.83	2.92	3.00	3.07	3.12
7	1.89	2.27	2.48	2.62	2.73	2.82	2.89	2.95	3.01
8	1.86	2.22	2.42	2.55	2.66	2.74	2.81	2.87	2.92
9	1.83	2.18	2.37	2.50	2.60	2.68	2.75	2.81	2.86
10	1.81	2.15	2.34	2.47	2.56	2.64	2.70	2.76	2.81
11	1.80	2.13	2.31	2.44	2.53	2.60	2.67	2.72	2.77
12	1.78	2.11	2.29	2.41	2.50	2.58	2.64	2.69	2.74
13	1.77	2.09	2.27	2.39	2.48	2.55	2.61	2.66	2.71
14	1.76	2.08	2.25	2.37	2.46	2.53	2.59	2.64	2.69
15	1.75	2.07	2.24	2.36	2.44	2.51	2.57	2.62	2.67
16	1.75	2.06	2.23	2.34	2.43	2.50	2.56	2.61	2.65
17	1.74	2.05	2.22	2.33	2.42	2.49	2.54	2.59	2.64
18	1.73	2.04	2.21	2.32	2.41	2.48	2.53	2.58	2.62
19	1.73	2.03	2.20	2.31	2.40	2.47	2.52	2.57	2.61
20	1.72	2.03	2.19	2.30	2.39	2.46	2.51	2.56	2.60
24	1.71	2.01	2.17	2.28	2.36	2.43	2.48	2.53	2.57
30	1.70	1.99	2.15	2.25	2.33	2.40	2.45	2.50	2.54
40	1.68	1.97	2.13	2.23	2.31	2.37	2.42	2.47	2.51
60	1.67	1.95	2.10	2.21	2.28	2.35	2.39	2.44	2.48
120	1.66	1.93	2.08	2.18	2.26	2.32	2.37	2.41	2.45
∞	1.64	1.92	2.06	2.16	2.23	2.29	2.34	2.38	2.42

(continued)

IX. Critical Values for Dunnett's Test for Comparing Treatments with a Control
$$d_{0.01}(a - 1, f)$$
One-Sided Comparisons (*continued*)

f	$a - 1 =$ Number of Treatment Means (excluding control)								
	1	2	3	4	5	6	7	8	9
5	3.37	3.90	4.21	4.43	4.60	4.73	4.85	4.94	5.03
6	3.14	3.61	3.88	4.07	4.21	4.33	4.43	4.51	4.59
7	3.00	3.42	3.66	3.83	3.96	4.07	4.15	4.23	4.30
8	2.90	3.29	3.51	3.67	3.79	3.88	3.96	4.03	4.09
9	2.82	3.19	3.40	3.55	3.66	3.75	3.82	3.89	3.94
10	2.76	3.11	3.31	3.45	3.56	3.64	3.71	3.78	3.83
11	2.72	3.06	3.25	3.38	3.48	3.56	3.63	3.69	3.74
12	2.68	3.01	3.19	3.32	3.42	3.50	3.56	3.62	3.67
13	2.65	2.97	3.15	3.27	3.37	3.44	3.51	3.56	3.61
14	2.62	2.94	3.11	3.23	3.32	3.40	3.46	3.51	3.56
15	2.60	2.91	3.08	3.20	3.29	3.36	3.42	3.47	3.52
16	2.58	2.88	3.05	3.17	3.26	3.33	3.39	3.44	3.48
17	2.57	2.86	3.03	3.14	3.23	3.30	3.36	3.41	3.45
18	2.55	2.84	3.01	3.12	3.21	3.27	3.33	3.38	3.42
19	2.54	2.83	2.99	3.10	3.18	3.25	3.31	3.36	3.40
20	2.53	2.81	2.97	3.08	3.17	3.23	3.29	3.34	3.38
24	2.49	2.77	2.92	3.03	3.11	3.17	3.22	3.27	3.31
30	2.46	2.72	2.87	2.97	3.05	3.11	3.16	3.21	3.24
40	2.42	2.68	2.82	2.92	2.99	3.05	3.10	3.14	3.18
60	2.39	2.64	2.78	2.87	2.94	3.00	3.04	3.08	3.12
120	2.36	2.60	2.73	2.82	2.89	2.94	2.99	3.03	3.06
∞	2.33	2.56	2.68	2.77	2.84	2.89	2.93	2.97	3.00

X. Coefficients of Orthogonal Polynomials[a]

n = 3

X_j	P_1	P_2
1	-1	1
2	0	-2
3	1	1
$\sum_{j=1}^{n}\{P_i(X_j)\}^2$	2	6
λ	1	3

n = 4

X_j	P_1	P_2	P_3
1	-3	1	-1
2	-1	-1	3
3	1	-1	-3
4	3	1	1
$\sum_{j=1}^{n}\{P_i(X_j)\}^2$	20	4	20
λ	2	1	$\frac{10}{3}$

n = 5

X_j	P_1	P_2	P_3	P_4
1	-2	2	-1	1
2	-1	-1	2	-4
3	0	-2	0	6
4	1	-1	-2	-4
5	2	2	1	1
$\sum_{j=1}^{n}\{P_i(X_j)\}^2$	10	14	10	70
λ	1	1	$\frac{5}{6}$	$\frac{35}{12}$

n = 6

X_j	P_1	P_2	P_3	P_4	P_5
1	-5	5	-5	1	-1
2	-3	-1	7	-3	5
3	-1	-4	4	2	-10
4	1	-4	-4	2	10
5	3	-1	-7	-3	-5
6	5	5	5	1	1
$\sum_{j=1}^{n}\{P_i(X_j)\}^2$	70	84	180	28	252
λ	2	$\frac{3}{2}$	$\frac{5}{3}$	$\frac{7}{12}$	$\frac{21}{10}$

n = 7

X_j	P_1	P_2	P_3	P_4	P_5	P_6
1	-3	5	-1	3	-1	1
2	-2	0	1	-7	4	-6
3	-1	-3	1	1	-5	15
4	0	-4	0	6	0	-20
5	1	-3	-1	1	5	15
6	2	0	-1	-7	-4	-6
7	3	5	1	3	1	1
$\sum_{j=1}^{n}\{P_i(X_j)\}^2$	28	84	6	154	84	924
λ	1	1	$\frac{1}{6}$	$\frac{7}{12}$	$\frac{7}{20}$	$\frac{77}{60}$

n = 8

X_j	P_1	P_2	P_3	P_4	P_5	P_6
1	-7	7	-7	7	-7	1
2	-5	1	5	-13	23	-5
3	-3	-3	7	-3	-17	9
4	-1	-5	3	9	-15	-5
5	1	-5	-3	9	15	-5
6	3	-3	-7	-3	17	9
7	5	1	-5	-13	-23	-5
8	7	7	7	7	7	1
$\sum_{j=1}^{n}\{P_i(X_j)\}^2$	168	168	264	616	2184	264
λ	2	1	$\frac{2}{3}$	$\frac{7}{12}$	$\frac{7}{10}$	$\frac{11}{60}$

n = 9

X_j	P_1	P_2	P_3	P_4	P_5	P_6
1	-4	28	-14	14	-4	4
2	-3	7	7	-21	11	-17
3	-2	-8	13	-11	-4	22
4	-1	-17	9	9	-9	1
5	0	-20	0	18	0	-20
6	1	-17	-9	9	9	1
7	2	-8	-13	-11	4	22
8	3	7	-7	-21	-11	-17
9	4	28	14	14	4	4
$\sum_{j=1}^{n}\{P_i(X_j)\}^2$	60	2772	990	2002	468	1980
λ	1	3	$\frac{5}{6}$	$\frac{7}{12}$	$\frac{3}{20}$	$\frac{11}{60}$

n = 10

X_j	P_1	P_2	P_3	P_4	P_5	P_6
1	-9	6	-42	18	-6	3
2	-7	2	14	-22	14	-11
3	-5	-1	35	-17	-1	10
4	-3	-3	31	3	-11	6
5	-1	-4	12	18	-6	-8
6	1	-4	-12	18	6	-8
7	3	-3	-31	3	11	6
8	5	-1	-35	-17	1	10
9	7	2	-14	-22	-14	-11
10	9	6	42	18	6	3
$\sum_{j=1}^{n}\{P_i(X_j)\}^2$	330	132	8580	2860	780	660
λ	2	$\frac{1}{2}$	$\frac{5}{3}$	$\frac{5}{12}$	$\frac{1}{10}$	$\frac{11}{240}$

[a] Adapted with permission from *Biometrika Tables for Statisticians*, Vol. 1, 3rd edition by E. S. Pearson and H. O. Hartley, Cambridge University Press, Cambridge, 1966.

XI.   Random Numbers[a]

10480	15011	01536	02011	87647	91646	69179	14194	62590
22368	46573	25595	85393	30995	89198	27982	53402	93965
24130	48360	22527	97265	76393	64809	15179	24830	49340
42167	93093	06243	61680	07856	16376	39440	53537	71341
37570	39975	81837	16656	06121	91782	60468	81305	49684
77921	06907	11008	42751	27756	53498	18602	70659	90655
99562	72905	56420	69994	98872	31016	71194	18738	44013
96301	91977	05463	07972	18876	20922	94595	56869	69014
89579	14342	63661	10281	17453	18103	57740	84378	25331
85475	36857	53342	53988	53060	59533	38867	62300	08158
28918	69578	88231	33276	70997	79936	56865	05859	90106
63553	40961	48235	03427	49626	69445	18663	72695	52180
09429	93969	52636	92737	88974	33488	36320	17617	30015
10365	61129	87529	85689	48237	52267	67689	93394	01511
07119	97336	71048	08178	77233	13976	47564	81056	97735
51085	12765	51821	51259	77452	16308	60756	92144	49442
02368	21382	52404	60268	89368	19885	55322	44819	01188
01011	54092	33362	94904	31273	04146	18594	29852	71585
52162	53916	46369	58586	23216	14513	83149	98736	23495
07056	97628	33787	09998	42698	06691	76988	13602	51851
48663	91245	85828	14346	09172	30168	90229	04734	59193
54164	58492	22421	74103	47070	25306	76468	26384	58151
32639	32363	05597	24200	13363	38005	94342	28728	35806
29334	27001	87637	87308	58731	00256	45834	15398	46557
02488	33062	28834	07351	19731	92420	60952	61280	50001
81525	72295	04839	96423	24878	82651	66566	14778	76797
29676	20591	68086	26432	46901	20849	89768	81536	86645
00742	57392	39064	66432	84673	40027	32832	61362	98947
05366	04213	25669	26422	44407	44048	37937	63904	45766
91921	26418	64117	94305	26766	25940	39972	22209	71500
00582	04711	87917	77341	42206	35126	74087	99547	81817
00725	69884	62797	56170	86324	88072	76222	36086	84637
69011	65795	95876	55293	18988	27354	26575	08625	40801
25976	57948	29888	88604	67917	48708	18912	82271	65424
09763	83473	73577	12908	30883	18317	28290	35797	05998
91567	42595	27958	30134	04024	86385	29880	99730	55536
17955	56349	90999	49127	20044	59931	06115	20542	18059
46503	18584	18845	49618	02304	51038	20655	58727	28168
92157	89634	94824	78171	84610	82834	09922	25417	44137
14577	62765	35605	81263	39667	47358	56873	56307	67607

[a] Reproduced with permission from *Probability and Statistics in Engineering and Management Science,* 3rd edition, by W. W. Hines and D. C. Montgomery, Wiley, New York, 1990.

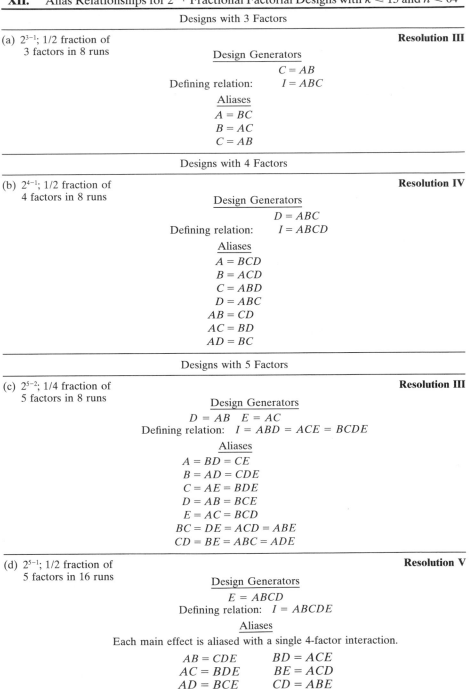

Designs with 3 Factors

(a) 2^{3-1}; 1/2 fraction of
3 factors in 8 runs

Resolution III

Design Generators

$$C = AB$$
Defining relation: $\quad I = ABC$

Aliases

$$A = BC$$
$$B = AC$$
$$C = AB$$

Designs with 4 Factors

(b) 2^{4-1}; 1/2 fraction of
4 factors in 8 runs

Resolution IV

Design Generators

$$D = ABC$$
Defining relation: $\quad I = ABCD$

Aliases

$$A = BCD$$
$$B = ACD$$
$$C = ABD$$
$$D = ABC$$
$$AB = CD$$
$$AC = BD$$
$$AD = BC$$

Designs with 5 Factors

(c) 2^{5-2}; 1/4 fraction of
5 factors in 8 runs

Resolution III

Design Generators

$$D = AB \quad E = AC$$
Defining relation: $\quad I = ABD = ACE = BCDE$

Aliases

$$A = BD = CE$$
$$B = AD = CDE$$
$$C = AE = BDE$$
$$D = AB = BCE$$
$$E = AC = BCD$$
$$BC = DE = ACD = ABE$$
$$CD = BE = ABC = ADE$$

(d) 2^{5-1}; 1/2 fraction of
5 factors in 16 runs

Resolution V

Design Generators

$$E = ABCD$$
Defining relation: $\quad I = ABCDE$

Aliases

Each main effect is aliased with a single 4-factor interaction.

$AB = CDE$	$BD = ACE$
$AC = BDE$	$BE = ACD$
$AD = BCE$	$CD = ABE$
$AE = BCD$	$CE = ABD$
$BC = ADE$	$DE = ABC$

2 blocks of 8: $AB = CDE$

(continued)

XII. Alias Relationships for 2^{k-p} Fractional Factorial Designs with $k \leq 15$ and $n \leq 64$ (*continued*)

Designs with 6 Factors

(e) 2^{6-3}; 1/8 fraction of 6 factors in 8 runs — Resolution III

Design Generators

$$D = AB \quad E = AC \quad F = BC$$

Defining relation: $I = ABD = ACE = BCF = ACDF = ABEF = DEF$

Aliases

$A = BD = CE = CDF = BEF \qquad E = AC = DF = BCD = ABF$
$B = AD = CF = CDE = AEF \qquad F = BC = DE = ACD = ABE$
$C = AE = BF = BDE = ADF \qquad CD = BE = AF = ABC = ADE = BDF = CEF$
$D = AB = EF = BCE = ACF$

(f) 2^{6-2}; 1/4 fraction of 6 factors in 16 runs — Resolution IV

Design Generators

$$E = ABC \quad F = BCD$$

Defining relation: $I = ABCE = BCDF = ADEF$

Aliases

$A = BCE = DEF \qquad\qquad AB = CE$
$B = ACE = CDF \qquad\qquad AC = BE$
$C = ABE = BDF \qquad\qquad AD = EF$
$D = BCF = AEF \qquad\qquad AE = BC = DF$
$E = ABC = ADF \qquad\qquad AF = DE$
$F = BCD = ADE \qquad\qquad BD = CF$
$ABD = CDE = ACF = BEF \qquad BF = CD$
$ACD = BDE = ABF = CEF$

2 blocks of 8: $ABD = CDE = ACF = BEF$

(g) 2^{6-1}; 1/2 fraction of 6 factors in 32 runs

Design Generators

$$F = ABCDE$$

Defining relation: $I = ABCDEF$

Aliases

Each main effect is aliased with a single 5-factor interaction.
Each 2-factor interaction is aliased with a single 4-factor interaction.

$ABC = DEF$	$ACE = BDF$
$ABD = CEF$	$ACF = BDE$
$ABE = CDF$	$ADE = BCF$
$ABF = CDE$	$ADF = BCE$
$ACD = BEF$	$AEF = BCD$

2 blocks of 16: $ABF = CDE$

4 blocks of 8: BC

$ABF = CDE$
$ACF = BDE$

Designs with 7 Factors

(h) 2^{7-4}, 1/16 fraction of 7 factors in 8 runs

Design Generators

$$D = AB \quad E = AC \quad F = BC \quad G = ABC$$

Defining relation: $I = ABD = ACE = BCDE = BCF = ACDF = ABEF = DEF = ABCG$
$= CDG = BEG = ADEG = AFG = BDFG = CEFG = ABCDEFG$

Aliases

$A = BD = CE = FG$	$E = AC = DF = BG$
$B = AD = CF = EG$	$F = BC = DE = AG$
$C = AE = BF = DG$	$G = CD = BE = AF$
$D = AB = EF = CG$	

(continued)

XII. Alias Relationships for 2^{k-p} Fractional Factorial Designs with $k \leq 15$ and $n \leq 64$ (*continued*)

Resolution IV

(i) 2^{7-3}; 1/8 fraction of 7 factors in 16 runs

Design Generators

$$E = ABC \quad F = BCD \quad G = ACD$$

Defining relation: $I = ABCE = BCDF = ADEF = ACDG = BDEG = ABFG = CEFG$

Aliases

$A = BCE = DEF = CDG = BFG$	$E = ABC = ADF = BDG = CFG$	
$B = ACE = CDF = DEG = AFG$	$F = BCD = ADE = ABG = CEG$	
$C = ABE = BDF = ADG = EFG$	$G = ACD = BDE = ABF = CEF$	
$D = BCF = AEF = ACG = BEG$		

$AB = CE = FG$	$AF = DE = BG$	
$AC = BE = DG$	$AG = CD = BF$	
$AD = EF = CG$	$BD = CF = EG$	
$AE = BC = DF$		

$ABD = CDE = ACF = BEF = BCG = AEG = DFG$

2 blocks of 8: $ABD = CDE = ACF = BEF = BCG = AEG = DFG$

(j) 2^{7-2}; 1/4 fraction of 7 factors in 32 runs

Resolution IV

Design Generators

$$F = ABCD \quad G = ABDE$$

Defining relation: $I = ABCDF = ABDEG = CEFG$

Aliases

$A =$	$AB = CDF = DEG$	$BC = ADF$	$CE = FG$	$ACE = AFG$
$B =$	$AC = BDF$	$BD = ACF = AEG$	$CF = ABD = EG$	$ACG = AEF$
$C = EFG$	$AD = BCF = BEG$	$BE = ADG$	$CG = EF$	$BCE = BFG$
$D =$	$AE = BDG$	$BF = ACD$	$DE = ABG$	$BCG = BEF$
$E = CFG$	$AF = BCD$	$BG = ADE$	$DF = ABC$	$CDE = DFG$
$F = CEG$	$AG = BDE$	$CD = ABF$	$DG = ABE$	$CDG = DEF$
$G = CEF$				

2 blocks of 16: $ACE = AFG$ 4 blocks of 8: $ACE = AFG$

$BCE = BFG$

$AB = CDF = DEG$

(k) 2^{7-1}; 1/2 fraction of 7 factors in 64 runs

Design Generators

$$G = ABCDEF$$

Defining relation: $I = ABCDEFG$

Aliases

Each main effect is aliased with a single 6-factor interaction.

Each 2-factor interaction is aliased with a single 5-factor interaction.
Each 3-factor interaction is aliased with a single 4-factor interaction.

2 blocks of 32: ABC	4 blocks of 16: ABC
	CEF
	CDG

Designs with 8 Factors

(l) 2^{8-4}; 1/16 fraction of 8 factors in 16 runs

Design Generators

$$E = BCD \quad F = ACD \quad G = ABC \quad H = ABD$$

Defining relation: $I = BCDE = ACDF = ABEF = ABCG = ADEG = BDFG = CEFG = ABDH$
$= ACEH = BCFH = DEFH = CDGH = BEGH = AFGH = ABCDEFGH$

Aliases

$A = CDF = BEF = BCG = DEG = BDH = CEH = FGH$	$AB = EF = CG = DH$
$B = CDE = AEF = ACG = DFG = ADH = CFH = EGH$	$AC = DF = BG = EH$
$C = BDE = ADF = ABG = EFG = AEH = BFH = DGH$	$AD = CF = EG = BH$
$D = BCE = ACF = AEG = BFG = ABH = EFH = CGH$	$AE = BF = DG = CH$
$E = BCD = ABF = ADG = CFG = ACH = DFH = BGH$	$AF = CD = BE = GH$
$F = ACD = ABE = BDG = CEG = BCH = DEH = AGH$	$AG = BC = DE = FH$
$G = ABC = ADE = BDF = CEF = CDH = BEH = AFH$	$AH = BD = CE = FG$
$H = ABD = ACE = BCF = DEF = CDG = BEG = AFG$	

2 blocks of 8: $AB = EF = CG = DH$

(continued)

XII. Alias Relationships for 2^{k-p} Fractional Factorial Designs with $k \leq 15$ and $n \leq 64$ (*continued*)

(m) 2^{8-3}; 1/8 fraction of 8 factors in 32 runs   **Resolution IV**

Design Generators

$$F = ABC \quad G = ABD \quad H = BCDE$$

Defining relation: $I = ABCF = ABDG = CDFG = BCDEH = ADEFH = ACEGH = BEFGH$

Aliases

$A = BCF = BDG$	$AE = DFH = CGH$	$DE = BCH = AFH$
$B = ACF = ADG$	$AF = BC = DEH$	$DH = BCE = AEF$
$C = ABF = DFG$	$AG = BD = CEH$	$EF = ADH = BGH$
$D = ABG = CFG$	$AH = DEF = CEG$	$EG = ACH = BFH$
$E =$	$BE = CDH = FGH$	$EH = BCD = ADF = ACG = BFG$
$F = ABC = CDG$	$BH = CDE = EFG$	$FH = ADE = BEG$
$G = ABD = CDF$	$CD = FG = BEH$	$GH = ACE = BEF$
$H =$	$CE = BDH = AGH$	$ABE = CEF = DEG$
$AB = CF = DG$	$CG = DF = AEH$	$ABH = CFH = DGH$
$AC = BF = EGH$	$CH = BDE = AEG$	$ACD = BDF = BCG = AFG$
$AD = BG = EFH$		

2 blocks of 16: $ABE = CEF = DEG$ 4 blocks of 8: $ABE = CEF = DEG$

                                                   $ABH = CFH = DGH$

                                                   $EH = BCD = ADF = ACG = BFG$

(n) 2^{8-2}; 1/4 fraction of 8 factors in 64 runs

<u>Design Generators</u>

$G = ABCD$ $H = ABEF$

Defining relation: $I = ABCDG = ABEFH = CDEFGH$

<u>Aliases</u>

AB = CDG = EFH	BG = ACD	EF = ABH	ADH =	BFG =
AC = BDG	BH = AEF	EG =	AEG =	BGH =
AD = BCG	CD = ABG	EH = ABF	AFG =	CDE = FGH
AE = BFH	CE =	FG =	AGH =	CDF = EGH
AF = BEH	CF =	FH = ABE	BCE =	CDH = EFG
AG = BCD	CG = ABD	GH =	BCF =	CEF = DGH
AH = BEF	CH =	ACE =	BCH =	CEG = DFH
BC = ADG	DE =	ACF =	BDE =	CEH = DFG
BD = ACG	DF =	ACH =	BDF =	CFG = DEH
BE = AFH	DG = ABC	ADE =	BDH =	CFH = DEG
BF = AEH	DH =	ADF =	BEG =	CGH = DEF

2 blocks of 32: CDE = FGH   4 blocks of 16: CDE = FGH
ACF
BDH

(continued)

XII. Alias Relationships for 2^{k-p} Fractional Factorial Designs with $k \leq 15$ and $n \leq 64$ (*continued*)

Designs with 9 Factors

(o) 2^{9-5}; 1/32 fraction of 9 factors in 16 runs   **Resolution III**

Design Generators

$E = ABC \quad F = BCD \quad G = ACD \quad H = ABD \quad J = ABCD$

Defining relation:

$I = ABCE = BCDF = ADEF = ACDG = BDEG = ABFG = CEFG = ABDH$
$\quad = CDEH = ACFH = BEFH = BCGH = AEGH = DFGH = ABCDEFGH = ABCDJ$
$\quad = DEJ = AFJ = BCEFJ = BGJ = ACEGJ = CDFGJ = ABDEFGJ = CHJ$
$\quad = ABEHJ = BDFHJ = ABCDEFHJ = ADGHJ = BCDEGHJ = ABCFGHJ = EFGHJ$

Aliases

$A = FJ$

$B = GJ$

$C = HJ$

$D = EJ$

$E = DJ$

$F = AJ$

$G = BJ$

$H = CJ$

$J = DE = AF = BG = CH$

$AB = CE = FG = DH$

$AC = BE = DG = FH$

$AD = EF = CG = BH$

$AE = BC = DF = GH$

$AG = CD = BF = EH$

$AH = BD = CF = EG$

2 blocks of 8: $AB = CE = FG = DH$

(p) 2^{9-4}; 1/16 fraction of 9 factors in 32 runs

Design Generators

$$F = BCDE \quad G = ACDE \quad H = ABDE \quad J = ABCE$$

Defining relation: $I = BCDEF = ACDEG = ABFG = ABDEH = ACFH = BCGH = DEFGH = ABCEJ$
$$= ADFJ = BDGJ = CEFGJ = CDHJ = BEFHJ = AEGHJ = ABCDFGHJ$$

Aliases

$A = BFG = CFH = DFJ$	$AD = CEG = BEH = FJ$	$BJ = ACE = DG = EFH$
$B = AFG = CGH = DGJ$	$AE = CDG = BDH = BCJ = GHJ$	$CD = BEF = AEG = HJ$
$C = AFH = BGH = DHJ$	$AF = BG = CH = DJ$	$CE = BDF = ADG = ABJ = FGJ$
$D = AFJ = BGJ = CHJ$	$AG = CDE = BF = EHJ$	$CJ = ABE = EFG = DH$
$E =$	$AH = BDE = CF = EGJ$	$DE = BCF = ACG = ABH = FGH$
$F = ABG = ACH = ADJ$	$AJ = BCE = DF = EGH$	$EF = BCD = DGH = CGJ = BHJ$
$G = ABF = BCH = BDJ$	$BC = DEF = GH = AEJ$	$EG = ACD = DFH = CFJ = AHJ$
$H = ACF = BCG = CDJ$	$BD = CEF = AEH = GJ$	$EH = ABD = DFG = BFJ = AGJ$
$J = ADF = BDG = CDH$	$BE = CDF = ADH = ACJ = FHJ$	$EJ = ABC = CFG = BFH = AGH$
$AB = FG = DEH = CEJ$	$BH = ADE = CG = EFJ$	$AEF = BEG = CEH = DEJ$
$AC = DEG = FH = BEJ$		

2 blocks of 16: $AEF = BEG = CEH = DEJ$ 4 blocks of 8: $AEF = BEG = CEH = DEJ$
$$AB = FG = DEH = CEJ$$
$$CD = BEF = AEG = HJ$$

(continued)

XII. Alias Relationships for 2^{k-p} Fractional Factorial Designs with $k \leqslant 15$ and $n \leqslant 64$ (*continued*)

(q) 2^{9-3}, 1/8 fraction of 9 factors in 64 runs   Resolution IV

Design Generators

$$G = ABCD \quad H = ACEF \quad J = CDEF$$

Defining relation: $I = ABCDG = ACEFH = BDEFGH = CDEFJ = ABEFGJ = ADHJ = BCGHJ$

Aliases

$A = DHJ$	$AC = BDG = EFH$	$BF =$
$B =$	$AD = BCG = HJ$	$BG = ACD = CHJ$
$C =$	$AE = CFH$	$BH = CGJ$
$D = AHJ$	$AF = CEH$	$BJ = CGH$
$E =$	$AG = BCD$	$CD = ABG = EFJ$
$F =$	$AH = CEF = DJ$	$CE = AFH = DFJ$
$G =$	$AJ = DH$	$CF = AEH = DEJ$
$H = ADJ$	$BC = ADG = GHJ$	$CG = ABD = BHJ$
$J = ADH$	$BD = ACG$	$CH = AEF = BGJ$
$AB = CDG$	$BE =$	$CJ = DEF = BGH$
$DE = CFJ$	$GJ = BCH$	$AFJ = BEG = DFH$
$DF = CEJ$	$ABE = FGJ$	$AGH = DGJ$
$DG = ABC$	$ABF = EGJ$	$AGJ = BEF = DGH$
$EF = ACH = CDJ$	$ABH = BDJ$	$BCE =$
$EG =$	$ABJ = EFG = BDH$	$BCF =$
$EH = ACF$	$ACJ = CDH$	$BDE = FGH$
$EJ = CDF$	$ADE = EHJ$	$BDF = EGH$
$FG =$	$ADF = FHJ$	$BEH = DFG$
$FH = ACE$	$AEG = BFJ$	$BFH = DEG$
$FJ = CDE$	$AEJ = BFG = DEH$	$CEG =$
$GH = BCJ$	$AFG = BEJ$	$CFG =$

2 blocks of 32: *CFG*   4 blocks of 16: *CFG*

$AGJ = BEF = DGH$

$ADE = EHJ$

Designs with 10 Factors

(r) 2^{10-6}, 1/64 fraction of 10 factors in 16 runs

Design Generators

$E = ABC \quad F = BCD \quad G = ACD \quad H = ABD \quad J = ABCD \quad K = AB$

Defining relation:

$I = ABCE = BCDF = ADEF = ACDG = BDEG = ABFG = CEFG = ABDH$
$= CDEH = ACFH = BEFH = BCGH = AEGH = DFGH = ABCDEFGH = ABCDJ$
$= DEJ = AFJ = BCEFJ = BGJ = ACEGJ = CDFGJ = ABDEFGJ = CHJ$
$= ABEHJ = BDFHJ = ACDEFHJ = ADGHJ = BCDEGHJ = ABCFGHJ = EFGHJ = ABK$
$= CEK = ACDFK = BDEFK = BCDGK = ADEGK = FGK = ABCEFGK = DHK$
$= ABCDEHK = BCFHK = AEFHK = ACGHK = BEGHK = ABDFGHK = CDEFGHK = CDJK$
$= ABDEJK = BFJK = ACEFJK = AGJK = BCEGJK = ABCDFGJK = DEFGJK = ABCHJK$
$= EHJK = ADFHJK = BCDEFHJK = BDGHJK = ACDEGHJK = CFGHJK = ABEFGHJK$

Aliases

$A = FJ = BK$	$J = DE = AF = BG = CH$
$B = GJ = AK$	$K = AB = CE = FG = DH$
$C = HJ = EK$	$AC = BE = DG = FH$
$D = EJ = HK$	$AD = EF = CG = BH$
$E = DJ = CK$	$AE = BC = DF = GH$
$F = AJ = GK$	$AG = CD = BF = EH = JK$
$G = BJ = FK$	$AH = BD = CF = EG$
$H = CJ = DK$	

2 blocks of 8: $AG = CD = BF = EH = JK$

(continued)

XII. Alias Relationships for 2^{k-p} Fractional Factorial Designs with $k \leqslant 15$ and $n \leqslant 64$ (*continued*)

(s) 2^{10-5}; 1/32 fraction of 10 factors in 32 runs

Design Generators

$$F = ABCD \quad G = ABCE \quad H = ABDE \quad J = ACDE \quad K = BCDE$$

Defining relation:
$I = ABCDF = ABCEG = DEFG = ABDEH = CEFH = CDGH = ABFGH = ACDEJ$
$= BEFJ = BDGJ = ACFGJ = BCHJ = ADFHJ = AEGHJ = BCDEFGHJ = BCDEK$
$= AEFK = ADGK = BCFGK = ACHK = BDFHK = BEGHK = ACDEFGHK = ABJK$
$= CDFJK = CEGJK = ABDEFGJK = DEHJK = ABCEFHJK = ABCDGHJK = FGHJK$

Aliases

$A = EFK = DGK = CHK = BJK$	$AH = BDE = BFG = DFJ = EGJ = CK$
$B = EFJ = DGJ = CHJ = AJK$	$AJ = CDE = CFG = DFH = EGH = BK$
$C = EFH = DGH = BHJ = AHK$	$AK = EF = DG = CH = BJ$
$D = EFG = CGH = BGJ = AGK$	$BC = ADF = AEG = HJ = DEK = FGK$
$E = DFG = CFH = BFJ = AFK$	$BD = ACF = AEH = GJ = CEK = FHK$
$F = DEG = CEH = BEJ = AEK$	$BE = ACG = ADH = FJ = CDK = GHK$
$G = DEF = CDH = BDJ = ADK$	$BF = ACD = AGH = EJ = CGK = DHK$
$H = CEF = CDG = BCJ = ACK$	$BG = ACE = AFH = DJ = CFK = EHK$
$J = BEF = BDG = BCH = ABK$	$BH = ADE = AFG = CJ = DFK = EGK$
$K = AEF = ADG = ACH = ABJ$	$CD = ABF = GH = AEJ = BEK = FJK$
$AB = CDF = CEG = DEH = FGH = JK$	$CE = ABG = FH = ADJ = BDK = GJK$
$AC = BDF = BEG = DEJ = FGJ = HK$	$CF = ABD = EH = AGJ = BGK = DJK$
$AD = BCF = BEH = CEJ = FHJ = GK$	$CG = ABE = DH = AFJ = BFK = EJK$
$AE = BCG = BDH = CDJ = GHJ = FK$	$DE = FG = ABH = ACJ = BCK = HJK$
$AF = BCD = BGH = CGJ = DHJ = EK$	$DF = ABC = EG = AHJ = BHK = CJK$
$AG = BCE = BFH = CFJ = EHJ = DK$	

$$2 \text{ blocks of } 16: AK = EF = DG = CH = BJ$$
$$4 \text{ blocks of } 8: AK = EF = DG = CH = BJ$$
$$AJ = CDE = CFG = DFH = EGH = BK$$
$$AB = CDF = CEG = DEH = FGH = JK$$

(t) 2^{10-4}; 1/16 fraction of 10 factors in 64 runs

Design Generators

$$G = BCDF \quad H = ACDF \quad J = ABDE \quad K = ABCE$$

Defining relation: $I = BCDFG = ACDFH = ABGH = ABDEJ = ACEFGJ = BCEFHJ = DEGHJ = ABCEK$
$= ADEFGK = BDEFHK = CEGHK = CDJK = BFGJK = AFHJK = ABCDGHJK$

Aliases

$A = BGH$	$AD = CFH = BEJ$	$BK = ACE = FGJ$
$B = AGH$	$AE = BDJ = BCK$	$CD = BFG = AFH = JK$
$C = DJK$	$AF = CDH = HJK$	$CE = ABK = GHK$
$D = CJK$	$AG = BH$	$CF = BDG = ADH$
$E =$	$AH = CDF = BG = FJK$	$CG = BDF = EHK$
$F =$	$AJ = BDE = FHK$	$CH = ADF = EGK$
$G = ABH$	$AK = BCE = FHJ$	$CJ = DK$
$H = ABG$	$BC = DFG = AEK$	$CK = ABE = EGH = DJ$
$J = CDK$	$BD = CFG = AEJ$	$DE = ABJ = GHJ$
$K = CDJ$	$BE = ADJ = ACK$	$DF = BCG = ACH$
$AB = GH = DEJ = CEK$	$BF = CDG = GJK$	$DG = BCF = EHJ$
$AC = DFH = BEK$	$BJ = ADE = FGK$	$DH = ACF = EGJ$
$EF =$	$GJ = DEH = BFK$	$AEG = BEH = CFJ = DFK$
$EG = DHJ = CHK$	$GK = CEH = BFJ$	$AEH = BEG$
$EH = DGJ = CGK$	$HJ = DEG = AFK$	$AFG = BFH = CEJ = DEK$
$EJ = ABD = DGH$	$HK = CEG = AFJ$	$AGJ = CEF = BHJ$
$EK = ABC = CGH$	$ABF = FGH$	$AGK = DEF = BHK$
$FG = BCD = BJK$	$ACG = BCH = EFJ$	$BCJ = EFH = BDK$
$FH = ACD = AJK$	$ACJ = EFG = ADK$	$BEF = CHJ = DHK$
$FJ = BGK = AHK$	$ADG = BDH = EFK$	$CDE = EJK$
$FK = BGJ = AHJ$	$AEF = CGJ = DGK$	$CFK = DFJ$

2 blocks of 32: $AGJ = CEF = BHJ$ 4 blocks of 16: $AGJ = CEF = BHJ$
$AGK = DEF = BHK$
$CD = BFG = AFH = JK$

(continued)

695

XII. Alias Relationships for 2^{k-p} Fractional Factorial Designs with $k \leq 15$ and $n \leq 64$ *(continued)*

Designs with 11 Factors

Resolution III

(u) 2^{11-7}; 1/128 fraction of 11 factors in 16 runs

Design Generators

Defining relation:

$I = ABC$ $F = BCD$ $G = ACD$ $H = ABD$ $J = ABCD$ $K = AB$ $L = AC$

$= ABCE = BCDF = ADEF = ACDG = BDEG = ABFG = CEFG = ABDH$
$= CDEH = ACFH = BEFH = BCGH = AEGH = DFGH = ABCDEFGH = ABCDJ$
$= DEJ = AFJ = BCEFJ = BGJ = ACEGJ = CDFGJ = ABDEFGJ = CHJ$
$= ABEHJ = BDFHJ = ACDEFHJ = ADGHJ = BCDEGHJ = ABCFGHJ = EFGHJ = ABK$
$= CEK = ACDFK = BDEFK = BCDGK = ADEGK = FGK = ABCEFGK = DHK$
$= ABCDEHK = BCFHK = AEFHK = ACGHK = BEGHK = ABDFGHK = CDEFGHK = CDJK$
$= ABDEJK = BFJK = ACEFJK = AGJK = BCEGJK = ABCDFGJK = DEFGJK = ABCHJK$
$= EHJK = ADFHJK = BCDEFHJK = BDGHJK = ACDEGHJK = CFGHJK = ABEFGHJK = ACL$
$= BEL = ABDFL = CDEFL = DGL = ABCDEGL = BCFGL = AEFGL = BCDHL$
$= ADEHL = FHL = ABCEFHL = ABGHL = CEGHL = ACDFGHL = BDEFGHL = BDJL$
$= ACDEJL = CFJL = ABEFJL = ABCGJL = EGJL = ADFGJL = BCDEFGJL = AHJL$
$= BCEHJL = ABCDFHJL = DEFHJL = CDGHJL = ABDEGHJL = BFGHJL = ACEFGHJL = BCKL$
$= AEKL = DFKL = ABCDEFKL = ABDGKL = CDEGKL = ACFGKL = BEFGKL = ACDHKL$
$= BDEHKL = ABFHKL = CEFHKL = GHKL = ABCEGHKL = BCDFGHKL = ADEFGHKL = ADJKL$
$= BCDEJKL = ABCFJKL = EFJKL = CGJKL = ABEGJKL = BDFGJKL = ACDEFGJKL = BHJKL$
$= ACEHJKL = CDFHJKL = ABDEFHJKL = ABCDGHJKL = DEGHJKL = AFGHJKL = BCEFGHJKL$

Aliases

$A = FJ = BK = CL$	$J = DE = AF = BG = CH$	
$B = GJ = AK = EL$	$K = AB = CE = FG = DH$	
$C = HJ = EK = AL$	$L = AC = BE = DG = FH$	
$D = EJ = HK = GL$	$AD = EF = CG = BH$	
$E = DJ = CK = BL$	$AE = BC = DF = GH = KL$	
$F = AJ = GK = HL$	$AG = CD = BF = EH = JK$	
$G = BJ = FK = DL$	$AH = BD = CF = EG = JL$	
$H = CJ = DK = FL$		

2 blocks of 8: $AE = BC = DF = GH = KL$

(v) 2^{11-6}, 1/64 fraction of 11 factors in 32 runs

Design Generators

Defining relation:

$F = ABC \quad G = BCD \quad H = CDE \quad J = ACD \quad K = ADE \quad L = BDE$

$I = ABCF = BCDG = ADFG = CDEH = ABDEFH = BEGH = ACEFGH = ACDJ = BDFJ = ABGJ = CFGJ$
$= AEHJ = BCEFHJ = ABCDEGHJ = DEFGHJ = ADEK = BDEFK = BCDEFK = ABCEGK = EFGK = ACHK = BFHK$
$= ABDGHK = CDFGHK = CDFGHJK = CEJK = ABEFJK = BDEGJK = ACDEFGJK = DHJK = ABCDFHJK = BCGHJK$
$= AFGHJK = BDEL = ACDEFL = CEGL = ABEFGL = BCHL = AFHL = DGHL = ABCDFGHL$
$= ABCEJL = EFJL = ADEGJL = BCDEFGJL = ABDHJL = CDFHJL = ACGHJL = BFGHJL = ABKL$
$= CFKL = ACDGKL = BDFGKL = ABCDEHKL = DEFHKL = AEGHKL = BCEFGHKL = BCDJKL$
$= ADFJKL = GJKL = ABCFGJKL = BEHJKL = ACEFHJKL = CDEGHJKL = ABDEFGHJKL.$

Aliases

$A = BCF = DFG = CDJ = BGJ = EHJ = DEK = CHK = FHL = BKL$
$B = ACF = CDG = EGH = DFJ = AGJ = FHK = DEL = CHL = AKL$
$C = ABF = BDG = DEH = ADJ = FGJ = AHK = EJK = EGL = BHL = FKL$
$D = BCG = AFG = CEH = ACJ = BFJ = AEK = HJK = BEL = GHL$
$E = CDH = BGH = AHJ = ADK = FGK = CJK = BDL = CGL = FJL$
$F = ABC = ADG = BDJ = CGJ = EGK = BHK = AHL = EJL = CKL$
$G = BCD = ADF = BEH = ABJ = CFJ = EFK = CEL = DHL = JKL$
$H = CDE = BEG = AEJ = ACK = BFK = DJK = BCL = AFL = DGL$
$J = ACD = BDF = ABG = CFG = AEH = CEK = DHK = EFL = GKL$
$K = ADE = EFG = ACH = BFH = CEJ = DHJ = ABL = CFL = GJL$
$L = BDE = CEG = BCH = AFH = DGH = EFJ = ABK = CFK = GJK$

$AB = CF = GJ = KL$	$AE = HJ = DK$	$AH = EJ = CK = FL$	$AL = FH = BK$	$BH = EG = CL = FK$
$AC = BF = DJ = HK$	$AF = BC = DG = HL$	$AJ = CD = BG = EH$	$BD = CG = FJ = EL$	$CE = DH = JK = GL$
$AD = FG = CJ = EK$	$AG = DF = BJ$	$AK = DE = CH = BL$	$BE = GH = DL$	$EF = GK = JL$

$ABD = CDF = ACG = BFG = EFH = BCJ = AFJ = DGJ = BEK = GHK = AEL = HJL = DKL$
$ABE = CEF = DFH = AGH = EGJ = BHJ = BDK = CGK = FJK = ADL = FGL = CJL = EKL$
$ABH = DEF = AEG = CFH = BEJ = GHJ = BCK = AFK = DGK = EHK = AJK = BFL = DJL = HKL$
$ACE = BEF = ADH = FGH = DEJ = CHJ = CDK = BGK = EHK = AJK = DFL = AGL = BJL$
$AEF = BCE = DEG = BDH = CGH = FHJ = DFK = AGK = BJK = CDL = BGL = EHL = AJL$

4 blocks of 8: $AB = CF = GJ = KL$
$AD = FG = CJ = EK$

2 blocks of 16: $AB = CF = GJ = KL$

$AD = FG = CJ = EK$
$BD = CG = FJ = EL$

(continued)

XII.   Alias Relationships for 2^{k-p} Fractional Factorial Designs
with $k \leq 15$ and $n \leq 64$ (*continued*)

Designs with 12 Factors

(w) 2^{12-8}; 1/256 fraction of   **Resolution III**
12 factors in 16 runs

Design Generators

$E = ABC \quad F = ABD \quad G = ACD \quad H = BCD$
$J = ABCD \quad K = AB \quad L = AC \quad M = AD$

Aliases

$A = HJ = BK = CL = DM$
$B = GJ = AK = EL = FM$
$C = FJ = EK = AL = GM$
$D = EJ = FK = GL = AM$
$E = DJ = CK = BL = HM$
$F = CJ = DK = HL = BM$
$G = BJ = HK = DL = CM$
$H = AJ = GK = FL = EM$
$J = DE = CF = BG = AH$
$K = AB = CE = DF = GH$
$L = AC = BE = DG = FH$
$M = AD = BF = CG = EH$

$AE = BC = FG = DH = KL = JM$
$AF = BD = EG = CH = JL = KM$
$AG = EF = CD = BH = JK = LM$

2 blocks of 8: $AE = BC = FG = DH = KL = JM$

Designs with 13 Factors

(x) 2^{13-9}; 1/512 fraction of   **Resolution III**
13 factors in 16 runs

Design Generators

$E = ABC \quad F = ABD \quad G = ACD \quad H = BCD$
$J = ABCD \quad K = AB \quad L = AC \quad M = AD \quad N = BC$

Aliases

$A = HJ = BK = CL = DM = EN$
$B = GJ = AK = EL = FM = CN$
$C = FJ = EK = AL = GM = BN$
$D = EJ = FK = GL = AM = HN$
$E = DJ = CK = BL = HM = AN$
$F = CJ = DK = HL = BM = GN$
$G = BJ = HK = DL = CM = FN$
$H = AJ = GK = FL = EM = DN$
$J = DE = CF = BG = AH = MN$
$K = AB = CE = DF = GH = LN$
$L = AC = BE = DG = FH = KN$
$M = AD = BF = CG = EH = JN$
$N = BC = AE = FG = DH = KL = JM$

$AF = BD = EG = CH = JL = KM$
$AG = EF = CD = BH = JK = LM$

2 blocks of 8: $AF = BD = EG = CH = JL = KM$

XII. Alias Relationships for 2^{k-p} Fractional Factorial Designs
with $k \leq 15$ and $n \leq 64$ (*continued*)

Designs with 14 Factors

(y) 2^{14-10}; 1/1024 fraction of   **Resolution III**
14 factors in 16 runs

Design Generators

$E = ABC$ $F = ABD$ $G = ACD$ $H = BCD$ $J = ABCD$
$K = AB$ $L = AC$ $M = AD$ $N = BC$ $O = BD$

Aliases

$A = HJ = BK = CL = DM = EN = FO$
$B = GJ = AK = EL = FM = CN = DO$
$C = FJ = EK = AL = GM = BN = HO$
$D = EJ = FK = GL = AM = HN = BO$
$E = DJ = CK = BL = HM = AN = GO$
$F = CJ = DK = HL = BM = GN = AO$
$G = BJ = HK = DL = CM = FN = EO$
$H = AJ = GK = FL = EM = DN = CO$
$J = DE = CF = BG = AH = MN = LO$
$K = AB = CE = DF = GH = LN = MO$
$L = AC = BE = DG = FH = KN = JO$
$M = AD = BF = CG = EH = JN = KO$
$N = BC = AE = FG = DH = KL = JM$
$O = BD = AF = EG = CH = JL = KM$

$AG = EF = CD = BH = JK = LM = NO$

2 blocks of 8: $AG = EF = CD = BH = JK = LM = NO$

Designs with 15 Factors

(z) 2^{15-11}; 1/2048 fraction of   **Resolution III**
15 factors in 16 runs

Design Generators

$E = ABC$ $F = ABD$ $G = ACD$ $H = BCD$ $J = ABCD$
$K = AB$ $L = AC$ $M = AD$ $N = BC$ $O = BD$ $P = CD$

Aliases

$A = HJ = BK = CL = DM = EN = FO = GP$
$B = GJ = AK = EL = FM = CN = DO = HP$
$C = FJ = EK = AL = GM = BN = HO = DP$
$D = EJ = FK = GL = AM = HN = BO = CP$
$E = DJ = CK = BL = HM = AN = GO = FP$
$F = CJ = DK = HL = BM = GN = AO = EP$
$G = BJ = HK = DL = CM = FN = EO = AP$
$H = AJ = GK = FL = EM = DN = CO = BP$
$J = DE = CF = BG = AH = MN = LO = KP$
$K = AB = CE = DF = GH = LN = MO = JP$
$L = AC = BE = DG = FH = KN = JO = MP$
$M = AD = BF = CG = EH = JN = KO = LP$
$N = BC = AE = FG = DH = KL = JM = OP$
$O = BD = AF = EG = CH = JL = KM = NP$
$P = CD = EF = AG = BH = JK = LM = NO$

Index